IFMBE Proceedings Volume 28

Series Editor: R. Magjarevic

Biomag 2010

The International Federation for Medical and Biological Engineering, IFMBE, is a federation of national and transnational organizations representing internationally the interests of medical and biological engineering and sciences. The IFMBE is a non-profit organization fostering the creation, dissemination and application of medical and biological engineering knowledge and the management of technology for improved health and quality of life. Its activities include participation in the formulation of public policy and the dissemination of information through publications and forums. Within the field of medical, clinical, and biological engineering, IFMBE's aims are to encourage research and the application of knowledge, and to disseminate information and promote collaboration. The objectives of the IFMBE are scientific, technological, literary, and educational.

The IFMBE is a WHO accredited NGO covering the full range of biomedical and clinical engineering, healthcare, healthcare technology and management. It is representing through its 58 member societies some 120.000 professionals involved in the various issues of improved health and health care delivery.

IFMBE Officers
President: Makoto Kikuchi, Vice-President: Herbert Voigt, Former-President: Joachim H. Nagel
Treasurer: Shankar M. Krishnan, Secretary-General: Ratko Magjarevic
http://www.ifmbe.org

Previous Editions:

IFMBE Proceedings BIOMAG2010, "17th International Conference on Biomagnetism Advances in Biomagnetism – Biomag2010", Vol. 28, 2010, Dubrovnik, Croatia, CD

IFMBE Proceedings ICDBME 2010, "The Third International Conference on the Development of Biomedical Engineering in Vietnam", Vol. 27, 2010, Ho Chi Minh City, Vietnam, CD

IFMBE Proceedings MEDITECH 2009, "International Conference on Advancements of Medicine and Health Care through Technology", Vol. 26, 2009, Cluj-Napoca, Romania, CD

IFMBE Proceedings WC 2009, "World Congress on Medical Physics and Biomedical Engineering", Vol. 25, 2009, Munich, Germany, CD

IFMBE Proceedings SBEC 2009, "25th Southern Biomedical Engineering Conference 2009", Vol. 24, 2009, Miami, FL, USA, CD

IFMBE Proceedings ICBME 2008, "13th International Conference on Biomedical Engineering" Vol. 23, 2008, Singapore, CD

IFMBE Proceedings ECIFMBE 2008 "4th European Conference of the International Federation for Medical and Biological Engineering", Vol. 22, 2008, Antwerp, Belgium, CD

IFMBE Proceedings BIOMED 2008 "4th Kuala Lumpur International Conference on Biomedical Engineering", Vol. 21, 2008, Kuala Lumpur, Malaysia, CD

IFMBE Proceedings NBC 2008 "14th Nordic-Baltic Conference on Biomedical Engineering and Medical Physics", Vol. 20, 2008, Riga, Latvia, CD

IFMBE Proceedings APCMBE 2008 "7th Asian-Pacific Conference on Medical and Biological Engineering", Vol. 19, 2008, Beijing, China, CD

IFMBE Proceedings CLAIB 2007 "IV Latin American Congress on Biomedical Engineering 2007, Bioengineering Solution for Latin America Health", Vol. 18, 2007, Margarita Island, Venezuela, CD

IFMBE Proceedings ICEBI 2007 "13th International Conference on Electrical Bioimpedance and the 8th Conference on Electrical Impedance Tomography", Vol. 17, 2007, Graz, Austria, CD

IFMBE Proceedings MEDICON 2007 "11th Mediterranean Conference on Medical and Biological Engineering and Computing 2007", Vol. 16, 2007, Ljubljana, Slovenia, CD

IFMBE Proceedings BIOMED 2006 "Kuala Lumpur International Conference on Biomedical Engineering", Vol. 15, 2004, Kuala Lumpur, Malaysia, CD

IFMBE Proceedings WC 2006 "World Congress on Medical Physics and Biomedical Engineering", Vol. 14, 2006, Seoul, Korea, DVD

IFMBE Proceedings BSN 2007 "4th International Workshop on Wearable and Implantable Body Sensor Networks", Vol. 13, 2006, Aachen, Germany

IFMBE Proceedings ICBMEC 2005 "The 12th International Conference on Biomedical Engineering", Vol. 12, 2005, Singapore, CD

IFMBE Proceedings Vol. 28

Selma Supek · Ana Sušac (Eds.)

17th International Conference on Biomagnetism Advances in Biomagnetism – Biomag2010

March 28 – April 1, 2010
Dubrovnik, Croatia

Springer

Editors

Selma Supek
University of Zagreb
Faculty of Science
Dept. of Physics
Bijenicka cesta 32
10000 Zagreb
Croatia
E-mail: selma@phy.hr

Ana Sušac
University of Zagreb
Faculty of Science
Dept. of Physics
Bijenicka cesta 32
10000 Zagreb
Croatia
E-mail: ana@phy.hr

ISSN 1680-0737
ISBN 978-3-642-12196-8 ISBN 978-3-642-12197-5 (eBook)

DOI 10.1007/978-3-642-12197-5

Library of Congress Control Number: 2010923826

The IFMBE Proceedings is an Official Publication of the International Federation for Medical and Biological Engineering (IFMBE)

Typesetting: Scientific Publishing Services Pvt. Ltd., Chennai, India.

Cover Design: deblik, Berlin

Printed on acid-free paper

9 8 7 6 5 4 3 2 1

springer.com

Foreword

40th anniversary of "medical uses of SQUID"

It is my great pleasure and honor to invite you to the *17th International Conference on Biomagnetism – Biomag2010* held in Dubrovnik, Croatia from Sunday, March 28 through Thursday, April 1, 2010.

The interdisciplinary field of biomagnetism includes dynamic and evolving SQUID-based technologies offering advanced real-time methods for noninvasive assessments of magnetic signals from the brain, heart and other organs as well as a range of modeling, mathematical and computational methods for functional source localization approaches. Excellent spatial resolution and unique, millisecond, temporal resolution of biomagnetic techniques allow insights into cortical neurodynamics and neurobiological basis of the human brain as well as assessment of heart and other organs functions in health and disease. Biomag2010 will be a great opportunity for an exchange of ideas and presentation of the latest developments in instrumentation, modeling approaches, basic and clinical biomedical studies. We are particularly proud to announce the celebration of the 40th anniversary of the first SQUID-based MCG measurements published on April 1, 1970. Since then "medical uses of SQUID" were dynamic and growing, including the most recent developments, in combination with a low field MRI, toward a "direct neuronal imaging".

Dubrovnik, the host city of the Biomag2010, a jewel on the Adriatic, will be a superb and stimulating setting for both scientific and social aspects of this meeting.

I am looking forward to hosting you in Dubrovnik, Croatia in spring of 2010.

Selma Supek
Conference Chair

Organization

Proceedings Editors

Selma Supek
Ana Sušac

Organizer

University of Zagreb, Faculty of Science, Department of Physics

Sponsor

Croatian Ministry of Science and Technology

Endoursed By

Croatian Biophysical Society
Croatian Cardiac Society
International Federation for Medical and Biological Engineering
Croatian Institute for Brain Research
Croatian Neurological Society
Croatian Physical Society
Croatian Society for Medical and Biological Engineering
Croatian Society for Neuroscience
Zagreb Epilepsy Center

Conference Chair

Selma Supek

Secretary General

Ana Sušac

International Advisory Board

Shinya Kuriki, Chair (Japan)
Kazuhiko Atsumi (Japan)
Douglas Cheyne (Canada)
David Cohen (USA)
Luder Decke (Austria)
Sergio Erne (Germany)
Eric Halgren (USA)
Jens Haueisen (Germany)

Toivo Katila (Finland)
Makoto Kotani (Japan)
Gian Luca Romani (Italy)
Selma Supek (Croatia)
Shoogo Ueno (Japan)
Harold Weinberg (Canada)
Chris Wood (USA)

Scientific Program Committee

Selma Supek, *Chair* (Croatia)
Jens Haueisen, *Co-chair* (Germany)
Robert Kraus, *Co-chair* (USA)
Cheryl J. Aine (USA)
Sylvain Baillet (USA)
Susan Bowyer (USA)
Line Garnero[†] (France)
Matti Hämäläinen (USA)
Riitta Hari (Finland)
Risto Ilmoniemi (Finland)
Ole Jensen (The Netherlands)
Shinya Kuriki (Japan)
Sari Levänen (USA)
Markku Mäkijärvi (Finland)
Ksenija Marinkovic (USA)
Maria Mody (USA)
Nobukazu Nakasato (Japan)
Yoshio Okada (USA)
Christo Pantev (Germany)
Andrew Papanicolaou (USA)
Gian Luca Romani (Italy)
Kensuke Sekihara (Japan)
Yoshinori Uchikawa (Japan)
Peter Van Leeuwen (Germany)

Local Organizing Committee

Hrvoje Hećimović
Sanja Josef-Golubić
Veljko Grilj
Ratko Magjarević
Marko Miljević
Sandra Požar
Ana Sušac
Nikolina Wolf
Marko Banušić, IT Manager
Branko Đurđević, IT Manager
Mia Vucic, Visual identity

National MEG Initiative Board – CroMEGin

Anto Bagić, *Coordinator* (USA)
Igor Antončić (Rijeka)
Vedran Deletis (Split)
Vida Demarin (Zagreb)
Dražen Domijan (Rijeka)
Sanja Hajnšek (Zagreb)
Hrvoje Hećimović (Zagreb)
Ivica Kostović (Zagreb)
Ratko Magjarević (Zagreb)
Joško Paladino (Zagreb)
Selma Supek (Zagreb)
Dinko Štimac (Osijek)

Reviewers

Cheryl J. Aine (USA)
Sylvain Baillet (USA)
Susan Bowyer (USA)
Markus Butz (Germany)
Chun Kee Chung (South Korea)
Maureen Clerc (France)
Paulo R. Fonseca (Italy)
Michael Funke (USA)
Matti Hämäläinen (USA)
Riitta Hari (Finland)
Richard Henson (UK)
Jens Haueisen (Germany)
Risto Ilmoniemi (Finland)
Sunao Iwaki (Japan)
Ole Jensen (The Netherlands)
Blake Johnson (Australia)
Sung Chan Jun (South Korea)
Robert Kraus (USA)
Shinya Kuriki (Japan)
Sari Levänen (USA)
Jo-Fu Lotus Lin (USA)
Burkhard Maess (Germany)
Markku Mäkijärvi (Finland)
Ksenija Marinkovic (USA)
Maria Mody (USA)
John C. Mosher (USA)
Nobukazu Nakasato (Japan)
Yoshio Okada (USA)
Christo Pantev (Germany)
Christos Papadelis (Italy)
Andrew Papanicolaou (USA)
Hubert Preissl (Germany)
Gian Luca Romani (Italy)
Kensuke Sekihara (Japan)
Peter Soros (USA)
Selma Supek (Croatia)
Ana Sušac (Croatia)
Norman Tepley (USA)
Yoshinori Uchikawa (Japan)
Peter Van Leeuwen (Germany)
Tony W. Wilson (USA)
Filippo Zappasodi (Italy)
Johanna Zumer (UK)

Table of Contents

Instrumentation: Multi-modal Integration - MEG, Low-Field MRI, EEG, fMRI, TMS, NIRS

MEG Modeling and Spatio-temporal Source Localization

Neural Signal Processing

MEG: Sensory Systems

MEG: Brain-Computer Interface

MEG: Clinical Applications

MCG: Modeling and Clinical Applications

Biosusceptometry, Nanoparticles and Effects of Magnetic Fields

Forty Years of Magnetocardiology

Gerhard Stroink

Physics Department, Dalhousie University, Halifax, Nova Scotia, Canada B3H-3J5

Abstract— **About 40 years ago the SQUID was introduced. Its impressive sensitivity made it possible to, relatively conveniently, measure the extreme small magnetic fields generated by the body. The first biomagnetic measurement with the SQUID was that of the heart. Since then the advantages and limitations of magnetocardiology have been investigated in many research labs, several of them in a clinical setting. In this article we will describe the growth of this field, particularly during the first pivotal fifteen years, highlight some of the research efforts over these forty years and indicate in what direction magnetocardiology is developing.**

Keywords— **magnetocardiology, cardiac magnetic field maps, arrhythmias, coronary artery disease, imaging.**

I. INTRODUCTION

Forty years ago the first SQUID measurements of the heart's magnetic field were published [1]. A series of technical achievements made this possible. David Cohen had just finished building a mu-metal shielded room at MIT, Jim Zimmerman and colleagues had developed stable point-contact SQUIDs and Edgar Edelsack, the third author on this paper, was the catalyst. These first measurements of the magnetocardiogram (MCG) with the SQUID were made during the last few days and nights of 1969. They were produced without time averaging and showed with excellent signal-to-noise ratio the main features of the electrical activity of the heart. Their experiment demonstrated the superior performance of the SQUID magnetometer for biomagnetic measurements and generated enthusiasm in the scientific community for this new, completely non-invasive method of measuring cardiac activity. In what follows we will illustrate some of the highlights of these forty years.

II. THE BEGINNING

The first recognizable magnetocardiograms were measured about six years earlier by Baule and McFee [2] as part of Baule's excellent theoretical and experimental PhD work. He measured the magnetic heart signals using a set of copper coils with several millions of windings in a gradiometer arrangement in an open field to avoid the urban magnetic interference. Similar feats were achieved somewhat later by Safonov et al [3] and David Cohen [4] with such room temperature coils in a shielded room. The MCGs measured with these induction coil magnetometers had poor signal-to-noise ratios that needed averaging to observe relevant details.

Cohen and Zimmerman demonstrated 40 years ago that the sensitivity of the SQUID in the, for the heart, relevant frequency range was unsurpassed and that consequently the SQUID would be *the* instrument for measuring MCGs. This was a major technical breakthrough. The superiority of the SQUID as a magnetic sensing device was not only realized by the three authors but also by others who then, with Jim Zimmerman formed, also 40 years ago, the SHE company to manufacture and market SQUID magnetometers. This California company made reliable SQUIDs and biomagnetic dewars accessible and was generous in distributing the knowledge needed to measure these biomagnetic signals. They stimulated discussions and helped sponsor the first Biomagnetism conferences. Zimmerman and Edelsack each wrote down their recollection of the, for Biomagnetism, pivotal 1969 events in the proceedings of the 1989 seventh Biomagnetism Conference in New York [5] to celebrate 20 years of SQUID biomagnetic measurements. David Cohen wrote down his recollections of the 1969 measurements in the proceedings of the fourteenth International Biomagnetism Conference in Boston [6].

The first SQUID measurements of the MCG signals predated the first MEG SQUID measurements by a few years because of the relatively large signals (peak signal about 50 pT compared to 1pT for the largest brain signals) and the, for noise suppression, more convenient higher frequency range used in MCG measurements. Because of these advantages, the readily available SQUID and perhaps the mystique of measuring non-invasively magnetic fields generated by the body, several research groups decided to take the plunge into Biomagnetism, focusing on MCG measurements.

III. THE SEVENTIES

A. *Experimental*

The seventies saw a steady expansion of MCG research. However, MCG measurements were not routine and were

S. Supek and A. Sušac (Eds.): Advances in Biomagnetism – BIOMAG2010, IFMBE Proceedings 28, pp. 1–8, 2010.
www.springerlink.com

generally performed in Physics departments or research institutions where one could combine the technical knowhow of low temperature research, SQUID technology, impedance matching and noise suppression. The initial MCG measurements were research intensive and it was clear that the future of MCG measurements would be tied to the ease with which they could be obtained, ideally in a clinical environment. Whereas Cohen's results demonstrated that MCG detection with a SQUID magnetometer produced excellent results these measurements also required expensive magnetic shielding. A major step forward was provided by Zimmerman and Frederick [7] who in 1971 showed that a properly balanced superconducting gradiometer connected to the SQUID would detect the MCG in a quiet unshielded laboratory environment. This created the possibility of measuring MCG's and getting involved in biomagnetism with modest amounts of research dollars, important for small groups.

In addition to the MCG measurements at Cohen's magnet lab [8] MCG's of normal subjects and patients were now routinely measured at a number of places. At the Helsinki University of Technology MCGs were obtained [9] with a SQUID based gradiometer in a wooden shed away from buildings [10]. They also showed that such measurements were possible in a hospital environment. As was the practice with ECG measurements, MCGs were typically obtained at several locations over the chest area and at each location compared with that for different cardiac patient groups. To account for the large spatial variations of MCGs over this area one typically measured sequentially at as many as 32 to 54 locations in a rectangular grid (normal component). At about the same time, MCG measurements were also obtained in a mu-metal shielded room at Stanford (Fairbank's group). They introduced the vector MCG, which measured the cardiac magnetic field in three orthogonal directions at a single position over the heart using a single SQUID and gradiometer. Based on theoretical arguments the 3D motion of the magnetic heart vector over the heart cycle represented, like the vector ECG that was regularly measured at that time, best the overall activity of the heart [11, 12].

More groups across the world became active in MCG with often their own flavor of gradiometers, measuring techniques and magnetic shielding. At the 3rd Biomagnetism conference in Berlin in 1980 (with 70 participants) Magnetocardiology was the most studied subsector of Biomagnetism [13]. So far the physics slanted cardiomagnetic research had made few inroads in the clinical world. Of particular clinical interest at that time were MCG measurement of the DC diastolic injury current by Cohen and Kaufman [14] and, somewhat later, the MCG signals of His bundle activity (see below). Both type of measurements aimed to demonstrate that MCG measurements could obtain clinical important information not readily available from ECG measurements.

Our involvement with biomagnetism started with discussions with Milan Horacek who also was at Dalhousie University and had worked on a digital model to predict the MCG [15]. We had experience with measuring small magnetic fields using a Clark junction (a simple but primitive DC SQUID, also known as the SLUG). Milan came from a group (Ratuharju) that had built a state-of-the-art 128 lead ECG system to obtain Body Surface Potential Maps (BSPM) that was supported by superior software with excellent algorithm to accurately average many ECG complexes to obtain high resolution data presented as voltage maps over the body surface. With a commercial SQUID and dewar, homemade 2nd order gradiometer, and supported by this software adapted to MCG data we presented our first MCG data in the late seventies, first measured in the open field, then in our laboratory within an eddy current shielded room especially built for MCG measurements and containing the 128 channel BSPM system [16].

B. Theoretical

The first MCG measurements stimulated theoretical work to calculate the observed MCGs from modeled current sources within the heart and the volume currents propagating through the conductive body compartments (the forward problem). It was recognized that such calculations would lead to a better understanding of MCGs and that the addition of MCG to the ECG could assist in the difficult problem of calculating from the MCGs and/or ECGs the cardiac current densities during the heart cycle (inverse problem). The basis for the theory of the magnetocardiogram was provided by Plonsey [17], Geselowitz [18] and their excellent students. The calculations to model the MCG or ECG also explored fundamental differences in their information content and how magnetocardiography may complement the ECG to provide valid diagnostic information where ECG showed a lack of sensitivity. The uniqueness question and possible advantages of one method over another was [19, 20] and still is [21] not trivial and was hotly debated in several good PhD theses, internal reports and articles that model different measuring methods [22], cardiac source configurations and torso properties [23].

IV. THE EARLY EIGHTIES

Typically, one would obtain simultaneously with the single channel MCG measurement at each of the grid points, the ECG of the limb leads. Utilizing these continuously recorded electrocardiograms as a time reference, one could then calculate for each probe position the magnetic field strength at a particular instance of the cardiac cycle. These magnetic field data points where then combined to produce a magnetic field map (MFM) for that moment. Repeating this process for each time point, sequential MFMs-typically 2 ms apart-were produced. During the early eighties one saw a transition from presenting data in the form of time traces at each measuring point to presenting magnetic field strength maps (iso-magnetic field maps) over the chest area [24, 25, 26, 27]. Such presentation was particularly promoted by MCG groups familiar with producing BSPMs maps similarly derived from multi-lead ECGs. They presented and analyzed both MFMs and BSPMs and thus the most complete information about the electrical activity of the heart that can be obtained non-invasively [25, 28, 29]. Analysis techniques then naturally focused on map features and allowed application of a variety of techniques for extracting relevant activities of clinical interest [30, 31]. Also, by presenting both the potentials and magnetic fields of the heart allowed one to compare and then characterize the unique information content of MFMs and consequently of MCG over ECGs [28, 29, 32]. MFMs also proved useful in comparing measured with simulated data, normally presented as maps, and as a starting point for calculating inverse solutions.

The superior quality of the Berlin shielded room built around 1980 suggested obtaining and exploring high resolution MCGs. This was at a time that the ECG community presented evidence that particular features in high resolution ECGs provided clinical useful information. This provided a challenge to the biomagnetism community to observe such features in MCG. Of particular interest were small deflections during the ST-segment and T-waves known as late potentials. High resolution surface ECG had revealed a definite association between the appearance of late potentials and potentially lethal, malignant ventricular arrhythmias. Catheter ablation of the arrhythmogenic area in selected cases resulted in the disappearance of both symptoms and late potentials. The first MCG of late magnetic fields associated with the late potentials were recorded in the Berlin shielded room by Erne and others [33]. Many MCG studies, also in relatively light shielded environment, have since confirmed that such delayed depolarization can be detected and even localized with non-invasive MCG.

Another area of interest at that time was detecting abnormalities in the conduction system using high resolution ECG. Particularly the non-invasive detecting of His-Purkinje signals (HPS) was of clinical interest. The signals are small and difficult to separate from the larger atrial repolarization signals. The first magnetic record of HPS was published by Farrell et al [34] and Fenici and co-workers [35, 36]. HPS maps provided a good testing ground for source localization methods because the small spatial extend of this activity suggested that the equivalent current dipole (ECD) was a reasonable approximation as the source model [37]. HPS studies provided a logical avenue towards studying accessory pathways present near the AV ring in patients with WPW syndrome and observed as a characteristic signal during the PR interval. Locating this accessory pathway with MCG in WPW patients was recognized as potentially useful to assist the clinician where the accessory pathway needed ablation [38]. Both, research in late potentials and WPW presented around 1980 were seen as promising avenues to be pursued by the MCG community for possible clinical applications.

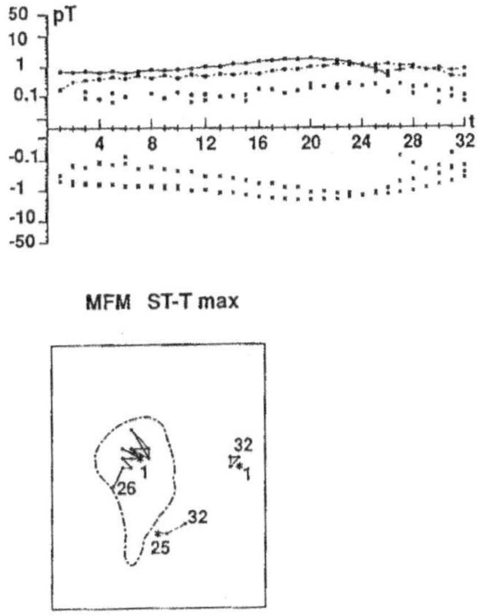

Fig. 1 Top: Variation over time of the extrema amplitudes (in pT) in a MFM of a VT patient during the ST-T interval. A normal subject shows a single positive and negative maximum. Bottom: The location of the fragmented extrema maximum over time on the MFM of the same patient during the same interval. Both frames show an abnormal fragmentation of map features during the ST-T interval [46]

Up to the mid eighties MCG research focused on the study of normal subjects, some patients, the development of instrumentation, general measuring and analysis techniques as well as some modeling at different levels of sophistication. Clearly, one of the main goals was to introduce MCG in the clinical environment. At the Vancouver Conference (1984) possible directions for clinical studies were formulated particularly by Fenici and Siltanen [39]. The measurements of clinical significant small features in the MCG in a shielded environment provided a challenge to observe similar features in an unshielded environment or ideally in a clinical unshielded environment. At the same Vancouver conference Fenici et al. indeed published MCG results obtained in the catherisation unit of his hospital in Rome for the PR interval demonstrating that MCG signals during this for the conduction system important segment could indeed be mapped and analyzed in an unshielded hospital environment [30].

In the eighties the MCG community experienced a dynamic expansion, but was still in the pioneering stage. MCG researchers met each other regularly as part of Biomagnetism conferences and workshops, first as a smaller group in Boston (1976) and Grenoble (1978), then larger in Berlin (1980), followed by Frascati (1982), Rome (1982), Vancouver (1984), Helsinki (1985), Turin (1986), Tokyo (1987) and New York (1989). We had stimulating exchanges of ideas, intensive discussions with lots of wine till late at night, exchanged data, set up collaborative programs and above all made many friends.

From a clinical viewpoint, there were several indications that MCG could replicate whatever was possible in terms of diagnostic and predictive value with ECG. There were some excellent studies pointing to some unique additional information that could be obtained with MCG and also that risk factors as expressed as predictive values obtained with ECG or BSPM could be further substantiated with enhances sensitivity or specificity using MCG and MFM as well. There was clearly a need for MFM studies involving larger and also better defined patient populations. With new noise suppressing technologies using additional SQUIDs, actively promoted by several companies, new imaging software and a more solid theoretical basis for inverse solutions a new era seemed nearby.

V. The multichannel era

Obtaining MFM was dramatically simplified with the production around 1990 of the first commercial multichannel system with first 31 then 37 SQUID channels [40]. Since then MCG multichannel systems (60 – 168 channels) produced by a number of companies have been installed at a number of places all over the world; some are located in a hospital setting. They provide conveniently and for the patient comfortable, detailed spatial-temporal information about the electrical activity of the heart in a matter of minutes. In tandem with this development in instrumentation substantial progress was made in mathematical techniques to use ECG and/or MCG mapping data with anatomical data to image the cardiac activation sequence or locate arrhythmogenic regions on the heart surface. In what follows we will discuss some of the more recent results obtained with multichannel systems. For a more complete review see Stroink et al [23] or Tavarozzi et al [41].

A. CAD Patients

Of the many diseases of the heart, ischemic heart disease (IHD) is the leading cause of death in adults in industrial countries. The routine clinical test to diagnose IHD is the 12 lead ECG and echocardiography before (rest) and immediate after or during exercise (stress). These 12 lead ECG tests are often used as the first evaluation to be followed by other clinical tests to determine subsequent clinical action.

IHD often manifests itself in ECG by changes in the ST segment and T wave morphology due to the onset of ischemia during the exercise. Several similar stress MCG studies also reported changes in ST segment and T wave pattern in such patients with coronary arterial disease (CAD). Saarinen et al [9] and later Cohen et al [42] were the first to explore ST shifts in MCG with exercise tests. The measurements by Cohen et al demonstrated that dc injury currents resulting from ischemia, virtually impossible to measure with ECG, could be measured with MCG.

Recently, Hänninen et al [43] showed that in a well documented group of 44 patients with induced ischemia the location of the most prominent of changes in the MFMs was dependent on the location of the stenosed vessel. This type of information obtained non-invasively is important in planning treatment of stenosed vessels with invasive means.

Morguet et al [44] studied features and parameters to determine MFM's ability to assess the viability of heart muscle (indicating the degree of blood perfusion and overall function) in patients with previous MI. The results were compared with PET and SPECT studies. The amplitude of the R and T waves were identified as parameters with the highest selectivity for determining myocardial viability. They correctly classified, in retrospect, all patients with regard to the extension of myocardial scar within the viable tissues. This study was followed up by imaging with MCG the viable myocardium using current density algorithms [45] and comparing it with the PET results. Even without exercise, changes in the MFM features due to CAD are readily identified using a variety of MCG parameters [46, 47]. Even if after infarction, the 12 lead ECG shows no ST-elevation (an indicator of ischemia), MCG can distinguish MI patients from normal subjects [48] confirming results by Goernig et al [49] that MFMs has clear advantages over conventional 12 lead ECG tests for identifying MI. These clinical studies exemplify more recent clinical research projects by several research groups with well characterized, relatively large groups of patients with CAD, to assess the diagnostic capabilities of MCG multichannel systems [23].

B. Risk Analysis

One of the main reasons for malignant, potentially lethal arrhythmias is myocardial damage and scarring, resulting from prior MI. Researchers have focused on a variety of different parameters in ECG, MCG, BSPM and MFM to identify patients at risk for ventricular arrhythmias. In such studies one compares, for instance, data of patients with MI and patients with MI and documented VT to analyze and quantify map differences. Such risk parameters are often directly related to the delayed ventricular activity through or around the damaged myocardial area resulting in additional delays in or fragmentation of map features of patients with VT [46, 47] (see Fig. 1). Multichannel systems and their extensive imaging software provide an excellent platform to analyze such features and have resulted in several studies identifying patients prone to VT after myocardial infarction. For instance, Korhonen and co-workers [50] demonstrated that time-domain and QRS fragmentation parameters in MFM correlate well with invasive electrograms obtained with an epicardial jacket and an endocardial balloon electrode array. The study involved 22 patients with old MI undergoing subendocardial resection to treat sustained VT. After the therapeutic intervention, the abnormalities in the MCG parameters mentioned above were reduced and in 20 of the 21 patients tested postoperatively, the VT became non-inducible providing further evidence for the direct association between these parameters and the delayed ventricular conduction leading to VT. In a later study [51], involving 158 patients with acute MI with left ventricular dysfunction they showed that increased QRS fragmentation over the QRS interval predicts arrhythmic events and mortality in such post-MI patients. Local and global spatial irregularities observed with multichannel MFM, as quantified by smoothness indices and QT variability, were used by Van Leeuwen et al. [52] and again other indices by Bydnyk et al [53] and others to successfully separate different patient groups including patients with VT. It is clear that such patient identification and classification is done conveniently and fast with multichannel MCG systems.

C. Localization

Stimulated by predictions that MCG would have advantages over ECG and following early successes in MEG, source localization using the equivalent current dipole (ECD) as source model for specific cardiac events was presented early at the first Biomagnetism conferences. However, cardiac inverse solutions are often complicated by spatial extended cardiac wavefronts, bounded by epicardial and endocardial surfaces and magnetic fields modified by anisotropic cardiac tissue and torso compartments with different conductivities. So an inverse solution that provides a dynamic image of the activation sequence during the heart cycle is extremely challenging. However when activity starts in a small focused area, like for some arrhythmias, simple ECD models provided good localization accuracies. To treat such arrhythmias a catheter is lead intravenously to the suspected region. By pacing this catheter and monitoring the induced activity by electrophysiological study (EPS) this catheter can be guided to near or at the arrhythmia site to ablate it. By the absence of further inducible arrhythmias this also validates its location as determined by the MCG inverse solution [39]. Over the years, several groups have used MCG assisted pacing to evaluate both the accuracy of a variety of inverse solutions and the use of MCG for ablation in a clinical environment. Comparing MCG and BSPM localization accuracies of a pacing catheter in the heart, Pesola et al. [54] found the localization accuracy of MCG to be superior to that of BSPM. From November 2002 till July 2006 Fenici's group, who introduced the multichannel in the pacing lab [55], collected more than 600 MCG maps in the catheter lab for assessment in patients with arrhythmias [56]. They showed that the non-invasive localization with MCG of the target area was successful in patients with WPW syndrome, atrial flutter or fibrillation and ventricular arrhythmias. They found that MCG localization with simultaneous catheter action potential recording aimed at the arrhythmogenic target is useful to define the arrhythmogenic substrate before intervention [57]. In general, the accurate three dimensional imaging of the onset of focal arrhythmias and the guidance of ablation catheter to this location has become a major goal in treating arrhythmias. Although competing technologies are proposed none is as risk free as the non-invasive MCG combined with the minimal invasive non-magnetic pacing catheter [57].

For more complex heart conditions and for a detailed knowledge of the electrophysiological function of the heart more detailed images of the propagation of the cardiac wave-fronts are required. In principle, such images can be obtained with voltage measurements directly on the epicardial surface. This however requires surgical intervention. The goal is clearly to derive at propagation sequences non-invasively. Progress has been made towards this goal using MFMs. In one such study Oostendorp and Pesola [58] used data obtained from patients undergoing open-chest surgery to treat arrhythmias. The authors calculated activation times from pre-surgical BSPM, MFM and combined BSPM/MFM datasets with as source model the uniform double layer (UDL) and as volume conductor a torso and heart geometry determined from MRI of that patient. These calculated activation times were then compared with the activation times measured on the epicardial surface with 102 leads on an electrode sock placed around the heart. The results were very encouraging.

Non-invasive activation time imaging or viability imaging, mentioned earlier, based on this method or current density imaging [59], calculated from either MFM, BSPM or both is a challenging but rewarding field of study with many potential applications. With many new imaging modalities entering the clinical environment the accurate, complete non-invasive imaging of the electrical activity of the heart and displayed on the actual (obtained from MRI) epicardial or endocardial surface would give magneto-cardiography a distinct advantage.

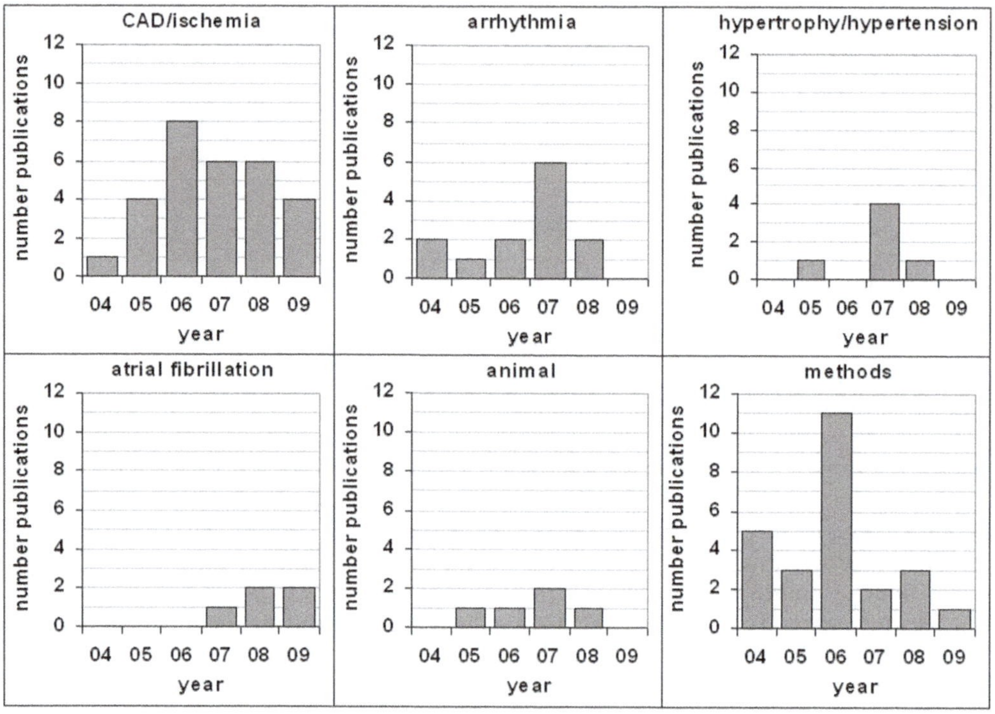

Fig. 2 The number of MCG publications since 2004 in some of the focus areas of research

It is interesting to compare the publication activity of MCG in some of the subfields described above. As can be seen from Fig. 2 the last five years most papers in peer-reviewed journals are in research on CAD and ischemia. “Methods” refers to data analysis methods and some modeling. Perusing the proceedings over the last 40 years one observes that there has been a shift from instrumentation papers, typical for a developing field, to manuscripts focusing on clinical research; an encouraging development.

VI. Conclusions

Cardiac research activity using MCG remains small compared to such non-invasive research using ECG or BSPM studies. Although clinical applications have been demonstrated and validated, few of the MCG systems have found, as yet, clinical acceptance as a method of choice to assess cardiac dysfunction. Magnetocardiology has grown over the last 40 years from a novelty, showcased by the occasional measurement on a few patients and by instrumentation development, to a field with well designed studies involving different patient groups addressing clinical issues. Results published in internationally respected cardiology journals indicate that MCG have been accepted as a research tool with potentially useful clinical applications. In particular the introduction of MCG in the catheter lab has shown to be beneficial for patient treatment and follow-up. The continued introduction of MCG multichannels in hospital environments to characterize and asses CAD patients is also a promising development. Both areas can benefit from advanced visualization techniques aimed at presenting detailed images of the electrophysiological function of the region of interest. Some of the recent results presented above point to successes in this area.

This article gave a condensed overview of what happened in the 40 years since the first SQUID MCG measurements. This review is far from complete with several areas ignored. It has not paid enough attention to areas such as the fusion of MCG with other imaging modalities, animals studies, cellular cardiac magnetic fields or the relatively large field of fetal magnetocardiology; areas where we see new developments and clinical relevant research opportunities.

Acknowledgement

We thank Peter van Leeuwen for valuable suggestions and providing Fig.2. This work was supported by NSERC.

References

1. Cohen D, Edelsack E A, Zimmerman J E (1970) Magnetocardiograms taken inside a shielded room with a superconducting point-contact magnetometer. Appl Phys Lett 16:278-280
2. Baule G M, McFee R (1963) Detection of the magnetic field of the heart. Am Heart J 55:95-96
3. Safonov Y D, Provotorov V M, Lube V M et al (1967) Method of recording the magnetic field of the heart. Bull Exp Bio Med 64:1022-1024
4. Cohen D (1967) Magnetic fields around the torso: production by electrical activity of the heart. Science 156:652-654
5. Williams S J, Hoke M, Stroink G, Kotani M (1989) Advances in Biomagnetism, Proc Seventh Internat Conf Biomagnetism, New York, USA, 1989, pp 67-86
6. Cohen D (2004) Boston and the history of biomagnetism. Neurology and Clinical Neurophysiology 30: 1-4
7. Zimmerman J E and Frederick N V (1971) Miniature ultrasensitive superconducting magnetic gradiometer and its use in cardiography and other applications. Appl Phys Lett 1:16-19
8. Cohen D, Lepeschkin E, Hosaka H et al (1976) Abnormal patterns and physiological variations in magnetocardiograms, J Electrocardiol 9:398-409
9. Saarinen M, Karp P J, Katila T E et al (1974) The magnetocardiogram in cardiac disorders. Cardiovascular Res 8:820-834
10. Katila T (1981) Instrumentation for medical applications. In: Biomagnetism, Eds: Erne S N, Hahlbohm H D and Lubbig H Proc. 3rd Biomagnetism Conf, Berlin, 1980, pp 3-31
11. Baule G and McFee R (1970) The magnetic heart vector. Am Heart J 79:223-236
12. Barry W H, Fairbank W M, Harrison D C, Lehrman K L, Malmivuo J A V and Wikswo J P (1977) Measurement of the human magnetic heart vector. Science 198:1159-1162
13. Karp P (1981) Cardiomagnetism. In: Biomagnetism, Eds. Erne S N, Hahlbohm H D and Lubbig H Proc. 3rd Biomagnetism Conf, Berlin, 1980, pp 219-258
14. Cohen D and Kaufman L A (1975) Magnetic determination of the relationship between the S-T segment shift and the injury current produced by coronary artery occlusion. Circulation Res36:414-424
15. Horacek M (1973) Digital model for studies in magnetocardiography, IEEE Trans Magn MAG-9:440-444
16. Stroink G, Horacek M and Blackford B L (1978) A simple gradiometer to measure magnetocardiograms. In: Computers in Cardiology. A Murray and S Swiryn, Eds. Piscataway, N.J.: IEEE Press, pp 21-24.
17. Plonsey R, Collin R E (1961) Principles and applications of electromagnetic fields. McGraw-Hill, New York
18. Geselowitz D B (1970) On the magnetic field generated outside an inhomogeneous volume conductor by internal current sources. IEEE Trans Mag MAG-6: 346-347
19. Wikswo J P and Barach J P (1982) Possible sources of new information in the magnetocardiogram. J Theoretical Biol 95:721-729
20. Plonsey R W (1971) Capability and limitations of electrocardiography and magnetocardiography. IEEE Trans Biomed Eng BME-19:239-244
21. Hren R, G Stroink and Horacek B M (1998) Spatial resolution of body surface potential maps and magnetic field maps: A model study applied to the identification of ventricular preexcitation sites. Med. & Biol. Eng. & Comp. 36:145-157.
22. Malmivuo J and Plonsey R (1995) Bioelectromagnetism. Oxford Univ Press, New York
23. Stroink G, Hailer B and Van Leeuwen P (2007) Cardiomagnetism. In: Magnetism in Medicine. Eds: Andrea W and Nowak H, 2nd edition, Wiley-VCH, Berlin, pp164-209.
24. Peters M J, van de Graaf A P and Van Oosterom A (1981). The influence of the torso on the magnetic field of a current dipole. In: Biomagnetism, Eds: Erne S N, Hahlbohm H D and Lubbig H Proc. 3rd Biomagnetism Conf, Berlin, 1980, pp 337-342
25. Nakaya Y, Sumi M, Saito K et al (1984) Analysis of current source of the heart using isomagnetic and vector arrow maps. Jpn Heart J 25:701-711
26. Erne S N, Fenici R R, Hahlbohm H D et al (1983) High-resolution isofield mapping in magnetocardiology. Il Nuovo Cimento 2D(2): 291-300
27. Stroink G, MacAuley C, Montague T J and Horacek B M (1985) Normal and abnormal components in magnetocardiographic maps of a subject with myocardial infarction. Med Biol Eng Comp 23(Suppl. 1): 61-62
28. Horacek B M and Stroink G (1981). Electrocardiogram and magnetocardiogram: independence testing. In: Proc. 34th Ann. Conf. of Eng. in Med. and Biol., Houston, Texas, pp. 430-432
29. Peters M J, Dunajski Z, Heringa A and Van Dam R T (1983) The mapping of the measured cardiac electric potential and magnetic field distribution. Il Nuovo Cimento, 2D(2):311-324
30. Fenici R R, Masselli M, Erne S N, Hahlbohm H D (1985) Magnetocardiographic mapping of the P-R interval phenomena in an unshielded hospital laboratory In: Weinberg H, Stroink G, & Katila T, (Eds.), *Biomagnetism*: Applications and Theory, Pergamon Press, New York, pp. 137-141
31. C MacAulay, G Stroink and BM Horacek (1985). "Analysis of magnetocardiographic spatial maps during the PR-interval". Med. & Biol. Eng. & Computing, 23, 1479-1480.
32. MacAulay C E, Stroink G and Horacek B M (1985) In: Weinberg H, Stroink G & Katila T (Eds.), *Biomagnetism*: Applications and Theory. Signal analysis of magnetocardiograms to test their independence, Pergamon Press, New York, 1985, pp. 115-120.

33. Erne S N, Fenici R R, Hahlbohm H D et al (1983) High-resolution magnetocardiographic recordings of the S-T segment in patients with electrical late potentials. Il Nuovo Cimento 2D(2): 340-345
34. Farrell D E, Tripp J H, Norgren R (1980) Magnetic study of the His-Purkinje conduction system in man. IEEE Trans Biomed Eng BME-27: 345-359
35. Barbanera S, Carelli P, Leoni R et al (1981) Magnetocardiographic study of some human cardiac electrophysiological phenomena. In: Biomagnetism, Eds: Erne S N, Hahlbohm H D and Lubbig H Proc. 3rd Biomagnetism Conf, Berlin, 1980, pp 283-290
36. Fenici R R, Masselli M, Lopez L et al. (1985) First simultaneous magnetocardiographic and invasive recordings of the P-R interval electrophysiological phenomena in man. Med Biol Eng Comp 23(Suppl. 2): 1483-1484
37. Fenici R R, Romani G L, Leoni R (1983). Magnetic measurements and modeling for the investigation of the human heart conduction system. Il Nuovo Cimento 2D(2): 280-290
38. Fenici R R, Romani G L, Erne S N (1983) High-resolution measurements of the human cardiac electrophysical events. Il Nuovo Cimento 2D(2): 231-247
39. Erne S N, Fenici R R, Katila T E et al (1985) Magnetocardiography: challenging clinical problems and promising analytic techniques In: Weinberg H, Stroink G & Katila T, (Eds.), *Biomagnetism*: Applications and Theory, Pergamon Press, New York, pp. 106-114
40. Schneider S, Hoenig E, Reichenberger H et al (1990) Multichannel biomagnetic system for study of the electrical activity of brain and heart. Radiology 176:825-830
41. Tavarozzi I, Comani S, Del Gratta C et al (2002) Magnetocardiography: current status and perspectives. Part II: Clinical applications. Ital Heart J 3:151-165
42. Cohen D, Savard P, Rifkin R D et al (1983) Magnetic measurement of S-T and T-Q segment shift in humans. Part II: exercise induced S-T segment depression. Circ Res 53:274-279
43. Hänninen H, Takala P, Korhonen P et al (2002) Features of ST segment and T wave in exercise induced myocardial ischemia evaluated with multichannel magnetocardiography. Ann Med 4:120-129
44. Morguet A J, Behrens S, Kosch O et al (2004) Myocardial viability evaluation using magnetocardiography in patients with coronary artery disease. Coronary Artery Dis 15(3):155-162
45. Goernig M, Haueisen J, Schreiber J et al. (2008) Comparison of current density viability imaging at rest with FDG-PET in patients after myocardial infarction. Comput Med Imaging Graph 33(1):1-6
46. Stroink G, Meeder R J J, Elliott P et al (1999) Arrhythmia vulnerability assessment using magnetic field maps and body surface potential maps. PACE 22:1718-1728
47. Van Leeuwen P, Hailer B, Lange S et al (2006) Identification of patients with coronary artery disease using magnetocardiographic signal analysis. Biomed Tech 51(2):83-88
48. Lim H K, Kim K, Lee Y and Chung N (2009) Detection of non-ST-elevation myocardial infarction using magnetocardiogram: new information from spatiotemporal electrical activation map. Ann Med 41(7):533-546
49. Goernig M, Liehr M, Tute C et al (2009). Magnetocardiography based spatiotemporal correlation is superior to conventional ECG analysis for identifying myocardial injury. Ann Biomed Eng 37(1):107-111
50. Korhonen P, Pesola K, Jarvinen A et al (2002) Relation of magnetocardiographic arrhythmia risk parameters to delayed ventricular conduction in postinfarction ventricular tachycardia. PACE, 25(9): 1339-1345
51. Korhonen P, Husa T, Tierala I et al (2006) Increased intra-QRS fragmentation in magnetocardiography as a predictor of arrhythmic events and mortality in patients with cardiac dysfunction after myocardial infarction. J Cardiovasc Electrophysiol 17:396-401
52. Van Leeuwen P, Hailer B, Lange S, Grönemeyer D (2003) Spatial distribution of repolarization times in patients with coronary artery disease. PACE 26:1706-1714
53. Budnyk M M, Kozlovsky V I, Stadnyuk L A et al (2004) Evaluation of magnetocardiography indices in patients with cardiac diseases. Neurology and Clinical Neurophysiol 111: 1-5
54. Pesola K, Nenonen J, Fenici R et al (1999) Bioelectromagnetic localization of a pacing catheter in the heart. Phys Med Biol 44:2565-2578
55. Fenici R, Brisinda D, Meloni A and Fenici P (2003) First 36-channel system for clinical magetocardiography in unshielded hospital laboratory for cardiac electrophysiology, Int J Bioelectromagn 1 (5):80-83
56. Fenici R and Brisinda (2007) Magnetocardiographic electroanatomical imaging for pre-interventional evaluation of arrhythmogenic substrates. In: New frontiers in biomagnetism. Proc 15th Int Conf on Biomagn, Vancouver 2006. Eds: Cheyne D, Ross B, Stroink G and Weinberg H. Elsevier Int Congress Series 1300: 476-479
57. Fenici R and Brisinda D (2007) Bridging noninvasive and interventional electroanatomical imaging: role of magnetocardiography. J Electrocard 40(1):S47-S52
58. Oostendorp and Nenonen J (2003) Non-invasive estimation of the activation sequence of the heart in the presence of old myocardial infarctions: comparison with invasive patient data. Int J Electromagnetism 5(1):221-224
59. Ferguson A S, Vardy D, Hren R and Stroink G (1995) A regularized minimum norm method for calculating distributions of source currents on epicardial surfaces". In: Biomagnetism: Fundamental Research and Clinical Applications. (Eds. Deecke, Baumgartner, Stroink and Williamson), Elsevier, Amsterdam, 676-679.

G. Stroink
Dalhousie University
Department of Physics
Halifax
Canada B3H-3J5

Highlights of 40 Years of SQUID-Based Brain Research and Clinical Applications

Cheryl J. Aine

Department of Radiology, University of New Mexico School of Medicine, Albuquerque, New Mexico, 87131, USA

Abstract— **A brief review of early MEG milestones, basic and clinical, conducted during the past 40 years is presented. Studies capturing the attention of the general audience via Science/Nature publications are highlighted as well as findings that could not have been discovered through the use of other functional neuroimaging techniques.**

Keywords— **Auditory, Motor, Somatosensory, Visual, Language.**

I. Introduction

Approximately 40 years have elapsed since the advent of the first magnetoencephalographic (MEG) recordings of fluctuating AC magnetic fields produced by alpha rhythm currents inside the human brain [1, 2]. These seminal studies (Figure 1) paved the way for subsequent basic research and clinical studies utilizing MEG. A synopsis of these memorable 40 years is presented herein. Given space limitations it is impossible to cover all studies. Therefore, three different approaches have been integrated for organizing this review: 1) the listing of 'first' (or early) sensory studies and clinical applications; 2) studies that Science and Nature journals found interesting; and 3) studies in neuroscience areas where MEG made a contribution that could not have been made (or was not yet made) using other noninvasive imaging techniques. The latter includes human studies that, for the first time, replicated results found through invasive work conducted in animals (e.g., tonotopy and retinotopy). Proceedings papers and edited book chapters are not included in this review. Some emphasis has been placed on Science and Nature articles since this is a good way to communicate with the rest of the world about the advantages of using MEG methods in general. Since this review primarily emphasizes early work, it may be particularly relevant to new investigators in this field.

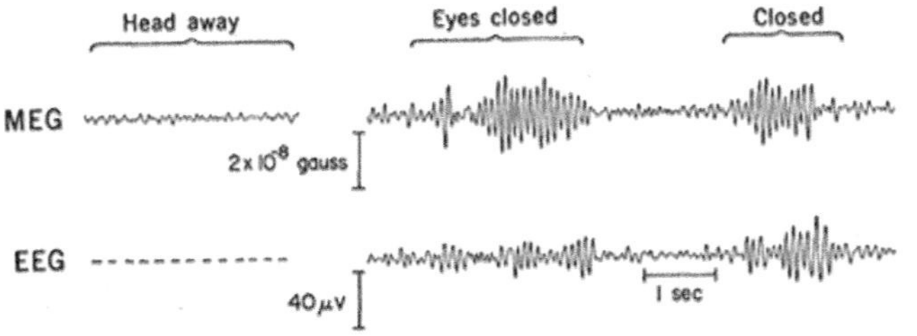

Fig. 1 Simultaneous MEG/EEG alpha-rhythm recordings from a normal subject who is closing and opening his eyes. Four subjects with large alpha rhythms were selected for this study, in addition to one subject with psychomotor epilepsy. The magnetometer was located at the left occipital region. Bandwidth for MEG/EEG was 4–15 Hz. Adapted from Cohen 1972

II. Auditory

The first recording of human magnetic auditory evoked fields (AEFs) was conducted by Reite [3]. Responses to 512 clicks were averaged to produce the fields shown in Figure 2A. Recordings were made by a second derivative SQUID gradiometer located in an aluminum shielded room. These recordings, along with recordings from other sensory modalities, revealed that in contrast to EEG, the MEG field distributions were more sharply focused over sensory cortices. Figure 2A demonstrates a systematic change in evoked response waveshape as the single-channel magnetometer was advanced from positions 1 through 5 across the left lateral surface. The amplitude increased as the magnetometer was placed closer to auditory cortex.

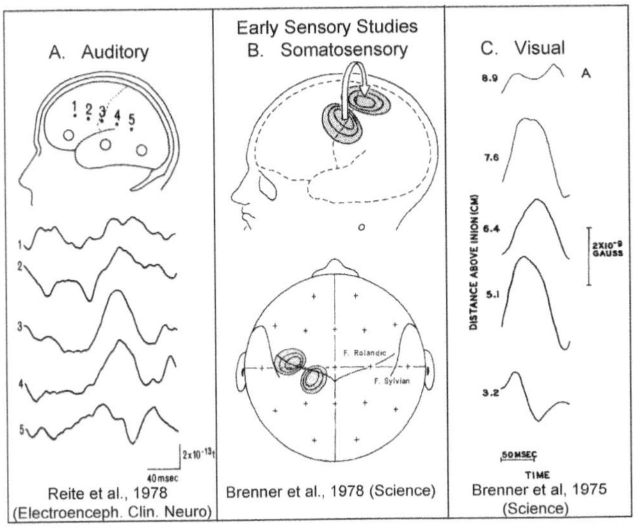

Fig. 2 Magnetic evoked fields recorded in the 1970s. A. Auditory responses to clicks. B. Steady-state somatosensory responses to electrical stimulation of little finger of right hand. C. Visual responses to vertical grating flickering at 10 Hz; average of 1000 responses. Recordings were made at different locations above the inion, at the back of the head

In the immediate years to follow the first demonstration of AEFs, issues such as reproducibility of the AEFs [4] and

S. Supek and A. Sušac (Eds.): Advances in Biomagnetism – BIOMAG2010, IFMBE Proceedings 28, pp. 9–34, 2010.
www.springerlink.com

locations of the neural generator(s), determined by variations in amplitudes and polarities across the surface, were of great interest. The 100 ms component and the sustained field evoked by long 1 kHz tones had a field distribution consistent with a generator within the Sylvian fissure [5, 6]. Without use of an invasive animal model for validating an imaging method, the next logical step was to demonstrate tonotopic mapping of auditory cortex, similar to that shown in invasive studies in nonhuman primates and cats [7, 8]. Tonotopy refers to the orderly projections in primary auditory cortex in which tones of various frequencies activate different neuronal populations. Romani, Williamson and Kaufman [9] presented in Science, the first evidence of tonotopic representation in auditory cortex of humans (Figure 3). The source of the evoked field systematically increased in depth with increasing frequency of the tone, which is similar to that found in animals. Pantev and colleagues [10] later reported in Science that the tonotopic

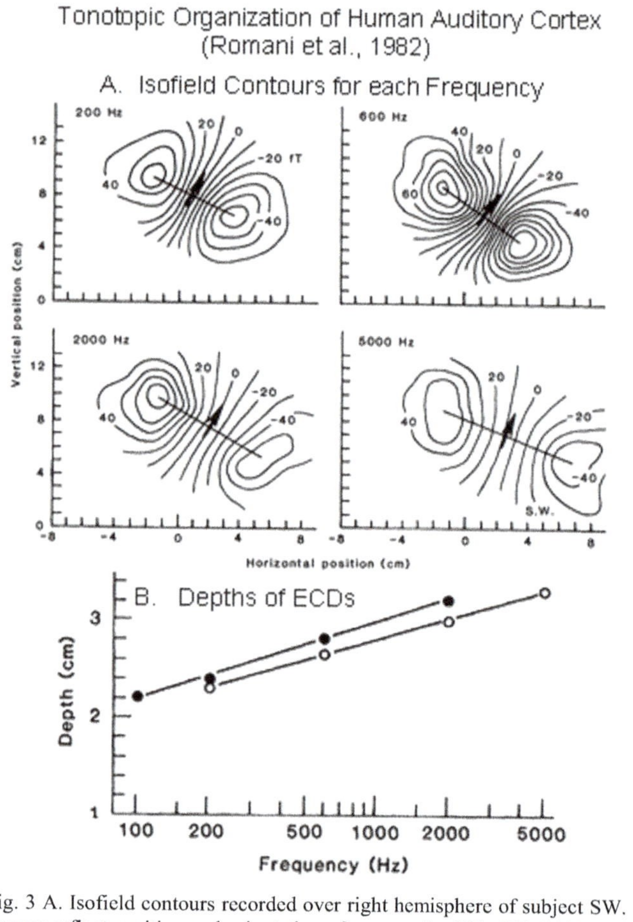

Fig. 3 A. Isofield contours recorded over right hemisphere of subject SW. Arrows reflect position and orientation of sources for 200, 600, 2000, and 5000 Hz tones. The origin is at the ear canal. B. Depths of equivalent current dipoles (ECDs) for 2 subjects as a function of the logarithm of the frequency. Adapted from Romani et al., 1982

organization of secondary auditory areas reflects the pitch rather than the frequency of the sound. This was true for the generators of the M100, rather than the M200 component.

Soon after Gray and colleagues [11] found neurons in cat visual cortex that oscillate in the range of 40–60 Hz, Pantev and colleagues [12] identified a 40 Hz transient oscillatory response evoked by the onset of auditory stimuli. Lately, interest in oscillatory activity has rekindled since many believe that synchronization of oscillatory responses of spatially distributed, feature-selective cells may be the mechanism for the ultimate binding together of different features into percepts [11, 13]. Pantev and colleagues found gamma band bursts consisting of 4 or more cycles during the 20–130 ms period after stimulus onset which were strongly locked in phase. The spectrum of the wideband response (0.1–95 Hz) revealed one peak near 10 Hz (reflecting predominantly the slow-wave components with peak latencies near 50, 90 and 170 ms) and another peak in the 30–40 Hz range (reflecting the gamma band response or GBR–see Figure 4). The equivalent current dipoles (ECDs) of both responses were in auditory cortex, although the locations differed somewhat. The authors suggested that perception in most or all sense modalities involves coherent

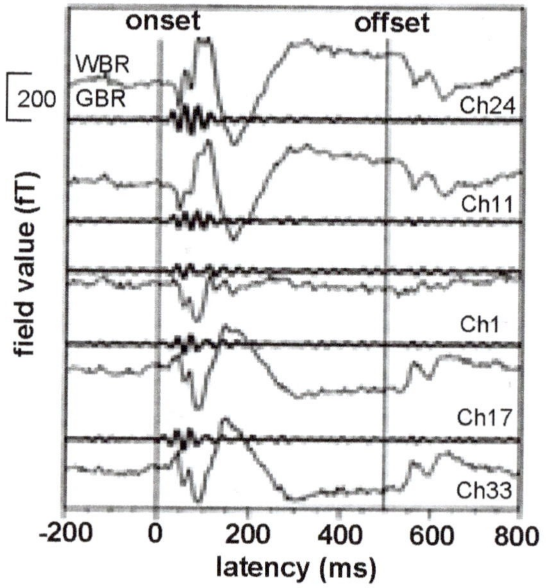

Fig. 4 Wide-band response with GBRs shown locked to the onset of the stimulus for 5 channels. Inversion of polarity of responses can be seen for both wideband responses and GBRs. Adapted from Pantev et al., 1991

rhythmic brain activity at gamma-band frequencies. When auditory cortex was driven by 40 Hz clicks, the gamma band response was also localized to auditory cortex [14]. The issue of oscillatory activity will be revisited in other sections below.

Another interesting area of study is the lifetime of neuronal activation traces in primary auditory and association cortices. Lu and colleagues [15, 16] found that the neuronal activation traces in auditory association cortex (they were established 100 ms after the onset of a probe stimulus) were several seconds longer than that in primary auditory cortex. In addition, their Science article suggested that echoic memory directly reflects the decay of the physiological activation trace in primary auditory cortex and that the lifetime for decay of the activation trace was found to predict the psychophysically determined duration of memory for the loudness of a tone.

A. Music

A number of studies examining how the brain processes music and studies comparing the cortical representations in auditory and somatosensory cortex between brains of musicians and non-musicians caught the attention of both Science and Nature as news features [17, 18]. Holden [17] notes that Thomas Carlyle referred to music as "the speech of angels." A Nature Neuroscience article by Maess and colleagues [19] suggest that language and music have a lot in common. Music-syntactic incongruities were set up by placing Neapolitan chords (that contain out-of-key notes with respect to the preceding harmonic context) in the 3rd or 5th position of chord sequences (each sequence consisted of 5 chords). Dipole modeling for Neapolitan chords at the 5th position of the chord sequences showed early effects (200 ms) in Broca's area and its right hemisphere counterpart. These frontal regions appear to process musical syntax, in the sense of determining harmonic relationships within a musical phrase. A Nature article by Patel and Balaban [20] examined the phase-time series for each sensor to demonstrate that some regions of the brain track the pitch contour of tone sequences (1-min long sequences) and that the accuracy of tracking increases as tone sequences become more predictable in structure.

Abbott's news feature in Nature [18] also focused on how the brain processes music by highlighting the works of Pantev and Elbert. In a Nature article, Pantev and colleagues [21] mapped the cortical representations of auditory cortex in highly skilled musicians and found greater dipole moments in auditory cortex (~25%) for piano tones, compared with pure tones of similar fundamental frequency that were matched in loudness, relative to a group of subjects who never played an instrument. In addition, dipole moment strength correlated with the age at which musicians began to practice. Elbert and colleagues [22] concluded in a Science article that somatosensory stimulation of the digits of the left hand of violin and string musicians resulted in enlarged activity (i.e., dipole moment was greater) in the cortical representation of left hand digits, compared to non-musicians. The center of cortical responsivity for tactile stimulation of the left hand digits was shifted in the musicians as well. Here too, a correlation was found between dipole moment strengths and the age at which the musicians began playing their instruments.

B. Basic Auditory Processes and Functional Reorganization

Many of the early auditory studies were concerned with localizing the generators of main deflections of ERPs found in studies of pre-attentive change detection processes and effects of selective attention. For example, Hari and colleagues [23] were the first to localize the mismatch negativity response (MMN) to supratemporal auditory cortex. Näätänen and colleagues [24] consider the MMN to reflect a mismatch process between the neuronal representation set up by a string of standard stimuli and the afferentation caused by a new stimulus that is deviant in some way from the string of standards (e.g., intensity, frequency, duration). The MMN is different from the well-known P300 response of the electrical potential since it occurs earlier in time (~200 ms or earlier) and does not require attention to be directed to the deviants in order to evoke it. Otherwise, the paradigms are essentially the same in that they both use standard stimuli that are presented frequently (e.g., 80% probability) and deviant stimuli that are presented infrequently (e.g., 20%). Pantev and colleagues [25, 26] have recently shown that the MMN may also be elicited by violations of regularities in sound sequences (patterns) and even by imagined melodies.

Sams and colleagues [27, 28] further suggested that there may be an additional source of the MMN in the frontal region while Alho and colleagues [29] differentiated between sources of the MMN and the magnetic counterpart of P300a. The electrical P300a is more closely associated with MMN in the sense that attention is not required to evoke it [30]. Alho and colleagues located the generator of the magnetic P300a to auditory cortex on the superior plane of the temporal lobe adjacent to and slightly anterior to the MMNm source. Before leaving this topic, it is important to point out a clinical application associated with the use of the MMN. Because the MMN does not require subjects to attend to stimuli this method is ideal for the assessment of sound and phonemic discrimination in fetuses, newborns and infants. The electrical MMN was used initially for these purposes (e.g., [31]) followed by the magnetic counterpart (e.g., [32]).

This section would be incomplete without acknowledging the many attempts to localize the generators of the electrical P300b. Unlike the electrical P300a, P300b requires attention to be directed to infrequent or novel stimuli. The

P300b is often associated with psychological constructs such as 'cognitive decision processes' since this component occurs relatively late in time (≥ 300 ms) and is not influenced by the stimulus parameters per se, but rather is influenced by the information processing demands of the task. Gordon and colleagues [33] were the first to localize the auditory magnetic P300b to auditory cortex. Shortly after, Rogers and colleagues [34] agreed that there is a source in auditory cortex and perhaps there is an initial deeper source (thalamus or hippocampus) contributing to P300b as well. Okada [35] was the first to suggest a hippocampal generator for the magnetic visual P300b.

While numerous studies using ERPs have investigated effects of selective attention, ERFs offer the distinct advantage of localizing these effects. Hari and colleagues [36] examined responses to either words or tones when participants were instructed to either attend (count the number of targets) or ignore the stimuli (reading condition). A dichotic listening task was conducted as well. In general, the initial transient responses, evoked in the attend and ignore conditions, did not differ in response amplitudes. The transient response was followed by a sustained field which was stronger in attend than ignore conditions beginning around 120-200 ms poststimulus and continuing for several hundred milliseconds. The source locations of both transient and sustained responses were localized to supratemporal auditory cortex. These results were supported soon after by Arthur and colleagues [37] when subjects either attended or ignored tone pitches. More recently, Pantev and colleagues [38, 39] have shown that attention improves auditory performance in noisy environments by either enhancing the processing of task-relevant stimuli ("gain"), suppressing task-irrelevant information ("sharpening"), or both.

Similar to visual studies, two streams of processing have been identified in the auditory system of monkeys associated with "where" (spatial) and "what" (object or pattern) streams of processing [40]. Kaiser and colleagues began identifying and characterizing the spatial [41] and pattern streams [42] by examining gamma-band responses evoked by a mismatch paradigm. They suggested that the human anterior temporal and ventrolateral prefrontal cortex is involved in the processing of changes associated with the spectral composition of complex sounds ("what" stream) while sound localization ("where" stream) appears to be mediated in posterior parieto-temporal regions. Brunetti and colleagues [43, 44] used fMRI and MEG to examine the functional and temporal dynamics of the spatial stream via a sound localization task. The stimuli used in both studies were audio recordings of a knife tapping on an empty glass (a natural sound) and subjects were instructed to localize the position of the incoming stimuli. In one condition the stimuli consisted of a random sequence of sounds projected from 5 locations while in other conditions the sounds were projected from the left, right, or central directions across the meridian. A network of regions was identified with Heschl's gyrus revealing a peak latency of ~140 ms poststimulus, followed by the superior temporal gyrus at ~155 ms, and then responses localized to the inferior parietal lobe and supramarginal gyrus peaked at ~160 ms. They also found a hemispheric difference where stronger activation was evident in the right hemisphere. Brunetti et al. [44] extended these results by examining the reorienting of attention toward regularly varying sound locations versus sounds presented randomly at different locations on the azimuthal plane. Two MEG sources were seeded in the right hemisphere (inferior parietal and prefrontal cortex) where fMRI routinely revealed activation. The fMRI data suggested that: 1) the supratemporal plane is modulated by variations of sound location; 2) the inferior parietal lobe is modulated by the cross-meridian effect; and 3) the inferior frontal cortex is engaged by the inhibition of a motor response. The MEG data provided the temporal dynamics of this network.

This section concludes with clinical research studies demonstrating how pathology of the auditory system can result in the functional reorganization of auditory cortex. In the first example, Tecchio and colleagues [45] compared the tonotopic organization of primary auditory cortex in otosclerotic patients (i.e., a progressive hearing deficit caused by the anomalous growth of the bone around the annular ligament that locks the stapes footplate to the oval window), pre- and post-surgery, with normal controls. In the pre-surgery recordings, AEF morphology, latency and strengths were comparable to normal control subjects but the tonotopic distributions were significantly different with the mean value of the ECD depth being little affected by stimulus frequency. In post-surgery recordings, the extent of the cortical tonotopic map (based on 4 tonal frequencies) was enlarged with respect to pre-operative measurements and had a tonotopic distribution similar to the control group. In addition, the cortical extent of the tonotopic map correlated with the duration of the post-surgery period. It was concluded that the initial restriction of the cortical tonotopic map, caused by long-term reduction of acoustic input, was followed by a reorganization of auditory cortex during recovery of function which can occur within a few weeks.

The second example focuses on tinnitus, characterized by auditory perception (e.g., ringing in the ear) in the absence of any external sound input. By examining the tonotopic organization of auditory cortex in patients with tonal tinnitus and healthy controls, Muhlnickel and colleagues [46] found a marked shift of the cortical representation of the tinnitus frequency into an area adjacent to the expected tonotopic location. These investigators suggested the possibility of developing a new therapeutic approach by having patients attend to and discriminate features of the acoustic stimuli that are close to the tinnitus frequency to drive cortical reorganization of the non-tinnitus frequencies into the tinnitus representation, in order to reduce the perception of tinnitus. Pantev and colleagues [47] showed that listening to spectrally "notched" music can reduce cortical activity corresponding to the notch center frequency. That is, the removal of a specific frequency from the acoustic environment results in rapid (within hours) reorganization of auditory cortex. Recently, a tinnitus treatment strategy was developed and tested using the spectrally "notched" music approach [48]. For 12 months subjects listened regularly to music notched in the frequency range of each individual's tinnitus frequency. The patient group showed significantly reduced tinnitus loudness.

C. Schizophrenia

Schizophrenia is included here since most of the studies, and most certainly the early studies, examined the functional organization of auditory cortex in schizophrenia patients. The earliest study by Reite and colleagues [49], examined the M100 response in schizophrenia patients. MEG activity was recorded from ~35 locations from each hemisphere while 1 kHz tone pips, at 90 dB sound pressure level, were delivered at the contralateral ear. Male patients with paranoid schizophrenia exhibited less interhemispheric asymmetry of the M100 response compared with normal males. That is, the source of M100 in normal male subjects is more anterior in the right hemisphere, compared with the left hemisphere, while the male schizophrenia patients revealed less of this asymmetry. These results were replicated later by the same group [50]. Several auditory studies found a general sex-specific interhemispheric asymmetry in healthy adults; the M100 source tends to be more anterior in the right hemisphere of males compared to females [51-53]. Since the M100 appears to be generated in or near Heschl's gyrus, Rojas and colleagues [54] quantified the volume, surface area, and 3-D location of Heschl's gyri on MRI in order to determine if there is a sensory cortical reorganization in schizophrenia patients. They concluded that some temporal lobe abnormalities in schizophrenia are sex-specific and that the anomalous lateralization of the auditory evoked field cannot be explained by a shift in the underlying anatomy since the anatomical substrate is lateralized in both healthy control subjects and patients of both sexes. Hajek and colleagues [55, 56] also noted sex-specific differences for M100 in their schizophrenia patients (i.e., right hemisphere alterations were ascribed to female patients and pronounced left hemisphere differences were seen in male patients).

Cañive and colleagues [57] began examining the spontaneous data obtained from schizophrenia patients, both medicated and unmedicated. At the group level, the data also revealed abnormal patterns of alpha activity with respect to both alpha power and alpha frequency. Within the group of schizophrenia patients, one unmedicated patient showed epileptiform sharp waves while four other unmedicated patients showed abnormal bitemporal slow waves. No gross abnormalities were seen in the medicated patients even though three of them did show slow waves when they were in the unmedicated group at one point in time. Interestingly, in a subsequent study that was conducted as part of a multicenter clinical trial to study the efficacy of aripiprazole [58], three patients revealed increased delta and theta activity along with paroxysmal bitemporal slow waves at washout. Cañive and colleagues then began examining the M50 response using an auditory paired-click paradigm (e.g., clicks separated by 500 ms). They replicated EEG findings showing that patient responses to the 2nd click is not as suppressed as it is for normal controls (e.g., [59-61]). This is a robust result that is referred to as a sensory gating deficit. Most important, they determined that the primary generator of the M50 response localized to the Superior Temporal Gyrus [60], that there are additional sources of the M50 response in patients with schizophrenia compared with control subjects [61], and that the M50 gating deficit is more pronounced in the left hemisphere [60] while the M100 gating deficit is bilateral [61].

III. Somatosensory

Following Brenner's first demonstration showing that median nerve stimulation produces activity in contralateral, primary somatosensory cortex (SI–see Figure 2B) and that the magnetic signals were ~2 cm more lateral for thumb versus little finger stimulation [62], Okada and colleagues [63] offered a quantitative demonstration of somatotopic organization of human somatosensory cortex while Hari and colleagues [64, 65] were the first to identify and characterize secondary somatosensory (SII) cortex. In Okada's study, single dipole modeling was used to demonstrate a medial shift of source location for stimulation of the thumb, index finger, little finger and ankle. The estimated depths of the sources placed them near the middle of the Rolandic fissure. Since the early studies, somatotopy maps have been repeatedly demonstrated which are generally consistent with the "homuncular" organization offered by Penfield and Jasper [66], showing a medial to lateral representation of the foot, trunk, hand, lips and tongue [67, 68]. In addition, source localization of contralateral SI activation, evoked by median nerve stimulation at the wrist, was compared using EEG,

MEG and electrocorticogram (ECoG) methods in epilepsy patients. Sutherling and colleagues [69] found good agreement between the different methods (i.e., MEG and EEG localization estimates were similar), with combined MEG and EEG estimates approaching the accuracy of the ECoG.

Several groups began to partially replicate invasive animal studies showing functional reorganization of primary somatosensory cortex following restricted deafferentation (e.g., when the median nerve was transected and ligated) [70]. The invasive animal studies revealed that within 2–9 months after transection, the representation of primary somatosensory cortex was topographically reorganized. The first experimental study examining the functional reorganization of somatosensory cortex was conducted by Rossini and colleagues [71]. They recorded somatosensory evoked fields (SEFs) from one finger of the left hand before, during, and after transient sensory deprivation (3 stages) of the other four digits of the left hand, produced by an anesthetic block. The nerve compound potential remained stable throughout the 3 stages, indicating that the afferent input did not change. The test digit (either the thumb, 3rd, or 5th finger) was stimulated with electric square wave pulses. The SEFs from the unanesthetized finger were characterized by an increase in strength and deepening for the middle finger, and by a shift on the coronal plane for the thumb and little finger, indicative of functional reorganization.

In a Nature article, Flor and colleagues [72] found phantom-limb pain to be a perceptual correlate of cortical reorganization following arm amputation. The amount of cortical reorganization occurring after unilateral upper-extremity amputation was positively correlated with the intensity of the phantom pain. Yang and colleagues [73] emphasize in their Nature article that noninvasive somatosensory pre-surgical mapping is especially important since massive reorganization of sensory maps can occur in the adult human brain. These investigators demonstrated considerable reorganization of somatosensory maps in two adult amputees. The surgical separation of webbed fingers provided a good opportunity to examine cortical reorganization of the hand area. Mogilner and colleagues [74] described the pre-surgical maps as being shrunken without somatotopic hand representations. Within weeks after surgery cortical reorganization for the digits occurred over distances of 3–9 mm. Stavrinou and colleagues [75] experimentally webbed four fingers of healthy subjects for a 5-hour period and monitored the Euclidean distance (ED) between cortical sources evoked by electrical stimulation of the index and small fingers during this time. Before webbing, the ED was in the range of 6.0–13.7 mm. After webbing there was an initial decrease in ED (30 min) before reaching the maximum 3-fold increase in ED (at 2.2 h) with respect to baseline values. After 2-3 hours, the ED began returning toward baseline values. These studies collectively provide ample evidence for cortical reorganization that occurs within hours, days, or weeks.

Similar to the auditory studies, studies of the somatosensory system began mapping networks within the somatosensory system. Hari and colleagues [64, 65] used electrical stimulation of the median nerve at the wrist and the peroneal nerve at the ankle and found contralateral and ipsilateral activation for SII. The field distributions of these responses were consistent with sources on the upper bank of the Sylvian fissure. Kawamura and colleagues [76] set out to determine whether the pre-central cortex contributes to the generation of the early components of somatosensory responses even though EEG studies have not found such a source. In 176 hemispheres the isofield maps indicated an anterior-oriented dipolar pattern for the first peak and a posterior-oriented dipolar pattern for the second peak. Dipoles fitted to the first peak localized on the posterior wall of the central sulcus and the second peak dipoles were localized medial and superior to the first peak dipoles (i.e., pre-central sulcus or motor cortex). This result stimulated some controversy. Although it is generally agreed that the generator of the P35 deflection is more medial and superior than the first deflection, which is generally located on the posterior bank of the central sulcus, it was suggested that an extended sheet of dipoles (i.e., the cortical representation of the median nerve may span several centimeters) fitted by a single dipole model may be an artifact of the modeling procedures [77]. Huang and colleagues [78], using multidipole modeling methods, verified the existence of a source found on the anterior bank of the central sulcus associated with median nerve stimulation and compared it with results from a motor task (i.e., index finger lift). Primary motor-cortex sources localized from an index-finger lift task were either in the same locations or slightly more medial than the locations found on the anterior bank of the central sulcus associated with median nerve stimulation (Figure 5A). It was further suggested that median nerve stimulation tasks may be used in clinical populations where it is difficult to perform motor tests, in order to test the integrity of tissue in motor cortex. Later, this approach was applied to a group of ischemic stroke patients to monitor recovery of sensory and motor function after stroke [79]. It was shown that viability of tissue in primary motor area, in addition to the adjacent primary somatosensory area, can be assessed by the passive median nerve MEG exam (Figure 5B). Changes in MEG source strength of primary motor and primary somatosensory sources significantly correlated with changes in neurological scores of sensorimotor functions, demonstrating the potential value of MEG for studying recovery of sensorimotor functions in stroke patients

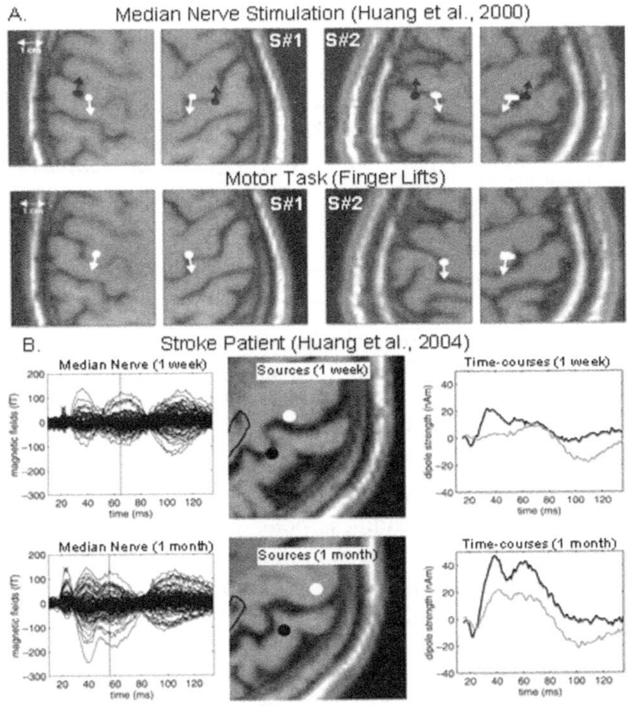

Fig. 5 A. Sources localized to posterior bank of central sulcus (black arrows) and anterior bank of central sulcus (white arrows) for median nerve stimulation. Immediately below, sources localized to motor cortex (white arrows) during a finger lift task are shown for the same two subjects. One hundred Monte Carlo analyses were conducted for all sources to obtain the uncertainty of the source locations. B. Waveforms (left panel), source locations (middle panel), and time-courses (right panel–black=SI and gray=M1) evoked by median nerve stimulation are shown for one stroke patient 1 week following stroke and 1 month following stroke. Adapted from Huang et al., 2000 and 2004

Since an example from a stroke study was presented above, it is worth highlighting that MEG is particularly useful for examining patient populations where the neurovascular coupling (i.e., the process by which neural activity influences hemodynamic properties of the surrounding vasculature) may be altered, compared with hemodynamic-based functional imaging techniques. A warning that the neurovascular coupling may be altered in aging and disease was offered from fMRI investigators (e.g., [80]). However, Rossini and colleagues [81] have clearly demonstrated uncoupling of neuronal activity from fMRI activation that is strongly related to altered vasomotor reactivity. Median nerve stimulation was conducted in stroke patients and control subjects while MEG and fMRI data were acquired. Transcranial Doppler (TCD) was also acquired for evaluating arterial cerebral blood flow. In the patient group, fMRI activation did not reach statistical significance in SI or in other sensory areas of the affected hemisphere in 5 out of 10 patients. In 3 of these 5, fMRI activation did not reach significant levels in the unaffected hemisphere either. In contrast, clear MEG signals were obtained for all patients. It was concluded that the uncoupling between the MEG and BOLD contrast was due to exhausted vasomotor reactivity, which reflects impaired ability of cerebral vessels to adapt to a vasodilatory stimulus.

A final interesting topic for this section on general studies of the somatosensory system is the high frequency oscillations (HFOs) found superimposed on the 20-ms response, evoked by median nerve stimulation [82, 83]. A wide recording passband (0.5-1500 Hz), 6 kHz sampling rate, and a high number of averages (n=4000) were used to identify these responses. A FFT analysis (Figure 6) reveals a major low amplitude peak (A) contributed by the N20m and a second, weaker peak (B) at 600 Hz. A digital high-pass filter allowed Curio and colleagues to isolate the high frequency burst (see Figure 6–right panel) and identify its source in the vicinity of the generator of the N20m response, with a similar somatotopic order as the N20m. Hashimoto and colleagues [84] found that HFOs decreased in amplitude during sleep while the N20m increased in amplitude. Sources of the HFOs and N20m were both in the primary somatosensory cortex (area 3b). The inverse relation between them, however, suggests different neural substrates. Hashimoto and colleagues speculate that the HFOs represent localized activity of the GABAergic inhibitory interneurons of layer 4. Activity underlying N20m is believed to represent activity of pyramidal neurons. Although the significance of HFOs has not yet been determined, this is clearly an area unique to MEG study since similar results are unlikely to be demonstrated by methods relying on the monitoring of hemodynamic responses, given the sluggishness of these responses.

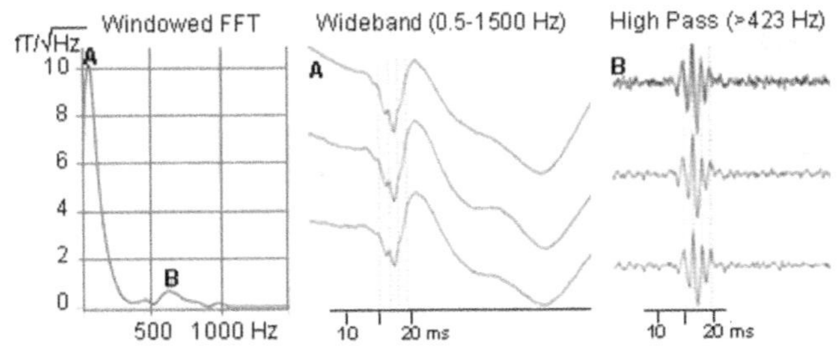

Fig. 6 Left. FFT analysis shows two peaks: (A) reflects large-amplitude N20m response while (B) reflects low-amplitude 600 Hz activity. Middle (A). Wideband response revealing N20m response. Right (B). Filtered response reveals 600 Hz burst of activity that was riding on the N20m (see vertical dashed lines in both plots). Adapted from Curio et al., 1994

A. Mapping Somatosensory Responses Following Painful Stimulation

A number of studies have examined the neural correlates of pain perception with the ultimate goal of differentiating between pathways mediated by thinly myelinated Aδ-fibers (faster conducting) and unmyelinated (more slowly conducting) C-fibers. The special problem in studies of acute pain is to apply stimuli that do not produce tactile-fiber contamination. Early studies in this area used different types of pain stimulation. The first study conducted by Hari and colleagues [85], for example, used painful dental stimulation (i.e., the upper central incisor was stimulated electrically) while others used: 1) painful CO_2 gas applied to the nasal mucosa [86]; 2) painful CO_2 laser beam applied to the skin of limbs [87]; and 3) painful electrical stimulation of the finger [88, 89]. The different studies by Hari and colleagues, Huttunen and colleagues and Kakigi and colleagues all reported SII activation while studies using electrical stimulation of the finger found SI activation for both painful and non-painful stimulation. Peak deflections around 100-150 ms and 250 ms, in contrast, were only evident for painful stimulation. Kitamura and colleagues [90] applied painful electrical stimulation to the lower limb and found similar effects.

Ploner and colleagues [91] used cutaneous laser stimulation and found that contralateral SI and bilateral SII were involved in pain processing and were active simultaneously after 130 ms. Therefore, they suggested a direct thalamo-cortical projection to SII (dual pathways rather than serial activation from SI to SII) thereby emphasizing the quickness of the SII response and its probable role in nociceptive learning and memory. Later, Ploner and colleagues [92] suggested that the sensation of first pain (e.g., typically brief pricking) is mediated by area SI (a brief response that is focal) while the second pain (a longer lasting burning sensation) is mediated by the anterior cingulate (a more generalized response). SII was involved in both pain sensations. This conclusion was reached by comparing timecourses of stimulus perception with the timecourses of stimulus-evoked MEG activity.

More recently attempts have been made to separate out responses to Aδ-fibers from responses to C-fibers since it is difficult to selectively stimulate the C-fiber system without activating Aδ-fibers. Kakigi, Tran and colleagues [93, 94] developed a method for selectively activating the C-fiber system and showed simultaneous activation of SI and SII in the hemisphere contralateral to the site of stimulation followed by ipsilateral SII activation. Forss and colleagues [95] used two different methods of pain stimulation to selectively activate the Aδ-fiber and C-fiber pathways in the same participants. The earliest cortical responses to Aδ-fiber stimulation peaked around 165 ms while responses to C-fiber stimulation occurred later (~800 ms). For both types of stimulation, bilateral activation was evident in SII. However these authors did not find SI activation but instead found activation of the posterior parietal cortex by both types of stimuli. Apparently, distinct nociceptive Aδ- and C-fiber systems mediated noxious input within different time windows in a common cortical network that included at least bilateral SII and posterior parietal cortex. This area of work continues to be active with ongoing debate and is another unique area of study for MEG, i.e., in patients suffering from chronic pain.

IV. Motor

Early reports on motor responses were initially related to the slow magnetic field shift seen before actual initiation of movement. In electrophysiology studies, this readiness potential was referred to as Bereitschaftspotential (BP) and was coined by Kornhuber and Deecke [96, 97]. Deecke and Weinberg and colleagues [98] soon identified the magnetic counterpart of the readiness potential for volitional finger flexions and called it Bereitschaftsmagnetfeld (BM). The BP is a slow bihemispherical negativity over precentral and parietal areas (and the midline) that begins about 1 s prior to the onset of muscle contraction (see EEG tracing in Figure 7). The BM has the same basic waveshape as the BP. A generator of the BM is in the rolandic area of the contralateral hemisphere.

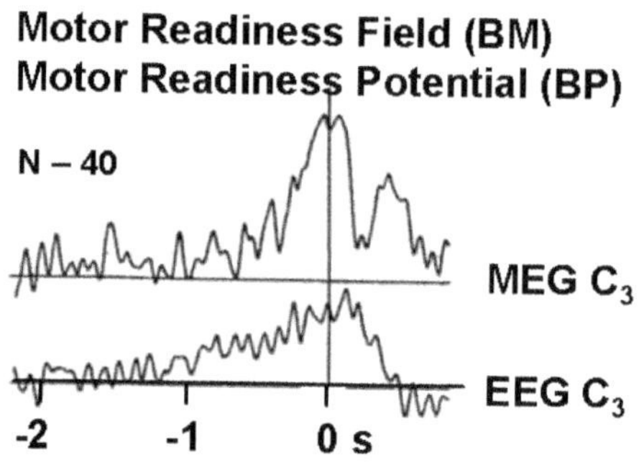

Fig. 7 Averaged MEG/EEG recordings to self-paced right finger flexions. Adapted from Deecke et al., 1982

BPs associated with hand and foot movements have different distributions. While the BP associated with hand movement typically reveals a distribution with maxima over the contralateral hemisphere, BPs associated with foot movement reveals a paradoxical distribution. That is, the BP distribution showed maxima over the ipsilateral hemisphere. In the same year Deecke and colleagues [99] and Hari and collegues [100] used MEG to examine the paradoxical distribution associated with foot movement, reasoning that the potential distributions were paradoxical due to the orientation of the mesial cortical source associated with foot movement. Indeed, the BM distribution revealed two sources in the contralateral hemisphere; rolandic and supplementary motor (SMA) sources.

As mentioned previously, characterization of spontaneous MEG rhythms or oscillatory responses is of great interest since it may represent an underlying mechanism mediating perception per se. Salmelin and Hari [101] localized sources of rhythmic activity in different frequency ranges within auditory, visual and somatosensory projection cortices. They found four main frequency ranges, two of which centered at 10 Hz and two at 20 Hz (see Figure 8) that were concentrated over the rolandic and parieto-occipital regions. Their findings support previous suggestions that the rolandic mu rhythm (a composite of oscillations in the 10 Hz and 20 Hz frequency ranges) is a normal resting rhythm. It was further suggested that during movement the 10 Hz rolandic rhythm seems to be mainly somatosensory in origin whereas the 20 Hz component receives its major contribution from the motor cortex.

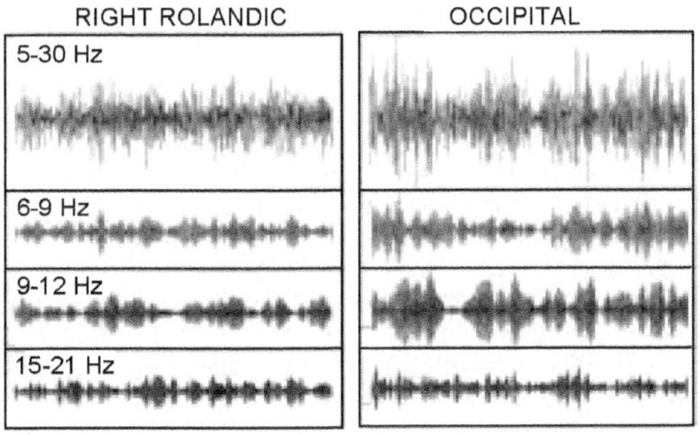

Fig. 8 Spontaneous activity in right rolandic area (left column) and occipital area (right column) of 1 subject. Composite 5–30 Hz signal is shown above with sub-components shown below. Adapted from Salmelin and Hari, 1994

Several investigators have used MEG to study the relationship between rhythmic activities in distributed sensorimotor regions and performance during motor tasks. Invasive monkey studies suggest that oscillatory activity in the range of 25–35 Hz may facilitate interactions between cells during exploratory and manipulative movements [102]. Conway and colleagues [103] and later Salenius and colleagues [104] examined the relationship between localized rhythmic cortical activity (MEG) and correlated motor-unit activity (EMG) during sustained or relaxed isometric contraction. Both of these studies found oscillatory activity (15–33 Hz) originating from the anterior bank of the central sulcus (primary motor cortex) during motor unit firing; however, Salenius and colleagues demonstrated that the cortical signals preceded motor unit firing by 12–53 ms. Salmelin and Hari [105] had already shown that the ~20-Hz component of the rolandic MEG rhythm localizes more anteriorly than the 10 Hz component, suggesting an origin in primary motor cortex. Salenius and colleagues concluded that oscillatory activity in motor cortex may drive the spinal motor neuron pool with a characteristic frequency during isometric contractions.

Hari and colleagues [106] began a series of studies aimed at examining mirror neurons in premotor cortex. Invasive monkey studies had shown that the ventral premotor cortex contains a specific set of neurons that discharge when it performs hand actions and when it observes another individual performing a similar hand action (e.g., [107]). Based on the MEG studies just mentioned above, rhythmic MEG oscillations of ~20 Hz were used as indicators of motor cortex activity. Hari and colleagues [105, 108] reported on 20-Hz oscillations showing suppression and subsequent transient rebound after median nerve stimulation. Suppression during sensorimotor tasks revealed activation of bilateral sensorimotor cortex, but the activation was predominantly reflected in the hemisphere contralateral to the hand performing the task. Three experimental conditions during alternating left and right stimulation of median nerves consisted of: 1) Resting; 2) Acting–subject manipulates a small object with the right hand; and 3) Action Viewing–subject views another person performing similar manipulations. The poststimulus rebound of the rolandic 15–25 Hz activity was strongly suppressed bilaterally during object manipulation (see arrow pointing to the Act condition in Figure 9). The rebound was also significantly suppressed during action observation (View, dotted lines). It was concluded that primary motor cortex is activated during observation as well as execution of motor tasks.

Studies of the mirror neuron system have been shown to generalize to tool use [109] and even when observers saw still pictures of a male face with three different kinds of lip forms [110]. For example, three of the lip forms showed the lip positions when pronouncing three different Japanese vowels. When subjects saw still pictures of lip forms, imitated them or executed them, a sequence of activation beginning in occipital cortex, superior temporal region, inferior parietal lobe, inferior frontal lobe and finally 50–140 ms later, primary motor cortex were activated. It was concluded that still pictures, implying motion, activate the mirror neuron system in a well-defined order. Since subjects with Asperger's syndrome (AS), an autism-spectrum disorder characterized by their limited nonverbal communication skills, are known to be impaired in social interaction and imitation, a group of AS and control subjects were studied in a version of the still pictures of lip form paradigm [111]. Although similar activation sites were found for control subjects and AS patients, activation

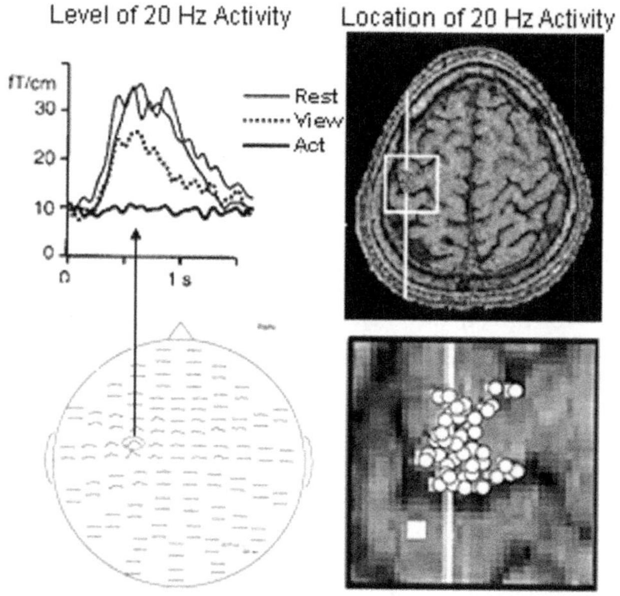

Fig. 9 Left. Level of 20 Hz activity for all conditions are superimposed. Arrow shows sensor region from which the waveforms were derived and points most directly at traces for the Act condition. Right. 20 Hz activity was localized to precentral motor cortex, just anterior to the central sulcus. Dots shown below represent single ECDs used to model the field pattern during single cycles of the 20-Hz oscillation. White square reflects location of source to right median nerve stimulation. Adapted from Hari et al., 1998

of inferior frontal lobe was delayed by 45–60 ms and activations in inferior frontal lobe and primary motor cortex were weaker in AS patients compared with control subjects. And finally, it was shown in another study that rebound occurs after both "seen" and "heard" motor actions (i.e., tapping a drum with right index finger) at about the same time that it occurs after self-performed actions [112]. The mirror neuron system seems to form a basis for understanding the interaction between one's own motor action and the motor actions of others.

V. Visual

The first published study on MEG visual evoked responses was Cohen's IEEE article [113]. In the same year, a Science paper was published by Brenner, Williamson and Kaufman [114] (Figure 2C), followed by another article from Cohen's laboratory [115]. The first two studies used stroboscopic flashes or vertical bar gratings, respectively. The earliest study examining the retinotopic organization of visual cortex (i.e., the systematic mapping between the placement of the stimulus in the visual field and the corresponding projections to primary visual cortex) was conducted by Maclin [116]; the depth of the source (inferred to be primary visual cortex) increased as the stimulus was placed further into the periphery, consistent with the cruciform model. Ahlfors and colleagues [117] also examined retinotopy using small checkerboard octants; they localized sources of activity using a single dipole model and minimum norm methods. Later, Aine, Supek and colleagues used very small stimuli and multidipole modeling to examine retinotopic organization of primary visual cortex [118] and of extrastriate sources [119].

Again, as an initial attempt to partially validate MEG methods in the absence of animal models, the early single-unit studies of the feline visual system served as a rich repository of findings to replicate noninvasively in humans. Studies of feline retinal ganglion cells suggested that the visual system contains two or more classes of neurons that differ in their receptive field (RF) and signal transmission characteristics [120, 121]; these studies laid the foundation for the now popular view of two processing streams in primate visual cortex (dorsal and ventral or the "where" and "what" processing streams), as originally described by Ungerleider and Mishkin [122]. The RF of a neuron is that region of visual space that excites the neuron when stimulated by light. Spatial frequency relates to the amount of detail the cells can process and is inversely related to cell size. Therefore, "Y" type cells in felines (magnocellular in monkeys) have larger RFs, faster conduction velocities, and respond best to lower spatial frequencies and faster temporal frequencies than "X" type cells. "X" type cells (parvocellular in monkeys) have smaller RFs, slower conduction velocities, and respond best to higher spatial frequencies and slower temporal frequencies.

Kaufman and Williamson [123] used MEG to examine the relationship between spatial and temporal frequency in humans. They measured the phase lag of the evoked neuromagnetic response to contrast reversal gratings when the spatial frequency and reversal rate of these gratings were varied. These data showed that for a grating of a particular spatial frequency, the phase of the response is proportional to the temporal frequency of presentation. Consequently, similar to the animal data, there was an increase in the latency of the neuromagnetic response with an increase in spatial frequency content.

Okada and colleagues [124] systematically examined peak-to-peak amplitudes of the visually evoked magnetic fields (VEFs) to varying spatial frequencies, temporal frequencies, and contrast levels using contrast reversal vertical gratings. MEG and EEG responses were recorded and threshold contrasts were determined psychophysically. The peak-to-peak means showed greater amplitude for higher spatial frequencies at low temporal frequencies and conversely, higher temporal frequencies revealed greater amplitudes to low spatial frequencies. The phase lag increased linearly with temporal frequency of the stimulus, similar to results found by Williamson and colleagues [125, 123].

Okada and colleagues also showed that the latency of the VEF increases with higher spatial frequency and decreases when contrast is increased. Although many of these results have been shown previously, this was the first study to quantitatively document the linear relation between the steady-state magnetic field and electrical potential for both phase and amplitude. These data suggested that the VEP and VEF were produced by a common source.

Although many investigators were aware that multiple generators contribute to the peaks in the MEG waveforms even as early as 130 ms [126, 117, 127, 118], a search for the neural origin of the electrical P100 visual response became a focus of several early MEG studies, using single dipole models. Seki and colleagues [128] used a single-dipole model to account for activity occurring within a 90−135 ms time window to pattern reversals of full-field, half-field, and quadrant-field stimulation. In each of these cases, the MEG correlate of P100 localized to the bottom of the calcarine fissure. Other MEG studies focused either on effects of check size on the P100 [129] or on localizing the sources of the different deflections of the visual ERPs; the latter studies showed various localizations around the calcarine fissure for the 1st through 3rd components [130, 131, 128, 132-134]. Similar to the ERP studies, source locations were often variable in these early studies, most likely because response peaks were analyzed using single-dipole models.

A. Dorsal and Ventral Processing Streams

Given the overwhelming evidence in monkeys for the existence of two processing streams (magnocellular/parvocellur or dorsal/ventral streams) [122, 135], many MEG investigators began examining the special functions (e.g., color versus motion) assigned to these streams (e.g., [136-149]) and the timing and properties of these different regions (e.g., [150-155]). Tasks involving face perception or face recognition became popular as a means to access ventral stream functions [156-159, 152, 160, 161]. While many of these studies indicate that the fusiform gyrus is active around 170 ms when discriminating between faces and objects, more recent studies began noting that even occipital cortex seems to differentiate between faces and control stimuli [162-165] and between happy, sad and neutral faces as early as 100−130 ms [166].

B. The Binding Problem and Oscillatory Activity

Contemporary views of higher cognitive functions suggest that information concerning the attributes of stimuli are not stored as a unified percept in a single cortical location; but rather, appear to be stored in a distributed cortical system in which information about specific features are stored close to the regions of cortex that mediate the perception of those features (e.g., [167-169]). Consequently, the binding problem or how features and attributes of stimuli become integrated across widespread cortical regions has been an issue of intense interest and debate in vision research. Recent evidence in animal studies now indicate that 1) local field potentials (LFPs), which provide a measure of mainly postsynaptic dendritic responses, show strong sub-threshold synchrony of ongoing fluctuations in the cell's membrane potentials [170] and 2) coordinated sub-threshold excitability changes have been demonstrated to modulate local networks [13]. From this perspective, induced oscillations are part of different cell assemblies that are activated to temporally bind different stimulus characteristics or to bind the activation of a system of distributed areas necessary for the task at hand. While the specific roles these rhythmic activities play are still debated (i.e., is it an epiphenomenon?), the existence of oscillatory activity is not [171, 172]. Jensen and Tesche [173] were among the first using a visual cognitive task to establish the behavioral relevance of oscillatory activity by showing that theta band activity over frontal regions increased parametrically with the number of items retained in working memory and there was stronger theta during the memory task compared with a control task. There was also a systematic increase in RT with increase in memory load. This study suggests a functional role for oscillatory activity, but the conditions under which gamma, theta, beta or alpha activity is involved as well as their specific roles is still unclear. There are many recent studies examining oscillatory activity in the visual system, too many to be discussed here (e.g., [174-179]). This is clearly a unique field of study for MEG that is capturing considerable interest again; here are several reviews on this topic [180-187].

C. Selective Attention

For the past four decades, behavioral and neurophysiological investigations of selective attention have attempted to specify the functional level in information-processing models in which information is selected for or rejected from further processing [188, 189]. "Early selection" theories (e.g., [190, 191]) have proposed that irrelevant information can be rejected before the semantic analysis of the stimulus, i.e., attention operates at the sensory or perceptual level. "Late selection" theories, in contrast, propose that selection occurs after both a physical and semantic analysis of all stimuli impinging on an organism (e.g., [192]). The question of "early" versus "late" selection in cognitive studies of selective attention translated into the question of how early in the ERP or MEG waveforms do effects of selective attention occur and more specifically, can selective attention influence very early levels of the visual system such as primary visual cortex (V1)? But, in light of recent monkey studies revealing 1) parallel processing streams throughout the visual system and 2) effects of feedback or efferent activity on lower-order regions of visual cortex, "late" processing could actually occur at "early" levels within the visual system (e.g., V1).

Aine, Supek and colleagues [127] demonstrated: 1) attention-related effects in area V1 that occurred later in time (~150 ms) and 2) the attention-related activity had a ~180° change in the net direction of current flow relative to the initial feedforward response around 80 ms (Figure 10A). These two conditions suggested that the attention-related effects were most likely due to feedback from higher-order areas into lower-order visual areas, consistent with a model of selective attention that emphasized the role of efferent activity [193]. Recent ERP and fMRI studies (see [194-196]) support these conclusions. Additionally, fMRI studies (e.g., [197, 198]) routinely show that attention can affect activity at the level of primary visual cortex and invasive studies show that it is via feedback influence which occurs later in time (e.g., [199-201]). This study was the first to suggest, according to S. Hillyard (personal communication), that attention can modulate V1 of humans but it occurs later in time (~130–150 ms) and is likely due to feedback from higher-order areas. MEG provides a unique tool for characterizing the spatio-temporal dynamics of neuronal activity with millisecond resolution that can even be used in certain circumstances (e.g., imagery) to determine feedforward activation sequences from feedback activity [202] (Figure 10B). Unlike many other imaging techniques, MEG can be used for determining net direction of current flow, which may aid in the interpretation of the data.

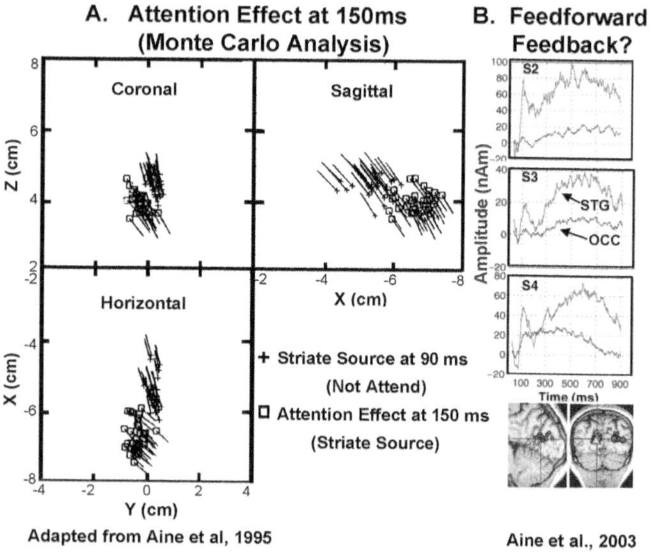

Fig. 10 A. Adapted from Aine et al., 1995. Monte Carlo analyses suggest adjacent locations in primary visual cortex (striate) for the 150 ms difference ECDs (Boxes: Attend – Not Attend) and the 90 ms "not attend" ECD (plus signs). The extent of scatter provides an estimate of error. Symbol "tails" represent direction of current flow. Negative x values represent the posterior half of the head (0 is head center, -10 is near the inion). Positive y values represent the left hemisphere and positive z values represent the top of the head. B. Adapted from Aine et al., 2003. Large amplitude responses represent timecourses from the superior temporal gyrus (STG) when subjects heard words representing common nouns. Smaller amplitude responses were localized to occipital cortex (OCC: see MRIs below) when they heard the nouns. A likely explanation for visual sources during an auditory task is that subjects imagined the objects as they heard the nouns, especially since they had to conduct a size discrimination task when they heard the list the first time. If this is the case, then the OCC activity reflects feedback activation. Putative feedback responses have a ramping behavior rather than revealing initial spike-like activity, as evident in STG sources

VI. Epilepsy

Hughes and colleagues [203] conducted the first study on epilepsy patients by comparing simultaneous EEG and MEG recordings in 10 patients with various types of EEG abnormalities, including epilepsy (three with petit mal seizures and two with probable seizure disorders). Their goal was to determine the clinical usefulness of MEG methods. They found many examples in which the abnormalities normally seen in EEG activity were poorly represented in MEG. In contrast, MEG was better at detecting slow waves associated with tumor, alpha activity and eye blink artifacts. Unfortunately, only one MEG channel was recorded and compared with 5–6 EEG channels in this study.

Barth and colleagues [204-206], reported in a series of Science papers, also compared EEG with MEG and used an animal model to determine the utility of MEG for localizing epileptiform spike activity in human brain. Because the magnetic field is 1) primarily influenced by intracellular axial currents confined to the region of discharging cortex and 2) transparent with respect to the electrically resistive skull, then interictal spikes should be more sharply focused in MEG. In addition, MEG is an absolute rather than differential measure which does not require a reference to a secondary scalp location. The single-channel system was moved over a rectangular matrix of 28 closely spaced points (2 cm) centered on the approximate EEG spike focus for two patients. Six channels of EEG were recorded. Magnetic field extrema were nicely captured using the rectangular recording matrix enabling these investigators to identify with precision the 3-D location of intracortical sources producing epileptiform discharges [204] (Figure 11). In a follow-up article where 17 patients were tested, Barth and colleagues [206] showed two sources, spatially distributed in 8 of these patients (i.e., the 2 sources differed in location and orientation). They were now able to display a fixed

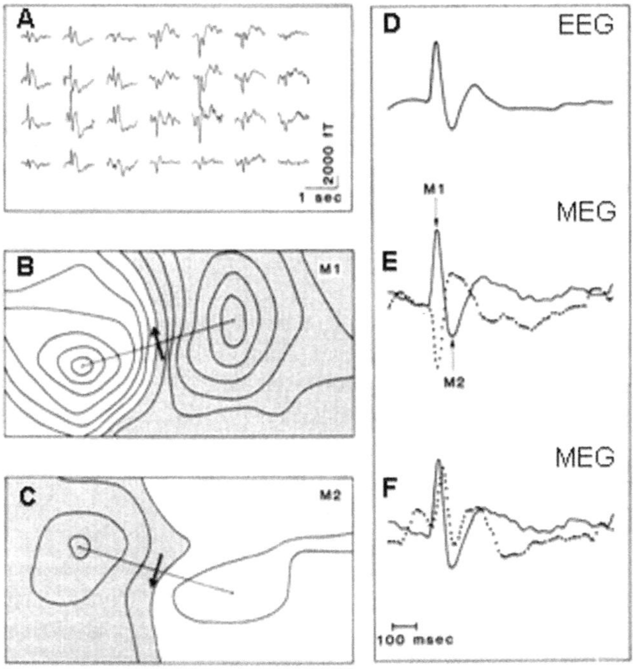

Fig. 11 A. Averaged magnetic spikes recorded from left hemisphere. B and C. Magnetic field distributions of magnetic component 1 (M1) and magnetic component 2 (M2) over left hemisphere. D. Averaged EEG spike recorded from T-4-T6 over right hemisphere. E. Averaged MEG spikes near extrema from left (solid) and right (dotted) hemispheres. F. Averaged MEG spikes from left (dotted) hemisphere is delayed by 20 ms when compared to averaged MEG spike from right hemisphere. Adapted from Barth et al., 1982

chronological sequence of discharge between two generators in the temporal lobe.

In order to obtain MEG measurements of the magnetic phenomena directly during seizure activity, an animal model was used where seizure activity was induced in rats and experimentally verified with ECoG [205]. The injection of penicillin into rat cerebral cortex produced the stereotypical focal epileptic spiking containing both fast and slow magnetic phenomena normally associated with focal epilepsy. In all animals the initial spike recorded by MEG (single channel) was similar to that of the ECoG but the later components differed in morphology, similar to that reported in humans. In addition to the fast spiking activity a steady magnetic DC shift from preseizure baseline was evident in all animals. The steady field shift began at the appearance of afterdischarge, reached a maximum at the onset of rapid seizure spiking, and dropped abruptly to baseline at the end of the seizure. This slow focal seizure phenomenon has not been mapped noninvasively in humans; but it has been seen during surgery. For example, a study by Ikeda and colleagues [207] recorded subdural and scalp EEG activity in three epilepsy patients in order to characterize ictal DC shifts in epilepsy patients. Subdural recordings revealed ictal DC shifts in 85% of all recorded seizures compared to scalp EEG recordings which showed ictal DC shifts in only 23% of the recorded seizures. In addition, scalp recorded DC shifts were identified only when the seizures were clinically intense. Therefore, the recording of DC magnetic shifts is another unique capability for MEG.

MEG localization of interictal epileptic foci continues to be an area where MEG can contribute clinically. As noted by an early review by Rose and colleagues in Science [208], other noninvasive methods such as EEG, PET, MRI, and CT often yield inadequate information for accurately determining the location of the epileptogenic focus within the brain while invasive diagnostic procedures (i.e., recordings from the surface of the brain via ECoG and use of depth electrodes) have risks associated with them and are costly and time consuming. If the presurgical candidate can have the epileptogenic region surgically removed then the patient may become seizure-free. A case study by Hari and colleagues [209] showed how MEG's good spatial localization accuracy combined with temporal precision was used to differentiate between primary and secondary epileptic foci in homotopic areas. As a final note, presurgical evaluation is especially important in children, where functional reorganization is expected to go hand-in-hand with development.

VII. Spreading cortical depression (SCD)

As mentioned above, Barth and colleagues [205] found a slow magnetic DC shift in addition to fast spiking activity in their animal model of epilepsy (rats), an effect that was not demonstrated before using noninvasive methods. Spreading cortical depression (SCD) is also evidenced as a slow DC shift and has been associated with migraine, along with similar phenomena (e.g., peri-infarct depolarization) found in cerebral ischemia, subarachnoid hemorrhage and cortical trauma). It has been proposed that SCD begins with a cellular efflux of K+, resulting in a wave of neuronal depolarization (i.e., cortical hyper-excitation) propagating away from the initiation site at a rate of 3-4 mm/minute, followed by suppression (see reviews [210-212]). At the scalp or MEG sensor level, the suppression is seen as a reduction of the fast activity of the EEG/MEG signals and a reduction in the cortical DC potential. Okada and colleagues [213] recorded SCD in turtle cerebellum in their animal model for migraines, while Gardner-Medwin et al. [214] witnessed SCD in their rabbit preparation and Barkley and colleagues [215] detected SCD in humans during migraine headaches. Recognizing the potential clinical utility of examining DC shifts commonly associated with epilepsy, migraines, stroke and head trauma, Tepley and colleagues [216, 217] fine-tuned their procedures for imaging DC shifts while Bowyer and colleagues [218, 219] validated the MEG measurements made in humans during migraines. That is, they successfully predicted the behavior of SCD using both rabbit and gyrencephalic swine models. MEG signals associated with SCD were small in amplitude when starting at the initiation site of induced SCD and were maximal when propagating within the fissural wall of cortex. At this point it stopped without crossing the longitudinal fissure, as predicted and supported by previous *in vitro* studies (Figure 12). The more complex brain model (swine) showed the spreading depression propagation slowed through the sulcal folds. Next, a whole head MEG system was used to demonstrate the detection of DC shifts similar to those detected in animal models during SCD [220]. And finally, Leistner and colleagues [221] demonstrated the feasibility of using DC-MEG in patients with stroke.

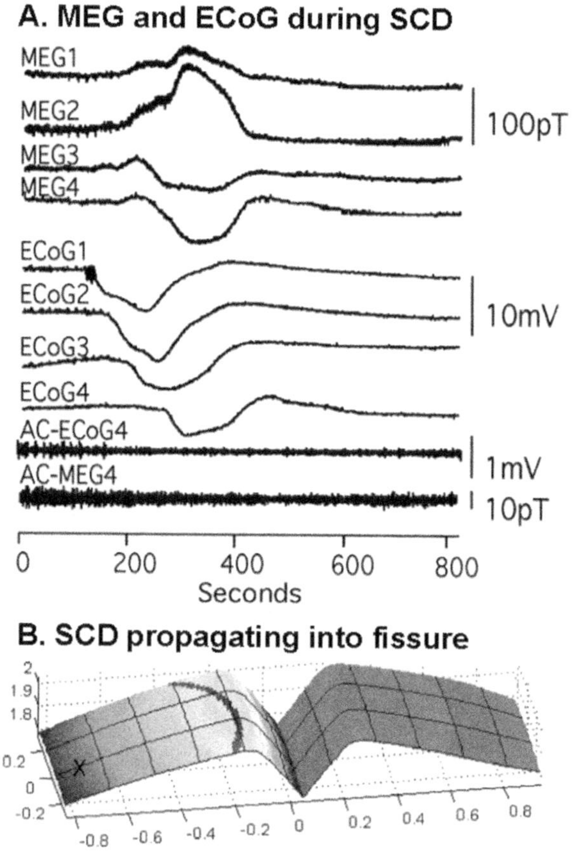

Fig. 12 A. MEG and ECoG traces for SCD entering longitudinal fissure from left side of rabbit cortex. Note polarity reversal across MEG channels. B. Computer model of SCD in cortical strip. Partial arc (black) represents SCD wave propagating into longitudinal fissure. Adapted from Boyer et al., 1999a

VIII. LANGUAGE STUDIES

Characterization of normal language functions, in addition to disabilities in language function, is another unique opportunity for MEG methods since: 1) there is considerable intersubject variability regarding the representation of language (e.g., [222]); 2) functional reorganization of cortical networks representing language functions may occur due to surgical or cognitive intervention techniques (e.g., [223, 224]); and 3) deficits in processing may be due to differences in timing between cortical regions acting as a functional circuit rather than frank differences in source locations. Salmelin and colleagues [225] began mapping the network of brain regions and the timing across these regions (Nature article) when subjects were engaged in a picture naming task, similar to the Boston Naming Task. They found similar activation sequences in both hemispheres, although left hemisphere sources were active earlier than right hemisphere sources. These general regions have been found to be active across many language tasks (e.g., occipital cortex, followed by temporo-parieto-occipital junction, auditory cortex and a region consistent with Broca's area). A couple areas of active research are discussed below.

A. Presurgical Mapping of Language Dominance

One noninvasive clinical application unique to MEG is the identification of sensory and language functions in the brain in order to facilitate surgical planning for removal of tumors or epileptogenic tissue and to reduce the risks of language deficits. This section pertains to efforts aimed at replacing the invasive Wada procedure used to determine hemispheric dominance for language with noninvasive MEG procedures. The Wada cannot provide information regarding the precise location and extent of language-specific cortex and it is an invasive procedure that has risks associated with it [226]. The goal is to map receptive language-specific cortex using simple language tasks (e.g., auditory and visual word recognition) and to compare the results with results from the Wada procedure, the "gold standard," to determine whether MEG can correctly identify the language-dominant hemisphere [227, 228]. The Wada procedure involves the sequential injection of barbiturates into the left and right internal carotid arteries causing temporary arrest of functions, which is characterized by administering simple language-related commands and tests (e.g., comprehension, naming objects/parts of objects, repetition of simple phrases) [229]. A hemisphere is classified as dominant when its injection results in the disruption of language in more than one test. Several studies have examined the reliability and validity of MEG methods for determining language dominance by comparing results with Wada tests in the same patients [230-232, 229] or through direct cortical stimulation mapping [226, 233]. According to Papanicolaou and colleagues [229], the MEG alternative to Wada procedures for determining the language dominant hemisphere was in complete agreement in 74 out of 85 cases (87%) with 11 discordant cases. But in the majority (7) of the discordant cases, MEG detected considerable activity in both hemispheres when the Wada suggested left-hemisphere language dominance. Maestú and colleagues [232] similarly found good agreement in 7 out of 8 patients. MEG results for the 8th patient suggested right-hemisphere dominance when the Wada procedure did not show clear hemispheric specialization for language. Papanicolaou cautions, however, that while MEG appears to be an acceptable substitute for the Wada for the purpose of establishing which hemisphere is language-dominant, the Wada procedure is also used to determine the relative importance of each hemisphere for memory function [229]. MEG offers a reasonable substitute in cases where memory is not an issue.

B. Dyslexia

Dyslexia is a specific disability in learning to read which cannot be attributed to diminished intellectual capacity or disadvantaged sociocultural opportunities. Early anatomic studies suggested alterations in the temporal bank of the left Sylvian fissure [234, 235], suggesting low-level auditory cortex involvement perhaps related to phonological processing. Suspecting that the N100m evoked by auditory speech sounds plays a role in phonological processing, Helenius and colleagues [236] assessed auditory cortical activation patterns to natural bisyllabic pseudowords, complex nonverbal sound pairs, and 1 kHz tones in dyslexic and non-reading-impaired control subjects. The N100m originating in the left supratemporal cortex was abnormally strong in dyslexics only for vowels in the beginning of pseudowords. These investigators concluded that the supratemporal auditory cortex appears to be coding stimulus-specific features critical for speech perception. Consistent with these results, Heim and colleagues [237, 238] also found differences between dyslexics and control subjects for the N100m source evoked by a passive oddball task using syllables. Their results showed the N100m to be asymmetrical in normal controls with a more anterior localization in the right perisylvian cortex, compared with left. In contrast, the dyslexic group revealed more symmetrical source localization across hemispheres. Therefore, they suggested that the lack of planum temporale asymmetry in dyslexics may be the basis for their problems. It was concluded that dyslexics and normal control children differ in the organization of their left hemisphere auditory cortex. Nagarajan and colleagues [239], using rapid presentations of pairs of tones consisting of two frequencies (high=1.2 kHz and low=800 Hz) and four possible pairings (high-high, high-low, low-high, and low-low), found differences in primary auditory cortex and its immediate surrounding areas between adults who are poor readers and normal controls. These differences correlated with task performance. Two of the physiological findings were: 1) the amplitude of responses to the second stimulus of paired stimuli were reduced for short ISIs (100 and 200 ms) when compared with normal controls; and 2) the average distributed response coherence evoked by the rapidly successive stimuli was weaker in the beta- and gamma-band frequency ranges (20-60 Hz) in poor readers, compared with controls. These investigators suggest that reading disabilities are correlated with the abnormal representation of brief and rapidly successive sensory inputs, at the entry level of the cortical auditory/aural speech representational system(s).

Salmelin and colleagues [240] examined the visual presentation of words (passive viewing of concrete/abstract words, pronounceable pseudowords and letter strings) and found differences between control and dyslexic adults 0–200 ms after word presentation in left inferior temporo-occipital cortex (word form area) followed by differences 200–400 ms in the left temporal lobe. Differences were also noted in the left inferior frontal lobe. It was concluded that the coherent response to letter strings observed in control subjects at ~180 ms after word presentation was absent or delayed in the left hemisphere of dyslexics. The source localized to the left temporal lobe (200–400 ms) was also reduced in strength in the dyslexics suggesting impairment in phonological processing as well. Instead, left inferior frontal lobe activation was seen during this 200–400 ms time interval in the dyslexics, possibly suggesting a compensatory top-down process.

Simos and colleagues [241] examined reading in dyslexic children but they used a word recognition task instead of the passive viewing of words, as used by Salmelin and colleagues. Initial left inferior temporal activation was essentially equivalent between dyslexic and control groups but differences were evident at the next stage of temporal-parietal area involvement. The left temporal-parietal area was active in controls while the right temporal-parietal area was active in dyslexics (i.e., there was reduced left temporal-parietal activation in dyslexics). They suggested that the functional connectivity between regions normally involved in reading was aberrant and that temporal-parietal regions are most likely involved in phonological decoding. When reading comprehension was compared between dyslexic adults and control subjects (sentences were presented one word at a time and ended with a word that was either semantically appropriate/inappropriate or expected/not expected), the same general regions of activation were found for the two groups but differences existed between timing and strength of the responses; in particular the onset latency of the N400m deflection was delayed in dyslexics compared with control subjects [236].

C. Autism Spectrum Disorders (ASD)

Autism is a neurodevelopmental syndrome that is defined by deficits in social interaction and communication. Distinctions among the ASDs can be made according to the degree of accompanying language deficit, general cognitive delay, or according to the severity of social or behavioral symptoms [242]. A recent review of MEG studies in this area is offered by Roberts and colleagues [243]. A brief discussion of Asperger's syndrome was included under the Motor section since there is no significant delay in language development for this syndrome. Instead there is a lack of nonverbal communication skills. Since approximately 30% of autistic patients also have an associated seizure disorder [242], Lewine and colleagues [244] compared ASD children with children diagnosed as Landau-Kleffner syndrome (LKS), a childhood neurological disorder characterized by sudden or gradual loss of the ability to understand and use spoken language and an abnormal EEG. The question was whether the same brain regions were involved in both disorders. MEG was used to evaluate patterns of epileptiform activity during state III sleep in 6 children with classic LKS and 50 children with regressive ASDs (16 with autism and 34 with pervasive developmental disorder (PDD) not otherwise specified). All children with LKS showed primary or secondary epileptiform involvement of the left intra/perisylvian region (and one child showed involvement of right sylvian region). MEG identified epileptiform activity in 82% of children with ASDs and in 85% of these cases the epileptiform activity was in the same intra/perisylvian regions seen in LKS children. Seventy-five percent of the ASD showing epileptiform activity also revealed additional nonsylvian zones of independent epileptiform activity, unlike the LKS children. A recent study replicated these findings using wakefulness MEG and EEG recordings. Muñoz-Yunta and colleagues [245] identified subclinical epileptiform activity in 86% of children with PDD, which was evident predominantly in the perisylvian regions, especially for ASD children. An earlier case study by Paetau and colleagues [246] suggested that unilateral discharges at or near the auditory cortex disrupts auditory discrimination in the affected hemisphere and leads to suppression of auditory information from the opposite hemisphere, thereby resulting in verbal auditory agnosia. All of these results are in good agreement in showing epileptiform activity in the same or similar brain regions associated with verbal auditory agnosia.

Later studies in this area tended to focus more on the auditory M100 evoked field to investigate maturational changes via measures of latency and frequency modulation in the cortical auditory systems in left and right hemispheres of children with autism disorder [247-249]. For the autism versus control groups, these investigators report generally longer M100 latencies in both hemispheres and a reduced dynamic range of frequency modulation in the right hemisphere but no differences have been found for the earlier M50 component between autism disorder and control groups. The transient auditory gamma band response (40 Hz) has also been examined in 11 adults with autism, 16 parents of children with autism, and 16 control participants [250]. Briefly, they found that induced gamma-power was significantly higher in 1st degree relatives and autism groups, compared with control subjects, while evoked gamma-band power was reduced. Finally, because ASD is characterized by impairments in reciprocal social communication, MEG has been used to compare neural responses to images of faces between ASD and control groups and face and gaze processing [251-253].

IX. Aging and Alzheimer's Disease (AD)

Early studies of aging examined the auditory evoked fields using tone pips and found that the amplitude of the P50m, unlike the N100m, was larger in older subjects [254]. In both groups, the peak latency of P50m and N100m were shorter over the contralateral hemisphere; however, in older subjects the interhemispheric latency difference of the N100m was increased. In AD patients, compared with control subjects, the peak latencies of P50m and N100m responses were significantly longer over ipsilateral but not contralateral auditory cortex [255]. Ahveninen and colleagues [256] used scopolamine in elderly subjects to examine 'preattentive' processing (i.e., P50m and N100m) since scopolamine is often used as a pharmacological model of dementia (i.e., it produces transient cognitive deficits similar to those observed in AD). The auditory evoked 40-Hz magnetic response was significantly larger after scopolamine administration compared to glycopyrrolate administration, suggesting modulation of preattentive processes. Osipova and colleagues [257] found that the steady-state response to 40-Hz stimulation is enhanced in AD compared with age-matched control subjects. One interpretation of their data is impaired adaptation of auditory neurons to repetitive stimulation. Pekkonen and colleagues [258] injected 9 non-demented elderly with scopolamine and glycopyrrolate to determine if it affected preattentive processing only or if it also affected the mismatch negativity (MMN) which has been associated with a later stage of auditory processing (i.e., short-lived sensory memory trace) and/or the P200m. Scopolamine affected the electric P50, electric and magnetic N100m but did not affect the subsequent MMN and P200 responses. The investigators suggest that the cholinergic system modulates preattentive auditory processing (stimulus detection) in aging and that the N100 response in particular offers an objective tool to monitor cholinergic activity in aging and AD.

Aine, Kovacevic and colleagues [259, 260] also noted an increase in amplitude for early activity in elderly (40–50 ms) for repetition of tones and words (i.e., in an auditory incidental verbal learning task). In an auditory oddball task, amplitudes at ~40 ms were enhanced to both rare and frequent tones in the elderly, compared with the young. As for word stimuli, there was a main effect of amplitude (~50 ms) for superior temporal gyrus sources in the elderly, relative to the young (i.e., both old and new words evoked larger amplitudes at 50 ms relative to the young). For the late activity (~100–800 ms) there was an age by task interaction which was specific to old and new words (i.e., the young showed reduced amplitudes throughout the entire interval when a list of old words was heard 3 times whereas the elderly did not). In contrast, when the list was presented the first time or when a list containing old/new words was presented and subjects heard the new words, the amplitudes of all response peaks (30–900 ms) did not differ between young and elderly. It is as if the responses in the elderly were not suppressed with repeated exposure as they were in the young. These results suggest that effects of stimulus repetition are sustained across a long interval of time representing a continuum of preattentive to attentive processes. However, it is still likely that different mechanisms mediate these different processes as suggested by Pekkonen and colleagues [258].

The first study to examine spontaneous activity in AD using MEG was Berendse and colleagues [261]. MEG was recorded and spectral power and coherence were calculated for four conditions: eyes open, eyes closed, recognition memory test for words and the Boston Naming test. They found a general increase in low frequency power coupled with a decrease in high frequency power values in AD patients. Mental task performance in AD patients failed to produce significant changes in MEG power. And finally, there was a generalized absolute reduction of coherences for both far and local coherences as well as frequencies ranging from 2 to 22 Hz.

Maestú, Fernandez and colleagues began a number of studies with a goal toward early diagnosis of Mild Cognitive Impairment (MCI) and probable AD. In a working memory study, Maestú and colleagues [262] used MEG and magnetic resonance spectroscopy (MRS) to examine 10 patients with probable AD and 10 control subjects. They suggested that the combined use of measures (e.g., reduced dipole density or DD over left temporoparietal areas at 400–800 ms and a myoinositol/N-acetyl-aspartate ratio) accounted for 65% of the variance noted on the Mini-mental State Examination. In another study, Maestú and colleagues [263] examined 15 healthy subjects but after a 2 year follow-up, 5 of these subjects converted to MCI; these individuals revealed lower DD in left medial temporal areas between 400–800 ms, compared with the group that didn't show cognitive decline. In another study [264] they found higher activity levels bilaterally in the ventral pathway in MCI patients relative to controls when equivalent performance levels were achieved across groups.

Fernandez and colleagues [265-267] took a different approach by recording 5 min of rest with eyes open and localizing slow waves using a single dipole model. They found that DD was enhanced in the AD group and that the slow wave activity was greater in temporoparietal regions of both hemispheres in the AD group. In addition, when left hippocampal volume was correlated with left temporal and parietal delta DD and left temporal theta DD, correct classification of AD was made in 87.1% of the cases [266]. In a similar study, the MCI group was divided into two groups, those with higher left parietal DD and those without. In a 2-year follow-up, they found that the group with higher DD in left parietal cortex was at a much greater risk for developing AD. Osipova and colleagues [268] applied Minimum Current Estimates (MCE; [269]) analysis to 2 min of resting eyes-closed data and showed a decrease of alpha band activity in the parieto-occipital sulcus in non-medicated AD patients and an increase in temporal regions (i.e., an anterior shift in sources). The increase in temporal regions was either associated with upper theta or lower alpha activity (i.e., 6-12.5 Hz) thus they considered their results to be in line with Fernandez and colleagues [265]. A subsequent study from this group did not find deficits in the alpha source distribution in MCI patients compared to healthy controls [270].

Aine, Stephen and colleagues [271, 272] also noted abnormally high amplitudes in anterior temporal regions for MCI/AD patients during an auditory word recognition task and for MCI patients in somatosensory cortex following median nerve stimulation. In the first case, several of the MCI/AD patients did not reveal activity in the anterior temporal lobe, a region where high functioning normal controls revealed activity. However, of the MCI/AD patients who did reveal activity in this region, the amplitudes of the time-courses were higher than for the control subjects. In addition, the high amplitude responses in anterior temporal lobe of MCI/AD patients correlated with the fact that many of the patients had abnormalities on their MRIs as reported by the neuroradiologist (i.e., either moderate to severe volume loss or moderate to severe white matter hyperintensities). The high amplitude responses and MRI abnormalities also correlated with poor task performance and poor performance on many of the neuropsychological tests. In the study examining SI responses to median nerve stimulation by Stephen et al. [272], MCI patients revealed abnormally high amplitudes compared with normal controls. AD patients, in contrast, revealed lower amplitude time-courses in SI compared with MCI and control subjects. It is as if some optimum level of activity is necessary which correlates with general well-being as evidenced by performance on neuropsychological tests whereas too much activation is negatively correlated with performance. Recently, this group has been trying to separate effects of normal aging, complete with MRI abnormalities associated with hypertension and type 2 diabetes, from healthy successful aging where memory deficits are not evident between young and elderly groups [273, 271].

X. Fetal, neonatal and infant sensory recordings

MEG provides a direct, noninvasive method for identifying and characterizing brain development intrauterine. In contrast, electrical measurements are typically made during labor after rupture of the fetal membranes while other imaging techniques such as fMRI, PET and SPECT pose potential hazards to the fetus. The first magnetoencephalographic recordings of the human fetus were made by Blum and colleagues using a single-channel magnetometer [274]. Three hundred auditory tone bursts (1 kHz, 100 ms duration, 40–100 dB HL) were transmitted through the mother's abdominal wall at 34 and 35 weeks gestation. Recordings were also made postnatal to help insure the deflection seen at 135 ms prenatally was a feature of the auditory evoked field. Postnatally, the 135 ms deflection was twice the size, compared with prenatal results. Wakai and colleagues [275, 276] measured fetal and neonatal auditory evoked fields and showed how difficult it is to make these measurements since the success rate of obtaining good fetal recordings was low (e.g., 29–54%) due to environmental interference (e.g., low signal amplitude, cardiac interference from both the fetus and the mother, fetal position relative to probe placement and fetal movement). Nevertheless, they found reliable triphasic waveforms with the largest deflection around 250 ms; the latency of this peak response decreased with gestational age. Their more recent study showed a higher percentage of good recordings and dipole patterns that were consistent with a source in the auditory cortex of the fetus. Several studies followed as investigators continued acquiring auditory evoked responses [277, 278], visual evoked responses [279] and spontaneous recordings upon which spectral analyses could be conducted [280, 281]. However, issues concerning technical difficulties in obtaining spontaneous recordings sparked some debate [282]. As a follow-up to a general finding that early peak deflections decrease in latency with an increase in age, Roberts and colleagues [283] recently demonstrated a negative correlation between the integrity of the white matter tracts (e.g., white matter diffusion anisotropy increases with age) and the latency of the auditory N100m (i.e., the latency decreases with age), in a group of children.

As noted by Lengle and colleagues [276], signal strengths in neonates were about 10 times greater than signals from the fetus. Several investigators began focusing on characterizing the maturation of the somatosensory system in neonates and infants. Gondo and colleagues [284] conducted the first MEG study of the somatosensory system of sedated infants, using tactile stimulation (thumb and ring finger). Three deflections were seen with the most prominent deflection having a latency of ~100 ms. The latency of the early deflection, which appeared to correspond to the N20m in adults, decreased with age (6–8 months vs. 11–21 months). In addition, the distances between ECDs of the thumb and ring finger were greater than that observed in adulthood, suggesting more extensive projection fields in infancy. MRI was obtained from one infant; ECDs for this infant were located along the postcentral gyrus. Interestingly, they recorded SEFs according to developmental milestones (e.g., palmar grasp versus pincers grasp stages) and found that the SEFs of the hands changed when the infants attained a new hand skill (i.e., functional reorganization).

Pihko and colleagues [285] examined the somatosensory evoked potentials (SEPs) and SEFs, evoked by tactile stimuli in newborns (1–3 days after birth) during active and quiet sleep cycles. A N1 deflection (~30 ms) was only seen in active sleep and only in a few subjects; therefore, this group emphasized the importance of monitoring the sleep stages in event-related responses in newborns. Lauronen and colleagues [286] used tactile stimulation (index finger) to examine SEFs in newborn, 6-mo old infants, and adults. Median nerve stimulation was also examined in newborns. When the median nerve responses in the newborns were compared with adult responses in the literature, the activation sequences differed markedly (e.g., in neonates, the direction of current did not reverse after the M30, a subsequent deflection of the same polarity, M60, appeared instead). In newborns and 6-mo old infants, all early tactile evoked-responses within the first 100 ms could be modeled by current sources pointing anteriorly which was clearly different from the tactile responses recorded in adults. They concluded that the cortical networks for processing somatosensory input by newborns and infants apparently function in a different way than in adults. Nevalainen and colleagues [287] provided new evidence concerning the network of cortical generators of the somatosensory system in newborn. As shown previously, the source of the M60 corresponds to the contralateral SI. In addition, the M200 is most likely associated with the contralateral SII, and ipsilateral activation of SI and SII was found in some babies. Pihko and colleagues [288] expanded the age groups to evaluate the maturation of tactile processing (e.g., newborns, 6 months, 12–18 months, 1.6–6 years, and adults). They found an orderly systematic change in the waveforms and ECD time-courses which suggested that adult-like responses for M30 and M50 components were present by 2 years of age. As emphasized by these authors, MEG has distinct advantages over EEG for studying cortical activity in infants since MEG is not affected as much by variations in skull thickness and conductivity or distortions introduced by fontanels and sutures in the skulls of infants.

XI. Concluding remarks

This review covers only a small fraction of all the basic and clinical research that has taken place over the past 40 years. The intent was to provide a brief overview of some of the historical research moments and popular research areas. In order to make the closing points it seems only fitting to highlight one more "Research News" article from Science entitled "Biomagnetism Attracts Diverse Crowd" [289] (Figure 13). This article begins with a historical note by announcing that 40 scientists gathered in 1976 at the First International Conference on Biomagnetism at MIT. This article was published following the 7th International Conference on Biomagnetism at NYU in 1989 so the intent was to show how much this field had grown during its first 20 years of existence [Currently, the meeting draws around 600 participants]. Some of the topics discussed at the NYU meeting were highlighted by Mr. Crease such as "Spreading cortical depression in migraine patients," by a team that included neurologists and physicists, and "The different responses of the visual cortex when processing visual and linguistic tasks of comparable mental load," by a team of psychologists and physicists. Crease was impressed that "these strange interdisciplinary bedfellows" could work together toward common goals in this field. This is precisely what makes this field enjoyable to work in; the diversity of thought, training, and its people.

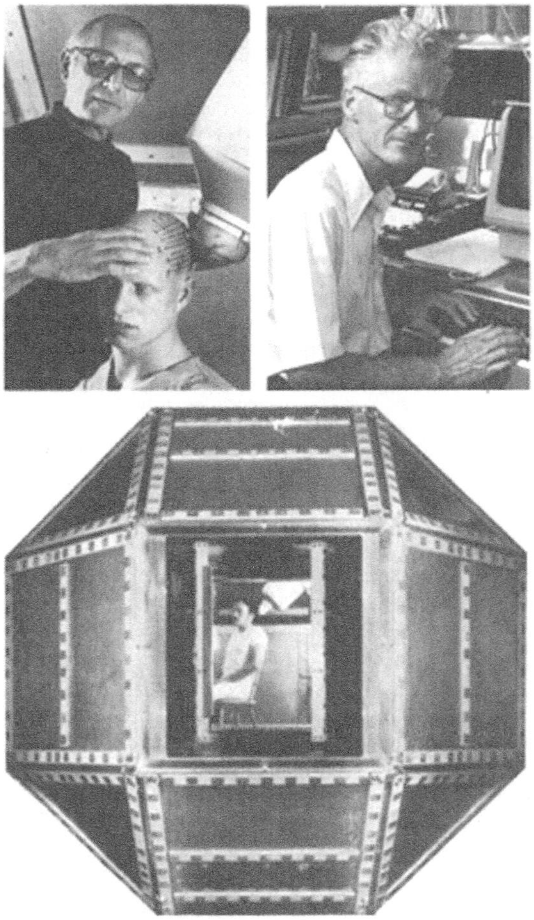

Fig. 13 Left. David Cohen. Right. James Zimmerman–SQUID developer. Bottom. First shielded room at MIT. Adapted from Crease, 1989

Acknowledgments

Special thanks go to Prof. Drs. Riitta Hari, Gian Luca Romani, Christo Pantev, and Selma Supek for their insightful comments and support throughout the revisions of this manuscript. Thanks go to Drs. Lori Sanfratello and Julia Stephen as well. This work was supported in part by NIH Awards R01 AG020302-05, R01 AG029495-02 and 1P20 RR021938-02.

References

1. Cohen D (1968) Magnetoencephalography: evidence of magnetic fields produced by alpha-rhythm currents. Science 161: 784-786.
2. Cohen D (1972) Magnetoencephalography: detection of the brain's electrical activity with a superconducting magnetometer. Science 175: 664-666.
3. Reite M, Edrich J, Zimmerman J T et al. (1978) Human magnetic auditory evoked fields. Electroencephalogr Clin Neurophysiol 45: 114-117.
4. Farrell D E, Tripp J H, Norgren R et al. (1980) A study of the auditory evoked magnetic field of the human brain. Electroencephalogr Clin Neurophysiol 49: 31-37.
5. Elberling C, Bak C, Kofoed B et al. (1980) Magnetic auditory responses from the human brain. A preliminary report. Scand Audiol 9: 185-190.
6. Hari R, Aittoniemi K, Järvinen M L et al. (1980) Auditory evoked transient and sustained magnetic fields of the human brain. Localization of neural generators. Exp Brain Res 40: 237-240.
7. Merzenich M M, Brugge J F (1973) Representation of the cochlear partition of the superior temporal plane of the macaque monkey. Brain Res 50: 275-296.

8. Merzenich M M, Knight P L, Roth G L (1975) Representation of cochlea within primary auditory cortex in the cat. J Neurophysiol 38: 231-249.
9. Romani G L, Williamson S J, Kaufman L (1982) Tonotopic organization of the human auditory cortex. Science 216: 1339-1340.
10. Pantev C, Hoke M, Lutkenhöner B et al. (1989) Tonotopic organization of the auditory cortex: Pitch versus frequency representation. Science 246: 486-488.
11. Gray C M, König P, Engel A K et al. (1989) Oscillatory responses in cat visual cortex exhibit inter-columnar synchronization which reflects global stimulus properties. Nature 338: 334-337.
12. Pantev C, Makeig S, Hoke M et al. (1991) Human auditory evoked gamma-band magnetic fields. Proc Natl Acad Sci U S A 88: 8996-9000.
13. Engel A K, Fries P, Singer W (2001) Dynamic predictions: oscillations and synchrony in top-down processing. Nat Rev Neurosci 2: 704-716.
14. Mäkelä J P, Hari R (1987) Evidence for cortical origin of the 40 Hz auditory evoked response in man. Electroencephalogr Clin Neurophysiol 66: 539-546.
15. Lu Z L, Williamson S J, Kaufman L (1992) Behavioral lifetime of human auditory sensory memory predicted by physiological measures. Science 258: 1668-1670.
16. Lu Z L, Williamson S J, Kaufman L (1992) Human auditory primary and association cortex have differing lifetimes for activation traces. Brain Res 572: 236-241.
17. Holden C (2001) How the brain understands music. Science 292: 623.
18. Abbott A (2002) Music, maestro, please! Nature 416: 12-14.
19. Maess B, Koelsch S, Gunter T C et al. (2001) Musical syntax is processed in Broca's area: an MEG study. Nat Neurosci 4: 540-545.
20. Patel A D, Balaban E (2000) Temporal patterns of human cortical activity reflect tone sequence structure. Nature 404: 80-84.
21. Pantev C, Oostenveld R, Engelien A et al. (1998) Increased auditory cortical representation in musicians. Nature 392: 811-814.
22. Elbert T, Pantev C, Wienbruch C et al. (1995) Increased cortical representation of the fingers of the left hand in string players. Science 270: 305-307.
23. Hari R, Hämäläinen M, Ilmoniemi R et al. (1984) Responses of the primary auditory cortex to pitch changes in a sequence of tone pips: neuromagnetic recordings in man. Neurosci Lett 50: 127-132.
24. Näätänen R, Paavilainen P, Alho K et al. (1987) The mismatch negativity to intensity changes in an auditory stimulus sequence. Electroencephalogr Clin Neurophysiol Suppl 40: 125-131.
25. Herholz S C, Lappe C, Knief A et al. (2008) Neural basis of music imagery and the effect of musical expertise. Eur J Neurosci 28: 2352-2360.
26. Herholz S C, Lappe C, Pantev C (2009) Looking for a pattern: an MEG study on the abstract mismatch negativity in musicians and nonmusicians. BMC Neurosci 10: 42.
27. Sams M, Hämäläinen M, Antervo A et al. (1985) Cerebral neuromagnetic responses evoked by short auditory stimuli. Electroencephalogr Clin Neurophysiol 61: 254-266.
28. Levänen S, Ahonen A, Hari R et al. (1996) Deviant auditory stimuli activate human left and right auditory cortex differently. Cereb Cortex 6: 288-296.
29. Alho K, Winkler I, Escera C et al. (1998) Processing of novel sounds and frequency changes in the human auditory cortex: magnetoencephalographic recordings. Psychophysiology 35: 211-224.
30. Squires N K, Squires K C, Hillyard S A (1975) Two varieties of long-latency positive waves evoked by unpredictable auditory stimuli in man. Electroencephalogr Clin Neurophysiol 38: 387-401.
31. Cheour M, Haapanen M L, Hukki J et al. (1997) The first neurophysiological evidence for cognitive brain dysfunctions in children with CATCH. Neuroreport 8: 1785-1787.
32. Draganova R, Eswaran H, Murphy P et al. (2005) Sound frequency change detection in fetuses and newborns, a magnetoencephalographic study. Neuroimage 28: 354-361.
33. Gordon E, Sloggett G, Harvey I et al. (1987) Magnetoencephalography: locating the source of P300 via magnetic field recording. Clin Exp Neurol 23: 101-110.
34. Rogers R L, Baumann S B, Papanicolaou A C et al. (1991) Localization of the P3 sources using magnetoencephalography and magnetic resonance imaging. Electroencephalogr Clin Neurophysiol 79: 308-321.
35. Okada Y C, Kaufman L, Williamson S J (1983) The hippocampal formation as a source of the slow endogenous potentials. Electroencephalogr Clin Neurophysiol 55: 417-426.
36. Hari R, Hämäläinen M, Kaukoranta E et al. (1989) Selective listening modifies activity of the human auditory cortex. Exp Brain Res 74: 463-470.
37. Arthur D L, Lewis P S, Medvick P A et al. (1991) A neuromagnetic study of selective auditory attention. Electroencephalogr Clin Neurophysiol 78: 348-360.
38. Okamoto H, Stracke H, Wolters C H et al. (2007) Attention improves population-level frequency tuning in human auditory cortex. J Neurosci 27: 10383-10390.
39. Stracke H, Okamoto H, Pantev C (2009) Interhemispheric support during demanding auditory signal-in-noise processing. Cereb Cortex 19: 1440-1447.
40. Rauschecker J P, Tian B (2000) Mechanisms and streams for processing of "what" and "where" in auditory cortex. Proc Natl Acad Sci U S A 97: 11800-11806.
41. Kaiser J, Lutzenberger W, Preissl H et al. (2000) Right-hemisphere dominance for the processing of sound-source lateralization. J Neurosci 20: 6631-6639.
42. Kaiser J, Lutzenberger W, Ackermann H et al. (2002) Dynamics of gamma-band activity induced by auditory pattern changes in humans. Cereb Cortex 12: 212-221.
43. Brunetti M, Belardinelli P, Caulo M et al. (2005) Human brain activation during passive listening to sounds from different locations: an fMRI and MEG study. Hum Brain Mapp 26: 251-261.
44. Brunetti M, Della Penna S, Ferretti A et al. (2008) A frontoparietal network for spatial attention reorienting in the auditory domain: a human fMRI/MEG study of functional and temporal dynamics. Cereb Cortex 18: 1139-1147.
45. Tecchio F, Bicciolo G, De Campora E et al. (2000) Tonotopic cortical changes following stapes substitution in otosclerotic patients: a magnetoencephalographic study. Hum Brain Mapp 10: 28-38.
46. Muhlnickel W, Elbert T, Taub E et al. (1998) Reorganization of auditory cortex in tinnitus. Proc Natl Acad Sci U S A 95: 10340-10343.
47. Pantev C, Wollbrink A, Roberts L E et al. (1999) Short-term plasticity of the human auditory cortex. Brain Res 842: 192-199.
48. Okamoto H, Stracke H, Stoll W et al. (2009) Listening to tailor-made notched music reduces tinnitus loudness and tinnitus-related auditory cortex activity. Proc Natl Acad Sci U S A.
49. Reite M, Teale P, Goldstein L et al. (1989) Late auditory magnetic sources may differ in the left hemisphere of schizophrenic patients. A preliminary report. Arch Gen Psychiatry 46: 565-572.
50. Reite M, Sheeder J, Teale P et al. (1997) Magnetic source imaging evidence of sex differences in cerebral lateralization in schizophrenia. Arch Gen Psychiatry 54: 433-440.
51. Baumann S B, Rogers R L, Guinto F C et al. (1991) Gender differences in source location for the N100 auditory evoked magnetic field. Electroencephalogr Clin Neurophysiol 80: 53-59.
52. Scheuneman D, Teale P, Linnville S et al. (1991) Magnetic auditory M100 source location in normal females. Brain Res Bull 26: 747-751.

53. Reite M, Sheeder J, Teale P et al. (1995) MEG based brain laterality: sex differences in normal adults. Neuropsychologia 33: 1607-1616.
54. Rojas D C, Teale P, Sheeder J et al. (1997) Sex-specific expression of Heschl's gyrus functional and structural abnormalities in paranoid schizophrenia. Am J Psychiatry 154: 1655-1662.
55. Hajek M, Boehle C, Huonker R et al. (1997) Abnormalities of auditory evoked magnetic fields in the right hemisphere of schizophrenic females. Schizophr Res 24: 329-332.
56. Hajek M, Huonker R, Boehle C et al. (1997) Abnormalities of auditory evoked magnetic fields and structural changes in the left hemisphere of male schizophrenics--a magnetoencephalographic-magnetic resonance imaging study. Biol Psychiatry 42: 609-616.
57. Cañive J M, Lewine J D, Edgar J C et al. (1996) Magnetoencephalographic assessment of spontaneous brain activity in schizophrenia. Psychopharmacol Bull 32: 741-750.
58. Cañive J M, Lewine J D, Edgar J C et al. (1998) Spontaneous brain magnetic activity in schizophrenia patients treated with aripiprazole. Psychopharmacol Bull 34: 101-105.
59. Edgar J C, Huang M X, Weisend M P et al. (2003) Interpreting abnormality: an EEG and MEG study of P50 and the auditory paired-stimulus paradigm. Biol Psychol 65: 1-20.
60. Thoma R J, Hanlon F M, Sanchez N et al. (2004) Auditory sensory gating deficit and cortical thickness in schizophrenia. Neurol Clin Neurophysiol 2004: 62.
61. Hanlon F M, Miller G A, Thoma R J et al. (2005) Distinct M50 and M100 auditory gating deficits in schizophrenia. Psychophysiology 42: 417-427.
62. Brenner D, Lipton J, Kaufman L et al. (1978) Somatically evoked magnetic fields of the human brain. Science 199: 81-83.
63. Okada Y C, Tanenbaum R, Williamson S J et al. (1984) Somatotopic organization of the human somatosensory cortex revealed by neuromagnetic measurements. Exp Brain Res 56: 197-205.
64. Hari R, Hämäläinen M, Kaukoranta E et al. (1983) Neuromagnetic responses from the second somatosensory cortex in man. Acta Neurol Scand 68: 207-212.
65. Hari R, Reinikainen K, Kaukoranta E et al. (1984) Somatosensory evoked cerebral magnetic fields from SI and SII in man. Electroencephalogr Clin Neurophysiol 57: 254-263.
66. Penfield W, Jasper H, Epilepsy and the functional anatomy of the human brain. 1954, Boston: Little, Brown & Co.
67. Narici L, Modena I, Opsomer R J et al. (1991) Neuromagnetic somatosensory homunculus: a non-invasive approach in humans. Neurosci Lett 121: 51-54.
68. Hari R, Forss N (1999) Magnetoencephalography in the study of human somatosensory cortical processing. Philos Trans R Soc Lond B Biol Sci 354: 1145-1154.
69. Sutherling W W, Crandall P H, Darcey T M et al. (1988) The magnetic and electric fields agree with intracranial localizations of somatosensory cortex. Neurology 38: 1705-1714.
70. Merzenich M M, Kaas J H, Wall J et al. (1983) Topographic reorganization of somatosensory cortical areas 3b and 1 in adult monkeys following restricted deafferentation. Neuroscience 8: 33-55.
71. Rossini P M, Martino G, Narici L et al. (1994) Short-term brain 'plasticity' in humans: transient finger representation changes in sensory cortex somatotopy following ischemic anesthesia. Brain Res 642: 169-177.
72. Flor H, Elbert T, Knecht S et al. (1995) Phantom-limb pain as a perceptual correlate of cortical reorganization following arm amputation. Nature 375: 482-484.
73. Yang T T, Gallen C, Schwartz B et al. (1994) Sensory maps in the human brain. Nature 368: 592-593.
74. Mogilner A, Grossman J A, Ribary U et al. (1993) Somatosensory cortical plasticity in adult humans revealed by magnetoencephalography. Proc Natl Acad Sci U S A 90: 3593-3597.
75. Stavrinou M L, Della Penna S, Pizzella V et al. (2007) Temporal dynamics of plastic changes in human primary somatosensory cortex after finger webbing. Cereb Cortex 17: 2134-2142.
76. Kawamura T, Nakasato N, Seki K et al. (1996) Neuromagnetic evidence of pre- and post-central cortical sources of somatosensory evoked responses. Electroencephalogr Clin Neurophysiol 100: 44-50.
77. Huttunen J (1997) Does the P35m SEF deflection really come from the motor cortex? Electroencephalogr Clin Neurophysiol 104: 101-102.
78. Huang M X, Aine C, Davis L et al. (2000) Sources on the anterior and posterior banks of the central sulcus identified from magnetic somatosensory evoked responses using multistart spatio-temporal localization. Hum Brain Mapp 11: 59-76.
79. Huang M, Davis L E, Aine C et al. (2004) MEG response to median nerve stimulation correlates with recovery of sensory and motor function after stroke. Clin Neurophysiol 115: 820-833.
80. D'Esposito M, Deouell L Y, Gazzaley A (2003) Alterations in the BOLD fMRI signal with ageing and disease: a challenge for neuroimaging. Nat Rev Neurosci 4: 863-872.
81. Rossini P M, Altamura C, Ferretti A et al. (2004) Does cerebrovascular disease affect the coupling between neuronal activity and local haemodynamics? Brain 127: 99-110.
82. Curio G, Mackert B M, Burghoff M et al. (1994) Localization of evoked neuromagnetic 600 Hz activity in the cerebral somatosensory system. Electroencephalogr Clin Neurophysiol 91: 483-487.
83. Curio G, Mackert B M, Burghoff M et al. (1997) Somatotopic source arrangement of 600 Hz oscillatory magnetic fields at the human primary somatosensory hand cortex. Neurosci Lett 234: 131-134.
84. Hashimoto I, Mashiko T, Imada T (1996) Somatic evoked high-frequency magnetic oscillations reflect activity of inhibitory interneurons in the human somatosensory cortex. Electroencephalogr Clin Neurophysiol 100: 189-203.
85. Hari R, Kaukoranta E, Reinikainen K et al. (1983) Neuromagnetic localization of cortical activity evoked by painful dental stimulation in man. Neurosci Lett 42: 77-82.
86. Huttunen J, Kobal G, Kaukoranta E et al. (1986) Cortical responses to painful CO2 stimulation of nasal mucosa; a magnetoencephalographic study in man. Electroencephalogr Clin Neurophysiol 64: 347-349.
87. Kakigi R, Koyama S, Hoshiyama M et al. (1995) Pain-related magnetic fields following painful CO2 laser stimulation in man. Neurosci Lett 192: 45-48.
88. Joseph J, Howland E W, Wakai R et al. (1991) Late pain-related magnetic fields and electric potentials evoked by intracutaneous electric finger stimulation. Electroencephalogr Clin Neurophysiol 80: 46-52.
89. Kitamura Y, Kakigi R, Hoshiyama M et al. (1995) Pain-related somatosensory evoked magnetic fields. Electroencephalogr Clin Neurophysiol 95: 463-474.
90. Kitamura Y, Kakigi R, Hoshiyama M et al. (1997) Pain-related somatosensory evoked magnetic fields following lower limb stimulation. J Neurol Sci 145: 187-194.
91. Ploner M, Schmitz F, Freund H J et al. (1999) Parallel activation of primary and secondary somatosensory cortices in human pain processing. J Neurophysiol 81: 3100-3104.
92. Ploner M, Gross J, Timmermann L et al. (2002) Cortical representation of first and second pain sensation in humans. Proc Natl Acad Sci U S A 99: 12444-12448.
93. Tran T D, Inui K, Hoshiyama M et al. (2002) Cerebral activation by the signals ascending through unmyelinated C-fibers in humans: a magnetoencephalographic study. Neuroscience 113: 375-386.

94. Kakigi R, Tran T D, Qiu Y et al. (2003) Cerebral responses following stimulation of unmyelinated C-fibers in humans: electro- and magneto-encephalographic study. Neurosci Res 45: 255-275.
95. Forss N, Raij T T, Seppä M et al. (2005) Common cortical network for first and second pain. Neuroimage 24: 132-142.
96. Kornhuber H H, Deecke L (1965) [Changes in the Brain Potential in Voluntary Movements and Passive Movements in Man: Readiness Potential and Reafferent Potentials.]. Pflugers Arch Gesamte Physiol Menschen Tiere 284: 1-17.
97. Deecke L, Grozinger B, Kornhuber H H (1976) Voluntary finger movement in man: cerebral potentials and theory. Biol Cybern 23: 99-119.
98. Deecke L, Weinberg H, Brickett P (1982) Magnetic fields of the human brain accompanying voluntary movement: Bereitschaftsmagnetfeld. Exp Brain Res 48: 144-148.
99. Deecke L, Boschert J, Weinberg H et al. (1983) Magnetic fields of the human brain (Bereitschaftsmagnetfeld) preceding voluntary foot and toe movements. Exp Brain Res 52: 81-86.
100. Hari R, Antervo A, Katila T et al. (1983) Cerebral magnetic fields associated with voluntary limb movements in man. Il Nuovo Cimento 2D: 484-494.
101. Salmelin R, Hari R (1994) Characterization of spontaneous MEG rhythms in healthy adults. Electroencephalogr Clin Neurophysiol 91: 237-248.
102. Murthy V N, Fetz E E (1992) Coherent 25- to 35-Hz oscillations in the sensorimotor cortex of awake behaving monkeys. Proc Natl Acad Sci U S A 89: 5670-5674.
103. Conway B A, Halliday D M, Farmer S F et al. (1995) Synchronization between motor cortex and spinal motoneuronal pool during the performance of a maintained motor task in man. J Physiol 489 (Pt 3): 917-924.
104. Salenius S, Portin K, Kajola M et al. (1997) Cortical control of human motoneuron firing during isometric contraction. J Neurophysiol 77: 3401-3405.
105. Salmelin R, Hari R (1994) Spatiotemporal characteristics of sensorimotor neuromagnetic rhythms related to thumb movement. Neuroscience 60: 537-550.
106. Hari R, Forss N, Avikainen S et al. (1998) Activation of human primary motor cortex during action observation: a neuromagnetic study. Proc Natl Acad Sci U S A 95: 15061-15065.
107. di Pellegrino G, Fadiga L, Fogassi L et al. (1992) Understanding motor events: a neurophysiological study. Exp Brain Res 91: 176-180.
108. Salenius S, Schnitzler A, Salmelin R et al. (1997) Modulation of human cortical rolandic rhythms during natural sensorimotor tasks. Neuroimage 5: 221-228.
109. Järveläinen J, Schurmann M, Hari R (2004) Activation of the human primary motor cortex during observation of tool use. Neuroimage 23: 187-192.
110. Nishitani N, Hari R (2002) Viewing lip forms: cortical dynamics. Neuron 36: 1211-1220.
111. Nishitani N, Avikainen S, Hari R (2004) Abnormal imitation-related cortical activation sequences in Asperger's syndrome. Ann Neurol 55: 558-562.
112. Caetano G, Jousmäki V, Hari R (2007) Actor's and observer's primary motor cortices stabilize similarly after seen or heard motor actions. Proc Natl Acad Sci U S A 104: 9058-9062.
113. Cohen D (1975) Measurements of the magnetic fields produced by the human heart, brain, and lungs. IEEE Trans Magnetics MAG-11: 694-700.
114. Brenner D, Williamson S J, Kaufman L (1975) Visually evoked magnetic fields of the human brain. Science 190: 480-482.
115. Teyler T J, Cuffin B N, Cohen D (1975) The visual evoked magnetoencephalogram. Life Sci 17: 683-691.
116. Maclin E, Okada, YC, Kaufman, L, Williamson, SJ (1983) Retinotopic map on the visual cortex for eccentrically placed patterns: first noninvasive measurement. Il Nuovo Cimento 2: 410-419.
117. Ahlfors S P, Ilmoniemi R J, Hämäläinen M S (1992) Estimates of visually evoked cortical currents. Electroencephalogr Clin Neurophysiol 82: 225-236.
118. Aine C J, Supek S, George J S et al. (1996) Retinotopic organization of human visual cortex: departures from the classical model. Cereb Cortex 6: 354-361.
119. Supek S, Aine C J, Ranken D et al. (1999) Single vs. paired visual stimulation: superposition of early neuromagnetic responses and retinotopy in extrastriate cortex in humans. Brain Res 830: 43-55.
120. Enroth-Cugell C, Robson J G (1966) The contrast sensitivity of retinal ganglion cells of the cat. J Physiol 187: 517-552.
121. Breitmeyer B G, Ganz L (1976) Implications of sustained and transient channels for theories of visual pattern masking, saccadic suppression, and information processing. Psychol Rev 83: 1-36.
122. Ungerleider L G, Mishkin M, Two cortical visual systems., in Analysis of Visual Behavior, D.J. Ingle, M.A. Goodale, R.J.W. Mansfield, Editors. 1982, MIT Press: Cambridge. p. 549-586.
123. Kaufman L, Williamson S J (1980) The evoked magnetic field of the human brain. Ann N Y Acad Sci 340: 45-65.
124. Okada Y C, Kaufman L, Brenner D et al. (1982) Modulation transfer functions of the human visual system revealed by magnetic field measurements. Vision Res 22: 319-333.
125. Williamson S J, Kaufman L, Brenner D (1978) Latency of the neuromagnetic response of the human visual cortex. Vision Res 18: 107-110.
126. Aine C J, Bodis-Wollner I, George J S (1990) Generators of visually evoked neuromagnetic responses. Spatial-frequency segregation and evidence for multiple sources. Adv Neurol 54: 141-155.
127. Aine C J, Supek S, George J S (1995) Temporal dynamics of visual-evoked neuromagnetic sources: effects of stimulus parameters and selective attention. Int J Neurosci 80: 79-104.
128. Seki K, Nakasato N, Fujita S et al. (1996) Neuromagnetic evidence that the P100 component of the pattern reversal visual evoked response originates in the bottom of the calcarine fissure. Electroencephalogr Clin Neurophysiol 100: 436-442.
129. Nakamura M, Kakigi R, Okusa T et al. (2000) Effects of check size on pattern reversal visual evoked magnetic field and potential. Brain Res 872: 77-86.
130. Harding G F, Janday B, Armstrong R A (1991) Topographic mapping and source localization of the pattern reversal visual evoked magnetic response. Brain Topogr 4: 47-55.
131. Harding G F, Degg C, Anderson S J et al. (1994) Topographic mapping of the pattern onset evoked magnetic response to stimulation of different portions of the visual field. Int J Psychophysiol 16: 175-183.
132. Shigeto H, Tobimatsu S, Yamamoto T et al. (1998) Visual evoked cortical magnetic responses to checkerboard pattern reversal stimulation: a study on the neural generators of N75, P100 and N145. J Neurol Sci 156: 186-194.
133. Hashimoto T, Kashii S, Kikuchi M et al. (1999) Temporal profile of visual evoked responses to pattern-reversal stimulation analyzed with a whole-head magnetometer. Exp Brain Res 125: 375-382.
134. Moradi F, Liu L C, Cheng K et al. (2003) Consistent and precise localization of brain activity in human primary visual cortex by MEG and fMRI. Neuroimage 18: 595-609.
135. De Yoe E A, Van Essen D C (1988) Concurrent processing streams in monkey visual cortex. Trends Neurosci 11: 219-226.
136. ffytche D H, Guy C N, Zeki S (1995) The parallel visual motion inputs into areas V1 and V5 of human cerebral cortex. Brain 118 (Pt 6): 1375-1394.

137. Anderson S J, Holliday I E, Singh K D et al. (1996) Localization and functional analysis of human cortical area V5 using magnetoencephalography. Proc Biol Sci 263: 423-431.
138. Patzwahl D R, Elbert T, Zanker J M et al. (1996) The cortical representation of object motion in man is interindividually variable. Neuroreport 7: 469-472.
139. Fylan F, Holliday I E, Singh K D et al. (1997) Magnetoencephalographic investigation of human cortical area V1 using color stimuli. Neuroimage 6: 47-57.
140. Holliday I E, Anderson S J, Harding G F (1997) Magnetoencephalographic evidence for non-geniculostriate visual input to human cortical area V5. Neuropsychologia 35: 1139-1146.
141. Kaneoke Y, Bundou M, Koyama S et al. (1997) Human cortical area responding to stimuli in apparent motion. Neuroreport 8: 677-682.
142. Kaneoke Y, Bundou M, Kakigi R (1998) Timing of motion representation in the human visual system. Brain Res 790: 195-201.
143. Ahlfors S P, Simpson G V, Dale A M et al. (1999) Spatiotemporal activity of a cortical network for processing visual motion revealed by MEG and fMRI. J Neurophysiol 82: 2545-2555.
144. Bundo M, Kaneoke Y, Inao S et al. (2000) Human visual motion areas determined individually by magnetoencephalography and 3D magnetic resonance imaging. Hum Brain Mapp 11: 33-45.
145. Kawakami O, Kaneoke Y, Kakigi R (2000) Perception of apparent motion is related to the neural activity in the human extrastriate cortex as measured by magnetoencephalography. Neurosci Lett 285: 135-138.
146. Lam K, Kaneoke Y, Gunji A et al. (2000) Magnetic response of human extrastriate cortex in the detection of coherent and incoherent motion. Neuroscience 97: 1-10.
147. Naito T, Kaneoke Y, Osaka N et al. (2000) Asymmetry of the human visual field in magnetic response to apparent motion. Brain Res 865: 221-226.
148. Kawakami O, Kaneoke Y, Maruyama K et al. (2002) Visual detection of motion speed in humans: spatiotemporal analysis by fMRI and MEG. Hum Brain Mapp 16: 104-118.
149. Singh K D, Barnes G R, Hillebrand A (2003) Group imaging of task-related changes in cortical synchronisation using nonparametric permutation testing. Neuroimage 19: 1589-1601.
150. Portin K, Salenius S, Salmelin R et al. (1998) Activation of the human occipital and parietal cortex by pattern and luminance stimuli: neuromagnetic measurements. Cereb Cortex 8: 253-260.
151. Portin K, Vanni S, Virsu V et al. (1999) Stronger occipital cortical activation to lower than upper visual field stimuli. Neuromagnetic recordings. Exp Brain Res 124: 287-294.
152. Sato N, Nakamura K, Nakamura A et al. (1999) Different time course between scene processing and face processing: a MEG study. Neuroreport 10: 3633-3637.
153. Vanni S, Tanskanen T, Seppa M et al. (2001) Coinciding early activation of the human primary visual cortex and anteromedial cuneus. Proc Natl Acad Sci U S A 98: 2776-2780.
154. Stephen J M, Aine C J, Christner R F et al. (2002) Central versus peripheral visual field stimulation results in timing differences in dorsal stream sources as measured with MEG. Vision Res 42: 3059-3074.
155. Stephen J M, Ranken D, Aine C J (2006) Frequency-following and connectivity of different visual areas in response to contrast-reversal stimulation. Brain Topogr 18: 257-272.
156. Lu S T, Hamalainen M S, Hari R et al. (1991) Seeing faces activates three separate areas outside the occipital visual cortex in man. Neuroscience 43: 287-290.
157. Sams M, Hietanen J K, Hari R et al. (1997) Face-specific responses from the human inferior occipito-temporal cortex. Neuroscience 77: 49-55.
158. Swithenby S J, Bailey A J, Brautigam S et al. (1998) Neural processing of human faces: a magnetoencephalographic study. Exp Brain Res 118: 501-510.
159. Liu L, Ioannides A A, Streit M (1999) Single trial analysis of neurophysiological correlates of the recognition of complex objects and facial expressions of emotion. Brain Topogr 11: 291-303.
160. Streit M, Ioannides A A, Liu L et al. (1999) Neurophysiological correlates of the recognition of facial expressions of emotion as revealed by magnetoencephalography. Brain Res Cogn Brain Res 7: 481-491.
161. Ioannides A A, Liu L C, Kwapien J et al. (2000) Coupling of regional activations in a human brain during an object and face affect recognition task. Hum Brain Mapp 11: 77-92.
162. Linkenkaer-Hansen K, Palva J M, Sams M et al. (1998) Face-selective processing in human extrastriate cortex around 120 ms after stimulus onset revealed by magneto- and electroencephalography. Neurosci Lett 253: 147-150.
163. Liu J, Harris A, Kanwisher N (2002) Stages of processing in face perception: an MEG study. Nat Neurosci 5: 910-916.
164. Itier R J, Herdman A T, George N et al. (2006) Inversion and contrast-reversal effects on face processing assessed by MEG. Brain Res 1115: 108-120.
165. Susac A, Ilmoniemi R J, Pihko E et al. (2009) Early dissociation of face and object processing: a magnetoencephalographic study. Hum Brain Mapp 30: 917-927.
166. Halgren E, Raij T, Marinkovic K et al. (2000) Cognitive response profile of the human fusiform face area as determined by MEG. Cereb Cortex 10: 69-81.
167. Goldman-Rakic P S (1988) Topography of cognition: parallel distributed networks in primate association cortex. Annu Rev Neurosci 11: 137-156.
168. Ungerleider L G (1995) Functional brain imaging studies of cortical mechanisms for memory. Science 270: 769-775.
169. Mesulam M M (1998) From sensation to cognition. Brain 121 (Pt 6): 1013-1052.
170. Lampl I, Reichova I, Ferster D (1999) Synchronous membrane potential fluctuations in neurons of the cat visual cortex. Neuron 22: 361-374.
171. Salinas E, Sejnowski T J (2001) Correlated neuronal activity and the flow of neural information. Nat Rev Neurosci 2: 539-550.
172. Tallon-Baudry C, Mandon S, Freiwald W A et al. (2004) Oscillatory synchrony in the monkey temporal lobe correlates with performance in a visual short-term memory task. Cereb Cortex 14: 713-720.
173. Jensen O, Tesche C D (2002) Frontal theta activity in humans increases with memory load in a working memory task. Eur J Neurosci 15: 1395-1399.
174. Sokolov A, Lutzenberger W, Pavlova M et al. (1999) Gamma-band MEG activity to coherent motion depends on task-driven attention. Neuroreport 10: 1997-2000.
175. Braeutigam S, Bailey A J, Swithenby S J (2001) Phase-locked gamma band responses to semantic violation stimuli. Brain Res Cogn Brain Res 10: 365-377.
176. Hall S D, Holliday I E, Hillebrand A et al. (2005) The missing link: analogous human and primate cortical gamma oscillations. Neuroimage 26: 13-17.
177. Hsiao F J, Lin Y Y, Hsieh J C et al. (2006) Oscillatory characteristics of face-evoked neuromagnetic responses. Int J Psychophysiol 61: 113-120.
178. Maratos F A, Anderson S J, Hillebrand A et al. (2007) The spatial distribution and temporal dynamics of brain regions activated during the perception of object and non-object patterns. Neuroimage 34: 371-383.
179. Osipova D, Hermes D, Jensen O (2008) Gamma power is phase-locked to posterior alpha activity. PLoS One 3: e3990.
180. Hari R, Salmelin R (1997) Human cortical oscillations: a neuromagnetic view through the skull. Trends Neurosci 20: 44-49.

181. Pfurtscheller G, Lopes da Silva F H (1999) Event-related EEG/MEG synchronization and desynchronization: basic principles. Clin Neurophysiol 110: 1842-1857.
182. Tallon-Baudry C, Bertrand O (1999) Oscillatory gamma activity in humans and its role in object representation. Trends Cogn Sci 3: 151-162.
183. Basar E, Basar-Eroglu C, Karakas S et al. (2000) Brain oscillations in perception and memory. Int J Psychophysiol 35: 95-124.
184. Schnitzler A, Gross J (2005) Normal and pathological oscillatory communication in the brain. Nat Rev Neurosci 6: 285-296.
185. Jensen O, Colgin L L (2007) Cross-frequency coupling between neuronal oscillations. Trends Cogn Sci 11: 267-269.
186. Jensen O, Kaiser J, Lachaux J P (2007) Human gamma-frequency oscillations associated with attention and memory. Trends Neurosci 30: 317-324.
187. Jerbi K, Ossandon T, Hamame C M et al. (2009) Task-related gamma-band dynamics from an intracerebral perspective: review and implications for surface EEG and MEG. Hum Brain Mapp 30: 1758-1771.
188. Kahneman D, Treisman A, Changing views of attention and automaticity., in Varieties of Attention, R. Parasuraman, D.R. Davies, Editors. 1984, Academic Press: London. p. 29-61.
189. Johnston W A, Dark V J (1986) Selective attention. Annu. Rev. Psychol. 37: 43-75.
190. Broadbent D E, Perception and Communication. 1958, London: Pergamon.
191. Treisman A M (1964) Selective Attention in Man. Br Med Bull 20: 12-16.
192. Deutsch J A, Deutsch D (1963) Attention: Some theoretical considerations. Psychol Rev 70: 80-90.
193. Harter M R, Aine C J, Brain mechanisms of visual selective attention, in Varieties of Attention, R. Parasuraman, D.R. Davies, Editors. 1984, Academic Press: London. p. 293-321.
194. Martinez A, Anllo-Vento L, Sereno M I et al. (1999) Involvement of striate and extrastriate visual cortical areas in spatial attention. Nat Neurosci 2: 364-369.
195. Noesselt T, Hillyard S A, Woldorff M G et al. (2002) Delayed striate cortical activation during spatial attention. Neuron 35: 575-587.
196. Di Russo F, Martinez A, Hillyard S A (2003) Source analysis of event-related cortical activity during visuo-spatial attention. Cereb Cortex 13: 486-499.
197. Tootell R B, Hadjikhani N, Hall E K et al. (1998) The retinotopy of visual spatial attention. Neuron 21: 1409-1422.
198. Brefczynski J A, DeYoe E A (1999) A physiological correlate of the 'spotlight' of visual attention. Nat Neurosci 2: 370-374.
199. Lamme V A, Roelfsema P R (2000) The distinct modes of vision offered by feedforward and recurrent processing. Trends Neurosci 23: 571-579.
200. Mehta A D, Ulbert I, Schroeder C E (2000) Intermodal selective attention in monkeys. II: physiological mechanisms of modulation. Cereb Cortex 10: 359-370.
201. Mehta A D, Ulbert I, Schroeder C E (2000) Intermodal selective attention in monkeys. I: distribution and timing of effects across visual areas. Cereb Cortex 10: 343-358.
202. Aine C J, Stephen J M, Christner R et al. (2003) Task relevance enhances early transient and late slow-wave activity of distributed cortical sources. J Comput Neurosci 15: 203-221.
203. Hughes J R, Cohen J, Mayman C I et al. (1977) Relationship of the magnetoencephalogram to abnormal activity in the electroencephalogram. J Neurol 217: 79-93.
204. Barth D S, Sutherling W, Engel J, Jr. et al. (1982) Neuromagnetic localization of epileptiform spike activity in the human brain. Science 218: 891-894.
205. Barth D S, Sutherling W, Beatty J (1984) Fast and slow magnetic phenomena in focal epileptic seizures. Science 226: 855-857.
206. Barth D S, Sutherling W, Engle J, Jr. et al. (1984) Neuromagnetic evidence of spatially distributed sources underlying epileptiform spikes in the human brain. Science 223: 293-296.
207. Ikeda A, Taki W, Kunieda T et al. (1999) Focal ictal direct current shifts in human epilepsy as studied by subdural and scalp recording. Brain 122 (Pt 5): 827-838.
208. Rose D F, Smith P D, Sato S (1987) Magnetoencephalography and epilepsy research. Science 238: 329-335.
209. Hari R, Ahonen A, Forss N et al. (1993) Parietal epileptic mirror focus detected with a whole-head neuromagnetometer. Neuroreport 5: 45-48.
210. James M F, Smith J M, Boniface S J et al. (2001) Cortical spreading depression and migraine: new insights from imaging? Trends Neurosci 24: 266-271.
211. Smith J M, James M F, Fraser J A et al. (2008) Translational imaging studies of cortical spreading depression in experimental models for migraine aura. Expert Rev. Neurotherapeutics 8: 759-768.
212. Charles A, Brennan K (2009) Cortical spreading depression-new insights and persistent questions. Cephalalgia 29: 1115-1124.
213. Okada Y C, Lauritzen M, Nicholson C (1988) Magnetic field associated with spreading depression: a model for the detection of migraine. Brain Res 442: 185-190.
214. Gardner-Medwin A R, Tepley N, Barkley G L et al. (1991) Magnetic fields associated with spreading depression in anaesthetised rabbits. Brain Res 540: 153-158.
215. Barkley G L, Tepley N, Nagel-Leiby S et al. (1990) Magnetoencephalographic studies of migraine. Headache 30: 428-434.
216. Barkley G L, Moran J E, Takanashi Y et al. (1991) Techniques for DC magnetoencephalography. J Clin Neurophysiol 8: 189-199.
217. Tepley N, Bowyer S, Clifton Y et al. (1996) A technique for sequential measurements of DC neuromagnetic fields. Electroencephalogr Clin Neurophysiol 99: 16-18.
218. Bowyer S M, Okada Y C, Papuashvili N et al. (1999) Analysis of MEG signals of spreading cortical depression with propagation constrained to a rectangular cortical strip. I. Lissencephalic rabbit model. Brain Res 843: 71-78.
219. Bowyer S M, Tepley N, Papuashvili N et al. (1999) Analysis of MEG signals of spreading cortical depression with propagation constrained to a rectangular cortical strip. II. Gyrencephalic swine model. Brain Res 843: 79-86.
220. Bowyer S M, Aurora K S, Moran J E et al. (2001) Magnetoencephalographic fields from patients with spontaneous and induced migraine aura. Ann Neurol 50: 582-587.
221. Leistner S, Sander T, Wachs M et al. (2009) Differential infraslow (<0.1 Hz) cortical activations in the affected and unaffected hemispheres from patients with subacute stroke demonstrated by noninvasive DC-magnetoencephalography. Stroke 40: 1683-1686.
222. Ojemann G, Ojemann J, Lettich E et al. (1989) Cortical language localization in left, dominant hemisphere. An electrical stimulation mapping investigation in 117 patients. J Neurosurg 71: 316-326.
223. Pataraia E, Billingsley-Marshall R L, Castillo E M et al. (2005) Organization of receptive language-specific cortex before and after left temporal lobectomy. Neurology 64: 481-487.
224. Simos P G, Fletcher J M, Sarkari S et al. (2007) Altering the brain circuits for reading through intervention: a magnetic source imaging study. Neuropsychology 21: 485-496.
225. Salmelin R, Hari R, Lounasmaa O V et al. (1994) Dynamics of brain activation during picture naming. Nature 368: 463-465.
226. Simos P G, Papanicolaou A C, Breier J I et al. (1999) Localization of language-specific cortex by using magnetic source imaging and electrical stimulation mapping. J Neurosurg 91: 787-796.
227. Simos P G, Breier J I, Zouridakis G et al. (1998) Assessment of functional cerebral laterality for language using magnetoencephalography. J Clin Neurophysiol 15: 364-372.

228. Papanicolaou A C, Simos P G, Breier J I et al. (1999) Magnetoencephalographic mapping of the language-specific cortex. J Neurosurg 90: 85-93.
229. Papanicolaou A C, Simos P G, Castillo E M et al. (2004) Magnetocephalography: a noninvasive alternative to the Wada procedure. J Neurosurg 100: 867-876.
230. Breier J I, Simos P G, Zouridakis G et al. (1999) Language dominance determined by magnetic source imaging: a comparison with the Wada procedure. Neurology 53: 938-945.
231. Breier J I, Simos P G, Wheless J W et al. (2001) Language dominance in children as determined by magnetic source imaging and the intracarotid amobarbital procedure: a comparison. J Child Neurol 16: 124-130.
232. Maestu F, Ortiz T, Fernandez A et al. (2002) Spanish language mapping using MEG: a validation study. Neuroimage 17: 1579-1586.
233. Castillo E M, Simos P G, Venkataraman V et al. (2001) Mapping of expressive language cortex using magnetic source imaging. Neurocase 7: 419-422.
234. Galaburda A M (1993) Neuroanatomic basis of developmental dyslexia. Neurol Clin 11: 161-173.
235. Steinmetz H (1996) Structure, function and cerebral asymmetry: In vivo morphometry of the planum temporale. Neuroscience and Biobehavioral Reviews 20: 587-591.
236. Helenius P, Salmelin R, Service E et al. (1999) Semantic cortical activation in dyslexic readers. J Cogn Neurosci 11: 535-550.
237. Heim S, Eulitz C, Elbert T (2003) Altered hemispheric asymmetry of auditory N100m in adults with developmental dyslexia. Neuroreport 14: 501-504.
238. Paul I, Bott C, Heim S et al. (2006) Reduced hemispheric asymmetry of the auditory N260m in dyslexia. Neuropsychologia 44: 785-794.
239. Nagarajan S, Mahncke H, Salz T et al. (1999) Cortical auditory signal processing in poor readers. Proc Natl Acad Sci U S A 96: 6483-6488.
240. Salmelin R, Service E, Kiesilä P et al. (1996) Impaired visual word processing in dyslexia revealed with magnetoencephalography. Ann Neurol 40: 157-162.
241. Simos P G, Breier J I, Fletcher J M et al. (2000) Cerebral mechanisms involved in word reading in dyslexic children: a magnetic source imaging approach. Cereb Cortex 10: 809-816.
242. Lord C, Cook E H, Leventhal B L et al. (2000) Autism spectrum disorders. Neuron 28: 355-363.
243. Roberts T P, Schmidt G L, Egeth M et al. (2008) Electrophysiological signatures: magnetoencephalographic studies of the neural correlates of language impairment in autism spectrum disorders. Int J Psychophysiol 68: 149-160.
244. Lewine J D, Andrews R, Chez M et al. (1999) Magnetoencephalographic patterns of epileptiform activity in children with regressive autism spectrum disorders. Pediatrics 104: 405-418.
245. Munoz-Yunta J A, Ortiz T, Palau-Baduell M et al. (2008) Magnetoencephalographic pattern of epileptiform activity in children with early-onset autism spectrum disorders. Clin Neurophysiol 119: 626-634.
246. Paetau R, Kajola M, Korkman M et al. (1991) Landau-Kleffner syndrome: epileptic activity in the auditory cortex. Neuroreport 2: 201-204.
247. Gage N M, Siegel B, Callen M et al. (2003) Cortical sound processing in children with autism disorder: an MEG investigation. Neuroreport 14: 2047-2051.
248. Gage N M, Siegel B, Roberts T P (2003) Cortical auditory system maturational abnormalities in children with autism disorder: an MEG investigation. Brain Res Dev Brain Res 144: 201-209.
249. Gage N, Roberts T P, Hickok G (2006) Temporal resolution properties of human auditory cortex: reflections in the neuromagnetic auditory evoked M100 component. Brain Res 1069: 166-171.
250. Rojas D C, Maharajh K, Teale P et al. (2008) Reduced neural synchronization of gamma-band MEG oscillations in first-degree relatives of children with autism. BMC Psychiatry 8: 66.
251. Bailey A J, Braeutigam S, Jousmaki V et al. (2005) Abnormal activation of face processing systems at early and intermediate latency in individuals with autism spectrum disorder: a magnetoencephalographic study. Eur J Neurosci 21: 2575-2585.
252. Kylliäinen A, Braeutigam S, Hietanen J K et al. (2006) Face and gaze processing in normally developing children: a magnetoencephalographic study. Eur J Neurosci 23: 801-810.
253. Braeutigam S, Swithenby S J, Bailey A J (2008) Contextual integration the unusual way: a magnetoencephalographic study of responses to semantic violation in individuals with autism spectrum disorders. Eur J Neurosci 27: 1026-1036.
254. Pekkonen E, Huotilainen M, Virtanen J et al. (1995) Age-related functional differences between auditory cortices: a whole-head MEG study. Neuroreport 6: 1803-1806.
255. Pekkonen E, Huotilainen M, Virtanen J et al. (1996) Alzheimer's disease affects parallel processing between the auditory cortices. Neuroreport 7: 1365-1368.
256. Ahveninen J, Jääskeläinen I P, Kaakkola S et al. (2002) Aging and cholinergic modulation of the transient magnetic 40-Hz auditory response. Neuroimage 15: 153-158.
257. Osipova D, Pekkonen E, Ahveninen J (2006) Enhanced magnetic auditory steady-state response in early Alzheimer's disease. Clin Neurophysiol 117: 1990-1995.
258. Pekkonen E, Jääskeläinen I P, Kaakkola S et al. (2005) Cholinergic modulation of preattentive auditory processing in aging. Neuroimage 27: 387-392.
259. Aine C, Adair J, Knoefel J et al. (2005) Temporal dynamics of age-related differences in auditory incidental verbal learning. Cognitive Brain Research 24: 1-18.
260. Kovacevic S, Qualls C, Adair J et al. (2005) Age-related effects on superior temmporal gyrus activity during an oddball task. Neuroreport 16: 1075-1079.
261. Berendse H W, Verbunt J, Scheltens P et al. (2000) Magnetoencephalographic analysis of cortical activity in Alzheimer's disease: a pilot study. Clinical Neurophysiology 111: 604-612.
262. Maestu F, Garcia-Segura J, Ortiz T et al. (2005) Evidence of biochemical and biomagnetic interactions in Alzheimer's disease: an MEG and MR spectroscopy study. Dement Geriatr Cogn Disord 20: 145-152.
263. Maestu F, Campo P, Gil-Gregorio P et al. (2006) Medial temporal lobe neuromagnetic hypoactivation and risk for developing cognitive decline in elderly population: a 2-year follow-up study. Neurobiol Aging 27: 32-37.
264. Maestu F, Campo P, Del Rio D et al. (2008) Increased biomagnetic activity in the ventral pathway in mild cognitive impairment. Clin Neurophysiol 119: 1320-1327.
265. Fernandez A, Maestu F, Amo C et al. (2002) Focal temporoparietal slow activity in Alzheimer's disease revealed by magnetoencephalography. Biol Psychiatry 52: 764-770.
266. Fernandez A, Arrazola J, Maestu F et al. (2003) Correlations of hippocampal atrophy and focal low-frequency magnetic activity in Alzheimer disease: volumetric MR imaging-magnetoencephalographic study. AJNR Am J Neuroradiol 24: 481-487.
267. Fernandez A, Hornero R, Mayo A et al. (2006) MEG spectral profile in Alzheimer's disease and mild cognitive impairment. Clin Neurophysiol 117: 306-314.
268. Osipova D, Ahveninen J, Jensen O et al. (2005) Altered generation of spontaneous oscillations in Alzheimer's disease. Neuroimage 27: 835-841.
269. Uutela K, Hämäläinen M, Somersalo E (1999) Visualization of magnetoencephalographic data using minimum current estimates. Neuroimage 10: 173-180.

270. Osipova D, Rantanen K, Ahveninen J et al. (2006) Source estimation of spontaneous MEG oscillations in mild cognitive impairment. Neurosci Lett 405: 57-61.
271. Aine C J, Bryant J E, Knoefel J E et al. (2010) Different strategies for auditory word recognition in healthy versus normal aging. Neuroimage 49: 3319-3330.
272. Stephen J M, Montano R, Donahue C H et al. (2010) Somatosensory responses in normal aging, mild cognitive impairment, and Alzheimer's disease. J Neural Transm 117: 217-225.
273. Aine C J, Woodruff C C, Knoefel J E et al. (2006) Aging: Compensation or Maturation? NeuroImage 32: 1891-1904.
274. Blum T, Saling E, Bauer R (1985) First magnetoencephalographic recordings of the brain activity of a human fetus. Br J Obstet Gynaecol 92: 1224-1229.
275. Wakai R T, Leuthold A C, Martin C B (1996) Fetal auditory evoked responses detected by magnetoencephalography. Am J Obstet Gynecol 174: 1484-1486.
276. Lengle J M, Chen M, Wakai R T (2001) Improved neuromagnetic detection of fetal and neonatal auditory evoked responses. Clin Neurophysiol 112: 785-792.
277. Schneider U, Schleussner E, Haueisen J et al. (2001) Signal analysis of auditory evoked cortical fields in fetal magnetoencephalography. Brain Topogr 14: 69-80.
278. Zappasodi F, Tecchio F, Pizzella V et al. (2001) Detection of fetal auditory evoked responses by means of magnetoencephalography. Brain Res 917: 167-173.
279. Eswaran H, Wilson J, Preissl H et al. (2002) Magnetoencephalographic recordings of visual evoked brain activity in the human fetus. Lancet 360: 779-780.
280. Kotini A, Koutlaki N, Anninos P et al. (2002) Fetal magnetoencephalography in intrauterine growth retarded pregnancies. Prenat Diagn 22: 1093-1097.
281. Rose D F, Eswaran H (2004) Spontaneous neuronal activity in fetuses and newborns. Exp Neurol 190 Suppl 1: S37-43.
282. Preissl H, Eswaran H, Pamela M et al. (2003) The effect of fetal heart signals on fetal magnetoencephalogram recordings: a possible explanation for increased spectral power in case of intrauterine growth-retarded pregnancies. Prenat Diagn 23: 603-605.
283. Roberts T P, Khan S Y, Blaskey L et al. (2009) Developmental correlation of diffusion anisotropy with auditory-evoked response. Neuroreport 20: 1586-1591.
284. Gondo K, Tobimatsu S, Kira R et al. (2001) A magnetoencephalographic study on development of the somatosensory cortex in infants. Neuroreport 12: 3227-3231.
285. Pihko E, Lauronen L, Wikström H et al. (2004) Somatosensory evoked potentials and magnetic fields elicited by tactile stimulation of the hand during active and quiet sleep in newborns. Clin Neurophysiol 115: 448-455.
286. Lauronen L, Nevalainen P, Wikström H et al. (2006) Immaturity of somatosensory cortical processing in human newborns. Neuroimage 33: 195-203.
287. Nevalainen P, Lauronen L, Sambeth A et al. (2008) Somatosensory evoked magnetic fields from the primary and secondary somatosensory cortices in healthy newborns. Neuroimage 40: 738-745.
288. Pihko E, Nevalainen P, Stephen J et al. (2009) Maturation of somatosensory cortical processing from birth to adulthood revealed by magnetoencephalography. Clin Neurophysiol 120: 1552-1561.
289. Crease R (1989) Biomagnetism attracts diverse crowd. Science 245: 1041-1043.

Author: Cheryl J. Aine
Institute: University of New Mexico School of Medicine
Street: Department of Radiology
City: Albuquerque, New Mexico
Country: USA
Email: aine@unm.edu

Development of a Whole-Head Child MEG System

Y. Adachi[1], M. Miyamoto[1], J. Kawai[1], M. Kawabata[1], M. Higuchi[1], D. Oyama[1], G. Uehara[1], H. Ogata[1], H. Kado[1], Y. Haruta[2], G. Tesan[3], and S. Crain[3]

[1] Applied Electronics Laboratory, Kanazawa Institute of Technology, Kanazawa, Japan
[2] Yokogawa Electric Corporation, Tokyo, Japan
[3] Macquarie Centre for Cognitive Science, Macquarie University, Sydney, Australia

Abstract— **A whole-head magnetoencephalography (MEG) system was developed to study cognitive processing in young children. The child MEG system has a helmet-shaped sensor array designed to fit child head sizes. The sensor array is composed of 64 LTS-SQUID axial-type gradiometric magnetometers with 50 mm of baseline length, arranged about 100 mm from the center of the child's head. The sensor array is installed in a helmet of a horizontal dewar with a head circumference of about 530 mm, which was determined on the basis of the preliminary investigation of the standard pre-school children's head. The liquid helium capacity of the dewar is roughly 100 liters, and the helium consumption rate is less than 6 liters/day. The sensors have been positioned in the dewar using a ship-in-a-bottle approach. To verify the performance of the child MEG system, an auditory evoked field measurement was taken of a healthy 4-year-old child subject. Large simultaneous magnetic field components corresponding to the P100m were successfully observed in the child over both the right and the left hemispheres. The latency of the effect was at around 130 ms, and two equivalent current dipoles were found in the temporal lobes in both hemispheres of the child subject.**

Keywords— **LTS-SQUID, MEG, child, whole-head, auditory evoked field.**

I. Introduction

We developed a whole-head magnetoencephalography (MEG) system to study cognitive processing in young children. This child MEG system contrasts with conventional systems that have a helmet-shaped sensor array designed to fit adult head sizes. When adult MEG systems are used with young child subjects, it is difficult to achieve a sufficient S/N ratio, because children's head sizes are much smaller than those of adults and the distance between the sensors and the magnetic sources in the brain is significant. Therefore, the use of conventional MEG system to study child cognitive processing requires the development of unsophisticated methods to acquire MEG signals from a pediatric population, such as positioning the child's head to one side of the helmet and to make two measurements for one hemisphere after another.

In this paper, the details of the developed child MEG system and preliminary measurements using the system for the acquisition of the whole head auditory evoked magnetic field (AEF) distribution are described. The necessity of the child MEG system is also discussed from the view point of S/N ratio of the measurement.

II. Instrumentation

A. System Configuration

Figure 1 shows the configuration of the child MEG system. SQUID sensors inside the dewar are connected to the SQUID driving electronics outside the magnetically shielded room (MSR). The SQUID signals are amplified and filtered by analog signal processing electronics, and then they are digitally recorded for visualization and further analysis.

The appearance of the child MEG system installed in a magnetically shielded room (MSR) at Macquarie Centre for Cognitive Science (MACCS) at Macquarie University is shown in Fig. 2. The MSR is made of three mu-metal layers and one RF shielding layer and it houses two MEG systems. The child MEG system is to the left and a conventional adult MEG system is to the right of the figure. The child MEG system has the gantry-free structure designed for supine-position subjects as is the adult MEG [1]. The supine-position is effective to suppress the movement of the subjects' heads during the measurement.

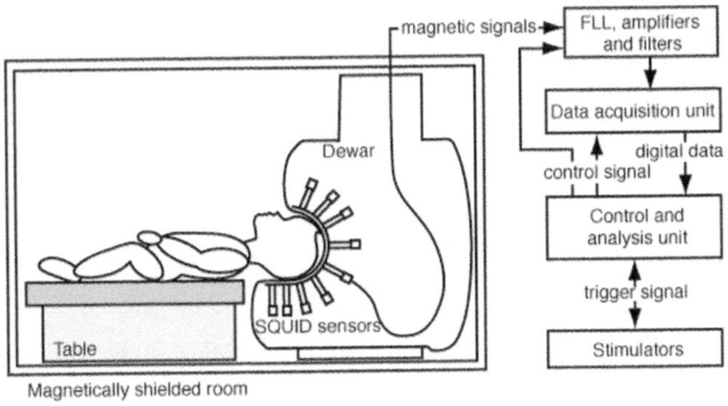

Fig. 1 System configuration

S. Supek and A. Sušac (Eds.): Advances in Biomagnetism – BIOMAG2010, IFMBE Proceedings 28, pp. 35–38, 2010.
www.springerlink.com

Fig. 2 Appearance of the child MEG system (left) and the conventional adult MEG system (right)

B. Sensor Array and Dewar

The sensor array consists of 64 LTS-SQUID axial-type first-order gradiometric magnetometers. Niobium-based LTS-SQUIDs known as Ketchen type [2][3] are used. The pick-up coil made of niobium thin wire with 15.5-mm diameter and 50-mm baseline length is wound around a bobbin and connected to the input coil of each LTS-SQUID mounted at the top of the bobbin. The typical noise characteristics of the sensor is less than 10 $fT/Hz^{1/2}$ at 1 Hz and less than 5 $fT/Hz^{1/2}$ at 100 Hz. The number of the SQUID sensors is expandable up to more than one hundred. The sensor array is helmet-shaped and its size is about 200 mm in diameter, which was determined on the basis of the preliminary investigation of the standard pre-school children's heads. This size was about 20 percent less than the size of the conventional adult MEG sensor array.

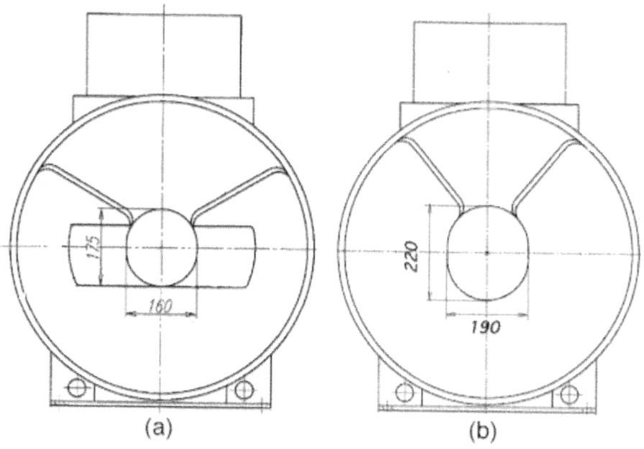

Fig. 3 Front view of the dewars. (a) for the child MEG; (b) for a conventional adult MEG

The dewar to keep liquid helium is a horizontal type optimized for the measurement in the supine mode. The sensor array described above is positioned inside of the dewar at the helmet side. The size of the helmet is shown in Fig. 3 in comparison with the conventional adult MEG. The aperture for the face was designed wider than that for the adult MEG in order to broaden the field of vision during the measurement. The liquid helium capacity of the dewar is roughly 100 liters, and the helium consumption rate is less than 6 liters/day. The sensor array is assembled using the *ship-in-a-bottle* approach [4], a method used to assemble our adult MEG systems. This technique enables to keep the diameter of the opening to access to the dewar inside smaller than the total size of the sensor array and as a result to reduce the liquid helium consumption rate.

The dewar and the sensor array is made of glass fiber reinforced plastic (GFRP). It is non-magnetic and effective in preventing magnetic artifacts. However, it is known that GFRP can be mechanically distorted when it is cooled in liquid helium and the position of a sensor can inevitably be dislocated from its originally designed value. The displacement is not predictable before cooling and it is not negligible in view of the accuracy of magnetic source analysis. Therefore, the positioning and calibration of each sensor is executed after cooling using a precisely machined coils array and standard electric current to produce standard magnetic fields [5].

C. Flux Locked Loop (FLL) and Data Acquisition Unit

A double integrator type FLL based on direct offset integration technique (DOIT) [6][7] is adopted to linearize the flux-voltage characteristics and to improve the dynamic range of the SQUID signal. The second feedback loop of the FLL provides an automatic offset adjustment. The effective frequency range of the FLL is 0.16 Hz – 5 kHz, consequently.

All the signals from the SQUID electronics are transferred to the data acquisition unit for digitization. The analog-digital conversion of the data acquisition unit has a resolution of 16 bits and a maximum sampling rate of 10 kHz. Sixty-four auxiliary analog inputs are reserved to monitor signals from electroencephalography and other peripheral instruments. The total number of the input channels is 128 and it is expandable when the number of the SQUID sensors is increased in the future.

III. AUDITORY EVOKED FIELD MEASUREMENT

In this section, the measurement of AEFs from a healthy 4-year-old male child is described. The child lay down on a bed and his head was placed inside the helmet of the dewar.

An online low pass filter of 200 Hz was applied to the signals from the 64 SQUID sensors. A high pass filter of 0.03 Hz was also applied to remove the residual baseline offset of the signals. The signals were digitally recorded at a sampling rate of 1 kHz. A 1-kHz tone burst stimulus was repeatedly delivered to the subject's ears binaurally through air tubes. The duration of the tone and inter stimulus interval (ISI) were 100 ms and 700 ms, respectively. One hundred responses synchronized with the sound stimulation were extracted and averaged after an offline digital band-pass filter of 1 – 100 Hz was applied. During the MEG measurement, the subject was instructed to watch a muted animation movie projected on a screen positioned in front of the subject. The animation didn't synchronize with the sound stimulation. This was an effective method to minimize the subject's head movement during the measurement.

Five marker coils were attached to the subject's head. Their positions, anatomical fiducial points of the head, and the shape of the subject's head were scanned with a stylus of a 3D digitizer system (Polhemus, Fastrak) before the AEF measurement. The source localization of the AEF was performed based on the scanned positional information.

The measurement lasted for 5 minutes including the preparation time. The head movement of the subject was monitored by measuring the location of the five marker coils before and after the auditory experiment. The displacement of each marker coils was 2.8 mm on average.

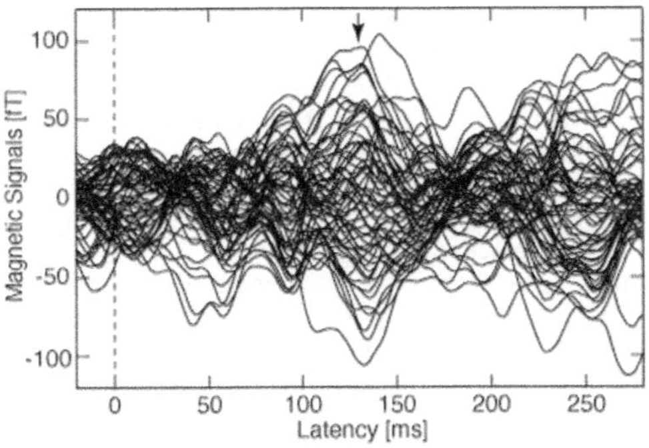

Fig. 4 Waveforms of AEF signals

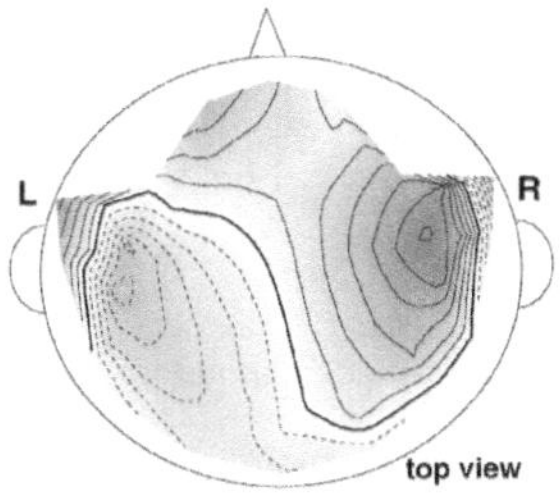

Fig. 5 Isofield contour map at 130 ms in latency

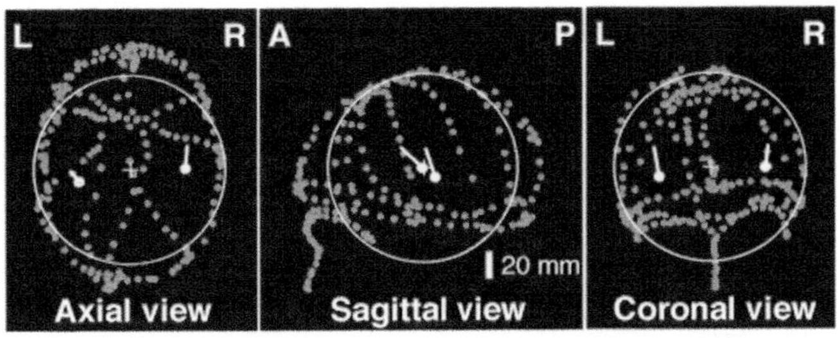

Fig. 6 The position of equivalent current dipoles

The overlapped 64 temporal waveforms of the response are displayed in Fig. 4. A large peak is clearly observed at about 130 ms in latency as indicated by the arrow. The isofield contour map at 130 ms in latency is shown in Fig. 5. Solid, dotted, and bold lines represent outward, inward and zero magnetic fields, respectively with a 15 fT step. The amplitude at the peak of the activity is of 104 fT.

For the magnetic source analysis, Sarvas's equation for the spherical conductor model [8] was used to fit equivalent current dipoles for the magnetic field distribution shown in Fig 5. The spherical model was defined based on the scanned head shape. Two dipoles were estimated in both hemispheres as a result. Their positions are shown together with the scanned head shape and the spherical model in Fig. 6. The intensities of the dipoles in the left and right hemispheres are 30.0 nAm and 30.7 nAm, respectively. The goodness of fit for the dipole estimation is 91.88%.

Each estimated current dipoles is located to an area reasonably corresponding to the temporal lobe. The large response at around 130 ms in latency is interpreted as the P100m response mentioned in a previous study [9]. This response shows a reverse polarity of the magnetic field distribution pattern compared to a typical adult M100 response. The 30 ms gap in latency is due to a delay introduced by the length of the air tubes used to deliver the stimuli.

IV. NUMERICAL EXPERIMENT

A numerical experiment was performed to estimate a hypothetical magnetic field distribution if the child AEF measurement were performed using a conventional adult MEG under the conditions described in the previous section. The purpose of this numerical experiment was to confirm whether the adult MEG could detect the AEF response of the child subject with a sufficient S/N ratio.

We assumed that our adult MEG system was applied to the child AEF measurement. Figure 7(a) shows the sensor arrangement of the adult MEG system superimposed to the child MEG sensor array in reference to the sphere brain model. Filled and unfilled circles indicate the position of the

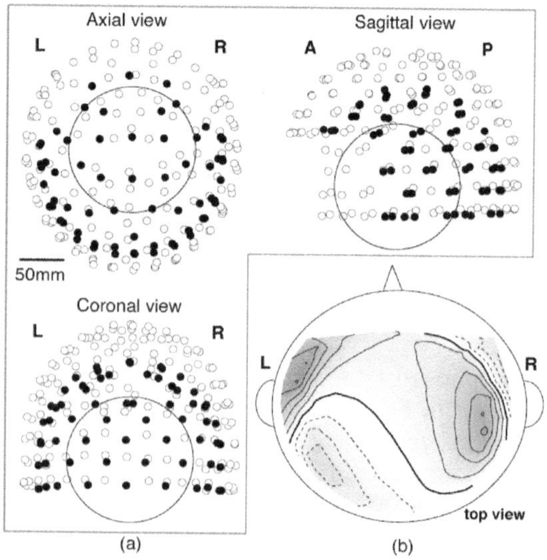

Fig. 7 (a) Sensor position relative to the sphere model; (b) an isofield contour map obtained in the numerical experiment

sensors of the child and of the sensors of the adult MEG system, respectively. The adult MEG sensor array was positioned so that the inferior positions of both adult and child MEG sensor arrays were aligned and concentric in the axial view. Two current dipoles, under the same conditions as stated in the previous section, were arranged in the spherical brain model. The distances between the sensor closest to each dipole and the center of the sphere for the child and adult MEG systems were 102.5±0.33 mm and 118.1±0.55 mm, respectively. The magnetic field distribution modeled by each sensor of the adult MEG was obtained using Sarvas's equation [8]. The resulting isofield contour map is shown in Fig. 7(b). The interval of contour lines was 15 fT as well as Fig. 5. The resulting magnetic field intensity was 77 fT at maximum. This indicates that the S/N ratio of the child AEF measurement by the adult MEG system becomes worse than one by the child MEG system.

To estimate how much this S/N ratio reduction effects on the magnetic source analysis, the equivalent current dipoles fitted to the calculated magnetic field distribution shown in Fig. 7(b) was inversely estimated. Before the dipole estimation, the magnetic field distribution was contaminated by the random values from a gaussian distribution (σ=13.37 fT) to include the realistic noise equivalent to the signal fluctuation in the pre-trigger period of the actual child AEF measurement shown in the previous section. The estimated dipoles were positioned 13.35±0.40 mm apart from their original positions determined from the child AEF measurement. The goodness of fit reduced to 83.36% from 91.88%.

According to the results of the numerical experiment, observation of the child AEF response might not be impossible even with an adult MEG system. However, the error in the source analysis will considerably increase because of the deterioration of the S/N ratio. It is necessary to increase the number of epochs to be averaged by almost twice (= $(104\ \text{fT} / 77\ \text{fT})^2$) in order to achieve the same S/N ratio achieved with the child MEG. This implies that the measurement time also has to be extended. A prolonged MEG experiment would be tedious and uncomfortable even for adult subjects. Therefore, it is beneficial for the child subjects and researchers to shorten the measurement time using the child MEG.

V. Conclusion

We developed a whole-head type MEG system for young children. To improve the fit of the MEG helmet to the size of a child's head, the size of the helmet-shaped sensor array was reduced by about 20 % when compared to a conventional adult MEG system. An auditory evoked field measurement was demonstrated and whole head distribution of the AEF was successfully acquired at a sufficient S/N ratio.

Acknowledgment

This study was partly supported by CLUSTER project, MEXT, Japan.

References

1. Kado H et al. (1999) Magnetoencephalogram systems developed at KIT, IEEE Trans Appl Supercond, 9: 4057–4062.
2. Jaycox JM, Ketchen MB (1981) Planar coupling scheme for ultra low noise DC SQUIDs, IEEE Trans Magn, M-17: 400–403.
3. Ketchen MB, Jaycox JM (1982) Ultra-low noise tunnel junction dc SQUID with a tightly coupled planar input coil, Appl Phys Lett, 40: 736–738.
4. Kado H, Method of assembling a magnetomeasuring apparatus, US patent, Patent number 5,896,645.
5. Higuchi M et al. (1989) The positioning of magnetic pickup coil in dewar by artificial signal source, Advance in Biomagnetism, Plenum Press, New York, pp. 701–704.
6. Drung D et al. (1990) Low-noise high-speed dc superconducting quantum interference device magnetometer with simplified feedback electronics, Appl Phys Lett, 57: 406–408.
7. Adachi Y et al (2007) A 75-ch SQUID biomagnetometer system for human cervical spinal cord evoked field, IEEE Trans Appl Supercond, 17: 3867–3873.
8. Sarvas J (1987) Basic mathematical and electromagnetic concepts of the biomagnetic inverse problem, Phys Med Biol, 32: 11–22.
9. Fujioka T et al (2006) One year of musical training affects development of auditory cortical-evoked fields in young children. Brain, 129: 2593–2608.

Author: Yoshiaki Adachi
Institute: Applied Electronics Laboratory, Kanazawa Institute of Technology
Street: 3 Amaike, Kanazawa
City: Ishikawa
Country: Japan
Email: adachi@ael.kanazawa-it.ac.jp

Characteristics of the Helium Circulation System on the 440 CH Vector Type MEG

Tsunehiro Takeda, Masayoshi Okamoto, and Keishi Katagiri

The University of Tokyo, Graduate School of Frontier Sciences, Kashiwa, Japan

Abstract— **We report the characteristics of the Helium Circulation System (HCS) mounted on the 440 CH vector type MEG of the University of Tokyo Japan, that re-liquefies all the evaporating helium gas. It collects warm helium gas about 300 K, cools it to about 40K and returns it to the neck tube of the Dewar of the MEG to keep it cold. It also collects helium gas just above the liquid helium surface while it is still cold, re-liquefies and returns it to the Dewar. A special transfer tube (TT) about 2 m length with 7 multi-concentric pipes was developed to allow the dual helium streams. It separates the HCS with a MEG to reduce magnetic noise. A refiner to collect the contaminating gases effectively by freezing the gases is developed. It has an electric heater to remove the frozen contamination in the form of gases into the air. A gas flow controller is also developed, which automatically control the heater to cleanup the contamination.**

The developed TT has very low heat inflow less than 0.1W/m to the liquid helium ensuring the efficient operation. The HCS can re-liquefy up to 35.5 l/D of liquid helium from the evaporated helium gas using two 1.5W@4.2K GM cryocoolers (SRDK-415D, Sumitomo Heavy Industries, Ltd.). Reduction of noise level induced by the HCS has been conducted and confirmed to be below 10 fT/Hz$^{1/2}$ for 2-40 Hz, below 30 fT/Hz$^{1/2}$ at 1 Hz, and 200 fT/Hz$^{1/2}$ at 50 Hz, which is the power supply frequency. Our MEG system has been used in real brain experiments without any problem over two years. Diameter of insert tube is reduced to the standard 1/2 inch. As the amount of liquid helium used decreases less than one percent, the maintenance cost of the MEG becomes less than one-tenth of the previous cost.

Keywords— **Helium, Reuse, Circulation, Condensation, MEG.**

I. Introduction

Magnetoencephalography (MEG) is becoming common throughout the world and, in the near future, will probably become indispensable for measuring brain functions [1]. However, they are very expensive to run because of their cooling system. They use about 10 litters per day (l/D) of liquid helium (LHe), and commonly waste all of it by letting it escape into the atmosphere, necessitating the troublesome task of refilling the Dewar with LHe once or twice per week, which must be done by a trained technician.

The most common and efficient LHe producing system uses a Collins-type liquefier [2]. As this system uses very high pressure, it is very large and its capacity far exceeds that required for MEGs. Although a cooling system that can achieve LHe temperatures by direct cooling with a small cryocooler has been developed, it is too noisy for MEGs [3]. Other small systems that collect the evaporating helium at room temperature and return it to the Dewar after liquefaction have also been developed (e.g., TRG-350D, TaiyoNissan Co., Ltd, Tokyo; HRT-K212, Sumitomo Heavy Industries, Ltd., Tokyo). They cool the collected evaporated helium to about 40 K using sub-cryocoolers, then to below 4.2 K using the main cryocoolers, and then return the liquefied helium to the cryostat. However, the liquefaction requires much electricity because the evaporated helium approaches room temperature before being cooled and because the specific heat capacity of helium is relatively high. Moreover, they are also still noisy for MEGs. Therefore, at present, there is no cooling system specifically suited to MEGs.

As MEG sensor coils must be placed near to the patient's head to detect their very weak magnetic fields, a liquid nitrogen heat shield, commonly used for Magnetic Resonance Imaging (MRI), cannot be used with MEGs. Hence, LHe is used to cool the Dewar as well as the Super Conducting Quantum Interference Devices (SQUIDs). In fact, because SQUIDs produce little heat, the LHe is mostly used for the former purpose, which is very inefficient. It would be more efficient to use relatively higher temperature helium gas (HeG) rather than LHe, to cool the Dewar [4-5].

In light of all the above serious drawbacks of the existing cooling systems, the aim of the present study was to develop an efficient small cooling system for MEGs that has a noise level low enough, does not need to have the helium frequently refilled, and is cheap to maintain.

II. Methods

The main principle of our system is to use relatively warm HeG to counter heat flowing into the Dewar from the surroundings, while using the LHe to cool the SQUID. The basic design is outlined in Fig. 1. In the Dewar, the HeG near the surface of the LHe will be much colder than the HeG near the top of the Dewar. The colder HeG goes through pipe B to the condenser at the second cooling stage where it is liquefied, and then the LHe flows under gravity back to the Dewar through pipe A. The outlet of pipe A is near the surface of the LHe and about 10 mm below the

S. Supek and A. Sušac (Eds.): Advances in Biomagnetism – BIOMAG2010, IFMBE Proceedings 28, pp. 39–42, 2010.
www.springerlink.com

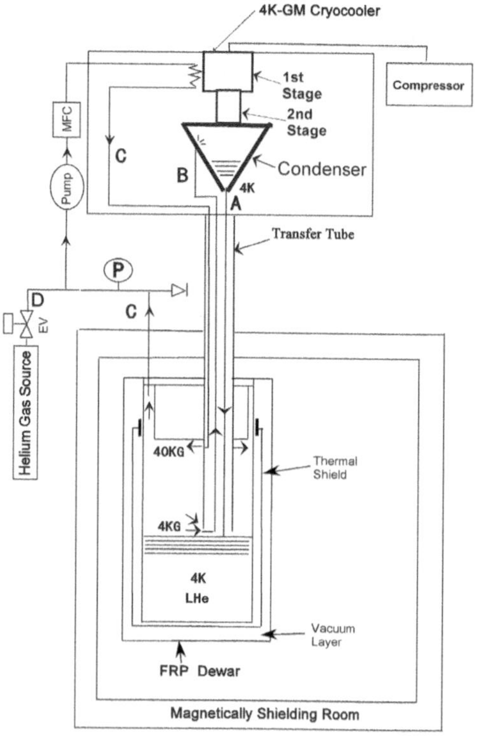

Fig. 1 Helium circulation system. A: pipe to lead LHe from condenser to Dewar, B: pipe to lead lower temperature HeG from Dewar to condenser, C: pipe to circulate higher temperature HeG, D: pipe to add pure HeG. EV: electric valve, MFC: mass flow controller

inlet of pipe B. The warmer HeG, which is at about 40 K at the outlet of pipe C, cools the Dewar as it passes through its neck. The warmed HeG is led to the first stage of the cryocoolers at a flow rate governed by a small pump and a mass flow controller (MFC). The outlet of pipe C is about 200 mm above the outlet of pipe A. Having the outlet of pipe C high above the LHe ensures that the temperature gradient between the outlet of pipe C and the LHe is relatively small, thus lowering the heat flow into the LHe. If the LHe level drops for some reason, the helium gas from a reservoir can be supplied through pipe D and liquefied, which recovers LHe level.

Since the amount of helium that evaporates, and thus the amount that has to be liquefied, depends on the ambient temperature, a large capacity cryocooler is essential to cope with a wide range of ambient temperature. On the other hand, if the cryocooler liquefies too much helium, the pressure inside the system may drop, causing air to flow into the system. Therefore, our system uses two 1.5W@4.2K GM (Gifford-McMahon) cryocoolers to ensure sufficient cooling capacity and, under feedback control from the pressure in the Dewar, a 3 W heater attached to the condenser to prevent overcooling.

A transfer tube is attached to the Dewar such that a vacuum separating pipes A, B and C, which are concentric, is a continuum of the vacuum in the wall of the cold chamber. The vacuum also separates pipe C from the ambient air. The heat flowing from the surroundings to the LHe through the transfer tube was estimated to be lower than 0.1 W/m.

III. Results

In a preliminary experimental set up used to measure the performance of the system, there was an additional heater at the bottom of an experimental Dewar which was used to measure the residual capacity of the cryocoolers to maintain the level of the LHe. The capacity was found to be 1.1 W. About 35.5 l/D of helium was evaporated from the Dewar, when the TT was not attached to the Dewar and 1.1 W of heat to the LHe was supplied. In the real situation, some more heat would be flowing from the inserted TT. Therefore, it was confirmed that at least 35.5 l/D of LHe could be re-liquefied in the experimental system. This amount is adequate for nearly all the existing MEG systems, which require about 10 l/D.

Fig. 2 is a photograph of the set up that we used to test the cooling performance of the HCS installed to real MEG. We fixed the TT to the wall of a Magnetically Shielded Room (MSR).

Fig. 3 shows how the level of LHe (LL) in the Dewar changed according to the flow rate (FR) of 40K helium gas to the neck tube of the Dewar for eight days. LL decreased, when FR was 6 or 5 l/m during the first two days. Then FR was set to 10 l/m from day three and was increased stepwise to 14 l/m. LL began to increase after the FR was set over 10 l/m at a rate of about 2.1 l/D on average (the diameter of the

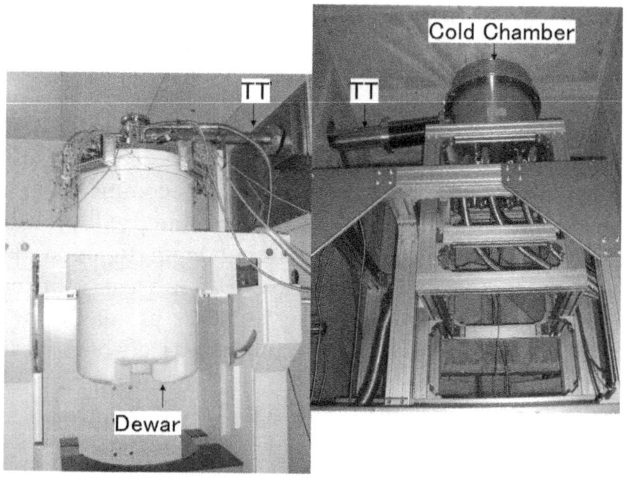

Fig. 2 MEG and HCS in operation

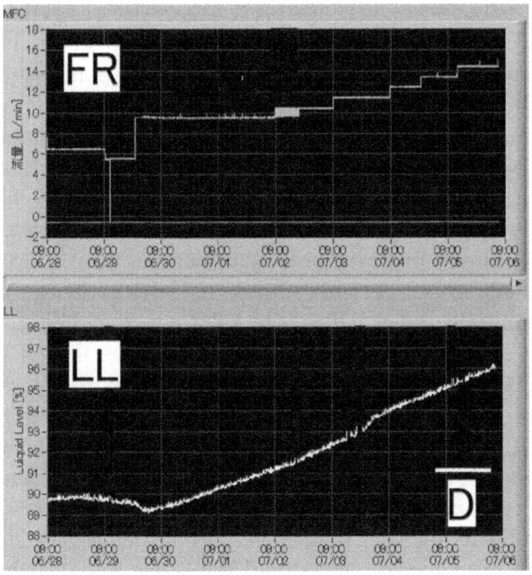

Fig. 3 The change of liquid helium level (LL: lower panel) in relation to flow rate (FR: higher panel)) of the 40K helium gas for 8 days

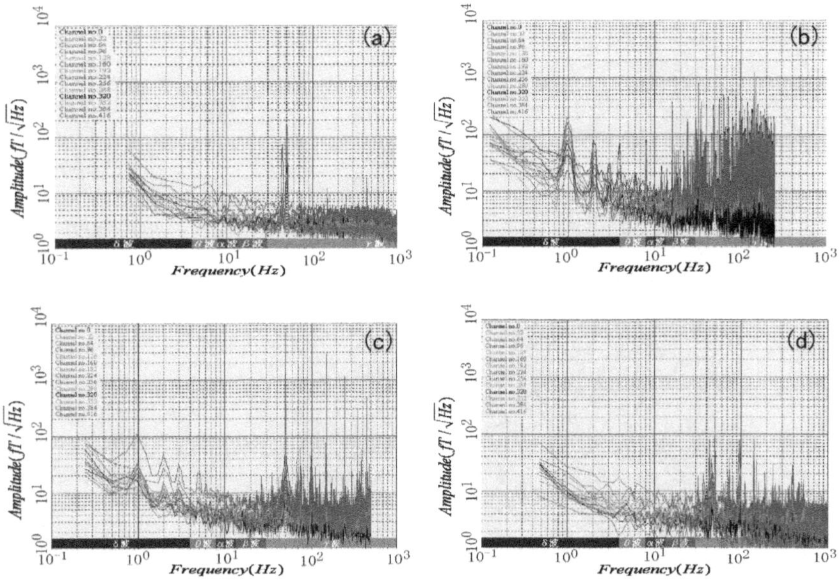

Fig. 4 Noise amplitude spectra of the selected 14 channels of the system, (a) before HCS installation, (b) just after HCS installation, (c) after stiffness increase of the HCS mount, and (d) after grounding improvements

Dewar was 488 mm and 1 % was roughly equal to 1 cm). Increase rate did not change proportional to FR, but was rather stationary. As the increase of the LL in the experimental Dewar was 5.5 l/D, it was estimated the Dewar of our MEG had more helium evaporation by 3.4 l/D. It is natural, because the MEG Dewar has 44CH SQUIDs and larger volume.

Acoustic noise in the hollow of the Dewar (see Fig. 2) measured using a sound level meter (Yokokawa Ltd., DT-805), was 34.7 dB, while the cryocoolers were running. As there is a hole (300x250) in our MSR to allow visual image projection by a liquid crystal projector, the acoustic noise level could be lowered to 33 dB if the hole was closed. This is small enough for MEG measurements.

Just after the HCS was installed to the MEG, magnetic noise was very large with about 1 pT@1Hz and 10 pT@50Hz. As the large noise mainly came from vibration induced by the GM cryocoolers, the rigidity of the base for the GM cryocoolers was increased. Various improving methods for the electric ground were also applied and the magnetic noise at 50Hz was decreased. Tentatively, the magnetic noise has been decreased to 40 fT@1Hz and 220 fT@50Hz. The magnetic noise level is lower than 10 fT between 2 and 35 Hz.

Fig. 4a depicts the amplitude spectrum of magnetic noise measured by the MEG before the HCS was installed. The noise levels are roughly below 10 $fT/Hz^{1/2}$ for 2-1000 Hz, similar to that in common MEGs. The amplitude spectrum has a typical 1/f noise pattern below 2 Hz. It is below 35 $fT/Hz^{1/2}$ at 1 Hz and 170 $fT/Hz^{1/2}$ at 50 Hz, which is the power supply frequency.

Fig.4b depicts the amplitude spectrum of magnetic noise measured just after the HCS was installed. It shows noise of about 200 $fT/Hz^{1/2}$ at 1 Hz, which was apparently produced by the vibration of the TT driven by the GM cryocoolers. The GM cryocoolers produce a strong 1 Hz vibration when the compressed helium gas expands adiabatically in the cryocoolers. The amplitude spectrum also has a very large noise spectrum at other frequencies, especially at the higher harmonics of 1 Hz and 50 Hz.

To reduce the noise levels, thick plates were placed between the aluminum frames where the cryocoolers are mounted to act as stiffeners. Fig. 4c shows the amplitude spectrum of magnetic noise measured after the modification of the cryocooler stage. The magnitude of the amplitude spectrum at 1 Hz was reduced considerably, to about 100 $fT/Hz^{1/2}$. However, there remained considerable noise caused by the HCS, especially at 50 Hz and its higher harmonics. The noise amplitude at 50 Hz was about 4 $pT/Hz^{1/2}$, which was more than ten times of that in Fig. 4a.

Various improvements were made to reduce the noise added by the HCS. As the details of those measures are not in the scope of this paper, they are not explained here. The most effective measure was to reduce current leak from the power supplier to the TT by various methods in the electrical grounding of the MEG and the HCS. Fig. 4d depicts the amplitude spectrum of magnetic noise measured after the improvements. The noise level at 1 Hz is reduced to less than 30 $fT/Hz^{1/2}$, which is a bit smaller than that in Fig. 4a. The noise level at 50 Hz is reduced to 200 $fT/Hz^{1/2}$, which is still a little bit bigger than before installation. There are also slight noise increases in the range of 30-50 Hz and at the higher harmonics of 50 Hz.

Fig. 5 Developed HCS with a 1/2 inch insert tube

The most serious problem remained to be improved was that the HCS had an insert tube with diameter of 3/2 inches. As all the existing MEGs have 1/2 inch insert hole to refill liquid helium, it required the extensive redesign or reconstruction of MEGs to use the developed HCS. Hence, we have tried to reduce the diameter of the insert tube to 1/2 inch. Though it was very hard to reduce the diameter of the insert tube to 1/3, we managed to achieve the goal as shown in Fig. 5.

IV. Discussion

MEGs would be much cheaper to run if high-Tc SQUIDs, cooled by liquid nitrogen, could be used. Unfortunately, high-Tc SQUIDs are currently too noisy, so costly LHe systems must be used. Our new LHe cooling system would reduce the running cost of MEGs because almost none of the He, a rare material, is allowed to escape wastefully into the atmosphere, eliminating the hazardous task of replacing lost LHe, which requires trainings. Also, although helium itself is not a major pollutant, because our system consumes less energy, it is indirectly more environment friendly.

The developed HCS is under modification to allow measurement posture using a new multiple tube flexible TT. Though we designed our system for MEGs, our system, with its advantages, can be used for other devices in the field of cryogenics, which is now on the way. It might even be suitable for magnetic resonance imaging (MRI): Though the reduction in running costs for MRI would not be so high, because they commonly use nitrogen thermal shields which reduces helium consumption, our system could be used to downsize the MRI Dewar because a large nitrogen container would no longer be necessary.

Our system can produce at least 35.5 l/D of LHe from the evaporated helium with two 1.5W@4.2K GM cryocoolers. The total cooling capacity of the system was estimated to be about 2.4 W by heating the condenser, because there was inevitable heat invasion at the cold chamber and the TT.

The magnetic noise level between 1 to 35 Hz is about the same with the noise level before the HCS was installed. As the noise level at 50 Hz was about 200 fT before the HCS installation, the noise level at the 50 Hz is a little bit larger. However we expect it will be reduced to the noise level before the HCS installation in the near future. Anyway, if an appropriate band pass filter or data averaging is adopted as usual, these noises have virtually no bad effect. Furthermore, if very strict measurement is required, the HCS could be stopped during measurement and the noise level is the same as that before HCS installation.

V. Conclusions

We developed a new helium circulation system (HCS) for MEGs. The HCS uses two 1.5W@4.2K GM cryocoolers, and has dual helium streams; one to collect evaporated HeG immediately and returns it as liquefied helium to the Dewar, and the second to use higher temperature HeG (about 40 K) to cool the Dewar. We developed a special transfer tube with seven concentric tubes to reduce the amount of heat flowing into the system. As the diameter of the TT insert tube is the standard 1/2 inch, no modification of MEG dewar itself is necessary. The HCS can liquefy at least 35.5 l/D of evaporated LHe. It can increase 2.1 l/D of liquid helium with our 440CH MEG system. The HCS adds virtually no acoustic and magnetic noises to the MEG.

References

1. Takeda T, Owaki T, Haruta Y, Uehara G (2004) Characteristics of a 440ch MEG system with vector sensors, *Biomag 2004*, 640-641
2. Kang Y *et. al.*(1998) Cryocooled SQUID system, *J. of Japan Biomagnetism and Bioelectromagnetics Society*, **11,** 1, 15 (in Japanese)
3. Takeda T, Takae T (2001) A liquid helium circulation system for biomagnetometers, *Biomag 2000*, 951-952
4. Takeda T, Okamoto M, Katagiri K (2008) An efficient helium circulation system with small GM cryocoolers, J. of Cryogenics, 48, 1, 6-1
5. Takeda T, Okamoto M, Katagiri K (2009) Performance of a helium circulation system for a MEG, J. of Cryogenics, 49, 2, 144-150

Author: Takeda T
Institute: The University of Tokyo
Street: 318 Kibanto, 5-1-5 Kashiwanoha
City: Kashiwa, Japan
Email: takeda@brain.k.u-tokyo.ac.jp

Correction of Disturbances Caused by Temporary and Continuous Head Movements

J. Nenonen[1], J. Nurminen[2], D. Kičić[2,4,6], R. Bikmullina[2], P. Lioumis[2], V. Jousmäki[5], S. Taulu[1], L. Parkkonen[1], M. Putaala[1], and S. Kähkönen[2,3]

[1] Elekta Oy, P.O. Box 34, FIN-00531 Helsinki, Finland
[2] BioMag Laboratory, HUSLAB, and [3] Pain Clinic, Hospital District of Helsinki and Uusimaa, Helsinki University Central Hospital, Finland
[4] Department of Biomedical Engineering and Computational Science, and [5] Low Temperature Laboratory, Helsinki University of Technology, Finland
[6] Current address: Viševica Komp d.o.o., Jordanovac 47, 10000 Zagreb, Croatia

Abstract— **This study was performed to assess the effectiveness of the spatiotemporal Signal Space Separation (tSSS) and head movement compensation (MC). Continuous MEG data were acquired from 20 adult and 10 child subjects in four conditions: (1) A reference with a static head position, (2) the subject changed the head position twice, (3) the subject moved the head constantly, and (4) two magnetized pieces were attached to the scalp and the subject moved the head. The recordings were repeated with two whole-head MEG systems.**

Auditory and tactile stimuli were applied. The tSSS and MC were performed on the recordings, and evoked fields were selectively averaged. Neuronal activity was modeled with equivalent current dipoles (ECDs) whose locations and strengths were fitted to the auditory and somatosensory evoked fields (AEF and SEF). ECD parameters were compared to those obtained from the reference responses.

The average differences of the dipole localizations from the reference recordings with the two systems were below 10 mm. In condition 1, the responses were unaffected by tSSS. In conditions 2-4, the head position could be tracked reliably, and tSSS suppressed the interference caused by magnetized impurities. In condition 2, the average dipole localization difference from the reference location was below 10 mm both for SEF and AEF. Dipole waveforms and amplitudes were similar to the reference results. In conditions 3 and 4, continuous and regular movements influenced the AEFs and triggered bursting rhythmic activity from the motor cortex and muscles moving the head. Consequently, the average AEF dipole localization differences were between 10 and 15 mm and the amplitudes were about 40% smaller than the references. The SEF results were, however, similar to the reference results.

In summary, tSSS and MC effectively suppressed disturbances caused by head movements and magnetized impurities.

Keywords— **Repeatability, Movement compensation, Signal space separation, AEF, SEF.**

I Introduction

Disturbances caused by head movements and magnetized impurities in/on the head have been problematic in MEG studies. The Signal Space Separation (SSS) [1] is a novel signal processing method capable of suppressing interference originating outside of the MEG sensor array. The spatiotemporal SSS (tSSS) method [2] improves the software shielding capability by removing artifacts caused by interference sources inside the sensor array, such as magnetized material on the head. In addition, the SSS method can be used for transforming data between different head positions and for head movement compensation (MC) [1].

The aim of this study was to assess the effectiveness of the tSSS and MC methods. We present the results obtained from MEG recordings of 30 subjects under four conditions, repeated with two MEG devices. Ref [3] presents the application of tSSS and MC on repeated MEG recordings in a single subject.

II. Methods

A. Recordings

MEG measurements were performed with the Elekta Neuromag® systems at the BioMag Laboratory of Helsinki University Central Hospital (BioMag), and at the facility of Elekta Oy (ENM), both in Helsinki, Finland. BioMag has a standard two-layer magnetically shielded room. ENM employs a lightweight, single-shell magnetically shielded room with an internal active shielding system (MaxShield™, Elekta Oy).

Two study populations were examined: (i) 20 healthy adult volunteers (age 23 – 38 years, 7 males and 13 females), (ii) 10 healthy children (5 children between 8 – 12 years and 5 children between 3 – 6 years). In the older children, the head sizes of the subjects are comparable with the adult population, while the children younger than 7 years have smaller heads. In this analysis, the head size (and thus the distance from the cortex to the MEG detectors) was considered more relevant than the age, because each subject served as his/her own reference. Thus, all subjects were pooled into one study population. Adult subjects and the parents of the children gave a written consent. The study protocol was approved by the ethical committee of the hospital.

Interleaved auditory and somatosensory stimuli were presented to the subject. The auditory stimulus was a 1-kHz sinusoidal signal, cut into a 50-ms pulse with a Hanning window. Presentation software (Neurobehavioral Systems, Inc., Albany, CA, USA) was used to produce a binaural stimulus tone every 2 seconds through a panel speaker mounted on the inner wall of the magnetically shielded room. The somatosensory stimuli were given by a tactile brush stimulator [4], operated by a nurse who was touching the root of the left thumb of the subject with the tactile stimulator once in the middle of each auditory stimulus interval. Both stimuli were repeated 150 times, and one recording lasted about 5 minutes.

At both sites recordings were made under four conditions:

1. **Ref1**: Reference measurement, the subject was asked to keep the head as still as possible. The spatial SSS was used to obtain the reference data.
2. **Dev2**: Temporary movement, the subject was first asked to keep the head still in the reference position and then move the head twice during the recording.
3. **Mov3**: Continuous movement, the subject was asked to move the head all the time inside the MEG helmet.
4. **Imp4**: Two small magnetized pieces were attached to subject's forehead and chin and the subject was asked to do continuous head and mouth movements.

The sampling rate and recording passband were, respectively, 600 Hz and 0.1–172 Hz at BioMag, and 1000 Hz and 0.1–330 Hz at ENM. The location of subject's head relative to the MEG sensor helmet was measured using four Head Position Indicator (HPI) coils attached to the scalp. Continuous HPI tracking was applied while raw data were recorded, and the head positions were calculated off-line [5].

B. Interference Suppression

Disturbances caused by *internal* interference arising from magnetized pieces on the subject's head and head movements were previously considered as exclusion criteria in MEG studies. Here, we utilized the tSSS method to suppress artifacts caused by the magnetized pieces, and SSS was then applied to transform data from the estimated head positions into a static reference position.

The tSSS and MC processing was performed with the MaxFilter™ software (Elekta Oy). In order to improve the signal to noise ratio, evoked fields were thereafter averaged in windows starting 200 ms before and ending 400 ms after stimulus. An epoch was rejected if a magnetometer signal magnitude exceeded 5000 fT, a gradiometer signal magnitude exceeded 3000 fT/cm, or the eye-movement EOG signal magnitude exceeded 200 μV.

C. Equivalent Current Dipole Analysis

The averaged response sets were analyzed by modelling the neuronal activity with equivalent current dipoles (ECDs) whose parameters (locations and strengths as a function of time) were fitted to the response data [6]. A homogeneous spherical head model was used, with the origin at (0, 0, 40 mm).

The baseline interval was set between -50 and 0 ms, and a low-pass filter with a 100-Hz corner and a 10-Hz transition band was applied. A single ECD was used to localize the source of the SEF peak response between 40 – 70 ms [4] in the somatosensory cortex.

For AEF, a low-pass filter with a 40-Hz corner and a 5-Hz transition band was applied. Two channel selections consisting of about 150 channels were applied one after the other over each primary auditory response area. A single dipole at the maximum N100m response amplitude was fitted for both hemispheres.

After finding the dipole locations, all 306 channels were used to fit the dipole amplitudes as a function of time [6]. Altogether, 120 SEF and 240 AEF dipoles were fitted at both sites. Dipole amplitudes as a function of time were saved both for reference and test sets. The two time series were compared by shifting the latter one by -20 ms to +20 ms in 1 ms steps with respect to the first one. Pearson correlation coefficients between the time series were calculated. The shift producing the maximum correlation coefficient represents a latency delay between the responses.

A data set was excluded from the comparisons if

- The subject had magnetized material other than the controlled impurities in condition Imp4. In such cases the reference results could not be obtained with SSS.
- The confidence volume of the dipole fit was over 2000 mm^3.
- The response was weak and non-dipolar.

In addition, some data sets were excluded for technical reasons:

- Some Dev2 or Mov3 recordings of small children were not performed to shorten the measurement session.
- Two raw data files were missing or corrupted.
- In two recordings, the continuous head position tracking was not switched on.
- If the landmark digitizations had a mismatch larger than 8 mm, the subject was omitted from Ref1 comparison between the two devices.

III. RESULTS

A. Visual Comparison

The processed response sets were overlaid with the reference set. Visual inspection was focused on the intervals of the baseline and primary responses: from -50 to +70 ms for SEF and from -50 to +150 ms for AEF responses. Both AEF and SEF peak polarities and latencies were similar within stimulus modality in all recordings. Notably, there were no MEG channels showing differing peak polarities.

B. Repeatability from Static Head Positions

The differences in SEF and AEF dipole fittings between the two recordings with a static head position are presented in Table 1. The 5 children younger than 7 years were excluded because of smaller head sizes.

Table 1 Differences of the ECD localizations between BioMag and ENM reference recordings in 20 adult subjects and 5 children over 7 years. *N* denotes to the number of accepted response sets

	SEF	AEF left	AEF right
Mean [mm]	8.4	8.5	8.3
Stdev [mm]	3.6	6.3	5.0
N	16	15	16

C. ECD Analysis

Tables 2, 3 and 4 present summaries of the ECD analysis results of somatosensory, auditory left and auditory right responses, respectively, over the whole subject population. The BioMag and ENM results were pooled in these tables.

Table 2 The SEF dipole analysis results, mean (1st line) and standard deviation (2nd line). *N* indicates the number of accepted sets, Diff means the 3D distance between the fitted and reference ECD localizations, Q is the average dipole moment strength, and correlation coefficient CC and shift indicate the similarity of the shape and delay of latencies between the test and reference ECD amplitude waveforms. Conf. denotes to the 95% confidence region of the dipole fit

Test	*N*	Diff [mm]	Q [nAm]	CC	Shift [ms]	Conf [mm^3]
Refl	54	1.5	31	0.997	0.0	58
tSSS		1.6	11	0.006	0.0	125
Dev2	42	6.6	32	0.972	0.2	33
tSSS+MC		3.1	12	0.031	0.5	38
Mov3	49	8.5	31	0.952	0.4	42
tSSS+MC		3.8	9	0.045	0.8	81
Imp4	52	9.1	29	0.949	0.4	69
tSSS+MC		4.6	11	0.059	1.2	138

Table 3 Summary of AEF left hemisphere dipole analysis results. The labels are as in Table 2

Test	N	Diff [mm]	Q [nAm]	CC	Shift [ms]	Conf [mm^3]
Refl tSSS	51	2.6	29	0.988	0.0	188
		2.8	14	0.040	0.0	199
Dev2	40	7.8	29	0.973	2.0	168
tSSS+MC		4.4	13	0.031	2.6	234
Mov3	36	12.2	22	0.919	5.7	457
tSSS+MC		6.4	16	0.063	3.7	557
Imp4	42	12.9	18	0.925	3.9	693
tSSS+MC		6.8	10	0.058	4.5	561

Table 4 Summary of AEF right hemisphere dipole analysis results. The labels are as in Table 2

Test	N	Diff [mm]	Q [nAm]	CC	Shift [ms]	Conf [mm^3]
Refl tSSS	49	2.2	30	0.992	0.3	185
		2.6	15	0.029	1.3	287
Dev2	39	7.9	32	0.973	1.8	108
tSSS+MC		4.8	14	0.040	2.4	123
Mov3	38	11.5	24	0.926	5.6	408
tSSS+MC		6.6	16	0.071	4.6	437
Imp4	35	12.0	20	0.929	4.3	466
tSSS+MC		6.1	11	0.057	3.4	441

IV. CONCLUSIONS

A. Environmental Interference

The two measurement sites exhibit very different environments for MEG recordings. The BioMag Laboratory represents a typical measurement site at an urban hospital where the environmental interference level is moderate. The Elekta office facility is in close proximity to heavy car traffic and a subway line, and shows ten times higher ambient noise levels at low frequencies. Internal active shielding reduces efficiently the environmental interference, and the SSS methods suppress the remaining interference. The remaining background noise levels were comparable at BioMag and at ENM.

B. Repeatability

Generally, differing head positions cause the largest inter- and intra-subject variations in waveforms and amplitudes at individual sensors. Main part of the data analysis focused on dipole modelling which expresses the locations of neuronal activity in a head-based coordinate system and is thus less sensitive to the differing head positions.

The differences between the two reference recordings in Table 1 are similar to those in earlier repeatability studies (see [3] and the references therein). Thus, the different environments and magnetical shielding methods did not cause differences in evoked fields.

Response latencies were very consistent across the repeated recordings. Dipole amplitude and localization exhibited more variability, likely due to factors such as inaccuracies in landmark digitization, variation in the way the brush stimulator was operated, and differences in the loudness of the auditory stimulus. Variability of the responses in children was generally larger than that in adults.

C. *Refl, tSSS*

The results in Tables 2–4 show that tSSS changes the ECD localization on the average less than 2 mm and does not affect the latency. Also the amplitude changes are small. Thus, it may be concluded that tSSS does not affect the brain signals.

D. *Dev2, tSSS and MC (Temporary Movement)*

The ECD localization differences are less than 10 mm and there are no amplitude or latency changes. The results show that tSSS and MC are very efficient in temporary head movements. The results of the condition Dev2 show smaller localization differences than the comparisons of Refl recordings between the two measurement sites.

E. *Mov3, tSSS and MC (Regular Movement)*

Tables 3 and 4 show that the AEF dipole moments are about 30% smaller than the ECD moments obtained from Ref1. Also, the localization differences are on the average 40% larger than the differences between the two reference recordings. Notably, the average SEF dipole moments and localizations in Table 2 are similar to the reference results.

Almost all subjects moved their head periodically combining left–right rotations, up–down nodding and tilting to left or right. A movement period was relatively fast, typically 2–4 seconds. Such continuous movements affect AEF responses in the following ways: i) they generate additional acoustic noise, e.g., due to hair rubbing against the helmet, ii) they decrease the auditory stimulus strength when the ear turns away from the speaker, iii) they generate bursts of rhythmic background activation in the motor cortex and in the muscles moving the head, and iv) direct the attention away from the stimuli.

Such background effects are inherent to the physiological changes during head motion and thus cannot be eliminated by SSS or tSSS. Despite this activity is not time-locked to stimuli, the amplitudes may be so strong that the disturbing background signals still contribute after averaging 100 epochs. Thus, decreased stimulus strength and physiological background activation are the most likely cause of lower AEF signal amplitudes and dipole moments, slightly delayed peak latencies and shifted dipole localizations. Notably, the background effects caused much less distortions in the SEF.

F. *Imp4, tSSS and MC (Magnetized Pieces and Movement)*

The AEF dipole localizations and amplitudes showed a similar degradation as in the condition Mov3. The SEF results were, however, similar to the reference results. The tSSS method was able to clean efficiently the interference caused by moving magnetized pieces. Thus, the AEF dipole analysis problems were due to the same factors as in the condition Mov3, because most subjects moved their heads in a similar regular manner.

G. *Summary*

The tSSS and MC methods facilitate analysis of MEG data contaminated by disturbances due to head movements and magnetized material on the head. Such suppression of interference originating from inside the sensor helmet is difficult to achieve by other analysis or noise reduction techniques.

References

1. Taulu S, Kajola M. (2005) Presentation of electromagnetic multichannel data: The signal space separation method. J Appl Phys 97: 124905 (1-10).
2. Taulu S, Simola J (2006) Spatiotemporal signal space separation method for rejecting nearby interference in MEG measurements. Phys Med Biol 51: 1759-1768.
3. Nenonen J, Parkkonen L, Taulu S, et al (2010) Repeatability of AEF and SEF from static and moving head positions. This volume.
4. Jousmäki V, Nishitani N, Hari R (2007) A brush stimulator for functional brain imaging. Clin Neurophysiol 118: 2620-2624.
5. Uutela K, Taulu S, Hämäläinen M (2001) Detecting and correcting for head movements in neuromagnetic measurements. NeuroImage 14: 1424-1431.
6. Hämäläinen M, Hari R, Ilmoniemi R, Knuutila J, Lounasmaa O (1993) Magnetoencephalography—theory, instrumentation, and applications to noninvasive studies of the working human brain. Rev Mod Phys 65: 413-497.

Author: Jukka Nenonen
Institute: Elekta Oy
Street: Siltasaarenkatu 18-20
City: Helsinki
Country: Finland
Email: jukka.nenonen@elekta.com

Video-MEG: Integration of Digital Video to MEG Epilepsy Recordings

Juha Wilenius[1], Andrey Zhdanov[1], Erik Larismaa[2], Lauri Parkkonen[2], Matti Kajola[2], Jaakko Lyytinen[2], Antti Ahonen[2], Ossi Miikkulainen[3], Jyrki P. Mäkelä[1], and Ritva Paetau[1]

[1] Biomag-laboratory, HUSLAB, Helsinki University Central Hospital, Helsinki, Finland
[2] Elekta Oy, Helsinki, Finland
[3] Helsinki University of Technology, Espoo, Finland

Abstract— **Video-EEG is the gold standard in clinical epilepsy diagnostics. MEG systems have traditionally lacked video capabilities. Here we describe the addition of a synchronized digital video recording (DVR) system to a standard Elekta Neuromag® 306-channel MEG device operated at BioMag laboratory of the Helsinki University Central Hospital. We also provide some insights obtained from using the system for ictal MEG recordings. Video and audio were recorded simultaneously with MEG, using a separate digital video recording station. Synchronization was achieved by timestamping the video, audio and MEG streams and synchronizing all parts of the system via Network Time Protocol (NTP). Video was recorded at 25 fps with an infrared-sensitive camera necessary to study patients that have seizures during sleep. MEG and video are synchronized to the maximum achievable accuracy of one video frame. At this time, five epileptic patients have undergone video-MEG recordings with the setup described above. In three patients, ictal recordings were obtained. In two cases the video considerably facilitated the exact temporal localization of the seizure onset, thus reducing the amount of time required for data analysis and improving the reliability of the spatial analysis.**

Keywords— **video-meg, meg, epilepsy, ictal, clinical.**

I. INTRODUCTION

Epilepsy source localization and mapping of the functional areas during preoperative assessment is a well established clinical application of magnetoencephalography [1]. Ictal recordings are now feasible due to new artifact rejection methods and continuous head position tracking [2].

Video-EEG has been for years a gold standard of epilepsy diagnostics [3]. Video has proved extremely helpful in identification of various artifacts, in documenting seizure behavior, and in timing of behavioral seizure onset relative to its electrophysiological correlates.

Commercially available MEG systems have traditionally lacked the ability to record and analyze video of the clinical manifestations of epileptic seizures. In this paper we describe a system modified to allow such video recordings and describe an example of data analysis.

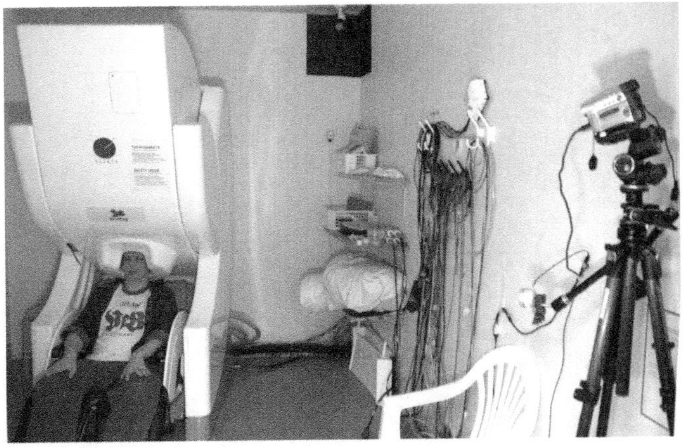

Fig. 1 Video-MEG setup inside the shielded room. One of the researchers seated in the MEG

II. MATERIALS AND METHODS

A. Data Acquisition and Analysis

MEG was recorded with a standard Elekta Neuromag® 306-channel MEG device operated at BioMag laboratory of Helsinki University Central Hospital.

Video and audio were recorded simultaneously with MEG using a separate digital video recording station by Mirasys Ltd., (Helsinki, Finland). Synchronization was achieved by inserting real-time stamps in the video, audio and MEG streams while keeping the real-time clocks of all parts of the system synchronized via Network Time Protocol (NTP). Additional timing information was provided by a regularly blinking LED visible in the video and recorded in a separate MEG channel. Video was recorded at 25 fps with an infrared-sensitive camera necessary to study patients that have seizures during sleep. Standard Elekta Neuromag Data Analysis software was modified by having the signal processor program "Graph" support bi-directional synchronization with the video. MEG and video are synchronized to the maximum achievable accuracy of one video frame.

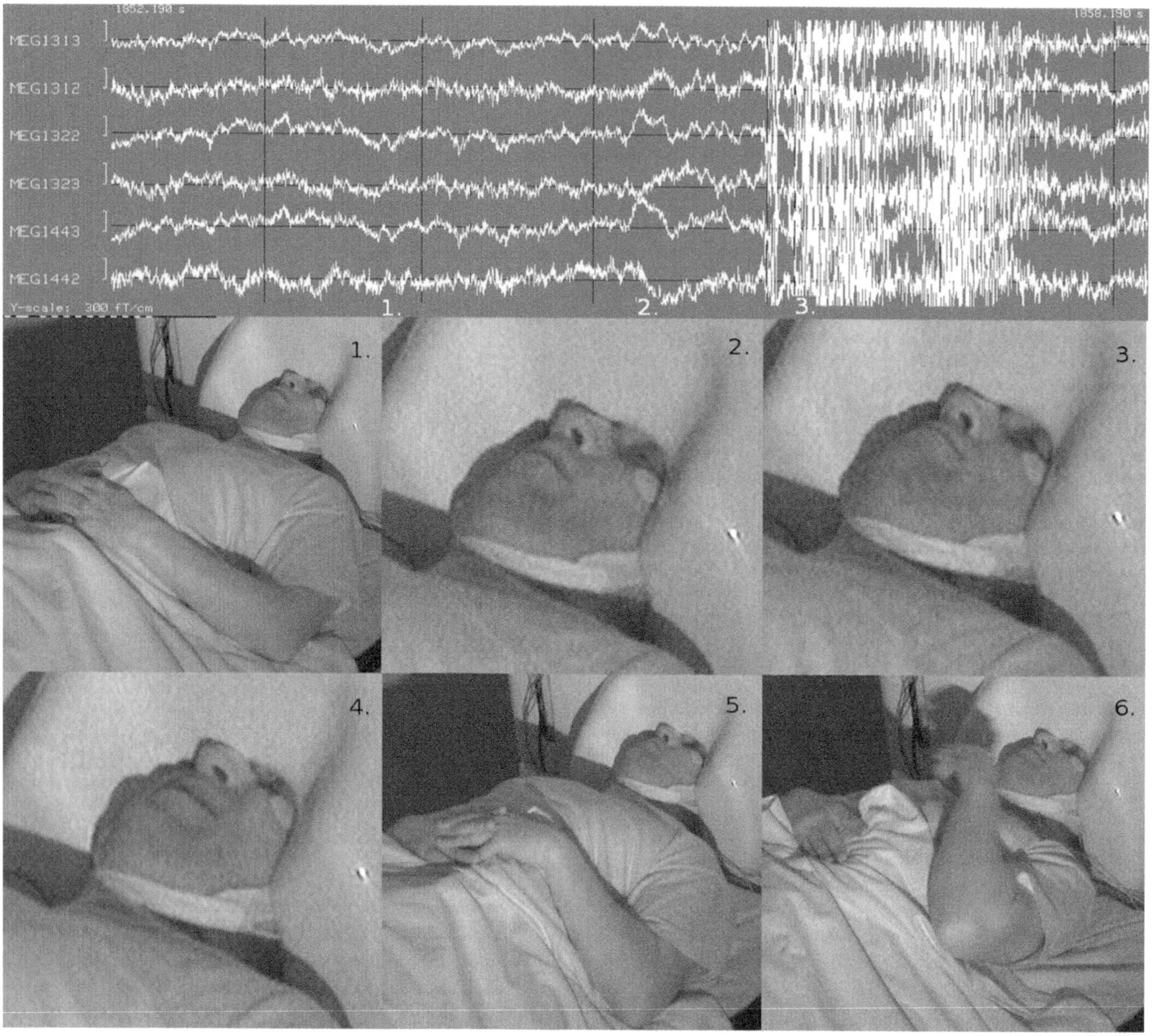

Fig. 2 Top: Six MEG channels from right fronto-temporal area. Numbers indicate the timepoints corresponding to the three still shots from the video (middle): 1. Patient is quietly sleeping, 2. Slow transients appears on MEG, no movement is seen, 3. The first visible movement of the jaw causes movement artifact on MEG. Pictures 4-6: Seizure continues with facial muscle contractions and tonic-clonic movement of the left arm

B. Utility of Video

Finding the ictal onset in MEG-EEG-data is not always straightforward. Here the video and audio recordings, synchronized with electrophysiological data, give additional information.

Movement artifacts on MEG may be more readily identified when correlated with video information. Short absence seizures that would otherwise be unnoticed may be documented during tasks. In some patients, no previous video recording of a typical seizure is available. In such cases the video alone may be of clinical importance.

So far five patients with epilepsy have undergone the simultaneous MEG and video recording. A typical seizure was registered in three of them. Presently, the analysis of two cases has been completed. In both of these cases the video recording had a significant impact on the analysis.

C. *An Example of Clinical Use*

Here we briefly describe one patient whose preoperative MEG result was affected by the video recording.

The patient is a 33-year old male with history of epilepsy since childhood. Antiepileptic medication was started at the age of 16 years. After testing several combinations, current medication consists of oxcarbazepine and lamotrigine. According to medical history, the typical seizure has been of psychomotor type (complex partial seizure with automatisms). Secondarily generalized seizures have been rare. Seizures most frequently occur during sleep. At the time of MEG recording seizures occurred once every 2 weeks.

Head MRI (1.5T) was normal. In video-EEG, ictal recording was not achieved. Frontal slow wave activity was observed in interictal data. The patient gave a written informed consent concerning the use of the MEG and video data in scientific publications.

Two seizures occurred during the measurement. One of these was recorded successfully by both MEG and video. The seizure started with repetitive facial muscle contractions and continued with tonic-clonic movements of the left arm. The seizure duration was about two minutes. The patient was mildly confused for a few minutes post-ictally and then fell asleep. After waking up he could not recall the seizure. *Fig. 2 illustrates combined video and MEG recorded during the seizure.*

In this patient, video-MEG helped to define the timing of ictal onset and understand the sources of movement artifacts. Moreover, it was also the first video recording of a typical seizure in this patient.

MEG source analysis by current dipole models points to the right temporal and frontal lobes as the epileptogenic area. Based on the clinical features, frontal aetiology seems the most likely.

III. Discussion

While technical details of setting up synchronized video-MEG recordings have already been described previously (for example [4]), no actual recordings in epilepsy patients have been reported before.

The installation of the video system has brought up many practical issues. One important conclusion from our initial experience with the system is that camera position is crucial for accurate detection of seizure manifestations. Optimal camera location is quite different for supine and seated patient positions. It is also often beneficial to have a separate video stream zooming in on a certain part of the body – usually face or limbs – where clinical manifestations of epilepsy are expected to be seen.

The capability of night recordings i.e. the IR-camera has already proved to be necessary as many of the patients have predominately nocturnal seizures.

The above conclusions need to be taken into consideration when designing future video-MEG systems.

IV. Conclusion

For the first time the benefit of synchronized video-MEG recordings has been demonstrated in real-life clinical setting. Further studies are needed to properly quantify the clinical value of video-MEG and establish standard procedures for clinical practice.

References

1. Mäkelä JP, Forss N, Jääskeläinen J, Kirveskari E, Korvenoja A, Paetau R (2006) Magnetoencephalography in neurosurgery. Neurosurgery 59:493-510.
2. Medvedovsky M, Taulu S, Bikmullina R, Paetau R (2007) Artifact and head movement compensation in MEG. Neurol Neurophysiol Neurosci 29:4
3. Engel J Jr, Troupin AS, Crandall PH, Sterman MB, Wasterlain CG (1982) Recent developments in the diagnosis and therapy of epilepsy. Ann Intern Med 97:584-98.
4. Burgess RC, Liu P, Wooledge GJ, Horning KE, Mosher JC (2009) Adaptation to MEG of Basic Capabilities Used in Long-Term Video-EEG Monitoring. Poster presentation at the American Clinical Neurophysiology Society Annual Meeting 2009.

Author: Juha Wilenius
Institute: Biomag-laboratory, HUSLAB, HUCH.
Street: P.O. Box 340
City: 00029 HUS
Country: FINLAND
Email: juha.wilenius@helsinki.fi

Demodulated Effect of SQUID to Magnetic Signal Outside Frequency Bandwidth

Daisuke Oyama, Masanori Higuchi, Jun Kawai, Yoshiaki Adachi, Gen Uehara, and Hisashi Kado

Applied Electronics Laboratory, Kanazawa Institute of Technology, Kanazawa, Japan

Abstract— **The authors proposed a new method for image formation of magnetic particles or electric currents by detecting magnetic signal modulated by sounds like ultrasound. This method is a combination of sonography and magnetometry, thus we named it "sonomagnetometry". The method enables the simultaneous observation of spatial distribution and time shift of magnetic source. To realize this technique, we need a SQUID magnetometer which can detect MHz-order frequency signal. Recently, some groups have realized bandwidths detection higher than MHz. In contrast to them, the authors take a different approach which is to detect MHz signal by a narrow bandwidth in basic type SQUID magnetometer system based on the idea similar to a diode demodulator. SQUID magnetometer with 8 kHz bandwidth which is normally used in MEG system was used for the examinations. When 1 MHz magnetic signal whose amplitude is modulated by 500 Hz signal applied to the SQUID, output voltage of 500 Hz was appeared. Its shape is very similar to the modulating 500 Hz signal. This result demonstrates that SQUID magnetometers have demodulation effect to magnetic signals whose frequency is outside the bandwidth. This effect can be applied to sonomagnetometry by measuring time delay from ultrasonic excitation and capturing the shape of the demodulated signals.**

Keywords— **sonomagnetometry, SQUID, ultrasound, amplitude modulation, outside frequency bandwidth.**

I. Introduction

Magnetoencepharography (MEG) and magnetocardiography have been practically used in laboratories and clinics for a few decades. They can measure activity of a brain or a heart noninvasively and passively. These are advantages of magnetic measurements compared with electric measurements.

We started to develop active measurement methods with SQUID magnetometers to expand applications of biomagnetic measurements. The aims of these methods are to get information of structures and functions simultaneously.

In this paper, we propose a new method to form images of magnetic particles or electric currents by detecting magnetic signals excited by sounds, for example, ultrasound. This method is a combination of sonography and magnetometry, so we named it "sonomagnetometry". The method enables the simultaneous observation spatial distribution and time shift of magnetic sources.

To realize this technique, we need a SQUID magnetometer which can detect MHz order frequency signal. We discuss the demodulation effect of a SQUID magnetometer useful to detect high-frequency magnetic signals outside the frequency bandwidth.

In section II, we describe the summary of sonomagnetometry. In section III, we discuss detection of high-frequency magnetic signals by demodulation effect of SQUID magnetometers.

II. Sonomagnetometry

Fig. 1 shows a schematic diagram of sonomagnetometry. An ultrasonic device provides ultrasonic pulse to a target inside a body, and then the target vibrates with the frequency of the ultrasound after time delay corresponding to the distance. Targets are magnetic particles or current sources, thus magnetic signal is generated by the vibration of the target. SQUID sensors measure the magnetic flux signals and the time delay $\triangle T/2$, and the ultrasonic device measures time delay of ultrasonic echo $\triangle T$. By integrating them, sonomagnetometry forms images of structures and functions.

Sonomagnetometry is an active sensing method by mechanical acoustic revitalization. The method realizes simultaneous measurements of spatial distribution and time shift by detecting acoustically excited magnetic signal. Synthesizing them to ultrasonic echo, sonomagnetometry can produce information of structures and functions at the same time.

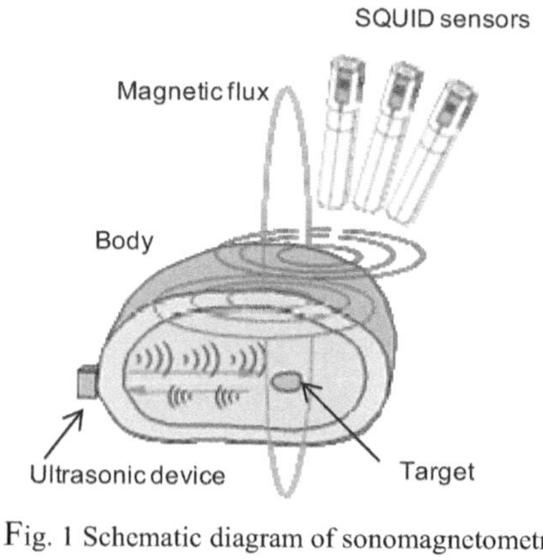

Fig. 1 Schematic diagram of sonomagnetometry

S. Supek and A. Sušac (Eds.): Advances in Biomagnetism – BIOMAG2010, IFMBE Proceedings 28, pp. 50–53, 2010.
www.springerlink.com

Previously, two similar methods were reported by other groups: one is the detection of electromagnetic radiation generated by sounds [1], and another is the detection of sounds generated by magnetic excitation [2]. Simultaneous detecting of two or more information is one of the characteristics of sonomagnetometry.

We project that sonomagnetometry brings forth new fields of biomagnetic measurements. Visualization of distributing magnetized medicines is one of the expected applications. This leads to much contribution to the basic medicine sciences, drug developments, life sciences, and so on.

It is difficult for ultrasonography to observe distribution of magnetic particles because the echo signals from dispersing particles are too small to detect. On the other hand, it may be possible to detect magnetic signals generated by vibration of the magnetic particles. Furthermore, it can be expected to identify density and magnetization of the particles by measuring the amplitude of the magnetic signal.

In addition, we use SQUID magnetometers which are employed for passive measurement like MEG and MCG. Integration of them and sonomagnetometry may bring new techniques to the imaging of the electrical activity of human body.

In this paper, we mainly describe a signal detection technique; examinations of the magnetic source and the sonic power are the topics of our further study.

III. Behavior in SQUID to High-Frequency Magnetic Signal

A. Bandwidth of SQUID Magnetometer

Mechanical vibration in several MHz is needed to achieve the resolution as high as that of ultrasonography. However, bandwidth of SQUID magnetometers which are used for MEG and MCG is generally about several kHz. In recent years, some groups have developed SQUID magnetometers with wide bandwidth over 1 MHz [3][4] and they may be applied to sonomagnetometry with ultrasounds. Their methods have achieved MHz bandwidth by high-speed flux-locked loop (FLL) electronics. In contrast to them, we propose a method to detect MHz signal by narrow-bandwidth basic type SQUID magnetometer system based on the idea similar to diode demodulator, in order to avoid the complexity of the circuits.

Fig. 2 shows a schematic diagram of direct readout FLL. The feedback system is composed of a SQUID, a preamp, an integrator, and a feedback coil. If the amplitude of the feedback signal is the same as the input signal, then the magnetic flux shift can be measured as output voltage.

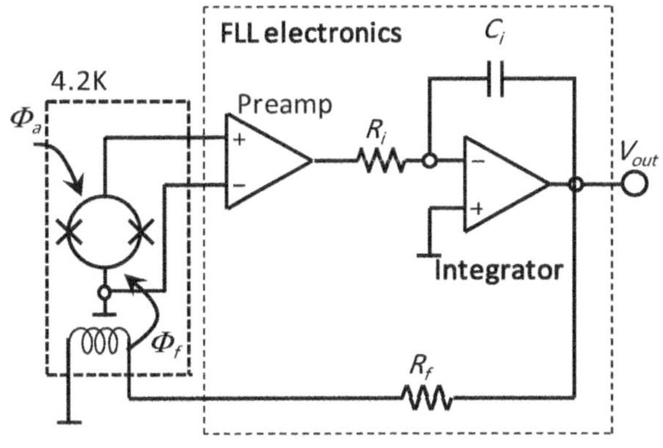

Fig. 2 Schematic diagram of FLL

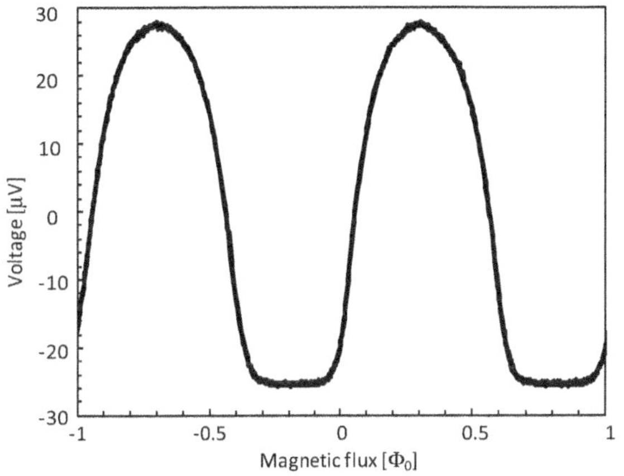

Fig. 3 Flux-voltage characteristic of a SQUID

It is well known that the bandwidth of SQUID magnetometer is limited by signal transmission lines, circuits and so on; though SQUIDs themselves usually have up to GHz response. The frequency response of SQUID magnetometers, which are generally used to biomagnetic measurements, is dc to several kHz. FLL cannot output feedback magnetic signal over its bandwidth. In consequence, we observe output voltage waveform via flux-voltage characteristic of SQUID and low-pass filter responses of the electronic components. It is similar to an open-loop measurement method.

SQUID magnetometers measure magnetic signal as linear response inside their bandwidths and as nonlinear response outside them.

B. Demodulation Effect in Nonlinear Characteristic of SQUID

We demonstrated demodulation effect in the nonlinear characteristic of a SQUID with open-loop read-out electronics. It can be considered that the FLL is the same as the open-loop outside its bandwidth.

Fig. 3 shows the flux-voltage characteristics of our SQUID used for this demonstration. It is a distorted sinusoidal shape with nonlinearity. Fig.4 shows (*a*) input magnetic flux and (*b*) output voltage of the SQUID measured by open-loop read-out electronics. The input signal was 100 Hz sine wave whose amplitude was modulated by 10 Hz. The output signal was distorted by nonlinearity of the SQUID's characteristic. Fig. 4 (*c*) and (*d*) show the spectra of input and output signal, respectively. In Fig. 4 (*c*), there were peaks at 100 Hz, 90 Hz and 110 Hz, because input is 100 Hz signal modulated by 10 Hz. As oppose to them, 10 Hz peak and its harmonics appeared in Fig. 4 (*d*). They were demodulated peaks generated by nonlinearity of the SQUID.

SQUID magnetometers can detect modulated high-frequency magnetic flux signal as demodulated low-frequency output voltage if the readout is open-loop electronics.

C. Demodulation Detection of Magnetic Field Outside the Bandwidth Using Nonlinearity of SQUID

We examined to detect magnetic signal outside the bandwidth of a SQUID magnetometer using demodulation effect described in the previous section. We used a direct readout FLL which had 8 kHz bandwidth. In this case, the carrier frequency of input signal was 50 kHz, and the modulation frequency was 1 kHz. Fig.5 shows the spectra (*a*) input magnetic flux and (*b*) output voltage of the SQUID magnetometer. It is equivalent to the open-loop detection because 50 kHz is outside the bandwidth of the FLL. As well as the case in Fig. 4, 1 kHz peak appeared by the demodulation effect.

SQUID magnetometers can detect high-frequency magnetic signals outside the bandwidth using nonlinearity of SQUIDs.

D. Detection of High-Frequency Magnetic Flux Signal

We demonstrated detections of high-frequency magnetic flux signal by the demodulation effect of SQUID magnetometers. The carrier frequency of input magnetic signal was 1 MHz, and the bandwidth of the SQUID magnetometer was 8 kHz. In this paper, we carried out two examinations in the number of the input pulse: the first was 1 pulse and the second was 500 pulses.

Fig. 6 (*a*) shows waveforms of input magnetic flux and output voltage of FLL. The sampling period of data acquisition was 0.01 μs. The number of input 1 MHz pulse was just 1. The onset of the pulse is indicated by a dotted line. It is considered that about 1 μs delay was caused by the delay of transmission lines and electronic components. The SQUID magnetometer, which had only 8 kHz bandwidth, could respond to 1 MHz magnetic signal. This result can be applied to sonomagnetometry. The distance between an ultrasonic device and a target

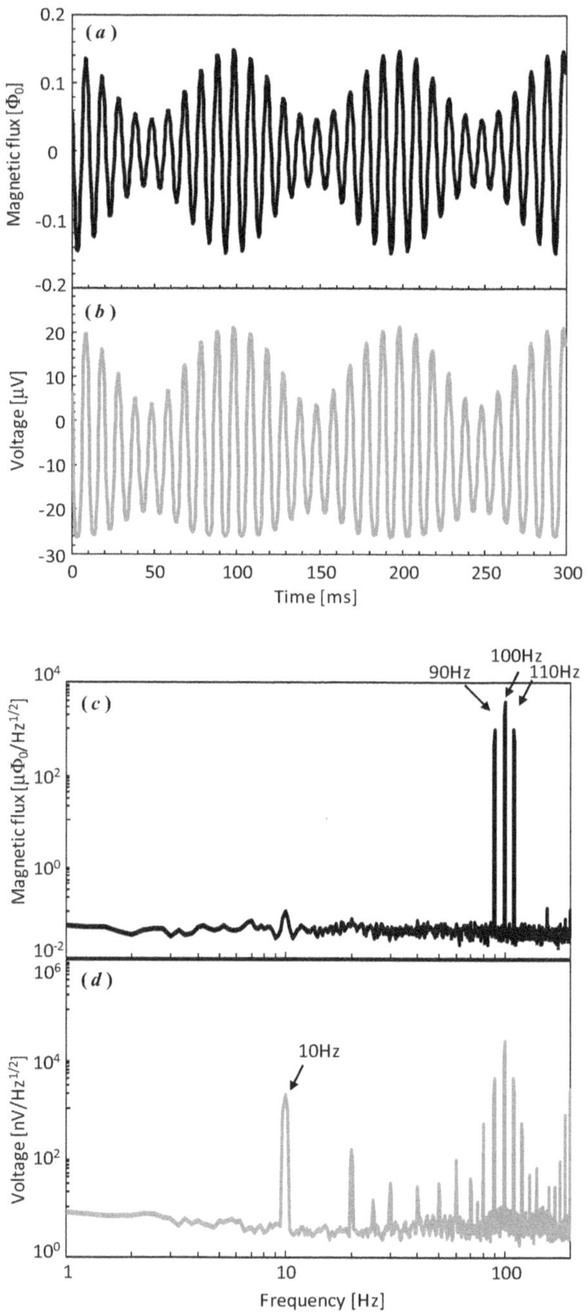

Fig. 4 Detection of amplitude-modulated (AM) signal by nonlinearity in flux-voltage characteristic of the SQUID. (*a*) and (*c*) show input AM signal. (*b*) and (*d*) show output demodulated signal

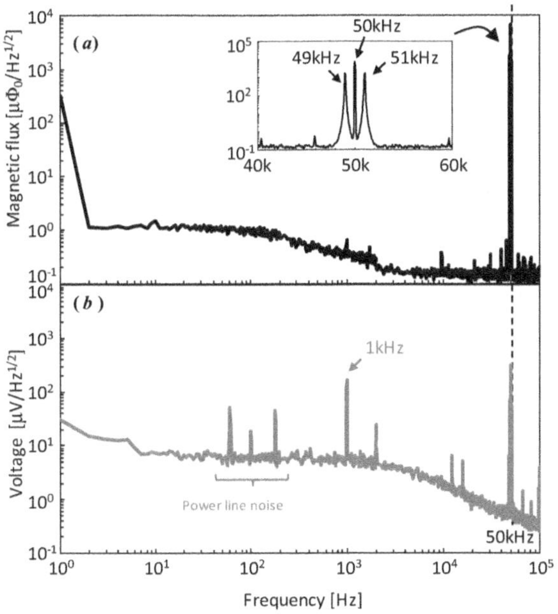

Fig. 5 Demodulation effect of SQUID to high-frequency magnetic signal. (*a*) shows spectrum of input AM signal, and (*b*) shows output voltage of the SQUID magnetometer. A peak at 1 kHz is appeared by demodulation effect to SQUID magnetometer

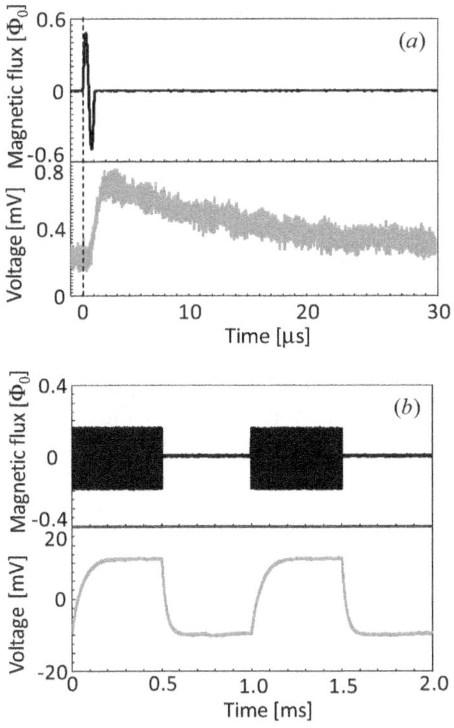

Fig. 6 Detection of high-frequency magnetic signal. They shows input and output signal respectively. The number of 1 MHz input pulse is 1 in (*a*) and 500 in (*b*), respectively

will be estimated by measuring time delay of output signal.

Fig. 6 (*b*) shows the result of the burst signal composed of consecutive 500 pulses. The period of the burst signal was 1 ms and the duty ratio of 50 %. The square wave with the period of 1 ms was detected by demodulation effect of the SQUID magnetometer. Roundness of squares was caused by the time constant of the FLL system. It is expected that SQUID magnetometers can receive some information included in the shapes of modulation signal and that these results can be applied to sonomagnetometry.

It is difficult to evaluate the amplitude of demodulated signal because the flux-voltage characteristic of SQUID is different according to the device. This fact requires further study.

IV. Conclusions

We proposed a new method to form magnetic source images by sonomagnetometry and examined behavior in a SQUID magnetometer to magnetic signal outside its bandwidth. The demodulation effect can be applied to sonomagnetometry.

Acknowledgment

This research was supported by Hokuriku Innovation Cluster for Health Science.

References

1. K. Ikushima, W. Watanuki and S. Komiyama (2006) Detection of Acoustically induced electromagnetic radiation, Appl. Phys. Lett., 89: 194103.
2. M. Abe, T. Ueda, T. Nakagawa, and H. Handa (2009) Sentinel-lympnode mapping for staging cancers by sonic wave emission by magnetically stimulated particles, Digest of the 33rd Annual Conference on Magnetics in Japan 2009: 82-83.
3. D. Drung, C. Aßmann, J. Beyer, M. Peters, F. Ruede, and Th. Schurig (2005) dc SQUID Readout Electronics With Up to 100 MHz Closed-Loop Bandwidth, IEEE Trans. on Appl. Supercond., 15, 2 : 777-780.
4. D. F. He and H. Itozaki (2006) High-Tc dc SQUID readout electronics with low noise and high bandwidth, Physica C, 445-448: 982-985.

Use macro [author address] to enter the address of the corresponding author:

Author: Daisuke Oyama
Institute: Applied Electronics Laboratory, Kanazawa Institute of Technology
Street: 3, Amaike
City: Ishikawa, Kanazawa
Country: Japan
Email: oyama@neptune.kanazawa-it.ac.jp

Asymmetric Three-Dimensional Finite Element Analysis of a Magnetically Shielded Room with Access Ports

Broc A. Burke[1,2] and Solomon G. Diamond[1]

[1] Thayer School of Engineering at Dartmouth, Hanover, NH, USA
[2] Dartmouth Medical School, Hanover, NH, USA

Abstract— **As part of a new initiative in multimodal functional neuroimaging at Dartmouth, a novel design for an economical magnetically shielded room (MSR) has been custom fabricated and installed on site. A three-dimensional finite element analysis (FEA) model was developed in concert with the MSR for evaluation of critical aspects of the design and then determination of degaussing effectiveness. The FEA model generates a full three-dimensional map of the shielding factor as a function of frequency and is sensitive to the number, thickness, and spacing of wall shielding, as well as access ports such as the door, vents and feed through panel. This simulation uses quadratic vector Lagrange elements to solve for the total electric and magnetic potentials using a time-harmonic analysis. With this model, estimation of the spatial variation in the shielding factor for a modestly shielded room is demonstrated. The Dartmouth MSR has internal dimensions of 2.35m x 2.35m x 2.16m and has two high conductivity eddy current layers and one mu metal layer (two additional layers, one eddy current and one mu metal layer will be added in the near future). The simulation results indicate that the shielding factor at the center of the MSR does not vary significantly with incident angle of the applied magnetic field. However, spatial inhomogeneities in the shielding factor are evident near MSR access ports. The results of this simulation will be used in future work, where the degaussing of the room will be validated against modeling predictions. Our future modeling work will add a virtual patient and virtual instruments, electroencephalography (EEG), and near-infrared spectroscopy (NIRS), for evaluation of achievable multimodal contrast to noise.**

Keywords— **MSR, FEA, FEM, CAD, modeling.**

I. Introduction

As part of a new multimodal neuroimaging initiative at the Thayer School of Engineering, the design, analysis and fabrication of a magnetically shielded room (MSR) have been undertaken. This MSR is being designed to accommodate simultaneous acquisition of magnetoencephalography (MEG), electroencephalography (EEG), and near-infrared spectroscopy (NIRS). In support of this effort, an asymmetric finite element model (FEM) of the MSR has been created for the purposes of economical room design, validation of the degaussing procedures, and the planning of appropriate placement of multimodal equipment and lines. The model results presented herein are of an intermediate step of the build process (3 of 5 eventual layers are modeled). The results show the localized effects of room asymmetries such as the door, vents, and feedthru panel, where material properties vary due to the lack of particular layers and thin low magnetic permeability conditions are applied due to imperfect continuation of the mu metal layer. The addition of thin film conditions and asymmetrical access ports extends the work of previous models [1-5] and extends the utility of MSR modeling.

II. Methods

A. *Computer-Aided Design and Meshing*

A three-dimensional representation of the MSR has been created using parametric computer-aided design (CAD) software, SolidWorks® (Dassault Systèmes SolidWorks Corp., Concord, MA, USA). This CAD model consists of three concentric square prismatic shielding layers, and two concentric square prismatic non-shielding layers. Additionally, the room has been modeled with the air that it contains, a sphere at the center of the room representative of a human head, and a sphere of air surrounding the room, allowing for variation of the incident angle of the applied magnetic field.

A compliant tetrahedral mesh was then generated from the CAD model using SolidWorks® Simulation. The generation of this mesh required regional mesh controls (Tables 1, 2 and 3) to produce a balance of adequate mesh density to represent the geometry and electromagnetic dynamics without exceeding available computer memory. The specific control parameters are the target element size, which is synonymous with the average edge length for tetrahedral

Table 1 Regional Tetrahedral Mesh Control Parameters

		Head	Inner Room Air	All Room Shells	Doorway Faces	Vent Faces	Feedthru Faces
Target Element Size	(m)	0.06	1.00	0.50	0.25	0.07	0.10
Maximum Element Growth Ratio	(~)	3.09	3.09	3.09	3.09	3.09	3.09

S. Supek and A. Sušac (Eds.): Advances in Biomagnetism – BIOMAG2010, IFMBE Proceedings 28, pp. 54–57, 2010.
www.springerlink.com

Table 2 Model Layer Geometric Properties, Mesh Statistics and Material Properties

Layer ID	Model Layer	Height (m)	Width (m)	Depth (m)	Layer Thickness (cm)	Tetrahedral Elements (#)	Element Volume Ratio	Minimum Element Quality	Relative Permeability	Relative Permittivity	Conductivity (S/m)
1	Surrounding Air	5.50	N/A	N/A	N/A	4068	9.60E-06	2.19E-01	1	1	0.00E+00
2	Galvanized Steel	2.16	2.35	2.35	0.05	2312	1.28E-04	1.99E-06	500[8]	1	7.04E+06
3	Kiln Dried Wood	2.16	2.35	2.35	1.14	1640	2.10E-03	4.20E-03	1	2	1.00E-16[9]
4	Aluminum	2.15	2.34	2.34	0.10	1780	4.99E-04	3.64E-05	1	1	3.77E+07
5	Air Gap	2.15	2.34	2.34	5.00	1748	1.30E-03	2.44E-02	1	1	0.00E+00
6	Mu Metal	2.15	2.34	2.34	0.10	1715	3.32E-04	8.03E-05	20000	1	1.82E+05
7	Inner Room Air	2.15	2.34	2.34	N/A	3839	3.66E-04	8.50E-02	1	1	0.00E+00
8	Central Sphere	0.11	N/A	N/A	N/A	171	1.44E-01	4.89E-01	1	1	0.00E+00

elements, and the maximum element growth ratio, which bounds the rate of volumetric change for adjacent mesh elements. These parameters were iteratively validated through visual inspection of the mesh, convergence of the FEM, and not exceeding 32GB of system RAM during finite element analysis (FEA). The resulting mesh consists of 3098 nodes and 4363 triangular boundary elements. Tables 2 and 3 also summarize mesh statistics per modeling layer. In particular, mesh quality is a measure of geometric anisotropy [6], where q is quality, *Volume* is the element volume, and h_i are the edge lengths of the tetrahedral element (1). The element volume ratio is defined as the minimum element volume divided by the maximum element volume in a given modeling layer and is useful for comparing the physical differences in the scale of the mesh within a layer.

$$q = \frac{72\sqrt{3}\times Volume}{\left(h_1^2 + h_2^2 + h_3^2 + h_4^2 + h_5^2 + h_6^2\right)^{3/2}} \tag{1}$$

B. Finite Element Analysis

Looking forward to future multimodal applications for this model, the FEA is being performed with COMSOL Multiphysics® (COMSOL Inc., Burlington, MA, USA). First, the mesh generated by SolidWorks® is imported into COMSOL® as quadratic Lagrange vector elements. Given the ratio of the largest physical dimension to shortest wavelength being simulated (0.06) is less than 0.10[6], a quasi-static formulation of Maxwell's equations is used (2 and 3). This formulation is applied across the quadratic Lagrange vector elements, where ω is angular frequency, σ is conductivity, ε_0 is permittivity in a vacuum, ε_r is relative permittivity, **A** is the magnetic vector potential, V is electric potential, **v** is the conductor velocity, $\mathbf{J}^e$ is an externally generated current density, μ_0 is magnetic permeability in a vacuum, and μ_r is relative magnetic permeability.

$$-\nabla\cdot\begin{pmatrix}\left(j\omega\sigma - \omega^2\varepsilon_0\varepsilon_r\right)\mathbf{A} - \sigma\mathbf{v}\times\left(\nabla\times\mathbf{A}\right)\\ +\left(\sigma + j\omega\varepsilon_0\varepsilon_r\right)\nabla V - \mathbf{J}^e\end{pmatrix} = 0 \tag{2}$$

$$\begin{gathered}\left(j\omega\sigma - \omega^2\varepsilon_0\varepsilon_r\right)\mathbf{A} + \nabla\times\left(\mu_0^{-1}\mu_r^{-1}\nabla\times\mathbf{A}\right)\\ -\sigma\mathbf{v}\times\left(\nabla\times\mathbf{A}\right) + \left(\sigma + j\omega\varepsilon_0\varepsilon_r\right)\nabla V = \mathbf{J}^e\end{gathered} \tag{3}$$

Once the global governing equations have been applied, the element boundary conditions are applied. This model invokes one of three boundary conditions [6]: magnetic continuity (4), magnetic field (5), or the thin low permeability condition (6 and 7), where **H** indicates a magnetic field, **J** is a current density, d is the thin gap (0.001 mm here), and **n** indicate the normal vector. The magnetic field condition is only applied to the outside of the external sphere; this is the condition that simulates the ambient magnetic field. The thin low permeability condition is applied at mu metal to mu metal boundaries on the model access ports listed in Table 3. Table 3 also provides an accounting of which layers are present for a given port (e.g. the galvanized steel and wood layers are not present for the vent or feed-thru). All other boundaries are governed by the continuity condition.

$$\mathbf{n}\times\left(\mathbf{H}_1 - \mathbf{H}_2\right) = 0 \tag{4}$$

$$\mathbf{n}\cdot\mathbf{J} = -\nabla\left(\mathbf{n}\times\mathbf{H}\right) = \nabla\mathbf{J}_s \tag{5}$$

$$\mathbf{n}\times\left(\mathbf{H}_1 - \mathbf{H}_2\right) = \nabla_t\times\left(d\mu_0^{-1}\mu_r^{-1}\nabla_t\times\mathbf{A}_t\right) \tag{6}$$

$$\mathbf{n}\times\left(\mathbf{J}_1 - \mathbf{J}_2\right) = 0 \tag{7}$$

Table 3 Model Access Port Characteristics

Access Port	Height (m)	Width (m)	Table 2 Layer ID	Tetrahedral Elements	Minimum Element Quality
Door	2.07	0.91	2-6	609	7.77E-06
Vent (each)	0.17	0.17	4-6	324	2.30E-03
Feedthru	0.38	0.38	4-6	365	3.20E-04

At this point, the material properties listed in Table 2 are assigned to all of the elements based upon their layer and access port if applicable. The resulting forward problem is then solved for the total potentials to a relative tolerance of 1E-06 using the PARDISO direct solver with nested dissection preordering and stiff springs constraint handling. The applied magnetic field is then swept over one hemisphere with 0.1π spacing and is also swept in frequency from 0.01Hz to 4 kHz in 12 logarithmic increments.

III. Results

Of over 120 simulation scenarios completed, one representative case was selected for graphical presentation. This particular case corresponds to an applied frequency of 40Hz due to its physiological relevance. The angle of incidence of the applied magnetic field is somewhat offset from vertical (offset angle is 0.1π) as a demonstration of the flexibility of this asymmetric model. Additionally, the particular incident angle that is being presented falls within one of the slice planes to facilitate visualization. Figure 1 depicts the L2 norm of the magnetic field in three orthogonal slice planes with an applied sinusoidal magnetic field of magnitude 1 and a frequency of 40 Hz.

Figures 1 and 2 additionally show the outline of the MSR CAD model and its access ports: the door on the left front face of the MSR, the feedthru panel on the right side of the right face, and the two air vents on the ceiling.

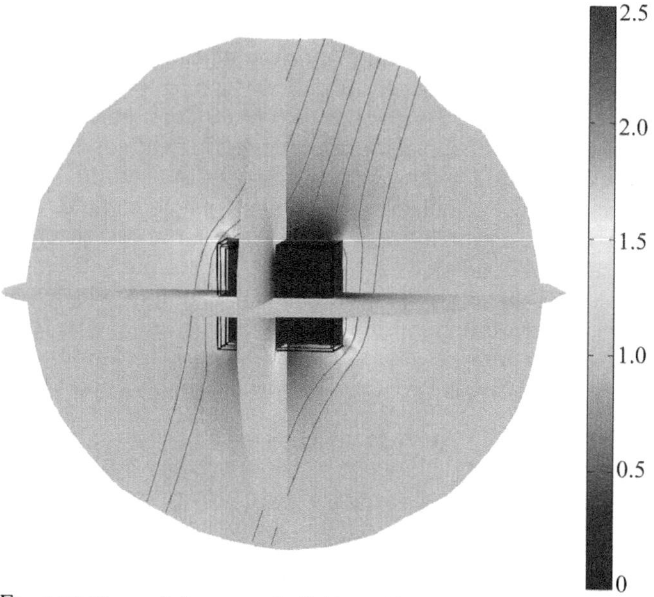

Fig. 1 L2 Norm of the magnetic field resulting from a 40 Hz sinusoidal applied field of magnitude 1 at phase=0: Red lines indicate magnetic field streamlines, while black lines indicate the contours of the MSR

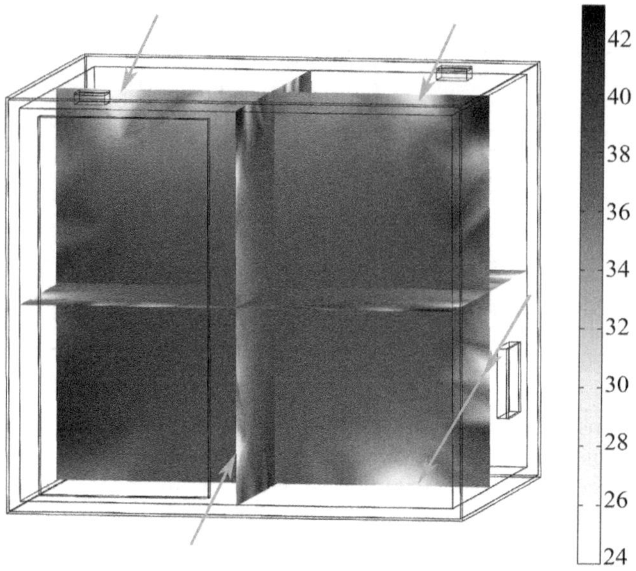

Fig. 2 Shielding effectiveness in decibels at 40 Hz: The black lines indicate MSR feature edges, and blue arrows highlight effects of the access ports

Figure 2 visualizes the shielding factor in decibels for the same non-axially oriented magnetic field shown in Figure 1. The perspective is also the same as that used in Figure 1; however, the view is zoomed in to better illustrate the potential effects of the access ports in the simulation.

Figure 3 shows the shielding factor in decibels at the MSR center as frequency is varied from 0.01 Hz to 4 kHz. This shielding factor corresponds to a traditional purely vertical mathematically applied magnetic field. The simulation of frequency sweeps at other incident angles (not

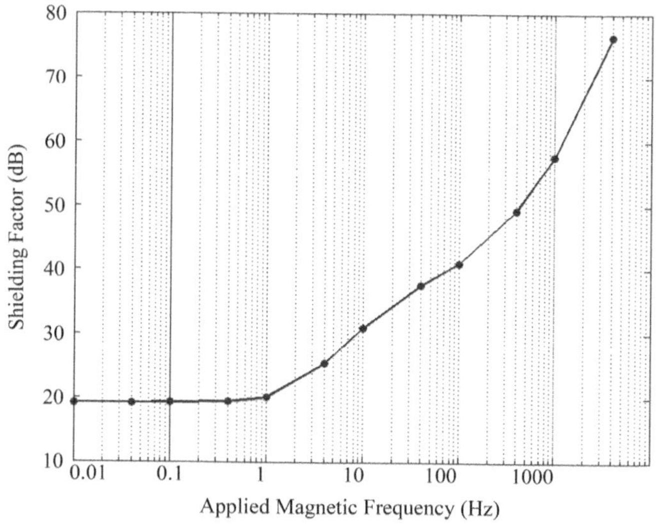

Fig. 3 Shielding effect in decibels at the center of the MSR as a function of applied magnetic frequency. The applied magnetic field illustrated here is purely vertical and closely matches a prior estimate at 0.01 Hz [4]

shown) were found to have little effect on the calculated shielding factor (less than 1%) at the MSR center.

IV. Discussion

The shielding factor estimates provided by the FEM (Figure 3) are in close agreement with previous computational work. For example, Yamazaki investigated the variation in shielding factor given different layering options [4]. One of the configurations investigated represents a MSR about 50% larger than the one considered here and had a comparable aluminum conductive layer but no galvanized steel layer. For this similar layering configuration, a center of room shielding factor of 17 dB at 0.1 Hz was calculated as compared with the 19dB shown in the present analysis. With the Dartmouth MSR under construction (steel and aluminum layers complete with the mu metal layer approximately 92% complete by surface area), a preliminary test of the DC shielding factor was made with a fluxgate magnetometer (FVM-400 MEDA Dulles, Virginia). The resulting shielding factor was 24 dB with an internal field of 2,500 nT as compared to the predicted internal field of 4,376 nT.

Figure 1 also provides insight into the magnetic field interaction with the edges of the MSR, where the simulation shows higher L2 norms of the magnetic field at the edges requiring a greater turning angle of the magnetic field lines (the upper right and lower left edges). This localized increase in the magnetic field magnitude can also be seen as a decreased shielding factor at the same edges in Figure 2.

A close inspection of Figure 2 also reveals spatial effects of MSR access ports (indicated by blue arrows). Reduced shielding can be seen near the vents, feedthru and door. These local variations in shielding factor may be attributed to the applied field and the MSR design choices. With adequate mesh refinement in specific regions of interest, informed decisions can be made regarding MSR design, auxiliary device placement, and MEG sensor head placement.

These estimates of local shielding variation will also inform those performing the degaussing of the MSR as to the success of the procedure. This will be achieved by using the model to take into account the local field effect resulting from the MSR geometry. Taking these estimates and subtracting them from empirical measurements of the field present in the room after degaussing will produce the residual field. This should allow for an improvement in the evaluation of the degaussing by providing a metric that previously was restricted to highly shielded rooms [7].

V. Conclusions

By eliminating the previous constraints of perfectly connected layers and symmetric systems, this computational model expands the potential applications of MSR simulations. Some immediately apparent new applications are

1. assessing the impact of MSR design for site specific, application specific scenarios,
2. providing a benchmark for the effectiveness of maintenance procedures, and
3. estimation of the impact of physical test configurations on the signal to noise of measurements.

Given the reasonable results provided by the FEM, the next step, in keeping with number 2 above, will be to validate the model and the degaussing procedure that will be used on the MSR. After degaussing the model will be enhanced to include the two additional shielding layers (4.8 mm of aluminum and 1.0 mm of mu metal). Following from number 3 above, the model will then be used to aid in the optimization of multimodal instrumentation design through the informed selection of instrument locations and the related effect on the magnetic potential inside of the MSR.

Acknowledgment

I would like to acknowledge the generous support provided by the Thayer School of Engineering, Dartmouth Medical School, and the United States Department of Education Grant P116Z080112. Additionally, this work would not have been possible without the technical and practical guidance provided by Peter Teale, and David Cohen.

References

1. Wills, A.P. (1899) On the magnetic shielding effect of trilamellar spherical and cylindrical shells. Physical Review (Series I) 9: 193
2. Mager, A. (1970) Magnetic shields. IEEE Transactions on Magnetics 6(1): 67-75.
3. Baum, E. and J. Bork (1991) Systematic design of magnetic shields. Journal of Magnetism and Magnetic Materials 101(1-3): 69-74
4. Yamazaki, K., K. Kato, and K. Fujiwara (2000) Effective combination of magnetic and conductive layers of magnetically shielded room. IEEE Transactions on Magnetics 36(5): 3649-3651 DOI 10.1109/tmag.2008.2002524
5. Yamazaki, K., et al. (2006) Optimal structure of magnetic and conductive layers of a magnetically shielded room. IEEE Transactions on Magnetics 42(10): 3524-3526 DOI 10.1109/tmag.2006.879156
6. COMSOL AB (2008) COMSOL Multiphysics AC/DC Module User's Guide
7. Thiel, F., et al. (2007) Demagnetization of magnetically shielded rooms. Review of Scientific Instruments 78(3) DOI 10.1063/1.2713433
8. Gale, W., et. al. (2004) Smithells Metals Reference Book: 20(4)
9. Simpson, W., TenWolde, A. (2001) Wood Handbook – Wood as an Engineering Material: 3(21)

Author: Broc A. Burke
Institute: Dartmouth College
Street: 8000 Cummings
City: Hanover, NH 03755
Country: USA
Email: broc@stanfordalumni.org

Mapping the Cardiomagnetic Field with 19 Room Temperature Second-Order Gradiometers

A. Weis[1], G. Bison[2], N. Castagna[1], S. Cook[3], A. Hofer[1], M. Kasprzak[1], P. Knowles[1], and J.-L. Schenker[1]

[1] Physics Department, University of Fribourg, Fribourg, Switzerland
[2] Neurology Department, Friedrich-Schiller-University, Jena, Germany
[3] Cardiology, Hôpital Cantonal, Fribourg, Switzerland

Abstract— **We present a laser pumped 19 channel second-order magnetometer system for measuring magnetic fields produced by the beating human heart. Each field sensor is a room-temperature paraffin-coated cesium vapor cell operated in a standard M_x magnetometer configuration based on optically detected magnetic resonance with circularly polarized D1 pumping on the $F=4 \to F'=3$ component of the D_1 transition. Dynamic magnetic field maps are derived from the simultaneous recording at 19 grid points. The system has been successfully used for recording (in less than 2 minutes) cardiac field maps of more than 30 healthy subjects. We also report a recording of the MCG of a patient with a nonspecific intraventricular conduction defect, which represents the first detection of a cardiac anomaly by an atomic magnetometer array.**

Keywords— **Magnetocardiography, atomic magnetometers, magnetic gradiometers, magnetic field mapping.**

I. Introduction

For many decades superconducting quantum interference devices (SQUIDs) have been the only magnetometers with a sensitivity sufficient for recording the faint magnetic fields of biological origin. In the past decade, rapid progress in the development of magnetometers based on laser-pumped paramagnetic atomic vapors has led to devices whose sensitivity approaches [1, 2] or surpasses [3] those of conventional low-temperature SQUIDs. For many years our team at the University of Fribourg (Switzerland) has been active in the development of laser-pumped atomic magnetometers (LOPMs) for applications in fundamental and applied fields [4]. The room temperature operation of LOPMs makes such magnetometers promising alternatives to SQUIDs by avoiding the complex logistics associated with their cryogenic cooling in view of clinical applications. Since each LOPM sensor requires only a few μW of light power, a single laser can operate a large number of individual sensors, a prominent feature which has not been exploited so far.

We have previously reported on the measurement of human cardiac magnetic field maps (cMFM) by stepping a single magneto-gradiometer [5, 6] over the chest. The recording times in excess of two hours excluded the use of that device with cardiac patients. Here we describe a system based on 25 individual LOPM sensors which allows the simultaneous recording of magnetocardiographic (MCG) signals at 19 points over the chest [7]. With a recording time of less than two minutes, this system has allowed us the first measurement of MCG signals and a cMFM of a patient with a cardiac disorder.

II. Principle

Atomic magnetometers rely on measuring the Larmor frequency

$$\omega_L = \frac{\vec{\mu} \cdot \vec{B}}{\hbar}, \tag{1}$$

where $\vec{\mu}$ is the atomic magnetic moment and $\vec{B}$ the total magnetic field that the atoms experience. The gyromagnetic ratio γ is $2\pi \times 3.5$ Hz/nT for the atoms (^{133}Cs) used in this work. From Eq. (1) one sees that the magnetometer is, in principle, a scalar magnetometer, sensitive only to the modulus of the magnetic field. However, the double resonance technique we use (described below) requires the magnetometer to be operated in an offset magnetic field $\vec{B}_0 = (0, 0, B_0)$. If an arbitrarily oriented (but much weaker) biomagnetic field, $\vec{b}$, adds to this offset field, the modulus of the total field is given, to first order in the components b_i of $\vec{b}$, by

$$|\vec{B}| = |\vec{B}_0 + \vec{b}| \approx B_0 + b_z. \tag{2}$$

For the biomagnetic field of interest, the system acts as an effective vector magnetometer, but one sensitive only to the biomagnetic field component along the offset field.

The magnetometric detection method is based on a double resonance process involving resonant laser excitation/detection and magnetic resonance in a room temperature cesium vapor. The device is operated in the so-called M_x configuration (see for example [1], and references therein), in which the offset magnetic field $\vec{B}_0$ is applied at 45° with respect to the laser light direction traversing the atomic

S. Supek and A. Sušac (Eds.): Advances in Biomagnetism – BIOMAG2010, IFMBE Proceedings 28, pp. 58–61, 2010.
www.springerlink.com

medium. The M_x magnetometer uses circularly polarized laser light, whose frequency is actively stabilized to the $F = 4 \rightarrow F' = 3$ hyperfine component of the $6S_{1/2}, \rightarrow 6P_{1/2}$ transition of Cs. A laser beam of a few μW traverses an evacuated glass cell containing Cs vapor at room temperature and spin-polarizes the sample by the transfer of angular momentum from the light to the atoms in multiple absorption-emission cycles. The magnetization associated with the spin polarization precesses at the Larmor frequency ω_L around the magnetic field $\vec{B}_0$. A weak magnetic field $\vec{B}_1$ — oscillating at frequency ω_{rf} along a direction perpendicular to $\vec{B}_0$ — drives the spin precession in a resonant manner (magnetic resonance process). Since the optical absorption coefficient of the atomic medium depends on the orientation and magnitude of its magnetization, the driven precession induces a periodic modulation of the transmitted optical power $P(t)$, doing so at the drive frequency ω_{rf}. The amplitude and phase of that power modulation depend in a resonant way on the magnetic resonance detuning ($\delta\omega = \omega_L - \omega_{rf}$). In particular, the phase φ between the drive $\vec{B}_1(t)$ and the response $P(t)$ obeys, up to a constant phase offset,

$$\varphi = \arctan \frac{\delta\omega}{\Gamma}, \qquad (3)$$

where Γ is the magnetic resonance linewidth. The phase φ is extracted using a digital phase detector implemented on a FPGA (field-programmable gate array) chip. The linear zero-crossing of $\varphi(\delta\omega)$ near resonance ($\delta\omega \approx 0$), can be used in feed-back loops for locking either ω_L to ω_{rf}, or vice-versa. The linewidth Γ is typically $2\pi \times 6$ Hz [2]. The magnetometric sensitivity of the device is proportional to the slope $d\varphi/d\omega_L$, which itself is proportional to the spin relaxation time $\tau = \Gamma^{-1}$. Long spin relaxation times are obtained by coating the inner cell walls with a layer of paraffin, which strongly suppresses spin depolarization in wall collisions [2].

III. EXPERIMENTAL APPARATUS

The apparatus is installed inside of a two layer aluminum shielding room ($3.3 \times 2.5 \times 2.3$ m^3) produced by IMEDCO [8, 9]. Three orthogonal pairs of square Helmholtz-like coils mounted outside the chamber are used to compensate the earth field and to produce the vertical offset field $\vec{B}_0$ (on the order of $B_0 \approx 10$ μT). The size of the coils guarantees the required field homogeneity over the volume spanned by the sensors. Additional gradient coils allow the fine tuning of the three transverse field gradients dB_0/dx, dB_0/dy, and dB_0/dz. The multichannel magnetometer system consists of

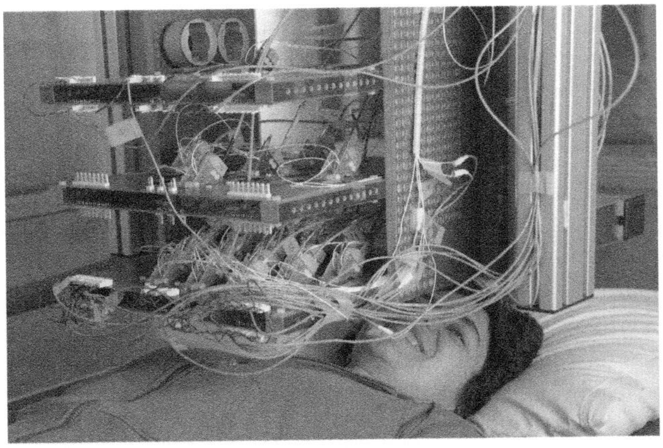

Fig. 1 Photograph of the 25-sensor 19-channel system with one of the authors (NC) in measurement position. The bottom plane holds the 19 measurement channels, while the mid plane and top plane hold sensors used for field and gradient fluctuation suppression

25 individual magnetic sensor heads that are arranged in three planes (vertically spaced by 10 cm) containing 19, 3, and 3 sensors, respectively, as shown in the photograph of Fig. 1. The 19 sensors of the lower plane are arranged in a hexagonal grid with a 4 cm spacing between sensors. They represent the actual measurement sensors, while the sensors in the upper two planes serve as reference sensors for realizing 19 second-order gradiometers.

Each sensor head is a compact block containing the Cs vapor cell, an optical fiber coupler with collimating and polarizing optics and a nonmagnetic photodiode that measures the light transmitted through the cell. The whole assembly is held together by two printed circuit boards that carry coils for generating the $\vec{B}_1$ field. The beam from the laser (mounted outside of the shielding room) is split into 25 equal intensity beams by means of a diffracting hologram. Each of those beams is coupled into a 400 μm multimode fiber that carries the light to the sensor head.

Each sensor plane consist of two large printed circuit boards (PCB) with a hexagonal grid of openings into which the individual sensor modules can be inserted (Fig. 1) and reconfigured with ease. Each opening in the PCB board is surrounded by a coil, so that every sensor has its own pair of feedback coils that allow the application of a small vertical magnetic field, individually controllable for each sensor.

The phase signals from the sensors in the middle plane are used to generate the drive frequency ω_{rf} by actively locking it to the Larmor frequency ω_L of the total magnetic field. In this way the system can handle relatively large drifts of the external field. Two further feedback loops whose error signals are derived from the phase signals in the upper and lower plane sensors, and which drive the feedback coils are then used to derive second order gradiometer signals for each of the sensors in the lower plane. The actual signal of interest is proportional to a combination of feedback currents in the lower and upper plane sensors as described in detail in [7].

The digital phase detectors and feedback controls are fully implemented in an FPGA. The 25 photodiode signals form the inputs to the electronics, which generates the common *rf* frequency for all sensors as well as the feedback currents. This digital control and acquisition system works with an internal sampling rate of 130 kHz. The 19 lower plane feedback current values are down-sampled to a rate of $\approx$ 5.5 kHz and are streamed to a data acquisition computer for offline analysis. The temporal resolution of the raw data is thus $\approx$ 2 ms.

Because of the close sensor packing, each sensor feels the field produced by the feedback coils of adjacent sensors. We have measured the matrix of the cross-talk signals and use it for off-line corrections of the data.

IV. Measurements

A. System Optimization

A non-metallic patient bed which allows the 3*D* positioning of lying patients with respect to the multichannel sensor system is installed in the center of the shielding room. For software development and system optimization we have used signals from a phantom heart consisting of a current dipole in a physiological saline solution contained in a spherical glass flask. The dipole was excited by a synthetic ECG signal yielding MCG signals with a 100 pT amplitude at the *R*-peak.

B. MCG Signals from Healthy Subjects

So far MCG signals have been recorded from more than 30 healthy subjects of different sex and age. Figure 2 shows the 19 MCG traces from a healthy subject. We typically record 70 heart beats and use a simultaneously recorded three-lead ECG trace for off-line slicing and averaging of the data. Because of the weak shielding condition provided by the aluminum chamber, the raw signal data are dominated by a 50 Hz line frequency oscillation with a peak-peak amplitude of 150 pT. Prior to averaging, the data are therefore filtered by a polynomial low-pass filter with a 50% cut-off frequency of 45 Hz [7]. The residual 16 2/3 Hz oscillations on the time traces originates from magnetic fields produced by a nearby (1 km) railway line.

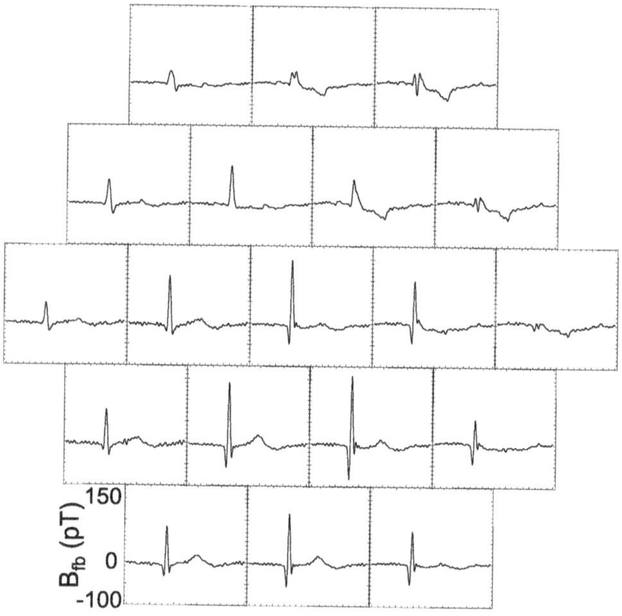

Fig. 2 Time series showing averages of 66 MCG traces of a healthy subject at each of the 19 sensor positions. Each trace is 1 s long and all graphs have the same (280 pT) vertical scale

C. MCG Signals of a Subject with a Heart Disease

After excluding morphological heart conditions, such as ischemic heart disease, left ventricular hypertrophy, cardiomyopathy, and valvulopathy, we selected one patient (a 38 years old healthy woman with no specific cardiac history) with a pure anomaly in the cardiac conduction system. The 12-lead ECG shows an increased QRS duration lasting more than 120 ms, indicating a slowed conduction in the ventricles. No ECG criteria for specific bundle branch or fascicular blocks were visible and the diagnosis remains that of a nonspecific intraventricular conduction defect (IVCD). This heart condition is stable so that a measurement could be performed without additional risk. Nevertheless, a cardiologist was present during the preparation and measurement procedures. Figure 3 shows the result. In this particular patient and in comparison with healthy subjects, a prominent r' activation was observed in the MCG time traces during the "*S*-slurring". The pathology is visible on most of the time trace data, but as yet, no analysis has been completed on the magnetic field maps.

For each given time, the 19 individual signals can be used to construct a magnetic field map by using a numerical interpolation algorithm. A sequence of such maps forms a movie showing the magnetic field dynamics. Two examples of magnetic field maps derived from the data in Fig. 3 are shown in Fig. 4. The left graph shows the out-of-chest component of the magnetic field at the time corresponding to the minimum of the S-dip, while the graph on the right shows the field distribution at the peak of the repolarization (T-) wave. The signals shown in Figs. 3 and 4 represent — to our knowledge — the first cardiac anomaly ever recorded with a multichannel array of atomic magnetometers.

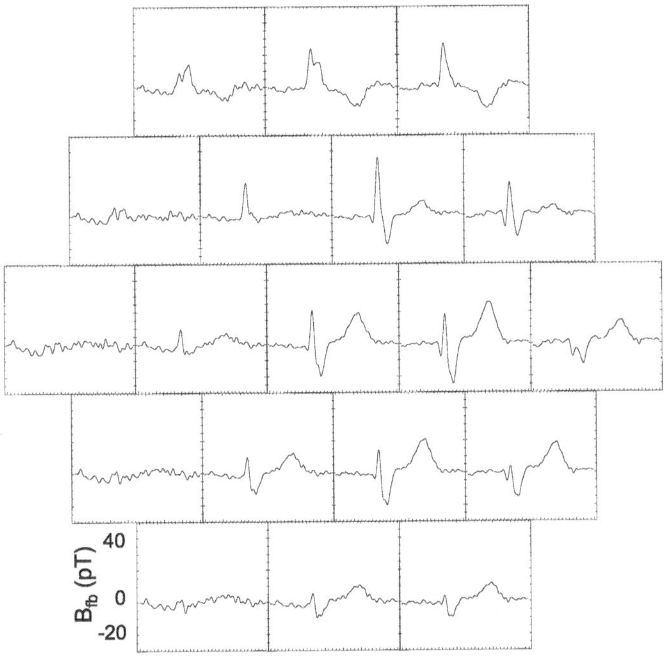

Fig. 3 Time series showing the average of 88 MCG traces of a cardiac patient with a nonspecific intraventricular conduction defect at each of the 19 sensor positions. Each trace is 0.7 s long and all graphs have the same (80 pT) vertical scale

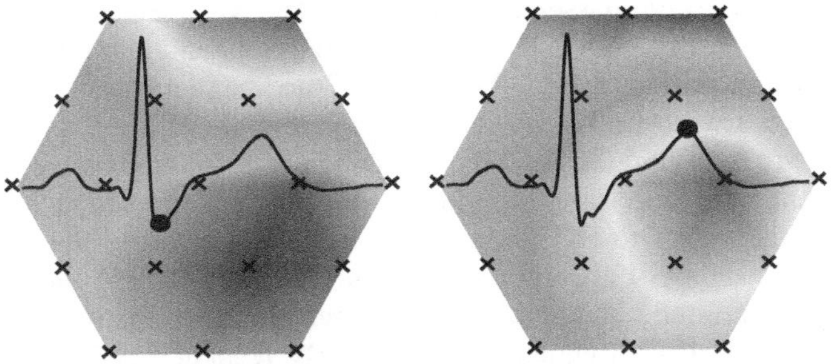

Fig. 4 Magnetic field map for the out-of-chest component of the cardiac field. Crosses show sensor positions, and the superimposed traces represent time-averaged ECG signals which identify, via black dots, the time at which the map was evaluated: S-dip on the left and T-wave at the right

V. Conclusion

We have reported MCG traces and magnetic field maps recorded by a multisensor array of laser pumped atomic magnetometers. The system allows the simultaneous recording of 19 second-order gradiometer signals at room temperature in a weakly shielded environment. So far we have measured cardiomagnetic field maps of more than 30 healthy subjects and (so far) a single subject with a known stable heart disease. The latter measurement represents the first observation of a cardiac pathology by an array of atomic magnetometers.

Acknowledgements

This work was supported by the Velux Foundation. We thank the laboratory technicians C. Macchione and H. Saudan for the production of the coated Cs cells and the mechanical workshop at the Physics Department for efficient support in the construction of the system.

References

1. Groeger S, Bison G, Schenker J-L, Wynands R, Weis A, A high-sensitivity laser-pumped M_x-magnetometer, *Eur. Phys. J. D.* 2006;38:239–247.
2. Castagna N, Bison G, Di Domenico G, et al, A large sample study of spin relaxation and magnetometric sensitivity of paraffin-coated Cs vapor cells, *Appl. Phys. B.* 2009;96:763–772.
3. Kominis I K, Kornack T W, Allred J C, Romalis M V, A subfemtotesla multichannel atomic magnetometer, *Nature.* 2003;422:596.
4. Groeger S, Bison G, Knowles P E, Wynands R, Weis A, Laser-pumped cesium magnetometers for high-resolution medical and fundamental research, *Sensors and Actuators A: Physical.* 2006;129:1–5. EMSA 2004 — Selected Papers from the 5th European Magnetic Sensors & Actuators Conference — EMSA 2004, Cardiff, UK, 4–6 July 2004.
5. Bison G, Wynands R, Weis A, A laser-pumped magnetometer for the mapping of human cardiomagnetic fields, *Appl. Phys. B.* 2003;76:325–328.
6. Bison G, Wynands R, Weis A, Dynamical mapping of the human cardiomagnetic field with a room-temperature, laser-optical sensor, *Opt. Express.* 2003;11:904–909.
7. Bison G, Castagna N, Hofer A, et al, A room temperature 19-channel magnetic field mapping device for cardiac signals, *Appl. Phys. Lett.* 2009;95:173701.
8. IMEDCO AG, `http://www.imedco.ch/`.
9. Bison G, Schläpfer U, Di Domenico G, Saudan H, Weis A, A new optical magnetometer for MCG measurements in a low-cost shielding room, *Int. Congr. Ser.* 2007;1300. New Frontiers in Biomagnetism. Proceedings of the 15th International Conference on Biomagnetism, Vancouver, BC, Canada, August 21–25, 2006.

Author: A. Weis
Institute: University of Fribourg
Street: Chemin du Musée 3
City: 1700-Fribourg
Country: Switzerland
Email: antoine.weis@unifr.ch

Towards Clinical MCG with Room-Temperature Sensors

Gertrud Lembke[1], Markus Stumpf[1], Sergio Nicola Erné[2], and Georg Bison[1]

[1] University Hospital Jena, Erlanger Allee 101, 07747 Jena, Germany
[2] BMDSys Jena, Wildenbruchstr. 15, 07745 Jena, Germany

Abstract— **OBJECTIVE: Laser pumped magnetometers (LOPM) have received growing attention over the past decade as they represent a room temperature alternative to SQUIDs. We present intermediate results of a project at the Jena University Hospital which develops a demonstration system for clinical MCG using LOPMs in collaboration with BMDSys. The system offers great potential for cost efficient MCG and is planned to be integrated into ongoing clinical studies. METHODS: The physical principle of the used sensors is identical to the 19 channel LOPM system developed at the University of Fribourg [1]. Since the new clinical demonstration system will have 57 primary sensors and two times 13 reference sensors, a new technology has been developed which allows a much closer packing of the sensor modules. Considerable effort went also in the design of a shielding room which is compatible with the clinical environments. The aluminum based shield was set up at the Jena University Hospital and was equipped with special coils needed to generate an offset magnetic field essential for LOPM operation. RESULTS: Laboratory prototypes of the new sensor technology are successfully tested. We present performance details in strong and weak shielding. CONCLUSION: We made important technological steps towards an affordable clinical MCG system. Development will continue until the system is ready to be included in clinical studies.**

Keywords— **optical magnetometers, room temperature sensor, clinical MCG.**

I. Introduction

One way to visualize the electrophysiological function of the human heart is the contact-free measurement of its magnetic field [2] (magnetocardiogram – MCG). To date the standard for MCG are magnetometers based on superconducting quantum interference devices (SQUIDs) [3], which need helium cooling to operate, making their handling sophisticated and expensive.

In the last years a novel class of devices based on a completely different method for measuring weak magnetic fields was developed: optically pumped magnetometers (OPMs). First used for geological measurements [4], these devices were improved further to become sensitive enough for the detection of the human heart's weak magnetic fields [5, 6].

The latest development of laser OPMs (LOPMs) resulted in the 19-channel mapping device installed at the University of Fribourg [1], allowing the recording of magnetic field maps of the beating human heart (cardiac magnetic field mapping – cMFM). The promising results encouraged the design of a 57-channel magnetometer that is planned to be installed as clinical demonstration system at the University Hospital in Jena. The system offers great potential for cost efficient MCG and is intended to be integrated into ongoing clinical studies.

II. Materials and Methods

A. Sensor Head

The sensor head will consist of 57 primary sensors placed in a grid with 32 mm spacing, directly above the chest of the patient. Above the primary sensors two planes are installed, each with 13 sensors providing reference signals for the compensation of magnetic stray fields.

The physical principle of the used sensors is identical to those of the 19-channel LOPM system developed at the University of Fribourg – optical-rf double resonance in the so-called M_x configuration [7]. A spherical Cs vapor glass cell, coated with paraffin [8] is positioned in a vertical magnetic field B_0. A circularly polarized laser beam, whose frequency is locked to the $F=4 \rightarrow F=3$ transition of the cesium D_1 line, passes at an angle of 45° with respect to B_0 through the cell and creates through optical pumping a magnetization inside the vapor. This magnetization precesses with the Larmor frequency $\omega_L=\gamma|B_0|$, with $\gamma\approx 2\pi\times 3.5$ Hz/nT, around the field B_0. It is coherently driven by a weak magnetic field $B_{rf}<<B_0$, with a frequency ω_{rf}, oriented at 90° with respect to B_0. The magnetization is related to the atoms' index of absorption. Thus the transmitted light yields an amplitude modulation at ω_{rf}, which is resonantly enhanced at the frequency $\omega_{rf} = \omega_L$. By feedback control of either B_0 or ω_{rf}, that resonance condition can be actively stabilized. Near resonance, small magnetic fields like the heart's field lead to a linear phase shift in the oscillating signal, allowing the measurement of the corresponding field.

Each sensor is designed as a compact module consisting of a plastic holder carrying the optical components to collimate and polarize the light (lens, linear polarizer, $\lambda/4$-plate), a Cs vapor cell (27 mm in diameter), and a photodiode to

S. Supek and A. Sušac (Eds.): Advances in Biomagnetism – BIOMAG2010, IFMBE Proceedings 28, pp. 62–65, 2010.
www.springerlink.com

detect the light transmitted through the cell. The modules can be arranged in two-dimensional arrays with 32 mm sensor separation. This is considerably more compact than the 50 mm sensor separation in the system installed in Fribourg [1]. The sensors will be mounted in a quadratic grid on large printed circuit boards (PCBs) that firstly serve as mechanical support structures positioning the sensors in horizontal planes and secondly carry the coils for the application of the magnetic fields (Fig. 1).

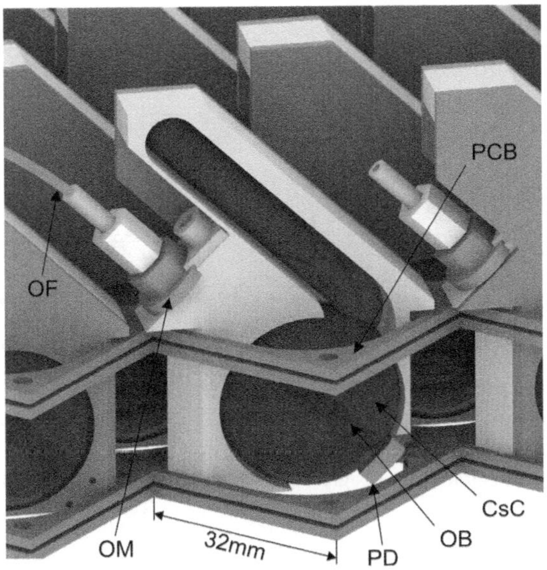

Fig. 1 Sketch of the sensor modules mounted in a quadratic grid on the large PCBs inside the sensor head. Inside the large PCBs the coils for applying the magnetic fields are implemented. OF: Optical fiber, OM: Optical module containing the components to collimate and polarize the light, PD: Photodiode, OB: Optical beam, CsC: Cesium cell

B. Magnetic Shielding Room

For the successful operation of any magnetometer, especially an OPM, the noise at 50 Hz present in any clinical environment, has to be reduced below its dynamic range. Thus a magnetic shielding room (MSR) with 12 mm thick aluminum walls and dimensions of 3.46 m × 2.46 m × 2.46 m (length × width × height) in modular construction technique was set up at the University Hospital of Jena for the 57-channel sensor system.

Although mu-metal is the material of choice for magnetic shielding, we opted for a purely aluminum based MSR for several reasons. First, it is much cheaper than mu-metal while providing sufficient shielding at 50 Hz. Since our project aims at a low priced MCG-system, the costs for the MSR play an important role.

Second, aluminum does not influence static currents in its direct vicinity, thus inside an aluminum room compensation coils for static and very slowly changing magnetic fields can be implemented directly inside the walls. Likewise the calculation of the coil geometry is much easier than in a mu-metal shield because no interaction between shielding material and coil has to be regarded. Thus a very compact design of the whole room without additional coils inside or outside is possible.

Inside our MRS we installed directly on the walls, the ceiling and the floor a coil system for the compensation of constant fields (Bx, By and Bz) and five field gradients: the three gradients of Bz (Bz/dx, Bz/dy, Bz/dz) and Bx/dx and By/dy. The coils compensate the ambient magnetic field such that the needed vertical field $B_0 \parallel z$ with an amplitude of approximately 1 μT remains. The Bz gradient coils are used to optimize the homogeneity of B_0 while Bx/dx and By/dy can be used to optimize the compensation of the transverse components x and y. For an optimal setup we calculated the expected field with Mathematica and Radia [9] and optimized the coil geometry to obtain ideal compensation properties at the sensor's position. A sketch of the MSR's interior with the coil system (only for the Bx-, By- and Bz- components of the magnetic field) and lowest sensor plane as it will be positioned above a patient is shown in Fig. 2.

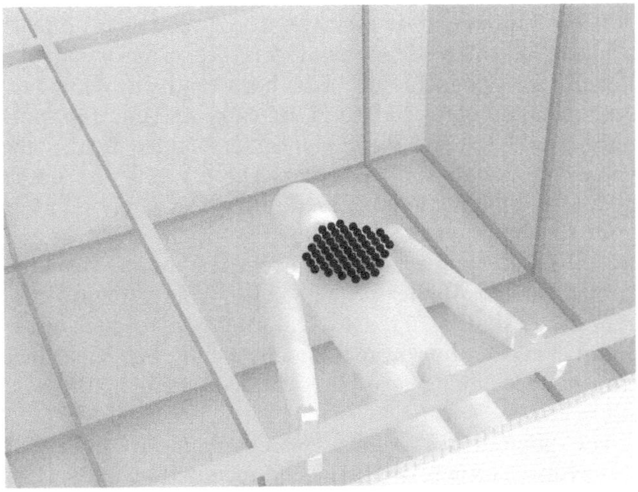

Fig. 2 Sketch of the MSR with the coils for the compensation of the Bx-, By- and Bz- components of the magnetic field, and the lowest sensor plane as it is positioned above a test person

III. RESULTS

A. Shielding Factor

The shielding factor of our MSR was measured by a fluxgate magnetometer installed in the middle of the room, with big driving coils around the chamber providing the magnetic field. Fig. 3 shows the shielding factor for frequencies up to 500 Hz. As expected, the shielding factor starts at 1 at low frequencies and increases to approximately 200 at 500 Hz. At 50 Hz we obtain a shielding factor of approximately 90, which is sufficient for the operation of our magnetometer.

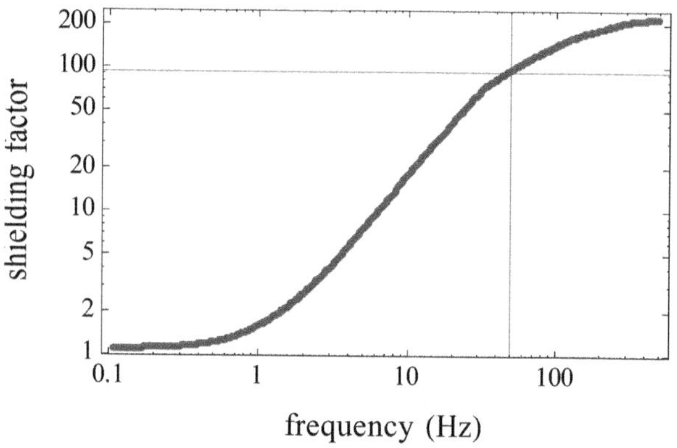

Fig. 3 Shielding factor of our MSR for frequencies of 0.1 Hz to 500 Hz. The gridlines show the shielding factor at 50 Hz

B. Noise Measurements of a Single Sensor

The performance of a single sensor module was tested inside a well shielded environment (3 layers mu-metal). The Cs-cell, the optics for the light beam and the photodiode were mounted between two PCBs carrying coils for applying the fields B_0 and B_{rf}, designed for optimal field homogeneity at the sensor's position. The Cs cell, the optical module and the photodiode used in this test setup are prototypes for the 57-channel system (Fig. 1). In order to study several noise sources of the current test setup we used a special version of the magnetometer. This version differs from the design in Fig. 1 by featuring two readout beams. Both beams can be recorded independently and yield independent measurements of the magnetic field in the same cell.

The photodiodes' signals were amplified by FEMTO Low Noise Current Amplifiers and recorded by a PC via an ADC card with a sampling rate of 1 MHz. Signal records of 10 seconds length were demodulated with a special software using a two phase lock-in algorithm which computes the phase of the signals.

Near resonance, the phase signal yields a linear dependence on the magnetic field with a zero crossing at the center of the resonance[1, 5]. When the frequency of B_{rf} is chosen close to the center of the resonance, small changes of the magnetic field cause a linear shift of the phase signal. The noise of the phase signal at the resonance determines the sensitivity of the magnetometer, thus its minimization is crucial for device optimization.

Effectively the two beam setup represents a gradiometer with zero base length. If the two signals are subtracted, the result becomes independent of magnetic field noise (Fig. 4, curve B) and can be used to optimize several system parameters. If the two signals are averaged, a single magnetic field dependant signal is obtained. The averaged signal (Fig. 4, curve A) is equivalent to a magnetometer with one beam where the beam can be twice as powerful as in the two beam setup.

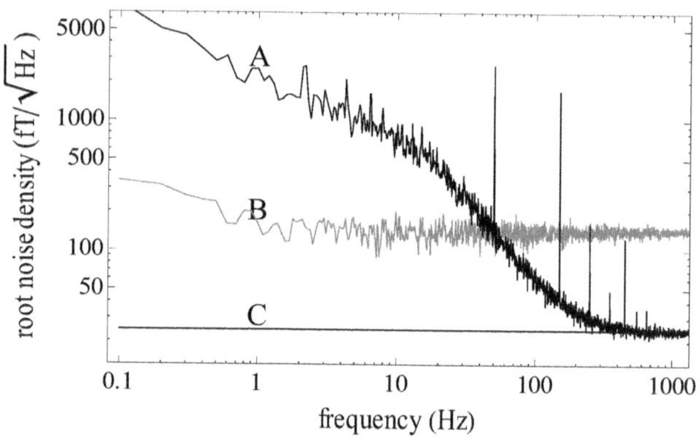

Fig. 4 Noise of a single sensor module inside a 3-layer mu-metal shield. A: Noise spectrum in a setup for heart measurements. B: Field independent gradiometer signal in a setup for system optimization. C: Intrinsic sensitivity

IV. Discussion

Figure 4 shows noise spectra of the signals discussed above which were calibrated in magnetic field units using the $d\phi/dB_0$ slope of the phase signal. At small frequencies the spectrum of the field dependant signal is dominated by magnetic field noise (Fig. 4, curve A). Towards high frequencies the noise decreases until reaching a constant level of 24.5 $fT/(Hz)^{1/2}$ which is limited by the fundamental quantum noise (shot noise) of the optical measurement (Fig. 4, line C). This decrease in noise is due to the limited bandwidth f_{bw} with which the spins of the Cs atoms react to magnetic field changes. In the simple phase-recording mode of operation used here, f_{bw} is limited by the transverse spin coherence time (T_2 time) of the atoms. Under the conditions used, f_{bw} was approximately 15 Hz and shows the characteristic of a first order low pass filter.

The measurement bandwidth can be significantly higher in other modes of OPM operation. In the 57-channel system we plan to use feedback techniques [1, 10] in order to achieve a bandwidth of at least 500 Hz. In the test experiment the limited bandwidth mode of operation was used intentionally in order to study the fundamental noise which is reached at approximately 400 Hz. This noise level is the intrinsic limitation of the magnetometer and is mainly given by the quality of the Cs cell. The level found coincides well with the theoretical levels calculated during cell production [8]. We expect that the noise falls to the intrinsic noise level at all frequencies if the magnetic noise in the test setup can be made small enough. In order to achieve such a demonstration we plan to measure in better shielding and to use gradiometric sensor configurations.

With the present experiment we can achieve a magnetic field independent signal if we subtract the two independent measurement channels. This signal, however, suffers from correlations in the light intensity and shows for that reason a noise level approximately one order of magnitude higher than the intrinsic noise. We will use this signal in the future to optimize the stability of the laser light. Currently it can only be interpreted as an upper limit of the magnetometer noise.

V. Conclusion

We made important technological steps towards an affordable clinical MCG system.

- A low-priced MSR with sufficient shielding factor for heart measurements with integrated coils for constant magnetic stray field and field gradient compensation was implemented.
- We designed a compact sensor module for a multi-channel sensor head with 57 primary channels, which improves the channel density by more than a factor of two compared to earlier designs [1]. The new sensor head will allow the recording of cardiac magnetic field maps with a spatial resolution similar to existing SQUID-based systems.
- The magnetometric sensitivity of the new compact sensor modules could be tested with a prototype setup. The intrinsic noise level could be confirmed at the 25 $fT/(Hz)^{1/2}$ level. This improvement in comparison to earlier designs [1, 6, 10, 11] lets us expect that the 57-channel system will be able to record diagnostically relevant information at least in the QRS-complex and the T-wave.

Development towards a well performing and user-friendly diagnosis system will continue until it is ready to be included in clinical studies.

Acknowledgment

This work was supported by the German Federal Ministry for Education and Research (#0315024 B).

References

1. G. Bison, N. Castagna et al. (2009) A room temperature 19-channel magnetic field mapping device for cardiac signals. Appl. Phys. Lett. 95:173701
2. W. Andrä, H. Nowak (1998) Magnetism in Medicine. Wiley-VCH Verlag GmbH, Berlin
3. D. Cohen, E. A. Edelsak, J. E. Zimmerman (1970) Magnetocardiograms taken inside a shielded room with a superconducting point-contact magnetometer. Appl. Phys. Lett. 16: 278-280
4. E. B. Alexandrow, V. A. Bonch-Bruevich (1992) Optically pumped atomic magnetometers after three decades. Opt. Eng. 31(4):711-717
5. G. Bison, R. Wynands, A. Weis (2003) A laser-pumped magnetometer for the mapping of human cardiomagnetic fields. Appl. Phys. B 76:325-328
6. G. Bison, R. Wynands, A. Weis (2003) Dynamical mapping of the human cardiomagnetic field with a room-temperature, laser-optical sensor. Optics Express 11:904-909
7. S. Groeger, J.-L. Schenker et al. (2006) A high-sensitivity laser-pumped Mx-magnetometer. Phys. J. D 38:239-247
8. N. Castagna, G. Bison et al. (2009) A large sample study of spin relaxation and magnetometric sensitivity of paraffin coated Cs vapor cells. Appl. Phys. B 96:763-772 DOI 10.1007/s00340-009-3464-5
9. Radia, 3D magnetostatics computation software (Mathematica Addon by ESRF), http://www.esrf.eu/Accelerators/Groups/InsertionDevices /Software/Radia
10. G. Bison, R. Wynands, and A.Weis (2004) Mapping of the cardiomagnetic field using a optical room-temperature magnetometer. Biomed. Tech. 48: Suppl. 2, 292-294
11. R. Fenici, G. Bison, R. Wynands, D. Brisinda, A. M. Meloni, A. Weis (2004) Comparison of Magnetocardiographic Mapping with SQUID-based and Laser-pumped Magnetometers in Normal Subjects. Biomed. Tech. 48: Suppl. 2, 192--194.

Author: Gertrud Lembke
Institute: University Hospital of Jena
Street: Erlanger Allee 101
City: 07747 Jena
Country: Germany
Email: Gertrud.Lembke@med.uni-jena.de

Improvement of Small-Channel MCG System for Unshielded Environment

Mykola M. Budnyk[1,2], Oleksandr V. Zakorcheny[2], Vitaly C. Koshelnyk[2], Vitaly M. Budnyk[1], Volodymyr V. Lukashyk[2], Yuriy D. Minov[1], Pavlo G. Sutkovyi[1], and Tetyana M. Ryzhenko[1]

[1] Glushkov Institute for Cybernetics/Department Sensors and Technologies for Non-contact Diagnostics, Kyiv, Ukraine
[2] Taras Shevchenko National University of Kyiv/Department Medical Radio-physics, Radio-physical Faculty, Kyiv, Ukraine

Abstract— **Some improving of small-channel system have been performed, including synthetic channels formed by two physical ones, Principal Component Analysis (PCA)-based algorithm for de-noising of input signal, and Lead Fields (LF) for 36 spatial channels taking into account the coverage diagram of gradiometers. Above improvements are directed to increase reliability and signal-to-noise ratio (SNR) at unshielded environment and also to overcome blurring of current density distribution (CDD) images of the heart electric activity.**

Keywords— **synthetic channels, PCA analysis, lead field (LF) of gradientometer.**

I. Introduction

Today in the world a few dozens biomagnetic laboratories actively work and conduct research based on SQUID-magnetometer systems. But, they, as a rule, need expensive magnetic shielded room (MSR), skilled service and persistent cooling by liquid helium [1]. All these factors limit application of such systems. So, spreading of promising magnetocardiographic (MCG) method is firstly limited by using MSR. However, at unshielded environment the strong urban noises lead to:

1) unstable operation (breakdowns);
2) poor signal-to-noise ratio (SNR) of input signal;
3) non-reliable current density distribution (CDD) images cause diagnostic errors.

To overcome above drawbacks, small-channel MCG systems utilize: a) the 2nd order axial gradiometers, b) reference vector magnetometer (RVM), and c) methods of inverse problem solution (IPS) using a priory data about nature of biomagnetic sources [2].

II. Problem Solved

Aim of study is to improve system operated without MSR by increasing stability, SNR, and reliability of CDD images. We propose to solve the above-mentioned in Introduction problems 1)-3) with the help of, respectively:

1) synthetic channels (higher-order effective gradiometer);
2) de-noising input data with help of PCA;
3) contrasting of channel LFs (including synthetic ones).

III. Comparison of Channels

To diagnose cardiologic diseases the 4-channel system was developed [2] based on pulse-relaxation SQUID-magnetometers [3]. Antenna is axial gradiometer of the 2nd order: graphite cylinder ∅20 mm, coils are wounded by Nb wire ∅50 mkm according to rule 1-2-1 [4]. Channel design and parameters were described in detail before [5], and took into account experiments [6] and calculations [7].

The 2-channel system is early developed for the study magnetic nanoparticles into the body of lab animals [8]. Two identical channels based on DC SQUIDs form a differential magnetometer. Antenna is similar gradientometer having textolite frame ∅8 mm and base 35 mm, winding rule 2-4-2 [9]. High balance $\leq 10^{-3}$ is achieved by manufacturing (accuracy of making, care of winding, fixing glue [10]) and the advanced mechanical balancing [11].

For comparison of channels (K1 and K4 – MCG, K2 and K3 – suseptometric), pair of channels of different types were installed into cryostat of the 4-channel system. Their calibration was conducted by the measure of magnetic induction (Certificate #185 from 01.03.2005) with the calibrated transfer function (TF) equal to 4.19 pT/mkA [12].

5 calibrated values of the field: 53.95 pT, 10.89 pT, 215.78 pT, 323.67 pT, and 431,57 pT were registered 30 times. TF of i-th channel Ki (dependence of output signal in ADC samples Ni from the input field Bi in pT) was determined based on linear regression [13].

$$N_i = K_i \text{ [samples/pT]} \times B_i \text{ [pT]} + N_{0i} \qquad (1)$$

where N_{0i} – “zero” shift for non-absolute sensor. As a result, such channel TF are measured: K1=7.62, K4=7.08, K2=5.28, K3=4,83 (Fig. 1).

The relative errors between 5 points were calculated [14] for estimation of the channel accuracies (tabl. 1). Average accuracy for MCG channels 1&4 is 2.1%, and for suseptometric channels 2&3 is in 1,62 times worse, i.e. 3.4%.

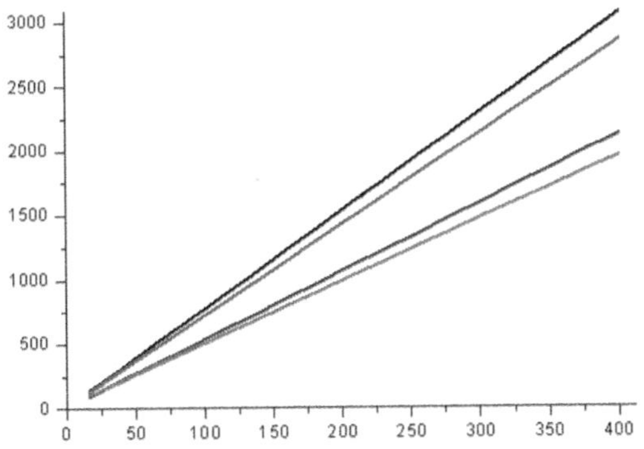

Fig. 1 Calibrating curves for SQUID channels (from top to bottom): 1, 4 (MCG) and 2,3 (susceptometric)

Table 1 Relative errors of channels

Field, pT	Channel number			
	1	2	3	4
53,95	1,77	3,16	3,65	2,41
107,89	0,88	1,81	2,13	1,47
215,78	0,40	0,79	0,79	0,44
323,68	0,57	0,62	0,68	0,53
431,57	1,06	0,58	0,61	0,23
Marginal	1,77	3,16	3,65	2,41

Stability (reliability) of channels operation was also estimated. For this purpose during calibrating procedure the amount of channel failures (out of dynamic range) was counted up. Information is resulted in tabl. 2, from which it is evidently, that stability in relation to strong disturbances in channels 1&4 in 74/4=18,5 times worse compare to channels 2&3. Solution at presence of strong disturbances consists in finding of place with the minimum level of disturbances that enables to decrease the noise by order [15].

Table 2 Quantity of the channel failures

Field, pT	Cnannel number			
	1	2	3	4
53,95	25	1	1	23
107,89	20	2	2	16
215,78	10	0	0	13
323,68	12	1	1	15
431,57	6	0	0	7
Total	73	4	4	74

IV. Synthesis of Logical Channels

Channel antennas are closely positioned into cryostat and form synthetic gradientometr registering the planar field gradient [16] in relation to far sources. Two synthetic gradientometer were created: 1st (2nd) is formed by channels 2&3 (1&4). But physical channels have different TF that is why logical channels should be synthesized as weighted differences of physical ones.

From (1) we can obtain expressions for the output signals B_{ik} and relative errors δ_{ik} of logical channels (i=3,4; k=2,1)

$$B_{41}=W_4N_4-W_1N_1+B_{041},\quad B_{32}=W_3N_3-W_2N_2+B_{032} \qquad (2)$$

$$B_{0ik}=W_iN_{0i}-W_kN_{0k},\quad W_i=1/K_i,\quad W_k=1/K_k \qquad (3)$$

$$\delta_{ik}=K_i{*}\,\delta_i+K_k{*}\,\delta_k, \qquad (4)$$

where $W_{i,k}$ and $\delta_{i,k}$ are weights and relative errors of physical channels, B_{0ik} is constant shift. Based on (2-4) and data from tabl.1 one can calculate TF W and relative errors δ of logical channels collected in Fig. 2 and tabl. 3.

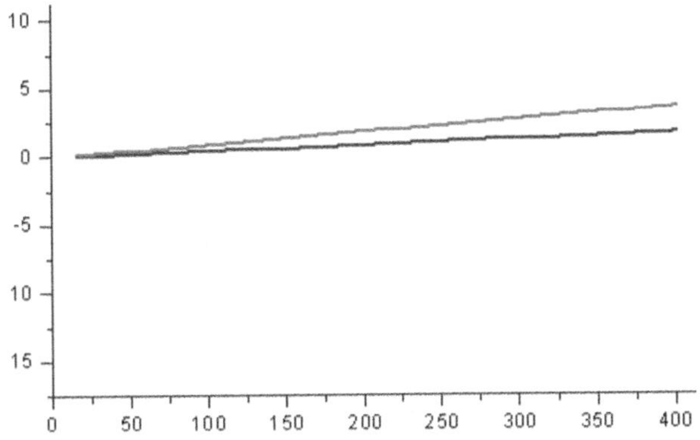

Fig. 2 Trunsfer functions for synthetic channels 3-2 (bottom) and 4-1 (top)

Table 3 Parameters of synthetic channels

Channel	$W_{4,3}$	$W_{1,2}$	B_0	δ, %	W
4-1	0.141	0.131	-1.89	2.26	0.009
3-2	0.207	0.189	-3.50	3.41	0.004

V. Signal De-noising by PCA Method

Purpose of PCA is to select useful information from a certain set of data. Data almost always contain an undesirable component – noise. What to be considered as noise and what is information? It is decided based on the nature of task. In results, it is obtained new presentation having too lower dimension (complexity).

It is known that filtration brings to non-linear amplitude-phase distortions. That is why it is needed the new methods for signal denoising, which do not distort a signal. 36 MKG signals were processed by MATLAB package. It was found denoised useful signal and 5 PCA noisy components – high-frequency (HF) uncorrelated fluctuations, 50 Hz, ½ and 1/3 power-line sub-harmonics, low-frequency (LF) trend [18]. MCG signal with minimum SNR is presented at Fig. 3.

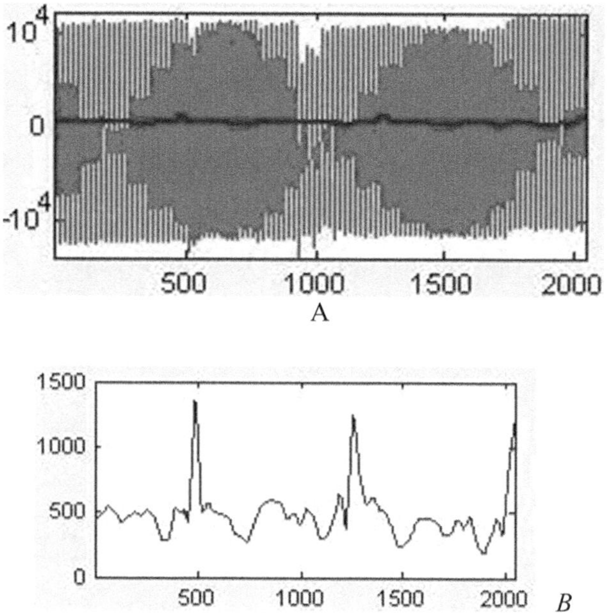

Fig. 3 View of MCG signal with SNR=3%: noisy (A), denoised (B)

VI. Calculation of Lead Fields (LF)

It is well-known that the view of MCG map depends on a measuring grid, that is why there is a problem of recalculating maps to the standard grid (i.e. theirs standardization), for example see [19].

Aim of study is calculation of LF for axial antenna taking into account the form of its coverage diagram (CoD) in order to visualize the heart activity. Radial CoD of magnetometric antenna has form (5) into plane, which is perpendicular to it axis, in the cylindrical co-ordinates [20]

$$F_{\rho} = \frac{3Z\rho}{2[\rho^2 + Z^2]^{5/2}} \tag{5}$$

where $r^2=\rho^2+Z^2$ is a radius-vector directed from the antenna center to observation point; Z – depth of point. LF for 36 spatial channels calculated within square 20×20 sm with a step 4 sm, i.e. totally 21×21 nodes. At Fig. 4,a the LF is represented for channel shifted from the map center on 6 sm to the right and downward (grid point 5-5).

From Fig. 4a it is shown that LF has axial symmetry and maximum on certain distance from the antenna center. Also one can see that map is very blurred, that lead to blurring of current reconstruction that is why the method of LF contrasting was developed. It is found, that at contrast degree equal to 25, LF distribution has "ring" form with diameter equal to depth of plane. It is in accordance with the calculations of CoD having the cover angle of 60^0 [7].

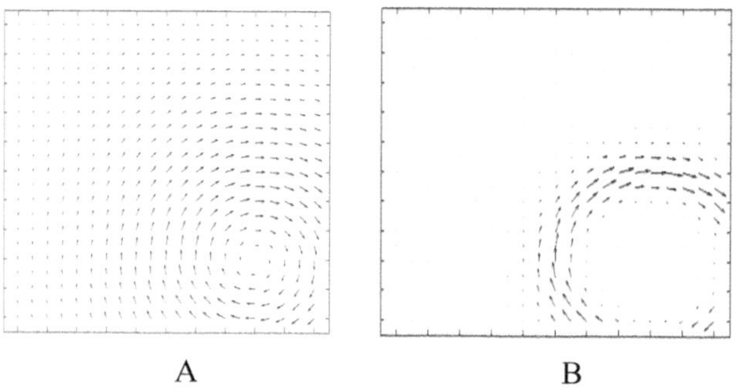

Fig. 4 LF of channel 5-5 at 12 sm depth: ordinary (a), contrasted (b)

It is also found, that the ring radius is increased with increasing of plane depth. Thus, it is needed to increase the degree of LF contrasting, in order to compensate the parasitic effect of increasing of blurring of CDD maps with increasing the plane depth. We will obtain total (integral) LF by summarizing LF of all 36 channels.

Integral LF has axial symmetry in relation to the map center, and its amplitude is decreased with depth increasing. Reason consists in that amplitude of the magnetic field is decreased with increasing of the source depth. It can be lead to error in determination of the source amplitude (for example, magnetic dipole), that is why the methods for LF normalization were developed [21].

VII. Conclusions

To improve features of 4-channel MCG system, it has been done:

1 Comparison of MCG and suseptometric channels have been performed, including:

a) Transfer functions, accuracy and operating stability of physical channels were experimentally determined at presence of industrial disturbances;

b) It is found that accuracy of channels 2&3 nearly in 1,5 times better than 1&4 ones. From other hand, failure statistics is clearly shown that channels 2&3 are more reliable against magnetic noises. Thus at the unshielded location, it is more optimal channels 2&3 which demonstrate high stability and insignificant worsening of accuracy.

c) Parameters of the logical software channels formed by 2 physical ones were calculated. Their TF on 2 orders of magnitude lower due to disturbances from far sources are suppressed more better. But for their application it is necessary the proper methods of inverse problem solution.

2 Adjusting and optimizing PCA-based algorithm for maximization of signal de-noising;

a) 5 kinds of disturbances were identified (50 Hz power line, its ½ (25 Hz) and 1/3 (17 Hz) sub-harmonics, zero drift and uncorrelated noise);
b) optimal parameters of PCA-analysis, which provide maximal de-noising, were found;
c) It is reached degree of de-noising allowing to reveal an useful signal on background disturbances with minimal SNR=3%.

3 Calculated LF for 36 spatial channels taking into account the coverage diagram of gradiometers and effect of blurring, including:

a) method of contrasting for visualization of «ring» which is in accordance with form of antenna coverage diagram [7];
b) integral LF of 36 MCG channels and its dependence vs. depth were calculated;
c) method of LF normalization was proposed.

Thus, above results demonstrate increasing noise-immunity (logical channels) and SNR (PCA-based de-noising), and are promising for CDD de-blurring (LF contrasting) for small-channel system operating at unshielded location.

Acknowledgment

This work was financed by the Science and Technology Center in Ukraine (STCU) within Project #4719 "Market investigation of new non-invasive imaging and analysis technology for early diagnosis of heart diseases" (manager – Dr Volodymyr Sosnytskyy). The authors would like to thank collaborators: Prof A. Hagege (Eur Hospital Pompidou, Paris), Prof B. Hailer (Catholical Hospital "Philippusstift", Essen), Dr med M. Blank (Universitatklinikum Essen), and Prof B. Schwabe (Margarethe von Wrangell Stiftung e.V.) for helpful support and cooperation.

References

1. Andra W, Nowak H (1998) Magnetism in Medicine. Wiley-VCH, Berlin
2. Budnyk M, Voytovych I, Minov Yu. et al. (2004) Low-cost 7-channel MCG system for unshielded environment. Neurol and Clin Neurophysiology 112:1-7
3. Budnyk M, Voytovych I, Gapelyuk A et al. (1995) Computer-Aided Biomagnetic Investigation System. Control System and Machines 3:31-46
4. Sosnytsky V, Voytovych I, Budnyk M et al. (1994) Pulse-relaxation oscillation SQUID magnetometer, Proc vol. III, XIII IMEKO World Congress, Torino, Italy, 1994, pp 2383-2387
5. Steinberg F, Budnyk M, Sosnytsky V (2002) Method for determining a diagnostically relevant parameter from the ECG and MCG data of the patient, Patent WO 0200108
6. Minov Yu, Budnyk M, Shpylyovyy P (2006) Sensor of superweak magnetic fields. Patent UA 75434
7. Budnyk M (1999) Coverage diagram of the axial superconductive gradiometer as receiving magnetostatic antenna. Med and Biol Eng and Comp 37(Suppl.2):404-406
8. Budnyk M, Dudchenko N, Minov Yu et al. (2009) SQUID-imaging technology to study magnetic nanocarriers for targeted magnetic transport of drugs. Materialwissenschaft und Werkstofftechnik 40:302-307
9. Budnyk M, Voytovych I, Minov Yu. et al. (2008) SQUID imaging system for studying magnetic nanoparticles, Bio-magnetism: Interdisciplinary Research and Exploration, Hokkaido Univ Press, Sapporo, 2008, pp. 30-32
10. Minov Yu, Budnyk M (2006) Superconductive gradientometer of magnetic field. Patent UA 16882
11. Budnyk M, Minov Yu (2007) Method for mechanical balancing of superconductive gradiometer. Patent UA 19997
12. Budnyk M, Budnyk S (2004) Development of measuring procedure for multi-channel magnetocardiograph, Proc vol.1, Int Conf "Metrology and measurement technique", Kharkiv, Ukraine, 2004, pp 332-334
13. Zakorcheny O, Budnyk V, Budnyk M (2009) Studying of SQUID-system for active&passive bio-magnetic researches at unshielded environment, Proc, 5th IEEE Workshop IDAACS, Rende (Cozenca), Italy, 2009, pp 77-81
14. Ostapchuk O, Zakorcheny O, Budnyk V (2009) Studying multi-channel SQUID-magnetometer at unshielded envi-ronment, Proc, 2nd summer school "Biomed IT in public health" (BMIT'2009), Kyiv, Ukraine, 2009, pp 93 - 98
15. Zakorcheny O, Minov Yu, Budnyk M (2007) Finding opti-mal design of SQUID-magnetometer antenna at unshielded environment. Com devices, networks&systems 6:47-50
16. Minov Yu, Budnyk M, Zakorcheny O (2008) 2-channel squid-magnetometer with changeable base, Pat UA 34263
17. Zakorcheny O, Budnyk M, Koronovska K (2009) Studying SQUID-magnetometer system at unshielded environment. Electronics and communications 4:164-168
18. Koshelnyk V, Budnyk M (2009) De-noising ECG signals by method of principal component analysis, Proc, 2nd summer school BMIT'2009, Kyiv, Ukraine, 2009, pp 125-128
19. Numminen J, Ahlfors S, Ilmoniemi R et al (1995) Transformation of multi-channel magnetocardiographic signal to standard grid from. IEEE Trans on BME 42: 75-79
20. Budnyk M (2007) Calculation of the magnetic antennas for moving the horizontal-oriented magnetic dipole into the horizontal plane. Electronics and Communication 3:39-42
21. Lukashyk V, Budnyk M (2009) Development of tools to visualize bioelectric activity based on lead fields, Proc, 2nd summer school BMIT'2009, Kyiv, Ukraine, pp 105-110

Address of the corresponding author:

Author: M.M. Budnyk
Institute: Glushkov Institute for Cybernetics/Department Sensors and Technologies for Non-contact Diagnostics
Street: 40 Glushkov Avenue
City: Kyiv
Country: Ukraine
Email: budnyk@meta.ua

Magnetoresistive Hybrid Sensors for Simultaneous Low-Field MRI and Biomagnetic Measurements

Natalia Sergeeva-Chollet, Hadrien Dyvorne, Hedwige Polovy, Myriam Pannetier-Lecoeur, and Claude Fermon

DSM/IRAMIS/SPEC-CEA Saclay, Gif-sur-Yvette, France

Abstract— **Recently developed magnetoresistive hybrid sensors can detect magnetic signals in the femtotesla range. This sensor is a combination of a Giant Magnetoresistive (GMR) field sensor and flux-to-field superconducting transformer [1]. Hybrid sensors are the good candidates for Low-Field Magnetic Resonance Imaging (LF-MRI) and neural signal detection. The primary advantages of these sensors are their robustness against external static fields and fast recovery after RF-pulses. We present the first the results obtained on low-field NMR with static fields up to 8 mT. MRI images obtained at LF-MRI without pre-polarization will be presented.**

Finally the combination of low-field MRI based on hybrid sensors with neural signal detection (MEG) will be discussed.

Keywords— **GMR, NMR, low-field MRI, MEG.**

I. Introduction

Low-field magnetic resonance imaging and biomagnetic measurements like neuronal signal detection (MEG) require sensitive magnetometers for detection of corresponding very weak magnetic signals of femtotesla range. Superconducting Quantum Interference devices (SQUID) are the most commonly used sensors for such applications [2]. However, recently developed new technologies have been proposed new devices such as atomic magnetometers [3] and hybrid sensors [1].

In this paper we present the results obtained with hybrid sensors which combine a field GMR sensor and a superconducting flux-to-field transformer. Such sensor is sensitive in the femtotesla range and is broadband. This gives the possibility of fast detection of low-field Nuclear Magnetic Resonance (NMR) signals.

II. Methods

A. Hybrid Sensor

Fabrication of hybrid sensors is based on thin film technology combining the GMR element and a superconducting loop which acts as a flux-to-field transformer. Two types of hybrid sensors were fabricated. The first combines the YBaCuO (YBCO) and GMR and can be cooled to liquid nitrogen temperature (77 K). The second one is based on a Nb loop with GMR and need to be cooled to liquid helium temperature (4 K). The GMR element is a spin valve and has a form of yoke to reduce the magnetic noise (Fig.1).

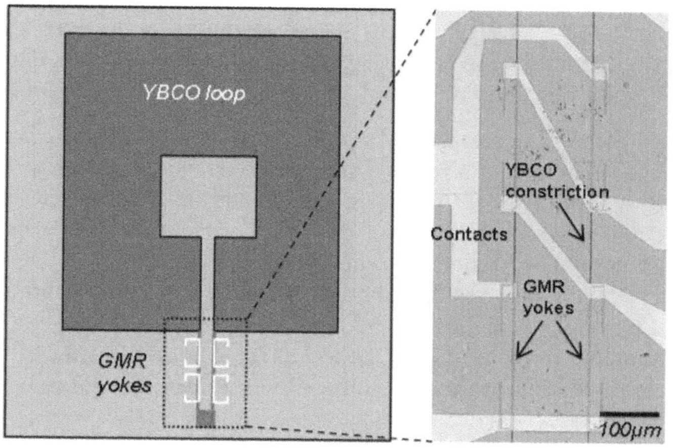

Fig. 1 Schematic (left) and micrograph (right) picture of a hybrid sensor with a YBCO superconducting loop with two constrictions and four GMR sensing elements mounted in Wheatstone bridge configuration

The typical response of a hybrid sensor to magnetic field applied perpendicular to the superconducting loop is shown in Fig. 2. The applied field generates a supercurrent that prevents the entrance of the flux. The supercurrent flowing through the constriction in the loop creates locally a high field that can be detected by GMR elements situated on top or under the constriction. Typically, for one square centimeter loop and 5 μm constriction, the local field detected by GMR is 500 times bigger than the field applied to the loop.

Hybrid sensor performance could be estimated from its gain and its noise characteristics. Sensor gain is the ratio between the slope of magnetoresistance when the loop is acting (field is applied perpendicular to the superconducting loop) and the slope of GMR (superconducting loop is not acting, field is applied to the GMR sensitive axis).

The field noise of a hybrid sensor (Fig. 3) is a function of frequency. The signal-to-noise ratio is limited by a low frequency 1/f noise which appears in kHz range. This contribution to noise is given by GMR elements of small volume. To reduce this contribution, a modulation technique is

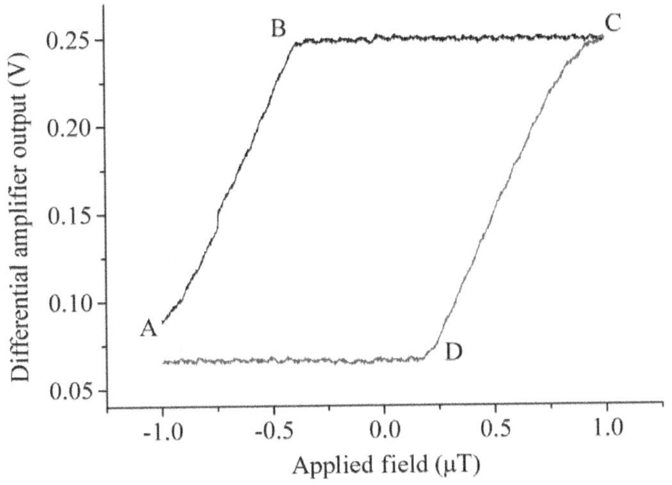

Fig. 2 Typical response of a hybrid sensor to a perpendicular applied field. From A to B, the output voltage increases as the supercurrent in constriction increases. At point B, the supercurrent reaches its critical value, and the local field does not increase anymore; when sweeping the field back (points C to D), the supercurrent decreases to minus critical current (point D) leading to change a magnetoresistance. The sensor is usually operated in the linear part of its response (A to B), well below the saturation

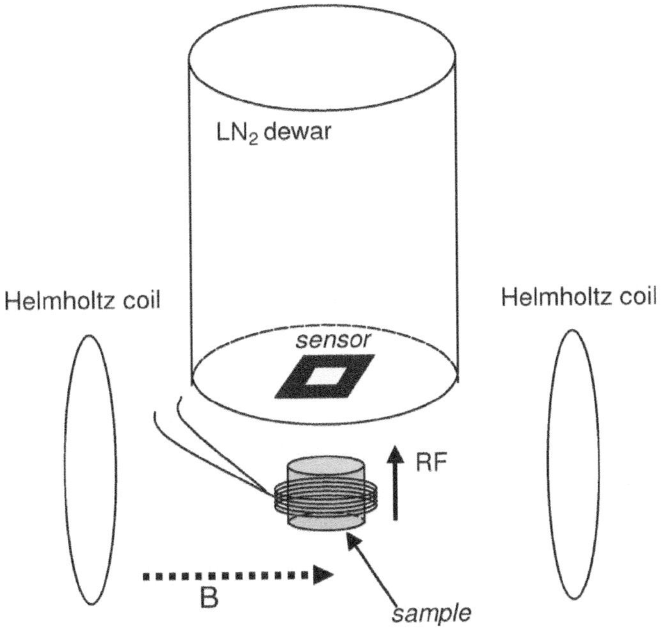

Fig. 4 Schematics of experimental setup. A main field (B) generated by Helmholtz coil is applied to the sample. An RF pulse is applied by coil wounded around the sample at corresponding frequency. The detection is done by a hybrid sensor placed into a liquid nitrogen dewar

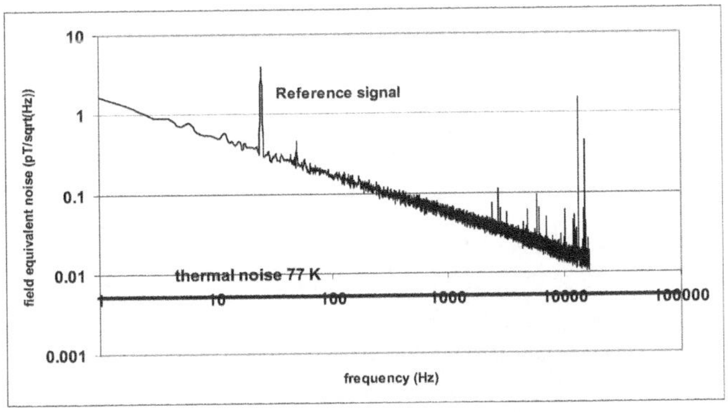

Fig. 3 Field equivalent noise as a function of frequency for a YBCO-based sensor. At 5pT/sqrt(Hz) reference signal could be clearly seen. Indication of thermal noise at 77 K is given by a straight line

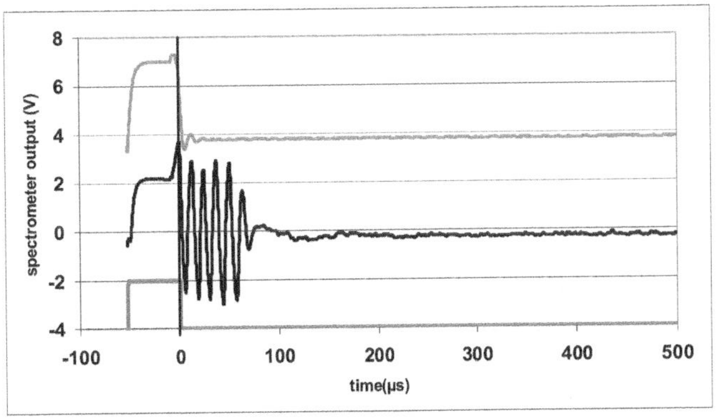

Fig. 5 Comparison of the response of a resonant coil (lower curve) and hybrid sensor (upper curve) to a 1 kW RF pulse of 50 μs at 800 kHz. The hybrid sensor being untuned, one does not observe the ringing as the coil response. The recovery is very fast and the small oscillation observed in the hybrid sensor response is already resonant signal

proposed which comprises the local heating of the constriction by driving the current in the local point of constriction [4]. This technique allows increasing the sensitivity in the 1/f regime which is very important for MEG applications.

B. Experimental Setup

The experimental setup shown in Fig. 4 consists of an excitation coil and a pair of Helmholtz coils generating permanent magnetic field of 8 mT (corresponding to 340 kHz). RF pulses are applied by a coil around the sample perpendicularly to the main field. The hybrid sensor used for the NMR measurements is the YBCO-based sensor with a gain of 716 and a transition temperature of T_C=80 K. The sensitivity of the sensor (20x20 mm loop and 5 μm constriction) is about 20 fT/sqrt(Hz). The hybrid sensor was placed in a liquid nitrogen dewar on top of room temperature sample. The distance between the sensor and the sample is about 30 mm.

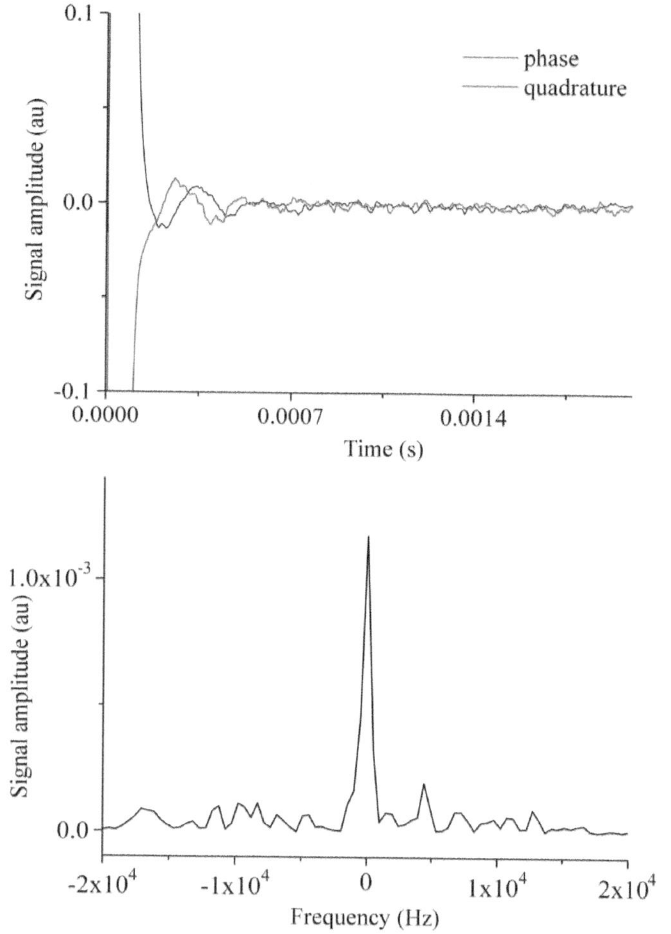

Fig. 6 NMR signal measured on doped water sample at 340 kHz with a hybrid sensor. Upper curve – FID and lower curve – resonance line

For signal amplification a low noise preamplifier is used and acquisitions are done with a homemade NMR spectrometer. First, detection of the NMR signal using a hybrid sensor has been performed to verify the robustness of the sensor and the possibility of using it for MRI.

The NMR signal is detected from a 3 ml water sample, doped with $CuSO_4$ at room temperature to reduce the relaxation time.

III. Results

A. Hybrid Sensor Properties in the Presence of RF-Signal

The RF pulse has been applied with the sensor on during 50 μs with repetition time 300 ms. During the pulses, saturation is reached in the sensor without destroying it. The sensor recovers in a microsecond range and is able to detect a signal just after the pulse (Fig.5). Therefore it can experience series of sequences without protection.

B. NMR Signal Detection

An NMR signal has been detected with a hybrid sensor. In that experiment, the homogeneity of applied field was rather poor inducing a large line width of the signal. However, hybrid sensors have the fast response which allows us obtaining free induction decay (FID) signal from nuclear spins. The applied RF pulse power is 1 kW. Fig. 6 shows the simple FID in-phase and quadrature signal detected with hybrid sensor in 5000 acquisitions [5].

C. MRI Images

We have developed a new setup with field homogeneity of 10 ppm. First the classical MRI was done to optimize imaging sequences. The detection is done with a coil placed perpendicular to the RF pulse coil. To perform imaging in low field, gradient coils and shimming coils are added. Applied external field is 8 mT.

The image of a peach was obtained with a gradient spin echo (Fig. 7) sequence. The selective excitation pulse has a corresponding bandwidth of 160 Hz. The spread in frequency due to the gradient during selection results in slices of order of 3 mm. The read-out gradient is included in a gradient spin echo sequence.

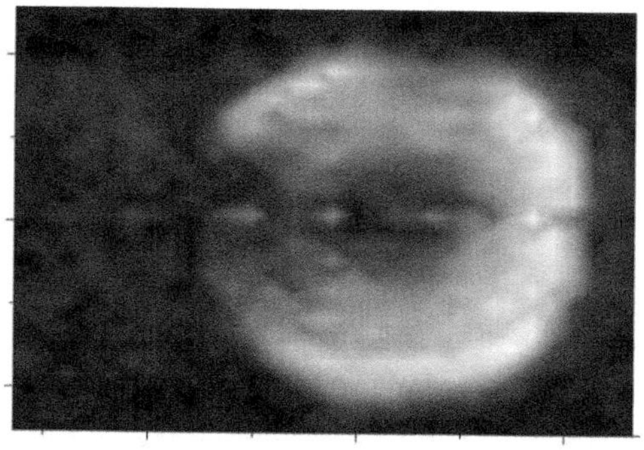

Fig. 7 MRI image of a peach

IV. Discussion

A. Detection Bandwidth

The fact that the hybrid sensors could operate without protection and without long recovery time is very important for low-field MRI imaging as a small field could be continuously applied to the sample in presence of sensor. The possibility of signal detection in microsecond range after RF pulse gives an opportunity to make multi-slice acquisitions like in high-field MRI.

B. Further Improvements for Low-Field MRI

The main advantages of the hybrid sensors are their robustness and wideband operation. These properties allow us to perform MRI sequences without prepolarization as these sensors could operate in mT-range static field.

The effective surface of hybrid sensor and the distance to sample play an important role in optimizing performance, which could be further enhanced by using of a flux transformer with a large pick-up loop [6].

For the moment the low frequency 1/f noise of GMR exhibits a corner in the kHz range which needs use of noise cancellation technique for MEG application.

V. Conclusions

Hybrid sensors are very promising devices for ultra-low signal detection. Their robustness against applied RF pulse is very important for high performance MRI imaging.

For MEG applications 1/f noise cancellation technique should be applied.

Combined system allowing simultaneous MEG and MRI acquisition are targeted with hybrid sensors.

Acknowledgment

This work was supported by the European project MEGMRI (HEALTH-F5-2008-200859).

References

1. Pannetier M., Fermon C., Legoff G., Simola J., Kerr E. (2004), Femtotesla magnetic field measurement with magnteoresistive sensors, Science 304:1648-1650
2. Clarke J., Braginskii A.I. (2004) The SQUID Handbook Wiley-vch
3. Kominis I.K., Kornack T.W., Allred J.C., Romalis M.V. (2003) A subfemtotesla multichannel atomic magnetometer Nature 422, 596
4. Pannetier-Lecoeur M., Fermon C., Dyvorne H., Jacquinot J.F., Polovy H., Walliang A.L. (2009) Magnetoresistive-superconducting mixed sensors for biomagnetic applications J. Magn. Magn. Mater. doi:10.1016/j.jmmm.2009.03.012
5. Dyvorne H., Fermon C., Pannetier-Lecoeur M., Polovy H., Walliang A.-L. (2009) NMR with superconducting-GMR mixed sensors IEEE Trans. Appl. Supercond. 19, 3:819-822
6. Dyvorne H., Scola J., Fermon C., Jacquinot J.F., Pannetier-Lecoeur M. (2008) Flux transformers made on commercial high critical temperature superconducting wires Rev.Sci.Instrum. 79, 025107-025111

Author: Natalia Sergeeva-Chollet

Institute: DSM/IRAMIS/SPEC-CEA Saclay
City: Gif-sur-Yvette
Country: France
Email: natalia.sergeeva-chollet@cea.fr

Safety in Simultaneous Ultra-Low-Field MRI and MEG: Forces Exerted on Magnetizable Objects by Magnetic Fields

Juhani Dabek, Fredrik Sannholm, Jaakko O. Nieminen, Panu T. Vesanen, and Risto J. Ilmoniemi

Department of Biomedical Engineering and Computational Science, Aalto University, Espoo, Finland

Abstract— **Ultra-low-field magnetic resonance imaging (ULF-MRI) and magnetoencephalography (MEG) can be combined to a unified imaging modality (MEG-MRI). To enhance the ULF-MR image quality in MEG-MRI, it is necessary to apply a prepolarization field of the order of 1–100 millitesla. When constructing the ULF-MRI apparatus, the effects of the prepolarization field and its gradients need to be considered as to what kinds of magnetizable objects are safe in their vicinity. The shape of the objects plays a role in the forces induced on them. Also, magnetized objects within the body may be subject to a torque that may press the tissue. These effects are investigated in this work.**

Keywords— **MEG-MRI, ultra-low-field MRI, prepolarization, safety.**

I. Introduction

In combined magnetoencephalography (MEG) and ultra-low-field magnetic resonance imaging (ULF-MRI), magnetic prepolarization of the sample is necessary [1]. In the following, for soft magnetic materials, it is verified in a field of the order of 100 mT both experimentally and theoretically that elongated objects experience a larger force than non-elongated ones. Also, the pressure effects of medical implants due to torque in the magnetic field is briefly inspected. First, necessary theoretical framework is presented.

II. Methods

The magnetic potential energy per unit volume of a magnetizable object in free space is

$$u = -\frac{1}{2}\mathbf{M}\cdot\mathbf{B}_0 \,, \tag{1}$$

where $\mathbf{B}_0$ is the strength of the external magnetic field density,

$$\mathbf{M} = \left(\frac{1}{\mu_0} - \frac{1}{\mu}\right)\mathbf{B} \tag{2}$$

is the magnetization of the object with permeability μ and magnetic induction $\mathbf{B}$ [2]. The permeability of free space is μ_0. The volume force is the gradient of this,

$$\mathbf{f} = -\frac{1}{2}\mathbf{M}\cdot(\nabla\mathbf{B}_0) \,, \tag{3}$$

where it has been assumed that $\mathbf{M}$ is constant. Assuming that $\mathbf{M} = M\hat{\mathbf{e}}_z$ and $\mathbf{B}_0 = B_0\hat{\mathbf{e}}_z$, where z belongs to the Cartesian coordinate system (x,y,z), this vector equation simplifies to a scalar form

$$f = -\frac{\partial u}{\partial z} = -\frac{1}{2}M\frac{\partial B_0}{\partial z} \,. \tag{4}$$

In the following subsections, the magnetic field is considered on the symmetry axis of a Helmholtz coil, and the magnetization is calculated for a prolate spheroidal object. Also, the pressure effects on tissue by torque are analyzed. For prolate spheroidal objects, it is possible to determine the magnetization analytically. An analytical solution for a cylinder, e.g., would be very difficult, if not impossible, to find. Hence, the elongated objects are assumed prolate spheroids, where necessary.

A. *Magnetic Field Due to a Helmholtz Coil*

A Helmholtz coil set-up, such as the 0.1-T MERIT MRI device at Aalto University used in this work, produces a homogeneous magnetic field in its center, by two identical short solenoids separated by the distance of their radii. The Helmholtz coil field on the symmetry axis is

$$B_\mathrm{H} = \frac{\mu_0 R^2 NI}{2}\left(\frac{1}{[z^2+R^2]^{3/2}} + \frac{1}{[(z+R)^2+R^2]^{3/2}}\right), \tag{5}$$

where R is the radius of the loop, I the current in the loop and N is the number of turns in both parts of the Helmholtz coil. The z derivative of the Helmholtz coil field is

$$\frac{\partial B_\mathrm{H}}{\partial z} = -3\frac{\mu_0 R^2 NI}{2}\left(\frac{z}{[z^2+R^2]^{5/2}} + \frac{z+R}{[(z+R)^2+R^2]^{5/2}}\right). \tag{6}$$

S. Supek and A. Sušac (Eds.): Advances in Biomagnetism – BIOMAG2010, IFMBE Proceedings 28, pp. 74–77, 2010.
www.springerlink.com

B. Magnetization of Prolate Spheroidal Objects

The prolate spheroidal coordinate system is presented in Fig. 1 [3]. The prolate spheroidal coordinates (η, ξ, φ) are related to the Cartesian coordinates as

$$x = \frac{d}{2}\sqrt{(1-\eta^2)(\xi^2-1)}\,\cos\varphi\,, \quad (7)$$

$$y = \frac{d}{2}\sqrt{(1-\eta^2)(\xi^2-1)}\,\sin\varphi \quad \text{and} \quad (8)$$

$$z = \frac{d}{2}\eta\xi\,, \quad (9)$$

where $-1 \le \eta \le 1$, $1 \le \xi < \infty$ and $0 \le \varphi < 2\pi$. The prolate spheroidal object is formed when ξ is constant. As $\xi \to 1$ the object becomes increasingly prolate. When $\xi \to \infty$, the object becomes increasingly spherical. The coordinate η can be associated with the deflection angle θ in spherical coordinates as $\eta = \cos\theta$. The azimuthal coordinate φ acts similarly as in the spherical coordinates.

The volume of a prolate spheroidal object is

$$V = \frac{\pi}{6}d^3\xi_0(\xi_0^2-1) = \frac{4}{3}\pi ab^2\,, \quad (10)$$

where a and b are half the spheroid's long and short axes lengths, respectively. For making the prolate spheroid similar to a cylinder of length l, it is assumed that the distance between the foci is $d = l$. A value of ξ_0 making their volumes the same can be solved from Eq. (10) as

$$\xi_0 = \begin{cases} \left(\frac{3V}{\pi l^3} - \sqrt{\frac{9V^2}{\pi^2 l^6} - \frac{1}{27}}\right)^{1/3} + \\ \left(\frac{3V}{\pi l^3} + \sqrt{\frac{9V^2}{\pi^2 l^6} - \frac{1}{27}}\right)^{1/3}, & \frac{3V}{\pi l^3} \ge \frac{1}{\sqrt{27}} \\ \frac{2}{\sqrt{3}}\cos\frac{\arctan\sqrt{\frac{\pi^2 l^6}{243V^2}-1}}{3}, & \frac{3V}{\pi l^3} < \frac{1}{\sqrt{27}}\,. \end{cases} \quad (11)$$

Starting from Maxwell's equations, the magnetization of a prolate spheroidal object can be calculated analytically in a homogeneous magnetic field, where there are no electrical effects. In this case, for a linear medium

$$\mathbf{B} = -\mu\,\nabla\Phi\,, \quad (12)$$

where the scalar potential Φ follows the Laplace equation $\nabla^2\Phi = 0$ that in the prolate spheroidal coordinate system is

$$\left[\frac{\partial}{\partial\eta}(1-\eta^2)\frac{\partial}{\partial\eta} + \frac{\partial}{\partial\xi}(\xi^2-1)\frac{\partial}{\partial\xi} + \frac{\xi^2-\eta^2}{(\xi^2-1)(1-\eta^2)}\frac{\partial^2}{\partial\varphi^2}\right]\Phi = 0\,. \quad (13)$$

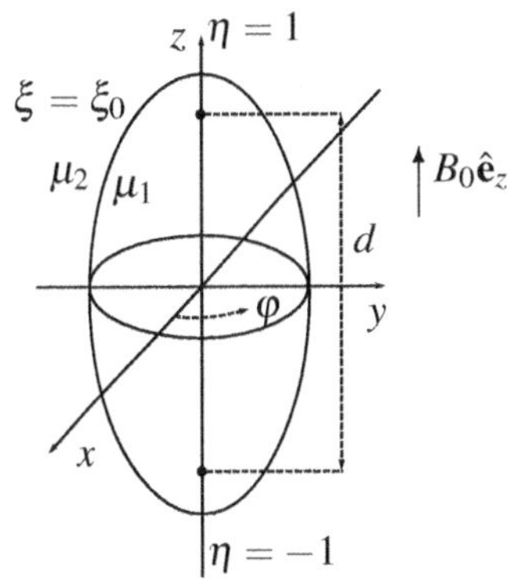

Fig. 1 The prolate spheroidal coordinate system with the Cartesian coordinate system. The inside of the prolate spheroid is volume 1 and the outside is volume 2 with permeabilities μ_1 and μ_2, respectively. The distance between the two foci is d. The external homogeneous field has strength B_0

Neglecting the φ dependence because of the azimuthal symmetry, the separation of variables leads to

$$\Phi(\eta, \xi) = \sum_{n=-\infty}^{\infty} S_n(\eta)R_n(\xi)\,, \quad (14)$$

where both S_n and R_n satisfy the same differential equation

$$\frac{d}{dz}\left[(1-z^2)\frac{d}{dz}f_n(z)\right] + n(n+1)f_n(z) = 0\,, \quad (15)$$

where $z = \eta$ and $f_n = S_n$, or, $z = \xi$ and $f_n = R_n$. This differential equation is satisfied by the associated Legendre functions of the first and second kind, $P_n(z)$ and $Q_n(z)$, respectively.

A solution to the original Laplace equation is now attempted by the separation method above. The inside of the prolate spheroidal object is denoted by $p = 1$ and the outside by $p = 2$. Thus, the scalar potentials become

$$\Phi_p(\eta, \xi) = \sum_{n=0}^{\infty}[\alpha_{n,p}P_n(\xi)P_n(\eta) + \beta_{n,p}Q_n(\xi)P_n(\eta)]\,, \quad p \in \{1,2\}\,, \quad (16)$$

where $Q_n(\eta)$ is excluded because it diverges. The boundary conditions are

$$\mathbf{B} \overset{\xi\to\infty}{\longrightarrow} B_0\hat{\mathbf{e}}_z \;\Rightarrow\; \Phi_2 \overset{\xi\to\infty}{\longrightarrow} -\frac{B_0}{\mu_2}z = -\frac{B_0}{\mu_2}\frac{d}{2}\eta\xi\,, \quad (17)$$

$$B_{1,\perp}(\eta, \xi_0) = B_{2,\perp}(\eta, \xi_0) \quad \text{and} \quad (18)$$

$$H_{1,\parallel}(\eta, \xi_0) = H_{2,\parallel}(\eta, \xi_0)\,, \quad (19)$$

where $B_{p,\perp}(\eta,\xi_0)$ means the normal component of the magnetic field on the prolate spheroid surface, and $H_{p,\parallel}(\eta,\xi_0)$ the tangential component of the auxiliary magnetic field. The auxiliary magnetic field and magnetic field are related as $\mathbf{B} = \mu\mathbf{H}$.

By recognizing that in volume $p = 1$, the potential must be finite, and from the boundary condition in volume $p = 2$ in Eq. (17), the simplifications

$$\begin{aligned} \Phi_1(\eta,\xi) &= \sum_{n=0}^{\infty} \alpha_{n,1} P_n(\xi) P_n(\eta) \quad \text{and} && (20) \\ \Phi_2(\eta,\xi) &= \alpha_{1,2} P_1(\xi) P_1(\eta) \\ &+ \sum_{n=0}^{\infty} \beta_{n,2} Q_n(\xi) P_n(\eta)\,, \quad \text{where} && (21) \\ \alpha_{1,2} &= -\frac{B_0}{\mu_2}\frac{d}{2}\,, && (22) \end{aligned}$$

can be made. Here, the constant terms have been set to 0. Developing Eqs. (18) and (19) to the forms

$$\begin{aligned} \mu_1 \left[\frac{\partial \Phi_1}{\partial \xi}\right]_{\xi=\xi_0} &= \mu_2 \left[\frac{\partial \Phi_2}{\partial \xi}\right]_{\xi=\xi_0} \quad \text{and} && (23) \\ \left[\frac{\partial \Phi_1}{\partial \eta}\right]_{\xi=\xi_0} &= \left[\frac{\partial \Phi_2}{\partial \eta}\right]_{\xi=\xi_0}\,, && (24) \end{aligned}$$

respectively, the remaining constants can be fixed to

$$\begin{aligned} \alpha_{1,1} &= \alpha_{1,2}\left[1 + \left(1 - \frac{\mu_1}{\mu_2}\right)(\xi_0^2 - 1)\right. \\ &\times \left.\left(1 - \frac{\xi_0}{2}\ln\frac{\xi_0+1}{\xi_0-1}\right)\right]^{-1} \quad \text{and} && (25) \\ \beta_{1,2} &= \alpha_{1,1}\left(1 - \frac{\mu_1}{\mu_2}\right)\xi_0(\xi_0^2 - 1)\,. && (26) \end{aligned}$$

Other constants vanish. Now, the potentials read

$$\begin{aligned} \Phi_1(\eta,\xi) &= \alpha_{1,1}\xi\eta \quad \text{and} && (27) \\ \Phi_2(\eta,\xi) &= \alpha_{1,2}\xi\eta + \beta_{1,2}\left(\frac{\xi}{2}\ln\frac{\xi+1}{\xi-1} - 1\right)\eta\,. && (28) \end{aligned}$$

Finally, assuming $\mu_1 \gg \mu_2 = \mu_0$ the magnetic field within the prolate spheroid is

$$B_1 = -\frac{2\mu_1}{d}\alpha_{1,1} \approx \frac{B_0}{(\xi_0^2 - 1)\left(\frac{\xi_0}{2}\ln\frac{\xi_0+1}{\xi_0-1} - 1\right)}\,, \qquad (29)$$

where the approximation is valid in the range $\xi_0 \in [1.001, 1.100]$, as for the spheroidally mimicked cylinders in the Results section. There, measurements with the cylinders can be compared to the analysis with prolate spheroids.

Using Eqs. (2) and (29), the spheroid's magnetization is

$$M_1 \approx \frac{B_0}{\mu_0(\xi_0^2 - 1)\left(\frac{\xi_0}{2}\ln\frac{\xi_0+1}{\xi_0-1} - 1\right)}\,. \qquad (30)$$

C. Pressure Effects of Torque on Tissue

The geometry of calculations of torque causing pressure is presented in Fig. 2. The shape of a box was chosen to simplify the calculations. As the box turns around one end, the displacement is

$$\Delta x = r\,\Delta\theta\,, \qquad (31)$$

which depends on the deflection angle $\Delta\theta$ and the distance r from the rotation axis. It is assumed that the pressure on the tissue follows Hooke's law, *i.e.*,

$$p(\Delta x) = p_0'\,\Delta x\,, \qquad (32)$$

where p_0' is a constant. For a small surface element $\Delta A = h\,\Delta r$, where h is the height of the box, the force due to the torque is

$$\begin{aligned} \Delta F(r\,\Delta\theta) &= p(r\,\Delta\theta)\,\Delta A && (33) \\ &= p_0'\,h\,r\,\Delta\theta\Delta r\,. && (34) \end{aligned}$$

A small increase in the torque can be assessed by using this force at a perpendicular distance r from the torque axis as

$$\Delta\tau = \Delta F(r\,\Delta\theta)\,r = p_0'\,h\,r^2\,\Delta\theta\,\Delta r\,. \qquad (35)$$

Making the distance change and the torque change infinitesimal, *i.e.*, $\Delta r \to dr$ and $\Delta\tau \to d\tau$, the total torque can be solved by integration as

$$\tau = \frac{p_0'\,h\,l^3\,\Delta\theta}{3}\,. \qquad (36)$$

Finally, from Eqs. (31), (32) and (36), the pressure profile of the object along the affected edge can be obtained as

$$p(r) = \frac{3\,\tau}{h\,l^3}r\,. \qquad (37)$$

With an angle α between a constant field and magnetization in volume V, the torque reads

$$\tau = V\,\mathbf{M}\times\mathbf{B} = V\,M\,B\,\sin\alpha \leq V\,M\,B\,. \qquad (38)$$

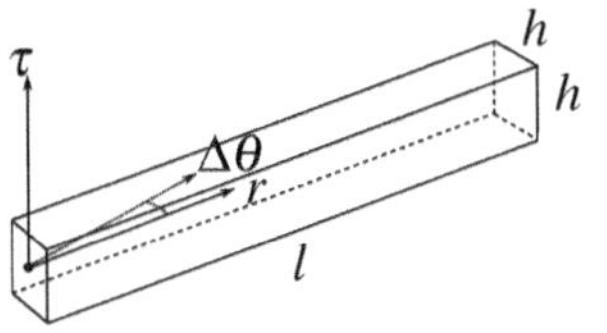

Fig. 2 The box experiencing the torque. See text for more details

III. Results

The key features of the 0.1 T MERIT MRI Helmholtz coil used in the experiments can be found in Table 1.

Table 1 Helmholtz coil properties. The coil's maximum field $B_{H,max}$ resides at the coil center. The theoretical maximum force magnitude of a magnetized object is reached at z_{max}

R (cm)	$\mu_0 R^2 NI/2$ (mT · m^3)	$B_{H,max}$ (mT)	z_{max} (cm)
44.4	5.78	94.5	12.8

Specifications of the four investigated reinforcing bars are presented in Table 2. Using the Helmholtz coil setup and approximating the bars by prolate spheroids and approximating $B_0 = B_H$ and $dB_0/dz = dB_H/dz$ as constants over the volume, the maximum force on the objects can be found at $z = z_{max}$ that is found in Table 1. The latter approximations are acceptable when $B_H \gg l \cdot dB_H/dz$. The force is calculated from Eqs. (4), (5), (6), (10) and (30) as

$$\max|F|_{calc.} \approx \left| \frac{\pi l^3 \xi_0 B_H \frac{dB_H}{dz}}{12\mu_0 \left(\frac{\xi_0}{2} \ln \frac{\xi_0+1}{\xi_0-1} - 1 \right)} \right|_{z=z_{max}} . \quad (39)$$

The results for the approximating prolate spheroids from this equation are presented in Table 2, together with the experimentally measured forces with the cylindrical bars in reality. From the former, it can be seen that when the volume of the spheroid becomes 7.5-fold, the maximum force grows 100-fold, and from the latter that the force grows 19-fold, respectively.

To evaluate the pressure effects of a magnetizable medical implant, the setup of Fig. 2 is crudely inspected with $2l =$ 50 mm and $h = 2$ mm. However, here the torque is applied in the middle of the object, whereby the pressure on tissue is halved compared to the situation with torque at one end. The corresponding spheroid has a $\xi_0(2l \rightarrow l) \approx 1.00152$ from Eq. (11) with $V = 2lh^2$. Placing the object in the center of the Helmholtz coil (see Table 1), and using Eqs. (30), (37) and (38), the maximum pressure of the implant on tissue is

$$p(r \leq l) \leq \frac{3h}{l} \cdot M_1(\xi_0, B_{H,max}) \cdot B_{H,max} \approx 215 \text{ kPa}. \quad (40)$$

Table 2 Reinforcing bar properties. Their diameter was approximately 1 cm. The length of the bar is represented by l and the mass by m. The maximum force magnitude on the bars, $\max|F|_{meas.}$, was interpolated from experimentally measurements on the symmetry axis of the Helmholtz coil described in Table 1. The spheroidal shape factors ξ_0 were computed with the latter definition of Eq. (11), and assuming a density of $\rho = 7860$ kg/m^3 for the material. The maximum force magnitude by calculations for an oblate spheroid approximation is given as $\max|F|_{calc}$

	Bar 1	Bar 2	Bar 3	Bar 4
l (cm)	3.2	8.1	15.1	22.4
m (g)	18	55	94	135
$\max\|F\|_{meas.}$ (N)	0.22	1.38	2.78	4.21
$(\xi_0 - 1) \cdot 10^{-3}$	61.0	12.3	3.3	1.5
$\max\|F\|_{calc.}$ (N)	0.10	0.81	3.72	10.26

IV. Conclusion

It was verified that an elongated object experiences a larger force than a non-elongated object. The theoretical maximum force in the Helmholtz coil field in the non-elongated case lags the measured one, whereas in the elongated case the situation is reversed. The shape differences of a prolate spheroid and a cylinder, and the constant-field-and-gradient assumption might explain this. The pressure calculations revealed that torque on a magnetizable implant may be dangerous.

As a conclusion, sharp elongated objects are the most dangerous ones in the vicinity of an MEG-MRI device if they are allowed to accelerate in the prepolarization field. This should be taken into account in the construction and operation of the device. Non-magnetizable medical implants are safer.

Acknowledgements

The research leading to these results has received funding from the European Community's Seventh Framework Programme (FP7/2007-2013) under grant agreement no. 200859.

References

1. Clarke J, Hatridge M, Mößle M. SQUID-Detected Magnetic Resonance Imaging in Microtesla Fields *Annu Rev Biomed Eng.* 2007;9:389–413.
2. Jackson JD. *Classical Electrodynamics.* USA: Wiley 1999.
3. Abramowitz M, Stegun IA. *Handbook of Mathematical Functions.* New York: Dover 1970.

Corresponding Author: Juhani Dabek
Institute: Aalto University
Street: Rakentajanaukio 2
City: 02150 Espoo
Country: Finland
Email: juhani.dabek@tkk.fi

Hybrid MEG-MRI: Geometry and Time Course of Magnetic Fields Inside a Magnetically Shielded Room

P.T. Vesanen, J.O. Nieminen, J. Dabek, and R.J. Ilmoniemi

Aalto University, Department of Biomedical Engineering and Computational Science, Espoo, Finland

Abstract— **In hybrid MEG-MRI, a magnetically shielded room (MSR) is often required to measure the weak magnetic fields (~0.1 pT) in MEG. On the other hand, to generate an MR image, relatively strong magnetic fields (<1 mT) need to be applied inside the MSR. These time-varying fields induce eddy currents and magnetize the walls of the room, possibly deteriorating MR image quality and hindering MEG data acquisition.**

In this study, static and dynamic magnetic fields are inspected inside a spherical MSR made of linear materials. An analytical expression is derived for the field generated by an arbitrary magnetic time dependent source inside the MSR, including the expressions for the amplitudes and time constants of the various eddy-current-associated fields. For the special case of a magnetic dipole in the centre of the room, it is shown that the response of the room is spatially homogeneous. The order of magnitude of the derived fields match those measured experimentally in a rectangular MSR.

Keywords— **MEG-MRI, magnetically shielded room, MSR.**

I. INTRODUCTION

Hybrid MEG-MRI is a novel technique that allows the direct combination of functional information acquired by MEG with the structural information produced by MRI [1,2]. For a successful MEG measurement, a magnetically shielded room (MSR) is vital. The walls of such a room are often composed of several shells, each including a μ-metal and an aluminum layer. The shielding properties of MSRs against external static and time-dependent magnetic sources are well-known [3,4]. However, the situation is dramatically changed in the case of an MEG-MRI device, which entails several static and time-dependent sources, *i.e.*, the polarization, measurement, and gradient coils, *inside* the MSR. The effects of such internal sources differ significantly from those caused by external sources.

In this paper, three distinct effects generated by magnetic sources inside an MSR are discussed. First, the fields will magnetize the μ-metal layers of the room, altering the desired spatial form of the MRI fields. Second, due to rapid switching of, *e.g.*, the polarization coil, eddy currents will be induced in both the μ-metal and aluminum layers, altering both the spatial form and time course of the fields. Depending on the different conductivities, permeabilities, and geometries of these layers, the time constants and amplitudes of these effects may differ widely.

Such MSR effects are expected to disturb the MEG-MRI measurement significantly, *e.g.*, by scrambling the phases of the spins or completely preventing data acquisition. The purpose of this paper is to study these effects by analytical calculations to find the parameters describing them. Such knowledge can be later used to understand the influence the effects have on MRI quality, or, if necessary, to find solutions to reduce the amplitude of the effects.

II. THEORY

Consider a spherical MSR of radius R as in Fig. 1. Suppose the walls of the room are thin and composed of an inner and an outer layer of linear and homogeneous materials with thicknesses $d_1 << R$ and $d_2 << R$, conductivities σ_1 and σ_2, and relative permeabilities μ_1 and μ_2, respectively. Suppose further that the inside and outside of the MSR have the properties of vacuum. Let us calculate the magnetic field inside the MSR in the case that the sources, *i.e.*, the MRI

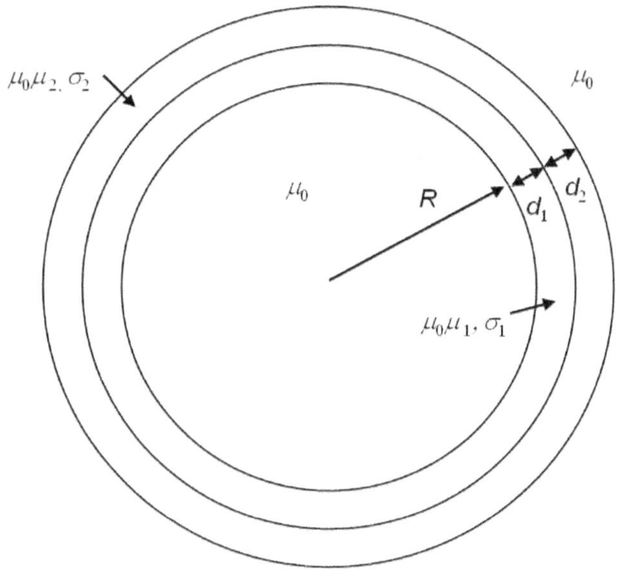

Fig. 1 Geometry of the spherical MSR model

S. Supek and A. Sušac (Eds.): Advances in Biomagnetism – BIOMAG2010, IFMBE Proceedings 28, pp. 78–81, 2010.
www.springerlink.com

coils, are modeled by a corresponding multipole in the center of the MSR. Let the source have a harmonic time dependence with a frequency low enough to justify the neglect of displacement currents. The neural currents creating the MEG signal are neglected as they are orders of magnitude lower than the currents flowing in the MRI coils. With these assumptions, the magnetic scalar potential V inside the room obeys the Laplace equation, $\nabla^2 V = \Delta V = 0$, whose solution in spherical coordinates is

$$V = \sum_{l,m} A_{l,m} r^l Y_l^m(\theta,\varphi) + \sum_{l,m} S_{l,m} r^{-l-1} Y_l^m(\theta,\varphi) = V_{\text{room}} + V_{\text{source}}, \tag{1}$$

where (r,θ,ϕ) are the spherical coordinates, $Y_l^m(\theta,\phi)$ are spherical harmonics, and $A_{l,m}$ and $S_{l,m}$ are constants to be identified later. The sum of terms diverging at the origin, V_{source}, is interpreted as the multipole expansion of the potential of the source in free space. The remaining terms in V_{room} are interpreted to represent the effects of the room.

In order to solve the coefficients $A_{l,m}$ characterizing the room response, the magnetic field is solved in the other regions as well. Outside the sphere, the magnetic scalar potential obeys the Laplace equation, and inside the layers the magnetic field **H** obeys the vector Helmholtz equation

$$\Delta \mathbf{H} = k_j^2 \mathbf{H}, \tag{2}$$

where $k_j^2 = i\sigma_j\mu_j\mu_0\omega$, σ_j and μ_j are the conductivity and relative permeability of the jth layer, ω is the angular frequency of the source, and i is the imaginary unit. By solving Eq. (2) and imposing the boundary conditions along all of the three boundaries, an explicit expression is obtained for coefficients $A_{l,m}(\omega)$. Because of the length of the expression, it is not written here.

So far, we have discussed a source with harmonic time dependence. The solution for an arbitrary time course of the source can be constructed from the harmonic solutions as a Fourier series

$$V_{\text{room}}(t) = \sum_{n=-\infty}^{\infty} c_n V_{\text{room}} e^{in\pi t/T} = \sum_{l,m} r^l Y_l^m(\theta,\varphi) f_{l,m}(t), \tag{3}$$

where c_n and $2T$ are the Fourier coefficients and the period of the source, respectively. The expression

$$f_{l,m}(t) = \sum_{n=-\infty}^{\infty} c_n A_{l,m}(n\pi/T) e^{in\pi t/T} \tag{4}$$

describes the time course of the room response for the multipole component (l,m). Here, we concentrate on the case of a square wave of period $2T$ with Fourier coefficients.

$$c_n = \frac{1}{2T}\int_0^T e^{-in\pi t/T} dt = \frac{i(e^{-in\pi}-1)}{2n\pi}. \tag{5}$$

The derived lengthy solution for $A_{l,m}(\omega)$ has the virtue of being accurate and general. However, due to its complexity, it is hard to interpret. In the next subsections, we concentrate on the case of the inner and outer layers made of μ-metal and aluminum, respectively. It turns out that in this case, the lengthy expression for $A_{l,m}(\omega)$ can be simplified under various justified approximations to reveal the magnetic field caused by the magnetization of the μ-metal layer and eddy currents both in the μ-metal and aluminum layers.

A. *Magnetization of the μ-Metal Layer and Eddy Currents in the Aluminum Layer*

In the following, approximations $\sigma_1 = 0$ and $\mu_2 d_2 << R$ are made. In words, the conductivity of μ-metal is neglected, and it is assumed that the magnetic width $\mu_2 d_2$ of the aluminum layer is small compared to R. It can be shown that the coefficients $A_{l,m}(\omega)$ can be simplified to

$$A_{l,m}(\omega) = A_{l,m}^s + A_{l,m}^\sigma \frac{i\omega}{1/\tau_l^\sigma + i\omega}, \tag{6}$$

where

$$A_{l,m}^s = -\frac{l(l+1)\mu_1 d_1 R^{-1-2l} S_{l,m}}{l(l+1)\mu_1 d_1 + (1+2l)R}, \tag{7}$$

$$A_{l,m}^\sigma = \frac{(l+1)(2l+1)R^{1-2l} S_{l,m}}{[(l+1)\mu_1 d_1 + R][(l+1)\mu_1 d_1 + (2l+1)R]}, \tag{8}$$

$$\tau_l^\sigma = \frac{(l+1)\mu_1 d_1 + R}{(l+1)\mu_1 d_1 + (2l+1)R}\mu_0 d_2 \sigma_2 R. \tag{9}$$

The meaning of these constants is clarified by substituting the simplified $A_{l,m}(\omega)$ from Eq. (6) in the Fourier series (4) to yield

$$f_{l,m}(t) = \begin{cases} -A_{l,m}^\sigma e^{-(t+T)/\tau_l^\sigma}, & -T \le t < 0 \\ A_{l,m}^s + A_{l,m}^\sigma e^{-t/\tau_l^\sigma}, & 0 \le t \le T \end{cases}. \tag{10}$$

Let us briefly explain the physical meaning of Eq. (10). First, a temporally constant term $A^s_{l,m}$ appears, describing the magnetization of the μ-metal layer during the time the source is switched on. Second, we see that the edges of the square wave pulsing of the source induce an exponentially decaying magnetic field due to eddy currents in the aluminum layer. The amplitude and time constant of the decay are $A^\sigma_{l,m}$ and

τ^{σ}_{l}, respectively. Inspecting Eq. (8) reveals that the amplitude of the eddy-current-associated magnetic field is essentially reduced as the square of the magnetic width of the µ-metal layer, $\mu_1 d_1$, reflecting the shielding effect of the µ-metal both when the field passes from the interior of the room through the µ-metal layer and when the field is reflected back from the aluminium layer.

B. *Eddy Currents in the μ-Metal Layer*

To derive an expression for the magnetic field generated by eddy currents in the µ-metal layer, different approximations for the general $A_{l,m}(\omega)$ have to be made. In this case, we approximate $\sigma_2 = 0$ and $\exp(d_1 k_1) >> 1$. The former approximation basically neglects the presence of the aluminum layer completely. With typical µ-metal parameters, the latter approximation is easily satisfied for frequencies above 10 Hz. Note that because of this approximation, this treatment does not give the right amplitude for the static magnetization of the µ-metal. However, here, we are interested only in the decaying effect, which justifies the approximation. With these approximations, $A_{l,m}(\omega)$ is simplified to

$$A_{l,m}(\omega) = A^{\mu}_{l,m}\left[1 - g_l(\omega)\right], \qquad (11)$$

where

$$g_l(\omega) = \frac{2l+1}{l+1+\sqrt{i\tau^{\mu}\omega}}, \qquad (12)$$

$$A^{\mu}_{l,m} = \frac{(l+1)S_{l,m}}{lR^{2l+1}}, \qquad (13)$$

$$\tau^{\mu} = \frac{\sigma_1 \mu_0 R^2}{\mu_1}. \qquad (14)$$

Unfortunately, substituting the simplified expression (11) in the Fourier series (4) does not give an explicit expression for the time dependence $f_{l,m}(t)$. Nevertheless, from Eqs. (11) and (12) we see that $A^{\mu}_{l,m}$ and τ^{μ} are the constants that describe the amplitude and the time scale of the magnetic field created by eddy currents in the µ-metal layer.

III. Results

As a simple application of the derived results, consider the case of a polarization coil in the centre of a magnetically shielded room of radius $R = 2$ m made of materials defined in Table 1. Suppose the polarization coil is approximated only by its dipole moment $m = 1000$ Am^2 pointing in the z-direction. In this case, the only nonzero term in the multipole expansion (1) is

Table 1 Typical material parameters of an MSR

Material	σ ($10^6/\Omega$m)	μ	d (mm)
Aluminum	36	1	10
µ-metal	2	30000	1

$$S_{1,0} = m/\sqrt{12\pi}. \qquad (15)$$

The amplitude of the magnetic field created by the room in response to a harmonically oscillating polarization coil is obtained from Eq. (1) yielding

$$\mathbf{B}_{\text{room}} = -\mu_0 \nabla V_{\text{room}} = -\mu_0 A_{1,0}(\omega)\mathbf{e}_z \qquad (16)$$

where $\mathbf{e}_z$ is the unit vector in the z-direction. For a square wave source waveform, the time course of the field is obtained by Fourier series in Eq. (4) with the help of approximations (6) or (11).

The magnetic field generated by the MSR in response to the square-wave oscillating dipole ($T = 3$ s) is plotted in Fig. 2. The amplitudes of the magnetic field due to the magnetization of MSR, eddy currents in the aluminum, and eddy currents in the µ-metal are, by Eqs. (7), (8), and (13), 12 µT, 80 nT, and 28 µT, respectively. The time constants in the aluminum and µ-metal layers are, by Eqs. (9) and (14), 850 ms and 0.3 ms, respectively.

IV. Conclusion

First, note that according to Eq. (16), the magnetic field of the room response, in the case of a dipolar source in the centre of the room, is spatially homogeneous everywhere inside the room. This simplifies the analysis of the various effects.

The magnetic field due to the magnetization of the MSR, 12 µT, in Fig. 2b, is buried under the field that a 1000 Am^2 coil itself typically produces. The amplitude of the magnetic field caused by eddy currents in aluminum, 80 nT, in Fig. 2a, corresponds to less than a pixel shift in a 128 x 128 pixel image with a typical measurement bandwidth of 50 µT, during a time period determined by the time constant 850 ms. Thus, no significant effect on MR image quality is expected. However, the effect on MEG data acquisition remains unknown, and the amplitude of this effect increases rapidly if materials with smaller μ_1 are used. The amplitude of the magnetic field caused by eddy currents in µ-metal, 28 µT, in Fig. 2c, is significant compared to the typical

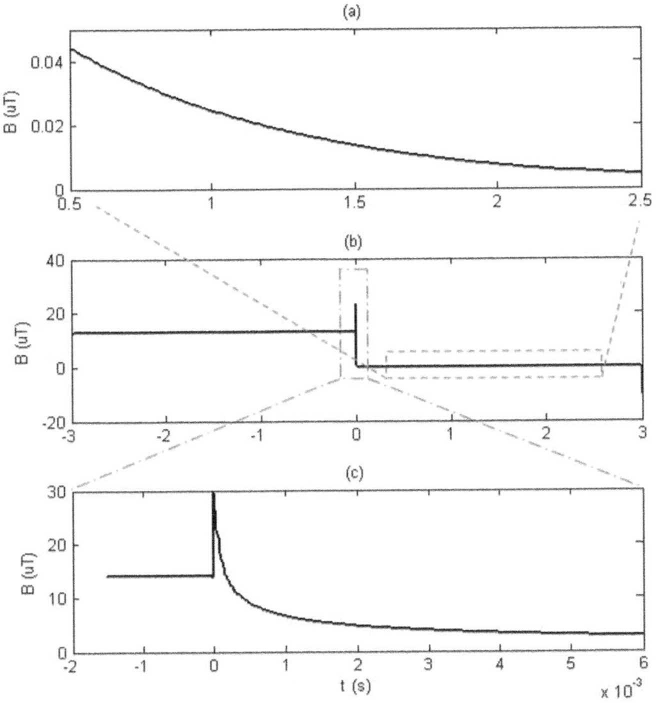

Fig. 2 In (b), the homogeneous magnetic field generated by the MSR in response to the oscillating dipole showing the magnetization of the μ-metal. In (a) and (c), plot (b) zoomed in to show the decaying magnetic fields created by eddy currents in the aluminum and μ -metal, respectively

strength 50 μT of an MEG-MRI measurement field. However, the effect vanishes rapidly, within a few milliseconds, such that it can be avoided simply by delaying the data acquisition by the respective time period.

V. Discussion

There are obviously several shortcomings in the model that allows the presented calculations. First, a typical MSR resembles more a cube than a sphere. Second, the permeability of the μ-metal is often highly nonlinear, inhomogeneous, and frequency-dependent. Third, there is typically more than one shell in an MSR. Therefore, the values presented here serve only as crude estimates. However, the model manages to capture the three important effects that sources internal to MSR produce: the magnetization and the shielding effect of the μ-metal, and eddy currents both in the μ-metal and aluminum.

To validate the analytical results, a simple experimental measurement mimicking the example in Fig. 2 was conducted. A polarization coil with m = 480 Am^2 was placed inside a 3x4x3 m^3 rectangular MSR manufactured by EuroShield Ltd. The room consisted of two shells both including μ-metal and aluminum with parameters around those in Table 1. The magnetic field was measured in the centre of the room with a Fluxgate sensor. In result, the magnetic field created by eddy currents in aluminum was measured to be decaying exponentially with time constant between 800 ms and 1100 ms depending on the field component. The respective time scale in the effect by μ-metal was measured to be between 1.5 and 3 ms. The measured and calculated amplitudes differ by a factor of 2 or 3, which is, however, understandable, taking into account the different geometries of the MSRs.

Note that while the dipole example in this article was very simple, Eq. (3) allows the determination of the room response for an arbitrary time-dependent source, *e.g.*, a gradient coil or a noncentrally positioned Helmholtz coil.

Acknowledgment

The research leading to these results has received funding from the European Community's Seventh Framework Programme (FP7/2007-2013) under grant agreement no HEALTH-F5-2008-200859.

References

1. Volegov P., Matlachov A.N., Espy M.A., et al. (2004) Simultaneous magnetoencephalography and SQUID detected nuclear MR in microtesla magnetic fields. Magn Reson Med 52:467-470
2. Mössle M., Myers W.R., Lee S.K., et al. (2005) SQUID-Detected in Vivo MRI at Microtesla Magnetic Fields. IEEE Trans Appl Supercond 15:757-760
3. Mager A.J. (1970) Magnetic Shields. IEEE Trans Magn MAG-6:67-75
4. Kelhä V., Pukki J., Ilmoniemi R., et al. (1982) Design, construction and performance of a large-volume magnetic shield. IEEE Trans Magn MAG-18:260-270

Corresponding author:

Author: Panu Vesanen
Institute: Aalto University, Department of Biomedical Engineering and Computational Science
Street: P.O. Box 2200, 02015 TKK
City: Espoo
Country: Finland
Email: panu.vesanen@tkk.fi

Magnetic Resonance Relaxometry at Low and Ultra Low Fields

P. Volegov, M. Flynn, R. Kraus, P. Magnelind, A. Matlashov, P. Nath, T. Owens, H. Sandin, I. Savukov, L. Schultz, A. Urbaitis, V. Zotev, and M. Espy

Applied Modern Physics, Los Alamos National Laboratory, Los Alamos, NM USA

Abstract— **Nuclear magnetic resonance (NMR) and magnetic resonance imaging (MRI) are ubiquitous tools in science and medicine. NMR provides powerful probes of local and macromolecular chemical structure and dynamics. Recently it has become possible and practical to perform MR at very low fields (from 1 μT to 1 mT), the so-called ultra-low field (ULF) regime. Pulsed pre-polarizing fields greatly enhance the signal strength and allow flexibility in signal acquisition sequences. Improvements in SQUID sensor technology allow ultra-sensitive detection in a pulsed field environment.**

In this regime the proton Larmor frequencies (1 Hz – 100 kHz) of ULF MR overlap (on a time scale of 10 ms to 100 ms) with "slow" molecular dynamic processes such as diffusion, intra-molecular motion, chemical reactions, and biological processes such as protein folding, catalysis and ligand binding. The frequency dependence of relaxation at ultra-low fields may provide a probe for biomolecular dynamics on the millisecond timescale (protein folding and aggregation, conformational motions of enzymes, binding and structural fluctuations of coupled domains in allosteric mechanisms) relevant to host-pathogen interactions, biofuels, and biomediation. Also this resonance-enhanced coupling at ULF can greatly enhance contrast in medical applications of ULF-MRI resulting in better diagnostic techniques.

We have developed a number of instruments and techniques to study relaxation vs. frequency at the ULF regime. Details of the techniques and results are presented.

Ultra-low field methods are already being applied at LANL in brain imaging, and detection of liquid explosives at airports. However, the potential power of ultra-low field MR remains to be fully exploited.

Keywords— **ultra-low field, nuclear magnetic resonance, relaxation.**

I. Introduction

Nuclear magnetic resonance (NMR) and magnetic resonance imaging (MRI) are ubiquitous tools in science and medicine. NMR provides powerful probes of local and macromolecular chemical structure and dynamics. In spite of that the main trend in NMR/MRI industry favors instruments with higher and higher magnetic fields - mainly due to higher signal-to-noise ratio at higher fields - in recent years several research groups all over the world have started or continued work in the area of NMR at low and ultra-low fields. Due to utilization of magnetically shielded enclosures, pre-polarizing magnetic field and improvements in SQUID sensor technology allowing ultra-sensitive detection in a pulsed field environment, it has become possible and practical to perform MR at very low fields (from 1 μT to 1 mT), the so-called ultra-low field (ULF) regime.

The ULF regime has a number of unique features which can be advantageously exploited in a number of applications. First, due to low measurement field strength, it is much easier to create a field with required absolute uniformity. For example, if we need to ensure that frequency spread over the sample is less than 1 Hz (protons) for a 1 T NMR system, we need to shim the coils to 0.02 ppm accuracy, while for the 100 μT system the shimming requirement is much less stringent ~ 200 ppm. This allows building rather simple and inexpensive magnetic systems with open access and convenience of measurement of large samples. Second, due to low Larmor frequency (~4.25 kHz at 100 μT for protons), it is possible to make NMR/MRI measurements of the samples with metal inclusions (implants, bullets, etc) and even completely enclosed by a metal container, such as a soda can. Third, and this is probably the most important and promising feature of the ULF regime, the Larmor frequency in this mode overlap (on a time scale of 10 μs to 100 ms) with the rate of "slow" molecular dynamic processes such as diffusion, intra-molecular motion, chemical reactions, and biological processes such as protein folding, catalysis and ligand binding. Although conventional NMR techniques can probe some of these processes, resonance-enhanced coupling at ULF can greatly enhance sensitivity of the relaxation measurements, thus yielding new information about these processes, and lead to the enhanced contrast.

Our team at Los Alamos has developed several applications of ULF MR including imaging of the human brain at 46 μT in combination with magnetoencephalography [1], classification of materials via relaxation properties measured at ULF relevant to numerous non-invasive inspection applications, and determination of uranium enrichment fraction via relaxation and/or J-coupling to address nuclear safeguards [2],[3]. Below we discuss the motivation, methods, and recent results in imaging and relaxometry of materials using the ULF MR approach.

S. Supek and A. Sušac (Eds.): Advances in Biomagnetism – BIOMAG2010, IFMBE Proceedings 28, pp. 82–87, 2010.
www.springerlink.com

II. METHODS AND INSTRUMENTS

The most widely used technique to measure NMR relaxation at low fields consists of polarizing the sample in a large field B_p, then, switching off B_p in a time short compared to T_1 and observing the free induction decay (FID) of the magnetization of the sample at a low field B_m. If the polarization field is switched off adiabatically ($dB/dt << \gamma B^2$, where γ is magnetogyric ratio) the bulk magnetization becomes aligned along the measurement field B_m and it is then necessary to apply a $\pi/2$ pulse to observe the precession of the magnetization. Alternatively, if B_p is switched off non-adiabatically ($dB_p/dt \geq \gamma B_m^2$) then the magnetization remains aligned along B_p and, if B_m is not parallel to B_p, the precession of the bulk magnetization is directly observable [4].

In our experiments the typical range of the polarization field is from 10 to 50 mT, while the range of the measurement field is from 5 to 200 μT. This protocol intrinsically allows measuring two very different relaxation parameters: 1) longitudinal relaxation (T_1) at relatively high field (Larmor frequency in MHz range for protons) by varying the polarization time, and 2) transverse relaxation (T_2) at low field (kHz range) by measuring decay of the signal with time. Such combination of the relaxation parameters proved to be very useful for characterizing materials. This can be qualitatively illustrated using the simplest model for NMR relaxation - an ensemble of spins which do not couple to one another but which couple to an external fluctuation field (Redfield theory, see for example [5]). According to this model the dependence of the relaxation times versus Larmor frequency ω_L can be expressed in terms of two microscopic quantities - an ensemble average of the local magnetic field $\overline{B_{loc}^2}$ and correlation time τ_0 :

$$\frac{1}{T_1} = \frac{2}{3}\gamma^2 \overline{B_{loc}^2} \frac{\tau_0}{1+(\omega_L \tau_0)^2}$$

$$\frac{1}{T_2} = \frac{1}{3}\gamma^2 \overline{B_{loc}^2} \left[\tau_0 + \frac{\tau_0}{1+(\omega_L \tau_0)^2} \right]$$

As the Larmor frequency during the polarization and the measurement intervals differs by 3 order of magnitude, T_1 and T_2 measurements in this case reflect different properties of the relaxation process or even different relaxation mechanisms.

In contrast to conventional NMR/MRI, relative homogeneity of the measurement field is not crucial in ULF MRI, because microtesla-range magnetic fields of even modest relative homogeneity are highly homogeneous on the absolute scale. In this case the signal attenuation by effects of the dephasing and diffusion due to field non-uniformity over the sample volume is negligible and this makes possible to measure transverse relaxation directly from the free inductance decay signal even for relatively large samples (~100 cc), avoiding use of spin-echo sequences, such as CPMG. First, this approach significantly simplifies the experimental apparatus - no spin flipping pulses, coils, synchronization, etc. Second, and this is probably more important, this approach also provides more data points (more dense sampling) of the transverse relaxation decay. It should be noted that the transverse relaxation data is typically a sum of exponentials associated to the different components of the sample (such as for example as water compartmentalization in biological tissue). Resolving the such data into individual components is very important for characterizing the sample. However, this problem is ill-conditioned and the results strongly depend on the quality of the initial relaxation curve measurement making the solutions difficult to interpret and validate [6]. In this context dense sampling of a relaxation curve improves the ability to resolve such data and extract the underlying characteristic relaxation times. At last, it should be noted that direct measurement of the relaxation curve preserves the information about local non-uniformity of the magnetic fields in the sample, such for example as the one associated with paramagnetic impurities or contrast agents, while spin-echo sequences mostly cancel these effects.

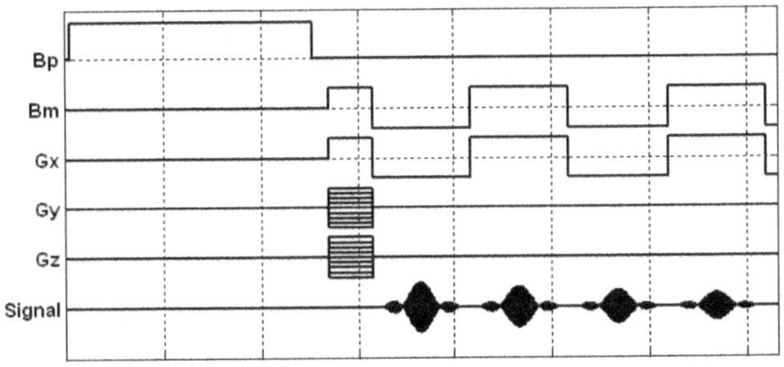

Fig 1 3D Fourier multiple gradient echo imaging sequence

The approach to measure T_1 at high field and measure T_2 at low field, can be combined with a MRI sequence to obtain spatial maps of the relaxation parameters. An example of such a sequence - 3D Fourier multiple gradient echo imaging - is shown in the figure 1. The sequence consists of the pre-polarization, phase encoding, and several echo intervals created by inverting both the measurement field and the readout gradient. The images are obtained for the each echo, and for several different polarization times. The spin density map $\rho(\mathbf{r})$ and the relaxation maps $T_1(\mathbf{r})$, $T_2^*(\mathbf{r})$ are obtained by fitting the parameters of the following model (see for example [7]): $I_{ij}(\mathbf{r}) \propto \rho(\mathbf{r}) \cdot \left(1 - e^{-T_p^{(j)}/T_1(\mathbf{r})}\right) \cdot e^{-T_e^{(i)}/T_2^*(\mathbf{r})}$, where $\left\{T_p^{(j)}\right\}_1^m$ are the polarization times, $\left\{T_e^{(i)}\right\}_1^n$ are the echo times, and $I_{ij}(\mathbf{r})$ is the image obtained for *i*-th echo and *j*-th polarization time. To compensate for the effects of any static residual fields, a uniform water phantom is used to calibrate the system. At last, one should mention here that the ratio of the MRI signal bandwidth, defined by the gradient strength and the size of the sample, to the Larmor frequency is typically larger in the ULF MRI regime comparing to its high field counterpart. This dictates that effects of concomitant gradients should be taken into account [8].

Rapidly switching off B_p inevitably results in the transient field which arises from the currents proportional to *dB/dt* induced in any conductive materials (e.g. other components of the experimental apparatus and surrounding equipment). While the transient fields are significantly smaller than B_m (~1 - 10μT), hence they do not significantly alter the evolution of the spins, they are much larger than the typical signal from bulk magnetization of the sample (~1..10 pT) thus impeding or even preventing measurement of the NMR signal and the associated NMR relaxation parameters. One approach to correct for the transient signals is to approximate the transients with a sum of exponents (typically 2-3) and subtract them in post processing. We have previously used this approach with good success. However this approach is rather limited to the extent it could cancel the transients, especially for shorter (~ 10 ms) periods. We designed a new technique to cancel transient-induced signals by acquiring interleaved NMR measurements at two different measurement fields (i.e. two different ω_L) and differencing these two signals. This method could cancel the transient signal by 3 orders of magnitude [9].

Over the last few years we developed an number of ULF NMR/MRI instruments ranging from a one channel NMR system designed to measure FID to multi-channel (7-channel) MRI system designed to image relaxation properties of multiple objects within one sample.

The detector of choice is a SQUID (a Superconducting QUantum Interference Device) [10]. This is due to the following two factors. First, the amplitude of the NMR signal at ULF is very low (for example a 1 cc sample of water polarized by 20 mT field generates ~2.3 pT field at the distance 3 cm from the sample), and, second, the Larmor frequency is also low, in the kHz range. The combination of these two factors makes it very difficult to detect the signal using a Faraday coil, but given the low intrinsic noise of the SQUID coupled with a untuned superconducting flux transformer, it is possible to achieve an acceptable signal-to-noise ratio [11]. The main challenge to use a SQUID detector in ULF NMR/MRI applications is to ensure that SQUID is in a working state after fast ramping down the polarization field. To protect the SQUID from the switching transients we use a fast cryogenic switch [12], which decouples the SQUID from the pick-up loops during the ramp down time, and then reconnects it within, typically, 5 μs. We also found it is very important to carefully isolate the SQUID chip itself from the ambient magnetic fields by a superconducting shield.

Another important and complex element of the systems is the pre-polarization magnet: a coil and power supply. Given a finite power supply, the design goal to generate as large polarization field as possible in the sample volume dictates using coils with relatively large number of turns (~1000) placed in the immediate vicinity of the sample. This presents several problems. First, the inductance of the polarization coils - typically for our instruments it is in the range from 25 mH to 500 mH - determines how fast the polarization field could be ramped down: $\Delta t = LI/U$, where L is the coil inductance, I is the polarization current, U is the voltage across the switching device. To address this problem we designed a current switching device based on optoMOS solid state relays [13]. The device is capable of switching on/off currents up to 80 A. The voltage across the switch is up to ~2.4 kV, which allows us to ramp down the polarization field in the time from 0.5 to 20 ms, depending on the inductance of the coils and the current. Second, the polarization coil effectively couples noise from the power supply to the SQUID, so we employ mechanical relays in addition to the solid state relays to completely decouple the polarization coils from the external circuits during the measurement. Third, the large amount of the conducting material near the SQUID pick-up loops is bound to generate strong noise due to Johnson currents. To overcome this effect we are using multi-stranded litz wire. Last, but not least, it should be noted that the coils generate significant amount of heat during the polarization period which should be dissipated without overheating the coils and the sample. Typically for our instruments it ranges from 0.5 to 4 kW. To remove the generated heat from the coils and sample volume we are using different cooling agents: air (convection and forced), liquid nitrogen (evaporation) [13], a fluorine-based industrial coolant (flow) [14].

III. Applications and results

Using the methods and instruments described in the previous section we have developed several applications of ULF MR designed for classification of materials via relaxation properties measured at ULF relevant to numerous non-invasive inspection applications.

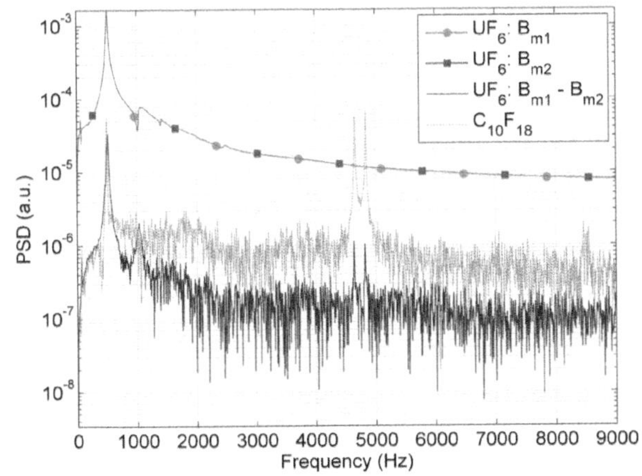

Fig. 2 Power spectral density of the FIDs from UF_6 and the reference liquid fluorine sample perfluorodecalin ($C_{10}F_{18}$). The effect of the two-frequency approach is shown as the PSDs from the odd and even epochs have a 100-fold higher noise level than the PSD of the differentiated FIDs

Uranium hexafluoride (UF_6) is used in the enrichment process of uranium. It is a solid at room temperature and atmospheric pressure and there exists a triple point at 64 ºC. At temperatures above the triple point the UF_6 consists of a mixture of liquid and vapor. A non-destructive ULF-NMR enrichment verification system could serve an important role in non-proliferation surveillance. We have developed an instrument to measure relaxation parameters of UF_6 in liquid phase and (possibly in gas phase) at ULF regime. As the transverse relaxation time of UF_6 is expected to be short (on the order of 10 ms for liquid and 1 ms for vapor) we employed the two frequency approach, described in the previous section, to minimize effects of the transients. FIDs have been recorded from a heated UF_6 sample at a mean temperature of 80 ºC. The results are shown in the figure 2. The fluorine peaks have been detected and match the frequencies of a reference liquid fluorine sample (perfluorodecalin; $C_{10}F_{18}$). The two-frequency approach was very effective in removing the transients and reducing the noise about 100 times. The FIDs have been analyzed using bootstrapping to find the corresponding spin-spin relaxation time T_2, a lognormal fit has been used to retrieve a mean of 24 ms and a variance of 13 ms.

To address the issue of measuring relaxation properties of several objects in an arbitrary configuration, we developed an ULF imaging relaxometer – MagViz [14]. The relaxometer is being applied to a homeland security application inspecting liquids in a check-point, and determining the presence of hazardous material. The instrument has also been used to distinguish liquids, such as wine, for quality. Fig. 3 presents results from the relaxometer. The pseudo-color image reflects relaxation properties of the objects: the intensity (the value component in hue-saturation-value scheme) is proportional to the spin-density, while the hue and saturation are functions of T_1 and T_2. The tests were performed with benign items from streams of commerce, and a threat material, hydrogen peroxide. The 2D analysis integrated over the depth of the bowl which was ~ 3 inches. The 3D analysis is shown with three slices. Circles indicate correctly identified objects found to contain threat material. In the upper images a foil-lined juice box was used, and the material inside still correctly identified. The unit was also used to analyze the difference between several different types of wine and vinegar.

We applied the ULF MR relaxation imaging approach to medical applications including imaging relaxation properties of the human brain at 46 μT. We obtained following T1 values at Br=46 μT relaxation field: 103±5 ms for gray matter, 75±2 ms for white matter, 344±9 ms for CSF, and 124±7 ms for scalp, and the following T2 values at the same field: 106±11 ms for gray matter, 79±11 ms for white matter, 355±15 ms for CSF, and 120±7 ms for scalp. The measurements confirm that at microtesla fields T_1-weighted contrast is similar to T_2 contrast for human brain tissues. This leads to the conclusion that the main advantage of ULF MRI is not the imaging contrast established at microtesla fields, but the unique opportunity to explore and utilize in vivo T_1-weighted contrast as a function of the relaxation field in a wide range of magnetic fields (from zero to 0.1 T and higher) [15].

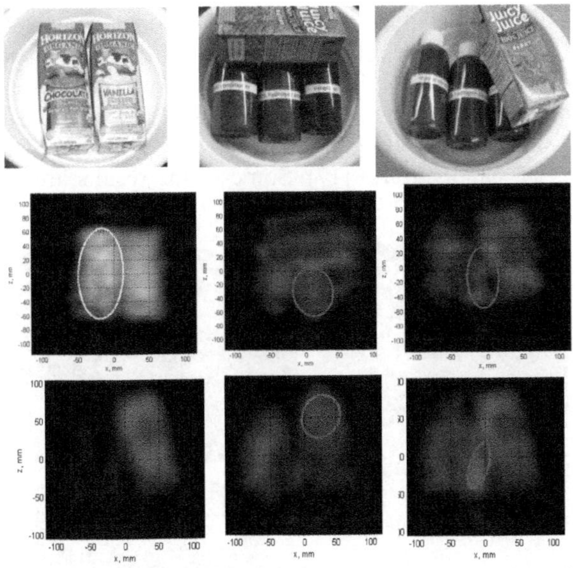

Fig. 3 Top: photographs of items. Middle: 2D images with threat detection. Bottom: 3D slices through the items at top right

One of the most interesting and important problems in neuroscience is to measure brain function. Two major techniques are widely used: a) magneto-encephalography (MEG), and b) functional MRI (fMRI). Each of these two techniques has its own advantages and drawbacks. As we have shown previously MEG can be combined with ULF

MRI. We would like to evaluate plausibility of using fMRI in the ULF regime. fMRI measures changes in the relaxation properties of spin population in blood produced by changes in blood oxygenation during neural activity. The main mechanism which changes the relaxation rate is a change in susceptibility. This effect is dependent on the strength of the measurement field and so would be unlikely to be present at ULF. However, it is possible that there are other relaxation mechanisms present at ULF that are dependent on blood oxygenation. To answer that question, we measured the changes in T_1 and T_2 in blood before and after deoxygenation at ULF. Deoxygenation was performed by evacuating the blood to 540 mm Hg. Measurements of T_1 at 18.8 mT and T_2 at 46 μT found T_1 of 351 ± 29 ms and T_2 of 194 ± 9 ms for the oxygenated blood and deoxygenated blood had a T_1 of 335 ± 82 ms and T_2 of 194 ± 9 ms. Overall, we found no statistically significant difference in T_1 or T_2 values when the level of oxygenation was changed. However, it may still be possible to perform fMRI in the ULF regime using other approaches based on changes in blood volume or flow.

The intrinsic MRI tissue contrast at ultra-low fields (ULF) has been shown to be higher than at high-fields [1]. However, further improvement of the contrast would be beneficial. To evaluate the performance of contrast agents at ULF we measured relaxivity of several contrast agents at different fields. We have chosen superparamagnetic iron oxide particles (SPIOs) for this study. Feridex® [16] is an FDA approved SPIO contrast agent. We have also studied different nanoparticles [17]. Some of the results are shown

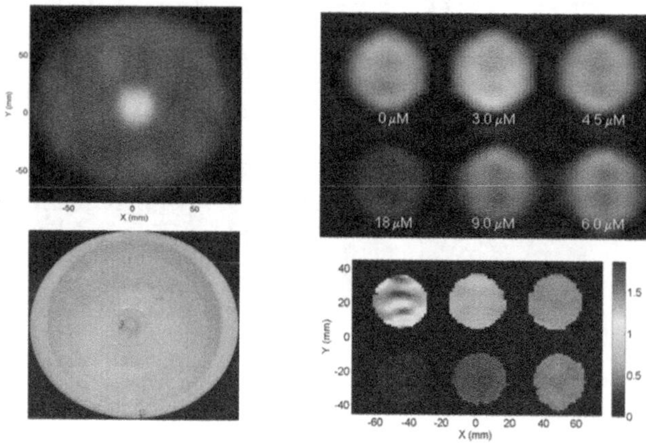

Fig. 4 Top left MRI of a bottle of *Feridex* contrast solution (1.6 μM) placed in a bowl of water, as shown in the lower left panel. Top right panel MRI retrieved in MagViz showing contrast induced by various concentrations of 20 nm nanoparticles dissolved in water. Bottom right panel shows R_2 map of the samples with the color indicating the R_2 values

in the figure 4. The R_1 relaxivities were measured at the prepolarization field (50 mT). With a different scheme, R_1 measurement could be performed at the measurement field. However, we observed no change in R_2 when reducing the polarization field from 50 mT to 25 mT. We found no consistent evidence of dispersion over the frequency range 0.8-7.3 kHz although there might be some shift at the lower part of the range. The main conclusion are a) iron oxide nanoparticles work as contrast agents at ultra-low fields, b) the T_1 contrast enhancement for Feridex is higher at low field (50 mT) than at typical MRI fields (≥1.5 T.

IV. Conclusions

Using ULR MR relaxometry we have made progress in numerous applications ranging from brain imaging to homeland security to industry. These applications illustrate the promise of the ULF relaxomentry approach to magnetic resonance as a scientific and industrial tool. However, the real potential of the method remains to be determined. While some relaxation mechanisms are already understood and experimentally verified at high field, ultra-low-field relaxation behavior has not been rigorously studied experimentally or theoretically. ULF MR provides a tool to study relaxation vs. frequency at ultra-low frequencies, to characterize relaxation behavior and hence the underlying molecular dynamics. One of the most promising applications of ULF MR is the study of the "slow" dynamic processes. Enhanced relaxation contrast in biological tissue, a unique benefit of imaging at ULF, is presumably the result of these complicated motions. Yet the theory of MR relaxation, which provides a window into these dynamics, has not been developed and validated below 10 kHz. The frequency dependence of relaxation at ULF may provide an exciting probe for molecular dynamics on the millisecond timescale (i.e. protein folding and aggregation, conformational motions of enzymes, binding and structural fluctuations of coupled domains in allosteric mechanisms) relevant to a host of applications yet to be discovered.

Acknowledgment

This work was done with support from the U.S. Department of Homeland Security, Department of Energy Office of Biological and Environmental Research, the National Institutes of Health R01 EB006456, and support from Los Alamos National Laboratory LDRD.

References

[1] V. S. Zotev, A. N. Matlashov, P. L. Volegov, I. M. Savukov, M. A. Espy, J. C. Mosher, J. J. Gomez, R. H. Kraus Jr., Microtesla MRI of the human brain combined with MEG, Journal of Magnetic Resonance 175 (2005) 103–113

[2] P. L. Volegov, A. N. Matlashov, and R. H. Kraus, Jr., "Ultra-low field NMR measurements of liquids and gases with short relaxation times," Journal of Magnetic Resonance, vol. 183, pp. 134-141, 2006.

[3] P. Magnelind, A. Matlashov, P. Volegov, and M. Espy, Ultra-Low Field NMR of UF6 for 235U Detection and Characterization, IEEE Transactions on Applied Superconductivity, vol. 19, no. 3, pp 816-818, June 2009

[4] A. Abragam, Principles of Nuclear Magnetism, Oxford University Press, 1961

[5] C.P. Slichter, Principles of Magnetic Resonance, Springer, 1990, pp 206-212

[6] L. Andrade, E. Micklander, I. Farhat, R. Bro, S. B. Engelsen, DOUBLESLICING: A non-iterative single profile multi-exponential curve resolution procedure Application to time-domain NMR transverse relaxation data, Journal of Magnetic Resonance 189 (2007) 286–292

[7] P.T. Callaghan, Principles of Nuclear Magnetic Resonance Microscopy, Oxford 2004

[8] P.L. Volegov, J.C. Mosher, M.A. Espy, R.H. Kraus Jr, On concomitant gradients in low-field MRI, Journal of Magnetic Resonance 175 (2005) 103–113

[9] P. L. Volegov, A. N. Matlashov, and R. H. Kraus, Jr., "Ultra-low field NMR measurements of liquids and gases with short relaxation times," Journal of Magnetic Resonance, vol. 183, pp. 134-141, 2006.

[10] John Clarke and Alex I. Braginski, The SQUID Handbook, vol.I, Wiley-VCH Verlag GmbH & Co. KGaA 2004

[11] John Clarke and Alex I. Braginski, The SQUID Handbook, vol.II, Wiley-VCH Verlag GmbH & Co. KGaA 2004

[12] "SQUID sensor Model CE2 blue and cryoswitch Model SW1 from Supracon AG," Jena, Germany [Online]. Available: www.supracon.com

[13] V. Zotev, A. Matlashov, P. Volegov, A. Urbaitis, M. Espy and R. Kraus Jr, SQUID-based instrumentation for ultralow-field MRI, Supercond. Sci. Technol. 20 (2007) S367–S373

[14] M. Espy, M. Flynn, J. Gomez, C. Hanson, R. Kraus, P. Magnelind, K. Maskaly, A. Matlashov, S. Newman, T. Owens, M. Peters, H. Sandin, I. Savukov, L. Schultz, A. Urbaitis, P. Volegov, V. Zotev, Ultra-low Field MRI for the Detection of Liquid Explosives Using SQUIDs, IEEE Transactions on Applied Superconductivity, vol. 19, no. 3, pp 835-838, June 2009

[15] V. S. Zotev, A. N. Matlashov, I. M. Savukov, T. Owens, P. L. Volegov, J. J. Gomez, and M. A. Espy, SQUID-Based Microtesla MRI for In Vivo Relaxometry of the Human Brain, IEEE Transactions on Applied Superconductivity, vol. 19, no. 3, pp 823-826, June 2009

[16] AMAG Pharmaceuticals, Inc., 100 Hayden Ave, Lexington, MA 02421, USA; http://www.amagpharma.com

[17] Ocean NanoTech, LLC, 2143 Worth Ln., Springdale, AR 72764, USA; http://www.oceannanotech.com

Characterizing Cerebral and Extracerebral Components in TMS-Evoked Near-Infrared Spectroscopy Signals

Hanna Mäki[1,2], Tiina Näsi[1,2], Kalle Kotilahti[1,2], Ilkka Nissilä[1], Pekka Meriläinen[1], and Risto J. Ilmoniemi[1,2]

[1] Department of Biomedical Engineering and Computational Science, Aalto University, Espoo, Finland
[2] BioMag Laboratory, HUSLAB, Helsinki University Central Hospital, Helsinki, Finland

Abstract— **Near-infrared spectroscopy (NIRS) can be used to study changes in blood volume and oxygenation level due to transcranial magnetic stimulation (TMS). In previous studies, no attention has been paid to the fact that TMS also activates superficial tissue, which may confound the analysis of the NIRS signal originating from the brain. In addition, stimulation-related changes in, e.g., blood pressure and heart rate may induce global interference in the NIRS signal. We delivered TMS trains to the left primary motor cortex of healthy subjects and recorded NIRS from both ipsi- and contralateral hemispheres. In addition, the shoulder of one subject was stimulated and NIRS was recorded simultaneously above the stimulation site. Extracerebral contribution was estimated from the shoulder stimulation data, using principal component analysis (PCA) and from NIRS data corresponding to short source–detector (SD) distances (multidistance method). Motor cortex stimulation resulted in pronounced reductions in the concentration of oxygenated hemoglobin (HbO_2) on the contralateral (non-stimulated) hemisphere. On the ipsilateral (stimulated) hemisphere, a less pronounced decrease of HbO_2 was observed. Also the shoulder stimulation resulted in reduction of HbO_2. Applying PCA resulted in smaller HbO_2 reductions on the contralateral hemisphere and increased signal-to-noise ratio on both hemispheres. The multidistance method also resulted in smaller HbO_2 reductions on the contralateral hemisphere. The present results suggest that some reductions of HbO_2 may be due to extracerebral effects. These effects have to be taken into account when analyzing and interpreting TMS-evoked NIRS signals.**

Keywords— **transcranial magnetic stimulation, near-infrared spectroscopy, cerebral signal, extracerebral component, systemic interference.**

I. Introduction

Transcranial magnetic stimulation (TMS) combined with hemodynamic-based neuroimaging enables the study of cortical excitability and connectivity in health and disease, can be used to elucidate the basic mechanisms of TMS and may be used to study neurovascular coupling. TMS has been combined with positron emission tomography (PET) [1], functional magnetic resonance imaging (fMRI) [2], and, in a few studies, also with near-infrared spectroscopy (NIRS) [3–10]. TMS–NIRS has several advantages over the other multimodal setups: the electromagnetic fields generated by TMS do not interfere with NIRS, the time resolution is better than in TMS–PET, and the subjects are not exposed to radiation. In addition, concentration changes in both oxygenated (HbO_2) and deoxygenated (HbR) hemoglobin can be calculated.

In previous studies, NIRS responses have been measured during TMS delivered to the primary motor cortex (M1) and the prefrontal cortex both near the stimulation site and from the contralateral hemisphere. Both increases [3, 4, 5] and decreases [6] of HbO_2 have been reported over the ipsilateral hemisphere. Over the contralateral hemisphere, both increases [7, 8] and decreases [8, 9, 10] of HbO_2 have been found as well. In most TMS–NIRS studies, changes in HbR have been either not reported or non-significant, but decreases on both ipsilateral [5] and contralateral [8] side have been shown.

In the previous TMS–NIRS studies, no attention has been paid to the fact that, in addition to activating the brain, TMS also has an effect on the superficial tissue such as nerves in the skin and scalp muscles. This activation of the superficial tissue is probably also reflected in the NIRS signal, since all absorption and scattering changes on the photon paths, not only the ones in brain tissue, contribute to the signal. In addition, task- or stimulation-related changes in systemic variables, such as blood pressure and heart rate, may induce some global interference in the NIRS signals [11].

In this study, we used several methods to characterize, identify and remove the extracerebral components from TMS-evoked NIRS signals. The results suggest that there is extracerebral contribution that should be taken into account.

II. Materials and methods

A. Subjects

We studied 14 subjects (1 female, 22–32 years old, mean age 27, SD 2.4) who gave their written informed consent before the experiment. Three subjects were excluded; one because of technical problems and two because of excessive

S. Supek and A. Sušac (Eds.): Advances in Biomagnetism – BIOMAG2010, IFMBE Proceedings 28, pp. 88–91, 2010.
www.springerlink.com

movement artifacts. The study was approved by the Ethics Committee of Helsinki University Central Hospital and was in compliance with the Declaration of Helsinki. The subjects sat comfortably on a reclining chair in a half-sitting position in a dimmed room and were instructed to stay relaxed and keep their eyes open.

B. Transcranial Magnetic Stimulation

TMS was delivered with a Nexstim eXimia TMS stimulator and its biphasic figure-of-8 coil (loop outer diameter 70 mm; Nexstim Ltd., Helsinki, Finland). MRI-guided navigated brain stimulation system Nexstim eXimia NBS was used to target the stimulation at the hand area of left M1. Motor evoked potentials were recorded from the right abductor pollicis brevis (APB) with a ME6000 EMG device and MegaWin software (Mega Electronics Ltd., Kuopio, Finland) to determine the resting motor threshold (MT). Stimulation intensity was adjusted to 75% of MT. Repetitive TMS (rTMS) trains lasting 8 s at 0.5, 1 or 2 Hz, 25 trains at each frequency, were delivered in random order with a randomized 28–38 s interval between them.

To qualitatively characterize the hemodynamic changes in the superficial tissue during and after TMS, also the shoulder – the proximal end of the humerus (armbone) – of one subject was stimulated. This shoulder stimulation might resemble stimulation of the superficial tissue above M1 as there are no muscles directly above the stimulation sites and TMS activates the temporal muscles during M1 stimulation and the deltoid muscle during shoulder stimulation to some extent.

As the TMS pulse is accompanied with a loud click, which may produce an auditory response in the NIRS signal, the subjects listened to masking white noise through in-ear headphones, which also damped external noise. The volume of the masking noise was adjusted so that the subjects did not hear the coil click.

C. Near-Infrared Spectroscopy

NIRS data were recorded with a frequency-domain instrument [12]. One NIRS probe was attached above each hemisphere, each probe comprising two source fibers and seven detector-fiber bundles at different source-to-detector (SD) distances (Fig. 1). Prism terminals were attached at the end of each fiber or fiber bundle to minimize the thickness of the probes. On the ipsilateral hemisphere, the central detector was positioned above the TMS target and, on the contralateral hemisphere, above M1 using the brain navigation system. During shoulder stimulation, the ipsilateral probe was positioned over the shoulder so that the central

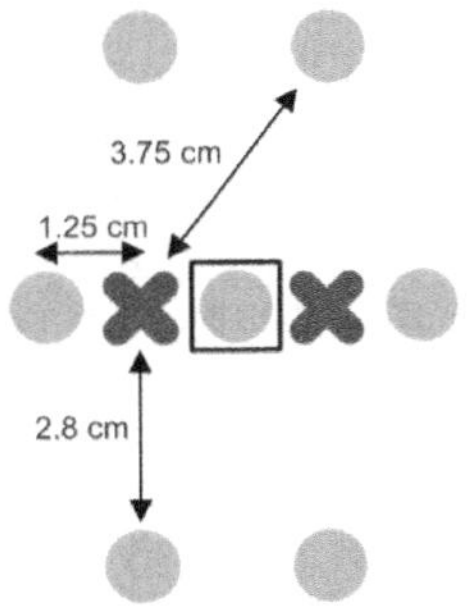

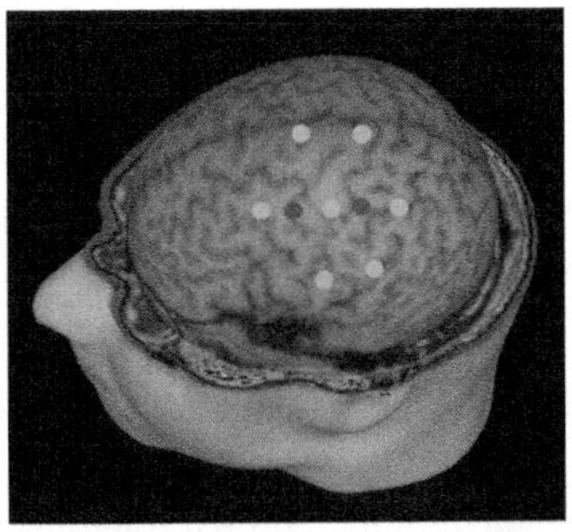

Fig. 1 Left: The NIRS probe. Sources are indicated with red crosses and detectors with yellow circles. The stimulation target is marked with a square. Right: The grid and its anatomical position. Digitized source (red circles) and detector (yellow circles) positions over the stimulated cortex in the MRI image of one subject. An identical grid was attached above the right hemisphere

detector was above the stimulation site and the sources anterior and posterior to it.

D. Analysis of NIRS Responses

Offline analysis of the NIRS data was performed using MATLAB (The Mathworks, Inc., Natick, Massachusetts, USA).

The signals were digitally filtered (passband 0.02–0.22 Hz) to attenuate systemic slow oscillations, artifacts caused by fiber contact variations, and high frequency noise. The NIRS amplitude signals were converted to HbO_2 and HbR values by using the modified Beer–Lambert law and an optical path length obtained from the calibrated NIRS phase signal [12, 13]. Epochs with artifacts were removed from further analysis and HbO_2 and HbR responses to rTMS were averaged time-locked to the stimuli (rTMS trains of different frequencies pooled). A baseline of 2 s before rTMS onset and a time window of 25 s after the stimulus onset were used. The NIRS responses were averaged over channels (short and long SD distances separately) and subjects.

E. Principal Component Analysis

Principal component analysis (PCA) can be used to present measured data as a linear combination of uncorrelated components. Each component is associated with an eigenvector and a corresponding eigenvalue. The component having the largest eigenvalue explains most of the variance in the data. The light measured with all NIRS channels has traversed through the surface tissue and is thus likely to have the largest eigenvalue. Deletion of the corresponding component has been used to remove systemic interference in NIRS data [15] and shown suitable also in applications with a small number of SD pairs [16].

F. Multidistance Method

To estimate extracerebral components of the NIRS signal, we utilized the fact that the light measured with a short SD distance has mostly traversed only through the surface tissue. By assuming that the extracerebral component is not correlated with the cerebral component, the brain signal can be obtained by subtracting appropriately weighted short SD distance signal from the long SD distance signal [14]. In our study, we subtracted an average of the 1.25 cm SD distance signals on each hemisphere from the long SD distance signals (SD distance > 1.25 cm) of this hemisphere.

III. Results

A. Characterization of Extracerebral Components

Reductions of HbO_2 were seen in the TMS-evoked NIRS measured above both ipsi- and contralateral hemispheres (see Fig. 2 A). Also, stimulation of the shoulder induced reduction of HbO_2 (see Fig. 2 D). This suggests that part of the HbO_2 reductions might be due to extracerebral effects. HbR was slightly reduced in the long SD distance channels after both brain and shoulder stimulation.

B. Removal of Extracerebral Components – PCA

Deletion of the first PCA component had two major effects on the data: increase of the signal-to-noise ratio, resulting in more clear reduction of HbO_2 in the long SD distance channels on the ipsilateral hemisphere and less pronounced reduction of HbO_2 on the contralateral hemisphere (see Fig. 2 B).

C. Removal of Extracerebral Components – Multidistance Method

Fig. 2 C shows the signals from long SD distance channels from which the estimated extracerebral component has been subtracted using the multidistance method. The procedure results in smaller HbO_2 reduction in the signal measured from the right hemisphere compared to the original signal.

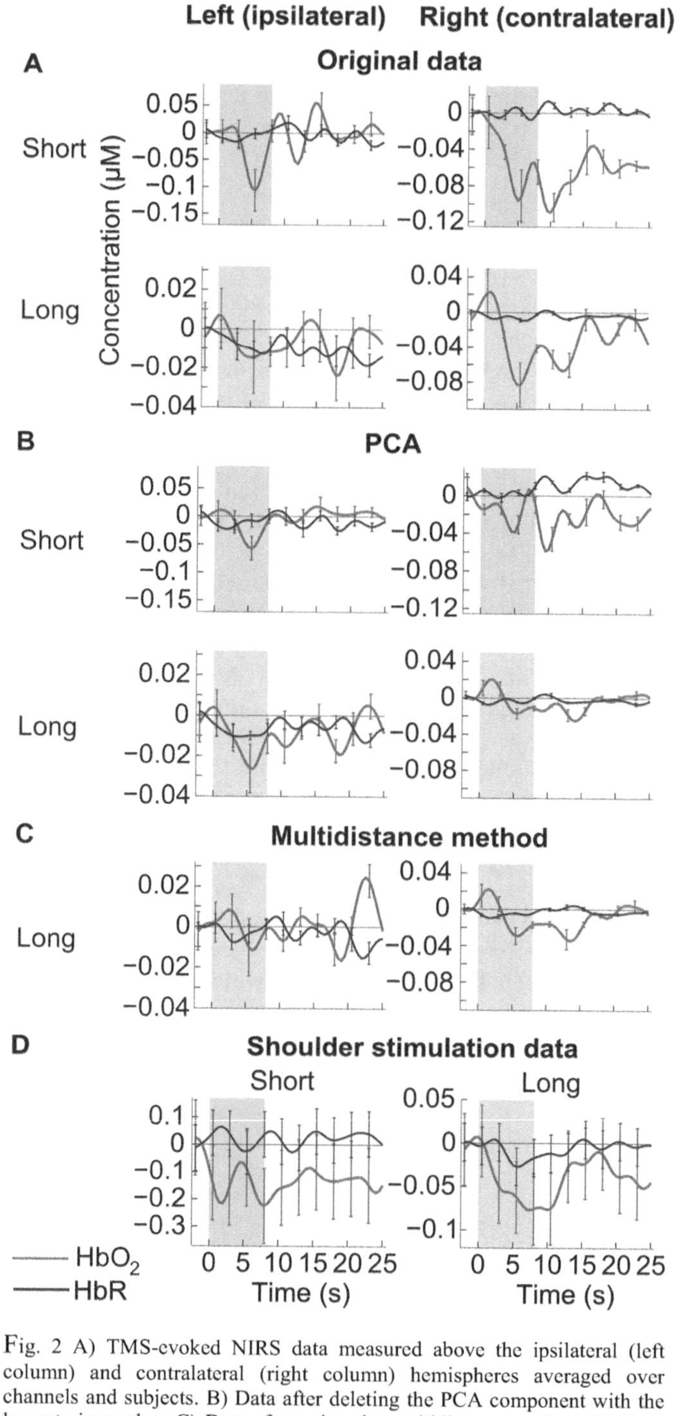

Fig. 2 A) TMS-evoked NIRS data measured above the ipsilateral (left column) and contralateral (right column) hemispheres averaged over channels and subjects. B) Data after deleting the PCA component with the largest eigenvalue. C) Data after using the multidistance method. D) Data measured during shoulder stimulation under the stimulation coil. The colored area shows the 8-s stimulation time (green: long SD distance channels, grey: short SD distance channels). Short and Long refer to SD distances. The errorbars in A–C show the SEM over subjects and in D the SD over epochs

IV. Discussion

The present study shows that TMS induces hemodynamic changes in the extracerebral tissue and that these are reflected in the TMS-evoked NIRS signal. This may confound the analysis of cerebral NIRS signals. When the shoulder of one subject was stimulated, it produced reduction in the

HbO_2 concentration around the stimulation site. The grand average NIRS response following left M1 stimulation showed similar reductions in HbO_2.

Muscle activity can lead to temporary reduction in HbO_2, whereas neural activation in the brain and the skin would be expected to lead to increases of blood flow and HbO_2 via the neurovascular coupling. A different pattern may be seen if TMS has an effect on the smooth muscles controlling blood vessel diameter. Also, if the net effect of the TMS train is inhibitory, resulting in a decreased net activation of both excitatory and inhibitory cells, reductions in HbO_2 could be expected. TMS delivered on cat visual cortex at frequencies (1–8 Hz) comparable to the present study resulted in increases in total hemoglobin concentration measured with intrinsic optical imaging, suggesting that the cerebral component would rather consist of increases in the sum of HbO_2 and HbR [17].

Both PCA and the multidistance method resulted in less pronounced reductions of HbO_2 on the contralateral hemisphere than in the original data. This part of the HbO_2 reduction probably reflects extracerebral effects. However, one should be cautious when interpreting results obtained with these methods. The argument stating that the PCA component with the largest eigenvalue corresponds to the surface component is somewhat heuristic. Problems with the multidistance method arise if cerebral signals are seen in the short SD data. If that is the case, also the cerebral signal is removed by the algorithm. This happens also if the cerebral and extracerebral signals correlate, which might be the case when both are time-locked to the stimulation. In addition, the processed signal contains more noise than the original, since noise is added by the subtraction procedure.

V. Conclusions

We show that TMS produces activation of the surface tissue, which may confound the analysis of the cerebral NIRS signal. PCA and a multidistance method using different SD separations can be used to reduce the extracerebral contribution, but the applicability of these methods for analyzing TMS-evoked NIRS data needs further study.

Acknowledgements

This study was supported by the Academy of Finland (grant N:o 121167 and International Graduate School in Biomedical Engineering and Medical Physics), the Finnish Cultural Foundation, and the Instrumentarium Science Foundation.

References

Paus T, Jech R, Thompson CJ et al. (1997) Transcranial magnetic stimulation during positron emission tomography: a new method for studying connectivity of the human cerebral cortex. J Neurosci 17:3178–3184

Bohning DE, Shastri A, Nahas Z et al. (1998) Echoplanar BOLD fMRI of brain activation induced by concurrent transcranial magnetic stimulation (TMS). Invest Radiol 33:336–340

Oliviero A, Di Lazzaro V, Plazza O et al. (1999) Cerebral blood flow and metabolic changes produced by repetitive magnetic brain stimulation. J Neurol 426:1164–1168

Noguchi Y, Watanabe E, Sakai KL (2003) An event-related optical topography study of cortical activation induced by single-pulse transcranial magnetic stimulation. NeuroImage 19:156–162

Mochizuki H, Ugawa Y, Terao Y et al (2006) Cortical hemoglobin-concentration changes under the coil induced by single-pulse TMS in humans: a simultaneous recording with near-infrared spectroscopy. Exp Brain Res 169:302–310

Hada Y, Abo M, Kaminaga T et al (2006) Detection of cerebral blood flow changes during repetitive transcranial magnetic stimulation by recording hemoglobin in the brain cortex, just beneath the stimulation coil, with near-infrared spectroscopy. NeuroImage 32:1226–1230

Nissilä I, Kotilahti K, Komssi S et al. (2002) Optical measurement of hemodynamic changes in the contralateral motor cortex induced by transcranial magnetic stimulation. Proc. of the 13th International Conference on Biomagnetism (BIOMAG 2002). Jena, Germany, 10-14 August 2002, pp 851–854

Hanaoka N, Aoyama Y, Kameyama M et al. (2007) Deactivation and activation of left frontal lobe during and after low-frequency repetitive transcranial magnetic stimulation over right prefrontal cortex: A near-infrared spectroscopy study. Neurosci Lett 414:99–104

Iwata NK, Ugawa Y, Watanabe E et al (2000) Interhemispheric connection between bilateral hand motor areas studied with near-infrared spectroscopic (NIRS) mapping and transcranial magnetic stimulation (TMS). NeuroImage 11:S520

Mochizuki H, Furubayashi T, Hanjima R et al (2007) Hemoglobin concentration changes in the contralateral hemisphere during and after theta burst stimulation of the human sensorimotor cortices. Exp Brain Res 180:667–675

Tachtsidis I, Leung TS, Chopra A et al (2009) False positives in functional near infrared topography. Adv Exp Med Biol 645: 307-314

Nissilä I, Noponen T, Kotilahti K et al (2005) Instrumentation and calibration methods for the multichannel measurement of phase and amplitude in optical tomography. Rev Sci Instrum 76:044302

Tarvainen T, Kolehmainen V, Vauhkonen M et al (2005) Computational calibration method for optical tomography. Applied Optics 44:1879–1888

Saager RB and Berger AJ (2005) Direct characterization and removal of interfering absorption trends in two-layer turbid media. J Opt Soc Am A 22:1874–1882

Zhang Y and Franceschini MA (2005) Eigenvector-based spatial filtering for reduction of physiological interference in diffuse optical imaging. J Biomed Opt 10: 011014

Virtanen J, Noponen T, Meriläinen P (2009) Comparison of principal and independent component analysis in removing extracerebral interference from near-infrared spectroscopy signals. J Biomed Opt 14: 054032

Allen EA, Pasley BN, Duong T et al (2007) Transcranial magnetic stimulation elicits coupled neural and hemodynamic consequences. Science 317:1918–1921

Author: Hanna Mäki
Institute: Aalto University School of Science and Technology
Street: P.O. Box 12200, 00076 Aalto
City: Espoo
Country: Finland
Email: hanna.maki@hut.fi

Real-Time Spatial Localization System of Brain Regions for TMS Application by Co-registration with fMRI

A.S.C. Peres[1], V.H.O. Souza[1], E.M. Rodrigues[1], C.E.G. Salmon[1], D.B. de Araujo[1,2], and O. Baffa[1]

[1] Universidade de São Paulo, Departamento de Física e Matemática, Ribeirão Preto, Brazil
[2] Universidade Federal do Rio Grande do Norte/IINN, Natal, Brazil

Abstract— **Most TMS applications still use subjective parameters to establish the brain area that is to be stimulated. An obvious way to improve TMS applications is to combine them with neuroimaging techniques. In order to accomplish this it is necessary to use complex techniques, i.e. neuronavigation, to perform the co-registration. Although there are commercial neuronavigation systems available, in general they are expensive and proprietary. This work aims to improve the precision of TMS applications by measuring the magnetic field produced by the coil and by using neuronavigation software based in an open architecture and free software, allowing the use of different positioning systems for the co-registration with the neuroimage.**

Keywords— **TMS, fMRI, Neuronavigation, Magnetic Field Mapping.**

I. Introduction

Transcranial magnetic stimulation (TMS) [1,2] has been extensively investigated for use as a clinical and research tool in neurophysiology, as evidenced by a strong increase in published papers in recent years. Since the magnetic pulses are applied externally, the operator has no means to know exactly which cortical area is being stimulated. Thus, usually pre-established fiducial points are used to guide TMS application, even though are large anatomical variations among individuals.

To know the regions that are being actually stimulated in a given subject it is necessary to rely on the subject's own neuroimage [3]. In order to co-register the positions of the scalp with the internal structures, it is necessary to use a neuronavigator [4], which is a complex system that integrates spatial tracking software with graphical and mathematical tools. Neuronavigators are usually based on a closed platform, making it impossible to implement the user-specific tools, and they are expensive. However, this scenario is changing due to the development of open source implementations, allowing multiple users to change the code according to their needs.

In this way, one of our goals is to implement an open source software available to groups that have no possibility to acquire a commercial system. As a second objective, we combined TMS with functional Magnetic Resonance Imaging (fMRI), in order to access the functional brain site that one which to stimulate [5,6,7]. Moreover, this platform should be versatile, and operate with different TMS coil geometry, and respective field map.

Although the geometry of most of the coils used is known and the magnetic field can be calculated, it is important to know as precisely as possible the field being applied. In order to do so, it is necessary to perform a field map of the coil to be used. The measurements of magnetic fields can be obtained with small search coils, hall sensors, magnetoresistive sensors or MRI techniques [8]. The advantage of using MRI for magnetic field mapping is the possibility to map a large volume with high spatial resolution relatively fast.

II. Matrials, methods and results

A. TMS Coil Studies

Using a 3 mm diameter coil, with acrylic and PVC core, as a search coil, the magnetic field time variation was analyzed. The electromotive force (emf) was measured as a function of time (Fig. 1).

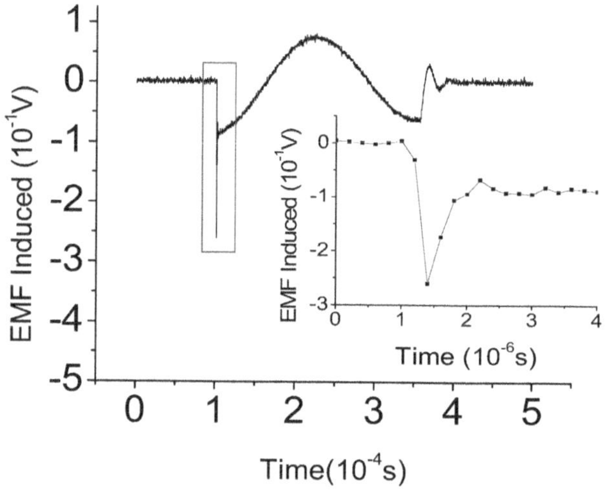

Fig. 1 Time variation of emf induced in the search coil to measure the magnetic flux produce by the TMS coil. The selected region in the box represents the initial ramp, detailed in the inset curve in the side

S. Supek and A. Sušac (Eds.): Advances in Biomagnetism – BIOMAG2010, IFMBE Proceedings 28, pp. 92–96, 2010.
www.springerlink.com

The signal starts with a large slope, of the order of 1 μs, and then becomes biphasic. The average rise time of the initial slope was 1.4 ± 0.2 μs and the total pulse duration was 307 ± 7 μs, both results obtained after an average of six measurements.

It is important to note that most the literature commonly report only the period of the biphasic wave, neglecting the initial slope. This can lead to an erroneous estimation of the intensity of the electric field induced, which causes depolarization of neurons.

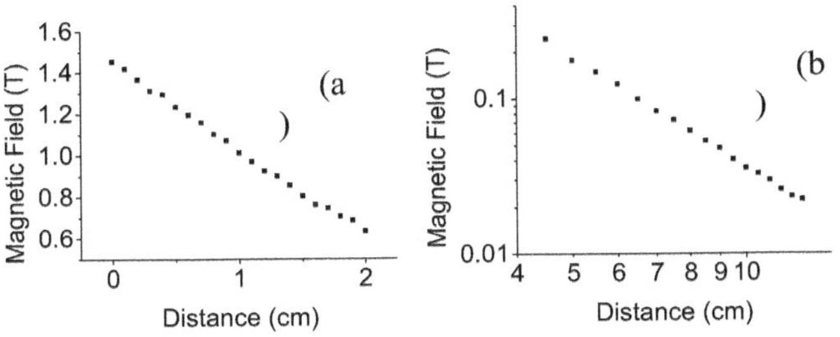

Fig. 2 a) Magnetic field decay at short distances to the TMS coil. b) Logarithmic representation to the intensity field decay, at 45 to 125 mm of distance to the TMS coil

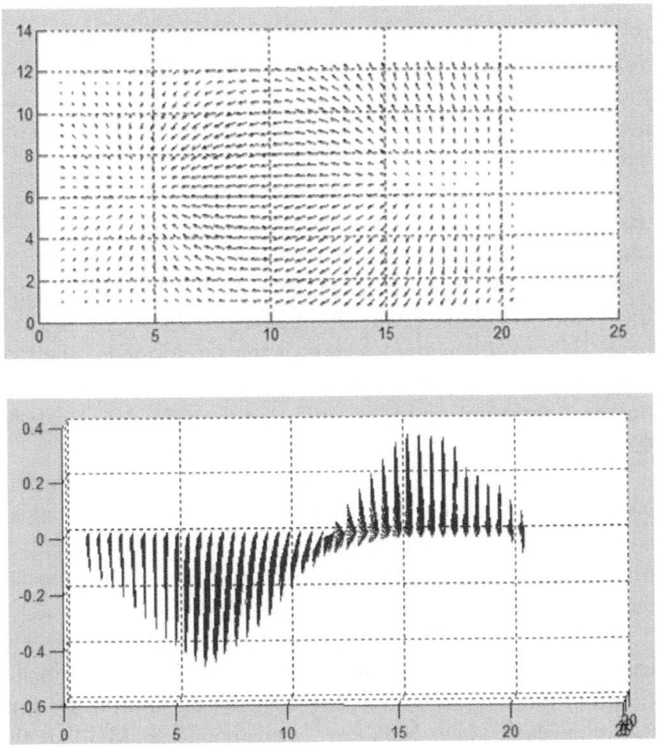

Fig. 3 Vector magnetic field distribution produced by butterfly coil shown in two perspectives

In order to know the magnetic field intensity dependence on a butterfly coil distance, we studied the field intensity at near points (20 mm) and at more distant points (45 to 125 mm). In the first case, a linear behavior was observed with slope -41.5 ± 0.5. This is expected, as the range of measurements is smaller than the outer radius of the coil. At distant points, it's possible to consider the butterfly coil as a point dipolar source, which equation is independent of coil size and falls with the third power of the distance.

A logarithmic plot of the measurements yielded the exponent of the distance dependence, which was equal to 2.34 ± 0.02. These results show that in this distance range the source size still influences the field intensity [9].

The results for the variation of magnetic field strength with distance from the source are shown in Figure 2.

Finally we use the phase images of gradient recalled echo technique to describe the spatial distribution of the butterfly coil magnetic field. In this experiment a 3T Magnetic Resonance Scanner (Philips Achieva, The Netherlands) and a phantom filled with $CuSO_4 \cdot 5H_2O$ solution were used. The magnetic field of the TMS coil was generated by a direct current of 1.12A. This was necessary since TMS pulses are very short, not changing the phase of the proton spins [10]. The results are shown in figure 3, presenting the vector distribution of the magnetic field.

B. The Neuronavigation Software

The neuronavigation software was built using the graphical user interface development environment (GUIDE) of Matlab. First, an algorithm to load medical images (DICOM, Analyze, REC/PAR, Nift, and Minc) was implemented. For brain segmentation we use the freeware software MRIcro [11]. A three-dimensional reconstruction algorithm for brain volume preview was also developed in this project.

After that, we implement an algorithm to co-register the patient (real coordinate) with his/her digital images (virtual coordinate).

To do so, a tracking system based on electromagnetic waves (Polhemus Isotrak II) [12] was used to acquire the positions of the following fiducial points: right and left pre-auricular and nasal cavity. Then the neuroimaging points corresponding to these reference points were also acquired. With these reference points, a basis associated with the Polhemus and another, based on the coordinated image, were built. Thus to take a point from the scanner basis to the image basis (pixel position) just one coordinate transformation is needed.

The precision of the whole system was analyzed using a head phantom. A precision of 6 mm was obtained mainly due to the size of the pointing device presently used with the Polhemus system. Use of the stylus pointing device will improve the precision. Software is under development, as well as the implementation of drivers that will allow the use of other tracking systems such as Claron (optical) [13] and Zebris (Ultra-sonic) [14].

C. fMRI Experiments

Ten healthy volunteers between 18 and 30 years, including 2 women and 8 men were used in the experiment. The project was approved by the Ethics Committee of *Hospital das Clínicas de Ribeirão Preto* (HCRP) according to HCRP No 6728/2007, and informed consent was obtained from all participants.

A 1.5 T scanner Siemens (Magneton Vision) with a quadrature coil transmitter/receiver head from HCRP was used in the experiment.

Echo Planar Images (EPI) of 66 blocks were acquired with 16 slices with 6 mm thickness, covering 9.6 cm, with the following parameters: ISI = 3540 ms, TE = 60 ms, flip angle = 90°, a matrix size = 128 x 128, FOV = 220 mm resulting in a voxel size = 1.72 x 1.72 x 6.00 mm.

As the region of interest in this study was the primary motor cortex, the slices were oriented approximately perpendicular to the axis of the central sulcus, thus covering a region beyond the lateral sulcus, ensuring the mapping of the entire pre- and post-central gyri.

For the high anatomical resolution images, a GRE, MPR, T1-weighed with 156 slices, TR of 9.7 ms, TE of 4 ms, flip angle of 12 degrees, matrix size of 256 x 256, FOV of 256 mm and a voxel size of 1 x 1 x 1 mm was used.

To verify that our neuronavigation system was viable, we decided to use a well-established block paradigm, known as finger tapping [15, 16], in which the volunteer alternates periods of rest and activity. Each period corresponds to 6 acquisitions in time. The volunteer started at rest and after a beep began the task until another signal was presented to end the task, thus totaling 6 periods of rest and 5 of activity.

The images were then processed with the software Brain Voyager QX, which performs statistical calculation algorithm using a General Linear Model (GLM) [17].

D. TMS and fMRI co-registration

Once the algorithm was tested with a phantom and the functional images were obtained, we set up a positioning system for co-registration with the fMRI images, consisting of a variable height chair and a table with a chin support.

To facilitate the positioning of the coil on the scalp, we developed a coordinate system for the skull of the volunteers. This system consists of a white nylon swim cap, on which we draw grid lines 1 cm apart horizontally and vertically.

An electromyography electrode was positioned on the thumb's abductor muscle to capture the MEP due to TMS in the motor cortex. This muscle was chosen due to the ease of positioning an electrode on it and also because of its large area of cortical representation.

The volunteer's head was positioned in the chin support, and the three points (real and virtual) for the co-registration were collected. Once co-registered, scans on the volunteers' scalp with the Polhemus's probe were made, until it was on the most probable of activity center found by the fMRI (hot spot). Besides the stimulation of this position, a set of points distant 1cm and 2cm from the left and right (longitudinal mapping), from the top and below (transversal mapping) and a distant point, totaling 10 points.

For the application of TMS, we first determined the motor threshold for each volunteer, that was found when at least 5 MEPs were elicited out of 10 pulses applied at the hot spot [18].

Once the motor threshold was determined, the intensity was adjusted to 120% of the threshold and for each point 8 stimulations separated by 5s were performed.

The data were then interpolated and plots of the MEP intensity versus position were made, summing 4 curves per volunteer, 2 for right hand and another 2 for left hand. Then the distance along the direction between the positions of the maximum activations obtained by fMRI and by TMS were determined, as shown in figures 4.

We analyzed first the behavior of all curves, computing the average and its deviation to the displacement of the peak for all curves, and obtained 7 ± 6 mm.

Thus we proceeded to a second analysis, which separated the values obtained in the stimulation of the right and left hand. The results show no significant variations for the two hemispheres, where we found 7 ± 6 mm for the stimulation on the right hand and 8 ± 7 mm for the stimulation of the left hand.

Finally a comparison between the sample into transversal curves and longitudinal curves was made. This result showed a significant difference in the orientation of the mapping, with the value (11±6) mm for the longitudinal curves and (4±3) mm for the transversal curves.

This discrepancy between distances obtained in the transverse and longitudinal direction is probably due to the fact that the longitudinal curve includes the motor and pre-motor cortex. Since both cortices can induce MEP in the thumb brevis, two peaks were found in the longitudinal curves in many volunteers, as can be seen in Figure 5d.

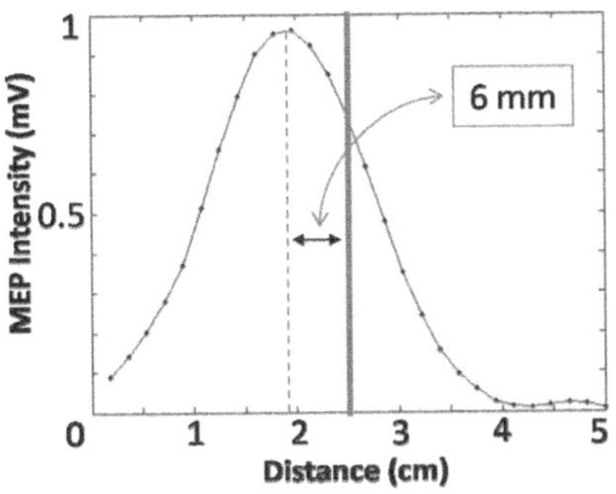

Fig. 4 MEP curve by the position of stimulation. The curve is represented by blue dots while the black line segment corresponds to the center of activity found by neuronavigator

As the value of maximum activity in fMRI was chosen visually by the experimenter, as explained above, when there were two points with high statistical significance, as shown in Figures 5a, b and c, the midpoint between them was chosen as the hot spot. The result can be seen in figure 5d, where the greatest fMRI activity was between 2 peaks of MEP, thus inducing an error in the measurement of distances. Therefore, only transverse curves must be used. The results obtained with the fMRI are very close to the precision of the neuronavigator, which are sufficient for the purpose for which it was designed, as in all volunteers we obtained the region of maximum activity defined by the neuronavigator. These results are similar to the ones obtained by Lotze, 2003 [19].

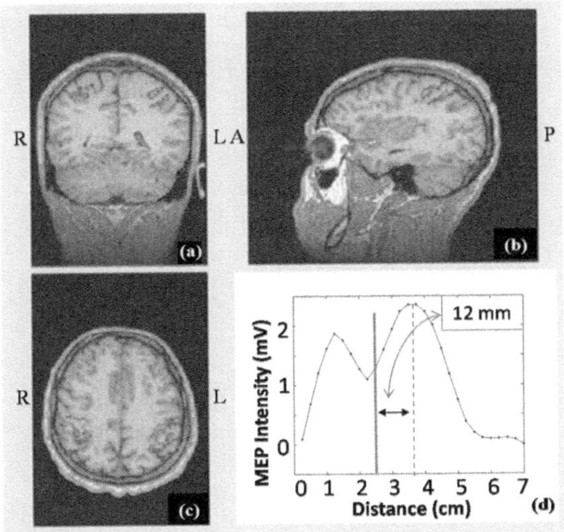

Fig. 5 fMRIs of a volunteer showing the motor area activated by the proposed protocol. (a) Coronal image; (b) Sagital image; (c) Axial image; (d) MEP intensity curve induced by TMS

III. Conclusion

We observed that 100% of the applications of TMS were in the center of activity found by neuronavigator were followed by MEP. Thus we may say that the neuronavigator is a good alternative to assist the localization of brain regions, not only reducing the application time of TMS, but also improving the accuracy of the test.

This technique, when allied with magnetic field maps, can predict with good precision what areas are being stimulated, and more than that, the intensity and direction of the magnetic field applied in a certain region.

Acknowledgments

The authors wish to thank the Brazilian agencies CAPES, CNPq, and FAPESP for partial financial support. Moreover, we are grateful to Mr. Lourenço Rocha for technical support. This work is part of the CINAPCE project.

References

1. Barker AT, Jalinous R, Freeston IL (1985) Non-invasive magnetic stimulation of human motor cortex. The Lancet 325:1106-1107.
2. Hallett M (2000) Transcranial magnetic stimulation and the human brain. Nature 406:147-150.
3. Siebner HR, Bergmann TO, Bestmann S, et al (2009). Consensuspaper: Combining transcranial stimulation with neuroimaging. Brain Stimulation 2:58-80.
4. Herwig U, Schönfeldt-Lecuona C, Wunderlich AP, et al. (2001). The navigation of transcranial magnetic stimulation. Psychiatry Research: Neuroimaging 108(2):123-131.
5. Kwong KK, Beliveau JW, Chesler DA, et al. (1992) Dynamic magnetic resonance imaging of human brain activity during primary sensory stimulation. Proceedings of the National Academy of Science 89:5675-5679.
6. Ogawa S, Tank DW, Menon R, et al. (1992) Intrinsic signal changes accompanying sensory stimulation: functional brain mapping with magnetic resonance imaging, Proceedings of the National Academy of Science 89(13):5951-5955.
7. Huettel SA, Song AW, MacCarthy G. Functional magnetic resonance imaging. Sinauer Associates, Inc., 1st ed. Cap 7, pp 159-184.
8. Tomasi D, Panepucci H (1999) Magnetic field mapping with the phase reference method. Mag Reson Imaging 17:157–160.
9. Wikswo JP Jr (1989). Biomagnetic Sources and Their Models. Advances in Biomagnetism. Proceedings of Seventh International Conference on Biomagnestism pp.1-18.
10. Peres ASC, Souza VHO, Maziero M, et al. (2009) Vector magnetic field mapping of a Transcranial Magnetic Stimulation coil using Magnetic Resonance Imaging: in vitro and in vivo experiments. Ifmbe Proceedings 25/9:571–574.
11. MRIcro at http://www.sph.sc.edu/comd/rorden/mricro.html
12. Polhemus at http://www.polhemus.com
13. Claron at http://www.clarontech.com/measurement.php
14. Zebris at http://www.zebris.de/english/medizin/medizin-3dmesssysteme.php?navanchor=1010025

15. de Araujo, DB, Araujo D, Rosset S, Wichert-Ana L, Baffa O, et al. (2004) FMRI in epilepsy. Medical Physics, 724, 3-7
16. Stewart H, Mostofsky SL, Rimrodt JGB, et al (2006). A typical Motor and Sensory Cortex Activation in Attention-Deficit/ Hyperactivity Disorder: A Functional Magnetic Resonance Imaging Study of Simple Sequential Finger Tapping. Biological Psychiatry 59:48-56.
17. Brain Voyager at (http://www.brainvoyager.com/)
18. Khol A, et all An integrative transcranial magnetic stimulation mapping, 2006.
19. Lotze M, Kaethner RJ, Erb M, et al. (2003). Comparison of representational maps using functional magnetic resonance imaging and transcranial magnetic stimulation. *Clinical Neurophysiology 114(2):306-312.*

Author: André Salles Cunha Peres
Institute: Dep. Física e Matemática – USP
Street: Av. Bandeirantes 3900
City: Ribeirão Preto
Country: Brazil
Email: andre@biomag.usp.br

Multimodal Integration: Constraining MEG Localization with EEG and fMRI

Richard N.A. Henson

MRC Cognition & Brain Sciences Unit, Cambridge, UK

Abstract— **I review recent methodological developments for multimodal integration of MEG, EEG and fMRI data within a Parametric Empirical Bayesian framework [1]. More specifically, I describe two ways to incorporate multimodal data during distributed MEG/EEG source reconstruction under linear Gaussian assumptions: 1) the simultaneous inversion of EEG and MEG data using a common generative model [2], and 2) the addition of spatial priors from fMRI data when inverting MEG or EEG data [3]. In the former, the addition of EEG data was shown to increase the conditional precision of source estimates relative to MEG alone; in the latter, the inclusion of each suprathreshold cluster in the fMRI data as a separate spatial prior was shown to increase the Bayesian model evidence for MEG and EEG reconstruction. The former is an example of "symmetric" integration, or "fusion", in which a single generative model of all data modalities is inverted; the latter is an example of "asymmetric" integration, in which the data from one modality is used to inform inversion of another. I will conclude by considering whether symmetric fusion of MEG/EEG and fMRI data is worthwhile.**

Keywords— **MEG, EEG, fMRI, multimodal, fusion.**

I. Introduction

There is much recent excitement in formal methods to integrate, or "fuse", data from different, non-invasive neuroimaging modalities, such as magnetoencephalography (MEG), electroencephalography (EEG) and functional magnetic resonance imaging (fMRI), in order to achieve more accurate spatiotemporal descriptions of human brain activity than is possible with any one modality alone. EEG/MEG data provide a relatively direct measure of (synchronous) neuronal local field potentials (LFP) and ensuing currents with millisecond (or higher) resolution, as recorded outside the scalp. Localizing this electrical activity within the brain is an ill-posed, inverse problem however [4]. Furthermore, EEG and MEG have different sensitivities to the orientation of electrical sources and to the different conductivities of skull and scalp [5]. fMRI on the other hand, normally relies on a blood oxygen level dependent (BOLD) signal that can be localized in the range of millimeters, but integrates over several seconds of neuronal activity.

Fusion of multimodal data can take several forms (Fig 1). A full integration, or "fusion", would entail inverting a single "generative" model that explains both types of data. This model must relate the same hypothetical neural causes (e.g., timecourses of neuronal activity in circumscribed brain regions) to each type of data, using modality-specific "forward models". In this framework, usually Bayesian, different data-types are treated symmetrically, and the priors on model parameters are not based upon the data in either modality [6]. This approach will be illustrated for simultaneous inversion of MEG and EEG data in Section III.

Developing such a generative model is the ultimate goal of multimodal fusion. In some cases however, these models can become complex and difficult to invert. For example, in order to integrate MEG/EEG data with fMRI data, one needs to relate the physiological causes of EEG/MEG signals to the physiological causes of the BOLD signal. While there is empirical evidence for a close relationship between LFPs and BOLD [7], accepted and accurate models do not yet exist. Furthermore, if the modalities are differentially sensitive to different dimensions of the neuronal causes, such as the temporal and spatial dimensions, it remains unclear whether the modalities can mutually constrain each other. In other words, the traditional complementarity of MEG/EEG and fMRI in terms of temporal and spatial resolution respectively makes the value of full fusion unclear (see Section V).

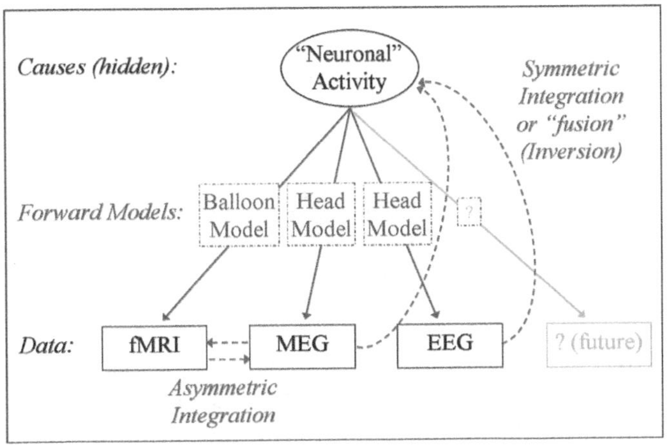

Fig. 1 Schematic illustrating distinction between symmetric and asymmetric multimodal integration within a generative model

An alternative approach therefore is to use data from one modality as a predictor (independent variable) for the data of another modality (dependent variable). For example, one might use some summary measure of EEG/MEG power at

S. Supek and A. Sušac (Eds.): Advances in Biomagnetism – BIOMAG2010, IFMBE Proceedings 28, pp. 97–100, 2010.
www.springerlink.com

each fMRI sample point as a regressor in a voxel-wise, classical statistical analysis of the fMRI data [8]. This type of "asymmetric" approach taken for MEG and fMRI integration in Section IV, except that we use partitions of the fMRI data as spatial priors on the localization of the sources of the EEG/MEG data (see also [9]).

II. A PARAMETRIC EMPIRICAL BAYESIAN FRAMEWORK

The approach to multimodal integration reviewed here benefits from a recent convergence of analysis methods for MEG/EEG and fMRI towards a common Variational Bayesian framework, specifically a Parametric Empirical Bayesian (PEB) framework [10]. The "Parametric" refers to the assumption of multivariate Gaussian distributions, e.g., for random error terms; while the "Empirical Bayesian" refers to hierarchical models in which the parameters at one level function as priors on those at lower levels. This means that the "hyper-parameters" that scale the Gaussian priors can be estimated directly from the data (e.g., through expectation-maximization, [10]).

In the context of distributed unimodal inversion of MEG (or EEG) data – i.e. estimation of the spatial distribution of activity over the brain that gives rise to the data recorded at the sensors – the PEB framework entails a two-level, linear, hierarchical model, where the first level represents the sensors and the second represents the sources [1]:

$$\begin{aligned} Y &= LJ + E^{(1)} \\ J &= 0 + E^{(2)} \end{aligned} \tag{1}$$

where Y is a n (sensors) by t (time points) matrix of sensor data; L is a n by p (sources) 'lead-field' matrix, or 'forward model', based on Maxwell's equations, and J is a p by t matrix of unknown dipole currents; i.e., the model parameters that we wish to estimate. These normally correspond to the amplitudes of several thousand dipoles distributed over a 2D tessellated surface of the neocortex (and are often, though not necessarily, assumed to be oriented normal to that surface). The fact that $p>>n$ means that inversion is ill-posed without additional constraints (i.e, regularization). These come from constraints on the random terms, E, which are assumed to be sampled from zero-mean, multivariate Gaussian distributions whose covariance factorizes into temporal components, V, and spatial components; $C^{(1)}$ and $C^{(2)}$ at the sensor and source level respectively.

The spatial covariance matrices are represented by a linear combination of covariance components:

$$C^{(i)} = \sum_{j=1}^{N} \lambda_j^{(i)} Q_j^{(i)} \tag{2}$$

where $\lambda_j^{(i)}$ is the 'hyperparameter' (regularization coefficient) for the j-th component of the i-th level. At the source-level, $C^{(2)}$ represents a spatial prior, and it can be shown that the standard 'minimum norm' solution corresponds to setting it to a p-by-p identity matrix [11]. Other assumptions are possible however, such as the use of multiple sparse priors (MSP), where each $Q_j^{(2)}$ represents a small "patch" of coherent activity [12]. For the sensor-level components, $Q_j^{(1)}$, one might assume white noise (an n-by-n identity matrix) or an empirical estimate of sensor noise (e.g., from MEG empty-room data, [2]).

The hyperparameters can be estimated using a Restricted Maximum Likelihood (ReML) algorithm. The objective function maximized by ReML is identical to the (negative) variational free-energy, F. For such linear models with Gaussian assumptions, the optimized free-energy provides a tight bound on the marginal log-likelihood of the generative model, M, or its "log-evidence" [10]:

$$\ln p(Y|M) = \ln \int p(Y, J \mid M) dJ \approx F \tag{3}$$

This log-evidence can be used to evaluate the advantage of asymmetric integration (Section IV). The hyperparameter estimates in turn allow Maximum A Posteriori (MAP) estimates of the source parameters (J), in addition to estimates of their posterior covariance, which quantifies the conditional precision of the source estimates. These conditional precisions can be used to evaluate the advantage of symmetric integration (Section III).

III. SYMMETRICAL FUSION OF MEG AND EEG

Several studies have shown that MEG and EEG data provide non-redundant information, both theoretically [13], and practically, in the sense that simultaneous inversion of both affords more accurate reconstructions than unimodal inversions (e.g., [14]). We recently reinforced these claims by outlining a new PEB method for full MEG-EEG fusion, and applying it to three types of evoked data recorded simultaneously [2]. These were two types of MEG data, from 1) 102 magnetometers and 2) 204 planar gradiometers, and 3) EEG data from 70 scalp electrodes.

Assuming that a lead-field matrix can be created for each of d sensor-types, then Eq. (1) can be extended as:

$$\begin{bmatrix} \tilde{Y}_1 \\ \tilde{Y}_2 \\ \vdots \\ \tilde{Y}_d \end{bmatrix} = \begin{bmatrix} \tilde{L}_1 \\ \tilde{L}_2 \\ \vdots \\ \tilde{L}_d \end{bmatrix} J + \begin{bmatrix} E_1^{(1)} \\ E_2^{(1)} \\ \vdots \\ E_d^{(1)} \end{bmatrix} \tag{4}$$

where the data of each type have been scaled (see below) and stacked into a single matrix; similarly for the modality-specific lead-field matrices. Note the sources (hidden neuronal causes) in J are common to all sensor-types. Likewise, the spatial covariance matrix of the sensor error, $C^{(1)}$, is formed by concatenating the covariance matrices for each sensor-type, which are themselves formed by linear combination of variance components, as in Eq. (2).

To accommodate different scaling and measurement units across the different sensor-types, the data and the forward model are re-scaled as follows:

$$\begin{aligned} \tilde{Y}_i &= \frac{Y_i}{\sqrt{\frac{1}{m_i} tr(Y_i Y_i^T)}} \\ \tilde{L}_i &= \frac{L_i}{\sqrt{\frac{1}{m_i} tr(L_i L_i^T)}} \end{aligned} \tag{5}$$

where tr is the trace of a matrix. This effectively normalizes the data so that the average variance (if the data are mean-corrected) is one for all sensor-types, and, in the absence of senor noise, the average variance expected under independent and identical sources with unit variance is one.

Note that these scalings are not based on any distinction between "signal" and "noise", the ratio of which (SNR) has been used previously to weight the relative contributions of MEG vs. EEG to source estimates (e.g., [9]). Such SNRs are often estimated from pre- versus post-stimulus periods; however, this confounds true "sensor noise" with endogenous "brain noise". In the present framework, the weightings (hyperparameters) of the sensor noise ($E^{(1)}$) for each modality, relative to the signal (J), are estimated automatically from the data, by optimizing the model-evidence, hence obviating the need for empirical SNR estimates.

Using this framework, Henson et al. [2] used statistical comparisons across twelve participants to show that the conditional precision of the source estimates (Section II) based on any one sensor-type generally improved with the addition of others. More specifically, while sensor-level error was greatest for EEG, the inclusion of EEG data increased the conditional precision of the underlying source estimates relative to MEG data alone. This is expected a priori from the ability of EEG to detect radial components of the electromagnetic field. The source reconstructions from fusion also had greater face-validity, in recovering both ventral and lateral posterior temporal activity related to the processing of faces (relative to scrambled faces); generators that are supported by independent data (such as fMRI and intracranial EEG; see [2] for further discussion).

IV. Asymmetrical fusion of MEG and fMRI

There have been several sophisticated attempts at symmetrical integration (fusion) of MEG (or EEG) data with fMRI data, using biologically-realistic generative models (e.g., [15,16]). We recently proposed a simpler, asymmetric approach, in which the partitions of the fMRI data function as separate spatial priors on the MEG (or EEG) inversion, using the PEB framework [3]. This way, we can estimate any temporal property of the MEG/EEG sources at any location in our solution space.

The use of priors in this context (rather than, say, fixing source activity at fMRI hotspots) is important in that MEG/EEG and fMRI data have different sensitivities to certain source configurations: For example, sources deep in the brain (far from the sensors) are likely to be represented only weakly in the MEG/EEG data, whereas very transient source activity may have minimal BOLD correlates. In other words, the fMRI data should impose "soft" rather than "hard" constraints on the MEG/EEG inversion [4].

Furthermore, the ability to use multiple spatial priors within the PEB framework allows each suprathreshold "cluster" from the fMRI data to form a separate prior (rather than entering all fMRI clusters, or even all voxels, as a single prior). Each suprathreshold cluster becomes a separate component, Q_j (Eq. (2)) for the source-level covariance, $C^{(2)}$, such that their relative weighting (hyperparameters) can differ (analogous to the relative weighting of the sensor-level covariances, $C^{(1)}$, for each sensor-type in Section III). This allows the "Automatic Relevance Detection" (ARD) behavior of our ReML algorithm to emphasize priors that are relevant, and de-emphasize priors that are not. In addition to possible discrepancies between the neuronal causes of the MEG/EEG and BOLD signals described above, irrelevant fMRI priors might reflect neuronal activity that occurs before or after the critical timewindow being localized (given the poor temporal resolution of BOLD).

Because the data (Y) are fixed (unlike for MEG+EEG fusion in Section III), and because the spatial priors are part of the generative model (M), the model evidence in Eq. (3) can be used to compare different types of fMRI prior. The log-evidence increases with the accuracy of the model (fit to the data), but decreases with complexity (favoring more parsimonious models). Using the same MEG+EEG dataset as in Section III, together with fMRI data on the same paradigm, Henson et al [3] showed that adding five suprathreshold fMRI regions as separate priors improved the log-evidence relative to no fMRI priors (just a minimum norm prior; Section II). This was true for all three sensor-types (magnetometers, gradiometers and electrodes), and was accompanied by more plausible source reconstructions for the face-evoked responses (deeper within the ventral temporal lobe).

Furthermore, this improvement in log-evidence was greater than when the suprathreshold clusters were entered as a single prior (despite the greater model complexity, and consistent with the down-weighting of some of the fMRI priors when entered separately, possibly because they reflected activity occurring outside the epoch of MEG/EEG data localized). Also important was the fact that adding invalid priors (by artificial displacement of the fMRI clusters) did not necessarily improve the model-evidence, particularly when added together with valid priors. This reinforces the ability of our PEB framework to ignore irrelevant (unhelpful) priors, and penalize overly-complex models. Finally, it was interesting to note that, for these data at least, the fMRI priors did not help so much when combined with multiple sparse priors rather than a single minimum norm prior (Section II), presumably due to the even greater flexibility of the former (see [3] for further discussion).

V. Conclusions

I have outlined a symmetric method for integrating (fusing) source reconstruction of MEG and EEG data, and an asymmetric method for integrating fMRI data as spatial priors on MEG/EEG source reconstruction. Both methods arise naturally within a Parametric Empirical Bayesian framework, in terms of multiple covariance components at either the sensor-level or source-level respectively.

The question remains as to whether symmetric (full) fusion of MEG/EEG and fMRI data is worthwhile. Unlike the fusion of MEG and EEG data, the parameters of generative models for hemodynamic and electromagnetic signals may not be shared. Put simply, if all one can estimate from fMRI data is "where" signals are coming from (i.e., spatial parameters) and all one can estimate from MEG/EEG data is "when" those signals are expressed (i.e., temporal parameters), then there is no point in using a common generative model. This is because multimodal fusion provides multiple constraints on the estimation of unknown causes generating data. If these quantities are constrained by only one modality, the conditional precision of their estimates will not be increased by adding another. The only way that fusion can work is if the spatial parameters estimated precisely by fMRI depend on the temporal parameters estimated precisely by MEG/EEG (or vice versa). Unfortunately, there is no principled reason to think that there will be strong dependencies of this sort, because the dynamics of electromagnetic sources are formally similar in different parts of the brain. If this argument turns out to be true, then the most powerful approaches may be asymmetric: i.e., using MEG/EEG as temporal constraints in whole-brain fMRI models, or using fMRI as spatial priors on the EEG/MEG inverse problem, as considered here.

Acknowledgment

This work is funded by the UK Medical Research Council (WBSE U.1055.05.012.00001.01).

References

1. Phillips C, Mattout J, Rugg M et al. (2005) An empirical Bayesian solution to the source reconstruction problem in EEG. NeuroImage 24:997-1011.
2. Henson R, Mouchlianitis E & Friston K (2009). MEG and EEG data fusion: Simultaneous localization of face-evoked responses. NeuroImage 47: 581-589.
3. Henson R, Flandin G, Friston K et al. (in press). A Parametric Empirical Bayesian framework for fMRI-constrained MEG/EEG source reconstruction. Human Brain Mapping.
4. Baillet S, Garnero L., Marin G. et al. (1999). Combined MEG and EEG source imaging by minimization of mutual information. IEEE Trans. biomed. eng. 46: 522-534.
5. Nunez P (1981). Electric fields of the brain: the neurophysics of EEG. New York: Oxford University Press.
6. Daunizeau J, Jbabdi S, Grova C et al. (2007). Symmetrical event-related EEG/fMRI information fusion in a variational Bayesian framework. NeuroImage 36:69-87.
7. Logothetis N, Pauls J, Augath M. et al. (2001). Neurophysiological investigation of the basis of the fMRI signal. Nature 412:150-157.
8. Debener S, Ullsperger M, Siegel M et al. (2006). Single-trial EEG-fMRI reveals the dynamics of cognitive function. TICS, 10:558-63.
9. Liu A, Belliveau J & Dale A (1998). Spatiotemporal imaging of human brain activity using functional MRI constrained magnetoencephalography data. Proc Natl Acad Sci U S A. 95:8945-50.
10. Friston K, Penny W, Phillips C et al (2002). Classical and Bayesian inference in neuroimaging: theory. NeuroImage 16:465–483.
11. Hauk O (2004) Keep it simple: a case for classical minimum norm estimation in analysis of EEG and MEG. NeuroImage 21:1612-21.
12. Friston K, Daunizeau J, Kiebel S et al. (2008) Multiple sparse priors for the M/EEG inverse problem. NeuroImage 39:1104-1120.
13. Molins A, Stufflebeam S, Brown E & Hamalainen M (2007) Quantification of the benefit from integrating MEG and EEG data in minimum L2-norm estimation. NeuroImage, 42:1069-1077.
14. Sharon D, Hamalainen M, Tootell R et al. (2007). The advantage of combining MEG and EEG: comparison to fMRI in focally stimulated visual cortex. NeuroImage, 36:1225-1235.
15. Riera J, Wan X et al. (2006). Nonlinear local electrovascular coupling I: A theoretical model. NeuroImage 27:896-914.
16. Sotero R & Trujillo-Barreto N (2008). Biophysical model for integrating neuronal activity, EEG, fMRI and metabolism. Neuroimage. 39:290-309.

Author: Richard Henson
Institute: MRC Cognition & Brain Sciences Unit
Street: 15 Chaucer Road
City: Cambridge
Country: England
Email: rik.henson@mrc-cbu.cam.ac.uk

Academic Software Toolboxes for the Analysis of MEG Data

S. Baillet[1], F. Tadel[2], R.M. Leahy[2], J.C. Mosher[3], A. Delorme[4,5], S. Makeig[5], R. Oostenveld[6], M. Hämäläinen[7], S.S. Dalal[8], J. Zumer[9], M. Clerc[10], C.H. Wolters[11], S. Kiebel[12], and O. Jensen[6,*]

[1] Department of Neurology, Medical College of Wisconsin, Milwaukee, USA
[2] Neuroimaing Group, University of Southern California, Los Angeles, USA
[3] Cleveland Clinic, Cleveland, USA
[4] CERCO, Universite Paul Sabatier/CNRS, Toulouse, France
[5] Swartz Center for Computational Neuroscience, Institute for Neural Computation, University of California San Diego, USA
[6] Donders Institute for Brain, Cognition and Behaviour, Radboud University Nijmegen, NL-6500 HB Nijmegen, The Netherlands
[7] Athinoula A. Martinos Center for Biomedical Imaging, Massachusetts General Hospital, Charlestown, MA, USA
[8] Dynamique Cérébrale et Cognition, INSERM Unité 821, Lyon, France
[9] Sir Peter Mansfield Magnetic Resonance Centre, School of Physics and Astronomy, University of Nottingham, University Park, Nottingham, NG7 2RD, UK
[10] INRIA Sophia Antipolis Méditerranée, 2004, route des Lucioles, BP 93, 06902 Sophia-Antipolis Cedex, France
[11] Institute for Biomagnetism and Biosignalanalysis, University of Muenster, 48149 Muenster, Germany
[12] Max Planck Institute for Human Cognitive and Brain Sciences, Leipzig, Germany
* Corresponding author.

Abstract— **Free MEG and EEG data analysis software packages springing from academic research are now widely used in published work. These toolboxes and applications are typically developed by or in close contact with researchers addressing cognitive or clinical neuroscience questions. Thus they often contain the latest methodological developments from the research community. It is therefore vital to educate MEG researchers and make them aware of the new possibilities offered by these toolboxes. The aim of this paper is to illustrate the characteristics and advantages of the various toolboxes to users and developers alike. We present each toolbox with their key features and target audience.**

Keywords— **MEG, source analysis, data analysis, imaging, mapping, connectivity, forward modeling.**

I. BRAINSTORM

BrainStorm is an open-source application developed using Matlab and dedicated to MEG and EEG data visualization, processing and cortical source estimation. For physicians and researchers, the interest of this software package resides in its rich and intuitive graphical interface, which does not require any scripting Matlab programming expertise. BrainStorm is an ongoing joint collaborative development between the University of Southern California (F. Tadel, D. Pantazis and R.M. Leahy), the Cleveland Clinic (J.C. Mosher) and the Medical College of Wisconsin (S. Baillet).

BrainStorm has the following key features:

1. Visualization and processing of EEG and MEG data:
   - Support for most of the common file formats for MEG, EEG and MRI;
   - Basic pre-processing: frequency filtering, resampling, epoching, baseline removal, averaging, etc;
   - Visualization of recordings: time series, 2D/3D topographies;
   - MRI/EEG/MEG co-registration;
   - Database and powerful database explorer (through subjects, studies, experimental conditions, etc.).
2. Source estimation
   - Forward modeling: spherical models, overlapping spheres, BEM;
   - Inverse modeling: efficient programming and data storage of weighted minimum-norm estimates, beamformers;
   - Visualization of sources: on 3D cortex surface, in original MRI volume (with or without maximum intensity projection glass brain techniques).
3. Data and source post-processing;
   - Definition of clusters (montages) at the sensor level;
   - Definition and easy management of regions of interests at the cortical level;
   - Group analysis, registration of individual brains on a common template through affine transformation
   - Statistical tests (contrasts): parametric and non-parametric (through permutation resampling) tests.

S. Supek and A. Sušac (Eds.): Advances in Biomagnetism – BIOMAG2010, IFMBE Proceedings 28, pp. 101–104, 2010.
www.springerlink.com

BrainStorm runs in Matlab. However, for users without a Matlab license, standalone binary executable versions are available. The toolbox can be downloaded from http://neuroimage.usc.edu/brainstorm

II. EEGLAB

The EEGLAB software environment is a widely used open source software resource for analysis and visualization of brain dynamics from electrophysiological data.

EEGLAB is designed to facilitate a major evolutionary step now taking place in the field of human electrophysiology. Human electroencephalographic (EEG), magnetoencephalographic (MEG), and invasive electrophysiological studies have long been dominated by simple time-locked event-related potential (ERP) and field (ERF) averaging methods. However, advances in modern signal processing can be applied to these data including time-frequency analysis, source localization onto magnetic resonance images of the head, three-dimensional, animated visualization methods, and spatial filtering including independent component analysis (ICA). These techniques are applied not only to averages but directly to the high-dimensional single-trial data. So far unexploited information about brain dynamics contained in human electrophysiological data is accessible to the methods provided by EEGLAB. Despite the ever-increasing affordability of high-speed computational resources that should allow every electrophysiological laboratory to perform sophisticated analyses on their data, the use of modern signal processing tools for these purposes is still rare, particularly for EEG analysis. The aim of EEGLAB is to make modern signal processing advances available to the users.

The freely available open source software toolbox EEGLAB runs under Matlab encompasses functions at three levels of user sophistication: a graphic user interface for exploratory signal processing, high-level and more low-level scripting methods for more experienced users, and a plug-in function facility that allows method innovators to distribute new or alternative analysis approaches to the EEG research community. EEGLAB can be freely download from http://sccn.ucsd.edu/eeglab

III. FIELDTRIP TOOLBOX FOR EEG/MEG ANALYSIS

The target users of FieldTrip are neuroscience researchers and method developers.

FieldTrip is a Free and Open-Source Matlab toolbox for MEG and EEG analysis that is since 2002 being developed at the Donders Institute for Brain, Cognition and Behaviour together with collaborating institutes. The toolbox provides functions for reading many EEG and all commercial MEG data formats, and for performing both simple and very advanced analyses methods, such as preprocssing, artifact removal, averaging, frequency and time-frequency decomposition using wavelets and multitapers, MEG and EEG source reconstruction using dipoles, distributed sources or beamformers, connectivity analysis using coherence or Granger connectivity, and statistical analysis on the channel- and source-level using parametric and non-parametric cluster-based randomization approaches. The toolbox includes many more functions which are all well structured with a clear end-user oriented interface. The toolbox does not have a graphical user interface, but instead works with well-structured scripts that are written by the end-user. These scripts facilitate standardized analysis pipelines, batch analyses of very large datasets and sharing of expertise within and across research groups. Furthermore the toolbox includes functionality for ad-hoc parallelization of the Matlab computations over a peer-to-peer cluster. The FieldTrip toolbox is currently used by an estimated 2500 people and has an active user community (email list and wiki). Besides for experimental neuroscientists, the toolbox is interesting for methods developers because of the clear structure of the Matlab code and existing support for many EEG, MEG and animal electrophysiology data formats. Parts of the functionality of the FieldTrip toolbox are shared and co-developped with the SPM and EEGLAB packages.

The FieldTrip toolbox requires Matlab version 7.1 or later, plus optionally a number of additional toolboxes for specialized functionality. It is tested and used on a daily basis by the developers on Windows, Linux and OSX. It is released under the GPL license. It can be found at http://fieldtrip.fcdonders.nl/

IV. MNE SOFTWARE

The target users of the MNE software are neuroscientists and clinicians.

The MNE software was developed at the Martinos Center of Massachusetts General Hospital to compute cortically-constrained L2 minimum-norm current estimates and associated dynamic statistical parametric maps from MEG and EEG data, optionally constrained by fMRI. This software includes MEG and EEG preprocessing tools, interactive and batch-mode modules for the forward and inverse calculations, as well as various data conditioning and data conversion utilities.

The architecture of MNE makes it ideal for batch processing in LINUX and Mac OSX environments. The interactive tools can be used to view the input data, averages, and analysis results while most of the analysis steps can be automated using standard UNIX scripting tools and run on a wide variety of hardware from laptops to multi-node computing clusters.

The MNE software also includes a Matlab toolbox composed entirely of m-files which facilitates access to the fif format data files and enables development of custom analysis tools.

MNE tools support analysis of data from all major MEG systems. The information regarding the signal-space projection (SSP) and signal-space separation (SSS) methods employed in Elekta-Neuromag systems and software gradient compensation using reference sensors for CTF and 4D Magnes data is extracted from the data files and taken into account in the analysis as appropriate. MNE software also includes both interactive and batch-mode to quickly create new SSP operators for suppression of environmental noise and biological artifacts.

The MNE tools have been written in C and Matlab. The binary executables available LINUX and Mac OS X. It can be downloaded from http://www.nmr.mgh.harvard.edu/martinos/userInfo/data/sofMNE.php

V. NUTMEG

NUTMEG is a source analysis toolbox geared towards cognitive neuroscience researchers using MEG or EEG. It is developed by the UCSF Biomagnetic Imaging Laboratory, its collaborators, and alumni diasporas in the UK, France, and Switzerland.

Evoked and single-trial data from a variety of MEG and EEG systems can be imported to the toolbox for source analysis in either the time or time-frequency domains. NUTMEG offers a choice of several beamformer variants, probabilistic models, and minimum-norm techniques to generate functional maps of spatiotemporal neural source activity. Lead fields can be calculated from single and multiple sphere head models or imported from other software. Group averages and statistics can be calculated as well. As a Matlab package, the end user may easily add customized functions or link with other toolboxes.

In addition to data analysis tools, NUTMEG provides a unique and intuitive graphical interface for visualization of results. Source analyses can be superimposed onto a structural MRI or headshape to provide a convenient visual correspondence to anatomy. These results can also be navigated interactively, with the spatial maps and source time series or spectrogram linked accordingly. Animations can be generated to view the evolution of neural activity over time. NUTMEG can also display brain renderings and perform spatial normalization of functional maps using SPM's engine.

The toolbox runs under Matlab and can be used with the Linux/UNIX, Mac OS X, and even Windows platforms. NUTMEG is available at http://nutmeg.berkeley.edu and is distributed under a BSD-style open source license.

VI. OPENMEEG

The OpenMEEG software is being developped at INRIA, France. Its target users are clinicians, and researchers in clinical and cognitive sciences, interested in inverse source localization from MEG and/or EEG. OpenMEEG implements the forward problem of MEG and EEG with the symmetric Boundary Element Method (state of the art in terms of accuracy).

The key features of the toolbox include

- Nested domains described by surfaces ;
- Combined discretization of electrical potential (piecewise linear) and normal current (piecewise constant) ;
- Very high performance in terms of accuracy;
- Can be used either standalone (command line interface, Python or Matlab scripting) or in integration with other packages (FieldTrip, ...)

Binary packages are available for GNU-Linux, Mac OS X and Windows. Open source code is distributed under the CeCILL B license. CMake/CPack are used for building and packaging the code. OpenMEEG can be found at http://openmeeg.gforge.inria.fr

VII. SIMBIO

The target users of SimBio are neuroscientists interested in EEG/MEG source analysis using realistic volume conductor modeling.

Within the Volkswagen-project CAUCHY (1992-1997), the EU-project SimBio (http://www.simbio.de, 2000-2003) and the DFG-projects FEMINVERS and EMOBEAMFEM (http://gepris.dfg.de/gepris/, search for C.H.Wolters), the freely available SimBio software (SimBio: A generic environment for bio-numerical simulations, https://www.mrt.uni-jena.de/simbio, 2009) was developed, and is currently used (and further developed) by more than 50 research groups worldwide. This cooperative work is enabled by means of two subversion repositories for both software and test data, by means of a wiki page (https://www.mrt.uni-jena.de/simbio) as well as an email-list (simbio-list@tu-ilmenau.de).

SimBio offers inverse algorithms such as dipole fitting, scanning methods such as goal function scan, MUSIC and SAM beamforming, current density reconstruction (CDR) approaches such as MNLS, LORETA, sLORETA and Spatio-Temporal (ST)-CDR on the brain surface or in its volume. The inverse EEG and MEG source analysis approaches can be combined with multi-compartment sphere, BEM or FEM volume conductor modeling. The modular structure of the toolbox enables an easy extension by new EEG/MEG forward or inverse methods, allowing an efficient comparison of different source analysis approaches. The key features of SimBio are the fast FEM volume conduction routines (SimBio-NeuroFEM). The software NeuroFEM-VGRID enables geometry-adapted hexahedra FE meshing, which, combined with an isoparametric FEM approach within NeuroFEM, leads to an accurate and efficient FEM forward modeling. SimBio is furthermore coupled to the constrained Delaunay tetrahedralization meshing code TETGEN and to the visualization environment SCIRun (http://www.sci.utah.edu/cibc/collaboration/technical/81-wolters.html), which also enables visualization also for the highest FE resolutions. SimBio is furthermore coupled to the commercial software packages ASA, BESA and CURRY.

The following platforms were successfully tested and detailed descriptions can be found on the internal webside https://www.mrt.uni-jena.de/simbio/index.php/Compiling_SimBio: Ubuntu 8.10 (Intrepid Ibex, 64-bit as well as 32-bit) ; Ubuntu 8.04 Hardy; Fedora 10; Gentoo (64-bit as well as 32-bit) ; Debian; Suse.

VIII. Statistical Parametric Mapping for MEG and EEG

The target users of this software package are neuroscientists. In its latest version, Statistical Parametric Mapping (SPM8) enables sophisticated analyses of MEG and EEG data. The functionality comprises classical analysis of sensor data, analysis of evoked and induced responses, Bayesian source reconstruction, statistical analysis of source reconstructed data, and the analysis of effective connectivity using Dynamic Causal Modelling (DCM). SPM uses both graphical user interfaces and scripts to allow for both step-by-step analysis and highly automated batch processing. The software is actively supported by an experienced developer team at the Wellcome Trust Centre for Neuroimaging, London, UK. An email list is used to resolve technical and analysis issues quickly, either by the developers or by experienced users. Novel developments are documented in both a manual and help texts, and reported in the peer-reviewed literature.

SPM8 runs under Matlab 7.xx on both Windows and Linux, 32 and 64bit, systems. It can be download from http://www.fil.ion.ucl.ac.uk/spm/

IX. Conclusions

This paper has documented the key features of the non-commercial software toolboxes available for the analysis of MEG data. Several of the toolboxes have been used to analyze MEG and EEG data published in high-ranking journals and it is clear that the importance of these tools continues to increase. The different toolboxes have different strengths and features and it is an advantage that they coexist. In the future we expect more cross fertilization and that the developers will share the strong features from their respective toolboxes.

The Adjoint Method for General EEG and MEG Sensor-Based Lead Field Equations

Théodore Papadopoulo, Sylvain Vallaghé, and Maureen Clerc

Odyssée Project-Team, INRIA Sophia Antipolis, France

Abstract— **The inverse source problem in electroencephalography (EEG) and magnetoencephalography (MEG) generally uses a lead field, which relates any source in the brain to its measurements at the sensors. For complex geometries, there is no analytical formula of the lead field. The common approach is to numerically compute the value of the lead field for a finite number of point sources (dipoles). There are several drawbacks: the model of the source space is fixed (a set of dipoles) and the computation can be expensive for as much as 10 000 dipoles. The common idea to bypass these problems is to compute the lead field from a sensor point of view. We show how the adjoint method can be used to derive general EEG and MEG sensor-based lead field equations. Within a simple framework, we provide a complete review of the explicit lead field equations, and we are able to extend these equations to non-point like sensors.**

Keywords— **Electroencephalography, Magnetoencephalography, Forward problem, Adjoint method.**

I. Introduction

Electroencephalography (EEG) and magnetoencephalography (MEG) inverse problems traditionally optimizes a source configuration by "comparing" actual measurements to quantities simulated using a forward model that predicts the electric potential or magnetic field. The EEG forward problem is governed by the following partial differential equation:

$$\begin{cases} \nabla\cdot(\sigma\nabla V) & = \nabla\cdot\mathbf{J}^{\mathrm{p}} \quad in \quad \Omega \\ \sigma\nabla V\cdot\mathbf{n} & = 0 \qquad\quad on \quad \partial\Omega \end{cases} \tag{1}$$

where V is the electric potential, σ is the conductivity, $\mathbf{J}^{\mathrm{p}}$ is the primary current vector representing brain electrical activity. Ω represents the head domain (an open bounded region of the ambient space $\mathbb{R}^3$). Similarly, the magnetic field is given by the Biot-Savart law:

$$\mathbf{B}(\mathbf{r}_i) = \mathbf{B}_0(\mathbf{r}_i) - \frac{\mu_0}{4\pi}\int_\Omega \sigma\nabla V \times \nabla\left(\frac{1}{R}\right) dr\,, \tag{2}$$

with $\mathbf{B}_0(\mathbf{r}_i) = \frac{\mu_0}{4\pi}\int_\Omega \mathbf{J}^{\mathrm{p}} \times \nabla\left(\frac{1}{R}\right) dr$ and $R = \|\mathbf{r}_i - \mathbf{r}\|$. It is obvious that both V and $\mathbf{B}$ are linear functions of $\mathbf{J}^{\mathrm{p}}$. Because of this property, the predictions of the forward problem at the sensor locations (the only information used by inverse problems) can be conveniently summarized by a leadfield, which is a linear operator $\mathbf{L}$ that maps source activations to simulated potentials of magnetic fields at sensor locations. Since it is linear, this operator can be written using the Riesz representation theorem:

$$\mathbf{L}(\mathbf{J}^{\mathrm{p}}) = < \frac{\partial\mathbf{L}}{\partial\mathbf{J}^{\mathrm{p}}}, \mathbf{J}^{\mathrm{p}} >\,, \tag{3}$$

where $<,>$ denotes for the inner product[1] of functions in $\mathbb{R}^3$:

$$< u, v > = \int_\Omega u(\mathbf{r})\cdot v(\mathbf{r}) dr\,.$$

Traditionally, the forward problem equations are discretized and the leadfield is represented by the matrix $\frac{\partial\mathbf{L}}{\partial\mathbf{J}^{\mathrm{p}}}$ whose number of lines equals the number of sensors and whose number of columns equals the number of parameters of the dipoles considered as possible sources (around 10 000 in distributed source models). Each column of this matrix gives the values of the electric or magnetic measurements at the sensor locations for a given dipole parameter. A simple way to compute the leadfield matrix is thus to solve as many forward problems as there are source parameters, which can be quite time consuming with detailed realistic geometries for the head. One solution to bypass this problem relies on some matrix manipulations on the discretized problem so as to estimate rows of the leadfield matrix (instead of columns) [1, 2]. This is quite interesting as there are generally much less sensors than source positions and as the resulting computation is very similar to the original forward problem. Another solution is to work in the continuous domain, and to use the Helmholtz reciprocity principle to express the leadfields as the solutions of different forward problems [1, 3, 4].

Both these approaches can been explained using the adjoint method [5], which is applied here in the continuous domain to both the EEG and MEG leadfields. As the approach is very general, it is quite easy to include within this framework new information such as sensor geometry.

[1] For scalar functions the inner product inside the integral becomes a simple product.

S. Supek and A. Sušac (Eds.): Advances in Biomagnetism – BIOMAG2010, IFMBE Proceedings 28, pp. 105–108, 2010.
www.springerlink.com

II. The adjoint method

Let p represent a parameter which can be a real scalar or vector function. For any p, we consider the (hopefully unique) function v which satisfies the state equation:

$$\mathbf{A}v = b(p) , \quad (4)$$

where $\mathbf{A}$ is a linear operator of $\mathbb{R}^3$ and b is a differentiable function of $\mathbb{R}^3$. To give some intuition, in our case the state function v is the electric potential $V(\mathbf{r})$, the parameter p is the primary source current vector $\mathbf{J}^\mathrm{p}(\mathbf{r})$, the operator $\mathbf{A}$ is the generalized Laplacian $\nabla \cdot (\sigma\nabla .)$ and b is the divergence operator.

Let us consider a "measurement function" g which is a linear functional of p and v. From the Riesz representation theorem, there exists m and $n \in E$ such that $g(p,v) = < m,v > + < n,p >$. To compute the differential of g with respect to p (knowing that v also depends linearly on p), we first build a Lagrangian $\mathscr{L}$ by adding to the measurement equation the inner product of the state equation with a Lagrange multiplier w:

$$\mathscr{L}(p,v,w) = < m,v > + < n,p > + < w, \mathbf{A}v - b(p) > .$$

Notice that $g(p,v(p)) = \mathscr{L}(p,v(p),w)$ for all w as $\mathbf{A}v(p) = b(p)$. Assuming that $\mathscr{L}$ is differentiable with respect to both v and p, the Lagrange multiplier w is chosen such that $\frac{\partial \mathscr{L}}{\partial v}(p,v(p),w)\delta v = < m, \delta v > + < w, \mathbf{A}\delta v > = 0$ for all δv . Introducing the adjoint operator $\mathbf{H}^*$ of a linear operator $\mathbf{H}$, such that $< x, \mathbf{H}y > = < \mathbf{H}^* x, y >$, the previous equation can be rewritten as $< m + \mathbf{A}^* w, \delta v > = 0$ for all δv, which is equivalent to:

$$\mathbf{A}^* w = -m . \quad (5)$$

This equation is called the adjoint equation. With this choice for w, the differential of g can be written as:

$$\delta g = \frac{\partial \mathscr{L}}{\partial p}(p,v(p),w)\delta p = < n - \frac{\partial b}{\partial p}^* w, \delta p > . \quad (6)$$

Since g is linear in p, $g(p) = < \frac{\partial g}{\partial p}, p >$. $g(p)$ being a measurement, $\frac{\partial g}{\partial p}$ is exactly the leadfield $\mathbf{L}$ and Eq. (6) exactly expresses this operator as a function of the adjoint variable w:

$$\mathbf{L} = n - \frac{\partial b}{\partial p}^* w . \quad (7)$$

III. Application to M/EEG leadfields

This section briefly details the application of the previous result to M/EEG leadfields. For more detailed calculations, the reader is referred to [6]. Using the previous result first requires the computation of the adjoint operators associated to our state equation (1) (i.e. $\mathbf{A}^*$ and $\frac{\partial b}{\partial p}^*$). In our case, v is the potential $V(\mathbf{r})$, p is the primary source $\mathbf{J}^\mathrm{p}(\mathbf{r})$ and b is the divergence operator, which is linear so that $\frac{\partial \nabla \cdot \mathbf{J}^\mathrm{p}}{\partial \mathbf{J}^\mathrm{p}} \mathbf{J}^\mathrm{p} = \nabla \cdot \mathbf{J}^\mathrm{p}$. Using the divergence theorem:

$$< w, \nabla \cdot \mathbf{J}^\mathrm{p} > = \int_\Omega w \nabla \cdot \mathbf{J}^\mathrm{p} dr = -\int_\Omega \mathbf{J}^\mathrm{p} \cdot \nabla w dr = < -\nabla w, \mathbf{J}^\mathrm{p} > ,$$

because sources are restricted to be strictly inside the head, so the boundary integrals vanish. Thus:

$$\frac{\partial b}{\partial p}^* w = \frac{\partial \nabla \cdot \mathbf{J}^\mathrm{p}}{\partial \mathbf{J}^\mathrm{p}}^* w = -\nabla w . \quad (8)$$

For $\mathbf{A}^*$, we start with $< w, \mathbf{A}v >$:

$$< w, \mathbf{A}v > = \int_\Omega w \nabla \cdot (\sigma \nabla V) dr .$$

Assuming that w is continuous on Ω, but that $\sigma \nabla w \cdot \mathbf{n}$ can be discontinuous at tissue interfaces, the divergence theorem applied twice leads to (using the facts that $\sigma \nabla V \cdot \mathbf{n} = 0$ on $\partial\Omega$ and that V and $\sigma \nabla V \cdot \mathbf{n}$ are continuous functions over Ω):

$$\begin{aligned} < A^* w, v > &= \int_\Omega w \nabla \cdot (\sigma \nabla V) dr \\ &= \int_\Omega V \nabla \cdot (\sigma \nabla w) dr - \sum_k \int_{S_k} V [\sigma \nabla w \cdot \mathbf{n}]_{S_k} ds , \end{aligned} \quad (9)$$

where S_k are the tissue interfaces and $[.]_{S_k}$ denotes the jump of a function on a given surface S_k.

A. EEG Leadfield

One line of the EEG leadfield is the potential of an electrode located at point $\mathbf{r}_i$ with respect to some reference electrode located at $\mathbf{r}_0$ (mean reference can also easily be handled with slightly more complicated formulae). Computing a leadfield for dipoles with fixed orientations can also be done easily following the same principle.

$$\begin{aligned} g_{EEG}(p,v) &= g_{EEG}(\mathbf{J}^\mathrm{p}, V) = V(\mathbf{r}_i) - V(\mathbf{r}_0) \\ &= \int_{\partial\Omega} V(\mathbf{r})(\delta(\mathbf{r}_i) - \delta(\mathbf{r}_0)) dr , \end{aligned}$$

where the notation $\delta(\mathbf{r})$ refers to the Dirac distribution at $\mathbf{r} \in \partial\Omega$. The integral is restricted to $\partial\Omega$ since all the EEG sensors are located on this surface. As $g(\mathbf{J}^\mathrm{p}, V)$ does not depend on $\mathbf{J}^\mathrm{p}$, this leads to $m_{EEG} = \delta(\mathbf{r}_i) - \delta(\mathbf{r}_0)$ and $n_{EEG} = 0$. Equating all the surfacic and volumic terms in the linear forms shows that $[\sigma_k \nabla w \cdot \mathbf{n}]_{S_k} = 0$ for all inner interfaces (as there are no other surfacic terms for those), hence only the last surface ($\partial\Omega$) is to be kept in the sum in Eq. (9):

$$< \mathbf{A}^* w, V >= \int_\Omega V \nabla \cdot (\sigma \nabla w) dr - \int_{\partial\Omega} V[\sigma \nabla w \cdot \mathbf{n}]_{\partial\Omega} ds \,.$$

From Eq. (5), w satisfies:

$$\left| \begin{cases} \nabla \cdot (\sigma \nabla w) & = 0 \quad\quad in \quad \Omega \\ \sigma \nabla w \cdot \mathbf{n} & = \delta_{\mathbf{r}_i} - \delta_{\mathbf{r}_0} \quad on \quad \partial\Omega \end{cases} \right| \qquad (10)$$

Combining Eq. (7) and Eq. (8), the EEG leadfield for punctual electrodes is:

$$\boxed{\mathbf{L}_{EEG} = \nabla w} \qquad (11)$$

B. MEG Leadfield

For MEG, the only difference is in the type of measurements: $g_{MEG}(p, v) = g_{MEG}(\mathbf{J}^p, V) = \mathbf{B}(\mathbf{r}_i) \cdot \mathbf{d}_i$, with $\mathbf{B}$ given by the Biot-Savart law (2). This law has two terms, $\mathbf{B}_0$ depending on $\mathbf{J}^p$ and the other term depending on V. Applying the adjoint state approach requires the computation of the corresponding m_{MEG} and n_{MEG}.

n_{MEG} corresponds to the $\mathbf{B}_0$ term:

$$\begin{aligned} \mathbf{B}_0(\mathbf{r}_i) \cdot \mathbf{d}_i &= \frac{\mu_0}{4\pi} \int_\Omega \left(\mathbf{J}^p \times \nabla \left(\frac{1}{R} \right) \right) \cdot \mathbf{d}_i \, dr \\ &= \frac{\mu_0}{4\pi} \int_\Omega \mathbf{J}^p \cdot \left(\nabla \left(\frac{1}{R} \right) \times \mathbf{d}_i \right) dr \,. \end{aligned}$$

Thus

$$n_{MEG} = \frac{\mu_0}{4\pi} \left(\nabla \left(\frac{1}{R} \right) \times \mathbf{d}_i \right) \,. \qquad (12)$$

m_{MEG} corresponds to the Ohmic current term:

$$\begin{aligned} < m_{MEG}, \mathbf{J}^p > &= -\frac{\mu_0}{4\pi} \int_\Omega \sigma \nabla V \cdot \left(\nabla \left(\frac{1}{R} \right) \times \mathbf{d}_i \right) dr \\ &= \frac{\mu_0}{4\pi} \int_\Omega V \nabla \cdot \sigma \left(\left(\frac{1}{R} \right) \times \mathbf{d}_i \right) dr - \\ &\quad \frac{\mu_0}{4\pi} \int_\Omega \nabla \cdot V \sigma \left(\left(\frac{1}{R} \right) \times \mathbf{d}_i \right) dr \\ &= \frac{\mu_0}{4\pi} \int_\Omega V \nabla \cdot \sigma \left(\left(\frac{1}{R} \right) \times \mathbf{d}_i \right) dr - \\ &\quad \frac{\mu_0}{4\pi} \sum_k \int_{S_k} V[\sigma]_{S_k} \left(\left(\frac{1}{R} \right) \times \mathbf{d}_i \right) \cdot \mathbf{n} ds \,. \end{aligned}$$

Thus m_{MEG} has both a volumic term and several surfacic terms. Equating all the surfacic and volumic terms in the linear forms yields the following equation for the adjoint:

$$\left| \begin{cases} \nabla \cdot (\sigma \nabla w) & = -\frac{\mu_0}{4\pi} \nabla \cdot \left(\sigma \nabla \left(\frac{1}{R} \right) \times \mathbf{d}_i \right) \quad in \ \Omega_k, \\ [\sigma \nabla w \cdot \mathbf{n}]_{S_k} & = -\frac{\mu_0}{4\pi} [\sigma]_{S_k} \nabla \left(\frac{1}{R} \right) \times \mathbf{d}_i \cdot \mathbf{n} \quad on \ S_k \end{cases} \right|$$

Note that if the conductivity σ is constant and isotropic in a domain Ω_k, then $\nabla \cdot \left(\sigma \nabla \left(\frac{1}{R} \right) \times \mathbf{d}_i \right) = 0$ and w is harmonic in each domain Ω_k. This result has been used previously for the numerical computation of the MEG leadfield [3, 4].

Combining Eq. (7), Eq.(12) and Eq. (8), the MEG leadfield for a punctual magnetometer is:

$$\left| \mathbf{L}_{MEG} = \nabla w + \frac{\mu_0}{4\pi} \left(\nabla \left(\frac{1}{R} \right) \times \mathbf{d}_i \right) \,. \right|$$

IV. Incorporating sensor geometry

A. MEG: Squids Geometry

A magnetometer measures the flux of the magnetic field through a small loop. For a magnetometer i, let M_i be the surface enclosed by the loop, and $\mathbf{d}_i$ the unitary vector normal to M_i. The function g can thus be written as $g(\mathbf{J}^p) = \int_{M_i} \mathbf{d}_i \cdot \mathbf{B}(\mathbf{r}')dr'$. Switching the integrals over M_i and Ω and since the only dependence of $\mathbf{B}(\mathbf{r}')$ on $\mathbf{r}'$ is in R, it is quite easy to show that following the lines of section III.B yields:

$$\left| \begin{cases} \nabla \cdot (\sigma \nabla w) & = -\frac{\mu_0}{4\pi} \int_{M_i} \nabla \cdot \left(\sigma \nabla \left(\frac{1}{R} \right) \times \mathbf{d}_i \right) dr' \quad in \ \Omega_k, \\ [\sigma \nabla w \cdot \mathbf{n}]_{S_k} & = -\frac{\mu_0}{4\pi} [\sigma]_{S_k} \nabla \left(\frac{1}{R} \right) \times \mathbf{d}_i \cdot \mathbf{n} \quad on \ S_k \end{cases} \right|$$

As in the point wise case, the term $\int_{M_i} \nabla \cdot \left(\sigma \nabla \left(\frac{1}{R} \right) \times \mathbf{d}_i \right) dr'$ vanishes if σ is constant and isotropic in a domain Ω_k. The resulting expression for the leadfield is:

$$\left| \mathbf{L}_{MEG} = \nabla w + \frac{\mu_0}{4\pi} \int_{M_i} \left(\nabla \left(\frac{1}{R} \right) \times \mathbf{d}_i \right) d\mathbf{r}' \,. \right|$$

This formulation is the same as the one given in [2]. Gradiometers can be treated similarly. Generally, MEG manufacturers give a set of positions and weights for each sensor, and the linear combination of the magnetic field at these positions using these weights is meant to recreate the measurement. But this requires as many adjoint problems as there are positions. The main advantage of the proposed approach is to require a single adjoint problem per sensor, reducing notably the needed for the leadfield computation.

B. EEG: Electrode Surfaces

Incorporating the surface of EEG electrodes in the model is more complicated as the electrode perturbs the electrical field due to the *shunt effect*: potential is constant at the surface of the electrodes due to the high conducting electrode (gel and metal). This effect has been modeled for EEG [7] and requires a modification of the boundary condition of the state equation (1), which becomes:

$$\begin{cases} V + z_k\sigma\nabla V\cdot\mathbf{n} & = v_k \quad on \quad e_k \\ \int_{e_k}\sigma\nabla V\cdot\mathbf{n}\,dr & = 0 \\ \sigma\nabla V\cdot\mathbf{n} & = 0 \quad on \quad S\backslash\cup e_k \end{cases} \tag{13}$$

where e_k is the k-th electrode, v_k is the constant value of the potential on the k-th electrode and z_k is the effective contact impedance which models the electrochemical effect at the skin-electrode interface. Following the adjoint approach yields the same leadfield equation as in the pointwise EEG case (Eq. (11)). However, the boundary condition of the adjoint state equation for w is transformed into:

$$\begin{cases} w + z_k\sigma\nabla w\cdot\mathbf{n} = W_k \quad on \quad e_k, \\ \int_{e_k}\sigma\nabla w\cdot\mathbf{n}\,dr = \begin{cases} 1 & k=i \\ -1 & k=0 \\ 0 & otherwise \end{cases}, \\ \sigma\nabla w\cdot\mathbf{n} = 0 \quad on \quad S\backslash\cup e_k\ , \end{cases} \tag{14}$$

where W_k is a constant.

V. Results

To illustrate the approach, lead field for magnetometers are computed for a standard three nested spheres model with radii of 0.87, 0.92, 1, meant to represent brain, skull and scalp tissues with respective conductivities of 1, 0.02, 1. While it is known that in a spherical geometry the magnetic field outside the conductor does not depend on the conductivities, we intentionally put different conductivities in our model to test that the numerical solution is actually similar to the case of a homogeneous sphere. Computations were made using a finite element method presented in [8] and results were compared with the ground truth given by the analytical formulation. The magnetometer is oriented in the x direction, positioned at a distance of 0.03 of the outermost sphere, and has a radius of 0.015. The differential equation of section IV A was solved, using our finite element method. The integral term over the magnetometer loop was computed after applying Stockes theorem with a Gauss-Kronod method using 61 points. Figure 1 shows the absolute error is similar and small for three dipole orientations. More results can be found in [6].

VI. Conclusion

The adjoint method is a powerful general tool to derive efficient leadfield computation for various sensors types.

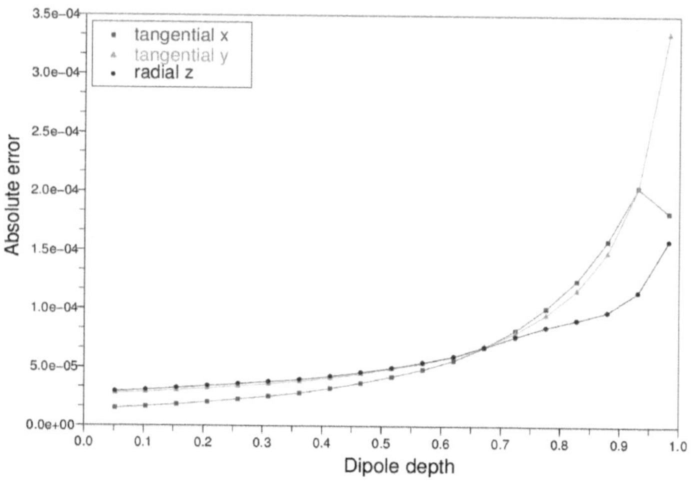

Fig. 1 Absolute error between numerical and analytical solution with respect to dipole depth

Acknowledgements

This work is partly supported by contract EADS 2118.

References

1. Weinstein D., Zhukov L., Johnson C.. Lead-field bases for electroencephalography source imaging *Annals of Biomedical Engineering.* 2000;28:1059-1164.
2. Wolters C. H., Grasedyck L., Hackbusch W.. Efficient computation of lead field bases and influence matrix for the FEM-based EEG and MEG inverse problem *Inverse Problems.* 2004;20:1099–1116.
3. Nolte G.. The magnetic lead field theorem in the quasi-static approximation and its use for magnetoencephalography forward calculation in realistic volume conductors *Physics in Medicine and Biology.* 2003;48:3637–3652.
4. Schimpf P. H.. Application of quasi-static magnetic reciprocity to finite element models of the MEG lead-field *IEEE Transactions on Biomedical Engineering.* 2007;54:2082-2088.
5. Lions J.L.. *Optimal control of systems governed by partial differential equations.* Springer 1971.
6. Vallaghé S., Papadopoulo T., Clerc M.. The adjoint method for general EEG and MEG sensor-based lead field equations *Physics in Medicine and Biology.* 2009;54:135–147.
7. Ollikainen Jorma O., Vauhkonen Marko, Karjalainen Pasi A., Kaipio Jari P.. Effects of electrode properties on EEG measurements and a related inverse problem *Medical Engineering & Physics.* 2000;22:535–545.
8. Papadopoulo Théodore, Vallaghé Sylvain. Implicit Meshing for Finite Element Methods using Levelsets in *Proceedings of MMBIA 07* 2007.

Author: Théo Papadopoulo
Institute: INRIA Sophia-Antipolis
Street: 2004 route des Lucioles, BP 93
City: Sophia-Antipolis, Cedex
Country: France
Email: Theodore.Papadopoulo@sophia.inria.fr

The Symmetric BEM: Bringing in More Variables for Better Accuracy

Maureen Clerc[1], Alexandre Gramfort[1,2], Emmanuel Olivi[1], and Theo Papadopoulo[1]

[1] INRIA Odyssée Project-Team, Sophia Antipolis, France
[2] INRIA Parietal Project Team, Saclay, France

Abstract— **Electrophysiological modeling of Magneto- and Electro-encephalography (MEG and EEG) rely on accurate forward solvers that relate source activities to sensor measurements. In comparing a Boundary Element (BEM) and a Finite Element Method (FEM) for forward electroencephalography, in our early numerical experiments, we found the FEM to have a better accuracy than the BEM. This triggered a quest to improve the accuracy of Boundary Element Methods and led us to study the extended Green representation theorem.**

A fundamental result in potential theory shows that, up to an additive constant, a harmonic function is determined within a domain from its value on the boundary (Dirichlet condition), or the value of its normal derivative (Neumann condition). The Green Representation Theorem has been used in forward EEG and MEG modeling, in deriving the Geselowitz BEM formulation, and the Isolated Problem Approach. The extended Green Representation Theorem provides a representation for the directional derivatives of a piecewise-harmonic function.

By introducing the normal current as an additional variable in the forward problem, we derive a new Boundary Element Method, which leads to a symmetric matrix structure: we hence call it the Symmetric BEM. Accuracy comparisons demonstrate the superiority of the Symmetric BEM to the FEM and to the classical BEM.

Keywords— **Boundary Element Method, Electroencephalography, Magnetoencephalography, Forward Problem.**

I. Introduction

Electrophysiological modeling of MEG and EEG rely on forward solvers that relate the source activity to the sensor measurements. In comparing a Boundary Element Method (BEM) and a Finite Element Method (FEM) for forward electroencephalography, in our early numerical experiments, we found a superior accuracy to the FEM [1]. This triggered a quest to improve the accuracy of Boundary Element Methods and led us to study the extended Green representation theorem [2]. We proposed a common formalism for the integral formulations of the forward EEG problem, and we derived three different Boundary Element Methods within the same framework [3]. In this paper we recall the mathematical background of Boundary Element Methods, and we present both the double-layer BEM, which is the most widespread method, and the Symmetric BEM, which is a new formulation. Numerical comparisons on a three-layer spherical model show the superiority of the Symmetric BEM in terms of accuracy, compared to the double-layer BEM.

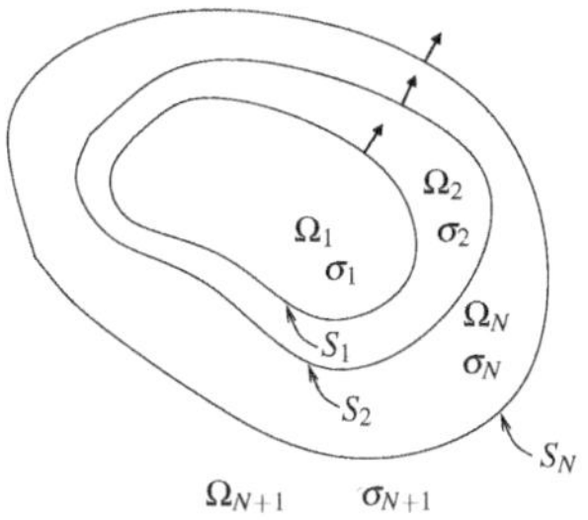

Fig. 1 Boundary Elements are well-suited for piecewise constant isotropic conductivity models. The head, denoted Ω, is here modeled as a set of nested domains Ω_i, $i = 1 \dots N$, each with a constant conductivity σ_i. The outer domain Ω_{N+1}, representing the air, is non-conductive ($\sigma_{N+1} = 0$)

II. Green Representation

A fundamental result in potential theory shows that, up to an additive constant, a harmonic function (i.e., such that $\Delta u = 0$) is uniquely determined within a domain Ω from its value on the boundary $\partial\Omega$ (Dirichlet condition), or the value of its normal derivative (Neumann condition). The Green Representation Theorem gives an explicit representation of a piecewise-harmonic function as a combination of boundary integrals of its jumps and the jumps of its normal derivative across interfaces. Before stating this theorem, some notation must be defined:

- The restriction of a function f to a surface S_j is indicated by f_{S_j}.
- The functions $f_{S_j}^-$ and $f_{S_j}^+$ represent the interior and exterior limits of f on S_j:

$$\text{for } \mathbf{r} \in S_j, \qquad f_{S_j}^{\pm}(\mathbf{r}) = \lim_{\alpha \to 0^{\pm}} f(\mathbf{r} + \alpha \mathbf{n}).$$

- The jump of a function f across S_j is denoted by:

$$[f]_{S_j} = f_{S_j}^- - f_{S_j}^+,$$

S. Supek and A. Sušac (Eds.): Advances in Biomagnetism – BIOMAG2010, IFMBE Proceedings 28, pp. 109–112, 2010.
www.springerlink.com

- $\partial_{\mathbf{n}} V = \mathbf{n} \cdot \nabla V$ denotes the partial derivative of V in the direction of a unit vector $\mathbf{n}$,
- The function $G(\mathbf{r}) = \frac{1}{4\pi\|\mathbf{r}\|}$ is the fundamental solution of the Laplacian in $\mathbb{R}^3$, such that $-\Delta G = \delta_0$.

Consider an open region Ω and a function u such that $\Delta u = 0$ in Ω and in $\mathbb{R}^3 \backslash \Omega$ (but not necessarily continuous across $\partial\Omega$). The Green Representation Theorem states that, for a point $\mathbf{r}$ belonging to $\partial\Omega$,

$$\frac{u^-(\mathbf{r})+u^+(\mathbf{r})}{2} = -\int_{\partial\Omega} [u]\, \partial_{\mathbf{n}'} G(\mathbf{r}-\mathbf{r}') ds(\mathbf{r}') + \int_{\partial\Omega} [\partial_{\mathbf{n}'} u]\, G(\mathbf{r}-\mathbf{r}') ds(\mathbf{r}') \,. \quad (1)$$

This representation also holds for the head model in Figure 1, when Ω is the union of disjoint open sets: $\Omega = \Omega_1 \cup \Omega_2 \cup \ldots \Omega_N$, with $\partial\Omega = S_1 \cup S_2 \cup \ldots S_N$. If u is harmonic in each Ω_i, for $\mathbf{r} \in S_i$,

$$\frac{u^-(\mathbf{r})+u^+(\mathbf{r})}{2} = -\sum_{j=1}^{N} \int_{S_j} [u]_{S_j}\, \partial_{\mathbf{n}'} G(\mathbf{r}-\mathbf{r}') ds(\mathbf{r}') + \int_{S_j} [\partial_{\mathbf{n}'} u]_{S_j}\, G(\mathbf{r}-\mathbf{r}') ds(\mathbf{r}') \quad (2)$$

The notation is simplified by introducing two integral operators, which map a scalar function f on $\partial\Omega$ to another scalar function on $\partial\Omega$: the "double-layer" operator $(\mathcal{D}f)(\mathbf{r}) = \int_{\partial\Omega} \partial_{\mathbf{n}'} f(\mathbf{r}')\, G(\mathbf{r}-\mathbf{r}')\, ds(\mathbf{r}')$, and the "single-layer" operators $(\mathcal{S}f)(\mathbf{r}) = \int_{\partial\Omega} f(\mathbf{r}') G(\mathbf{r}-\mathbf{r}')\, ds(\mathbf{r}')$. The restriction $\mathcal{D}_{ij}$ of an operator $\mathcal{D}$ maps a function of S_j to a function of S_i.

An extension of the Green Representation Theorem represents the directional derivative of a harmonic function as a combination of boundary integrals of higher order. This requires two more integral operators: the adjoint $\mathcal{D}^*$ of the double-layer operator, such that

$$\int_S \mathcal{D}^* f(\mathbf{r}) g(\mathbf{r}) ds(\mathbf{r}) = \int f(\mathbf{r}) \mathcal{D} g(\mathbf{r}) ds(\mathbf{r}) \,,$$

and a hyper-singular operator $\mathcal{N}$ defined by $(\mathcal{N}f)(\mathbf{r}) = \int_{\partial\Omega} f(\mathbf{r}') \partial^2_{\mathbf{n},\mathbf{n}'} G(\mathbf{r}-\mathbf{r}')\, ds(\mathbf{r}')$. For a point $\mathbf{r}$ belonging to $\partial\Omega$,

$$-\frac{\partial_{\mathbf{n}} u^-(\mathbf{r}) + \partial_{\mathbf{n}} u^+(\mathbf{r})}{2} = +\mathcal{N}[u] - \mathcal{D}^*[\partial_{\mathbf{n}} u]$$

and for a nested head model, if $\mathbf{r}$ is a point of S_i,

$$-\frac{\partial_{\mathbf{n}} u^-(\mathbf{r}) + \partial_{\mathbf{n}} u^+(\mathbf{r})}{2} = \sum_{j=1}^{N} \mathcal{N}_{ij} [u]_{S_j} - \sum_{j=1}^{N} \mathcal{D}^*_{ij} [\partial_{\mathbf{n}} u]_{S_j} \quad (3)$$

As we shall see in the next section, the Geselowitz formula exploits only the first boundary integral representation equation (2), while it is possible to exploit both (2) and (3). Thus, three Boundary Element Methods can be derived within a unified setting: a BEM involving only single-layer potentials, a BEM involving only double-layer potentials, and a BEM combining single- and double-layer potentials [3]. We concentrate hereforth on the latter two BEMs.

III. Forward EEG and MEG

To apply the representation theorem to the forward problem of EEG, a harmonic function must be produced, which relates the potential and the sources.

A. The Double-Layer BEM

Decomposing the source term as $f = \sum_i f_i$ where the support of each f_i lies inside Ω_i, consider v_{Ω_i} such that $\Delta v_{\Omega_i} = f_i$ holds in all $\mathbb{R}^3$. The function $v_d = \sum_{i=1}^{N} v_{\Omega_i}$ satisfies $\Delta v_d = f$ and is continuous across each surface S_i, as well as its normal derivative $\partial_{\mathbf{n}} v_d$. The function $u = \sigma V - v_d$ is a harmonic function in Ω, to which (2) can be applied. Since $[u]_{S_i} = (\sigma_i - \sigma_{i+1}) V_j$ and $[\partial_{\mathbf{n}} u] = 0$, we obtain, on each surface S_i,

$$\frac{\sigma_i + \sigma_{i+1}}{2} V_j + \sum_{j=1}^{N} (\sigma_j - \sigma_{j+1}) \mathcal{D}_{ij} V_j = v_d \,. \quad (4)$$

The above formula was established by Geselowitz [4], and was the only one used to model electroencephalography or electrocardiography, until recently, when [3] showed the diversity of BEMs that can be derived. This classical BEM is called a double-layer BEM because it only involves the double-layer operator $\mathcal{D}$.

As such, the double-layer BEM shows poor accuracy when there are large variations in conductivity between neighboring regions (such as the skull and the scalp). For this, a modified formulation, called the Isolated Problem Approach (or alternatively, Isolated Skull Approach), has been proposed [5]. This formulation is the most widespread in the forward BEM, and will be used in Numerical Results Section IV.

B. The Symmetric BEM

The originality of the Symmetric Boundary Element Method is to consider a different piecewise harmonic function for each domain Ω_i: a function u_{Ω_i} equal to $V - \frac{v_{\Omega_i}}{\sigma_i}$

within Ω_i and to $-\frac{v_{\Omega_i}}{\sigma_i}$ outside of Ω_i. This u_{Ω_i} is indeed harmonic in $\mathbb{R}^3 \backslash \partial\Omega_i$, and the representation equations (2) and (3) can be applied, leading to a system of integral equations involving two types of unknowns: the potential V_{S_i} and the normal current $p_{S_i} = (\sigma \partial_{\mathbf{n}} V)_i$ on each interface.

Evaluating the internal limit of u_{Ω_i} on S_i and the external limit of $_{\Omega_{i+1}}$ on S_i via the extended Representation Theorem, and subtracting the two representations gives, for $i = 1, \ldots, N$:

$$\sigma_{i+1}^{-1}(v_{\Omega_{i+1}})_{S_i} - \sigma_i^{-1}(v_{\Omega_i})_{S_i} = \mathcal{D}_{i,i-1}V_{S_{i-1}} - 2\mathcal{D}_{ii}V_{S_i} + \mathcal{D}_{i,i+1}V_{S_{i+1}} - \sigma_i^{-1}\mathcal{S}_{i,i-1}p_{S_{i-1}} + (\sigma_i^{-1} + \sigma_{i+1}^{-1})\mathcal{S}_{ii}p_{S_i} - \sigma_{i+1}^{-1}\mathcal{S}_{i,i+1}p_{S_{i+1}} \,. \quad (5)$$

Using the same approach, the quantities $\left(\sigma_i \partial_{\mathbf{n}} u_{\Omega_i}\right)_{S_i}^{-} = \left(p - \partial_{\mathbf{n}} v_{\Omega_i}\right)_{S_i}^{-}$ and $\left(\sigma_{i+1} \partial_{\mathbf{n}} u_{\Omega_{i+1}}\right)_{S_i}^{+} = \left(p - \partial_{\mathbf{n}} v_{\Omega_{i+1}}\right)_{S_i}^{+}$ can be evaluated using the extended Representation Theorem. Subtracting the two representations yields

$$(\partial_{\mathbf{n}} v_{\Omega_{i+1}})_{S_i} - (\partial_{\mathbf{n}} v_{\Omega_i})_{S_i} = \sigma_i \mathcal{N}_{i,i-1} V_{S_{i-1}} - (\sigma_i + \sigma_{i+1}) \mathcal{N}_{ii} V_{S_i} + \sigma_{i+1} \mathcal{N}_{i,i+1} V_{S_{i+1}} - \mathcal{D}^*_{i,i-1} p_{S_{i-1}} + 2\mathcal{D}^*_{ii} p_{S_i} - \mathcal{D}^*_{i,i+1} p_{S_{i+1}} \,, \quad (6)$$

C. Discretization

The surfaces are represented by triangular meshes. To fix ideas, consider a three-layer model for the head. Conductivities of each domain are respectively denoted σ_1, σ_2 and σ_3. The surfaces enclosing these homogeneous conductivity regions are denoted S_1 (inner skull boundary), S_2 (skull-scalp interface) and S_3 (scalp-air interface).

The Symmetric BEM is discretized with with mixed P1-P0 boundary elements. Surfacic P1 functions are continuous over their support and piecewise linear: they are linear on each triangle. Surfacic P0 functions are discontinuous over their support and piecewise constant: they are constant on each triangle. We denote $\psi_i^{(k)}$ the P0 boundary element associated to triangle i on surface S_k (equal to 1 on triangle i and to 0 elsewhere), and $\phi_j^{(l)}$ the P1 boundary element associated to node j on surface S_l (equal to 1 at node j, linear on all triangles that touch node j and equal to zero elsewhere). The potential V on surface S_k is approximated by $V_{S_k}(\mathbf{r}) = \sum_i x_i^{(k)} \phi_i^{(k)}(\mathbf{r})$, while $p = \sigma \partial_{\mathbf{n}} V$ on surface S_k is approximated by $p_{S_k}(\mathbf{r}) = \sum_i y_i^{(k)} \psi_i^{(k)}(\mathbf{r})$. The reason for discretizing $\sigma \partial_{\mathbf{n}} V$ with P0 elements while V is discretized with P1 elements is that the gradient of a piecewise-linear function is piecewise constant.

As an illustration, consider the source term to reside in the brain compartment Ω_1. After discretizing the equations (6) and (5), the variables $\left(\mathbf{x}_k\right)_i = x_i^{(k)}$ and $\left(\mathbf{y}_k\right)_i = y_i^{(k)}$ satisfy the linear system:

$$\begin{bmatrix} (\sigma_1+\sigma_2)\mathsf{N}_{11} & -2\mathsf{D}^*_{11} & -\sigma_2\mathsf{N}_{12} & \mathsf{D}^*_{12} & 0 \\ -2\mathsf{D}_{11} & (\sigma_1^{-1}+\sigma_2^{-1})\mathsf{S}_{11} & \mathsf{D}_{12} & -\sigma_2^{-1}\mathsf{S}_{12} & 0 \\ -\sigma_2\mathsf{N}_{21} & \mathsf{D}^*_{21} & (\sigma_2+\sigma_3)\mathsf{N}_{22} & -2\mathsf{D}^*_{22} & -\sigma_3\mathsf{N}_{23} \\ \mathsf{D}_{21} & -\sigma_2^{-1}\mathsf{S}_{21} & -2\mathsf{D}_{22} & (\sigma_2^{-1}+\sigma_3^{-1})\mathsf{S}_{22} & \mathsf{D}_{23} \\ 0 & 0 & -\sigma_3\mathsf{N}_{32} & \mathsf{D}^*_{32} & \sigma_3\mathsf{N}_{33} \end{bmatrix} \begin{bmatrix} \mathbf{x}_1 \\ \mathbf{y}_1 \\ \mathbf{x}_2 \\ \mathbf{y}_2 \\ \mathbf{x}_3 \end{bmatrix} = \begin{bmatrix} \mathbf{b}_1 \\ \mathbf{c}_1 \\ 0 \\ 0 \\ 0 \end{bmatrix}$$

where $\mathbf{b}_1$ and $\mathbf{c}_1$ are the coefficients of the P0 (resp. P1) boundary element decomposition of the source term $\partial_{\mathbf{n}} v_{\Omega_1}$ (resp. $-\sigma_1^{-1} v_{\Omega_1}$).

Blocks N_{ij} and D_{ij} map a potential V_j on S_j to a function defined on S_i. Block S_{ij} maps a normal current p_j on S_j to a function defined on S_i. The resulting matrix is block-diagonal, and symmetric, whence the name "Symmetric BEM".

The Symmetric BEM introduces an additional unknown into the problem: the normal current, and uses an additional set of representation equations linking the normal current and the potential. The Symmetric BEM departs from the double-layer BEM in several ways:

- the normal current to each surface is explicitely modeled;
- only the surfaces which bound a common compartment have an interaction (whence the blocks of zeros in the matrix);
- only the surfaces which bound a compartment containing sources have a source term (whence the zeros in the right-hand side vector);
- the design matrix is symmetric;
- the design matrix is larger than that of the double-layer BEM for a given head model, because there are more unknowns.

The magnetic field is computed from the electric field and the primary source distribution using the Biot and Savart equation, as proposed by Ferguson, Zhang and Stroink [6]. Two extensions of the Symmetric BEM have been proposed: the Symmetric BEM has been extended to non-nested geometries [8], and Fast Multipoles have been introduced in order to accelerate matrix-vector products for the inversion of the design matrix [7].

IV. Numerical results

The accuracy of a numerical method is measured by comparison with analytical solution, by means of the Relative

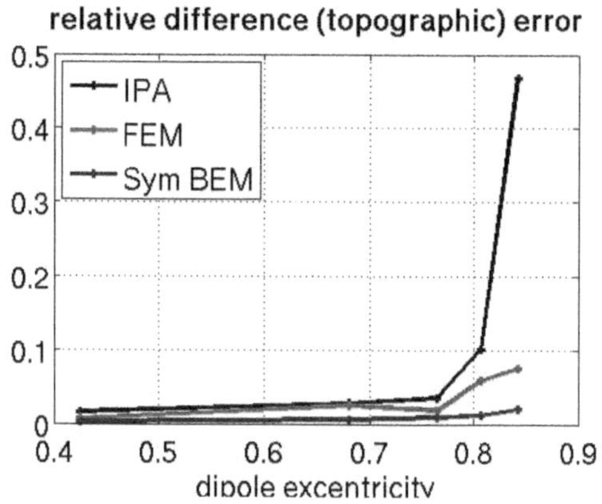

Fig. 2 Relative Difference Measure (RDM)

Difference Measure (RDM), defined by

$$RDM(g_{\text{numeric}}, g_{\text{analytic}}) = \left\| \frac{g_{\text{numeric}}}{\|g_{\text{numeric}}\|} - \frac{g_{\text{analytic}}}{\|g_{\text{analytic}}\|} \right\| \in [0\ ,\ 2].$$

The analytical solution was computed, for a classical, three-shell spherical head model [9]. The spheres had radii 0.87, 0.92 and 1 and were each triangulated with 642 points. The conductivities of the three domains was set to 1, 1/80 and 1. The volumes within the spheres were meshed with tetrahedra, which were designed to share the same vertices on the spheres as the ones used for the triangulations.

Figure 2 compares the accuracy of the Symmetric BEM (in blue) to a P1 tetrahedric Finite Element method (in red), and to a double-layer, IPA-corrected double-layer BEM (in black). The accuracy is here displayed for 5 dipoles on the z-axis with eccentricities 0.425, 0.68, 0.765, 0.8075 and 0.8415 and moment (1,0,1). Though the accuracy of the Finite Element method is higher than that of the IPA-corrected double-layer BEM, as initially shown in [1], the Symmetric BEM outperforms both methods.

V. Dissemination

The Symmetric BEM is made available through the open-source code OpenMEEG (http://openmeeg.gforge.inria.fr). This C++ code can used either standalone, through a command-line interface or via Python, and its integration into standard EEG and MEG packages is under way (e.g., Fieldtrip).

VI. Conclusion

By studying the extended Green Representation Theorem, several Boundary Element Methods for the forward EEG and MEG problems have been proposed. The most accurate among these is the Symmetric BEM. This increased accuracy for a given head model comes with the price of having more boundary integral operators to compute, and a larger design matrix to assemble. Interested readers may refer to [3] for more details on the method, and to http://openmeeg.gforge.inria.fr where an opensource implementation is available for download.

References

1. Clerc M., Dervieux A., Faugeras O., Keriven R., Kybic J., Papadopoulo T.. Comparison of BEM and FEM methods for the E/MEG problem in *Proceedings of BIOMAG 2002* 2002.
2. Nédélec Jean-Claude. *Acoustic and Electromagnetic Equations*. Springer Verlag 2001.
3. Kybic J., Clerc M., Abboud T., Faugeras O., Keriven R., Papadopoulo T.. A Common Formalism for the Integral Formulations of the Forward EEG Problem *IEEE Transactions on Medical Imaging*. 2005;24:12–28.
4. Geselowitz D. B.. On bioelectric potentials in an homogeneous volume conductor *Biophysics Journal*. 1967;7:1–11.
5. Hämäläinen M. S., Sarvas J.. Realistic Conductivity Geometry Model of the Human Head for Interpretation of Neuromagnetic Data *IEEE Trans. Biomed. Eng.*. 1989;36:165–171.
6. Ferguson A. S., Zhang X., Stroink G.. A Complete Linear Discretization for Calculating the Magnetic Field Using the Boundary Element Method *IEEE Trans. Biomed. Eng.*. 1994;41:455–459.
7. Kybic Jan, Clerc Maureen, Faugeras Olivier, Keriven Renaud, Papadopoulo Théo. Fast Multipole Acceleration of the MEG/EEG Boundary Element Method *Physics in Medicine and Biology*. 2005;50:4695–4710.
8. Kybic J., Clerc M., Faugeras O., Keriven R., Papadopoulo T.. Generalized head models for MEG/EEG: boundary element method beyond nested volumes *Phys. Med. Biol.*. 2006;51:1333-1346.
9. Zhang Zhi. A fast method to compute surface potentials generated by dipoles within multilayer anisotropic spheres *Phys. Med. Biol.*. 1995;40:335–349.

Author: Maureen Clerc
Institute: INRIA
Street: 2004 route des Lucioles, BP93
City: Sophia Antipolis Cedex
Country: France
Email: Maureen.Clerc@sophia.inria.fr

From Classical to Bayesian Estimators in the Interpretation of MEG and EEG

Jukka Sarvas and Risto J. Ilmoniemi

Dept. of Biomedical Eng. and Comput. Science, Aalto University, Espoo, Finland

Abstract— **The interpretation of MEG and EEG measurements often leads to solving a noisy, ill-conditioned matrix equation for the unknown source; the solution becomes an estimator of the true source. In practice, we need to know how to choose that estimator and what its properties are. In this paper, we present how a search for better estimators leads from the classical least-squares estimators through their regularized versions to Bayesian estimators. Though these estimators are rather well-known, the presented 'evolutionary' path is less known and not easily found in the literature.**

Keywords— **Interpretation of MEG and EEG, least-squares estimators, regularization, Bayesian estimators.**

I. Introduction

MEG and EEG measurements can be modeled by the linear observation equation

$$y = Ax + \eta\,, \tag{1}$$

where y is the m-component data vector, A is an $m \times n$ *matrix*, x is the *unknown* vector of n source parameters and η is the *measurement noise* [1], [2]. Interpreting the measurement amounts to solving (1) for x, but because of the noise it can never be solved exactly, and the solution $\hat{x}(y)$ becomes an *estimator* of the true x. A practical question is how to choose the solving method to get a good estimator $\hat{x}(y)$. In this paper, we show how a search for better estimators leads from the classical least-squares (LS) estimators to Bayesian ones.

The usual first choice is the LS estimator, but if A is ill-conditioned, it needs regularization. This leads to the *truncated singular value estimator*, the *Tikhonov estimator*, or the *Wiener estimator*. Usually we, however, have some prior knowledge of the unknown x, which can be used to form better-informed estimators than the LS ones. We show how the classical estimators can be optimized to become linear *minimum mean square error* (MMSE) estimators, which make full use of both the data y and the prior knowledge of x.

This raises the question whether one could find even better estimators than the linear ones. A complete answer is given by the Bayesian approach: the Bayesian *posterior mean* is always the best possible MMSE estimator, and it includes the MMSE Tikhonov and Wiener estimators as special cases. We illustrate the various estimators by numerical examples.

II. Notation and basic concepts

We present the $m \times n$ matrix A also by its *columns* $u_1, \ldots, u_n$ as $A = [u_1, \ldots, u_n]$. The *transpose* of A is denoted by A^{T}. The identity matrix is denoted by I, and $\|x\|$ is the *norm* of vector x. An $m \times n$ diagonal matrix with diagonal elements $d_1, \ldots, d_p$, $p = \min(m,n)$, is denoted by $\mathrm{diag}\,(d_1, \ldots, d_p)$.

If x is an n-variate *random variable* with the *density function* $p(x)$, then $\mathrm{E}(x) = \int x p(x)\,dx$ is the *mean* of x and $\mathrm{cov}(x) = \mathrm{E}\Big(\big(x - \mathrm{E}(x)\big)\big(x - \mathrm{E}(x)\big)^{\mathrm{T}}\Big)$ is the *covariance matrix*. The diagonal of $\mathrm{cov}(A)$ consists of variances s_j^2 of the components x_j of x, and we call $\mathrm{trace}\big(\mathrm{cov}(x)\big) = s_1^2 + \cdots + s_n^2$ the *total variance* of x.

III. Linear observation model and classical estimators

A. MSE of an Estimator

Let $\hat{x}(y)$ be an estimator of x in (1). If it depends linearly on y, i.e., $\hat{x}(y) = My$, where M is an $n \times m$ matrix, the estimator is called *linear*, otherwise *non-linear*. It is called *unbiased* if $E\left(\hat{x}(y)\right) = x$, otherwise *biased*, and $E\left(\hat{x}(y)\right) - x$ is the *bias*.

A natural measure of the goodness of an estimator is its *mean square error* (MSE) defined as $\mathrm{MSE}\,(\hat{x}) = E\left(\|\hat{x}(y) - x\|^2\right)$. It follows that $\mathrm{MSE}\,(\hat{x}) = \mathrm{trace}\Big(\mathrm{cov}\left(\hat{x}(y)\right)\Big) + \|\mathrm{bias}\|^2$, where $\mathrm{trace}\Big(\mathrm{cov}\left(\hat{x}(y)\right)\Big)$.

B. Least-Squares Estimator

A popular way to find an estimator $\hat{x}$ of the unknown x in (1) is to solve the equation

$$A\hat{x} = y, \tag{2}$$

for $\hat{x}$ in the sense of the least-squares, where $y = Ax + \eta$ is the data vector in (1), *i.e.*, we seek for a solution $\hat{x}$ that minimizes $\|A\hat{x} - y\|$ and has the minimum norm. This solution is called

S. Supek and A. Sušac (Eds.): Advances in Biomagnetism – BIOMAG2010, IFMBE Proceedings 28, pp. 113–116, 2010.
www.springerlink.com

the *minimum norm least-squares solution* and denoted by $\hat{x}_{LS}$. It is given by

$$\hat{x}_{LS}(y) = A^{\dagger} y = \sum_{j=1}^{r} \frac{1}{d_j} u_j^{\mathrm{T}} y v_j \qquad (3)$$

where $r = \mathrm{rank}(A)$, and $A = UDV^{\mathrm{T}}$ is the *singular value decomposition* (SVD) of A, where U and V are orthogonal matrices with columns $u_1, \ldots, u_m$ and $v_1, \ldots, v_n$, respectively, $D = \mathrm{diag}(d_1, \ldots, d_r, 0, \ldots, 0)$ and $d_1 \geq d_2 \geq d_{\min(m,n)}$, are the *singular values* of A. The matrix $A^{\dagger} = V \mathrm{diag}(d_1^{-1}, \ldots, d_r^{-1}, 0, \ldots, 0) U^{\mathrm{T}}$ is the *Moore–Penrose pseudoinverse* of A.

We call $\hat{x}_{LS}(y)$ the *least-squares estimator* of x in (1). If $m \geq n = r$, then $\hat{x}_{LS}(y)$ gets a simpler form as $\hat{x}_{LS}(y) = (A^{\mathrm{T}}A)^{-1}A^{\mathrm{T}}y$, and likewise, if $n \geq m = r$, then $\hat{x}_{LS}(y) = A^{\mathrm{T}}(AA^{\mathrm{T}})^{-1}y$.

If $m \geq n = r$ and the noise covariance $\mathrm{cov}(\eta) = \sigma^2 I$, then $\hat{x}_{LS}(y)$ has the well-known property that it is the *best linear unbiased estimator* (BLUE) of x; 'best' means that if we have another linear unbiased estimator of x, then its MSE is greater than that of $\hat{x}_{LS}(y)$, [3]. If $\mathrm{cov}(\eta) = \Gamma$ is of a more general form, then $\hat{x}_{LS}^{\Gamma}(y) = (A^{\mathrm{T}}\Gamma^{-1}A)^{-1}A^{\mathrm{T}}\Gamma^{-1}y$ is the BLUE estimator.

Though the LS estimator has the minimum MSE, that can, however, be very high, and make the estimator useless. This happens, *e.g.*, if the singular values d_j of A tend very rapidly to zero as j increases. As a remedy, we must *abandon the unbiasness* of the estimator and modify, or *regularize*, it so that the MSE decreases to an acceptable level without letting the bias to grow too much. Next we present two classical regularized estimators.

C. Truncated Singular Value Estimator

If the singular values d_j tend rapidly to zero and we truncate the sum in (3) after some $k < r$, then the total variance of $\hat{x}_{LS}(y)$ decreases, and we get the *k-truncated SVD estimator*

$$\hat{x}^{(k)}(y) = \sum_{j=1}^{k} \frac{1}{d_j} u_j^{\mathrm{T}} y v_j \,. \qquad (4)$$

On the other hand, with decreasing k the norm of the bias of $\hat{x}^{(k)}(y)$ increases, but usually much more slowly than the total variance. So, by lowering k we can *regularize* the estimator.

In practice, we need to know how k must be selected. One way to choose k is *Morozov's principle*, which considers $\hat{x}^{(k)}(y)$ as an acceptable estimate of the true x if the residual $\left\| A\hat{x}^{(k)}(y) - y \right\|$ is about the size of the noise. Accordingly, we choose k to be the largest integer $k \leq r$ so that

$$\left\| A\hat{x}^{(k)}(y) - y \right\|^2 = \sum_{j=k+1}^{m} (u_j^{\mathrm{T}} y)^2 \geq s^2 \,, \qquad (5)$$

where s is about the size $\|\eta\|$ of the noise. If $\mathrm{cov}(\eta) = \sigma^2 I$, then $s^2 = \mathrm{E}(\|\eta\|^2) = m\sigma^2$ is a good choice for s. Note that Morozov's principle does not control the bias of $\hat{x}^{(k)}(y)$, but in practice the principle usually offers a better estimator in the case of an ill-conditioned A than the LS estimator.

D. Tikhonov's Regularization

In this method, instead of Eq. (2), we solve its 'regularized' version $(A^{\mathrm{T}}A + \lambda I)\hat{x} = A^{\mathrm{T}}y$, where $\lambda > 0$ is the *regularization parameter*. The solution

$$\hat{x}_{\lambda}(y) = (A^{\mathrm{T}}A + \lambda I)^{-1}A^{\mathrm{T}}y = \sum_{j=1}^{r} \frac{d_j}{d_j^2 + \lambda} u_j^{\mathrm{T}} y v_j \qquad (6)$$

is the *Tikhonov estimator* of the unknown x in (1); the latter equation above is obtained by the SVD of A.

As in the case of the truncated SVD estimator we see that if λ increases in (6), then the total variance of $\hat{x}_{\lambda}(y)$ decreases and the bias of $\hat{x}_{\lambda}(y)$ increases, but usually more slowly. So, by an appropriate choice of $\lambda > 0$ we can regularize the estimator. Again the choice can be done with *Morozov's principle*: choose $\lambda > 0$ so that $\|A\hat{x}_{\lambda}(y) - y\| = s$, where s is as in (5).

Also here, Morozov's principle does not control the bias of $\hat{x}_{\lambda}(y)$, but it usually offers a better estimator in the case of an ill-conditioned A than the LS estimator.

IV. Minimum MSE estimators with prior knowledge

A. Observation Model with Prior Knowledge

We often in (1) have some *prior knowledge* of x, which can be used in forming estimators based on better information than the LS ones and their regularized versions. For that aim, the prior knowledge is presented in terms of the *prior (probability) distribution* of x with the density function $p(x)$. This makes x an n-variate random variable, which we assume to be statistically independent of η.

For notational convenience, we from now on also assume that the mean $\mathrm{E}(x) = 0$. Note that if $\hat{x}(y) = My$ is a linear estimator of x in (1) and $\mathrm{E}(x) = 0$, then $\hat{x}$ *is automatically unbiased*, because $\mathrm{E}(\hat{x}(y)) = \mathrm{E}(M(Ax + \eta)) = M(A\mathrm{E}(x) + \mathrm{E}(\eta)) = 0 = \mathrm{E}(x)$.

Usual prior distributions are the *uniform* and *normal* distributions. If our only prior knowledge of x is that its components x_j are bounded as $a_j \leq x_j \leq b_j$, $j = 1,\ldots,n$, then an appropriate prior distribution of x is the uniform one with density function $p(x)$ equal to $1/V$, in the n-box $a_j \leq x_j \leq b_j$, $j = 1,\ldots,n$, and vanishing outside it, where $V = (b_1 - a_1)\cdots(b_n - a_n)$.

The n-variate x is normally distributed, or *Gaussian*, with the *mean* x_0 and the *covariance matrix* Γ, if its density function is as $p(x) = c\exp\big(-\frac{1}{2}(x-x_0)^{\mathrm{T}}\Gamma^{-1}(x-x_0)\big)$, where $c = 1/[(2\pi)^{n/2}(\det\Gamma)^{1/2}]$.

B. Finding Minimum MSE Estimators

The MSE of an estimator $\hat{x}(y)$ of x with the prior density is still defined by $\mathrm{E}\big(\|\hat{x}(y) - x\|^2\big)$, but now the the mean is taken with the joint density function $p(x,y)$ of x and y. We want to find MSE's of the truncated SVD and Tikhonov estimators. To that end we first note that they are special cases of the (diagonal) *Wiener estimator* $\hat{x}_w(y)$ defined in terms of the SVD of A as

$$\hat{x}_w(y) = \sum_{j=1}^{r} w_j u_j^{\mathrm{T}} y v_j, \qquad (7)$$

where $w_1,\ldots,w_r$ are the *estimator coefficients* and $r = \operatorname{rank}(A)$, [5]. Clearly $\hat{x}_w(y)$ agrees with the truncated SVD estimator (4) with $w_j = 1/d_j$ and with the Tikhonov estimator (6) with $w_j = d_j/(d_j^2 + \lambda)$.

Let Γ_{pr} and Γ_{noise} be the covariance matrices of the prior and noise distributions in (1). Because x and η are independent and $\mathrm{E}(x) = \mathrm{E}(\eta) = 0$, we can, by the usual properties of mean and covariance, find the MSE of $\hat{x}_w(y)$ to be

$$\mathrm{MSE} = \sum_{j=1}^{r} \big[B(j,j)(w_j d_j - 1)^2 + w_j^2 C(j,j)\big] + R, \qquad (8)$$

where $R = \sum_{j=r+1}^{n} B(j,j)$, $B = V^{\mathrm{T}}\Gamma_{\mathrm{pr}}V$ and $C = U^{\mathrm{T}}\Gamma_{\mathrm{noise}}U$, and $A = UDV^{\mathrm{T}}$ is the SVD of A.

We can now minimize (8) with respect to w_j and obtain the coefficients yielding the *minimum* MSE *Wiener estimator* as

$$w_j = \frac{d_j B(j,j)}{d_j^2 B(j,j) + C(j,j)} \qquad (9)$$

and if $\Gamma_{\mathrm{pr}} = \kappa^2 I$ and $\Gamma_{\mathrm{noise}} = \sigma^2 I$, then $w_j = d_j/(d_j^2 + \frac{\sigma^2}{\kappa^2})$.

Because the truncated SVD estimator $\hat{x}^{(k)}$ and Tikhonov estimator $\hat{x}_\lambda$ are special cases of the Wiener estimator, we readily get the MSE's of those estimators by (8) for $\Gamma_{\mathrm{pr}} = \kappa^2 I$ and $\Gamma_{\mathrm{noise}} = \sigma^2 I$.

Namely, by minimizing the MSE of $\hat{x}^{(k)}$ with respect to k, we easily see that *the k yielding the minimum MSE truncated SVD estimator $\hat{x}^{(k)}(y)$ is the largest $k \leq r$ with $1/d_k \leq \kappa/\sigma$.*

By minimizing the MSE of $\hat{x}_\lambda$ with respect to $\lambda > 0$, we likewise get that *the λ yielding the minimum MSE Tikhonov estimator $\hat{x}_\lambda(y)$ is $\lambda = \sigma^2/\kappa^2$.*

Note that because the SVD and Tikhonov estimators are special cases of the Wiener estimator, they are never in the MSE sense better than the MMSE Wiener estimator, though there may be other reasons to use them.

After having found the minimum MSE Wiener estimator it remains to search for the unbiased, *general linear minimum MSE estimator* $\hat{x}(y) = My$, where M is an $n \times m$ matrix. To find M we must minimize $\mathrm{E}\big(\|My - x\|^2\big)$, the MSE of the estimator My. By observing that we can minimize the terms $\mathrm{E}\big((x_j - M_j y)^2\big)$ independently, where M_j, $j = 1,\ldots,n$, are the rows of M, we get that the minimizing matrix $M = \mathrm{E}(xy^{\mathrm{T}})\mathrm{E}(yy^{\mathrm{T}})^{-1}$. Applying that equation to (1) yields the wanted *general linear minimum MSE estimator* as

$$\hat{x}(y) = \Gamma_{\mathrm{pr}} A^{\mathrm{T}} (A\Gamma_{\mathrm{pr}} A^{\mathrm{T}} + \Gamma_{\mathrm{noise}})^{-1} y. \qquad (10)$$

Viewing the obtained results, we observe that if $\Gamma_{pr} = \kappa^2 I$ and $\Gamma_{\mathrm{noise}} = \sigma^2 I$, then Tikhonov and Wiener estimators agree with the general linear estimator (10) and $\hat{x}_\lambda(y) = (AA^{\mathrm{T}} + \frac{\sigma^2}{\kappa^2} I)^{-1} A^{\mathrm{T}} y$ *is the minimum* MSE *estimator among all linear estimators* of x in (1).

V. Bayesian approach to estimators

By using the Bayesian approach, we finally want to find the unbiased estimator which has the minimum MSE among all linear or *non-linear* estimators of x in (1) with given prior and noise distributions $p_{\mathrm{pr}}(x)$ and $p_{\mathrm{noise}}(\eta)$.

According to the the Bayesian approach all information on x after the measurement is contained in the *posterior distribution* $p(x|y)$ of x. The statistical inference then consists of appropriate summaries on $p(x|y)$ like its mean, peak values, the confidence sets, etc. By Bayes' formula for the posterior distributionis as $p(x|y) = c^{-1} p_{\mathrm{noise}}(y - Ax)\, p_{\mathrm{pr}}(x)$ for the observation model (1), where $c = \int p_{\mathrm{noise}}(y - Ax)\, p_{\mathrm{pr}}(x)\, dx$.

Using the properties of mean, one now can derive the main result [4]: *the unbiased minimum MSE estimator among all estimators of x in (1) is the posterior mean $\hat{x}_{\mathrm{post}}(y)$ given by*

$$\hat{x}_{\mathrm{post}}(y) = \mathrm{E}(x|y) = \int x\, p(x|y)\, dx. \qquad (11)$$

Furthermore, if $p_{pr}(x)$ and $p_{noise}(\eta)$ are *Gaussian* with zero mean and covariance matrices Γ_{pr} and Γ_{noise}, then one can show that $p(x|y)$ is also Gaussian with the mean $\hat{x}_{post}(y)$ and covariance matrix Γ_{post} given by

$$\Gamma_{post} = (A^T\Gamma_{noise}^{-1}A + \Gamma_{pr}^{-1})^{-1}, \quad (12)$$

$$\hat{x}_{post}(y) = \Gamma_{pr}A^T(A\Gamma_{pr}A^T + \Gamma_{noise})^{-1}y. \quad (13)$$

We observe that (13) agrees with (10), as it should do, because $\hat{x}_{post}(y)$ is a linear estimator by (13). Accordingly, (13) essentially tells us that in the Gaussian case $\hat{x}_{post}(y)$ is a linear estimator, but otherwise it may be non-linear and (11) must be used as its formula.

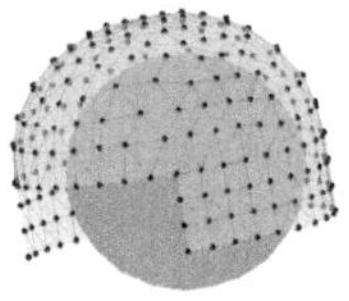

Fig. 1 Spherical head model and sensor positions

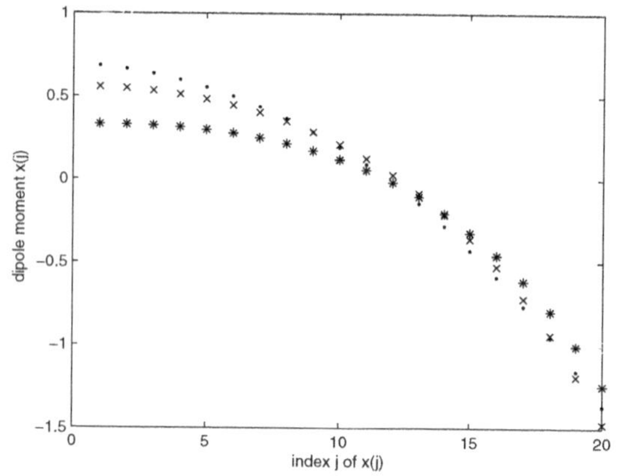

Fig. 2 The true x is depicted by circles (o), estimators $\hat{x}_\lambda$ by asterisks (*), $\hat{x}_w$ by crosses (x) and $\hat{x}_{post}$ by dots (·)

VI. Example with simulated MEG data

We use a spherical head model with radius 9 cm, the origin at the center of the sphere, the z-axis pointing upright and the x-axis to the nose. The source is $n = 20$ dipoles at points $P_j = (0,0,z_j)$, $z_j = 4 + (j-1)4.5/19$ cm, and with moments $(x_j,0,0)$, $x_j = \cos((j-1)\pi/19)$, $j = 1,\ldots,20$; max $|x_j|$ is normalized to one. On a MEG helmet, see Fig.(1), at the points R_j, $j = 1,\ldots,m = 260$, the normal components B_j of the magnetic field **B** are measured with noise η so that $\mathrm{cov}(\eta) = \sigma^2 I$, $\sigma = 0.1\max(|B_j|)$. The element $A(i,j)$ of A is the normal component of **B** at R_i due to a unit dipole $(1,0,0)$ at P_j, $1 \le i \le m$, $1 \le j \le n$, and computed with the analytic formula for **B** outside the head [6]. So, the measurement data is as $y = Ax + \eta$ with $x = (x_1,\ldots,x_n)$ being the true x given above.

We computed 4 estimators. The LS estimator $\hat{x}_{LS}$ had too big MSE and was abandoned. The Tikhonov estimator $\hat{x}_\lambda$ was adjusted with Morozov's principle. The Wiener estimator $\hat{x}_w$ was computed with the uniform prior in the box $|x_j| \le 1$, $j = 1,\ldots,m$, yielding $\Gamma_{pr} = 3^{-1}I$. Finally, the Bayesian estimator $\hat{x}_{post}$ was formed in two steps by using the posterior of the first measurement (assumed Gaussian with $\Gamma_{pr} = 3^{-1}I$) as the prior of the second measurement; applying (12) and (13) twice yields, $\hat{x}_{post} = GA^T(AGA^T + \sigma^2 I)^{-1}(y - A\hat{x}_w) + \hat{x}_w$ with $G = (\sigma^{-2}A^TA + 3^{-1}I)^{-1}$. The estimators are presented in Fig.(2) showing that $\hat{x}_{post}$ yields the best and $\hat{x}_\lambda$ the worst estimates, as expected.

VII. Conclusion

We have shown how the need for better estimators leads from the least-squares estimator and its regularized versions to Bayesian estimators. The various estimators were also illustrated with a numerical example with simulated MEG data.

References

1. Mosher J, Leahy R, Lewis P. EEG and MEG: Forward Solutions for Inverse Problems *IEEE Trans. on Biomed. Eng.* 46 3 1999 245-259.
2. Schmidt D, George J, Wood C. Bayesian inference for neural electromagnetic source localization: analysis of MEG visual evoked activity *Proc. SPIE*, 3661, 408 (1999).
3. Sorenson H W. *Parameter estimation*. New York and Basel: Marcell Dekker Inc 1980.
4. Gelman A, Carlin J, Stern H, Rubin D. *Bayesian data analysis*. Boca Raton, London and New York: Chapman & Hall/CRC 2004.
5. Numminen J, Ilmoniemi R J. Regularization of ill-conditioned linear systems: application to biomagnetism in *Proc. XXVI Ann.Conf Finnish Phys.Soc.*, TKK-F-A6973 (Lahti, Finland): p. 17:11 1992.
6. Sarvas J. Basic mathematical and electromagnetic concepts of the biomagnetic inverse problem *Phys.Med.Biol.* 32 1 (1987) 11-22.

Authors: Jukka Sarvas, Risto J Ilmoniemi
Institute: Dept. of Biomed. Eng. and Comp. Science, Aalto University
City: Espoo
Country: Finland
Email: jukka.sarvas@tkk.fi, risto.ilmoniemi@tkk.fi

Skull Thickness versus Mesh Density — A Study on BEM Discretization Error

Pascal Werner[1,2] and Matti Stenroos[2]

[1] Technische Universität Ilmenau, Institute for Biomedical Engineering and Informatics, Ilmenau, Germany
[2] Helsinki University of Technology, Department of Biomedical Engineering and Computational Science, Helsinki, Finland

Abstract— **In solving the EEG or MEG forward problem, the electric potential is commonly modeled with the boundary element method (BEM). It is known that the BEM has trouble dealing with the skull, and that the discretization error of the BEM solution increases, when the source is brought close to the skull. In addition, the BEM error depends also on the thickness of the skull. In this study, we assess the effect of the skull thickness on the discretization error in terms of the element side length of the inner sphere (ESL). The computations were carried out in a three-layer model using the linear Galerkin BEM, and the results were validated against analytically computed potentials. The error was measured in terms of the correlation coefficient. The conductivity ratio between the brain and the skull was varied between 20 and 80. Current-dipole sources were placed at different depths. Results showed that the conductivity ratio has has no big influence on the discretization error. For models with skull thickness of more than 0.6 ESL, source model can extend up to 0.5 ESL below the surface of the brain without introducing considerable error. For smaller skull thicknesses, the sources need to be place further away from the skull.**

Keywords— **electroencephalography, magnetoencephalography, boundary element method, forward problem, skull thickness.**

I. Introduction

In solving the EEG or MEG forward problem, the electric potential is commonly modeled with the boundary element method (BEM). Solving of the forward problem plays an essential part in the inverse problem, too. Thus an accurate solution to the inverse problem can be obtained only with help of an appropriate solution of the forward problem. It is known that the BEM has trouble dealing with the skull, and that the discretization error of the BEM solution increases, when the source is brought close to the skull. While the error due to bringing sources too close to a boundary surface is well known, there is, to the authors knowledge, no study that discusses the error caused by coarse discretization of thin structures such as the skull. The aim of this study is to better understand the influence of thin layers on the discretization error of the BEM.

II. Methods

A. Forward Problem

In boundary element modeling of a volume conductor, the electric potential is discretized on each conductivity interface with help of basis functions. In this study, the potential was approximated piece-wise linear on the boundary element mesh. This approximation causes errors, and these errors depend on the density of the element mesh. The side-length of a mesh-element is thus a good choice for the length measure.

The discretization error is minimized with respect to some weight functions. In this study, we used Galerkin weighting: the weight functions were piece-wise linear like the basis functions [1, 2]. The given values are related to the element side length of the inner layer of the used mesh (ESL). The BEM transfer matrices were computed as described in the appendix of [3]. An equivalent current dipole was used as the source model.

B. Head Model

The computations were carried out in a spherical three-layer model consisting of a brain, skull and scalp layer. Each layer was discretized with 2562 nodes and 5120 flat triangular elements. The radius of the brain was fixed and set to 90 mm, yielding the ESL of 6.8 mm. The radii of the two other layers depended on a specified thickness, ranging from 0.1 ESL to 2 ESL, in a step of 0.1 ESL. The radius of the skull (scalp) was set to the radius of the brain plus one (two) times the thickness. The conductivity of the brain was chosen to 0.25 S/m and the conductivity of the scalp to 0.35 S/m. The conductivity of the skull was computed according to a ratio from the conductivity of the brain over the conductivity of the skull. For this, the values 20, 40 and 80 were used, resulting in the conductivity values of 0.013 S/m, 0.006 S/m and 0.003 S/m.

C. Validations

The numerically computed potentials were validated against values computed analytically with the layered-sphere model. The BEM results were compared to the analytically obtained ones with the help of correlation coefficient (CC). We chose CC as error measure, because we were primarily interested in morphological error. The behavior of the more commonly used relative error (RE) measure was, however, rather similar to that of the CC.

D. Simulation Setup

For the simulations, a total of 4000 dipoles per model were used. The radial distance of the dipoles from the origin was varied between 0.85 to 1 times the radius of the innermost sphere (up to 2 ESL away from the boundary), resulting in a range from 76.5 mm to 90 mm. In this range, 20 different source depths were applied. For each depth, 100 random positions were chosen. In all these test dipole positions, potentials due to radial and tangential dipoles were computed.

The computations were done in Matlab 7.8.0 environment; the BEM programs and scripts were developed from those used in [2, 3].

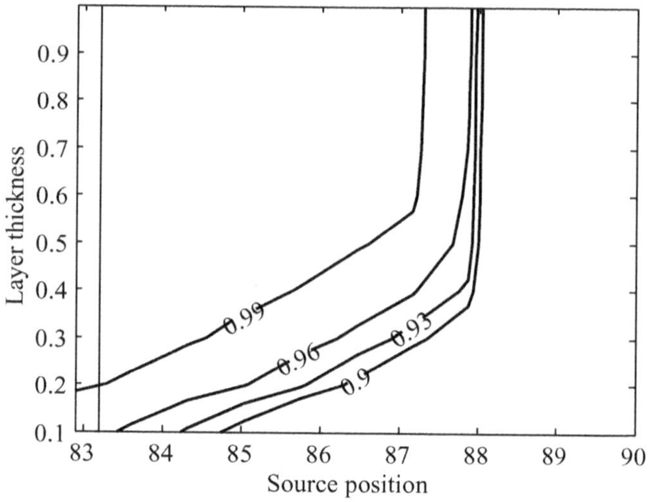

Fig. 1 Contour plot of the CC for a conductivity ratio of 20, scale of x axis in mm, scale of y axis in ESL

III. Results

From all simulations, the lowest CC for each source depths and model scenario was chosen and plotted in Figures 1–3. In the Figures, only the interesting part is plotted, meaning 1 ESL as maximum for the thickness, and for the source position an interval of 1 ESL below the surface of the brain. For bigger thicknesses, the contour lines stayed at the same source position as for a thickness of 1 ESL.

The figures 1-3 show contour plots of CC for different conductivity ratios. The x axis shows the position of the source in absolute values, ranging from 1 ESL below the brain surface to the radius of the brain (90 mm). The y axis shows the thickness of the skull and scalp layer, ranging from 0.1 ESL to 1 ESL. Only this part of the computed data was chosen, because the other data contained no further information. The left upper line shows the contour for a CC of 0.99. The region on the left and upper side to it shows bigger values. The right lower line shows the contour for a CC of 0.90, and the lines in between are linearly distributed between these two values. The CC of the region on the right and lower side of the 0.9-contour dropped significantly to unusable values, when bringing the source closer to the surface. The solid line on the left side each indicates a depth of 1 ESL below the brain surface.

IV. Discussion

The results for different conductivity ratios are morphologically similar; see curve shapes in Figs. 1–3. Overall, the conductivity ratio had only a small effect on this computations, especially for values nowadays used. We also did some simulations with the traditionally used Collocation BEM; those results showed a stronger dependency between the conductivity ratio and the discretization error. In recent studies, values for the brain–skull conductivity ratio between about 15 and 50 have been regarded as appropriate, while in former studies, conductivity ratios of about 80 were frequently used [4]. Hence, the following rules applies to the conductivity ratios of 20 and 40, but, with minor changes, it should also be valid for a conductivity ratio of 80.

For head models with a conductivity ratio of 20 and a thickness of more than 0.6 ESL, sources can be placed as close to the surface as 0.4 ESL. For situations, in which the position of the sources is estimated to be in a depth of 1 ESL related to the brain, the thickness can be set to 0.2 ESL. In between, there was a linear relation. The thinner the layer, the deeper must be the estimated position of the sources, in order to get reasonable results. For a conductivity ratio of 40, the behaviour is similar. To be able to use layers with a thickness of 0.6 ESL, the minimal depth for the sources must be set to 0.5 ESL. The linear behaviour for thinner meshes is similar to models with a conductivity ratio of 20.

With the used mesh generator, the spherical meshes were generated by subdividing an icosahedron. Thus it was not possible to have arbitrary number of nodes on each mesh, and we needed to adjust the thickness of the layers by changing the geometry of the model. This is the main limitation of this study; we are currently working on this issue.

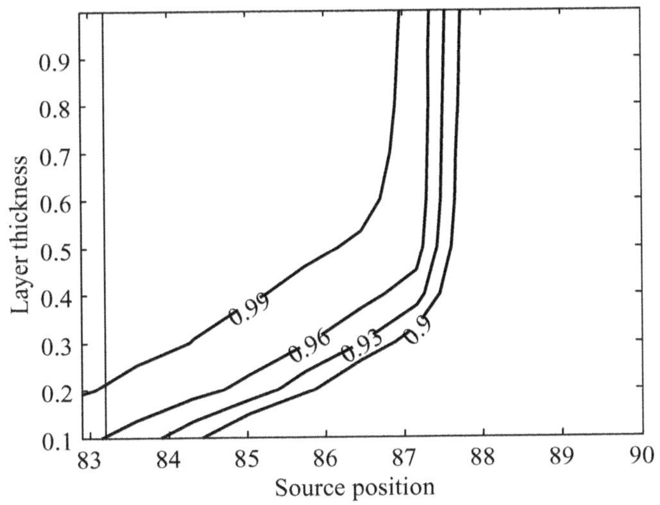

Fig. 2 Contour plot of the CC for a conductivity ratio of 40, scale of x axis in mm, scale of y axis in ESL

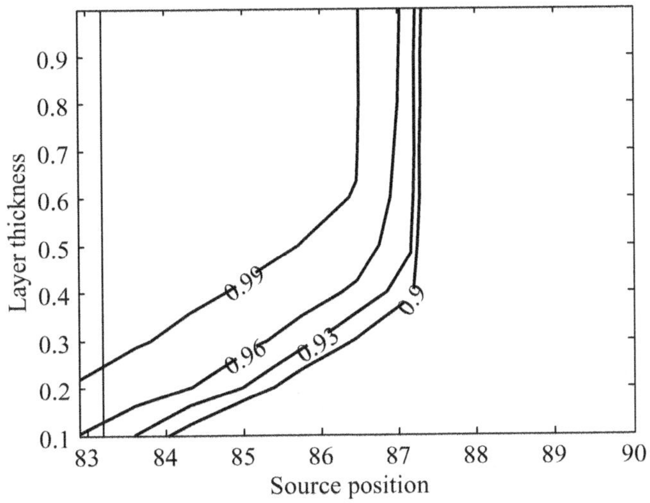

Fig. 3 Contour plot of the CC for a conductivity ratio of 80, scale of x axis in mm, scale of y axis in ESL

V. Conclusions

We investigated the discretization error of the BEM, applying thin meshes and bringing the sources close to the surface of the brain. For brain–skull conductivity ratios like 20 or 40 and a CC of 0.99, we can conclude that:

1. The conductivity ratio has no big influence on the discretization error.
2. For models with a thickness of more than 0.6 ESL, estimated source depth of 0.5 ESL below the surface of the brain can be used.
3. When it is clear, that the depth of the sources is bigger than 1 ESL, models with a thickness down to 0.2 ESL can be used.
4. When the sources are thought to be closer to the surface of the brain than 1 ESL and the thickness of the layers is smaller than 0.6 ESL, the Figures 1-2 show the appropriate source positions for a specific layer thickness.

Acknowledgments

We would like to thank J. Sarvas for providing an implementation of the analytical formula of the spherical head model.

References

1. Mosher J.C., Leahy R.M., Lewis P.S.. EEG and MEG: forward solutions for inverse methods *Biomedical Engineering, IEEE Transactions on.* 1999;46:245-259.
2. Stenroos M, Haueisen J. Boundary Element Computations in the Forward and Inverse Problem of Electrocardiography: Comparison of Collocation and Galerkin Weightings *IEEE Transactions on Biomedical Engineering.* 2008;55:2124–2133.
3. Stenroos M. The transfer matrix for epicardial potential in a piece-wise homogeneous thorax model: the boundary element formulation *Physics in Medicine and Biology.* 2008;54:5443–5455.
4. Oostendorp T.F., Delbeke J., Stegeman D.F.. The conductivity of the human skull: results of in vivo and in vitro measurements *Biomedical Engineering, IEEE Transactions on.* 2000;47:1487-1492.

corresponding Author: Pascal Werner
University: Technische Universität Ilmenau
Institute: Institute for Biomedical Engineering and Informatics
Street: Gustav-Kirchhoff Str. 2
City: 98693 Ilmenau
Country: Germany
Email: pascal.werner@tu-ilmenau.de

Domain Decomposition for Coupling Finite and Boundary Element Methods in EEG

Emmanuel Olivi, Maureen Clerc, and Théodore Papadopoulo

INRIA Odyssée Project-Team, Sophia Antipolis, France

Abstract— **The forward problem in electroencephalography aims to simulate on the scalp the potential V of an electromagnetic field generated by a simulated source. It must fit precisely with the electromagnetic propagation in the patient head. Yet, the skull anisotropy happens to be highly anisotropic, and must then be modeled. Although boundary element methods cannot deal with anisotropy like finite element methods, the symmetric BEM offers a higher accuracy than FEM wherever the conductivity can be considered as constant (*i.e.* for the brain and the scalp). A domain decomposition (DD) framework allows to split the global system into several ones with smaller computational domains. Then, one method (BEM or FEM) can be used per volume. This work presents such a coupling formulation of a 3-DD method solving iteratively a BEM for the brain, a FEM for the skull layer, and finally a BEM for the scalp.**

Keywords— **EEG, BEM, FEM, Coupling, Domain Decomposition.**

I. Introduction

The problem of finding the sources in the brain that are responsible for a measured EEG signal is an inverse problem of localization whose resolution requires many resolutions of the forward problem. The latter computes the potential V on the scalp of an electromagnetic field generated by a simulated source (in our case a dipole J_p). This propagation through the head is governed by the Maxwell equations in their quasi-static approximation. The forward problem resolution is achieved solving for V the Poisson equation:

$$\begin{cases} \nabla\cdot(\Sigma\nabla V) = \nabla\cdot J_p & \text{in } \Omega \\ (\Sigma\,\nabla V)\cdot n = 0 & \text{on } \Gamma \end{cases}, \qquad (1)$$

with Σ the head tissues' conductivity and Γ the Scalp/Air interface (outflow current through the neck is neglected). The forward model must fit precisely with the electromagnetic propagation in the patient head. Although the white matter makes the brain anisotropic, the skull anisotropy has a strong influence on the results, and must therefore be taken into account. Indeed, the skull conducts electricity ten to eighty times stronger in tangential directions than in the radial one. Finite Element Methods (FEM:s) can deal with this anisotropy, but offer a poor accuracy when compared to the symmetric Boundary Element Method (sBEM) in homogenous media [4] [5]. Therefore the coupling of both should yield better accuracy; using the BEM wherever the conductivity can be considered as constant (*i.e.* for the scalp and the brain when we neglect white matter anisotropy), and the FEM for the skull.

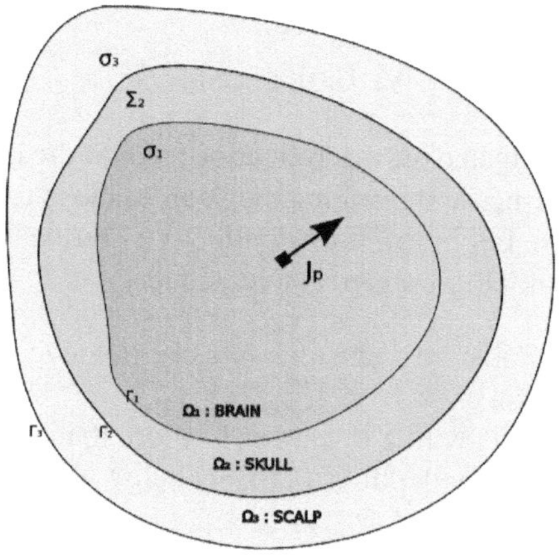

Fig. 1 A three-layer head model with nested regions

II. The symmetric BEM, the implicit FEM and coupling formulations

A. The Symmetric BEM, and the Implicit FEM

Wherever conductivity is considered as constant in a set of nested regions (see Fig. 1) then the first equation in Eq. 1 becomes $\sigma\Delta V = \nabla\cdot J_p$ and can be solved using integral formulations. Instead of the classical formulation that involves double-layer potentials, we use a formulation that involves both double-layer and single-layer potentials, to yield a symmetric system of equations. The symmetric BEM [1] involves both the potential and the normal current as unknowns on triangulated surfaces describing the different regions.

FEM:s usually require to mesh the entire volume to spread the information from the source through the computational

S. Supek and A. Sušac (Eds.): Advances in Biomagnetism – BIOMAG2010, IFMBE Proceedings 28, pp. 120–123, 2010.
www.springerlink.com

domain. The geometry used to solve the forward problem in EEG comes from anatomical MRI:s. These 3D-images are then segmented to obtain a levelset description of the regions to model. Generation of the (most often tetrahedral) meshes can be achieved but is usually expensive computationally and time consuming. The implicit FEM [3] bypasses this mesh generation step, going directly from the levelsets of the interfaces separating the various domains to the matrix associated to the FEM. It works on Cartesian grids with the potential V at the node locations as unknown, described with Q1 finite elements.

The sBEM developped in the open-source code OpenMEEG, turns out to be much more accurate than other BEM:s, and also than the implicit FEM (iFEM).

B. A BEM-FEM Coupling

Domain decomposition is a widely used numerical technique that allows to split the computational domain into regions of constant conductivities on the one hand and the skull on the other hand. We will then be able to use the appropriate method (BEM or FEM) per volume. Boundary conditions ensure the communication between the sub-problems. Several iterations are needed to solve the global system, and a relaxation at the interfaces is compulsory to ensure convergence.

We first propose a Neumann-Dirichlet approach which iteratively solves a BEM for the brain region, and a FEM for the skull and the scalp region. At iteration k, we first solve a BEM with Neumann boundary condition, to obtain V at the brain/skull interface. Secondly, a mixed Dirichlet - homogeneous Neumann problem is solved with the FEM using the newly computed V at the brain/skull interface as boundary condition. This gives V on the scalp:

$$\begin{cases} \sigma_1 \Delta V_1^k & = \nabla \cdot J_p \quad \text{in } \Omega_1, \\ \sigma_1 \partial_n V_1^k & = \lambda^k \quad \text{on } \Gamma_1, \\ \\ \nabla \cdot (\Sigma \nabla V_{2,3}^k) & = 0 \quad \text{in } \Omega_2 \cup \Omega_3, \\ V_{2,3}^k & = V_1^k \quad \text{on } \Gamma_1, \\ \sigma_3 \partial_n V_{2,3}^k & = 0 \quad \text{on } \Gamma_3, \end{cases} \tag{2}$$

Finally, λ^{k+1} is updated by combining the normal current computed by the FEM on the brain/skull interface and the previous λ^k with a relaxation parameter ω:

$$\lambda^{k+1} = (1-\omega)\lambda^k + \omega(\Sigma_2 \nabla V_2^k) \cdot n \text{ on } \Gamma_1 \text{ , } \omega > 0$$

We also present a Neumann-Dirichlet-Neumann coupling procedure, restricting the FEM to the skull only (see Fig. 2), and completing the computations with a BEM for the scalp:

$$\begin{cases} \sigma_1 \Delta V_1^k & = \nabla \cdot J_p \quad \text{in } \Omega_1, \\ \sigma_1 \partial_n V_1^k & = \lambda_1^k \quad \text{on } \Gamma_1, \\ \\ \nabla \cdot (\Sigma_2 \nabla V_2^k) & = 0 \quad \text{in } \Omega_2, \\ V_2^k & = V_1^k \quad \text{on } \Gamma_1, \\ V_2^k & = V_3^{k-1} \quad \text{on } \Gamma_2, \\ \\ \sigma_3 \Delta V_3^k & = 0 \quad \text{in } \Omega_3, \\ \sigma_3 \partial_n V_3^k & = \lambda_2^k \quad \text{on } \Gamma_2, \\ \sigma_3 \partial_n V_3^k & = 0 \quad \text{on } \Gamma_3, \end{cases} \tag{3}$$

With relaxations made on the Neumann's parameters λ_1 and λ_2:

$$\begin{aligned} \lambda_1^{k+1} &= (1-\omega_1)\lambda_1^k + \omega_1(\Sigma_2 \nabla V_2^k) \cdot n \quad \text{on } \Gamma_1 \text{ , } \omega_1 > 0 \\ \lambda_2^{k+1} &= (1-\omega_2)\lambda_2^k + \omega_2(\Sigma_2 \nabla V_2^k) \cdot n \quad \text{on } \Gamma_2 \text{ , } \omega_2 > 0 \end{aligned} \tag{4}$$

Initial condition for this coupling scheme is: $V_3^0 = 0$, with the relaxation parameters: $\omega_1 = \omega_2 = 1$ for $k = 1$. For $k > 1$, relaxation parameters are set manually to $\omega_1 = \omega_2 = 0.7$; a higher relaxation parameter would make the scheme diverge.

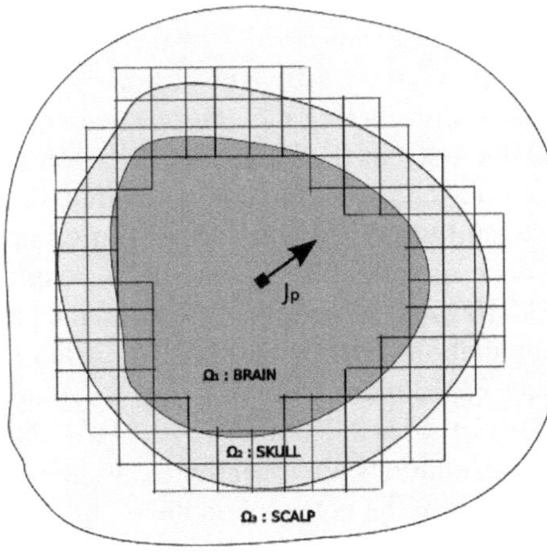

Fig. 2 The three-layer head model with the FEM's grid for the skull

Special care must be observed for the Neumann problem, indeed no accumulation of current should appear, therefore numerically we ensure that:

$$\begin{aligned} \int_{\Omega_1} \nabla \cdot J_p + \int_{\Gamma_1} \lambda_1^k &= 0 \quad , \forall k \\ \int_{\Gamma_2} \lambda_2^k &= 0 \quad , \forall k \end{aligned} \tag{5}$$

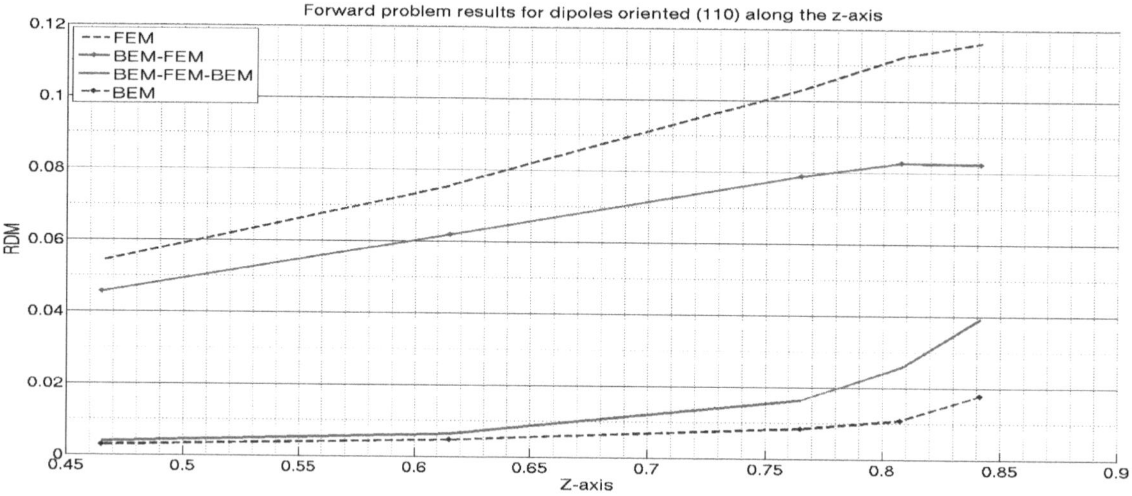

Fig. 3 RDM for the **isotropic** case, for pure and coupled methods

III. Numerical results

In order to validate results of the forward problem, analytical solutions are computed on a three-layer concentric sphere model [2], for both isotropic and anisotropic skull layer. Radii of the spheres and conductivities of the different layers are respectively $\{0.87, 0.92, 1.0\}$ and $\{1.0, 0.0125, 1.0\}$ for the isotropic case. Considering the anisotropic one, we have set up the conductivity in the tangential direction to ten times the normal one. The sBEM is computed on 642 point mesh per surface, and the iFEM considers a Cartesian grid of 90 points in each direction. The BEM-FEM coupling as well as the BEM-FEM-BEM coupling use the same previous grid sizes. Computations have been done for 5 dipoles oriented in Cartesian coordinates: $(1, 1, 0)$, and locations along the Z-axis: $\{0.465, 0.615, 0.765, 0.8075, 0.8415\}$. Accuracies of the numerical solutions are given by the Relative Difference Measure (RDM) of the potential on the scalp:

$$RDM = \left\| \frac{V_{analytic}}{\|V_{analytic}\|} - \frac{V_{computed}}{\|V_{computed}\|} \right\| \tag{6}$$

Fig. 3 shows the RDM for both pure and coupled methods in the isotropic case. The iFEM method in blue (dashed) is the less precise due to the poor grid definition. We see that the sBEM is the most accurate with a RDM always below the 2%. The BEM-FEM coupling in red (full with diamonds) shows a slight improvement compared to the accuracy of the FEM, whereas the BEM-FEM-BEM coupling really inherits its accuracy from the BEM. Coupling process have been stopped at 40 iterations. At this point the relative residuals computed as: $\frac{\left\| (\Sigma_2 \nabla V_2^{k+1})\cdot n|_{\Gamma_i} - \lambda_i^k \right\|}{\|\lambda_i^k\|}$, for $i = 1, 2$ were all below 6.10^{-5}. These results do not show the full benefit of the coupling, because the coupling takes its sense for anisotropic conductivities. As BEM cannot handle anisotropy, they must model the skull as isotropic and therefore their high accuracy is partly wasted because of the coarse model approximation. On Fig. 4, we have plotted the result of the BEM with an isotropic skull whose conductivity equals 0.0125 compared with the anisotropic analytical solution with conductivities $(radial, tangential) = (0.0125, 0.125)$. The iFEM as well as the BEM-FEM-BEM coupling have been computed with an anisotropic skull. This time, results have been compared on 15 dipoles: the five previous locations for three different orientations: $(1, 0, 1)$ in blue, $(1, 1, 0)$ in green and $(0, 0, 1)$ in red color. One can see that the coupling (full line) has a better precision than the BEM (dashed line with diamonds) for all dipoles expected the two located at $(0, 0, 0.8415)$ with orientations $(1, 0, 1)$ and $(0, 0, 1)$. As it comes closer to the skull, the singularity of the dipole makes the approximation due to the iFEM's grid worse. Using a higher resolution FEM's grid, this error should go diminishing. Results on the blue curve for the coupling are not gratifying; there are some convergence problems with our BEM-FEM-BEM algorithm, for some dipole orientations, that we have not yet solved.

IV. Conclusion

The domain decomposition framework really allows to take advantage of both the versatility of the FEM in modeling

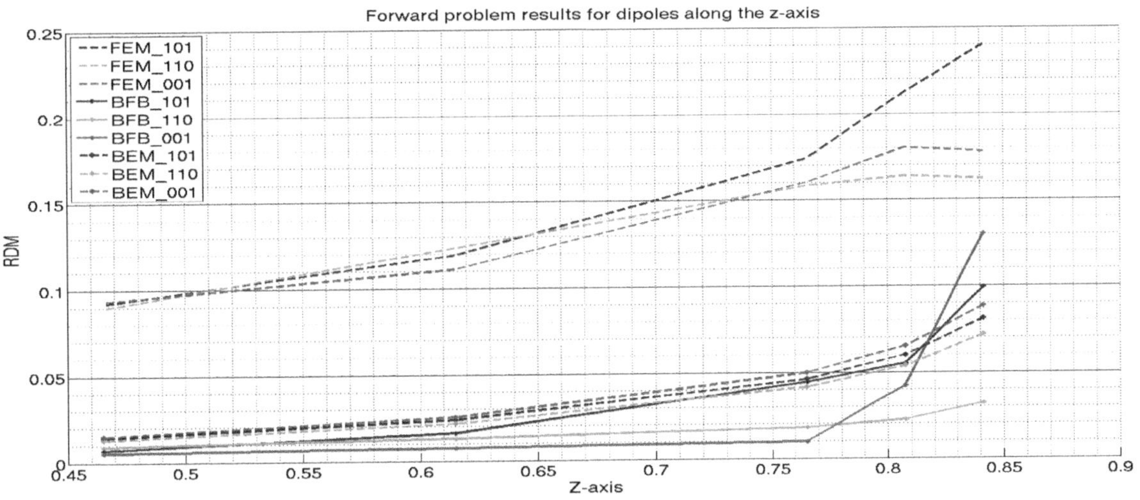

Fig. 4 RDM for the **anisotropic** case, for pure and coupled methods

conductivity and the accuracy of the BEM. It gives hope for providing accurate solutions of the forward problem without resorting to diminishing the mesh size in finite element methods. One could try to find a relaxation parameter changing through iterations to faster convergence. Parallelism is often used in the domain decomposition community, but it was not pertinent in our case since the sBEM has a direct solver, and only requires a matrix vector product to solve a new Neumann problem for each iteration ; most of the time is spent in the iterative solver of the FEM.

Acknowledgements

This work was supported by the Région Provence-Alpes-Côte d'Azur.

References

1. J. Kybic, M. Clerc, T. Abboud, O. Faugeras, R. Keriven, and T. Papadopoulo. A common formalism for the integral formulations of the forward EEG problem. *IEEE Transactions on Medical Imaging*, 24:12–28, jan 2005.
2. Zhi Zhang. A fast method to compute surface potentials generated by dipoles within multilayer anisotropic spheres Phys. Med. Biol. 1995
3. Théodore Papadopoulo and Sylvain Vallaghé. Implicit meshing for finite element methods using levelsets. In *Proceedings of MMBIA*, 2007.
4. G. Adde and M. Clerc and O. Faugeras and R. Keriven and J. Kybic and T. Papadopoulo. Symmetric BEM formulation for the M/EEG forward problem. In *Proceedings of IPMI*, 2003. ftp://ftp-sop.inria.fr/odyssee/Publications/2003/adde-clerc-etal:03.pdf
5. Emmanuel Olivi, Mariette Yvinec, Maureen Clerc, Théodore Papadopoulo. Quantitative comparisons of forward problems in MEEG. INRIA Research Report, 2007.

Author: Emmanuel Olivi
Institute: INRIA
Street: 2004 route des lucioles BP93
City: Sophia-Antipolis Cedex
Country: France
Email: Emmanuel.Olivi@sophia.inria.fr

Multi-condition M/EEG Inverse Modeling with Sparsity Assumptions: How to Estimate What Is Common and What Is Specific in Multiple Experimental Conditions

Alexandre Gramfort[1,2]

[1] INRIA, Parietal Project Team, Saclay, France
[2] INRIA, Odyssee Project Team, Sophia-Antipolis, France

Abstract— **M/EEG inverse modeling with distributed dipolar source models and penalizations with sparsity inducing norms (e.g. ℓ_1 with MCE [1], ℓ_0 with FOCUSS [2], ℓ_2-ℓ_1 [3]) offer a way to select a set of active dipoles. Indeed, sparsity inducing norms lead to solutions where most of the sources are set to zero and the remaining non zero sources form the set of estimated active dipoles. When running cognitive studies multiple experimental conditions are usually involved and cognitive hypothesis classically consist in quantifying the difference between these conditions. The problem is that when a sparse inverse solver is used independently for each experimental condition, it happens that the selection of dipolar sources is not consistent across conditions, thus limiting further analysis. Even if all conditions share a common dipolar source, due to noise, it can happen that such solvers do not select exactly the same dipole but two neighboring ones. To circumvent this limitation, we propose in this contribution to run the inverse computation with all the experimental conditions simultaneously. We use a penalization that achieves a joint selection of active dipoles while estimating two parts in the reconstructed current distributions: a part that is common to all the different conditions and a part that is specific to each condition. The penalization used in the inverse problem is based on groups of ℓ_2-ℓ_1 norms. The optimization is achieved with iterative least squares (iterative ℓ_2 Minimum Norm) making the solver tractable on large datasets. The method is illustrated on toy data and validated on synthetic MEG data reproducing activations appearing for somesthesic finger stimulations. We call our solver SMC (Sparse Multi-Condition).**

Keywords— **MEG, EEG, Inverse Problem, Sparse prior, IRLS.**

I. Introduction

Distributed source models in Magnetoencephalography and Electroencephalography (collectively M/EEG) use the individual anatomy derived from high resolution anatomical Magnetic Resonance Images (MRI) [4]. They consist in sampling the automatically segmented cortical ribbon with a high number of equivalent current dipoles (ECD). Each dipole adds linearly its contribution to the measured signal leading to a linear solution to the forward problem. The measurements $M \in \mathbb{R}^{d_m \times d_t}$ (d_m number of sensors and d_t number of time instants) are obtained by multiplying the current source amplitudes $X \in \mathbb{R}^{d_x \times d_t}$ (d_x number of dipoles) by a forward operator $G \in \mathbb{R}^{d_m \times d_x}$, called the lead field matrix, i.e., $M = GX$.

Solving the forward problem consists in computing G taking into account the electromagnetic properties of the head [5], while solving the inverse problem consists in estimating the neural currents $\hat{X}$ that explain the observed measurements. However, this latter problem is strongly ill-posed. It implies that $\hat{X}$ can only be computed if priors are set on the solution. Standard priors assume that a weighted ℓ_2 norm of $\hat{X}$, denoted $\|\hat{X}\|_{\mathbf{w};F}$ (Frobenius norm), is small. The estimated distribution of cortical currents $\hat{X}$ is obtained by solving:

$$\hat{X} = \arg\min_{X} \|M - GX\|_F^2 + \lambda \phi(X) \ , \lambda \in \mathbb{R}_+ \qquad (1)$$

with $\phi(X) = \|X\|_{\mathbf{w};F}^2$ and $\|X\|_{\mathbf{w};F}^2 = \sum_{t=1}^{d_t} \sum_{i=1}^{d_x} w_i x_{it}^2$, $\mathbf{w} = (w_i)_i \in \mathbb{R}_{+,*}^{d_x}$. Such priors provide the grounds of what are called in the M/EEG literature Minimum Norm (MN) inverse solvers [6, 5] and the noise normalized variants, *e.g.*, dSPM [7]. Such standard solvers are fast to compute since ℓ_2 priors lead to linear inversion. Indeed $\hat{X}$ is obtained by multiplication M with a matrix $H_{\mathbf{w}}$:

$$\hat{X} = W^{-1} G^T (G W^{-1} G^T + \lambda Id)^{-1} M = H_{\mathbf{w}} M \ , \qquad (2)$$

where Id stands for the identity matrix and W is a diagonal matrix whose diagonal elements are the weights w_i. However, there are shortcomings of such simple solvers. The main critic is that they tend to smear the estimated current distributions over widely extended cortical regions, limiting the spatial resolution of the reconstructed sources. To circumvent this limitation and obtain more focal activations, alternative strategies based on general ℓ_p norms have been proposed. With ℓ_p norms, a value of $p \leq 1$ induces *"sparsity"*, *i.e.*, a small number of sources with non-zero amplitudes, while

S. Supek and A. Sušac (Eds.): Advances in Biomagnetism – BIOMAG2010, IFMBE Proceedings 28, pp. 124–127, 2010.
www.springerlink.com

with ℓ_2 all sources have non-zero amplitudes. This observation led to the development of FOCUSS [2] and MCE [1] that work instant by instant, and more recently to a promising spatiotemporal inverse method based on a mixed ℓ_2-ℓ_1 norm [3]. We call such solvers *sparse solvers* and the set of estimated active dipoles the *active set*.

During an experiment, a subject is generally asked to perform different cognitive tasks or to respond to various external stimulations. They are referred to as different experimental conditions. Neuroscience questions ofter require to quantify the difference between these conditions, *i.e.*, to assess what are the different and the common active brain regions involved in the different conditions. Let us denote by M^k the measurements for condition k and d_k the number of conditions. With linear inverse solvers, a contrast map between condition k and condition l can be obtained simply by computing: $\hat{X}^k - \hat{X}^l = HM^k - HM^l = H(M^k - M^l)$.

With a sparse solver that provides better spatial resolution, the contrast $\hat{X}^k - \hat{X}^l$ can also be computed but it is likely to fail. Because of noise, the active set for condition k is likely to be different than the active set for condition l even if both conditions share a common active region. The aim of the solver detailed in this contribution is to circumvent this limitation. To do so, we propose a solver based on ℓ_2-ℓ_1 mixed norms that inverts all the conditions simultaneously as in [8] for the problem of functional mapping with M/EEG. By doing so, we want to estimate what are the common and specific active regions in all conditions.

The rest of this contribution consists of two parts. Section II, introduces the solver and briefly sketches the implementation details. Section III, presents some simulation results on two datasets: toy data and synthetic MEG data reproducing activations appearing for somesthesic stimulations of different fingers. Quantitative validation is provided on the realistic dataset.

II. Method

In [3], the problem solved corresponds to (1) where $\phi(X)$ equals to $\|X\|_{21}$. This ℓ_{21} mixed norm is defined for a matrix $X \in \mathbb{R}^{d_x \times d_t}$ by $\|X\|_{21} = \sum_{i=1}^{d_x} \|X_{i\cdot}\|_2 = \sum_{i=1}^{d_x} \sqrt{\sum_{t=1}^{d_t} x_{it}^2}$.

When penalizing the inverse problem with the ℓ_{21} norm the solution $\hat{X}_{21}$ has a sparse structure where only a few rows have non-zero coefficients. These rows form the active set $\mathscr{A}(\hat{X}_{21})$ of the ℓ_{21} solution: $\mathscr{A}(\hat{X}_{21}) = \{i \text{ s.t. } \hat{x}_{it} \neq 0 \text{ for all } t\}$. The reason for this sparse structure is that the ℓ_{21} norm groups row coefficients in different ℓ_2 norms: $\|X_{i\cdot}\|_2$.

When considering different experimental conditions, indexed by k, the ℓ_{21} solver provides the solutions $\hat{X}_{21}^k$ whose active sets are $\mathscr{A}(\hat{X}_{21}^k)$. Due to noise, is it likely that the two estimated active sets $\mathscr{A}(\hat{X}_{21}^k)$ and $\mathscr{A}(\hat{X}_{21}^l)$ are not exactly equal, even if the generators in both conditions are exactly the same. One way to go around this issue consists in computing the inverse problem on all conditions simultaneously by solving the problem:

$$\hat{X} = \arg\min_{X} \frac{1}{2} \left\| \sum_{k=1}^{d_k} M^k - GX^k \right\|_F^2 + \lambda \|\underline{X}\|_{21} , \lambda \in \mathbb{R}_+ \quad (3)$$

where the matrix $\underline{X} \in \mathbb{R}^{d_x \times d_k d_t}$ is obtained by concatenating horizontally all the X^k. By doing so, one forces $\mathscr{A}(\hat{X}_{21}^k)$ to be the same for all k.

In practice, this latter constraint is too strong as different conditions necessarily have specific current generators in addition to the common ones. This observation suggests that one could estimate each $\hat{X}^k$ in two components: a component $\hat{X}_c^k$ whose active set is common between all conditions and a component $\hat{X}_s^k$ that is specific. The problem addressed becomes:

$$(\hat{X}_c^k, \hat{X}_s^k)_k = \arg\min_{(X_c^k, X_s^k)_k} \frac{1}{2} \left\| \sum_{k=1}^{d_k} M^k - GX_c^k - GX_s^k \right\|_F^2 + \lambda\rho \|\underline{X}_c\|_{21} + \lambda(1-\rho) \sum_{k=1}^{d_k} \|X_s^k\|_{21} \quad (4)$$

where $\lambda \in \mathbb{R}_+$ and $\rho \in [0,1]$. Once $(\hat{X}_c^k, \hat{X}_s^k)_k$ are estimated one can get access to the components of the measurements that are common and specific, $(\hat{M}_c^k, \hat{M}_s^k)_k$, defined by $\hat{M}_c^k = G\hat{X}_c^k$ and $\hat{M}_s^k = G\hat{X}_s^k$.

Implementation. We solve the optimization problem in (4) using iterative Weighted Minimum Norm computations (2) where the weights are updated at each iteration. The approach is similar to the FOCUSS solver and is usually referred to as IRLS (Iterative Reweighted Least Squares). Some details can be found in section 4.2.1 in [9]. Once the active set is estimated, the time series are obtained by running an ordinary least square constrained to the active set, which is no longer an ill-posed problem as the active set size is smaller than the number of sensors.

III. Simulation study

To illustrate the method, we designed a toy dataset with 3 conditions, 1 common generator and 1 specific generator in each condition. The time series for each source is simulated to be smooth and last for 100 ms with positive and negative deflections similar to standard evoked responses measured by M/EEG. The toy dataset has $d_m = 10$ sensors and $d_x = 50$

possible sources. The matrix G is random with normalized columns. The noiseless time series, used later as ground truth, $(\check{M}_c^k, \check{M}_s^k)_k$ with $\check{M}^k = \check{M}_c^k + \check{M}_s^k$ are presented in figure 1. Since activations are not necessarily time locked with the same delay across conditions, activation time series are perturbed with random time jitters. This is illustrated in figure 2 where the source time series are plotted. The matrices X_c^k and X_s^k can be visualized as images like in figure 3(a) or by looking at the active sets like in figure 3(b). One can observe the row structure of the sparsity pattern. Starting from $(\check{M}_c^k, \check{M}_s^k)_k$ to which has been added a Gaussian white noise (SNR=3), the sources $(\hat{X}_c^k, \hat{X}_s^k)_k$ are estimated taking as parameters $\lambda = 0.01$ and $\rho = 0.4$. It can be observed that we recover the true sparsity pattern, *i.e.*, the good active sources. The associated estimated time series presented in figure 2(b) match with the ground truth presented in figure 2(a). The parameters were set empirically to maximize the fit with the ground truth. However, an optimal value for $\rho = 1/(1+\sqrt{d_k}) \approx 0.4$ can be analytically derived in the case where the time series of all generators have the same ℓ_2 norm. If this does not hold, the parameter ρ should be adapted.

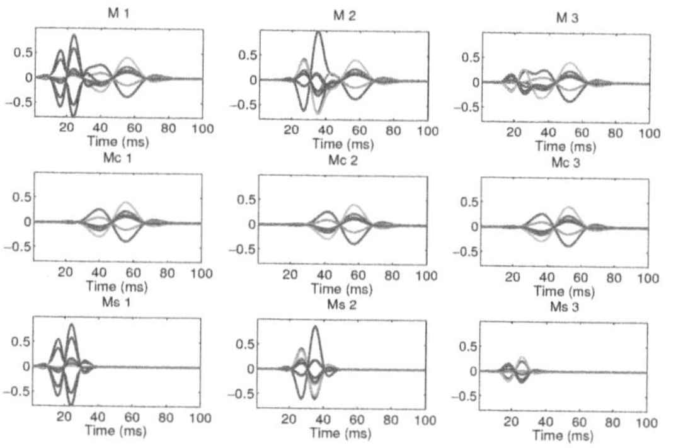

Fig. 1 Noiseless toy dataset with 3 conditions, 1 common generator and 1 specific generator in each condition. Each column corresponds to a condition. The 1st row corresponds to $\check{M}^k$, the 2nd to $\check{M}_c^k$ and the 3rd to $\check{M}_s^k$

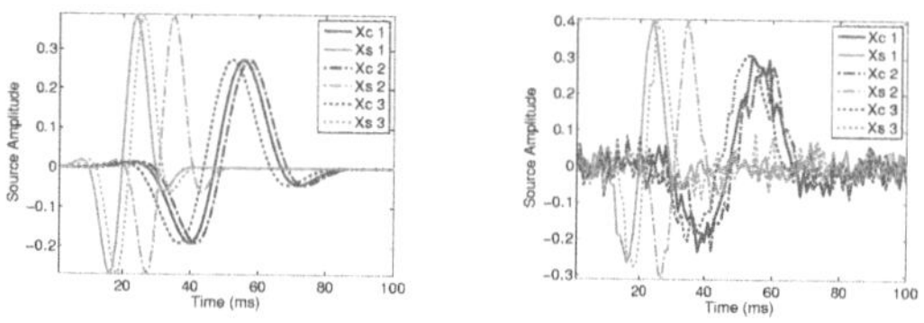

(a) Ground truth (b) Estimated

Fig. 2 Source time series. In green are the time series for the specific sources and in red the time series for the common generator. (a) Series used for simulation before corrupting the signal with an additive Gaussian noise (b) Estimated time series (SNR=3, $\lambda = 0.01$ and $\rho = 0.4$)

In order to quantitatively assess the performance of the method, we have conducted a simulation study with a realistic configuration. The matrix G is obtained by solving an MEG

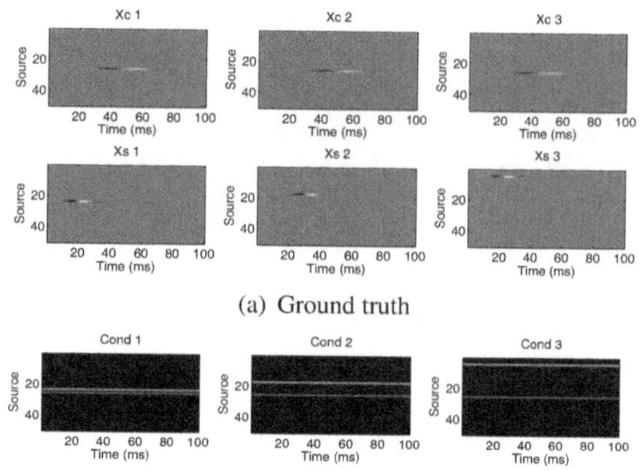

(a) Ground truth

(b) Active Sets for all conditions

Fig. 3 (a) Source amplitudes used for the toy data in all 3 conditions (b) Illustration of the active sets. In red is the active set common to all the X_c^k and in green is the active set specific to the different X_s^k

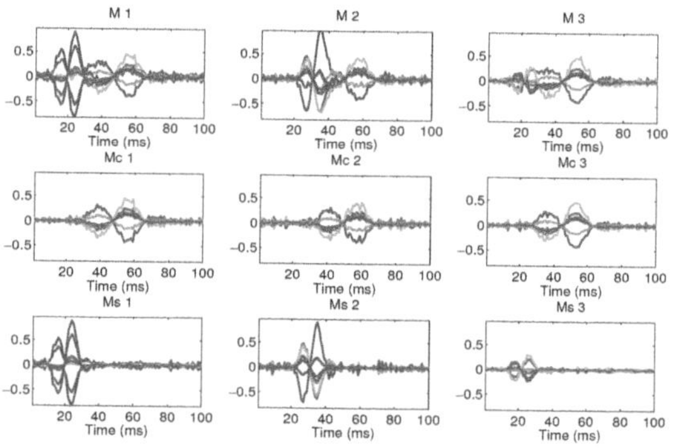

Fig. 4 Estimated times series $(\hat{M}_c^k, \hat{M}_s^k)_k$. Each column corresponds to a condition. The 1st row corresponds to $\hat{M}^k$, the 2nd to $\hat{M}_c^k$ and the 3rd to $\hat{M}_s^k$. One can observe a good match with the ground truth presented in figure 1

forward problem with a spherical headmodel. The sources are sampled over a cortically constrained domain consisting of a triangulated mesh of the cortical ribbon. In order to schematically reproduce activations appearing for somesthesic stimulations of two different right hand fingers, two specific generators have been positioned in the left primary somatosensory cortex (in green in figure 5) and one common generator has been positioned in the left parietal cortex. The time series of activations of these different generators are presented in figure 6(a). The computation has been carried out 10 times for multiple SNR values. One result of estimated source times series with SNR equals 1 is presented in figure 6(b). For each estimation, we chose to quantify the error using $(\hat{M}_c^k, \hat{M}_s^k)_k$ and the ground truth $(\check{M}_c^k, \check{M}_s^k)_k$. The errors E_c and E_s are defined by $E_a = \|\hat{\underline{M}}_a - \check{\underline{M}}_a\|_F / (\|\hat{\underline{M}}_a\|_F + \|\check{\underline{M}}_a\|_F) \in [0,1]$ where $\hat{\underline{M}}_a$ (resp. $\check{\underline{M}}_a$) is obtained by concatenating horizontally the $\hat{M}_a^k$ (resp. $\check{M}_a^k$) and "a" is set to "c" or "s".

The results of the simulations are presented in figure 7. One can observe that the method provides unbiased results at very high SNR. The errors obtained with SNR equal to 20 are almost 0. As the SNR drops down, the errors increase but stay small until the SNR gets negative, which is acceptable.

In order to demonstrate that the portion of the active set that is common between conditions is more consistently recovered with the SMC solver, we have compared our results with the single condition spatiotemporal sparse solver proposed in [3]. We ran this solver on each condition successively and defined the common active sets as the intersection between all the $\mathscr{A}(\hat{X}^k_{21})$. The d_k specific portions of the active set for the ℓ_{21} solver were obtained by subtracting from each $\mathscr{A}(\hat{X}^k_{21})$ the common portion. For both methods, the estimation error was quantified as the number of differences between the active sets obtained and the ground truth. Results are presented in figure 8. One can observe that the estimation of $\mathscr{A}_c$ is improved thanks to the SMC solver, especially at low SNR. This was the principal motivation for the development of this solver.

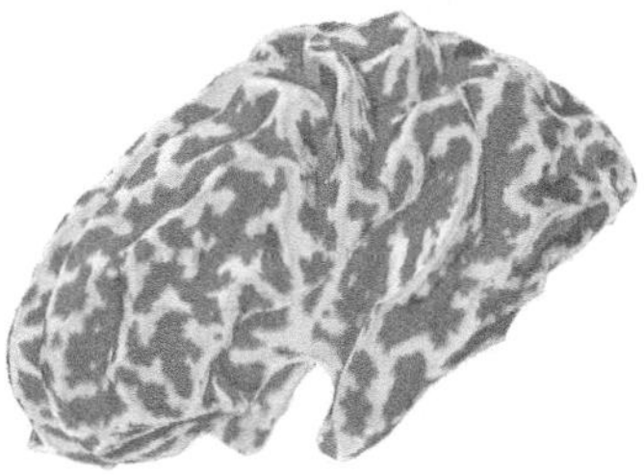

Fig. 5 Synthetic source configuration with 2 specific generators in the primary somatosensory cortex (green sphere end green pyramid) and 1 common generator in the parietal cortex (red sphere)

IV. Conclusion

In this contribution, we explained how sparse solvers can be used in M/EEG to go beyond simple localization by providing a way to estimate what are the specific and what are the common current generators involved in different cognitive tasks. By doing so we provide a principled and robust way to compare different conditions when using non-linear sparse inverse solvers.

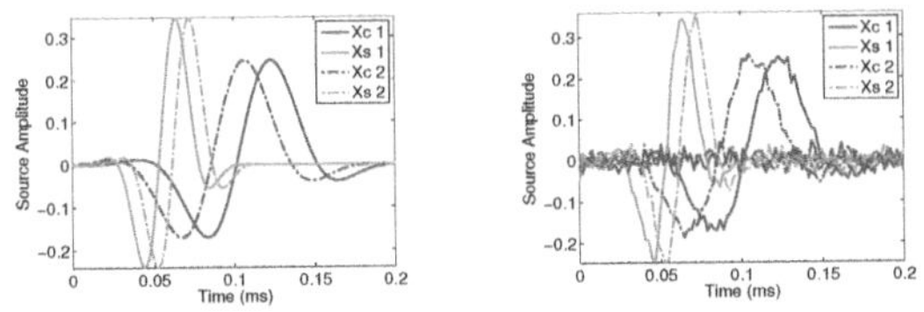

(a) Ground truth (b) Estimated

Fig. 6 Source time series. In green are the time series for the specific sources and in red the time series for the common generator. (a) Series used for simulation before corrupting the signal with an additive Gaussian noise (b) Estimated time series (SNR=1, $\lambda = 5$ and $\rho = 0.4$)

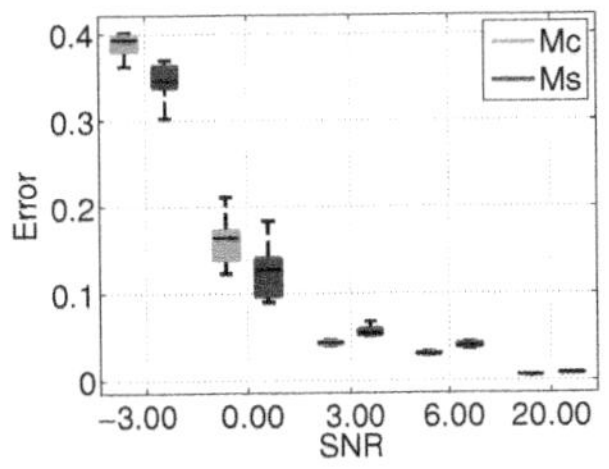

Fig. 7 Estimation errors for different values of SNR (see text). Errors are provided for 10 repetitions of the experiment. The method has no estimation bias for high SNRs and presents low errors even at low SNRs

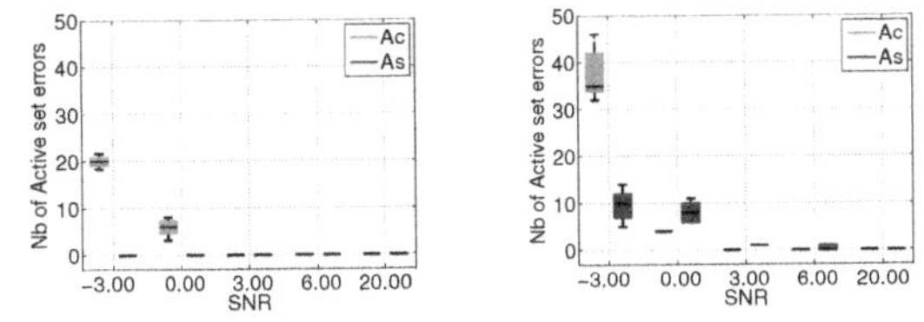

(a) SMC (b) Single condition ℓ_{21}

Fig. 8 Number of errors in the estimated active sets for both solvers: the ℓ_{21} solver [3] and the SMC solver. Results are presented for 10 repetitions of the experiment. It can be observed that the estimation of $\mathscr{A}_c$ is improved by the SMC solver especially at low SNR

References

1. Matsuura K., Okabe Y.. Selective minimum-norm solution of the biomagnetic inverse problem. *IEEE Trans Biomed Eng.* 1995;42:608–615.
2. Gorodnitsky I, George J, Rao B. Neuromagnetic source imaging with FOCUSS: a recursive weighted minimum norm algorithm *Electroencephalography and clinical Neurophysiology.* 1995.
3. Ou W, Hämäläinen M, Golland P. A Distributed Spatio-Temporal EEG/MEG Inverse Solver *Neuroimage.* 2009;44:932-946.
4. Dale A, Sereno M. Improved Localization of Cortical Activity By Combining EEG and MEG with MRI Cortical Surface Reconstruction *Journal of Cognitive Neuroscience.* 1993.
5. Hämäläinen M, Ilmoniemi R. Interpreting magnetic fields of the brain: minimum norm estimates *Medical and Biological Engineering and Computing.* 1994;32:35–42.
6. Wang J.-Z., Williamson S.J., Kaufman L.. Magnetic source images determined by a lead-field analysis: the unique minimum-norm least-squares estimation *Biomedical Engineering, IEEE Transactions on.* 1992;39:665-675.
7. Dale A, Liu A, Fischl B, Buckner R. Dynamic statistical parametric neurotechnique mapping: combining fMRI and MEG for high-resolution imaging of cortical activity *Neuron.* 2000;26:55-67.
8. Gramfort A, Kowalski M. Improving M/EEG source localization with an inter-condition sparse prior in *Proceedings ISBI* 2009.
9. Gramfort A. *Mapping, timing and tracking cortical activations with MEG and EEG: Methods and application to human vision.* PhD thesis 2009.

Author: Alexandre Gramfort
Institute: INRIA NeuroSpin CEA
Street: Bat. 145, PC 156
City: Gif-sur-Yvette
Country: France
Email: alexandre.gramfort@inria.fr

MEG-Clinic: A Comprehensive Software Solution for Routine MEG Analysis

Elizabeth Bock and Sylvain Baillet

MEG Program, Department of Neurology, Medical College of Wisconsin, Milwaukee USA

Abstract— **As the use of Magnetoencephalography (MEG) becomes more prevalent in the clinical setting, the need for an optimized workflow has become imperative. A comprehensive solution will relieve the reliance on the scientist and physician and move the majority of the workload to automated algorithms and a trained technician. The solution will include a user interface that will guide a technician through the entire workflow, while automating steps to reduce the time to reporting. The current workflow includes data acquisition, data conditioning, data analysis and report generation. MEG-Clinic has been developed as an integrated solution using Java and Matlab. This application manages the patient workflow by organizing and displaying those files associated with each patient, while providing a "wrapper" to several existing software packages that provide for data conditioning and analysis. MEG-Clinic lays out the entire workflow, step by step. As the technician completes each step, either by launching a manual process or reviewing the results of an automated process, new data are available for analysis and the next steps can be completed. MEG-Clinic is currently being developed for the Elekta-Neuromag system but could be extended to other MEG systems using appropriate file format conversion. Supported software packages include MaxFilter, MNE, Brainvisa, BrainStorm and Prism. Appropriate functions from these packages are called from MEG-Clinic when needed. If the process is automated (MEG-Clinic runs the process in the background and provides an output), the technician will be required to review the outputs. MEG-Clinic automates several steps of data conditioning, including MaxFilter (signal-space separation), artifact removal (ECG, EOG, etc), data averaging for functional protocols and Elekta FDA Source Analysis. In addition to this data conditioning, MEG-Clinic currently interacts with BrainStorm for final analysis and Prism for final reporting. The ongoing development will include integration of additional analysis programs.**

Keywords— **MEG, software, analysis, routine, clinical.**

I. Introduction

A typical bottleneck of routine clinical MEG is the considerable number of consecutive steps required before clinicians may access their data at an interpretable level. This is particularly critical in the clinical environment where results have to be made promptly available, and where data analysis requires some form of protocol standardization across patients. From an operational standpoint, state-of-the-art MEG source imaging consists of a multi-step procedure involving multiple processes and software applications and therefore, considerable user interventions and expertise.

MEG-Clinic is a proposition from the MEG Program of Froedtert Hospital and the Medical College of Wisconsin to make the routine of clinical MEG using advanced signal processing and source imaging techniques become reality. This article reports on the strategy adopted by our group in that matter and presents some preliminary results in terms of software implementation.

II. Methods

A. MEG Data Analysis as a Multistep Procedure

The analysis of large, individual MEG recordings includes mandatory pre-processing steps such as: 1) filtering and noise-rejection processes, 2) review for quality assessment, 3) artifact detection/correction/rejection, 4) event detection and classification. State-of-the-art source imaging techniques require further steps in the analysis pipeline and namely, the integration with anatomical data: 1) processing of MRI volume data (tissue segmentation and tessellation), 2) geometrical registration with MEG coordinate system, 3) forward modeling of distributed MEG sources [1]. Finally, MEG source analysis is contingent on the clinical objective and may take multiple faces: 1) functional mapping of essential systems (somatosensory, motor, language, etc.), 2) evaluation of ongoing brain activity in the spectral domain, 3) stimulus-induced oscillatory bursts.

The requirements of clinical data management add further constraints to the picture in terms of databasing (archiving and retrieval), workflow (identified operators and their respective expertise) and reporting. Clinical MEG centers have usually approached these issues with ad hoc solutions, which may result in inconsistent data management across the clinical-MEG community or even within a given MEG program as the workflow might be very much operator-dependent. For instance, each reviewer can impart his/her own data management standards and reporting, which eventually impacts the reproducibility of results. While this may not be a systematic issue in the context of a research program – though this certainly impedes productivity and adds considerable staffing loads on training and education of new

S. Supek and A. Sušac (Eds.): Advances in Biomagnetism – BIOMAG2010, IFMBE Proceedings 28, pp. 128–131, 2010.
www.springerlink.com

users – a consistent, standardized workflow for the MEG laboratory is crucial to the success of a clinical program.

B. Example of a Routine Data Analysis Pipeline

Our MEG Program uses an Elekta-Neuromag 306-channel system for data collection. Some of the steps and software elements described here are specific for this instrument but have equivalent counterparts in other MEG systems. Though the hardware and software used for data acquisition is system-specific, the rest of the software solutions available to users for data analysis is generally system-independent. Here we illustrate how advanced source analysis may be achieved using a series of software elements that will be integrated into MEG-Clinic.

Raw data is first processed using Elekta-Neuromag's MaxFilter software for attenuation of perturbations of far-field sources using a spatial-filtering approach (signal-source separation, SSS) [2, 3]. This software is a mandatory pre-processing step for a posteriori data rectification in order to benefit from the Maxshield active shielding solution provided by Elekta when a light magnetically-shielded room (MSR) is used. As a side note, light MSR may be considered as an adequate alternative to heavier and bulkier MSR in a clinical environment where space is generally problematic.

Raw data then needs to be reviewed for quality control and event marking. Typical artifacts caused by heartbeats, eye blinks and movements may be marked by a qualified operator or detected automatically. Data segments with artifacts may be either discarded or further processed for correction. We suggest heartbeat artifacts are detected using concurrent electrocardiographic recordings (ECG). Events are automatically defined at the R peak of the ECG QRS complex. A Signal Space Projection (SSP) vector is then defined using principal component analysis (PCA) of [10, 30]Hz bandpassed raw data about each heartbeat event ([-80, 80]ms time window). Further SSP projections may be defined manually about electro-oculographic (EOG) events or other transient artifact occurrences. The MNE software (http://www.nmr.mgh.harvard.edu/martinos/userInfo/data/sofMNE.php) conveys an efficient and user-friendly interface to proceed to artifact compensation and raw data reviewing.

Once the data is cleaned, off-line averaging about marked events may be readily performed using MNE or in-house software. Alternatively, the analysis of segments of raw data – for e.g., spectral analysis – may also be completed. We encourage testing for the performance of artifact compensation by generating the raw data about e.g., heartbeat events before and after the SSP correction to assess for improvement in artifact attenuation and possibly expand signal projections to supplementary dimensions (e.g., the heartbeat's T wave artifact may be more prevalent in some younger patients).

In the clinical context, the clean data may be further reviewed by an electrophysiologist for identifying events of interest (e.g. interictal epileptic spikes or other paroxystic events) which will be additional targets for source localization.

Source modeling may be handled using a variety of commercial and academic software. US clinical MEG programs are required to run FDA-approved software to complete source analysis. This latter is provided with the MEG instrument data analysis software suite. Our experience is that this step requires considerable user-intervention in terms of data retrieval, specification of head and source model parameters, estimation per se and the generation of figures and statistical summaries. Some software packages are poorly equipped with source modeling applied to long, spontaneous data analysis which may be of interest to investigators and clinicians. One of our objectives has been to develop a software wrapper to address this issue and generate scripting commands to pilot the software through large amounts of data with minimal user intervention. Our approach is further detailed below. Academic software solutions may offer complementary alternatives to commercial packages. In that respect, MNE offers a fully-integrated solution (available under Linux and MacOS only). Our group has been co-developing alternative approaches through the BrainStorm application for a number of years (http://neuroimage.usc.edu/brainstorm, available on all software platforms running MATLAB), which we propose to access through MEG-Clinic using scripting tools.

The final step of the MEG analysis consists in generating summary figures and measures that eventually are included in a formal report and may ultimately be archived in a picture archiving and communication system (PACS). We shall now describe the integrated solution provided by MEG-Clinic.

C. Automated Processes

MEG-Clinic is designed to create a consistent workflow through the entire process from data collection to reporting (Fig. 1). It is comprised of several parts, including automated processes, workflow visualization and control and report generation. MEG-Clinic can be used with a graphical user interface (GUI) to step through the workflow, configure processes, re-run processes and interact with images. MEG-Clinic can also be run from the command line for automatic SSS, heartbeat artifact removal (SSP) and offline averaging of evoked data sets. More specifically, MEG-Clinic monitors whenever a new recording, or run, is

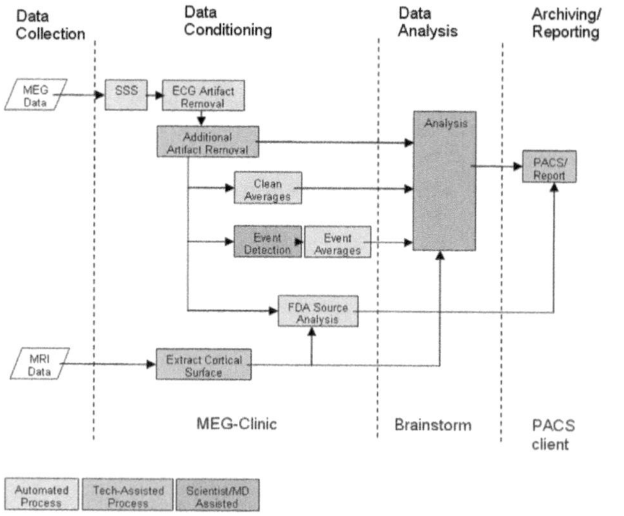

Fig. 1 Overview of the MEG data collection and analysis workflow using MEG-Clinic

recorded and saved. It then automatically launches MaxFilter to perform SSS.

When data acquisition is complete, MEG-Clinic launches automatic heartbeat artifact removal. The ECG channel is identified and heartbeats are detected and marked. MNE is then used to create an SSP projector from the cardiac events as described above and the cleaned file is saved. MEG-Clinic then determines if the data run contains functional event markers (as defined in a given stimulus paradigm). Offline averaging is used to re-average the evoked set using the cleaned data. After these steps, MEG-Clinic creates new entries into the Brainstorm database. Complementary to the FDA-approved source analysis solutions from Elekta, advanced head and source modeling are performed using the individual MRI volume data processed through brainVISA (http://brainvisa.info). MEG-Clinic automatically creates an entry in brainVISA's database and the segmentation and tessellation of brain tissues proceeds in parallel to MEG data pre-processing steps. All of these automated processes can be reviewed, configured and re-run using MEG-Clinic's GUI.

With minimal user interaction, MEG-Clinic aids the user in importing MRI and MEG data into BrainStorm. By capitalizing on the fact that many of the processes involved may be scripted, the time that an operator spends interacting with the software packages involved can be drastically reduced. Once the result files are created, they can be accessed from MEG-Clinic, stored with the patient record, or rendered and reviewed using BrainStorm.

The result files can be automatically imported into a multimodal review software and report generator software such as Prism View (http://www.prismclinical.com/). In a hospital environment, the report is stored with the patient record and image snapshots of MEG source analyses may be stored in the PACS archives.

D. User Interaction

MEG-Clinic is designed as a guide to allow the user to step through each task in the workflow. The GUI display is organized into two major sections, the workflow and the data file tree.

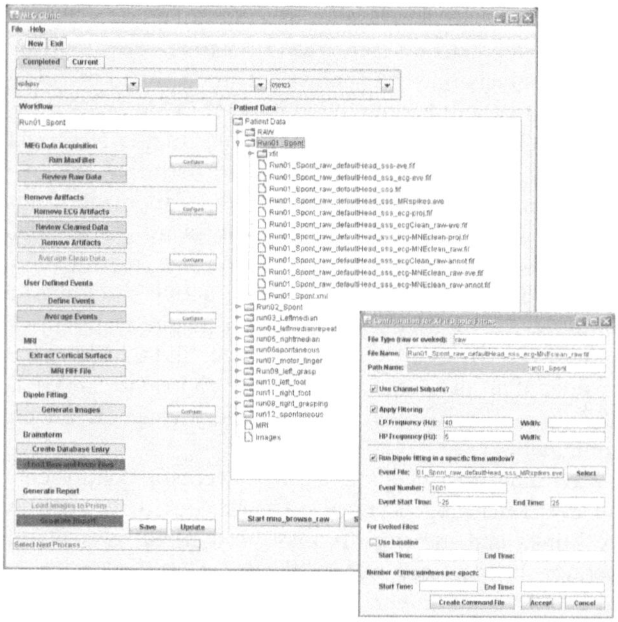

Fig. 2 The MEG-Clinic user-interface

Fig. 2 displays the MEG-Clinic GUI environment. The workflow is on the left side of the GUI and is itemized through an ordered series of buttons representing the chronological steps that need to be taken by the operator. MEG-Clinic manages the flow of data through the workflow by determining which file types are present in the dataset. The buttons are initially all disabled, but become enabled when the previous steps are performed, and corresponding file types are created. The buttons also change colors according to their status. A red button indicates a step is ready to be performed and a green button indicates that a process is complete. The orange color on a button indicates that the step is optional. For example, orange is used for "review" of the data. Selecting an orange button renders the data using the adequate software tool, so the user can confirm the quality of the automated process. Each recording has a separate workflow status which is saved in XML format so it can be recalled whenever a user interacts with the dataset.

Once all of the automatic processes are complete, the user can choose to review the results by selecting the review buttons or start the appropriate software for rendering a file. The user may also configure and re-run some or all of the processes by selecting the configure button available to the right of each process. This opens a configuration window containing the adjustable parameters for the process, allows the user to change these parameters and then starts running the process with the new configuration (Fig. 3).

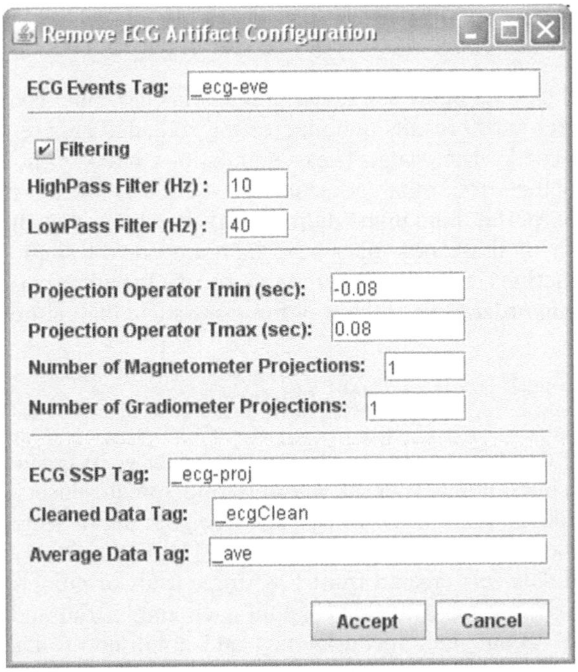

Fig. 3 An example of the configuration window used in MEG-Clinic to provide flexibility to the user

The user can choose to overwrite existing results files or create new ones but can never overwrite or delete an original raw file.

E. Additional Workflow Management

In addition to the visual workflow management, several other key features make MEG-Clinic particularly useful in a clinical setting. One feature is a "current" list that identifies new records and keeps track of those records that are currently being analyzed. Likewise it maintains a list of records for which analysis is complete. This helps the technician recall which records need to be completed and alert them to newly added records. These lists are stored as files in Extensible Markup Language (XML) format. Another feature includes flexible default configurations. These configurations are stored in XML files that can easily be added to or edited allowing changes in the default configurations of the processes.

F. Software Architecture

MEG-Clinic is designed as a Java/MATLAB application and is therefore operating-system (OS) independent. Note however that some of the tools called by MEG-Clinic may not be available under multiple OS. Elekta's suite for instance is only available under Linux and HP/UX; MNE is available under MacOS and Linux. The GUI is designed to run without MATLAB, but there will be limitations in the processing. The Java portion of the application is responsible for providing the GUI and the interactions with other software such as Elekta-Neuromag's and MNE. MATLAB is used to run the heartbeat detection algorithms and for running BrainStorm. These two modalities work together to give MEG-Clinic an easy-to-use and easy-to-customize feel.

III. CONCLUSION

Rather than suggesting yet another software solution, MEG-Clinic capitalizes on the wealth of existing software packages in the MEG community to propose an integrated solution with efficiency and minimal user intervention as a primary objective. The software prototype is in its first 6 months of development and is not yet available for distribution. We anticipate MEG-Clinic may be distributed in the close future through MEG vendor software packages and through academic collaborative projects. Approval from FDA in the United States and equivalent agencies elsewhere may be considered in partnership with a commercial or academic sponsor.

REFERENCES

1. Baillet, S.; Mosher, J. & Leahy, R. (2001) Electromagnetic brain mapping, *IEEE Signal Processing Magazine18:*14-30
2. Taulu, S.; Kajola, M. & Simola, J. (2004) Suppression of interference and artifacts by the Signal Space Separation Method. *Brain Topogr,16:* 269-275
3. Taulu, S. & Hari, R. (2009) Removal of magnetoencephalographic artifacts with temporal signal-space separation: demonstration with single-trial auditory-evoked responses. *Human Brain Mapp.*, 30: 1524-1534

Author: Elizabeth Bock
Institute: MEG Program, Department of Neurology
Street: 9200 W. Wisconsin Avenue
City: Milwaukee, WI 53226
Country: USA
Email: ebock@mcw.edu

MEG-SIM Portal: Reconstructions from Realistic Simulations of Sensory and Cognitive Processing

Lori Sanfratello[1,2], Julia M. Stephen[2], Douglas Ranken[3], Elaine Best[2], Theodore Wallace[1], Jason MacArthur[2], Katie Gilliam[2], and Cheryl J. Aine[1,2]

[1] UNM HSC, Dept. of Radiology, Albuquerque, USA
[2] The Mind Research Network, Albuquerque, USA
[3] Los Alamos National Lab, Los Alamos, USA

Abstract— **To enable and encourage comparisons between MEG analysis algorithms, a partnership was formed between 5 research institutions. MEG/MRI data was acquired from 5 subjects at 3 of these institutions using 3 different MEG systems and paradigms from 3 sensory modalities. A grant from NIMH (R21MH080141) allowed us to create realistic simulated data from the data acquired through this partnership. Visual simulated data sets, along with tools to read these data, have been placed on a collaborative portal established and maintained at MRN (http://portal.mind.unm.edu/megsim). Additional simulated data sets are added as they are developed. The present study reports on the construction of three different types of realistic simulated data sets and results using a semi-automated, multidipole, spatio-temporal (CSST) algorithm for source localization. Our most recent data sets supply a variety of source configurations in multiple sensory modalities, as well as simulating functional connectivity. The results show that CSST accurately reconstructs both temporal and spatial characteristics of the simulated data sets.**

Keywords— **Simulations, single trials, working memory, modeling, oscillatory.**

I. Introduction

Due to differences in hardware/software between MEG systems and the wide variety of MEG analysis algorithms currently in use, it has historically been difficult to meaningfully compare and combine data across MEG research institutions. In order to begin to address this issue a partnership was formed between the Mind Research Network (MRN), Massachusetts General Hospital (MGH), University of Minnesota/Veterans Affairs (UMN/VA) in Minneapolis, University of New Mexico (UNM), and Los Alamos National Laboratory (LANL). MEG/MRI data were acquired from 5 subjects using 3 different MEG systems (VSM MedTech 275, Elekta-Neuromag 306, 4-D Neuroimaging 3600) and 3 different paradigms (visual, auditory and somatosensory). A grant from NIMH (R21MH080141) allowed us to then create realistic simulated data derived from the real noise contained in the collected data. Here we present recent results in both creating and analyzing realistic simulated MEG data. These simulations have been made available to other researchers via a web portal (http://portal.mind.unm.edu/megsim). We hope that the addition of these new data sets, including novel single trial simulations, will foster comparisons of algorithm performance in order to pave the way for cross-site collaborations.

II. Methods

Three different types of simulated data were created: 1) realistic simulated visual, auditory and somatosensory data sets developed from 100 averaged noise trials in which the sources of activity were embedded, 2) realistic visual simulated data sets created from 128 single trials of noise where the sources are embedded within each single trial and "jittered" about their mean latency and amplitude (all single trials are available via the web portal), and 3) simulated data sets similar to those described in 2) with the addition of oscillatory activity, reflecting functional connectivity of different brain areas. For this discussion we will confine ourselves to the CTF MEG system, but work is under way to also create the same simulations for the other two MEG systems.

To produce the averaged noise trial for the realistic simulated somatosensory, auditory, and visual data (data set #1) MNE (http://www.nmr.mgh.harvard.edu/martinos/userInfo/data/sofMNE.php) was used for initial display, rejection of bad channels, and averaging of 100 1s noise trials. This single noise file was then imported into MRIVIEW from within IDL. MRIVIEW [1] is a software tool (http://www.lanl.gov/p/p21 /mriview.shtml) for integrating volumetric MRI head data with functional information (e.g.,EEG, MEG, fMRI). A Forward Simulator is included for creating multiple MEG focal or distributed-source regions of arbitrary size and orientation. User-specified, ellipsoidal regions of gray matter (GM) and white matter (WM) boundary can be labeled and used to create simulated

S. Supek and A. Sušac (Eds.): Advances in Biomagnetism – BIOMAG2010, IFMBE Proceedings 28, pp. 132–135, 2010.
www.springerlink.com

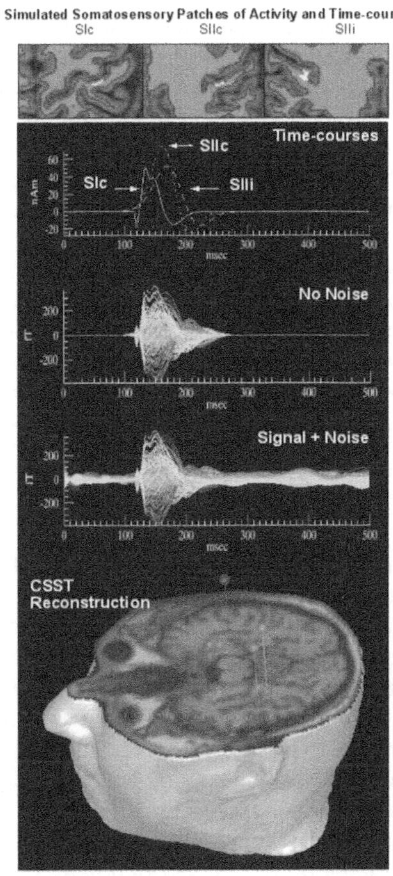

Fig. 1 Somatosenory realistic simulated data. Shown are cortical patches, time-courses, and CSST reconstruction of the source locations

regions of activity. The activation time-courses can then be modeled using multiple Gaussians or a sinusoid. Sensor geometries are obtained from a sister software package MEGAN (E. Best, http://portal.mind.unm.edu/megsim/tools/MEGAN/MEGAN.html), also programmed in IDL, for generating the forward fields. The simulated sensor measurements are obtained by summing the forward fields for all of the measurements. Once the averaged noise was brought into MRIVIEW, time-courses of simulated evoked activity were embedded within the noise to create the simulated data. These time-courses were based on physiologically plausible signals assigned to be emitted from cortical patches. The time-courses were created using multiple Gaussians (usually 3) in the MRIVIEW Forward Simulator. The locations of the cortical patches were based on our previous studies [2-5]; selected areas of activation depended on both the type of task being simulated and the particular subjects' cortical geometry (i.e., MRI).

Figure 1 shows a realistic simulated data set for a somatosensory task. The upper panel identifies the locations and extent of the cortical patches (yellow) for SI contralateral (SIc), SII contralateral (SIIc), and SII ipsilateral (SIIi) sources while the time-courses provided to the cortical patches are shown beneath. Below the time-courses, waveforms as seen across the 275 CTF MEG channels are shown both without noise and with noise embedded within 128 averaged noise trials. Lastly, the output of our source localization software CSST [1] is shown at bottom, producing a reconstruction with an average localization error of .32 cm for the 3 sources. Simulated data sets for an auditory task and a visual task were created in the same manner.

Next, a realistic simulated visual working memory data set was created (Figure 2), consisting of 128 single trials, where late activity (e.g., 400-600 ms) was synchronous across four cortical sites, as often seen in working memory studies (data set #2): 1) primary visual cortex (V1); 2) inferior lateral occipital gyrus (I.LOG); 3) intraparietal sulcus (IPS); and 4) dorsolateral prefrontal cortex (DLPFC). This data set was generated by "jittering" the amplitudes and peak latencies of the time-courses in every noise trial, using the MRIVIEW Forward Simulator. That is, 128 single trials were generated which varied from each other similar to that seen in normal human data sets. Because we have all the single trials in addition to the averaged waveforms, other investigators can use single-trial methods on these data, e.g., time-frequency

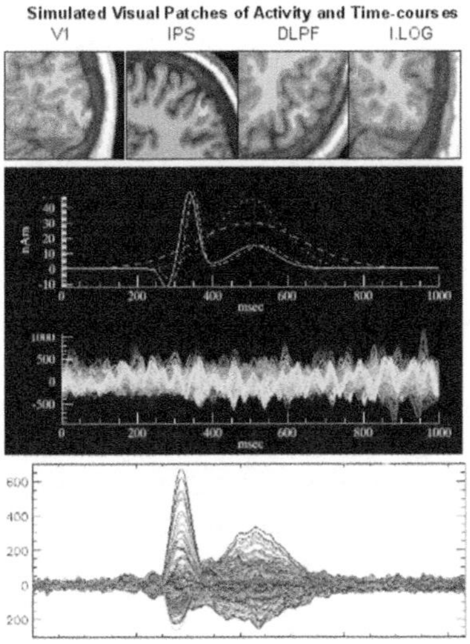

Fig. 2 Visual realistic simulated data

analysis or analyses literally using a single-trial, which is the ultimate goal for inverse methods in general.

For this data set, patch current strengths were initially assigned values similar to those we observe in visual working memory studies and were then randomly jittered about their peak values by up to +/- 50% across the single trials. Peak latencies were also jittered across each trial by a randomly selected value up to +/- the full width at half maximum (FWHM) divided by 2. FWHM equals 2.35 x the standard deviation (STD). The jittering of amplitudes and latencies across trials is necessary if one wishes, for example, to attempt Independent Component Analysis (ICA) on these data. At minimum, this more closely approximates what occurs in real data collection. Our goal is to generate data sets that are amenable to most MEG/EEG analysis techniques in order to permit direct comparisons between methods where the truth is known. Finally, to analyze these data using CSST the 128 single trials were then averaged together.

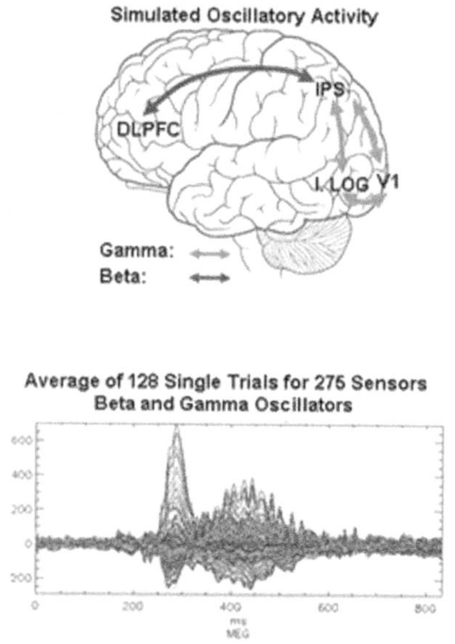

Fig. 3 Realistic simulated visual data with oscillations added

For the final data set created we again made 128 single trials; however, in this case oscillatory activity was added to the time-courses in data set #2, (generating data set #3). This data set was created in order to reflect functional connectivity of different brain areas. For the time-locked oscillatory activity shown in Figure 3, V1, I. LOG, and IPS oscillated between 30-60 Hz (Gamma band) across the 128 trials while IPS and DLPFC oscillated between 14-28 Hz (Beta band). Oscillatory activity for DLPFC was delayed by 20 ms relative to IPS and IPS gamma activity was delayed by 10 ms relative to IPS beta activity. The delays were meant to reflect normal time delays between visual areas. Gamma activity mimicked local circuitry activity between V1, I. LOG, and IPS while beta activity mimicked long-range connections between IPS and DLPFC.

For both beta and gamma oscillations, the average amplitudes were set at 10 nAm and were then jittered between 5-15 nAm across the 128 trials. Again the time-courses were constructed within MRIVIEW, but they had to be constructed independently at this point in time; i.e., one time-course contained the evoked response plus real noise while the other time-course contained the oscillations without noise. The two time-courses were then combined into one using Matlab. Again, to analyze these data using CSST the 128 single trials were then averaged together.

III. Results

Once the 3 data sets were created we tested the CSST algorithm for accuracy in source localization and time-course reconstruction. CSST uses the Nelder-Mead non-linear downhill simplex procedure to perform a spatial search and utilizes information based on singular value decomposition (SVD) of the data matrix for determining an estimate of the number of sources to be localized. The algorithm runs multiple instances of the nonlinear simplex search from random combinations of MRI-derived cortical starting locations or from a gridded head volume. We used a gridded head volume for the simulation results reported here. CSST does not refine solutions that are clearly incorrect, in a reduced chi-square sense, therefore this limited search of the head volume is able to be conducted quickly on a LINUX computer cluster located at the MRN. Furthermore, the solutions are completely unconstrained. CSST inverse and forward modeling procedures are launched from within MRIVIEW.

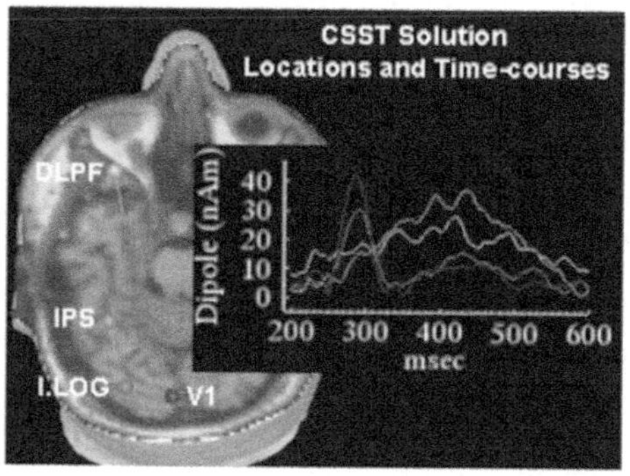

Fig. 4 Source and time-course reconstruction output from CSST algorithm

Figure 4 shows the output of the CSST algorithm for data set #2, consisting of an average of 128 jittered single trials. CSST outputs both the locations of the dipoles and the reconstructed time-courses of activity. Table 1 contains the results of this analysis for all 3 visual/working memory data sets that were created. Our results show that CSST accurately reconstructs both temporal and spatial characteristics of the simulated data sets, even with noisy and oscillating sources.

Table 1 CSST Results for Visual Sources, 4 Dipole Fit

Source	Distance from True Location (cm)	Difference from True Latency of First Peak (ms)	Difference from True Amplitude of First Peak (nAm)
With Averaged Noise:			
v1	.47	6	20
I.LOG	1.09	3	15.8
IPS	.57	1	5.8
DLPF	.79	7	10.1
Jittered Single Trials Averaged:			
v1	.15	6	2.5
I.LOG	.94	4	3.2
IPS	.37	3	7.9
DLPF	.89	13	6.8
Jittered Single Trials with Oscillations Added:			
v1	.47	6	9.8
I.LOG	.97	1	4.8
IPS	.7	1	11.2
DLPF	.49	16	2.4

IV. Discussion

With the creation of these simulated data sets a wider variety of MEG analysis tools can be compared. Construction of single trials that mimic the differences between epochs of real data allow analysis techniques such as ICA to be compared with analysis techniques such as CSST, where the analysis is conducted on the averaged data. With the addition of oscillations to the simulated data sets, analyses of functional connectivity between various brain areas can be investigated. This is important since it is incomplete to know, for example, that certain brain locations are active without information about which areas onset first, their durations, and whether cross-frequency oscillatory activity is evident across multiple disparate brain regions or not. Furthermore, the data will be made available in a variety of file formats, including netCDF and Curry (Compumedics, Neuroscan). Hopefully, the creation of these new data sets and formats, including novel single trial simulations, will foster algorithm performance comparisons to facilitate cross-site collaborations. All data sets and single trials discussed herein are currently available at the web portal (http://portal.mind.unm.edu/megsim).

Acknowledgment

This work was partially funded by NIH grants R21MH080141 and 1P20 RR021938-02. This work was also supported in part by the Department of Energy under Award Number DE-FG02-99#$62764 to the Mind Research Network. We thank M. Weisend, S. Ahlfors, M. Hämäläinen, J. Mosher, A. Leuthold, A. Georgopoulos, C. Donahue, F. Hanlon and R. Montaño for their help when the initial partnership between institutions was established.

References

1. Ranken, D., Best, E., Schmidt, D.M., George, J.S., Wood, C.C., Huang, M., 2002. MEG/EEG forward and inverse modeling using MRIVIEW. In: Nowak, H., Jaucisen, J., Giebler, F., Huonker, R., (Eds.). Proceedings of the 13th International Conference on Biomagnetism. pp. 785-787.
2. Stephen, JM, Aine, CJ, Christner, RF, Ranken, D, Huang, M, Best, E. Central versus peripheral visual field stimulation results in timing differences in dorsal stream sources as measured with MEG. Vision Research, 42:3059-3074, 2002.
3. Kovacevic, S, Qualls, C, Adair, JC, Hudson, D, Woodruff, CC, Knoefel, J, Lee, RL, Stephen, JM, Aine, CJ. Age-related effects on superior temporal gyrus activity during an oddball task. NeuroReport, 16:1075-1079, 2005.
4. Stephen, JM, Ranken, D, Best, E, Adair, J, Knoefel, J, Kovacevic, S, Padilla, D, Hart, B, Aine, CJ. Aging changes and gender differences in response to median nerve stimulation measured with MEG, Clinical Neurophysiology, 117(1): 131-143, 2006.
5. Aine, CJ, Woodruff, CC, Knoefel, JE, Adair, JC, Hudson, D, Qualls, C, Bockholt, J, Best, E, Kovacevic, S, Cobb, W, Padilla, D, Hart, B, Stephen, JM. Aging: Compensation or Maturation? NeuroImage, 32: 1891-1904, 2006.

Author: Lori Sanfratello
Institute: University of New Mexico
Street: 1 University of New Mexico
City: Albuquerque, NM 87131
Country: USA
Email: lsanfratello@mrn.org

Time-Frequency Source Estimation from MEG Data

R.E. Greenblatt[†,1], A. Ossadtchi[†,2], L. Kurelowech[3], D. Lawson[1], and J. Criado[3]

[1] Source Signal Imaging, Inc., San Diego CA USA
[2] St. Petersburg State University, St. Petersburg Russia
[3] Scripps Clinic, La Jolla CA USA

Abstract— **We extend adaptive beamformers (DICS) [1] and RAP MUSIC [2] to solve the MEG inverse problem in the time-frequency domain for induced and phase-locked components. Using event-related data, we estimate the complex covariances from time-frequency transformed single trials for total power, phase locked, and induced components [3]. We obtain the phase-locked DICS inverse as the difference between the total power and induced solutions. RAP-MUSIC was adapted to constrain the real and imaginary parts of the source space solution. We have verified these methods using simulated data, and validated them with somatosensory and motor MEG data.**

Keywords— **magnetoencephalography, source estimation, time-frequency, DICS, RAP MUSIC.**

I. Introduction

Interest in encephalographic oscillatory events has seen recent renewed attention with experimental and theoretical studies linking coherent multicellular rhythms with cognitive events. Time-frequency analysis overcomes some of the limitations of conventional FFT's, permitting simultaneous time-based and frequency-based analysis. Wavelets have well developed mathematical foundations [4] and have become the method of choice for electromagnetic encephalography (EMEG) time/frequency analysis. Tallon-Baudry et al. [3] introduced the useful distinction between phase-locked and induced components of the averaged oscillatory response. Nonlinearities in estimating the averages obtained from time-frequency transformed data have been an obstacle to the application of conventional source estimation methods to time-frequency transformed data. Greenblatt and Osman [5] and Lin et al. [6] have proposed methods to overcome the nonlinearity obstacle by computing single trial distributed source estimates and then averaging in source space. This approach, however, often has unacceptable noise sensitivity for typical experimental conditions, especially at higher frequencies.

Adaptive beamformers have been used to estimate source activity at specified locations for extracranial EMEG data in both the time [7] and the frequency [8] domains. They are adapted to the data via signal space covariance estimates. Gross et al., [8] proposed the use of an adaptive beamformer to solve the inverse problem in the frequency domain, and later for total power in the time-frequency domain, supporting both source power and coherence estimation [1]. Here, we generalize the adaptive beamformer method in the time-frequency domain, permitting the estimation of inverse solutions to both induced and phase-locked components of the event-related time frequency transformed data. We also show how dipole fitting, using the RAP MUSIC approach [2], may be adapted to all components in the time-frequency domain, extending the results of Sekihara et al [9].

† These authors contributed equally to the work in this report.

II. Methods

A. Time-Frequency Wavelet Transforms

For a single channel time series $s(t)$, we can obtain an estimate of the time-varying frequency component at time t and center frequency f as $z(t,f) = s(t) * w(t,f)$, where '*' represents the convolution operator and $w(t,f)$ is the complex wavelet function [10]. By successive applications at differing center frequencies, a time-frequency map may be constructed, averaged over a set of trials to extract phase-locked and total power components. The induced (non-phase-locked) component may then be obtained as the difference. Due to the linearity of the convolution operation, the resultant complex-valued transform coefficients may be used for source estimation.

B. Covariance Matrices

For each trial $1 \leq n \leq N$ and each channel $1 \leq k \leq K$ we compute a set of wavelet coefficients $z_{kn}(t,f)$ parametrized by time t, $t_{min} \leq t \leq t_{max}$ and frequency f, $f_{min} \leq t \leq f_{max}$ For each time-frequency bin (t,f) and for each n-th trial we form vectors $z_n(t,f) = [z_{n1}(t,f), z_{n2}(t,f), \ldots, z_{nK}(t,f)]^T$ by collecting the corresponding wavelet coefficients from all channels and use them in computing a set of covariance matrices.

We distinguish three different correlation matrices representing the spatial structure of total power, phase locked

S. Supek and A. Sušac (Eds.): Advances in Biomagnetism – BIOMAG2010, IFMBE Proceedings 28, pp. 136–139, 2010.
www.springerlink.com

and induced components. Estimate the total power correlation matrix as

$$\hat{\mathbf{C}}_{sig}^{tot}(t,f) \approx \frac{1}{N}\sum_{n=1}^{N}\mathbf{z}_n(t,f)\mathbf{z}_n^H(t,f) \quad (1)$$

and the phase-locked correlation matrix as

$$\hat{\mathbf{C}}_{sig}^{lock}(t,f) \approx \left(\frac{1}{N}\sum_{n=1}^{N}\mathbf{z}_n(t,f)\right)\left(\frac{1}{N}\sum_{n=1}^{N}\mathbf{z}_n(t,f)\right)^H \quad (2)$$

Under the assumption of independence of phase-locked and induced components in the data, the correlation for the induced component is the difference between the total power and phase-locked component correlations, or

$$\hat{\mathbf{C}}_{sig}^{ind}(t,f) = \hat{\mathbf{C}}_{sig}^{tot}(t,f) - \hat{\mathbf{C}}_{sig}^{lock}(t,f) \quad (3)$$

We define averaged correlation matrices as

$$\hat{\mathbf{C}}_{sig}^{tot,lock,ind} = \frac{\sum_{t=t_{min}}^{t_{max}}\sum_{t=f_{min}}^{f_{max}}\hat{\mathbf{C}}_{sig}^{tot,lock,ind}(t,f)}{(t_{max}-t_{min})(f_{max}-f_{min})}$$

Note, that in that case $\hat{\mathbf{C}}_{sig}^{lock}$ has no longer rank of unity. However, because of likely similarity of correlation matrices corresponding to the t-f bins in the region of interest $\hat{\mathbf{C}}_{sig}^{lock}$ is still significantly rank deficient.

C. Source Estimation with Scanning Beamformers

We propose minor extension of the DICS approach [1] to account for the induced and phase-locked, as well as the total power components by computing invertible data correlation matrices in order to build an estimator for each spatial location for each component separately.

For a specified target location, $\mathbf{r}$, an adaptive linearly constrained distortionless spatial filter [11] may be constructed as

$$\mathbf{W}_{\mathbf{r}}^T = \left[\mathbf{L}_{\mathbf{r}}^T\hat{\mathbf{C}}_{sig}^{-1}\mathbf{L}_{\mathbf{r}}\right]^{-1}\mathbf{L}_{\mathbf{r}}^T\hat{\mathbf{C}}_{sig}^{-1} \quad (4)$$

$\mathbf{L}_{\mathbf{r}}$ is the forward matrix for the dipole at location $\mathbf{r}$ and two orthogonal orientations in the tangent plane [12]. Note that due to the quasi-static approximation generally used in EMEG source estimation [13], the forward matrix is real. $\hat{\mathbf{C}}_{sig}$ is one of the signal space covariance matrices defined by equations 1-3, or an average of several of the same type over a selected time-frequency neighborhood. By combining the lead field vectors of all locations and orientations, we obtain an inverse operator $\mathbf{W} = \{\mathbf{w}_{\mathbf{r}}^T\}$ in which each two columns correspond to a location representing two orthogonal orientations at location $\mathbf{r}$. Then given $\mathbf{W}$ and the corresponding $\hat{\mathbf{C}}_{sig}$ we may estimate the source space covariance as

$$\hat{\mathbf{C}}_{src} = \mathbf{W}^T\hat{\mathbf{C}}_{sig}\mathbf{W}$$

The 2×2 diagonal submatrices of $\hat{\mathbf{C}}_{src}$ may be used as an estimate of the voxel-wise source space power, either by computing the trace or using the largest eigenvalue. The results may be presented either directly or as the voxel-wise ratio of estimated activations between two conditions (e.g., pre- vs. post-stimulus) [8].

Equation 4 may be applied directly for the total power and induced covariances. However, the phase-locked covariance is significantly rank-deficient, and therefore non-invertible. We propose to resolve this by estimating the source distribution underlying the the phase-locked component as the difference between total power and induced source space estimates. To do so, for each voxel we first compute $\hat{\mathbf{C}}_{src}^{tot}$ and $\hat{\mathbf{C}}_{src}^{ind}$ using the corresponding covariance matrices. Then, we compute source estimate for the phase-locked component as the larger eigenvalue of the diagonal submatrices of the difference matrix $\hat{\mathbf{C}}_{src}^{lock} = \hat{\mathbf{C}}_{src}^{tot} - \hat{\mathbf{C}}_{src}^{ind}$.

Taken together, both the diagonal and the off diagonal 2×2 submatrices of $\hat{\mathbf{C}}_{src}^{locked}, \hat{\mathbf{C}}_{src}^{tot}, \hat{\mathbf{C}}_{src}^{ind}$ may be used to estimate source space coherence between the corresponding locations.

D. Time-Frequency RAP MUSIC

RAP MUSIC as applied to neurophysiological data was developed originally to fit dipoles to real-valued covariance matrices [2]. MEG-MUSIC applied to total power was described in [9]. Although the original MUSIC algorithm was described for complex-valued matrices obtained from e.g., multisensor radar signals, the physical basis of the MEG signal differs from the radar case in a significant way. The complexity in the radar case arises from the necessity to model finite propagation delays. In EEG/MEG source modeling we use the quasi-static approximation [13] that implies the absence of delays in propagation of the electromagnetic field from neuronal generators to sensors. Real and imaginary parts of the wavelet coefficients carry complimentary information about the *temporal* structure of source activations and should be used concurrently in order to estimate the *spatial* parameters of the source.

To find the source dipoles we first perform SVD of the complex covariance matrix corresponding to the component of interest, i.e. $\mathbf{C}_{sig}^{tot,lock,ind} = \mathbf{USV}^T$, yielding complex left singular matrix $\mathbf{U}$ and real diagonal matrix $\mathbf{S} = diag(s_i), 1 \leq i \leq K$. Based on user specified threshold on percentage of unexplained variance, we find the effective rank R of the averaged covariance matrix. Since the spatial information in the wavelet-transformed signal is contained in both real and imaginary parts of the covariance matrix we turn R complex columns of U into a new $K \times 2R$ quasi-data matrix $\boldsymbol{D}$ by filling its columns with weighted by the corresponding singular values real and imaginary parts of columns of $\mathbf{U}$, i.e. $\mathbf{D} = [s_i \Re(\mathbf{u}_i) s_i \Im(\mathbf{u}_i)] \forall 1 \leq i \leq R$. We then use this newly created quasi-data matrix as an input to RAP-MUSIC algorithm with user-specified signal subspace rank parameter that we increase by a factor of 2 to account for the complexity of the original covariance matrices.

E. Simulations

To verify the methods, we simulated two dipoles in a virtual spherical head model, using sensor locations corresponding to a 148 channel first order magnetometer system. Both of the dipoles were active at 75 Hz frequency following an event marker, but differed in that the dipole on the left side was phase-locked to the event, while the dipole on the right side was phase-randomized with respect to the event. 70 trials were averaged in the time-frequency domain. The total power, phase locked and induced components were analyzed using both DICS and RAP MUSIC. The results are shown in Figure 1.

F. Experimental Data

We used magnetic field data from simple sensory and motor paradigms to evaluate the source estimation methods, since the paradigms we selected make strong predictions for the anatomical regions of interest [14]. In the case of the motor protocol, we predicted that we would observe beta desynchronization arising from contralateral primary motor cortex [15]. For the sensory evoked field paradigm, we predicted that we would observe a phase-locked activity shortly after the stimulus arising from contralateral primary sensory cortex. In addition, we felt confident that we could identify central sulcus anatomically from the MRI for this subject, and thus infer that primary motor cortex lay anterior to and primary sensory cortex lay posterior to the central sulcus.

Data acquisition: Magnetic field data were acquired from a healthy adult male volunteer using a 148 channel first order magnetometer system (Magnes 2500, 4D Neuroimaging, Inc, San Diego CA). The data were acquired in two separate sessions, once to acquire somatosensory evoked field data, and on another occasion to acquire voluntary motor data. A volumetric set of T1-weighted structural MRIs were also available for this subject. Protocols were approved by the Scripps Clinic Institutional Review Board. Magnetometer sensor locations were determined with respect to head fiducial points using 4D Neuroimaging software and a Polhemus (Colchester VT) 3D digitizer. The sensors were coregistered to the structural MRI using EMSE Suite v5.3 (Source Signal Imaging, Inc., San Diego CA). EMSE Suite software was used for all subsequent analysis steps, as well.

Motor paradigm: In the voluntary motor paradigm, the subject was instructed to squeeze his left hand into a fist, then extend his hand back to the resting position, approximately once every 3 seconds. A pair of electrodes were placed on the skin over the flexor muscles of the left forearm and the EMG signal was recorded in parallel with the MEG. A total of 42 successive squeezes were used for subsequent analysis. In order to determine movement onset, the EMG was Hilbert transformed to obtain the envelop. The onset was determined by visual inspection of the filtered EMG signal as the earliest time at which the envelop exceeded the background. The results of analysis are presented in Figure 2.

Somatosensory paradigm: Airpuff somatosensory stimuli were presented every 330 msec. to the left index finger of an adult male. Data were time-frequency transformed and averaged to obtain the phase-locked component of the evoked field. A RAP MUSIC dipole fit to the phase locked average was calculated for the interval [0.04-0.6 sec, 25-50 Hz], and the results are shown in Figure 3, where the single dipole fit is near or slightly posterior to right central sulcus.

III. RESULTS

Simulations: Simulations: The results for DICS are shown in Figure 1, and described here for both DICS and RAP MUSIC. For the total power case, both DICS and RAP MUSIC correctly identified both dipole locations and orientations, within the 1cm grid tolerance for DICS, and with approximately 1mm rms location error for RAP MUSIC. For the

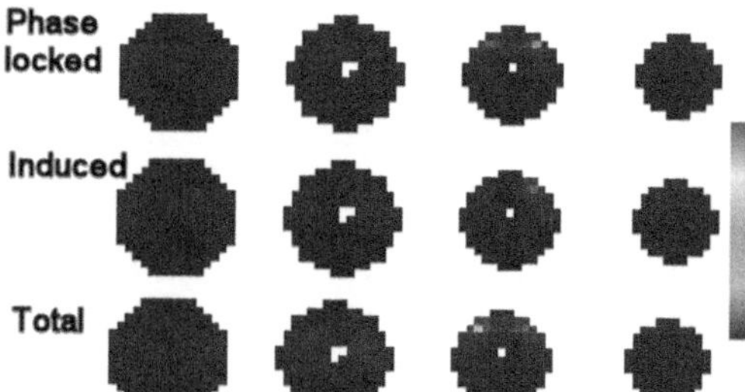

Fig. 1 Source estimates from simulated data are shown using DICS for the three components. DICS found both components for the total power case at the correct dipole locations. For the phase-locked case, DICS finds the correct dipole, but also sees a contribution from the non-phase-locked location. In the induced case, DICS finds a single dipole at the correct location. DICS maps computed as pSPM and plotted on axial slices, 1cm separation

phase-locked case, RAP MUSIC finds one dipole at the correct location for high (0.99) subspace correlation threshold, and two dipoles (at both simulated locations) for lower (0.95) threshold. Similarly DICS finds two locations in the phase-locked case, with a stronger source estimate for the (correct) phase-locked dipole. The appearance of two sources in the phase-locked estimates may be attributed to incomplete averaging, with the phase-randomized signal leaving a residual in the phase-locked transform. For the induced case, both methods correctly identified the (single) phase-randomized dipole source. In this case, RAP MUSIC obtains the same result at either high or low threshold. *Experimental data*: DICS localized primary motor cortex from the β-desynchronization induced correlation, shown in Figure 2. RAP MUSIC localized primary somatosensory cortex from phase-locked SEF data, shown in Figure 3.

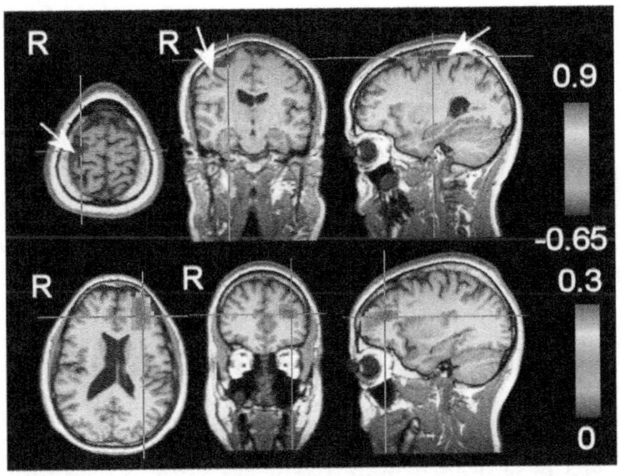

Fig. 2 (top) Induced DICS log(pSPM) inverse is shown for selected interval (0.05-0.18 sec, 17-24 Hz). The right central sulcus (arrow) was identified as the first sulcus anterior to the ascending branch of the right cingulate sulcus, which may be identified unambiguously in successive sagittal slices. (bottom) A brain voxel was chosen near the maximum of the event related desynchronization in right primary motor cortex, and coherence was estimated for the remaining brain voxels. The maximum coherence was observed in left dorsolateral prefrontal cortex

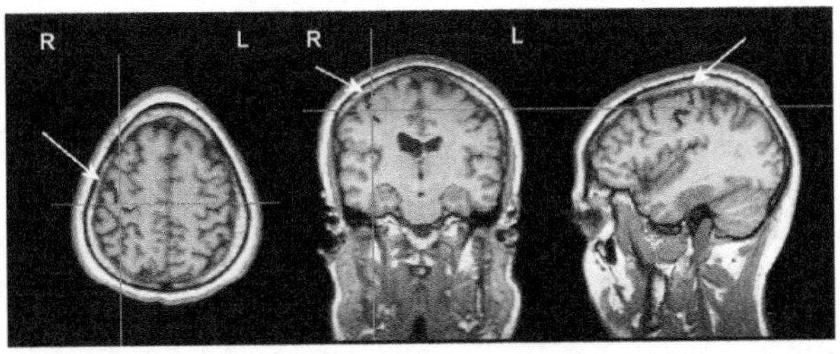

Fig. 3 RAP MUSIC fit a dipole (signal space correlation threshold 0.95, rank = 2) to SEF data . One dipole was found at the location shown. The right central sulcus (arrow, sulcus shown in blue) was identified as the first sulcus anterior to the ascending branch of the right cingulate sulcus, which may be identified unambiguously in successive sagittal slices

IV. Conclusions

We report a modest, but potentially useful, extension to existing MEG source estimation methods. By estimating distributed (DICS beamformer) and discrete (RAP MUSIC) methods to total power, phase-locked, and induced components of time-frequency transformed MEG data, we can address a broader range of questions than previously, for example, attributing brain sources to event-related synchronization, while minimizing contributions from phase-locked components.

Our results support the conclusion that time-frequency source estimation from MEG data is feasible for all components, using either distributed or discrete methods. They also suggest a potential clinical application in the localization of primary motor cortex using DICS-based source estimation of the β-desynchronization obtained during a voluntary motor task. Additional studies will be required to validate the clinical utility of this approach.

References

1. J Gross, F Schmitz, I Schnitzler, et al. Modulation of long-range neural synchrony reflects temporal limitations of visual attention in humans *Proc Natl Acad Sci (US).* 2004;101:13050-13055.
2. JC Mosher, RM Leahy. Source localization using recursively applied and projected (RAP)MUSIC *IEEE Trans Sig Proc.* 1999;47:332-340.
3. C Tallon-Baudry, O Betrand, C Delpuech, J Pernier. Stimulus Specificity of Phase-Locked and Non-Phase-Locked 40 Hz Visual Responses in Human *J. Neuroscience.* 1996;16:4240-4249.
4. Y Meyer. *Wavelets: algorithms and applications.* Philadelphia: SIAM 1993.
5. RE Greenblatt, A Osman. Wavelet-based Neuroelectromagnetic Source Estimation *NeuroImage.* 2003;19:CD-ROM.
6. F-H Lin, T Witze, MH Hämäläinen, AM Dale, JW Belliveau, SM Stufflebeam. Spectral spatiotemporal imaging of cortical oscillations and interactions in the human brain *NeuroImage.* 2004;23.
7. SE Robinson, DF Rose. *Current source image estimation by spatially filtered MEG*:761-765. Amsterdam: Excerpta Medica 1992.
8. J Gross, J Kujala, M Hamalainen, L Timmermann, A Schnitzler, R Salmelin. Dynamic imaging of coherent sources: Studying neural interactions in the human brain *Proc Natl Acad Sci USA.* 2001;98:694-699.
9. K Sekihara, S Nagarajan, D Poeppel, Y Miyashita. Time-Frequency MEG-MUSIC Algorithm *IEEE Trans Med Imaging.* 1999;18:92-97.
10. R Kronland-Martinet, J Morlet, A Grossmann. Analysis of sound patterns through wavelet transforms *Intl. J. Pattern Recog. Artificial Intelligence.* 1987;1:273-302.
11. RE Greenblatt, A Ossadtchi, ME Pflieger. Local Linear Estimators for the Bioelectromagnetic Inverse Problem *IEEE Trans Signal Proc.* 2005;53:3403-3412.
12. JC Mosher, RM Leahy, PS Lewis. EEG and MEG: Forward Solutions for Inverse Methods *IEEE Trans BME.* 1999;46:245-259.
13. Lütkenhöner B. Frequency-domain localization of intracerebral dipolar sources *EEG. Clin. Neurophysiol.* 1992;82:112-118.
14. A Papanicolaou. *Clinical Magnetoencephalography and Magnetic Source Imaging.* Cambridge: Cambridge Univ. Press 2009.
15. G Pfurtscheller, Silva FH Lopes. *Event-Related Desynchronization.* Amsterdam: Elsevier 1999.

Array-Gain Constraint Minimum-Norm Spatial Filter with Recursively Updated Gram Matrix for Biomagnetic Source Reconstruction

Kensuke Sekihara and Isamu Kumihashi

Department of Systems Design and Engineering, Tokyo Metropolitan University, Tokyo, Japan

Abstract— **This paper proposes a novel spatial filter for biomagnetic source imaging. The proposed spatial filter is derived based on a modified version of the minimum-norm spatial filter, and is designed to have a performance close to that of the adaptive minimum-variance spatial filter. In this method, the theoretical form of the measurement covariance matrix is estimated as an updated gram matrix in a recursive procedure. The proposed method does not require a large number of time samples, since the sample covariance matrix is not used. It can even be applied to single time-sample data, and is robust to source correlation. We have validated the method's effectiveness by our computer simulations as well as through experiments using auditory-evoked MEG data.**

Keywords— **biomagnetic source imaging, minimum-norm method, spatial filter, minimum-variance spatial filter, magnetoencephalography.**

I. Introduction

This paper proposes a novel spatial filter based on a modified version of the minimum-norm filter, called the array-gain constrained minimum-norm filter. The proposed method is designed to estimate the theoretical form of the measurement covariance matrix through recursively updating the gram matrix, and the method is able to provide three-dimensional volume reconstruction with a spatial resolution considerably higher than that of existing minimum-norm-based methods. Nonetheless, the method is free of the weaknesses occurring with adaptive spatial filters, because it does not depend on the sample covariance matrix. The method is also robust to the source correlation, and this robustness is demonstrated in our experiments using auditory-evoked magnetoencephalographic (MEG) data.

II. Spatial filter formulation

A. Definitions

Let us define the biomagnetic field measured by the mth sensor at time t as $b_m(t)$, and the measured data as the column vector $b(t) = [b_1(t), b_2(t), \cdots, b_M(t)]^T$. Here, M is the total number of sensors and the superscript T indicates the matrix transpose. A spatial location is represented by a three-dimensional vector r, as $r = (x, y, z)$. A source vector at r and time t is defined as a three-dimensional column vector $s(r,t)$, where $s(r,t) = [s_x(r,t), s_y(r,t), s_z(r,t)]^T$, and $s_x(r,t)$, $s_y(r,t)$, and $s_z(r,t)$ are the x, y, and z components of the source vector $s(r,t)$.

We denote the outputs of the mth sensor as $l_m^x(r)$, $l_m^y(r)$, and $l_m^z(r)$ when the unit-magnitude source at r is directed in the x, y, and z directions, respectively. Then, the column vectors $l_x(r)$, $l_y(r)$, and $l_z(r)$ are defined as $l_i(r) = [l_1^i(r), l_2^i(r), \cdots, l_M^i(r)]^T$, where $i = x, y$ or z. The vectors, $l_x(r)$, $l_y(r)$, and $l_z(r)$, express the sensor array sensitivity for a source at r directed in the x, y, and z directions. The lead-field matrix is expressed as the $M \times 3$ matrix: $L(r) = [l_x(r), l_y(r), l_z(r)]$, which represents the sensitivity of the whole sensor array for a source at r.

Using the measurement vector $b(t)$, the spatial filter reconstructs the source vector $s(r,t)$ using

$$\widehat{s}(r,t) = W^T(r)b(t). \tag{1}$$

where $\widehat{s}(r,t)$ is the estimated source vector at r and time t. In Eq. (1), the weight matrix $W(r)$ is an $M \times 3$ matrix defined as

$$W(r) = [w_x(r), w_y(r), w_z(r)], \tag{2}$$

where $M \times 1$ weight vectors $w_x(r)$, $w_y(r)$, and $w_z(r)$, respectively detect $\widehat{s_x}(r,t)$, $\widehat{s_y}(r,t)$, and $\widehat{s_z}(r,t)$.

B. Minimum-Variance Spatial Filter

The minimum-variance spatial filter is a representative adaptive spatial filter[1, 2], which is originally formulated using the unit-gain constraint, $W^T(r)L(r) = I$, where I is the identity matrix. However, for biomagnetic imaging, the norm of the lead-field $\|L(r)\|$ is spatially nonuniform and the minimum-variance spatial filter derived from the unit-gain constraint is affected by this nonuniformity of the lead field norm. When $\|L(r)\|$ is spatially nonuniform, instead of using the unit-gain constraint, it is more reasonable to use the constraint $W^T(r)L(r) = \Lambda$, which is called the array-gain constraint[2]. Here, $\Lambda = \mathrm{diag}[\|l_x(r)\|, \|l_y(r)\|, \|l_y(r)\|]$ and $\mathrm{diag}[\cdot]$ indicates the diagonal matrix in which diagonal

S. Supek and A. Sušac (Eds.): Advances in Biomagnetism – BIOMAG2010, IFMBE Proceedings 28, pp. 140–143, 2010.
www.springerlink.com

terms are equal to those between the square brackets. In this case, the weight matrix is expressed as

$$W(r) = R^{-1}\widetilde{L}(r)[\widetilde{L}^T(r)R^{-1}\widetilde{L}(r)]^{-1}, \quad (3)$$

where R is the measurement covariance matrix and $\widetilde{L}(r)$ is the lead field matrix consisting of the normalized columns, i.e.,

$$\widetilde{L}(r) = \left[\frac{l_x(r)}{\|l_x(r)\|}, \frac{l_y(r)}{\|l_y(r)\|}, \frac{l_z(r)}{\|l_z(r)\|}\right]. \quad (4)$$

This minimum-variance spatial filter is known to have significantly higher spatial resolution and much smaller source location bias, compared to the minimum-norm-based spatial filters[2, 3].

The weight matrix of the minimum-variance filter in (3) contain the covariance matrix R. When computing the weight matrix, a sample covariance estimate $\widehat{R}$ is used for R, and a large number of time samples are generally needed to obtain an accurate sample covariance matrix. However, in many bioelectromagnetic applications, it is difficult to obtain a large number of time samples, and this is one of the limitations of the minimum-variance filter.

The theoretical form of the covariance matrix is expressed as

$$R = \int_\Omega L(r)\langle s(r,t)s^T(r,t)\rangle L^T(r)dr, \quad (5)$$

where $\langle\cdot\rangle$ indicates averaging. The above expression for R is used in deriving the spatial filter proposed in the next section.

III. Proposed spatial filter

The proposed spatial filter is based on the array-gain-constraint minimum-norm filter, which is derived using the following optimization:

$$W(r) = \underset{W(r)}{\arg\min}\, \mathscr{F}(W), \quad \text{subject to} \quad W^T(r)L(r) = \Lambda, \quad (6)$$

where the cost function $\mathscr{F}(W)$ is defined as

$$\mathscr{F}(W) = \int_\Omega \mathrm{tr}\{\left[W^T(r)L(r') - \delta(r-r')\Lambda\right] P(r')\left[W^T(r)L(r') - \delta(r-r')\Lambda\right]^T\}d^3r', \quad (7)$$

where $P(r')$ is a 3×3 positive semi-definite matrix. This $P(r')$ is a matrix weighting function that controls location-dependent weighting when the resolution kernel is optimized. The resultant spatial filter is derived as

$$W(r) = \bar{G}^{-1}\widetilde{L}(r)[\widetilde{L}^T(r)\bar{G}^{-1}\widetilde{L}(r)]^{-1}, \quad (8)$$

where

$$\bar{G} = \int_\Omega L(r)P(r)L^T(r)d^3r. \quad (9)$$

If we compare Eq. (8) with Eq. (3), which is the weight equation for the minimum-variance filter. The only difference between Eqs. (8) and (3) is the use of $\bar{G}$, instead of R. We may then compare $\bar{G}$ (Eq. (9)) and R (Eq. (5)), and find that these two matrices are very similar and differ only in the manner of weighting in the integral. That is, the covariance matrix R is expressed as the integral of $L(r)L^T(r)$, with the weighting of the source power matrix $\langle s(r,t)s^T(r,t)\rangle$. On the other hand, the gram matrix $\bar{G}$ is given by the integral of $L(r)L^T(r)$, with the weighting of an arbitrary matrix function $P(r)$.

Therefore, if we could use a $P(r)$ that is similar to the source power matrix $\langle s(r,t)s^T(r,t)\rangle$, the gram matrix $\bar{G}$ could approximate the covariance matrix R, and consequently the performance of the array-gain minimum-norm filter would approach the performance of the minimum-variance filter. However, since the true source distribution $s(r)$ is unknown, we propose using the estimated source distribution $\widehat{s}(r)$ for computing $P(r)$.

The proposed recursive algorithm is described in the following manner: (i) By setting an initial $P(r)$ to be the identity matrix I, the weight $W(r)$ is derived using Eqs. (8) and (9). (ii) The estimated source distribution $\widehat{s}(r,t)$ is obtained using Eq. (1). (iii) This $\widehat{s}(r,t)$ is then used to compute $\bar{G}$ at each time point such that

$$\bar{G} = \int_\Omega L(r)\left[\widehat{s}(r,t)\widehat{s}^T(r,t)\right]L^T(r)d^3r, \quad (10)$$

whereupon the weight matrix is updated using Eq. (8). (iv) Finally, the updated source distribution is obtained using the updated weight matrix with Eq. (1). This procedure is repeated until a stopping criterion is satisfied.

IV. Experiments

A. Computer Simulation

A computer simulation was conducted to check the performance of the proposed spatial filter. In this computer simulation, we used a sensor alignment of the Magnes 2500™ (4D Neuroimaging Inc., San Diego, CA) whole-head sensor array, in which 148 sensors were arranged on a helmet-shaped surface. Three sources were assumed to exist on the vertical single plane: ($x = 0$ cm); their (y,z) coordinates chosen as $(-3.5,-6.5)$ cm, $(1.0,-5.0)$ cm, and $(4.0,-8.75)$ cm, respectively. The source-sensor configuration and the coordinate system are depicted in Fig. 1. The spherical homogeneous conductor model, with the sphere origin set at $(0,0,-12)$ cm, was used for the forward calculation. The

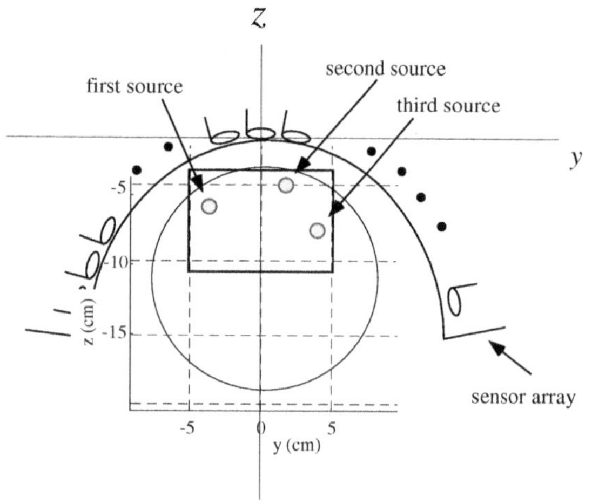

Fig. 1 The source-sensor configuration used in the computer simulation. The small circles show the locations of the three sources. The square shows the region for displaying the reconstruction results in Fig. 2

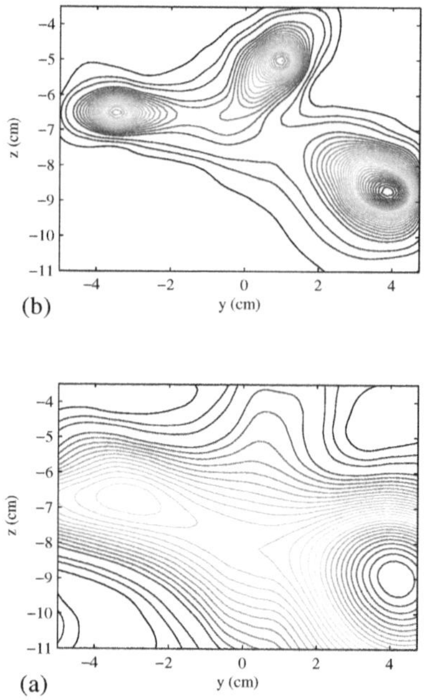

Fig. 2 The results of source reconstruction for single time-sample data. (a) The results when the sLORETA spatial filter was applied. (b) The results when the proposed spatial filter was applied

sensor data at a single time sample was calculated. Here, the intensities of the three sources were set equal in the sensor-domain. Gaussian noise was added, and the SNR of the data was set to 16 where the SNR was defined as the ratio of the norm of the signal vector to the norm of the noise vector. Note that this simulation compares the algorithm performances in considerably high SNR conditions.

The source reconstruction was performed using sLORETA [4], and the proposed spatial filter. Here, sLORETA was chosen for comparison, because it is a method representative of existing non-adaptive methods and is able to perform three-dimensional volume reconstruction. The source space Ω was defined as the three-dimensional region between $-4 \leq x \leq 4$ cm, $-5 \leq y \leq 5$ cm, and $-11 \leq z \leq -3$ cm. The three-dimensional reconstruction was performed on Ω with a 1-cm voxel interval. The squared source intensity $|\widehat{s}(r)|^2$ was plotted on the plane $x = 0$ cm on which the three sources were presumed to exist. The results from sLORETA and the proposed spatial filter are shown in Figs. 2(a) and (b). It can clearly be seen that the proposed spatial filter reconstructs the three sources at approximately correct locations. On the contrary, sLORETA produces blurred results, and fails to reconstruct the first and the second sources.

B. Experiments Using Auditory-Evoked MEG Data

The effectiveness of the proposed spatial filter was further tested using auditory-evoked MEG. Measurements were performed using the 160-channel MEG-VISION™ whole-cortex biomagnetometer (Yokogawa Electric Inc., Tokyo, Japan). Four hundred trials were measured and averaged. The averaged waveforms of the auditory evoked field are shown in Fig. 3(a).

A single time-sample data at the latency of 120 ms was used for the reconstruction experiments. This time point, indicated by a vertical broken line in Fig. 3(a), is close to the peak vertex of the M100, and the auditory cortices in the left and right hemispheres are expected to be simultaneously active at this instant. This is confirmed by the contour plots (Fig. 3(b)) of the sensor data at this time point, because the dipolar patterns are visible above the left and right hemispheres.

The results of the minimum-variance filter are shown in Fig. 4(a). To obtain these results, the sample covariance matrix was computed using a time window between 0 and 300 ms, and the source power distribution (obtained by averaging the squared source intensity over this time window) is shown in Fig. 4(a). In these results, the minimum-variance spatial filter fails to reconstruct the sources of the bilateral activation of auditory cortices, probably because the activities in these cortices are highly correlated[5]. The results from sLORETA are shown in Fig. 4(b) and the results from the proposed spatial filter are shown in Fig. 4(c). Those in Figs. 4(b) and (c) were obtained using the single time-sample data at 120 ms. The results of sLORETA are significantly blurred and biased. On the contrary, the proposed spatial filter can

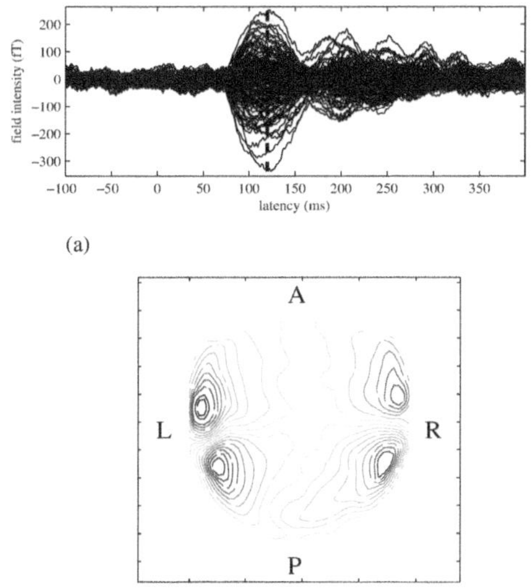

(a)

(b)

Fig. 3 (a) The averaged waveforms of the auditory evoked field used for the reconstruction experiments. The vertical broken line indicates the latency at 120 ms. (b) The sensor-field map at a latency of 120 ms

reconstruct the bilateral activities of the auditory cortices with significantly high spatial resolution.

In summary, we have proposed a novel spatial filter that outperforms existing nonadaptive, minimum-norm-based spatial filters. This paper presents our computer simulation that tests the algorithm performance in a relatively high SNR situation, as well as the results of the application to auditory-evoked MEG. The algorithm performance for low SNR data, stability of the solutions, and stopping criterion for the iterations will be reported in our forthcoming publication[6].

Acknowledgment

We would like to thank Mr. Hiroaki Tanaka for his help in providing auditory-evoked MEG data.

References

1. Veen B. D., Drongelen W., Yuchtman M., Suzuki A.. Localization of Brain Electrical activity via linearly constrained minimum variance spatial filtering *IEEE Trans. Biomed. Eng.*. 1997;44:867-880.
2. Sekihara K., Nagarajan S. S.. *Adaptive spatial filters for electromagnetic brain imaging*. Berlin, Heidelber: Springer-Verlag 2008.

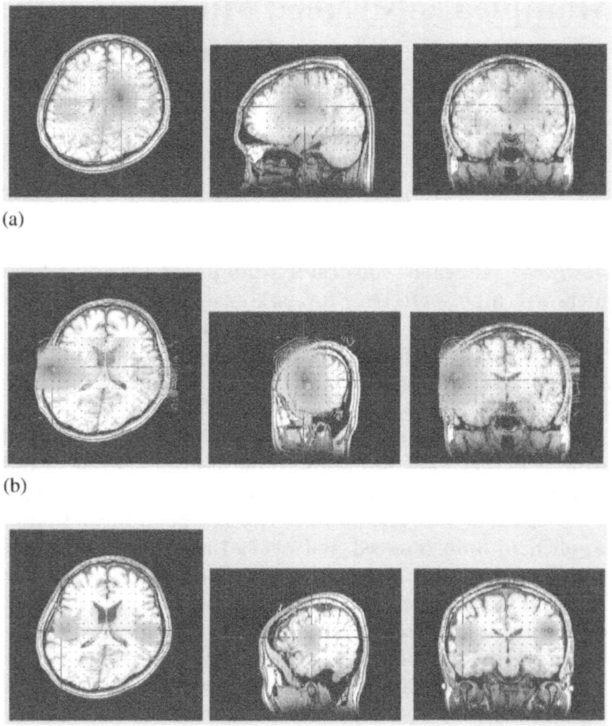

(a)

(b)

(c)

Fig. 4 The results of the source reconstruction for auditory-evoked field. (a) The results obtained from the minimum-variance spatial filter. (b) The results obtained from the sLORETA spatial filter. (c) The results obtained from the proposed spatial filter

3. Sekihara K., Sahani M., Nagarajan S. S.. Location bias and spatial resolution of adaptive and non-adaptive spatial filters for MEG source reconstruction *NeuroImage*. 2005;25:1056-1067.
4. Pascual-Marqui R. D.. Standardized low resolution brain electromagnetic tomography (sLORETA): technical details *Methods and Findings in Experimental and Clinical Pharmacology*. 2002;24:5-12.
5. Dalal S. S., Sekihara K., Nagarajan S. S.. Modified beamformers for coherent source region suppression *IEEE Trans. Biomed. Eng.*. 2006;53:1357-63.
6. Kumihashi I., Sekihara K.. Array-gain constraint minimum-norm spatial filter with recursively updated gram matrix for biomagnetic source imaging *IEEE Trans. Biomed. Eng.*. accepted for publication.

Multiple Constrained Minimum-Variance Beamformers for Reconstruction of Induced and Evoked Neural Activity

Alexander Moiseev

Down Syndrome Research Foundation, Vancouver, Canada

Abstract— **Linearly constrained minimum variance beamformers are highly effective for analysis of weakly correlated brain activity. When correlations become significant, their performance drastically degrades. One solution to this problem is to use a multiple constrained minimum variance (MCMV) beamformer which however requires *a priori* information about the correlated sources locations. Besides the question whether unbiased estimates of source positions can be obtained remained unanswered. In this work, we derive MCMV-based source localizers of two kinds: a) those that can be applied to both induced and evoked activity; and b) those targeting only the evoked component. We show that for arbitrary noise covariance both yield simultaneous unbiased estimates of multiple source positions and orientations and remain bounded at singular points. We also derive a search algorithm that may be used to find multiple *unknown* correlated sources. Simulations and analyses of real MEG data demonstrate that in situations where a single-source beamformer is unable to detect coherent sources, the MCMV-based ones are able to find either most or all of them, depending on the SNR.**

Keywords— **magnetoencephalography (MEG), inverse solutions, beamformers, correlated sources.**

I. Introduction

Minimum variance spatial filters [1, 2] are widely used for bioelectromagnetic source reconstruction. One of the fundamental assumptions needed for their derivation is that time courses of the brain sources are uncorrelated or statistically orthogonal. This assumption may often be violated. Indeed, we should expect the brain regions involved in certain tasks to be connected and to exhibit correlated activity. When correlations are significant they result in partial or complete cancellation of reconstructed sources and in distortion of their time courses [3]. A number of approaches have been suggested to overcome this problem [4, 5]. One solution is to use a multiple constrained minimum variance (MCMV) beamformer [6].

With the MCMV beamformer source cancellation does not occur because the filter weights for each source impose null-gain constraints on all other sources. To achieve this *a priori* knowledge about source locations and directions is required. This information is usually not available. The question whether the MCMV filter may be used to find multiple *unknown* correlated sources and whether unbiased estimates of source locations can be obtained remain unanswered. The goal of the current work is to address these issues.

II. Methods

A. MCMV Beamformer Formulation

Let $\boldsymbol{b}(t)$ denote the M-dimensional column vector of the MEG sensor readings at time t, where M is the number of sensors. Assume that $\boldsymbol{b}(t)$ is generated by $n<M$ point dipolar sources $s_i(\boldsymbol{\theta}^i,t), i=1,...,n$ plus noise, where s_i are the source amplitudes. Vectors $\boldsymbol{\theta}^i=\{\mathbf{r}^i,\mathbf{u}^i\}$ define source positions $\mathbf{r}^i$ and orientations $\mathbf{u}^i$ ($\|\mathbf{u}^i\|=1$) in the three-dimensional space. In the multi-source linear filter approach, an estimate $\hat{\mathbf{s}}(\boldsymbol{\theta},t)=\{\hat{s}_1(\boldsymbol{\theta}^1,t),...,\hat{s}_n(\boldsymbol{\theta}^n,t)\}^T$ of source amplitudes is found as a weighted sum of the array signals: $\hat{\mathbf{s}}(\boldsymbol{\theta},t)=W^T\mathbf{b}(t)$. Here $W=\{\mathbf{w}^1,...,\mathbf{w}^n\}$ is an $(M\,x\,n)$ matrix of the filter weights; $\boldsymbol{\theta}=\{\boldsymbol{\theta}^1,...,\boldsymbol{\theta}^n\}$; superscript "$T$" denotes the transposed matrix. The MCMV expression for the weights is [6]: $W=R^{-1}H\left(H^TR^{-1}H\right)^{-1}$ where H is an $(M\,x\,n)$ matrix composed of individual source forward solutions $\mathbf{h}^i(\boldsymbol{\theta}^i)$: $H=\{\mathbf{h}^1(\boldsymbol{\theta}^1),...,\mathbf{h}^n(\boldsymbol{\theta}^n)\}$. R is the measured field covariance matrix: $R=<\mathbf{b}\mathbf{b}^T>-\overline{\mathbf{b}}\overline{\mathbf{b}}^T$; $\overline{\mathbf{b}}\equiv<\mathbf{b}>$ and angular brackets denote statistical averaging. It holds that $W^TH=I_n$, where I_n is an n-dimensional identity matrix. In particular, $\mathbf{w}^{iT}\mathbf{h}^j=0$ when $i\neq j$ ("the null-gain constrain"). Consequently activity in location $\boldsymbol{\theta}^j$ does not contribute to reconstructed source amplitude $\hat{s}_i(\boldsymbol{\theta}^i)$, which explains why the MCMV filter is insensitive to the source correlations. However to calculate $\hat{s}_i(\boldsymbol{\theta}^i)$ one needs to know coordinates of all other sources $\boldsymbol{\theta}^j$.

B. Source Localizers

When source coordinates are not known, one can search for a global maximum of the total power

S. Supek and A. Sušac (Eds.): Advances in Biomagnetism – BIOMAG2010, IFMBE Proceedings 28, pp. 144–147, 2010.
www.springerlink.com

$P(\boldsymbol{\theta}) = <\|\hat{\mathbf{s}}(\boldsymbol{\theta},t)\|^2> = Tr((H^T R^{-1} H)^{-1})$ in the *5n*-dimensional source space, assuming that corresponding $\boldsymbol{\theta}$ determines the real source coordinates $\boldsymbol{\theta}_0$. However $P(\boldsymbol{\theta})$ tends to infinity when sources become close, that is when $\boldsymbol{\theta}^i \to \boldsymbol{\theta}^j$. Besides, even in a single source case (*n=1*) the peak of the *P* distribution is shifted relative to the true dipole position [7]. Therefore, other source localizers should be found.

When only one source exists and noise covariance matrix *N* has a form $N = \sigma^2 I_M$ (the diagonal white noise) several unbiased localizer functions can be identified [7]. In this work we introduce four such functions for a *multi-source* case and for an arbitrary noise covariance matrix *N*. Corresponding formulas are summarized in Tables 1 and 2.

The first two localizers listed in Table 2 are power-based and may be used for both evoked (phase-locked) and induced (oscillatory) activity. The other two apply when the evoked component of the source activity $\bar{\mathbf{s}}(t)$ (and therefore the averaged field $\bar{\mathbf{b}}(t)$) are not equal to zero.

The logic leading to expressions in Table 2 may be demonstrated with the MPZ beamformer. For *n=1* pseudo-Z is defined as a ratio of the signal plus noise power to the projected noise [2]: $\mathcal{Z} = (\mathbf{w}^T R\mathbf{w})/(\mathbf{w}^T N\mathbf{w})$ or (see Table 1) $\mathcal{Z} = (\mathbf{h}^T R^{-1}\mathbf{h})/(\mathbf{h}^T R^{-1} N R^{-1}\mathbf{h}) = S/T$. To get a pure signal to noise ratio we subtract the projected noise (*T*) in the numerator. For *n>1* we convert the result to a matrix form as: $\mathcal{Z} = T^{-1/2} S T^{-1/2} - I_n$. Finally, a trace of *matrix* $\mathcal{Z}$ gives us a *scalar* function of source coordinates $\boldsymbol{\theta}$ to be maximized to determine source locations and directions.

For *event*-related beamformers, for *n=1* we form a ratio of the power of the phase-locked response $\hat{\bar{\mathbf{s}}}(t) = \mathbf{w}^T \bar{\mathbf{b}}(t)$ to the projected noise: $\mathcal{Z}_{ER} = (\mathbf{w}^T \bar{\mathbf{b}}\bar{\mathbf{b}}^T \mathbf{w})/(\mathbf{w}^T N\mathbf{w}) = E/T$ [8]. Note that $\mathcal{Z}_{ER}$ is already a true SNR so no subtraction of the projected noise in the numerator is needed. To obtain more statistically robust results, $\mathcal{Z}_{ER}$ may be additionally averaged over a time interval [9] (denoted as $<\ldots>_t$ in the Table 1). Generalizing this approach to a multi source case leads to MER expressions in Table 2. The "reduced" version rMER of the localizer is obtained by replacing the noise covariance *N* with the full field covariance matrix *R*.

Table 1 Common matrix definitions

	Expression	Matrix size; comments
S	$H^T R^{-1} H$	(n × n)
G	$H^T N^{-1} H$	(n × n)
G_0	Matrix G when $\boldsymbol{\theta} = \boldsymbol{\theta}_0$	(n × n); value of G for true source coordinates
T	$H^T R^{-1} N R^{-1} H$	(n × n)
$\bar{R}$	$<\bar{\mathbf{b}}\bar{\mathbf{b}}^T>_t$	(M × M); 2nd moments matrix of the averaged field
E	$H^T R^{-1} \bar{R} R^{-1} H$	(n × n)
C	$<\mathbf{s}\mathbf{s}^T> - \bar{\mathbf{s}}\bar{\mathbf{s}}^T;\ \bar{\mathbf{s}} \equiv <\mathbf{s}>$	(n × n) ; the true source covariance matrix
$\bar{C}$	$<\bar{\mathbf{s}}\bar{\mathbf{s}}^T>_t$	(n × n); 2nd moments of the averaged source amplitudes

All localizers P_{**} in Table 2 (where P_{**} refers to any of $P_{MAI}, P_{MPZ}, P_{MER}, P_{rMER}$) have common properties, which we list here without a proof because of the space limitations.

1. P_{**} is a non-negative function of the source coordinates $\boldsymbol{\theta} = \{\boldsymbol{\theta}^1,\ldots,\boldsymbol{\theta}^n\}$. $P_{**}(\boldsymbol{\theta})$ remain finite at singular points where $\boldsymbol{\theta}^i \to \boldsymbol{\theta}^j$ $(i \neq j)$
2. $P_{**}(\boldsymbol{\theta})$ reaches a global maximum when the beamformer order becomes equal to the actual number of sources *n*, *and* coordinates $\boldsymbol{\theta}$ match those of the true sources $\boldsymbol{\theta}_0 : \max_{n,\boldsymbol{\theta}} P_{**}(\boldsymbol{\theta}) = P_{**}(\boldsymbol{\theta}_0) \equiv P_{**}^{\max}$. In other words, $P_{**}(\boldsymbol{\theta})$ provides an *unbiased* estimate of the true source locations and orientations.
3. Saturation: once the global maximum is reached, further increase of the beamformer order *does not change the value of the localizer*: for any $\tilde{\boldsymbol{\theta}} = \{\boldsymbol{\theta}_0^1,\ldots\boldsymbol{\theta}_0^n,\boldsymbol{\theta}^{n+1},\ldots\}$ holds $P_{**}(\tilde{\boldsymbol{\theta}}) = P_{**}(\boldsymbol{\theta}_0) \equiv P_{**}^{\max}$

For *n = 1* the MAI localizer is equal to Van Veen's neural activity index less 1; if additionally $N = \sigma^2 I_M$ it is equivalent to a scalar beamformer with a normalized lead field [7]. Comparing P_{MAI} to the P_{MPZ} beamformer, one can

Table 2 MCMV-based source localizers

Localizer name	Base matrix expression ("ME")	Localizer function (*Tr*("ME"))	Global maximum value
Multi-source activity index (MAI)	$G^{1/2} S^{-1} G^{1/2} - I_n$	$P_{MAI} = Tr(GS^{-1}) - n$	$P_{MAI}^{\max} = Tr(CG_0)$
Multi-source pseudo-Z (MPZ)	$T^{-1/2} S T^{-1/2} - I_n$	$P_{MPZ} = Tr(ST^{-1}) - n$	$P_{MPZ}^{\max} = Tr(CG_0)$
Multi-source event related beamformer (MER)	$T^{-1/2} E T^{-1/2}$	$P_{MER} = Tr(ET^{-1})$	$P_{MER}^{\max} = Tr(\bar{C}G_0)$
Reduced multi-source event related beamformer (rMER)	$S^{-1/2} E S^{-1/2}$	$P_{rMER} = Tr(ES^{-1})$	$P_{rMER}^{\max} = Tr(\bar{C}(G_0^{-1} + C)^{-1})$

prove that for any n, $\boldsymbol{\theta}$ holds $P_{MPZ} \le P_{MAI}$. At the same time, $P_{MPZ}^{\max} = P_{MAI}^{\max}$. Therefore the MPZ localizer has a higher spatial resolution (in $5n$-D space) than MAI. The bigger is the SNR, the larger is the difference in resolution. At the same time both P_{MAI} and P_{MPZ} are unbiased estimators providing alternative means of source reconstruction.

For event-related localizers it can also be shown that MER has a superior spatial resolution compared to rMER. The main advantage of the latter is that it is only based on matrices R and $\overline{R}$ that are easy to measure. Knowledge of the noise covariance matrix N whose estimation may be tricky, is *not required*.

C. *Source Search Procedure*

Without *a priori* information about sources locating a global maximum of $P_{**}(\boldsymbol{\theta})$ becomes a difficult problem. A search algorithm presented here in many cases allows detection of the sources (at least approximately) with a reasonable computational effort (about 1 - 2 minutes per source on a typical office desktop computer running MATLAB).

In practice even strongly correlated sources are unlikely to cancel each other completely [3]. For *full* cancellation sources should have a) approximately *equal* and relatively *high* SNR and b) very high correlation (≥ 0.8). More often than not at least one of those conditions is not met. Therefore the algorithm starts with a single-source beamformer. The largest activation detected is designated as a first source. On step k of the algorithm we fix coordinates of previously found *(k-1)* sources and label these as "references". The H-matrix may be written as $H = \{H^R, \mathbf{h}^k(\boldsymbol{\theta}^k)\}$, $H^R = const$ where columns of a "reference" matrix H^R are forward solutions of already found sources: $H^R = \{\mathbf{h}^1, ..., \mathbf{h}^{k-1}\}$. For $\mathbf{h}^k(\mathbf{r}^k, \mathbf{u}^k)$ we have $\mathbf{h}^k = H^k(\mathbf{r}^k)\mathbf{u}^k$ where matrix $H^k = \{\mathbf{h}_x^k, \mathbf{h}_y^k, \mathbf{h}_z^k\}$ is composed of forward solutions for sources at $\mathbf{r}^k$, oriented along axes x, y and z. To find the k-th source, vector $\mathbf{r}^k$ is run over the entire ROI. We leave it without proof that the dipole orientation $\mathbf{u}^k(\mathbf{r}^k)$ maximizing the localizer value at any location $\mathbf{r}^k$ can be found as an eigenvector of an eigenvalue problem: $D_{**}(\mathbf{r}^k)\mathbf{u}^k = \lambda F_{**}(\mathbf{r}^k)\mathbf{u}^k$, corresponding to the largest eigenvalue λ. Thus, the 2D search over the source orientations is not required. Generic expressions for the three dimensional square matrices D_{**}, F_{**} are: $D_{**} = A^{kR}(A^{RR})^{-1}B^{RR}(A^{RR})^{-1}A^{Rk} - A^{kR}(A^{RR})^{-1}B^{Rk} - B^{kR}(A^{RR})^{-1}A^{Rk} + B^{kk}$; $F_{**} = A^{kk} - A^{kR}(A^{RR})^{-1}A^{Rk}$. Here matrices A, B should be substituted with S, G for the MAI localizer, with T, S for the MPZ, with T, E for the MER or with S, E - for the rMER localizers, respectively. Matrices S, G, T, E with upper indices coincide with those in Table 1, with H replaced by either H^R or H^k. For example, $S^{kR} = (H^k)^T R^{-1} H^R$, etc. The largest maximum of $P_{**}(\mathbf{r}^k)$ determines the coordinates of the source number k. Note that on step k weights $\mathbf{w}^i, i < k$ of the sources found *earlier* are also updated. As a result, source cancellation effects are removed and previously hidden correlated activity is revealed.

III. Numerical examples

In both examples below no *a priori* assumptions about source locations were made. Therefore a whole head search was performed using the method described above. Fig. 1 shows results of a simulation with four parallel sources located in the same horizontal plane. Source 1 is uncorrelated to the other three, but sources 2, 3 and 4 are all highly correlated (the correlation coefficient is 0.8). The P_{MPZ} localizer was used. The first step of the algorithm corresponds to a conventional SAM beamformer [2]. The source #1 (uncorrelated) is found with correct amplitude and coordinates; source #2 is also located correctly but its amplitude is 41.4 nA*m (about 60% of the true value). Source #3 is missed and source #4 is found 1 cm off its true position with approximately the correct amplitude. In the following two steps a low contrast image is produced because a null-gain condition is imposed on source #1, while all remaining sources are highly correlated and tend to cancel each other. Finally, in step #4 all sources are found in their correct locations and with reasonably accurate amplitude values (23, 71, 61 and 23 nA*m). When P_{MAI} localizer was applied in the same simulation, sources 2, 3, 4 were found with location errors of about 1 cm as a consequence of MAI's lower spatial resolution compared to MPZ.

In a 2nd example the MER localizer was applied to find sources in an auditory steady state response experiment with stimulus frequency 40 Hz (fig. 2), which was conducted using a 151-channel CTF/VSM MEG system. 1230 s of data (about 1200 pulses of 0.5 s with varied ISI) were collected at 200 samples/s. The "true" source locations were identified using a spatio-temporal dipole fit. Measured brain noise covariance was used for matrix N. The single source beamformer (*n=1*) found sources 1 and 2, with source 1 (12 nA*m, the red dot) corresponding to the right auditory dipole fit location, and source 2 (8 nA*m, the blue dot) missing the left auditory location by ~1.6 cm. In the 2nd step (*n=2*), source 2 moved to position 3 (the red dot), which was within one voxel (~7 mm) of the "true" left auditory

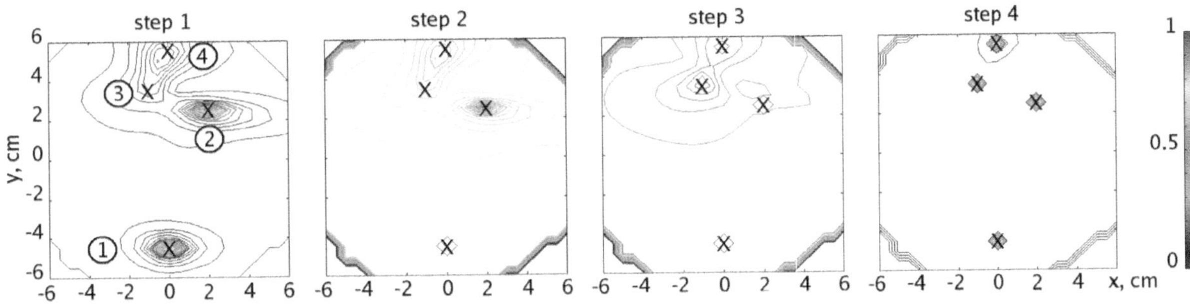

Fig. 1 Application of the multi-source pseudo-Z localizer for the case of three correlated and one uncorrelated sources

The $P_{MPZ}(\mathbf{r}^k)$ distributions for $z = 8$ cm in steps $k = 1,...,4$ of the algorithm, normalized on their max values excluding the reference locations. The source positions (marked with crosses) are (0, -4.5, 8), (2, 2.5, 8), (-1, 3.5, 8) and (0, 5.5, 8) cm; the directions are (1, 0, 0). Source amplitudes are 22, 70, 59 and 22 nA*m. Covariances R and N were calculated analytically assuming 20 fT/Hz$^{1/2}$ white diagonal sensor noise in the 0.5 – 50 Hz band. The correlation coefficient between sources 2, 3, 4 equals to 0.8.

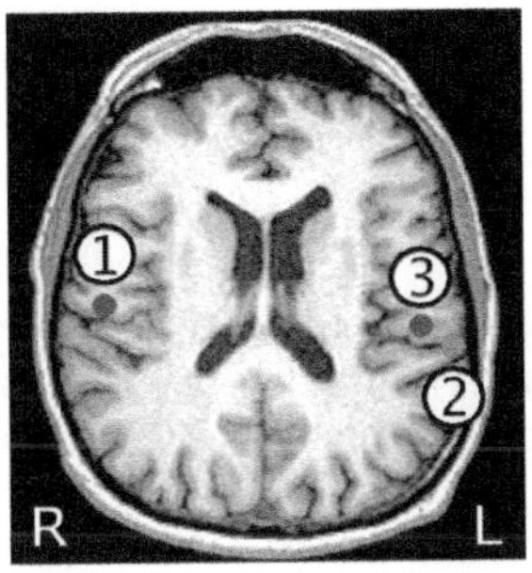

Fig. 2 Auditory sources found in an ASSR experiment. Red dots (1, 3)– final locations; blue dot (2) – a shifted left auditory source with reduced amplitude found by a single source beamformer (step #1)

location. Its amplitude increased to 15 nA*m because source cancellation effects were removed. Note that in this experiment signals are weak and covariances N and R do not differ much. As a result, rMER beamformer produced almost the same result as MER, while power-based localizers failed completely.

IV. Results and Conclusions

In this work, we introduced four new unbiased multiple source localizers based on the MCMV spatial filter. For $n = 1$, our results show that the no bias property of the pseudo-Z localizer [7] holds for an arbitrary noise covariance. The same is true for the neural activity index and for the event-related beamformers, for any n. To our knowledge, the location bias of these filters has not previously been studied. We also established the *saturation* property of the multi-source localizers which gives a principal possibility to determine the true number of sources.

An iterative algorithm presented here makes it possible to find approximate source coordinates $\boldsymbol{\theta}$ without *a priori* assumptions about their location or degree of correlation. The algorithm was shown to yield good results in simulations and with real MEG data, where correlations between the sources are significant.

References

1. Van Veen, B., van Drongelen, W. et al. (1997) Localization of brain electrical activity via linearly constrained minimum variance spatial filtering. *IEEE Trans. Biomed. Eng.*, vol. 44, no. 9, pp.867–880
2. Robinson, S. & Vrba, J. (1999) Functional neuroimaging by synthetic aperture magnetometry (SAM). *Recent Advances in Biomagnetism,* Sendai, Japan, Tohoku Univ. Press, pp. 302–305
3. Sekihara, K., Nagarajan, S. et al. (2002) Performance of an MEG adaptive-beamformer technique in the presence of correlated neural activities: Effects on signal intensity and time-course estimates. *IEEE Trans. Biomed. Eng.*, vol. 49, no. 12, pp. 1534–1546
4. Brookes, M., Stevenson, C. et al. (2007) Beamformer reconstruction of correlated sources using a modified source model. NeuroImage, vol. 34, pp. 1454-1465
5. Dalal, S., Sekihara, K. et al. (2006) Modified beamformers for coherent source region suppression. *IEEE Trans. Biomed. Eng.*, vol. 53, no. 7, pp. 1357–1363
6. Frost, O. (1972) An algorithm for linearly constrained adaptive array processing. Proc. of IEEE, vol. 60, no. 8, pp. 926-935
7. Sekihara, K, Sahani, M. et al. (2005) Localization bias and spatial resolution of adaptive and non-adaptive spatial filters for MEG source reconstruction. NeuroImage, vol. 25, pp. 1056-1067.
8. Cheyne, D., Bostan, A., et al. (2007) Event-related beamforming: a robust method for presurgical functional mapping using MEG. Clin. Neurophysiol., vol. 118, pp. 1691-1704.
9. Robinson, S. (2004) Localization of event-related activity by SAM(erf). Neurol Clin Neurophysiol, vol. 2004, p.109.

Author: Alexander Moiseev
Institute: Down Syndrome Research Foundation
Street: 1409 Sperling Avenue
City: Burnaby, British Columbia, V5B 4J8
Country: Canada
Email: amoiseev@dsrf.org

Speeding-Up MEG Beamforming Source Imaging by Correlation between Measurement and Lead-Field Vector

Jun Hee Hong and Sung Chan Jun

Department of Information and Communications, Gwangju Institute of Science and Technology, Gwangju, South Korea

Abstract— **Beamforming technique is promising in that it can easily deal with spatiotemporal multi-dipole problems and is robust to noise. One of weak points is to scan the whole source region, thereby requiring some amount of computing time. We introduce the method to speed-up beamforming source imaging. In order to speed-up localization as well as to keep good spatial resolution, we propose the way to reduce the number of scanning points by using the correlation between measurement and lead-field vector. It is found that this idea improves beamforming speed up to about 63% in computation time without sacrificing accuracy even though computing correlation requires some computing time. Through Monte Carlo simulation, correlation threshold depending on signal-to-noise ratio (SNR) is reasonably estimated in a least square sense.**

Keywords— **MEG/EEG, spatiotemporal source localization, beamformer, correlation, single-trial analysis.**

I. Introduction

Magnetoencephalography (MEG) or electroencephalography (EEG) source localization is to estimate current source in the brain from the spatiotemporal measurements. Among numerous source localization methods, beamforming techniques have recently paid more attention since it deals easily with spatiotemporal MEG/EEG problem and is robust to noise. Among many variants of beamformers, minimum-variance beamformer has been widely used and applied to solve MEG/EEG source localization problem [1].

Recently, non-invasive brain-computer interface (BCI) based on MEG/EEG system is one of hot research areas. BCI system enables persons to command a computer without spontaneous limb movement by measuring the brain signal such as MEG/EEG, and interpreting the intention of users in real-time basis. To develop high performance BCI system, extraction of good discriminative features in source space as well as in sensor space is of the utmost importance. For this purpose, real-time computation as well as good spatial resolution should be achieved in spatiotemporal source imaging. From that point of view, we proposed the method to speed-up beamforming source imaging in this work. We call this method fast-beamformer throughout this paper.

II. Methods

A. Adaptive Beamformer

In this work, minimum-variance beamformer is chosen as a source imaging technique [1]. In beamforming technique, to estimate the source activity the spatial filter $\mathbf{w}(\mathbf{r})$ is applied to the measurement data as follows:

$$\hat{Q}(\mathbf{r},t) = \mathbf{w}^T(\mathbf{r})\mathbf{m}(t), \tag{1}$$

where, $\hat{Q}(\mathbf{r},t)$ and $\mathbf{m}(t)$ are the estimated magnitude of the source activity at beamforming scanning source point $\mathbf{r}$ and at time t, and spatiotemporal measurement vector at time t. Spatial filter $\mathbf{w}(\mathbf{r})$ is defined in the following optimization problem:

$$\mathbf{w}(\mathbf{r}) = \arg\min \mathbf{w}^T(\mathbf{r})\mathbf{C}\mathbf{w}(\mathbf{r}), \text{ subject to } \mathbf{w}^T(\mathbf{r})\mathbf{l}(\mathbf{r}) = 1, \tag{2}$$

where $\mathbf{C}$ is the covariance matrix of the measurement defined as $\mathbf{C} = \langle \mathbf{m}(t)\mathbf{m}^T(t)\rangle_t$ and $\langle\cdot\rangle_t$ indicates the time average. $\mathbf{l}(\mathbf{r})$ denotes a lead-field vector at source point. This constraint minimization problem can be solved using the method of the Lagrange multiplier and explicit form of the weight spatial filter can be represented by

$$\mathbf{w}(\mathbf{r}) = \frac{\mathbf{C}^{-1}\mathbf{l}(\mathbf{r})}{\mathbf{l}^T(\mathbf{r})\mathbf{C}^{-1}\mathbf{l}(\mathbf{r})}. \tag{3}$$

Using this weight spatial filter, the time averaged source output power is obtained below:

$$\left\langle \hat{Q}(\mathbf{r},t)^2 \right\rangle_t = \mathbf{w}^T(\mathbf{r})\mathbf{C}\mathbf{w}(\mathbf{r}) = \frac{1}{\mathbf{l}^T(\mathbf{r})\mathbf{C}^{-1}\mathbf{l}(\mathbf{r})}. \tag{4}$$

When a spherical homogeneous conductor head model is introduced in forward computation, the false intensity may increase around the center of the sphere in the source imaging. To avoid this difficulty it is known that the constraint $\mathbf{w}^T(\mathbf{r})\mathbf{l}(\mathbf{r}) = \|\mathbf{l}(\mathbf{r})\|$ in place of $\mathbf{w}^T(\mathbf{r})\mathbf{l}(\mathbf{r}) = 1$ is applied. This is called array-gain minimum variance beamformer. Its output power is expressed as

S. Supek and A. Sušac (Eds.): Advances in Biomagnetism – BIOMAG2010, IFMBE Proceedings 28, pp. 148–151, 2010.
www.springerlink.com

$$\left\langle \hat{Q}(\mathbf{r},t)^2 \right\rangle_t = \frac{1}{\tilde{\mathbf{l}}^T(\mathbf{r})\mathbf{C}^{-1}\tilde{\mathbf{l}}(\mathbf{r})} = \frac{\mathbf{l}^T(\mathbf{r})\mathbf{l}(\mathbf{r})}{\mathbf{l}^T(\mathbf{r})\mathbf{C}^{-1}\mathbf{l}(\mathbf{r})}. \quad (5)$$

The vector type beamformer can be defined by weight matrix $\mathbf{W}(\mathbf{r}) = [\mathbf{w}_x(\mathbf{r}), \mathbf{w}_y(\mathbf{r}), \mathbf{w}_z(\mathbf{r})]$. The output power of unit gain vector type beamformer is represented below:

$$\left\langle \hat{Q}(\mathbf{r},t)^2 \right\rangle_t = \left[\mathbf{L}^T(\mathbf{r})\mathbf{C}^{-1}\mathbf{L}(\mathbf{r})\right]^{-1}. \quad (6)$$

Similarly, for array-gain vector type beamformer, its output power is expressed as

$$\left\langle \hat{Q}(\mathbf{r},t)^2 \right\rangle_t = \|\mathbf{l}(\mathbf{r})\|^2 \left[\mathbf{L}^T(\mathbf{r})\mathbf{C}^{-1}\mathbf{L}(\mathbf{r})\right]^{-1}. \quad (7)$$

Here $\mathbf{L}(\mathbf{r})$ denotes a lead-field matrix.

B. Proposed Method (Fast-Beamformer)

Scanning the whole brain region is one of main time dominant factors in beamforming technique. In order to speed-up beamforming localization as well as to keep good spatial resolution, we propose the idea to reduce the number of scanning points by using the correlation between measurement and lead-field vector, which is defined by

$$p(\mathbf{r},t) = \frac{\mathbf{m}(t)\cdot\mathbf{l}(\mathbf{r})}{\|\mathbf{m}(t)\|\|\mathbf{l}(\mathbf{r})\|}. \quad (8)$$

At each source point, root mean square correlation over time is computed as follows:

$$p_{rms}(\mathbf{r}) = \sqrt{\sum_t p(\mathbf{r},t)^2 / T}.$$

Here T is the number of time points. We observed that root mean square correlations at active points are more likely to have higher value than those at inactive points. From this observation, our idea is as follows:

- Correlation values $p(\mathbf{r},t)$ at each source point and each time point are computed. From all computed correlation values $p(\mathbf{r},t)$, root mean square correlation value over time $p_{rms}(\mathbf{r})$ at each source point $\mathbf{r}$ is computed.
- Reduce the scanning region by selecting points with root mean square correlation above the given correlation threshold. Correlation threshold is empirically estimated through Monte Carlo simulation for various SNRs. Its criterion is formulated by a least square sense optimization in the following section.
- Do beamforming scanning at the reduced region.

For significantly many time points, computing root mean square correlations over the whole source space takes significant time. To overcome this difficulty, two strategies are proposed. One strategy is called eigen-temporal-window (ETW). Using the projection of measurement matrix onto eigen-temporal space, the number of time points is dramatically reduced, thereby reducing correlation computing time. It is formulated as follows: from singular value decomposed measurement matrix (S×T) $\mathbf{m} = \mathbf{U\Sigma V}^T$, projection matrix $\mathbf{V}_{etw}$ consists of significant temporal-eigenvectors $\mathbf{v}_i$ (corresponding significant eigenvalues) of $\mathbf{m}$. Finally, projected measurement matrix is computed below:

$$\mathbf{m}_{etw} := \mathbf{m}[\mathbf{v}_1\ \mathbf{v}_2 \dots \mathbf{v}_J] = \mathbf{m}\mathbf{V}_{etw} \quad (9)$$

Other strategy called ESTW (eigen-spatio-temporal window) is in a similar way to EST. Measurement matrix is projected onto eigen-temporal space and eigen-spatial space. Eigen-spatial projection matrix $\mathbf{U}_{esw}$ consists of significant spatial-eigenvectors $\mathbf{u}_i$ (corresponding significant eigenvalues) of gram matrix $\mathbf{G}$ (defined in Equation (10)) having geometry information of source space:

$$\mathbf{G} = \int_\Omega \mathbf{L}(\mathbf{r})\mathbf{L}^T(\mathbf{r}) d^3 r, \text{ in source space } \Omega. \quad (10)$$

So projected measurement matrix $\mathbf{m}_{estw}$ is computed below:

$$\mathbf{m}_{estw} := [\mathbf{u}_1\ \mathbf{u}_2 \dots \mathbf{u}_I]^T \mathbf{m}[\mathbf{v}_1\ \mathbf{v}_2 \dots \mathbf{v}_J] = \mathbf{U}_{esw}^T \mathbf{m}\mathbf{V}_{etw} \quad (11)$$

Since ESTW can reduce the number of sensors additionally, ESTW is more efficient in computing correlations while EST can compute more accurate correlation distribution. The effectiveness of these strategies is demonstrated in the following result section.

III. RESULTS

A. Numerical Simulation

Numerical experiments were done in PC (Intel Core 2 Duo CPU 2.4GHz, 32bit OS and 4GB RAM). A spherical homogeneous conductor model of radius 8.5cm was used for forward computing and 160-sensor gradiometer array was applied. Two sources were assumed to locate on a box shaped region (include whole brain region) defined as $-8.5 \le x \le 8.5$, $-8.5 \le y \le 8.5$ and $-2.0 \le z \le 8.5$(cm) and the scanning points (total 10000 points) within the box are equally spaced lattices. The location of two sources and both orientations were set to (-3.7, -3.6, 5.0), (3.3, 4.1, 4.3)cm and (1, 0, 0). We generated synthetic data by adding calculated sensor values to white Gaussian noise. Synthetic magnetic field with 500 time samples was generated with SNR (defined by $\|\text{measurement matrix}\|_F^2 / \|\text{noise matrix}\|_F^2$) of 1. Each time course is damping sinusoidal with different wavelengths.

In our proposed way to reduce scanning points in beamforming, reasonable correlation threshold should be determined. We formulate the correlation threshold $c_{th}(SNR)$ depending on SNR as follows:

$$c_{th}(SNR) = \overline{m} + \overline{\sigma} \cdot \Gamma(SNR)$$
$$\overline{m} = mean(p_{rms}(\cdot)),\ \overline{\sigma} = \text{var}(p_{rms}(\cdot)) \quad (12)$$

To estimate reasonable Γ(SNR), 1000 simulated single dipole problems with different SNRs were generated. We computed the correlation value c_{or} at the original source dipole location, and $\overline{m}, \overline{\sigma}$ for each problem. Then for each problem, Γ(SNR) can be estimated from Equation (12) by using $\overline{m}, \overline{\sigma}$, c_{or} and replacing $c_{th}(SNR)$ into c_{or}. Then averaged Γ(SNR) was estimated for each given SNR. From these averaged Γ(SNR) values, a concise formulation was fitted in the least square sense, as shown in Figure 1.

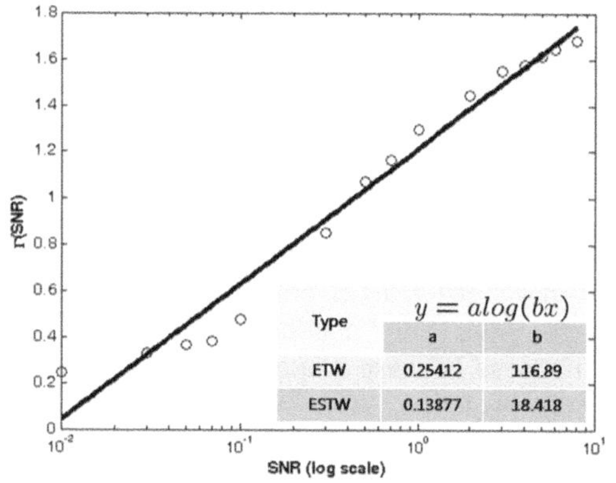

Fig. 1 Γ(SNR) was fitted in the model Γ(SNR)= a log(b·SNR) for ETW strategy. Dots indicate the result of Monte Carlo simulation at different SNRs. Lower right table represents the optimal coefficients of fitting model

Finally, in order for original source point to have far higher correlation value than correlation threshold, we use the following correlation formulation:

$$c_{th}(SNR) = \overline{m} + \overline{\sigma} \cdot [a \log(b \cdot SNR)]/2 \quad (13)$$

Figure 2 shows the results of fast-beamformer using ETW strategy. Correlation values are computed for each point **r** and we computed root mean squared correlation value over time. The correlation threshold is determined by Equation (13). In ETW strategy, we used significant eigenvectors explaining information of measurement matrix **m** up to 70%.

A proper correlation threshold value can reduce scanning region and keep the original source location, thereby reducing beamforming computation time without sacrificing accuracy, as illustrated in Figure 2. The source reconstruction result of our proposed method is identical to the conventional beamforming imaging scanning the whole region. Comparative computing time on various type of sources between ETW and ESTW strategies along with vector type minimum-variance are tabulated in Table 1.

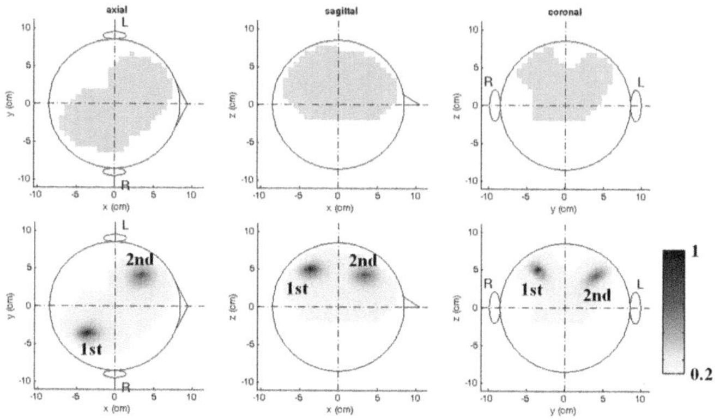

Fig. 2 Beamformer source reconstruction results using ETW strategy. Top rows represent the reduced scanning region having the higher correlation than given threshold value in Eq. (13). Bottom rows are beamforming source imaging on the reduced scanning region

Table 1 Computing time (total 10000 points and 500 time samples)

Beam-former type	Source type		Computing time (correlation + beamforming)	Improvement (%)
Conventional	single		**2.2** (0.0+2.2)	**0**
	double	near		
		far		
	triple			
Fast (ETW)	single		**1.20** (0.7+0.50)	**45.5**
	double	near	**1.22** (0.7+0.53)	**44.5**
		far	**1.29** (0.7+0.59)	**41.4**
	triple		**1.26** (0.7+0.56)	**42.7**
Fast (ESTW)	single		**0.73** (0.2+0.53)	**66.8**
	double	near	**0.75** (0.2+0.55)	**65.9**
		far	**0.84** (0.2+0.64)	**61.8**
	triple		**0.81** (0.2+0.61)	**63.2**

B. Empirical Experiment

We applied our proposed method to the empirical MEG data. Median nerve on the right hand was stimulated through small electric shock. Data were digitized at 2kHz, 500Hz low pass filtered, and post-processing digital filter was applied 1-100Hz except 50Hz. Total 399 single-trial sets were averaged and we analyzed the 5-30msec window. 160 gradiometers (MEGVISON Yokogawa electric Inc., Tokyo, Japan) were used to collect the data. Figure 3 illustrates the empirical MEG data for right-hand median nerve stimulation.

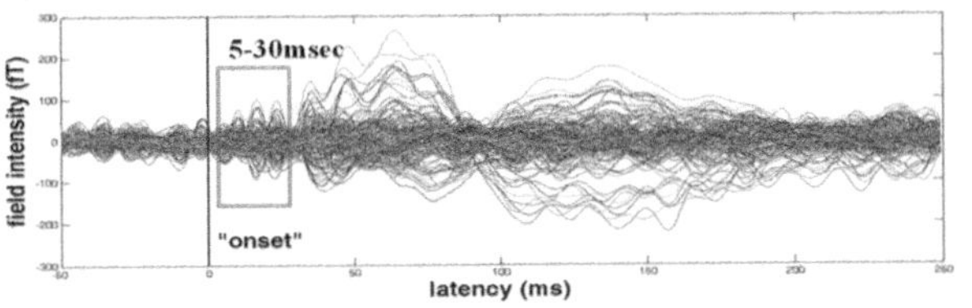

Fig. 3 160-channel overplotted empirical MEG data for right-hand median nerve stimulation. From 5-30msec were analyzed and vertical line indicates onset time point

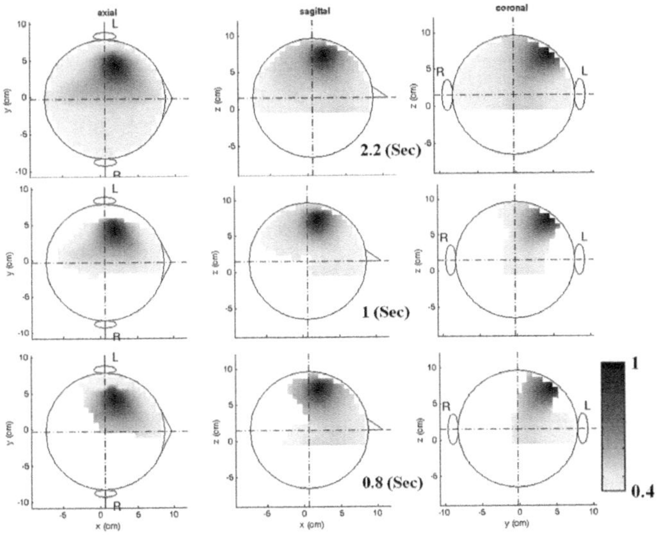

Fig. 4 Right-hand median nerve stimulation experiment: Conventional beamformer source reconstruction result (top), source reconstruction result of beamformer using ETW (middle), source reconstruction result of beamformer using ESTW (bottom)

Experiments are performed to verify the effectiveness of the proposed method. We determine the correlation threshold according to Equation (13). SNR was estimated as around 1. To apply ETSW strategy, eigen-temporal projection matrix V consists of eigenvectors explaining 70% information of measurement matrix and eigen-spatial projecttion matrix U consists of eigenvectors explaining 99% information of gram matrix. Figure 4 (top) represents the result of conventional beamformer (vector type minimum-variance) source reconstruction scanning the whole region. One source is estimated by the conventional beamformer. Our proposed methods using ETW or ESTW localized one source without sacrificing accuracy as shown in Figure 4 (middle, bottom). It means that reduction of the number of scanning points by ETW or ESTW can speed up beamforming source imaging. Total computing time of ETW and ESTW were about 1sec (54.5% improved) and 0.8sec (63.6% improved), respectively.

IV. Conclusions

We proposed the way to reduce the scanning region by correlation distribution, thereby speeding up beamformer source imaging. Efficient strategies to compute correlation distribution were discussed. The criterion to determine correlation threshold was empirically determined. For simulated data and empirical data, its efficacy was verified. Finally, we believe this method could be applicable for real-time brain-computer interface.

Acknowledgment

This work is supported by KRF-2008-531-C00012, the BioImaging Research Center at GIST, KOSEF/JSPS researcher exchange program, and the National IT Industry Promotion Agency (NIPA-2009-C1090-0902-0008).

References

1. K. Sekihara and S. S. Nagarajan (2008) Adaptive spatial filters for electromagnetic brain imaging. Berlin, Heidelber: Springer-Verlag

Author: S.C. Jun
Institute: Gwangju Institute of Science and Technology (GIST)
Street: 261 Cheomdan-gwagiro(Oryong-dong), Buk-gu
City: Gwangju
Country: South Korea
Email: scjun@gist.ac.kr

A Method for MEG Data That Obtains Linearly-Constrained Minimum-Variance Beamformer Solution by Minimum-Norm Least-Squares Method

Toshiaki Imada

Institute for Learning & Brain Sciences, University of Washington, Seattle, U.S.A.

Abstract— **A method to obtain the linearly-constrained minimum-variance (LCMV) beamformer solution simply by applying the minimum-norm least-squares (MNLS) method to the recorded brain magnetic fields is theoretically shown. First, apply the prewhitening process to the recorded data using the data variance-covariance matrix. Then, apply the MNLS method to its outcome. The obtained solution is mathematically proved to be equivalent to the one obtained by the original LCMV beamformer. This method works only when the number of independent neural currents (equivalent to double of the number of specified neural current locations if a spherical head model is used) in the brain to be calculated is not more than the number of magnetic sensors. This is because linear independence of the neural currents is one of prerequisites for obtaining the LCMV beamformer weight matrices. When it is more, the solution currents at different locations are not linearly independent, meaning that the obtained neural currents are mathematically, rather than physiologically, determined by the beamformer weight matrices or the MNLS pseudoinverse matrix, which should be avoided for further analyses on the obtained neural currents.**

Keywords— **minimum-norm least-squares method, linearly-constrained minimum-variance beamformer, prewhitening.**

I. Introduction

A minimum-norm least-squares (MNLS) method, which has been generally applied to construct a theoretical model from measured data by minimizing the differences between them, can also model the neural currents in the brain based on the recorded magnetoencephalography (MEG) data [1]. A linearly-constrained minimum-variance (LCMV) beamformer, originally developed to find a target in the specified direction in the radar array system [2], has been also applied to reconstruct a neural current activity in the brain from the data recorded by MEG or electroencephalography (EEG) [3]. The MNLS method could be called as a data-independent beamformer because it incorporates only physical configuration of the sensors and the brain to reconstruct the neural currents from the MEG data, whereas the LCMV beamformer (LCMVb) is a typical data-dependent beamformer that relies on the statistical features of the actual data in addition. The solutions from these two methods have been considered to be very different each other, but in this paper, it is theoretically shown that the solution by LCMVb is easily obtained by the MNLS method simply by introducing the data variance-covariance matrix. It is also shown that the neural current solutions from these two methods are equivalent except the noise; the solution by the LCMVb is always contaminated by the noise, but if the noise is estimated and subtracted, the solutions from these two methods are exactly the same.

II. Linearly-constrained minimum-variance beamformer

When we assume M magnetic sensors for the measurement and N neural current locations, $\mathbf{r}_1, \mathbf{r}_2, \cdots, \mathbf{r}_N$, in the brain, the measured brain magnetic field $\mathbf{y}$ is modeled by Eq. (1), where $\mathbf{L}_N$ is an M by JN lead field matrix, $\mathbf{x}$ is an JN by 1 neural current vector, and $\mathbf{e}$ is an M by 1 magnetic noise vector that is not modeled by the lead field matrix. J is the number of independent currents at each current location $\mathbf{r}_i$. For example, when the spherical head model is employed, J is equal to 2, representing x and y current components.

$$\mathbf{y} = \mathbf{L}_N \mathbf{x} + \mathbf{e} \tag{1}$$

Note that $\mathbf{x}$ may include any electric current that can produce magnetic fields modeled by the lead field, resulting in containing not only the brain signals of our interest but also the possible noise current, such as brain background activity currents.

Before calculating the variance-covariance matrix $\mathbf{C}$, we assume, in the same manner as in the original LCMVb [3], that (a1) the current source $\mathbf{x}$ is uncorrelated, namely, current sources, $\mathbf{x}(\mathbf{r}_i)$ and $\mathbf{x}(\mathbf{r}_i)$, at any of two different locations $\mathbf{r}_i$ and $\mathbf{r}_i$, are uncorrelated each other, where $\mathbf{x}(\mathbf{r}_i)$ is a J by 1 current vector at $\mathbf{r}_i$, (a2) the noise $\mathbf{e}$ at any two sensors are also uncorrelated each other, with mean value $\mathbf{0}$, and (a3) the magnetic field $\mathbf{L}_N\mathbf{x}$ derived from the current sources $\mathbf{x}$ and the noise magnetic field $\mathbf{e}$ are uncorrelated.

In the original LCMVb [3], a part of the assumption (a2), namely 'uncorrelated noise $\mathbf{e}$', was not included and this assumption is only implicitly used in this paper. In fact, in all of the MEG measurements, the MEG data could include

S. Supek and A. Sušac (Eds.): Advances in Biomagnetism – BIOMAG2010, IFMBE Proceedings 28, pp. 152–154, 2010.
www.springerlink.com

the correlated ambient noise. However, using a state-of-the-art noise-reduction method, such as signal space separation [4], the remaining noise in the MEG data could be only uncorrelated sensor noise, which makes the assumptions (a2) and (a3) realistic.

Based on these assumptions, the variance-covariance matrix $\mathbf{C}$ from the measured data is calculated as follows.

$$\begin{aligned}\mathbf{C} &= E(\mathbf{y}\mathbf{y}^T) = E((\mathbf{L}_N\mathbf{x}+\mathbf{e})(\mathbf{L}_N\mathbf{x}+\mathbf{e})^T) \\ &= \mathbf{L}_N E(\mathbf{x}\mathbf{x}^T)\mathbf{L}_N^{\ T} + E(\mathbf{e}\mathbf{e}^T)\end{aligned} \tag{2}$$

$\mathbf{A}^T$ is a transpose of matrix $\mathbf{A}$ and $E(\mathbf{z})$ is an ensemble mean of $\mathbf{z}$. We assume that (a4) $\mathrm{rank}(\mathbf{L}(\mathbf{r}_i)^T\mathbf{C}^{-1}\mathbf{L}(\mathbf{r}_i)) = \mathrm{J}$ ($i = 1\cdots\mathrm{N}$) and (a5) $\mathrm{rank}(\mathbf{C}) = \mathrm{M}$ so that we are able to define the LCMVb weight matrix.

From these assumptions (a1)-(a5) and Eq. (2), the transpose of the LCMVb weight matrix $\mathbf{W}(\mathbf{r}_i)^T$ is calculated as follows.

$$\mathbf{W}(\mathbf{r}_i)^T = (\mathbf{L}(\mathbf{r}_i)^T\mathbf{C}^{-1}\mathbf{L}(\mathbf{r}_i))^{-1}\mathbf{L}(\mathbf{r}_i)^T\mathbf{C}^{-1} \quad (i = 1\cdots\mathrm{N}) \tag{3}$$

Using these weight matrices, we obtain the LCMVb solution $\mathbf{x}_B(\mathbf{r}_i)$ at each scanned location $\mathbf{r}_i$ as shown in Eq. (4), where $\mathbf{x}_B(\mathbf{r}_i)$ is a J by 1 current vector ($i = 1\cdots\mathrm{N}$).

$$\mathbf{x}_B(\mathbf{r}_i) = \mathbf{W}(\mathbf{r}_i)^T\mathbf{y} = (\mathbf{L}(\mathbf{r}_i)^T\mathbf{C}^{-1}\mathbf{L}(\mathbf{r}_i))^{-1}\mathbf{L}(\mathbf{r}_i)^T\mathbf{C}^{-1}\mathbf{y} \quad (i = 1\cdots\mathrm{N}) \tag{4}$$

Since the LCMVb weight matrix $\mathbf{W}(\mathbf{r}_i)$ determines the neural current at each location $\mathbf{r}_i$, we call it as a local LCMVb weight matrix in this paper. By vertically concatenating $\mathbf{x}_B(\mathbf{r}_i)$ and local LCMVb weight matrices $\mathbf{W}(\mathbf{r}_i)^T$ ($i = 1\cdots\mathrm{N}$) in Eq. (4), we obtain the concatenated local weight matrix $\mathbf{W}_N^{\ T} \equiv [\mathbf{W}(\mathbf{r}_1)\ \mathbf{W}(\mathbf{r}_2) \cdots \mathbf{W}(\mathbf{r}_N)]^T$, as shown in Eq. (5).

$$\mathbf{x}_B \equiv \begin{bmatrix}\mathbf{x}_B(\mathbf{r}_1)\\ \mathbf{x}_B(\mathbf{r}_2)\\ \vdots \\ \mathbf{x}_B(\mathbf{r}_N)\end{bmatrix} = \begin{bmatrix}\mathbf{W}(\mathbf{r}_1)^T\\ \mathbf{W}(\mathbf{r}_2)^T\\ \vdots \\ \mathbf{W}(\mathbf{r}_N)^T\end{bmatrix}\mathbf{y} \equiv \mathbf{W}_N^{\ T}\mathbf{y} \tag{5}$$

From Eq. (5), to obtain uncorrelated current solution $\mathbf{x}_B$, which is assumed in the assumptions (a1), it is obvious that the local weight matrix, $\mathbf{W}(\mathbf{r}_i)^T$ ($i = 1\cdots\mathrm{N}$), should be linearly independent each other [5]. From this constraint, it is evident that JN is not greater than M. Moreover, from Eq. (3) and linear independency of $\mathbf{W}(\mathbf{r}_i)$, $\mathbf{L}(\mathbf{r}_i)$ ($i = 1\cdots\mathrm{N}$) should be also linearly independent each other. From these, we assume (a6) $\mathrm{JN} \le \mathrm{M}$ and (a6') $\mathrm{rank}(\mathbf{L}_N) = \mathrm{J\,N} \le \mathrm{M}$.

When we obtain an LCMVb solution with more than M/J brain locations, it is necessary to reduce the number of locations, by eliminating linearly dependent locations out of more than M/J locations, so that the solution could be used for further analyses.

Similarly to the LCMVb local weight matrix $\mathbf{W}(\mathbf{r}_i)^T$, we define a (JN) by M expanded or global LCMVb weight matrix $\mathbf{\Omega}_N^{\ T}$ simply by replacing the local lead field matrix $\mathbf{L}(\mathbf{r}_i)$ with the global lead field matrix $\mathbf{L}_N$ in Eq. (3), as shown in Eq. (6).

$$\Omega_N^{\ T} \equiv (\mathbf{L}_N^{\ T}\mathbf{C}^{-1}\mathbf{L}_N)^{-1}\mathbf{L}_N^{\ T}\mathbf{C}^{-1} \quad (\mathrm{M} \ge \mathrm{J\,N}) \tag{6}$$

Since from the assumption (a1), the LCMVb output, $\mathbf{x}_B(\mathbf{r}_i)$ ($i = 1\cdots\mathrm{N}$), is uncorrelated each other, the global weight matrix $\mathbf{\Omega}_N^{\ T}$ is equivalent to the concatenated local LCMVb weight matrix $\mathbf{W}_N^{\ T} \equiv [\mathbf{W}(\mathbf{r}_1)\ \mathbf{W}(\mathbf{r}_2) \cdots \mathbf{W}(\mathbf{r}_N)]^T$.

$$\begin{aligned}\Omega_N^{\ T} &\equiv (\mathbf{L}_N^{\ T}\mathbf{C}^{-1}\mathbf{L}_N)^{-1}\mathbf{L}_N^{\ T}\mathbf{C}^{-1} \\ &= ([\mathbf{L}(\mathbf{r}_1)\ \mathbf{L}(\mathbf{r}_2) \cdots \mathbf{L}(\mathbf{r}_N)]^T\mathbf{C}^{-1}[\mathbf{L}(\mathbf{r}_1)\ \mathbf{L}(\mathbf{r}_2) \cdots \mathbf{L}(\mathbf{r}_N)])^{-1} \bullet \\ &\quad [\mathbf{L}(\mathbf{r}_1)\ \mathbf{L}(\mathbf{r}_2) \cdots \mathbf{L}(\mathbf{r}_N)]^T\mathbf{C}^{-1} \\ &= [\mathbf{W}(\mathbf{r}_1)\ \mathbf{W}(\mathbf{r}_2) \cdots \mathbf{W}(\mathbf{r}_N)]^T \equiv \mathbf{W}_N^{\ T}\end{aligned} \tag{7}$$

From Eqs. (5) and (7), the following Eq. (8) holds.

$$\begin{aligned}\mathbf{x}_B &\equiv \begin{bmatrix}\mathbf{x}_B(\mathbf{r}_1)\\ \mathbf{x}_B(\mathbf{r}_2)\\ \vdots \\ \mathbf{x}_B(\mathbf{r}_N)\end{bmatrix} = \begin{bmatrix}\mathbf{W}(\mathbf{r}_1)^T\\ \mathbf{W}(\mathbf{r}_2)^T\\ \vdots \\ \mathbf{W}(\mathbf{r}_N)^T\end{bmatrix}\mathbf{y} \equiv \mathbf{W}_N^{\ T}\mathbf{y} \\ &= \Omega_N^{\ T}\mathbf{y} = (\mathbf{L}_N^{\ T}\mathbf{C}^{-1}\mathbf{L}_N)^{-1}\mathbf{L}_N^{\ T}\mathbf{C}^{-1}\mathbf{y}\end{aligned} \tag{8}$$

III. MINIMUM-NORM LEAST-SQUARES BEAMFORMER

To obtain the neural current solution equivalent to the LCMVb solution by the MNLS method or beamformer (MNLSb), first we prewhiten the recorded magnetic fields $\mathbf{y}$ by the data variance-covariance matrix $\mathbf{C}$, as shown below.

$$\mathbf{D}\mathbf{L}_N\mathbf{x} = \mathbf{D}(\mathbf{y} - \mathbf{e}) \tag{9}$$

The M by M matrix $\mathbf{D}$ is obtained by factoring the data variance-covariance matrix $\mathbf{C}$ as follows.

$$\mathbf{C}^{-1} = \mathbf{D}^T\mathbf{D} \tag{10}$$

Applying the MNLSb to Eq. (9), which is equivalent to multiplying $(\mathbf{D}\mathbf{L}_N)^+$, a generalized or pseudo inverse matrix of $\mathbf{D}\mathbf{L}_N$, from the left of Eq. (12), we obtain the solution $\mathbf{x}_{DM}$ as shown in Eq. (11), where $\Pi_N = (\mathbf{D}\mathbf{L}_N)^+\mathbf{D}$.

$$\mathbf{x}_{DM} = (\mathbf{D}\mathbf{L}_N)^+\mathbf{D}(\mathbf{y} - \mathbf{e}) \equiv \Pi_N(\mathbf{y} - \mathbf{e}) \tag{11}$$

It is shown that, by applying the factorization of the pseudo-inverse matrix [6], the matrix Π_N is equivalent to the matrix Ω_N^T, as shown in Eq. (12).

$$\Pi_N \equiv (\mathbf{DL}_N)^{+}\mathbf{D} = [(\mathbf{DL}_N)^T(\mathbf{DL}_N)]^{-1}(\mathbf{DL}_N)^T\mathbf{D}$$
$$= (\mathbf{L}_N{}^T\mathbf{C}^{-1}\mathbf{L}_N)^{-1}\mathbf{L}_N{}^T\mathbf{C}^{-1} = \Omega_N^T \quad (12)$$

From Eqs. (8), (11), and (12), it is mathematically proved that the data-prewhitened MNLSb solution $\mathbf{x}_{DM}$ is equivalent to the LCMVb solution $\mathbf{x}_B$ when the original recorded magnetic field data $\mathbf{y}$ is used instead of the data after subtracting the noise $\mathbf{e}$, namely $(\mathbf{y} - \mathbf{e})$ in Eq. (11). Needless to say, when the uncorrelated noise $\mathbf{e}$ equals to $\mathbf{0}$, these two solutions, $\mathbf{x}_{DM}$ and $\mathbf{x}_B$, are equivalent. When $\mathbf{e} = \mathbf{0}$, Mosher et al. [7] discussed about the LCMV solution from different point of view.

If the noise $\mathbf{e}$ can be estimated and is subtracted from the recorded data $\mathbf{y}$, the data-prewhitened MNLSb solution $\mathbf{x}_{DM}$ shows the better estimation of the neural current than $\mathbf{x}_B$.

The solution by the modified version of LCMVb that removes the noise $\mathbf{e}$ based on the eigen-space of the data variance-covariance matrix $\mathbf{C}$ [8] is exactly the same as $\mathbf{x}_{DM}$ if the same noise $\mathbf{e}$ is subtracted in MNLSb.

Finally, note that this method of obtaining the LCMVb solution by the MNLSb is valid when $JN \leq M$ is satisfied. If $JN > M$, however, the neural current at each brain location obtained by the solution $\mathbf{x}_B$ is correlated each other, which contradicts the assumption (a1).

IV. Conclusions

The method to obtain the LCMVb solution simply by applying the data-prewhitened MNLSb to the recorded magnetic fields is theoretically shown. This method is mathematically valid only when the number of neural current locations in the brain to be calculated is *not* more than the number of magnetic sensors. When it is more, however, the solution currents at different locations are not linearly independent, meaning that the obtained neural currents are mathematically, rather than physiologically, determined by the beamformer weight matrices, which should be avoided for further analyses of the obtained neural currents.

Acknowledgment

I thank Dr. P. K. Kuhl for the support. This research was funded by Washington State's Life Science Discovery Fund grant, by a Science of Learning Center grant to the University of Washington LIFE Center from the National Science Foundation, by University of Washington's Institute for Learning and Brain Sciences.

References

1. Hämäläinen MS, Hari R, Ilmoniemi RJ, Knuutila J, Lounasmaa OV (1993) Magnetoencephalography—theory, instrumentation, and applications to noninvasive studies of the working human brain. Rev Mod Phys 65:413–497
2. Frost III OL (1972) An algorithm for linearly constrained adaptive array processing. Proc IEEE 60:926-935
3. van Veen, BD, van Drongelen W, Yuchtman M, Suzuki A (1997) Localization of brain electrical activity via linearly constrained minimum variance spatial filtering. IEEE Trans Biomed Eng 44:867–880
4. Taulu S, Kajola M (2005) Presentation of electromagnetic multichannel data: The signal space separation method. J Appl Physics, 97: 124905.
5. Imada T (2008) Local and global weight matrices of linearly-constrained minimum-variance beamformer. In: Kakigi R, Yokosawa S, Kuriki S (Eds) Biomagnetism – Interdisciplinary Research and Exploration (Proc 16th International Conference on Biomagnetism, 2008), pp 65-67
6. Strang G, (1988) Linear algebra and its applications. 3rd ed. Orlando, FL, Harcourt Brace & Company, p 452
7. Mosher JC, Baillet S, Leahy RM (2003) Equivalence of linear approaches in bioelectromagnetic inverse solutions. 2003 IEEE Workshop on Statistical Signal Processing, pp 294-297
8. Yu J-L, Yeh C-C (1995) Generalized eigenspace-based beamformer. IEEE Trans on Signal Processing, 43:2453-2461.

Author: Toshiaki Imada
Institute: University of Washington
Street: 206 Old Fisheries
City: Seattle
Country: U.S.A.
Email: imada "at" u.washington.edu

Current Dipole Estimation in MEG by Spatial Interpolation of Magnetic Sensors

Hiroshi Fukuda[1], Masato Odagaki[1], Atsushi Kodabashi[2], Toshiro Fujimoto[2], and Osamu Hiwaki[1]

[1] Graduate School of Information Sciences, Hiroshima City University, Hiroshima, Japan
[2] Fujimoto Hayasuzu Hospital, Miyakonojo, Japan

Abstract— **In this study, a method for current dipole estimation with spatial interpolation of measured magnetic fields in magnetoencephalogram (MEG) is proposed. This method applies virtual magnetic (M_v) sensors uniformly interpolated in the surface comprised of MEG (M_r) sensors. Orientations and output signals of the virtual magnetic sensors are calculated by a spatial linear interpolation based on those of real magnetic sensors. For determining an optimal value of regularization parameter (α) in minimum norm inverse solutions, M_v sensors are divided into some groups randomly. Quantities of magnetic field (b_k) at the M_r sensors are calculated by estimated dipoles with each group of M_v sensors. α is determined by minimization of the total sum of differences between b_k and the output signals measured by M_r sensors. In order to confirm the validity of the proposed method, we estimated source distribution of N20m components of unilateral and bilateral somatosensory evoked fields (SEFs) measured by a 160-channel axial SQUID gradiometer system, when subject's right and/or left median nerve was electrically stimulated. The proposed method was able to estimate source distribution in the primary somatosensory cortex (S1). Furthermore, the source distribution estimated by the proposed method tended to be more confined than that estimated by a cross validation error method. These results indicate that the virtually improved spatial resolution of MEG data can estimate more confined source distribution.**

Keywords— **Magnetoencephalogram, Current Dipole Estimation, Inverse Problem, Somatosensory Evoked Field.**

I. Introduction

Magnetoencephalography (MEG) can measure human brain activity noninvasively. MEG has advantage in time resolution as compared with other noninvasive techniques including functional magnetic resonance imaging (fMRI) and near infrared spectroscopy (NIRS).

To estimate brain activity from MEG data, biomagnetic inverse problems must be solved. In MEG inverse problems, many candidates of current dipoles are discretely distributed in the whole brain and forward biomagnetic problems are represented by a linear function $b = Lq$, where b is an $M \times 1$ vector of magnetic fields, q is a $3N \times 1$ vector of moments of dipole candidates, L is an $M \times 3N$ lead field matrix based on the Biot-Savart's law, M is the number of magnetic sensors and N is the number of the candidates of current dipoles.

Generally, the inverse problems are ill-posed because the number of instantaneous MEG data measured with the magnetic sensors M is much less than the number of the candidates of current dipoles N. Conventionally, regularization has been used to determine a unique solution of the ill-posed inverse problems. In Tikhonov regularization, norm of moments of the dipole candidates is used as an objective function. This regularization method estimates distribution of current dipoles to minimize

$$\phi(q) = \left\| b^* - Lq \right\|^2 + \alpha \left\| q \right\|^2 \tag{1}$$

where α is a regularization parameter and b^* is an $M \times 1$ vector of measured magnetic fields which contains some measurement noises δ. An approximate solution $\hat{q}$ is calculated by

$$\hat{q} = L^{\mathrm{T}} \left(LL^{\mathrm{T}} + \alpha I_{\mathrm{M}} \right)^{-1} b^* \tag{2}$$

where I_{M} is an $M \times M$ identity matrix and the superscript "T" denotes transpose. This method and its solution $\hat{q}$ are called minimum norm method and minimum norm solution, respectively.

In the minimum norm method, the regularization parameter α must be determined to find a good compromise between the residual norm $\|b^* - Lq\|^2$ and the norm of the solution $\|q\|^2$. One of the useful methods to determine the optimal regularization parameter is the cross validation error (CVE) method [1, 2]. However, this method does not necessarily ensure the best compromise between the residual norm and the norm of the solution. A worse compromise causes large activity in undesired cortical sites such as outsides of the primary somatosensory area (S1) for the somatosensory evoked fields (SEF).

The two-source resolvability and single-source localization accuracy monotonically improve with increasing number of magnetic sensors when the data is analyzed with beamformers [3]. Hence, MEG system with higher dense magnetic sensors is expected to improve accuracy of dipole localization. However, it is difficult to manufacture the MEG system with actual high dense magnetic sensors.

In this paper, we demonstrated that virtually interpolated magnetic sensors comprised of real magnetic sensors are useful as an alternative method.

S. Supek and A. Sušac (Eds.): Advances in Biomagnetism – BIOMAG2010, IFMBE Proceedings 28, pp. 155–158, 2010.
www.springerlink.com

II. Methods

A. Spatial Interpolation of Magnetic Sensors

An example of arrangement of real magnetic sensors is show in Fig. 1 (a). Surfaces comprised of the real magnetic sensors are Delaunay triangles. A sphere is assumed inside of the Delaunay triangles. The radius of the assumed sphere is 95 % of the average of distances from the origin of the MEG coordinate system to the real magnetic sensors. Points are arranged on the surface of the assumed sphere with 5 mm of arc as shown in Fig. 1 (b). The virtual magnetic sensors are put onto Delaunay triangles, which are on the lines through the origin of the MEG coordinate system and the points on the assumed sphere as shown in Fig. 1 (c). An example of arrangement of the virtual magnetic sensors on the Delaunay triangles is shown in Fig. 1 (d).

Orientations of the virtual magnetic sensors are linearly interpolated by orientations of three real magnetic sensors at the vertices of the Delaunay triangles. Outputs of the virtual magnetic sensors are interpolated by the strengths of the magnetic field at the vertices of Delaunay triangles.

B. Determination of Regularization Parameter

It is assumed that the numbers of the real and virtual magnetic sensors are M_r and M_v, respectively. M_r virtual magnetic sensors are randomly selected from M_v virtual magnetic sensors and grouped. A total of K groups of M_r virtual magnetic sensors are prepared. For the k-th group of the virtual magnetic sensors, current dipole distribution of N dipole candidates $\hat{q}_k$ ($3N \times 1$) for an arbitrary α is calculated by

$$\hat{q}_k(\alpha) = L_k^T \left(L_k L_k^T + \alpha I_M \right)^{-1} b_k^* \tag{3}$$

where L_k is an $M_r \times 3N$ lead field matrix and b_k^* is an $M_r \times 1$ vector of magnetic fields corresponded to the virtual magnetic sensors of the k-th group. An optimal value of the regularization parameter α is determined so as to minimize the equation

$$f(\alpha) = \sum_{k=1}^{K} \left\| b^* - L_k \hat{q}_k(\alpha) \right\|^2 \tag{4}$$

where b^* is an $M_r \times 1$ vector of magnetic fields measured by the real magnetic sensors.

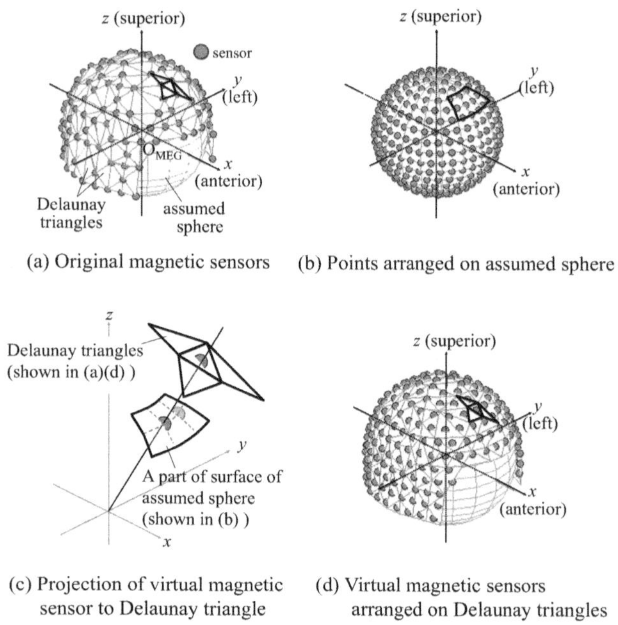

(a) Original magnetic sensors (b) Points arranged on assumed sphere

(c) Projection of virtual magnetic sensor to Delaunay triangle (d) Virtual magnetic sensors arranged on Delaunay triangles

Fig. 1 Spatial interpolation of magnetic sensors

C. MEG and MRI Measurements

Five subjects participated in the measurements of the somatosensory evoked fields (SEFs). The measurements of SEFs were performed using a 160-channel axial SQUID gradiometer system (PQ1160C, Yokogawa Electric Corp., Japan). The median nerves at right and left wrists were unilaterally or bilaterally stimulated by electric stimuli (labeled "right SEF", "left SEF" and "bilateral SEF", respectively). The duration, interval and strength of the electric stimuli were 0.1 ms, 0.5 s and 6-10 mA. SEFs were obtained at a sampling frequency of 2 kHz in an acquisition frame of 150 ms. The signals of the SEFs were filtered by a 1-500 Hz bandpass filter and then averaged 400 times. The baseline was determined by the averaged signals for 50 ms before the stimulus.

MR images of the subjects were scanned using a 1.5-T MR system (GEMS0372, GENESIS SIGNA) with a field of view of 250 mm × 250 mm and an image size of 256 pixels × 256 pixels. A total of 124 axial slices was obtained with a slice thickness of 2 mm.

D. Examination of Validity of Proposed Method

Current source distribution of the N20m components of SEFs was estimated by the proposed method and a CVE method which is one of the most useful methods to determine the regularization parameter.

In both methods, candidates of current dipole were arranged on the cerebral cortex at 4 mm spatial resolution based on the MR images of the subjects. Details on the arrangement of dipole candidates are described in our previous study [4]. For the five subjects, totals of 4022, 5120, 3585, 3852 and 3961 dipole candidates were arranged. The number of the real magnetic sensors M_r was 160. In the proposed method, the numbers of the virtual magnetic sensors M_v were 972 for four subjects and 993 for one subject. Eight hundreds virtual sensors were randomly selected and then they were randomly separated into five groups of 160 sensors. A total of 960 sensors (160 real and 800 virtual sensors) was used in the proposed method.

Comparison of current source distribution between the proposed method and the CVE method were performed as follows. An optimal regularization parameter was determined by eq. (3) and eq. (4). Dipoles in S1 in the both hemisphere of the cerebrum were evaluated. Magnetic fields produced by all dipoles (b_{all}) and those produced by the dipoles in the contralateral S1 (b_{S1}) were calculated and then goodness to fit (GOF) between b_{all} and b_{S1} was calculated. To compare GOF in the proposed method (GOF_P) and GOF in the CVE method (GOF_C), the relative error of GOF_P for GOF_C was calculated by

$$E_R = (GOF_P - GOF_C)/GOF_C \times 100\ \%. \tag{5}$$

A positive value of E_R indicates that current source distribution estimated with the proposed method is more confined to S1 than that with the CVE method.

All computations were performed by MATLAB (Math Works Inc., U.S.).

III. Results and discussion

An example of interpolation of the magnetic sensors is shown in Fig.2. The N20m component of right SEF measured by 160 real sensors (shown in Fig.2 (a)) was spatially interpolated with 993 virtual magnetic sensors (shown in Fig. 2(b)). The source and sink of the magnetic field appeared clearly by the interpolation.

Examples of curves of the cross validation error and the error function defined by eq. (4) with respect to regularization

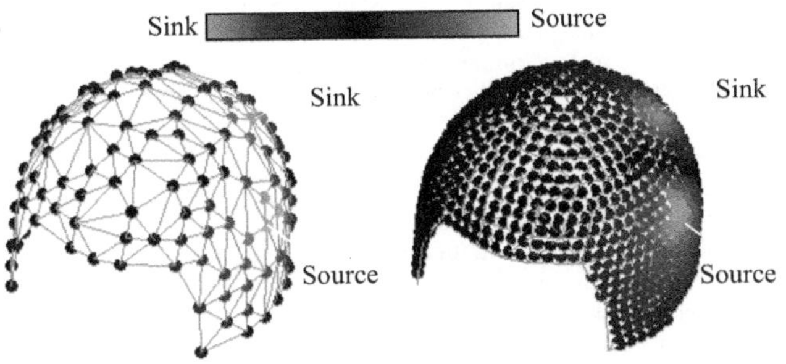

(a) Original magnetic sensors (b) Interpolated magnetic sensors

Fig. 2 An example of interpolation of magnetic sensors in right SEF

parameter α in a representative subject are shown in Fig. 3. The optimal value of α in the proposed method was different from that in the CVE method. In this case, the residual norm $||b^* - Lq||^2$ was more weighted in the proposed method than that in the CVE method.

The estimated current dipole distributions in the right SEF, left SEF and bilateral SEF in the representative subject are shown in Fig. 4 (a), (b) and (c), respectively. In this subject, the number of the estimated current dipoles at outside of S1in the proposed method was less than that in the CVE method.

The relative errors (E_R) in all stimulus conditions (right, left and bilateral SEFs) and all subjects are shown in Fig. 5. The dipoles with over 70 % of maximum activity are shown as white spots. E_R was positive in ten out of fifteen conditions. The average of E_R across all stimulus conditions and all subjects was 4.7 % (±7.2 % S.D.). The average of E_R was significantly greater than 0 % ($p < 0.05$). This result indicates that the estimated source distribution of the N20m component of SEFs was more confined to S1 in the proposed method.

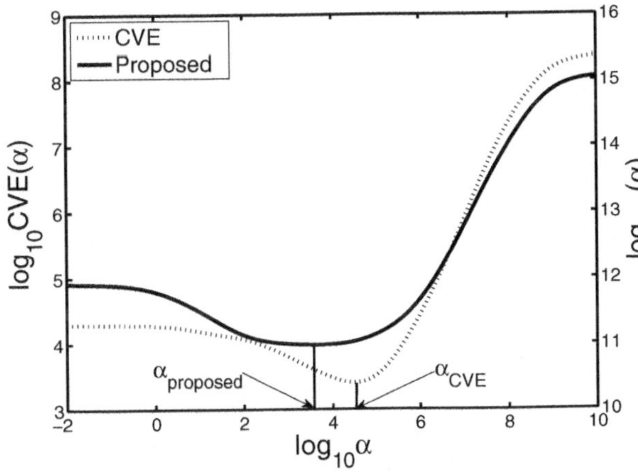

Fig. 3 Comparison of objective function with respect to α between the proposed method and the CVE method

IV. Conclusions

In this study, a method for current source estimation with spatial interpolation of measured magnetic fields in MEG was proposed. The proposed method was able to estimate source distribution at S1 in SEF measurements. Furthermore, the source distribution estimated by the proposed method was more confined to S1 than that estimated by the CVE method.

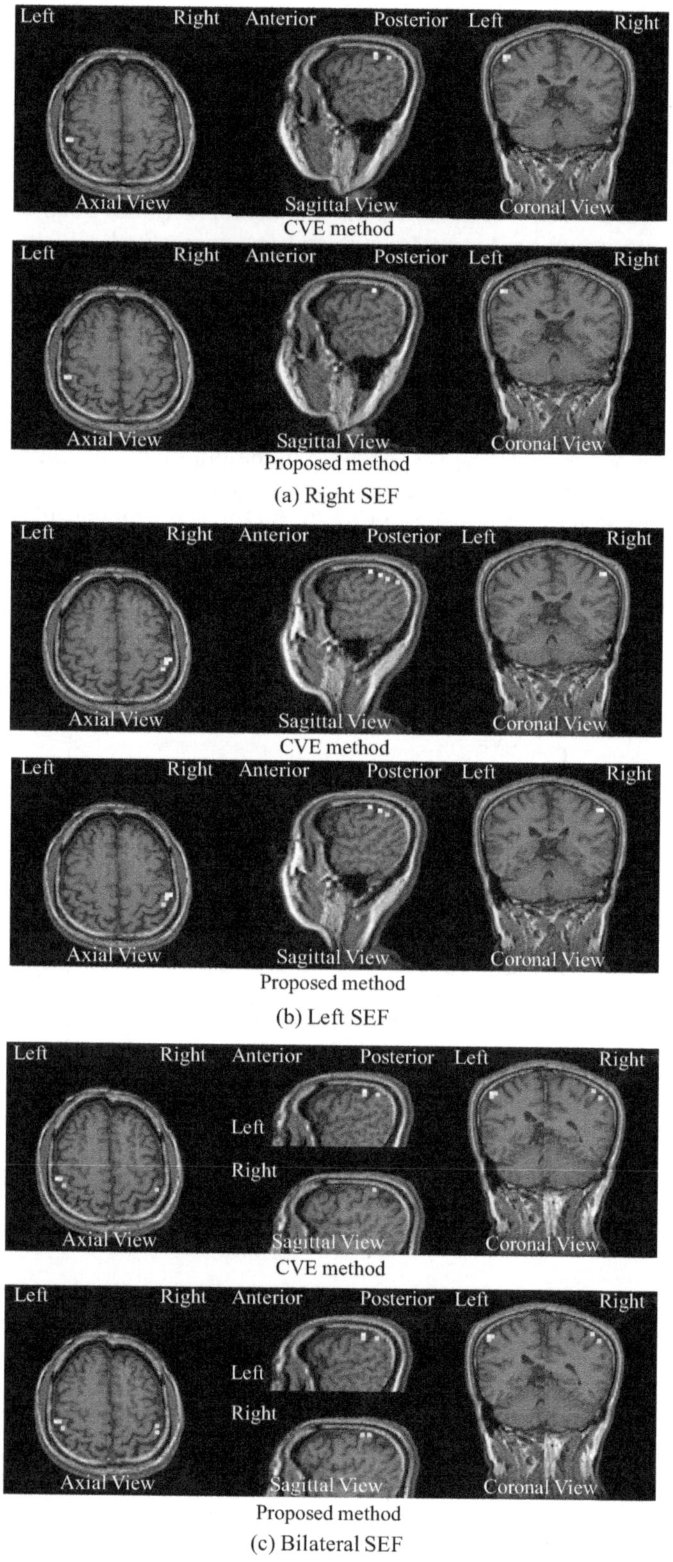

Fig. 4 Estimated source distribution in a representative subject. The dipoles with over 70 % of maximum activity are shown as white spots

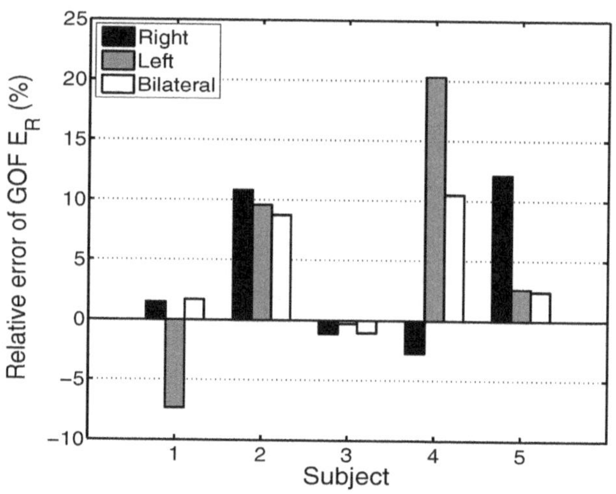

Fig. 5 Relative error of GOF between proposed method and CVE method. The abscissa indicates the subjects and the ordinate indicates the relative error of GOF between the proposed method and the CVE method

Acknowledgment

This work was supported by SCOPE (Contract No. 082108003) from the Ministry of Internal Affairs and Communications of Japan and Grant-in-Aid for Young Scientists (B) Contract No. 20760269 and Scientific Research (C) Contract No. 20560400 from the Ministry of Education, Science, Sports and Culture of Japan.

References

1. Pascual-Marqui RD (2002) Standardized low resolution brain electromagnetic tomography (sLORETA): technical details. Method Find. Exp. Clin. Pharmacol. 24D:5-12
2. Ding L, He B (2005) 3-dimensional brain source imaging by means of Laplacian weighted minimum norm estimate in a realistic geometry head model. Proc.IEEE Eng. Med. Biol. 5:4480–4483
3. Vrba J, Robinson SE, McCubbin J (2004) How many channels are needed for MEG? Neurol. Clin. Neurophysiol.99
4. Fukuda H, Odagaki M, Hiwaki O, Kodabashi A., Fujimoto T. (2008) Estimation method on multiple sources of MEG based on the columnar structure of the cerebral cortex. IEEE Trans. Mag. 44:4425-4431

Author: Osamu Hiwaki
Institute: Hiroshima City University
Street: 3-4-1, Ozuka-higashi, Asaminami-ku
City: Hiroshima
Country: Japan
Email: hiwaki@hiroshima-cu.ac.jp

The Compressible Estimate (CE) of MEG Current Sources

Wei-Tang Chang[1] and Fa-Hsuan Lin[1,2]

[1] Institute of Biomedical Engineering, National Taiwan University, Taipei, Taiwan
[2] MGH-HST Athinoula A. Martinos Center for Biomedical Imaging, Charlestown, MA, USA

Abstract— **Magnetoencephalography (MEG) non-invasively detects weak extracranial magnetic fields during tasks using super-conducting quantum interference devices (SQUIDs). The MEG inverse problem is to quantitatively model the underlying neuronal current sources by estimating their locations and orientations using MEG measurements. Many of the methods involve explicit preferences toward either spatially focal or diffused estimates of neuronal currents. However, considering a scenario that both focal and distributed current sources are simultaneously presented in the measurement, neither focal nor diffused bias is appropriate to reveal the underlying current sources. To address this challenge, we propose the Compressible Estimate (CE) based on the hypothesis that the current sources are sparse after a spatial transformation. CE promotes the compressibility of current sources by minimizing the ℓ_1-norm of transformed estimated current sources and avoids overly diffused source estimates by minimizing the ℓ_1-norm of estimated current sources. In this study we select the Laplacian and spherical wavelet as the spatial transformation. As demonstrated by the simulations and *in vivo* somatosensory and auditory experiments, CE provides the improved accuracy of the source localization involving both spatially focal and diffused neuronal currents.**

Keywords— **inverse problem, ℓ_1-norm, convex optimization, MEG, EEG, compressed sensing, sparsity, regularization.**

I. Introduction

Magnetoencephalography (MEG) allows non-invasive monitoring weak extracranial magnetic fields elicited by clusters of neurons using super-conducting quantum interference devices (SQUIDs) [1]. However, due to the ill-posed nature [2], MEG/EEG inverse problem has no unique solution unless additional constraints are imposed. Mostly such constraints involve explicit preferences toward either spatially focal or diffused estimates of neuronal currents, such as minimum current estimate (MCE) [3] or minimum norm estimate (MNE) [4]. However, considering a scenario that both focal and distributed current sources are simultaneously presented in the measurement, neither minimizing the ℓ_1-norm nor the ℓ_2-norm of estimated current is an appropriate constraint to reveal the underlying current sources.

The aim of this study is to develop a source reconstruction technique without preferring either focal or distributed current source *a priori*. We hypothesize that the spatial distribution of current sources are sparse after a spatial transformation. This hypothesis is inspired by the observation that natural images are mostly "compressible" [5]. And the sparsity can be quantified by the ℓ_1-norm of the sparsity transformed current estimates. Here we specifically focus on two sparsity transformations: second-order spatial derivatives (Laplacian) and spherical wavelet transforms in our proposed Compressible Estimate (CE). CE promotes the compressibility of current sources by minimizing the ℓ_1-norm of transformed estimated current sources and avoids overly diffused source estimates by minimizing the ℓ_1-norm of estimated current sources. As demonstrated by the following simulations and *in vivo* somatosensory and auditory experiments, CEs provide the improved accuracy of the source localization involving both spatially focal and diffused neuronal currents using MEG measurements.

II. Methods

A. Forward Model

The forward problem of MEG described the relationship between the measured MEG signals and the underlying neuronal currents, which can be modeled as a linear equation:

$$\mathbf{y}(t)=\mathbf{A}\mathbf{s}(t)+\mathbf{n}(t), \tag{1}$$

where $\mathbf{y}(t)$ is an m-dimensional vector containing the observed MEG data from m sensors at time t, $\mathbf{A}$ is a m-by-$3n$ forward matrix, $\mathbf{s}(t)$ is a $3n$-dimensional vector representing the unknown current sources, and $\mathbf{n}(t)$ is the additive noise. n is the number of current sources. Each column of $\mathbf{A}$ describes the signals in the MEG sensors that would be generated by a unit-strength source current element at a particular location and orientation in the source space. Accordingly, a dipole source at each source location can be expressed, via stacking up dipole components at all the points in source space, as $\mathbf{s}(t)=[\mathbf{s}_1(t)^T\ \mathbf{s}_2(t)^T \dots \mathbf{s}_n(t)^T]^T$ where $\mathbf{s}_i(t)=[s_{x,i}(t),\ s_{y,i}(t),\ s_{z,i}(t)]^T$, where the subscript x, and y describe two mutually orthogonal current sources tangential to the local cortical surface, and the subscript z describes a current source locally normal to the cortical surface. Without the loss of generality, we assume that Eq. (1) describes spatially whitened data such that the

S. Supek and A. Sušac (Eds.): Advances in Biomagnetism – BIOMAG2010, IFMBE Proceedings 28, pp. 159–162, 2010.
www.springerlink.com

data vector, the forward matrix, and the noise vector have been multiplied by $\mathbf{C}^{-1/2}$, where $\mathbf{C}$ is the estimated noise covariance matrix of the measurements.

B. Compressible Estimates

Instead of a direct preference toward spatially focal source estimates, we propose the Compressible Estimate (CE) which promotes highly "compressible" localizations results. CE can be mathematically described by the minimization of a cost function:

$$\min_{\mathbf{s}(t)} \left\| \mathbf{M}\mathbf{s}^{\ell 2}(t) \right\|_1 \quad \text{s.t. } \left\| \mathbf{y}(t) - \mathbf{A}\mathbf{s}(t) \right\|_2^2 < \varepsilon \tag{2}$$

where $\mathbf{M}$ is the sparsity transformation matrix to "compress" current estimates, $\mathbf{s}^{\ell_2}(t)=[s_1^{\ell_2}(t), s_2^{\ell_2}(t), \ldots s_n^{\ell_2}(t)]^T$ is the vector consisted of the ℓ_2-norms of orientation constrained dipole moments at each source location. $\mathbf{s}_i^{\ell_2}(t)$ $(1 \le i \le n)$ is the local ℓ_2-norm at the ith source location and is defined as

$$\mathbf{s}_i^{\ell 2}(t) = \left\| \left[w_{x,i}s_{x,i}(t), w_{y,i}s_{y,i}(t), w_{z,i}s_{z,i}(t) \right]^T \right\|_2 \tag{3}$$

where $[w_{x,i}, w_{y,i}, w_{z,i}]=[1/\sin\theta_i, 1/\sin\theta_i, 1]$ and θ_i ranges between 0° and 90° in order to control the orientation constraint. Note that as θ_i is small, the estimated current source at location i will be preferably normal to the local cortical surface. ε is the specified residual variance controlling the consistency between the modeled measurements using source estimates and the empirical measurements. It has been suggested that ε can be chosen such that the probability of $\|\mathbf{y}(t)-\mathbf{A}\mathbf{s}(t)\|_2^2 \ge \varepsilon$ is small [6]. Using the whitened measurements with variance σ^2, we have $\|\mathbf{y}(t)-\mathbf{A}\mathbf{s}(t)\|_2^2/\sigma^2 \sim \chi_m^2$, where χ_m^2 is the Chi-square distribution with m degrees of freedom. In practice, after whitening $\sigma^2=1$ and ε can be selected so that the probability of $\|\mathbf{y}(t)-\mathbf{A}\mathbf{s}(t)\|_2^2$ integrated within the interval $[0, \varepsilon]$ is high, e.g. 0.99.

In this study, we investigated Laplacian and wavelet transforms on cortical manifolds. However, according to our simulation, employing such compression matrices alone can prefer spatially blurred source estimates. Thus we included an additional ℓ_1-norm constraint on the source itself to counter-balance between spatially focal and diffused solutions tuned by a parameter α. The cost function becomes:

$$\left\| \mathbf{M}\mathbf{s}^{\ell 2}(t) \right\|_1 + \alpha \left\| \mathbf{s}^{\ell 2}(t) \right\|_1 \tag{4}$$

An α with a higher value favors a more focal solution. In our study, we selected α=0.4 according to the simulation results. Eq. (2) with the cost function replaced by Eq. (4) can be solved using the convex optimization technique [7]. For n=3793 and m=306, our current implementation takes about 40 s per sample point with a standard PC (1.6 GHz CPU and 2G bytes RAM).

C. Discrete Laplacian

From a physiological point of view, the neighboring neuronal populations are more likely to be active synchronously [8, 9]. This physiological consideration can be used in our model by applying the discrete surface Laplacian [10] on the source space. This Laplacian computes the second-order spatial derivatives:

$$\mathbf{M}_{uv}^{L} = \begin{cases} 1, & u = v \\ -\dfrac{1}{\sqrt{d_u d_v}}, & (u,v) \in \mathbf{E} \\ 0, & \text{otherwise} \end{cases} \tag{5}$$

where $\mathbf{M}_{uv}^{L}$ denotes the element at the uth column and v^{th} row in the matrix $\mathbf{M}^L$, and d_i denotes the number of immediate neighbors of current dipole i. $\mathbf{M}^L$ is an n-by-n matrix. $\mathbf{E}$ denotes the set of edges of the triangulated cortical model. The solution CE^L is calculated from the optimization problem:

$$\text{CE}^{L} \equiv \min_{\mathbf{s}(t)} \left\| \mathbf{M}^{L}\mathbf{s}^{\ell 2}(t) \right\|_1 + \alpha \left\| \mathbf{s}^{\ell 2}(t) \right\|_1 \quad \text{s.t. } \left\| \mathbf{y}(t) - \mathbf{A}\mathbf{s}(t) \right\|_2^2 < \varepsilon \tag{6}$$

D. Spherical Wavelet Operator

Wavelets have been proven to be powerful bases for signal compression. The implementation of spherical wavelet operator is based on the lifted Butterfly bases [11]. The wavelet coefficients can be computed using the lifting scheme. The vertices in a Butterfly basis are shown in Figure 1. The same lattice structure is used in different spatial scales of the semi-regular multi-resolution cortical model. In contrast to the Laplacian, vertices in a Butterfly basis are located in vicinity but they are not immediate neighbors. The whole wavelet transformation can be written as one gigantic sparse matrix multiplication for the convenience of solving a convex optimization.

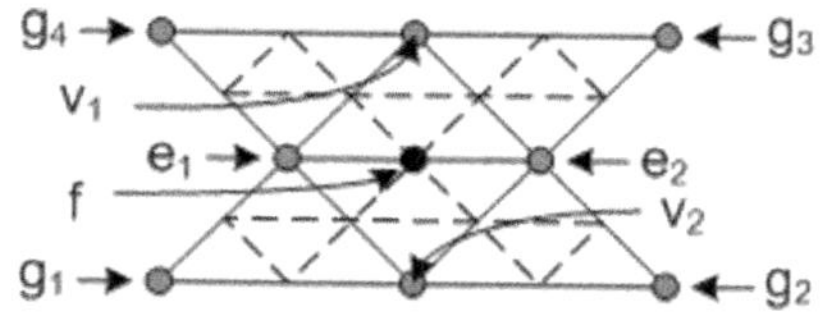

Fig. 1 The vertices in Butterfly bases. f is the vertex in the finer resolution level. e_1, e_2, v_1, v_2, g_1, g_2, g_3 and g_4 are the members of the neighbors used in the bases

The solution CE^W is calculated from the optimization problem:

$$\mathrm{CE}^{\mathrm{W}} \equiv \underset{\mathbf{s}(t)}{\arg\min} \left\| \mathbf{M}^{\mathrm{W}} \mathbf{s}^{\ell 2}(t) \right\|_1 + \alpha \left\| \mathbf{s}^{\ell 2}(t) \right\|_1$$
$$\text{s.t. } \left\| \mathbf{y}(t) - \mathbf{A}\mathbf{s}(t) \right\|_2^2 < \varepsilon \,. \tag{7}$$

Similar to CE^L, CE^W is globally focal and locally smooth. However, CE^W can separate close but not directly connected current sources than $\hat{\mathbf{s}}^{L}$, as demonstrated in our results.

III. Results

A. Simulations

We compare CE^L and CE^W with MNE and previously reported sparse source imaging (SSI) approach , where **M** is an identity matrix (see Figure 2). In this simulation, we created two sources of either small or large extent. The simulated MEG responses are generated by the cortical sources shown in the top row in Figure 2. The SNR of the simulated data was set to 2. Here SNR is defined as $\|\mathbf{y}(t)\|_2 / \sqrt{m}\sigma$. All pair-wise combinations of the simulation sources were tested.

The SSI generated globally focal source estimates regardless of the underlying source configurations due to the nature of the ℓ_1-norm constraint on the estimated source strengths. On the other hand, MNE generates overly diffused source estimates because of the ℓ_2-norm constraint. The source estimates of CE^L and CE^W can more correctly reflect the spatial pattern of the true simulation sources in four separate simulations. As two simulation sources were focal/diffused, both CE^L and CE^W estimated focal/diffused sources (see Figure 2).

Since the lead fields are similar for sources across a sulcus, it is quite common to obtain source estimates across one sulcus even though the true current source is located at only one side of the sulcus in simulations. Furthermore, it is difficult to discern two sources in the simulation as true sources are spatially adjacent to each other. These localization difficulties were further investigated in the following two cases. In the first case, a group of simulated current dipoles are allocated at left secondary somatosensory area (SII) (Figure 3). As the results shown in the case 1 in Figure 3, CE^L and CE^W correctly estimated the source at left SII. However, CE^W also reported a weak source on the superior temporal gyrus (indicated by a blue arrow in Figure 3). The constraint of spatial continuity of the Laplacian encouraged sources estimates of immediate neighbors in CE^L, which consequently is less prone to produce source estimates on the wrong side of the fissure. In case 2, we generated two separated but spatially close sources on the post central gyrus in order to test the ability of CE^L and CE^W on distinguishing two sources. CE^W demonstrated better separation of two sources than CE^L in Figure 3. Since many vertices in a wavelet basis are not immediate neighbors, spatial discontinuity is allowed and less penalized in CE^W than in CE^L.

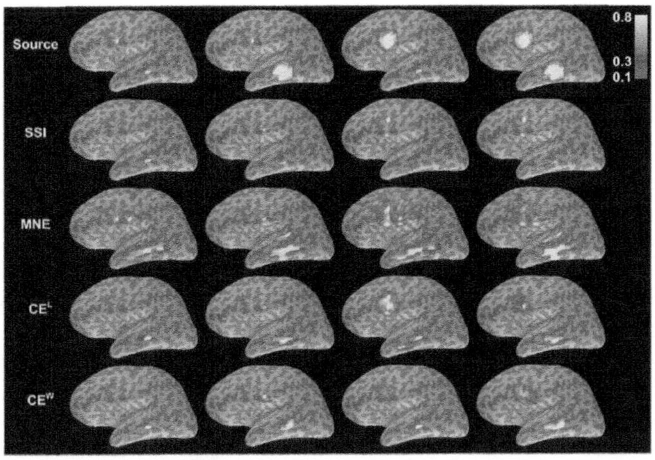

Fig. 2 Cortical maps of different source configurations and the corresponding localization results using different methods. All pair-wise combinations between focal and diffused sources were studied. From the second to the bottom rows are the source reconstructions using SSI, MNE, CE^L, and CE^W. The activation strength is color-coded as illustrated by the color bar

B. Somatosensory and Auditory MEG Experiments

In Figure 4(a), we show the source estimates using SSI, MNE, CE^L and CE^W in the right median nerve stimulation experiment. We used blue arrows to indicate the source locations inconsistent with the literatures. At 35 ms, all methods estimated the left SI sources on the posterior wall of the central sulcus. The MNE, however, produced the current source at the SII area. Consulting previous literatures, we found these sources are physiologically unlikely at this early timing [12]. At 82 ms, all the methods estimated activations at the contralateral SII. But, SSI and MNE also reported current sources at the superior temporal gyrus. These sources were unlikely activated in the median nerve stimulation experiment [12].

For the auditory experiment using 1KHz pure tone, Figure 4(b) shows the SSI, MNE, CE^L and CE^W at 88 ms after the stimulus onset. Previous studies have shown that the stimulus presented to one ear can evoke bilateral responses on primary auditory cortices at the superior temporal gyrus [13]. SSI and MNE solutions revealed sources at not only the primary auditory cortex but also middle temporal gyrus and inferior parietal gyrus, which have not been reported in the literatures. On the contrary, CE^L and CE^W

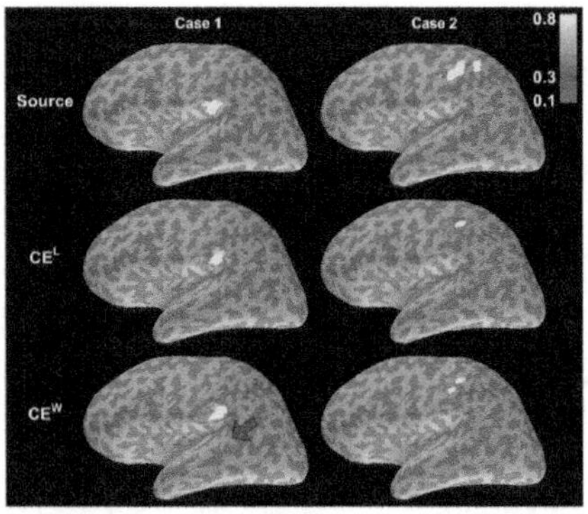

Fig. 3 Reconstructions of CE^L and CE^W using two different source configurations. The top row represents the simulated sources. The middle and bottom rows are the reconstructed signals of CE^L and CE^W, respectively. The activation strength is linearly scaled between 1 and 0 and is color-coded as illustrated by the color bar

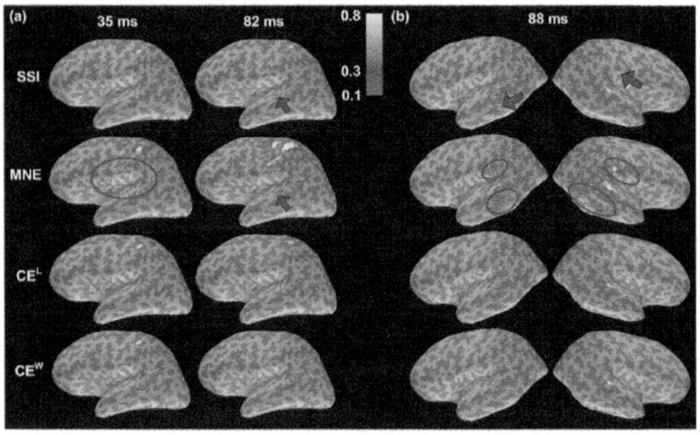

Fig. 4 (a) Activation maps for somatosensory experiment. From left to right are presented the activation maps at 35 and 82 ms after stimulus onset. (b) Activation maps for auditory experiment. All of them are snapshots at 88 ms after stimulus onset. The source estimates of SSI, MNE, CE^L and CE^W are listed from top to bottom. The activation strength is linearly scaled between 1 and 0 and is color-coded as illustrated by the color bar. The blue arrows/ellipsoid indicate the locations where the reported neuronal activities were inconsistent with the literatures

detected prominent source estimates on superior temporal gyrus in both hemispheres.

IV. Conclusions

In this paper we propose to incorporate compressibility as a constraint in solving the ill-posed MEG source localization problem. Specifically, our method is different from other approaches by 1) the explicit quantification of the compressibility of the source estimates using a sparsifying transformation matrix ($\mathbf{M}^L$ or $\mathbf{M}^W$), 2) the cost function is balanced between compressible and spatially focal source estimates, and 3) the constraint of the forward problem is flexibly bounded by a specified variance rather than a fixed residual variance. As demonstrated in our simulation and experimental results, CE^L and CE^W surpass sparse source imaging (SSI) and the Minimum norm estimates (MNE) by showing the capability of adaptively estimating the spatially focal and diffused sources. The CE approach can be used in both neuroscientific studies and clinical applications.

Acknowledgement

This work was supported by National Institutes of Health Grants by R01DA14178, R01HD040712, R01NS037462, P41 RR14075, R01EB006847, RO1EB000790, R21EB007298, National Center for Research Resources, NSC 97-2320-B-002-058-MY3, NSC 98-2320-B-002-004-MY3 (National Science Council, Taiwan), and NHRI-EX98-9715EC (National Health Research Institute, Taiwan).

References

1. Hamalainen, M., et al., *Magnetoencephalography - Theory, Instrumentation, and Applications to Noninvasive Studies of the Working Human Brain.* Reviews of Modern Physics, 1993. **65**(2): p. 413-497.
2. Helmholtz, H., *Ueber einige Gesetze der Vertheilung elektrischer Strome in korperlichen Leitern, mit Anwendung auf die thierisch-elektrischen Versuche.* Ann. Phys. Chem., 1853. **89**: p. 211-233, 353-377.
3. Uutela, K., M. Hamalainen, and E. Somersalo, *Visualization of magnetoencephalographic data using minimum current estimates.* Neuroimage, 1999. **10**(2): p. 173-80.
4. Hamalainen, M. and R. Ilmoniemi, *Interpreting measured magnetic fields of the brain: estimates of curent distributions.* Helsinki University of Technology, Helsinki, Finland, 1984.
5. Taubman, D.S. and M.W. Marcellin, *JPEG2000 : image compression fundamentals, standards, and practice.* Kluwer international series in engineering and computer science SECS 642. 2002, Boston: Kluwer Academic Publishers. xix, 773 p.
6. Ding, L. and B. He, *Sparse source imaging in electroencephalography with accurate field modeling.* Hum Brain Mapp, 2008. **29**(9): p. 1053-67.
7. Boyd, S. and M. Grant, *CVX.* 2009, Stephen Boyd: San Francisco. p. a Matlab-based modeling system for convex optimization.
8. Engel, A.K., et al., *Stimulus-Dependent Neuronal Oscillations in Cat Visual Cortex: Inter-Columnar Interaction as Determined by Cross-Correlation Analysis.* Eur J Neurosci, 1990. **2**(7): p. 588-606.
9. Silva, L.R., Y. Amitai, and B.W. Connors, *Intrinsic oscillations of neocortex generated by layer 5 pyramidal neurons.* Science, 1991. **251**(4992): p. 432-5.
10. Bradshaw, L.A. and J.P. Wikswo, Jr., *Spatial filter approach for evaluation of the surface Laplacian of the electroencephalogram and magnetoencephalogram.* Ann Biomed Eng, 2001. **29**(3): p. 202-13.
11. Schroder, P. and W. Sweldens. *SphericalWavelets: Efficiently Representing Functions on the Sphere.* in *Computer Graphics (SIGGRAPH '95 Proceedings).* 1995.
12. Hari, R. and N. Forss, *Magnetoencephalography in the study of human somatosensory cortical processing.* Philos Trans R Soc Lond B Biol Sci, 1999. **354**(1387): p. 1145-54.
13. Reite, M., J.T. Zimmerman, and J.E. Zimmerman, *MEG and EEG auditory responses to tone, click and white noise stimuli.* Electroencephalogr Clin Neurophysiol, 1982. **53**(6): p. 643-51.

Single-Trial Analysis for Empirical MEG Data

Jun Hee Hong, Minkyu Ahn, and Sung Chan Jun

Department of Information & Communications, Gwangju Institute of Science and Technology, Gwangju, South Korea

Abstract— **MEG/EEG single-trial data usually has low signal-to-noise ratio (SNR) because it is highly contaminated by background noise. Thus, it is too hard to detect sources and is hardly to distinguish between sources of our interest and sources of noise. To overcome this difficulty, time-locked averaged signal over many trials is used for source localization. However, averaging effect is more likely to filter out source activities which are slightly varied over trial. These varying sources may impose good and critical information in understanding brain dynamics. For this reason, we make attempt to analyze many single-trial data to investigate how brain activities over trial are going. We used conventional minimum-variance beamformer for source localization. We adopt bootstrap resampling technique to do various localization analysis between original single-trial analysis and fully averaged analysis. It is found from median nerve stimulation that some unseen sources in averaged data were frequently detected in a specific area. We believe it is evident that single-trial analysis enables to reveal more different dynamical behaviors than averaged one.**

Keywords— **single-trial analysis, spatiotemporal source localization, beamformer, bootstrap, MEG/EEG.**

I. Introduction

MEG/EEG single-trial data usually has far low SNR because it is highly contaminated by background noise. Thus, it is too hard to detect sources and is hardly to distinguish between sources of our interest and sources of noise. In order to overcome this difficulty, time-locked averaged signal over many trials is used in detecting sources. This seems to be relevant to evoked response signals. However, averaging effect is more likely to filter out source activities which are slightly varied over trial. These varying sources would be crucial in understanding brain dynamics. For this reason, we try to attempt to analyze many single-trial data to investigate how brain activities over trial are going on.

In this work, we attempt single-trial analysis by eigen-based minimum variance beamformer [1]. We used empirical MEG median nerve stimulation data consisting of about 400 single trials. Bootstrap resample [2] is adopted to generate various re-sampled averaged data by doing p (given number of samples) sampling with replacement from original single trials and by averaging chosen p samples. Sources are reconstructed from these re-sampled averaged data. We attempted this approach for different p values such as 1, 2, 10, 50, 100 and so on. Higher p value yields re-sampled averaged data with higher SNR which is more likely to get consistent localization results. Particularly, the case for p=1 means the original single-trial analysis. Localization results for various p are compared with those of total averaged data.

II. Methods

In this work, minimum-variance beamformer is chosen as a source imaging technique [3]. In beamforming technique, to estimate the source activity the spatial filter $\mathbf{w}(\mathbf{r})$ is applied to the measurement data as follows:

$$\hat{Q}(\mathbf{r},t) = \mathbf{w}^T(\mathbf{r})\mathbf{m}(t), \quad (1)$$

where, $\hat{Q}(\mathbf{r},t)$ and $\mathbf{m}$(t) are the estimated magnitude of the source activity at beamforming scanning source point $\mathbf{r}$ and at time t, and spatiotemporal measurement vector at time t, respectively. Spatial filter $\mathbf{w}(\mathbf{r})$ is defined in the following optimization problem:

$$\mathbf{w}(\mathbf{r}) = \arg\min \mathbf{w}^T(\mathbf{r})\mathbf{C}\mathbf{w}(\mathbf{r}), \text{ subject to } \mathbf{w}^T(\mathbf{r})\mathbf{l}(\mathbf{r}) = 1, \quad (2)$$

where $\mathbf{C}$ is the covariance matrix of the measurement defined as $\mathbf{C} = \langle \mathbf{m}(t)\mathbf{m}^T(t) \rangle_t$ and $\langle \cdot \rangle_t$ indicates the time average. $\mathbf{l}(\mathbf{r})$ denotes a lead-field vector at source point. This constraint minimization problem can be solved using the method of the Lagrange multiplier and explicit form of the weight spatial filter can be represented by

$$\mathbf{w}(\mathbf{r}) = \frac{\mathbf{C}^{-1}\mathbf{l}(\mathbf{r})}{\mathbf{l}^T(\mathbf{r})\mathbf{C}^{-1}\mathbf{l}(\mathbf{r})}. \quad (3)$$

Using this weight spatial filter, the time averaged source output power is obtained below:

$$\left\langle \hat{Q}(\mathbf{r},t)^2 \right\rangle_t = \mathbf{w}^T(\mathbf{r})\mathbf{C}\mathbf{w}(\mathbf{r}) = \frac{1}{\mathbf{l}^T(\mathbf{r})\mathbf{C}^{-1}\mathbf{l}(\mathbf{r})}. \quad (4)$$

When a spherical homogeneous conductor head model is introduced in forward computation, the false intensity may increase around the center of the sphere in the source imaging. To avoid this difficulty it is known that the constraint

S. Supek and A. Sušac (Eds.): Advances in Biomagnetism – BIOMAG2010, IFMBE Proceedings 28, pp. 163–166, 2010.
www.springerlink.com

$\mathbf{w}^T(\mathbf{r})\mathbf{l}(\mathbf{r}) = \|\mathbf{l}(\mathbf{r})\|$ in place of $\mathbf{w}^T(\mathbf{r})\mathbf{l}(\mathbf{r}) = 1$ is applied. This is called array-gain minimum variance beamformer. Its output power is expressed as

$$\left\langle \hat{Q}(\mathbf{r},t)^2 \right\rangle_t = \frac{1}{\tilde{\mathbf{l}}^T(\mathbf{r})\mathbf{C}^{-1}\tilde{\mathbf{l}}(\mathbf{r})} = \frac{\mathbf{l}^T(\mathbf{r})\mathbf{l}(\mathbf{r})}{\mathbf{l}^T(\mathbf{r})\mathbf{C}^{-1}\mathbf{l}(\mathbf{r})}. \quad (5)$$

In general, the estimated source activity obtained from minimum-variance beamformer may include source imaging distortion or noise source. Eigenspace-projection spatial filter enables to consider reducing the noise effect by the projection onto the signal subspace [1]. The projection matrix $\mathbf{E}_S$ can be obtained by singular value decomposition (SVD) of the covariance matrix $\mathbf{C}$ such that

$$\mathbf{C} = \mathbf{E}_S \mathbf{\Lambda}_S \mathbf{E}_S^T + \mathbf{E}_N \mathbf{\Lambda}_N \mathbf{E}_N^T \quad (6)$$

Where Λ_S is a diagonal matrix consisting of significant eigenvalues and Λ_N is a diagonal matrix consisting of negligibly small eigenvalues (remaining eigenvalues after excluding significant eigenvalues). Those corresponding eigenvectors are $\mathbf{E}_S$ and $\mathbf{E}_N$, respectively. When we consider about between lead-field vector $\mathbf{l}(\mathbf{r})$ and covariance $\mathbf{C}$, the term $\mathbf{C}^{-1}$ $\mathbf{l}(\mathbf{r})$ is orthogonal to each other in ideal conditions. In reality, the first term of Eq.(6) is used to calculate weight vector because the noise term is likely to weaken the orthogonality. To avoid this, eigenspace-based spatial filter is proposed below:

$$\mathbf{w}_{ES}(\mathbf{r}) = \mathbf{E}_S \mathbf{E}_S^T \mathbf{w}(\mathbf{r}). \quad (7)$$

We assume that MEG/EEG single trial data is expressed as:

$$\mathbf{X} = \{\mathbf{X}_1, \mathbf{X}_2, \ldots, \mathbf{X}_n\} \quad (8)$$

where n represents the total number of single-trial sets and $\mathbf{X}_i$, i=1,…,n indicates the original (not re-sampled) spatio-temporal single-trial matrix. Bootstrap resample is adopted to generate various re-sampled averaged data by doing p (given number of samples) sampling with replacement from original single trials and by averaging chosen p samples such that

$$\mathbf{D} = \{\mathbf{D}_1, \mathbf{D}_2, \ldots, \mathbf{D}_m\}$$
$$\mathbf{D}_j = \sum_{i=1}^{p} \mathbf{X}_{n_i} \Big/ p \quad (9)$$

where n_i is index of chosen trial $\mathbf{D}_j$, j=1,…,m indicates re-sampled averaged spatiotemporal single-trial matrix (i.e., p randomly chosen original single-trial sets are averaged for one re-sample) and here m is equal to about 1000 in this work. Sources are reconstructed from these re-sampled averaged data. We attempted this approach for various different p values such as 1, 2, 10, 50, 100 and so on.

III. Experiments

Empirical median nerve stimulation MEG data were acquired from the following experiment:

- Healthy male volunteer participated in the MEG measurement after appropriate informed consent. Median nerve of right or left hand was stimulated and all measurements were done in a magnetically-shielded room. A 2.9(right)/3.5mA(left) current intensity and 2Hz sampling was applied with 0.3msec current pulse. These different current intensity values were determined as threshold intensity. That is, for stronger stimulation than this threshold intensity subject started to feel pain. Interestingly, these thresholds are different between right/left hands. Data were digitized at 2kHz, 500Hz low pass filtered, and post-processing digital filter was applied 1-100Hz except 50Hz. Total 399 (right) / 418(left) single trials were acquired. Interstimulus was randomly chosen. Data were collected from a single subject on a whole-head gradiometer system with 160 channels (MEGVISON Yokogawa electric Inc., Tokyo, Japan) with eye closed.

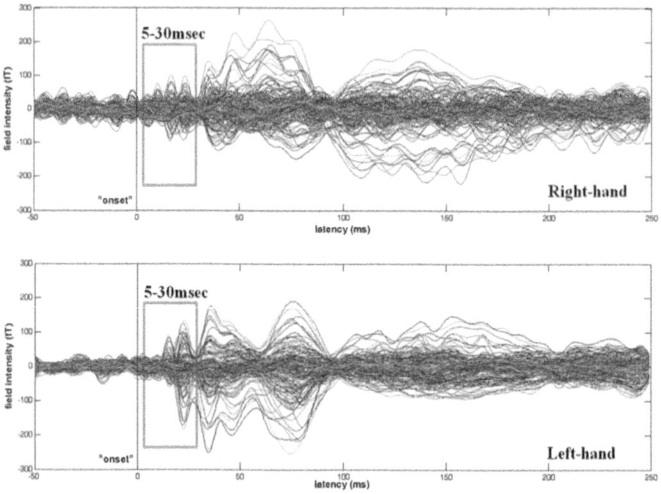

Fig. 1 160-channel overplotted empirical MEG data for right-hand (top)/left-hand (bottom) median nerve stimulation. From 5-30msec were analyzed and vertical line indicates onset time point

Figure 1 shows that time-locked total averaged empirical MEG data is overplotted. We found the fitted sphere model of radius 8.04cm through MRI imaging. Considering the sphere center as origin, three axes are roughly determined as follows: +x axis is direction from origin to the nasion of subject, +y axis is direction from origin to the left pre-auricular point, and +z axis is direction perpendicular to x and y axes.

We analyzed right/left-hand median nerve stimulation averaged (up to 399/418 trials) data (5-30msec time window), as illustrated in Figure 1. In Figure 2, source imaging results of eigenspace-based beamforming technique are depicted.

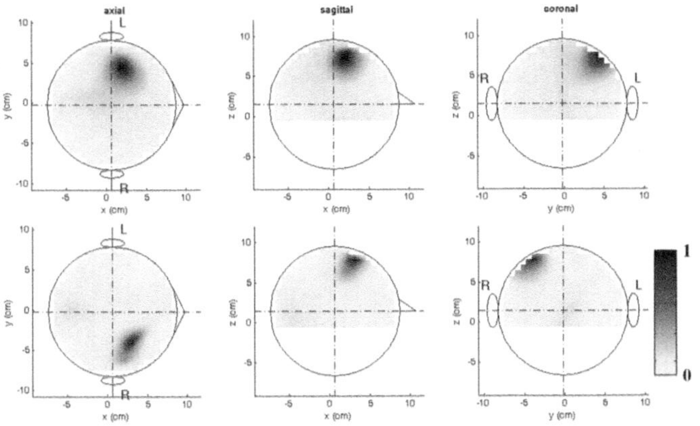

Fig. 2 Source reconstruction results using eigenspace-based beamformer. Top rows stand for the result of right-hand median nerve stimulation. The other rows represent the result of left-hand case. In both right and left case, one source was well reconstructed

Figure 2 shows that one contralateral source was well reconstructed in each right/left case. These localized sources seem to be relevant to results reported in many literatures. Next, in place of total averaged data we attempted to analyze MEG single-trials of median nerve stimulation. Bootstrap resample was adopted to generate various re-sampled averaged data by doing *p* (given number of samples) sampling with replacement from original single trials and by averaging chosen *p* samples. This method is well suited to the analysis of event-related MEG data since the experiments are repeated tens or even hundreds of times and averaged to achieve acceptable SNRs [2]. We also used eigenspace-based beamformer for source imaging.

Sources are reconstructed from these re-sampled averaged data. Figure 3 presents localization results for right-hand median nerve stimulation (n=399, m=1000, p=10) and Figure 4 shows the result of left-hand median nerve stimulation single-trial analysis (n=418, m=1000, p=10). Similarly, we observed other sources than primary sources found in total averaged data were frequently detected in a specific area, which is described as black rectangle in Figure 3 and Figure 4. This specific area is located in the same hemisphere to stimulated hand and thus newly detected sources seem to be physiologically related to secondary sources in median nerve stimulation. For its concrete conclusion, it is under investigation.

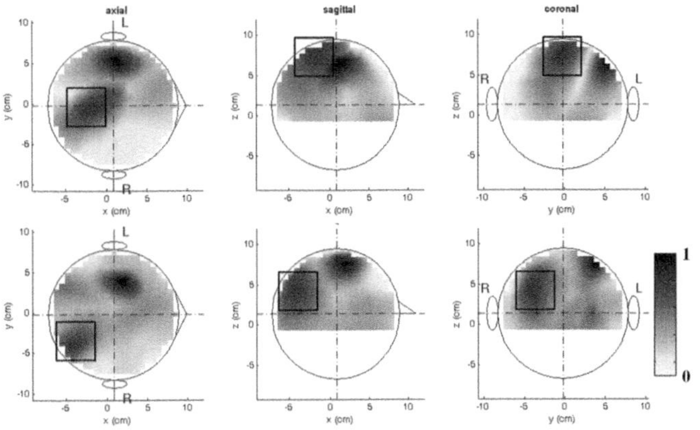

Fig. 3 Right-hand median nerve stimulation single-trial analysis: Source reconstruction results using eigenspace-based beamformer. Re-sampled data was obtained by doing p=10 sampling. The total number of re-sampled data is equal to 1000. Here, the rectangle means the specific area where the secondary sources were frequently detected

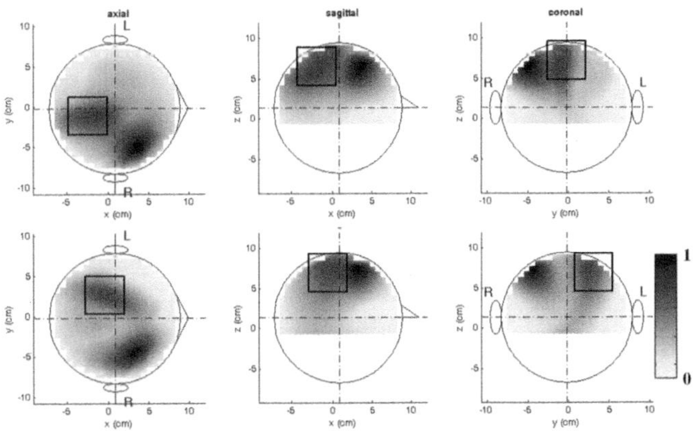

Fig. 4 Left-hand median nerve stimulation single-trial analysis: Source reconstruction results using eigenspace-based beamformer. Re-sampled data was obtained by doing p=10 sampling. The total number of re-sampled data is equal to 1000. Here, the rectangle means the specific area where the secondary sources were detected frequently

IV. Conclusions

We made attempt to analyze median nerve stimulation single-trial data through eigenspace-based minimum variance beamformer along with bootstrap method. It is found that other sources than those in averaged data were frequently detected in a specific area. They seem to be related to secondary source which is well known to be localized in the lateral area. We believe it is evident that single-trial analysis enables to reveal different dynamical behaviors than averaged one. Further investigation on this regard would be conducted to make sure this kind of approach is useful in understanding brain dynamics.

Acknowledgment

This work is supported by KRF-2008-531-C00012, the BioImaging Research Center at GIST, KOSEF/JSPS researcher exchange program, and the National IT Industry Promotion Agency (NIPA-2009-C1090-0902-0008). We appreciate Prof. C.K. Chun at SUNH MEG center and Dr. Haruta at Yokogawa Electric Corp. to his assistance for MEG/EEG data acquisition.

References

1. K. Sekihara et al (2002) Application of an MEG eigenspace beamformer to reconstructing spatio-temporal activities of neural sources, Human Brain Mapping, vol. 15, pp.199-215
2. F. Darvas et al (2005) Investigations of dipole localization accuracy in MEG using the bootstrap, Neuroimage, 25, 355-368
3. K. Sekihara and S. S. Nagarajan (2008) Adaptive spatial filters for electromagnetic brain imaging. Berlin, Heidelber: Springer-Verlag
4. Sarvas J (1987) Basic mathematical and electromagnetic concepts of the biomagnetic inverse problem. Phys. Med. Biol., 32(1) 11-22

Author: S.C. Jun
Institute: Gwangju Institute of Science and Technology (GIST)
Street: 261 Cheomdan-gwagiro(Oryong-dong), Buk-gu
City: Gwangju
Country: South Korea
Email: scjun@gist.ac.kr

Development of Human Brainwave Simulating Device for Magnetoencephalography and the Corresponding Dipole Localization Study

Jeff Liu[1], Yoshinao Kishimoto[2], Teresa Cheung[3], Ash M. Parameswaran[1], and Kenji Amaya[2]

[1] School of Engineering Science, Simon Fraser University, Burnaby, BC, Canada
[2] Graduate School of Information Science and Engineering, Tokyo Institute of Technology, Meguro-ku, Tokyo, Japan
[3] Department of Physics, Simon Fraser University, Burnaby, BC, Canada

Abstract— **Magnetoencephalography (MEG) has gained its importance in the real-time detection of brain neuronal activities, based on the magnetic flux generated by ion exchange in brain neurons [1]. However, mapping of the actual magnetic dipole locations from the magnetic fields detected by MEG still remains a challenge [2], [3]. This study presents the method of developing a device which physically simulates human brainwave due to neuronal activities. Instead of using a human test subject, in which the information of the exact locations of dipoles is difficult to obtain, a brain phantom, which consists of coils controlled by waveform-generating circuits and mounted on different locations of a head-shaped object, is constructed. The magnetic flux density, which is proportional to the current in a coil, is controlled by the electronic unit attached to it. This study has also developed a technique using inverse analysis to localize the corresponding dipoles from the data collected by an MEG machine [4]. We have conducted experiments to verify the proposed brain phantom. The accuracy of the dipole localization technique is also verified using the brain phantom device developed, as the magnitudes and locations of the dipoles created by these coils are carefully controlled. The proposed localization technique has demonstrated close estimation of the dipoles from the raw MEG data [5]. Future research of the brain mapping project includes up-scaling the number of individually controlled coils and generating magnetic field pattern more closely resembling the actual human brainwave.**

Keywords— **Magnetoencephalography, Human brain phatom, Brainwave simulation, Dipole localization, Inverse problem.**

I. Introduction

Magnetoencephalography (MEG) has been widely used, in conjunction with Electroencephalography (EEG), to provide real-time monitoring of brain neuronal activities. Both methods share the advantage of high temporal resolutions and are capable of detecting precisely the timing of neuronal activities in a brain cognition process, as opposed to methods such as MRI, in which the time required for the resulting image to be processed is substantially longer.

As human brain activities can be represented by the combination of multiple current dipoles, the challenge commonly faced in the mapping of these activities using MEG data is the inability to obtain the information about the actual locations of dipoles, if a human test subject is used. The present study features the development of a human brainwave simulating device which physically generates time-variant magnetic fields. Such a device is subsequently used for characterizing the method of brain current dipole localization also developed in the present study.

II. Methods

A. Brainwave Simulating Device

The pattern of magnetic fields generated by brain neuronal activities in a given cognition process can be modeled as the combination of individual current dipoles located at active neurons in which ion exchange occurs. More realistic modeling can be achieved by increasing the number of individual dipoles and decreasing the size of each of them. For the initial stage of the development, a coil is used to physically simulate the aggregated effect of a group of brain neurons. The coil is attached to an electronic control unit. Together the whole device is termed human brain phantom.

According to Ampere's Law, the magnetic field strength generated by a current-carrying coil is proportional to the amount of current carried by the coil. The control unit is designed to allow real-time controlling of the amplitude and frequency of the output waveform. The functional blocks of the device developed are illustrated in Figure 1.

The process flow of the design is illustrated in Figure 2. The device operates in two modes, namely the data sampling and the waveform generating modes. Upon initialization, the device operates in the data sampling mode, and the input amplitude and frequency values are sampled through analog-to-digital converters. After processed by the microcontroller, the values are updated and displayed. The process then loops back to start another sampling cycle, if the waveform generating function is not initiated. The device leaves the data sampling mode if the waveform generating mode is selected. Upon entering the waveform generating mode, the stored frequency value is used for computing the time interval between two successive data transfers between the microcontroller and the output digital-to-analog

S. Supek and A. Sušac (Eds.): Advances in Biomagnetism – BIOMAG2010, IFMBE Proceedings 28, pp. 167–170, 2010.
www.springerlink.com

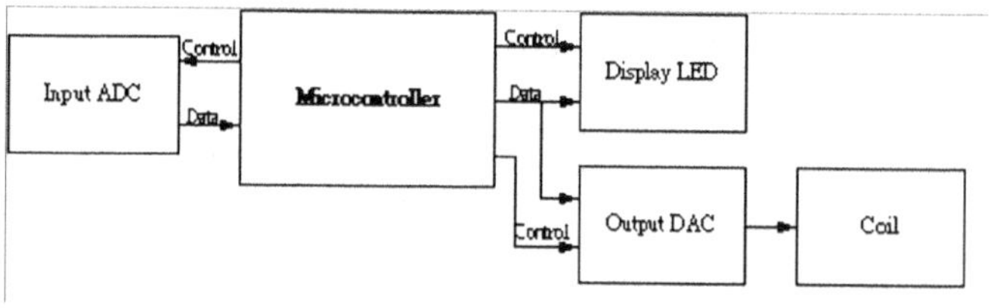

Fig. 1 Functional blocks of the brainwave simulating device developed

converter. Each data transfer corresponds to an indexed position in the waveform. The device keeps looping through the waveform generating sequence until it is switched back to the data sampling mode.

A digital low-pass filter is integrated to the output of the DAC unit to largely suppress the harmonics that are parasitic to the AC source powering the device.

B. Localization of Dipoles

In our research, we have developed multiple phase inverse analysis technique for localization of dipoles [4], [5]. This proposed technique consists of two stages, namely extracting and grouping phases. Figure 3 outlines the procedure. Its summary is written as follows.

Extracting Phase

When the locations of dipoles are known, the relationship between magnitudes of dipoles and magnetic flux density measured by MEG is linear. In the extracting phase, the locations of dipoles are pre-selected and magnitudes of the dipoles are estimated by TSVD. The order of the truncated singular value is determined with AICC.

Now the cutoff magnitude can be predetermined. Dipoles with the estimated magnitude less than the cutoff magnitude are discarded. Using the remaining dipoles, TSVD and AICC are applied iteratively to eliminate the invalid dipoles.

The remaining dipoles are the active units for our analysis. Depending on the required resolution these iterative steps have to be continued until the final threshold value.

Grouping Phase

The estimated distributions of dipoles by the extracting phase are grouped into different sets of dipoles by cluster analysis. Using these sets as initial values, the optimal locations and magnitudes of the dipoles which minimize the residual error between the measured MEG data and the data predicted using the mathematical model are calculated by downhill simplex method.

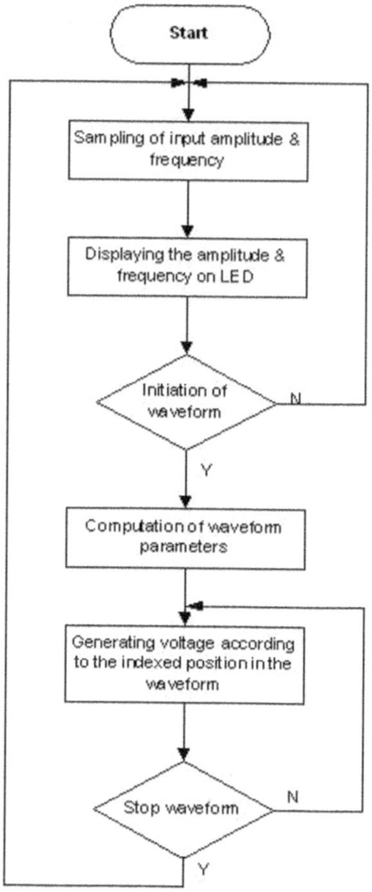

Fig. 2 Control process of the brainwave simulating device

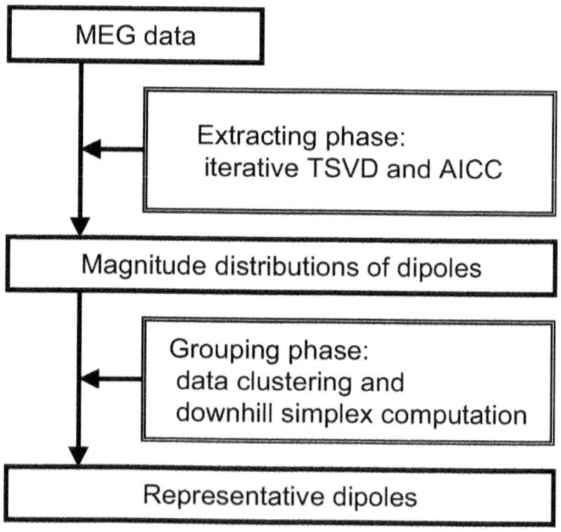

Fig. 3 Procedural flow of the localization technique

III. Experimental setup

Figure 4 shows the experimental setup for the experimental verification of the brainwave simulating device i.e. the human brain phantom we have developed. In this experiment, the brain phantom has 4 coils as magnetic dipoles and serves several voltage waveforms into the coils. The coils are placed in air, as opposed to liquid conducting medium. The major disadvantage of using liquid medium instead of air is the interference of ion exchange across different coils when multiple coils are in presence in the same medium.

To be used in a MEG machine, the coils are attached to a plastic cylinder as the base, as shown in Figure 5. Reference coils supplied by MEG manufacturer are attached to the base for pre-defining spatial coordinates. The locations of the magnetic dipoles generated by the coils are expressed in terms of their relative positions to the reference coils.

The magnetic field flux data generated by the dipoles are collected by the MEG machine in a magnetic-noise free environment.

IV. Results

A. MEG Measurement Results with the Device

A typical MEG measurement with the phantom is illustrated in Figure 6. This pattern represents the output of sinusoidal waveform signals in the coils. It can be observed that the magnetic flux density emitted by the coils closely resembles the input signal waveforms. Figure 7 shows the spatial mapping of the MEG data collected at time equal to 0.1 s. The localization technique is applied on these MEG data.

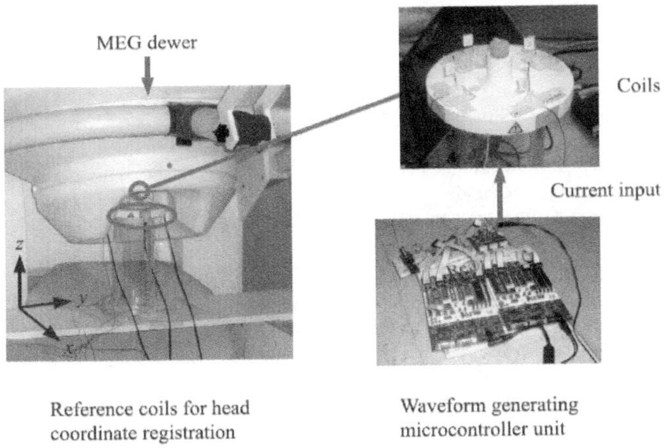

Fig. 4 Experimental setup for MEG measurement using the device developed

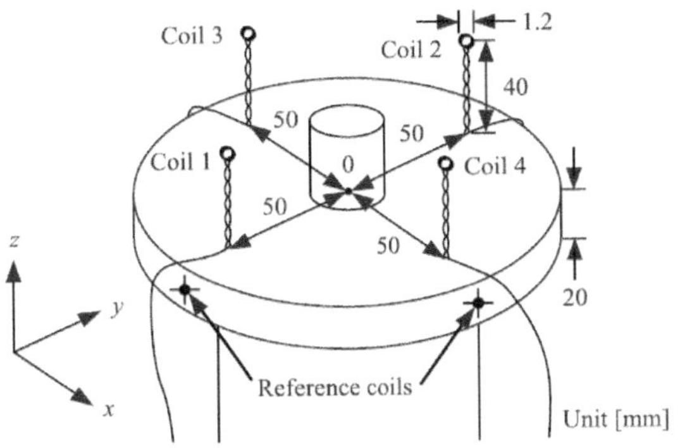

Fig. 5 Configurations of the coils

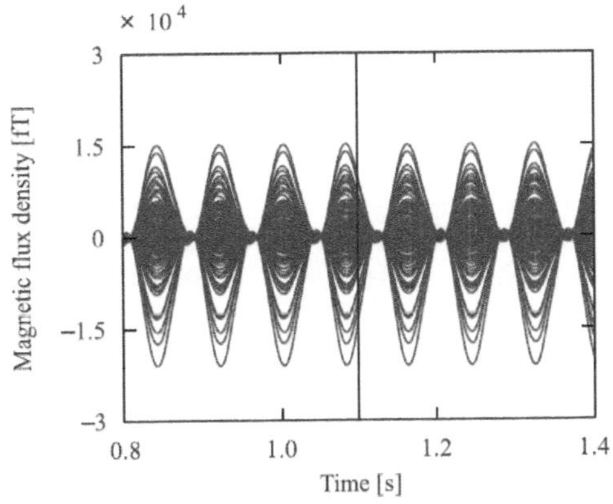

Fig. 6 Temporal magnetic flux density emitted by the brainwave simulating device

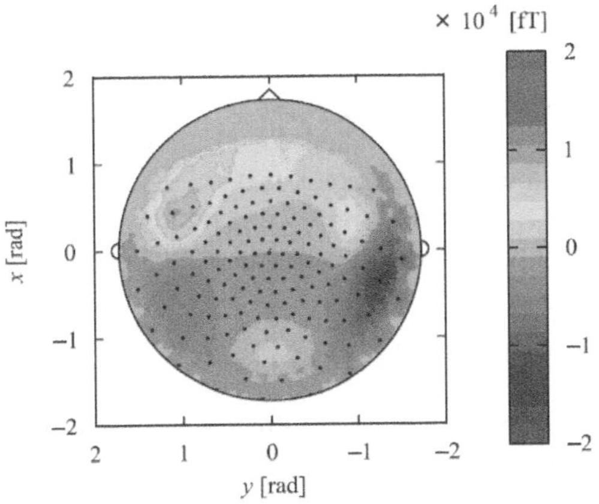

Fig. 7 Spatial distribution map of MEG data at time equal to 0.1s in Figure 6

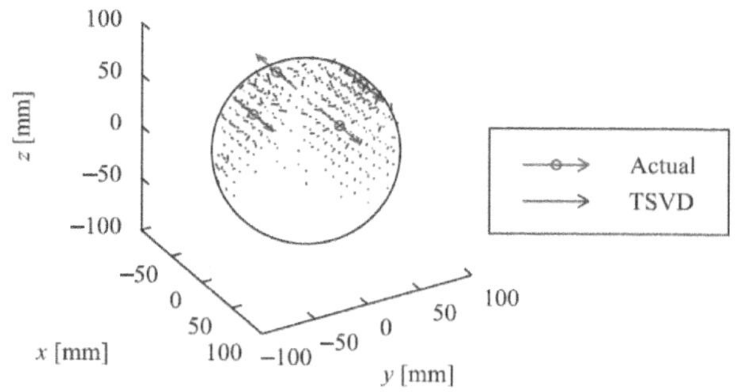

Fig. 8 Localization results on the data shown in Figure 7 after the extracting phase

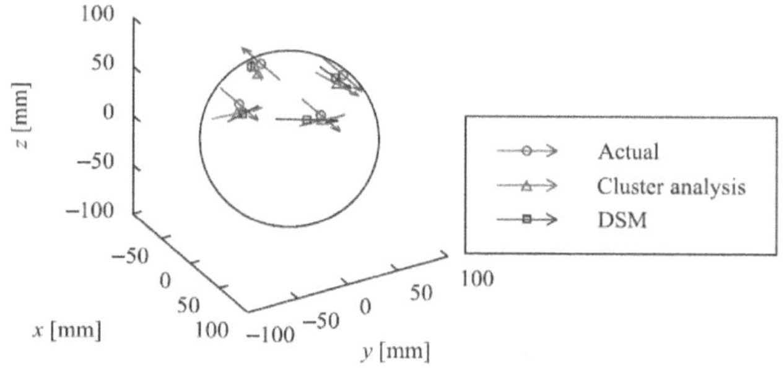

Fig. 9 Final localization results on the data shown in Figure 8 after the grouping phase

B. *Localization of Dipoles with MEG Measurements Results*

Figure 8 shows the localization result of TSVD, AICC and the cutoff process applied on the collected MEG data shown in Figure 7 iteratively. The cutoff magnitude used is 10% of the maximum magnitude of the estimated dipoles. The expected dipoles near the actual positions eventually remain after this extracting phase.

Figure 9 shows the final localization result of cluster analysis and downhill simplex method using the data shown in Figure 8. The positions and magnitudes of the estimated dipoles are in good agreement with the actual values.

V. Conclusions

We have developed a human brainwave simulating device for MEG application. This device consists of coils connected to waveform generating units. Each coil simulates an individual magnetic dipole. MEG data showed that the brain waveform device successfully induced magnetic flux densities resembling the input waveform. The device was used for verifying the localization technique developed in the present study. The technique was demonstrated to have successfully localized the dipoles when applied to the MEG data collected.

We propose to develop a brain phantom capable of more closely simulating brain neuronal activities using micromachining technology. The device will be proven useful for diagnostic and clinical applications.

Acknowledgment

We gratefully acknowledge DSRF for the giving access to MEG. This research was funded by NSERC.

References

1. Baillet, S., et al., (2001), Electromagnetic Brain Mapping, *IEEE Signal Processing Magazine*, Vol.18, No.6, pp.14-30.
2. McVeigh, P., et al., (2007), Study of Conducting Volume Boundary Influence on Source Localization using a Realistic MEG Phantom, Proceedings of the 15th Biomag, Vol.1300, pp.153-156.
3. Uehara, G., et al., (2007), An Error in Magnetic Field of Isosceles-Triangle Coil as a Representation of Equivalent Current Dipole, *Proceedings of the 15th Biomag*, Vol.1300, pp.607-610.
4. Kishimoto, Y., et al., (2009), Brain Current Dipole Localization Studies for Magnetoencephalography using Multiple Phase Inverse Analysis, *Proceedings of the 15th OHBM*, No.210.
5. Kishimoto, Y., et al., (2009), Magnetic Dipole Localization Studies for Magnetoencephalography using Multiple Phase Inverse Analysis and the Experimental Verification, *Proceedings of the 32nd CMBES*, No.09-32.

Co-author: Jeff Liu
Institute: Simon Fraser University
Street: 8888 University Drive
City: Burnaby
Country: Canada
Email: jliun@sfu.ca

Co-author: Yoshinao Kishimoto
Institute: Tokyo Institute of Technology
Street: 2-12-1 Ookayama, Meguro-ku
City: Tokyo
Country: Japan
Email: kishimoto.y.aa@m.titech.ac.jp

MEG Virtual Channel Methods for Movement Prediction and Training

T. Yeager[1,2], T.D. Ard[2], F.W. Carver[2], T. Holroyd[2], and R. Coppola

[1] Department of Bioengineering, University of Pittsburgh, Pittsburgh, PA
[2] Magnetoencephalography Core Facility, NIMH, NIH, Bethesda, MD

Abstract— **Using a whole-head MEG system, we investigated the possibility of distinguishing between directions of hand movement in the mu-beta (8-32 Hz) and high gamma (76-100 Hz) frequency bands in the hand knob and surrounding prefrontal gyrus. We found that four cardinal directions (up, down, left, right) in center-out tasks could be distinguished to a statistically significant extent in the beta band. While no statistically significant difference was found in the gamma band, it was possible to discriminate between hand movement and rest based on hand knob-localized event-related synchronization. Analysis of hand movement in the beta band could lead to finer coordination of brain computer interfaces.**

Keywords— **MEG, BCI, motor movement, beta band.**

I. Introduction

The search for a suitable modality for a brain machine interface (BMI) has become the focus of many studies in the past decade (Mason et al., 2007). The use of intracranial electrocorticographic (ECoG) brain activity recordings has demonstrated the ability to target individual neurons, yielding relatively stable signals that are capable of directly controlling prosthetic robotic limbs in primate models (Velliste et al., 2008). However, the invasive nature of ECoG, coupled with the limited biocompatibility of current cortical surface electrodes (Yeager et al., 2008), make such procedures unsuitable for human subjects who have not already undergone craniotomies for other reasons.

Extracranial electroencephalography (EEG) has decreased the need for intracranial electrode placement, instead relying on the measurement of electrical currents passing from the cortical surface through the surrounding tissue. Due to the variable conductivity of the dura, skull, and skin, however, EEG possesses limited spatial resolution compared to ECoG (Yeager et al., 2008).

Recently, magnetoencephalography (MEG), which uses superconducting quantum interference device (SQUID) to measure the oscillating weak magnetic fields generated by cortical currents, has been gaining acceptance as a viable research tool. Unlike EEG, MEG recordings are largely unaffected by the intervening tissues between the cortical surface and the SQUID sensors.

Decreases in MEG signal power are interpreted as desynchronization of oscillatory magnetic fields within neurons, while increases in signal power are representative of increased synchronization of magnetic field oscillation.

The relatively small number of SQUID sensors utilized in early MEG arrays led to it being combined with EEG to provide significantly higher spatial localization accuracy (Babiloni et al., 2001), however developments in whole-head MEG systems with over 200 SQUIDs have allowed standalone MEG analysis to provide spatial resolutions of 3-5 mm.

A previous MEG study has noted that regions of the motor cortex showed event-related desynchronization at 15-35 Hz during the formation of oro-facial gestures (Muthukumaraswamy et al., 2004). Additionally, the presence of finger movement in both hands has been detected using whole-head MEG at 20 Hz (Kauhanen et al., 2006).

Of greater interest to BMI applications is the ability to differentiate between directions of movement, rather than simply detecting the presence of movement. A recent MEG study focused on determining the direction of hand movement by analyzing low-passed MEG data with 3 Hz and 7 Hz cutoffs, within the delta frequency band. It was shown that both MEG and EEG could be used to predict movement direction prior to the onset of physical movement based on cortical activity in the primary motor cortex (Waldert et al., 2007; Waldert et al., 2008).

However, due to the inherently low temporal resolution afforded by analysis of delta oscillations, it becomes difficult to provide a smooth machine response to intended hand motion. We are therefore interested in the possibility of analyzing hand movement direction in the beta frequency band (15-30 Hz), as it is logical to assume that monitoring of higher frequency oscillations would allow for faster determination of intended movement directions, as well as smoother natural movement. Furthermore, the previous study (Waldert) addressed hand direction analysis on a case by case basis; it is our aim to generalize hand movement directionality across multiple subjects.

II. Methods

A. *Subjects and Tasks*

Data were collected from seven healthy volunteer subjects. In order to study simple movements of the wrist, an

S. Supek and A. Sušac (Eds.): Advances in Biomagnetism – BIOMAG2010, IFMBE Proceedings 28, pp. 171–174, 2010.
www.springerlink.com

optical joystick system (Mag Design and Engineering, Redwood City, CA) was used to record the physical motions of a test subject in response to visual cues on a projection screen, as shown in Figure 1A. The joystick operates by modulating the light in optical fibers; the only electrical components necessary were located in the receiving device located outside of the magnetically shielded room. The *x*- and *y*- outputs of the joystick were voltages that were recorded as a part of MEG datasets. Subjects were directed to move the joystick, at approximately 2-second intervals, from the center position to any of four outward positions (left, right, up, down) before returning to the center position, as illustrated in Figure 1B.

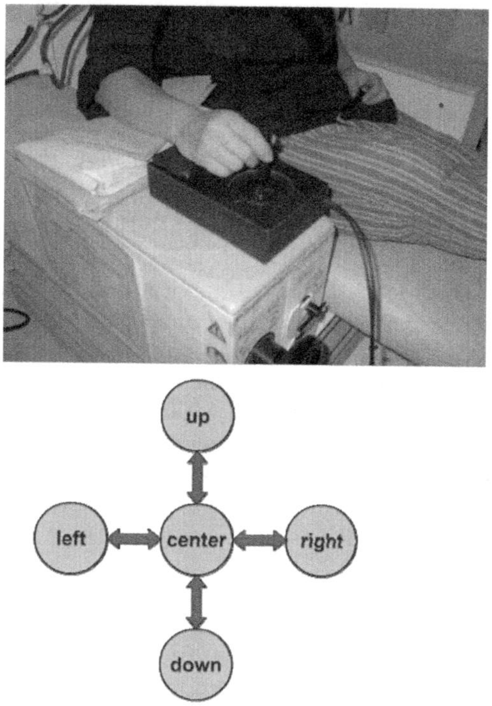

Fig. 1 (A) Experimental setup: subjects sat upright in MEG while operating an optical joystick using movement of the right wrist only. (B) Center-out joystick task setup

B. *Neuromagnetic Recordings*

Magnetic monitoring of subjects during joystick operation was accomplished by a CTF 275 MEG system (CTF Systems, Inc., Coquitlam, BC, Canada), composed of a whole-head array of 275 radial 1st order gradiometer/SQUID channels housed in a magnetically shielded room (Vacuumschmelze, Germany). Synthetic 3rd gradient balancing was used to remove background noise on-line. Three fiducial electrodes adhered to the skin were utilized to provide constant tracking of the orientation of the head. Neuromagnetic data was recorded at a sampling rate of 1200 Hz. Structural MRI scans were subsequently performed on all subjects, using the fiducial markers as physiological reference markers for later spatial correlation between localized MEG signals and physiological landmarks on subject brains.

C. *Event Markers*

Custom functions written for MATLAB (The Mathworks, Inc., Natick, MA) were used to read the joystick position data to mark the datasets. Markers were placed at time points in each dataset wherever changes in joystick output voltage were detected above a predetermined threshold, signifying movement. Individual markers were classified based on the direction of the detected movement (up, down, left, or right), and whether the movement was towards an outer position or the center position. From these markers, analysis could be performed to compare neuromagnetic activity at rest against the corresponding movement event related activity.

D. *Synthetic Aperture Magnetometry (SAM) Analysis*

Following data acquisition, CTF software (CTF Systems, Inc., Coquitlam, BC, Canada) was used to conduct synthetic aperture magnetometry (SAM) analysis on the MEG datasets. SAM analysis estimates the signal power, or variance of magnetic fields, for a group of voxels in the brain. This is accomplished by creating a covariance matrix that contrasts the theoretically expected signal power of individual voxels with measured signal power. A minimum variance beamformer, or spatial filter, is created for each voxel to retain the theoretical signal while blocking signals from elsewhere in the brain. The relationships between voxels can then be used to localize areas of the brain where magnetic field signal power is changing. Increases or decreases in signal power from MEG can be interpreted as synchronization or desynchronization events, respectively, as it is assumed that interactions between oscillating magnetic fields lead to variations in the amplitude of signals recorded by the MEG.

SAM analysis was conducted within 500 ms before and after each joystick movement marker within frequency bands of 8-32 Hz and 76-100 Hz, at a spatial resolution of 7.5 mm. The two frequency bands were chosen to avoid signal contamination from 60 Hz line noise and its harmonics; additionally, we sought to replicate hand movement results generated in an earlier study using ECoG (Miller et al., 2007). Due to right hand movement being correlated to activity within the hand knob, analysis was focused primarily on changes in activity within the hand knob and the surrounding precentral gyrus of the left hemisphere of the brain.

E. Statistical Analysis

Using AFNI software (Cox, 1996), individual SAM volumes were converted to standard Talairach coordinates, in effect normalizing neuromagnetic data across subjects. AFNI was then used to conduct student t-test statistical analysis across all subjects. It was then possible to visualize regions of significantly different neuromagnetic signal power within frequency bands of 8-32 Hz and 76-100 Hz.

III. Results

All statistically significant SAM analysis results are displayed on a Tailarached brain model, using the color-coding shown in Figure 2 to signify regions of increased synchronization or desynchronization.

Fig. 2 Decreases in MEG signal power are interpreted as desynchronization of oscillatory magnetic fields within neurons, while increases in signal power are representative of increased synchronization of magnetic field oscillation

Although localized synchronization was observed in the 76-100 Hz frequency band during wrist movement, as shown in Figure 3B, comparison between movement directions did not yield statistically significant differences in signal power, so results for 76-100 Hz are not discussed further.

It is evident from Figure 3 that the region of significant activity during wrist movement occupies the precentral gyrus and is roughly centered on the hand knob. Comparison of left and right movements at 8-32 Hz, shown in Figure 4A, indicates that a portion of the precentral gyrus inferior to the hand knob experiences greater synchrony during left movements than in right movements. A similar region shows greater synchrony during upward movements than in downward movements, as shown in Figure 4B. Figure 5 shows greater synchrony in an inferior portion of the precentral gyrus during upward movements compared to right or left movements, which suggests that upward movement induces the greatest synchrony of any movement directions in that area. Conversely, downward movements showed significantly greater desynchrony within the hand knob when compared to left or right movements. All activity shown was found to be significant, with $p<0.05$ (corrected) for each comparison.

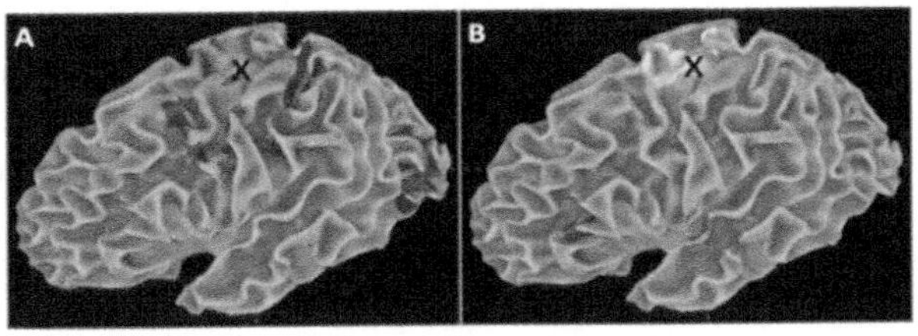

Fig. 3 Wrist movement is characterized by broad desynchronization over the hand knob in the 8-32 Hz band (A) and more focused synchronization in the 76-100 Hz band (B), regardless of movement direction. The hand knob is indicated by a black X mark

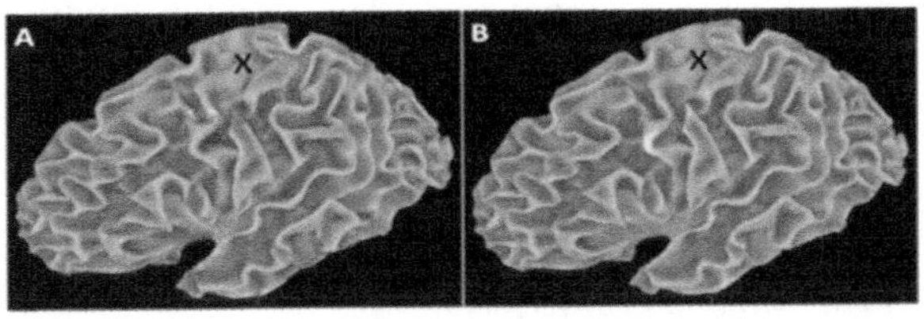

Fig. 4 (A) T-test comparison of left and right movements at 8-32 Hz, showing greater synchrony during movement to the left; $p<0.05$. (B) T-test comparison of up and down movements at 8-32 Hz indicates greater synchrony during upward movements; $p<0.01$

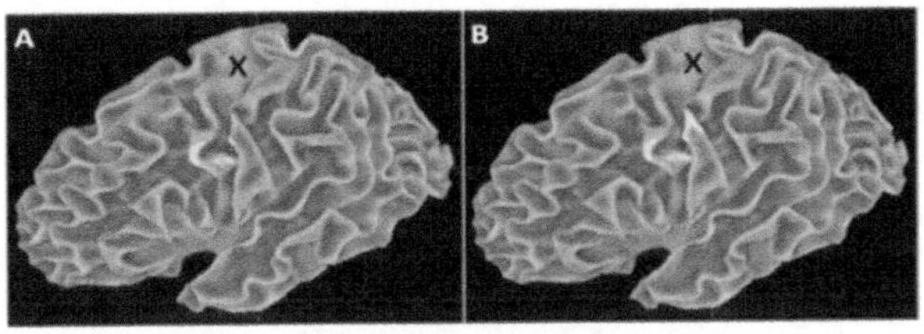

Fig. 5 (A) T-test comparison of up and left movements at 8-32 Hz shows greater synchrony during upward movements; $p<0.01$. (B) T-test comparison of up and right movements at 8-32 Hz also reveals greater synchrony during upward movements; $p<0.01$

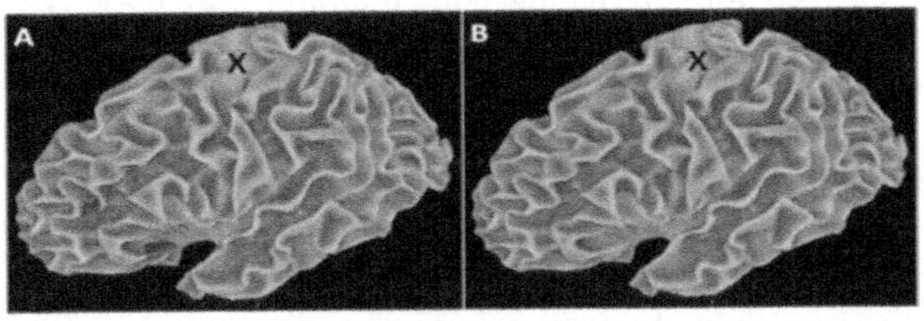

Fig. 6 (A) T-test comparison of down and left movements at 8-32 Hz indicates greater desynchrony during downward movements; $p<0.05$. (B) T-test comparison of down and right movements at 8-32 Hz shows greater desynchrony during downward movements; $p<0.05$

IV. Discussion

A. Correlating Movement Direction and Cortical Activity

The observation of statistically significant ($p<0.05$) patterns of cortical activity in the precentral gyrus during movement of the wrist in distinct directions suggests that directional movements can be mapped to discrete regions of decreased neuromagnetic signal power in and surrounding the hand knob.

Although much of the spatial difference in desynchronization between movement directions was observed in the precentral gyrus but not the hand knob itself, this could be a result of using a SAM spatial resolution of 7.5 mm to reduce processing time. The MEG system used in this study is capable of 3-5 mm spatial resolution, so future studies may be able to detect significant differences in neuromagnetic desynchrony within the hand knob, allowing MEG analysis to be focused on a smaller region of interest and thereby reducing processing requirements.

B. Brain Computer Interface

The overall goal of this project is to identify neuromagnetic signals with the potential for use in controlling man-made devices via brain computer interfaces (BCIs). The statistically distinct magnetic signatures observed in the hand knob and surrounding prefrontal cortex for movement of the wrist in different directions indicate that these signals could be used to accurately control a BCI in a two dimensional plane of movement. For example, a cursor on a screen could be directed toward a target in any direction, and the demonstrated presence of a beta desynchronization during wrist movement would likewise allow movement commands to be easily distinguished.

A particularly interesting feature of the directional neuromagnetic signals observed in this study is the fact that they were statistically significant across subjects. This indicates that the same core set of directional commands could be utilized for interfacing with multiple BCI users, potentially reducing the training time required for a BCI to learn the command signals of individual users.

This study analyzed neurological synchronization and desynchronization events using MEG, which possesses high temporal and spatial resolution, but is unsuitable for portable use with current technology. However, EEG, which relies on relatively more robust detection equipment, could be used to detect the same neurological signals in a viable BCI system.

C. Examining Other Neurological Regions and Frequency Ranges

The mu-beta and gamma frequency ranges analyzed in this study were chosen primarily to examine the previously reported event-related desynchronization in the beta band during motor movement, as well as the less well-documented event-releated synchronization in the gamma band during similar movement. The frequency ranges were constrained to avoid signal contamination from 60 Hz line noise and its harmonics, as well as to match and attempt to replicate ECoG results obtained by Miller and colleages (Miller et al., 2007). While this large frequency range provided statistically relevant results, Fourier analysis of signal amplitude distribution could lead to a more focused bandwidth. This could possibly allow for the discovery of significant differences in signal power within the hand knob region, perhaps within the gamma band, and also reduce signal processing requirements.

References

1. Babiloni F, Carducci F, Cincotti F, Del Gratta C, Pizzella V, Romani GL, Rossini PM, Tecchio F, Babiloni C (2001) Linear inverse source estimate of combined EEG and MEG data related to voluntary movements. Hum Brain Mapp 14:197-209.
2. Cox RW (1996) AFNI: software for analysis and visualization of functional magnetic resonance neuroimages. Comput Biomed Res 29:162-173.
3. Kauhanen L, Nykopp T, Sams M (2006) Classification of single MEG trials related to left and right index finger movements. Clin Neurophysiol 117:430-439.
4. Mason SG, Bashashati A, Fatourechi M, Navarro KF, Birch GE (2007) A comprehensive survey of brain interface technology designs. Ann Biomed Eng 35:137-169.
5. Miller KJ, Leuthardt EC, Schalk G, Rao RP, Anderson NR, Moran DW, Miller JW, Ojemann JG (2007) Spectral changes in cortical surface potentials during motor movement. J Neurosci 27:2424-2432.
6. Muthukumaraswamy SD, Johnson BW, Gaetz WC, Cheyne DO (2004) Modulation of neuromagnetic oscillatory activity during the observation of oro-facial movements. Neurol Clin Neurophysiol 2004:2.
7. Velliste M, Perel S, Spalding MC, Whitford AS, Schwartz AB (2008) Cortical control of a prosthetic arm for self-feeding. Nature 453:1098-1101.
8. Waldert S, Braun C, Preissl H, Birbaumer N, Aertsen A, Mehring C (2007) Decoding performance for hand movements: EEG vs. MEG. Conf Proc IEEE Eng Med Biol Soc 2007:5346-5348.
9. Waldert S, Preissl H, Demandt E, Braun C, Birbaumer N, Aertsen A, Mehring C (2008) Hand movement direction decoded from MEG and EEG. J Neurosci 28:1000-1008.
10. Yeager JD, Phillips DJ, Rector DM, Bahr DF (2008) Characterization of flexible ECoG electrode arrays for chronic recording in awake rats. J Neurosci Methods 173:279-285.

An Algebraic Method for Eye Blink Artifacts Detection in Single Channel EEG Recordings

Zoran Tiganj[1], Mamadou Mboup[2], Christophe Pouzat[3], and Lotfi Belkoura[4]

[1] Project ALIEN, INRIA Lille - Nord Europe, Villeneuve d'Ascq, France
[2] CReSTIC. Université de Reims Champagne Ardenne, Reims, France
[3] Laboratoire de Physiologie Cérébral, Université Paris Descartes, Paris, France
[4] LAGIS, Université des Sciences et Technologies de Lille, Lille, France

Abstract— **Single channel EEG systems are very useful in EEG based applications where real time processing, low computational complexity and low cumbersomeness are critical constrains. These include brain-computer interface and biofeedback devices and also some clinical applications such as EEG recording on babies or Alzheimer's disease recognition. In this paper we address the problem of eye blink artifacts detection in such systems. We study an algebraic approach based on numerical differentiation, which is recently introduced from operational calculus. The occurrence of an artifact is modeled as an irregularity which appears explicitly in the time (generalized) derivative of the EEG signal as a delay. Manipulating such delay is easy with the operational calculus and it leads to a simple joint detection and localization algorithm. While the algorithm is devised based on continuous-time arguments, the final implementation step is fully realized in a discrete-time context, using very classical discrete-time FIR filters. The proposed approach is compared with three other approaches: (1) the very basic threshold approach, (2) the approach that combines the use of median filter, matched filter and nonlinear energy operator (NEO) and (3) the wavelet based approach. Comparison is done on: (a) the artificially created signal where the eye activity is synthesized from real EEG recordings and (b) the real single channel EEG recordings from 32 different brain locations. Results are presented with Receiver Operating Characteristics curves. The results show that the proposed approach compares to the other approaches better or as good as, while having lower computational complexity with simple real time implementation. Comparison of the results on artificially created and real signal leads to conclusions that with detection techniques based on derivative estimation we are able to detect not only eye blink artifacts, but also any spike shaped artifact, even if it is very low in amplitude.**

Keywords— **Electroencephalography, eye blink artifacts, differential algebra, operational calculus, parameter estimation.**

I. Problem description

In electroencephalographic (EEG) recordings electrical activity of the levator muscle (which controls the upper eyelid and is responsible for eye blinks), projects on vectors pitched by the EEG electrodes causing artifacts. Eye blink artifacts affect more then 10% of an average human EEG (human in average blinks once in $5s$, and the duration of an artifact is more then $0.5s$), thus they are a very important problem in EEG signal processing.

The amplitude of these artifacts depends on the amplitude of the levator muscle activity and on the brain region where the EEG is recorded. Typically, the amplitude has the same order of magnitude as the recorded brain activity. Thus, during the eye blink EEG recording is useless for most of applications. On the other hand, the artifacts can not be detected only by setting a threshold.

Common approach for eye blinks detection is placing extra electrodes close to the eye. Signal from these electrodes is amplified with differential amplifier, then setting an appropriate threshold is enough for the detection. Moreover, if several EEG channels are recorded, blind source separation is often used, not only to detect artifacts, but also to reject them from the EEG data.

However, placing extra electrodes close to the eye or using EEG with several channels is not suitable in many applications. Upcoming consumer application based on EEG such as brain-computer interface [1] or biofeedback devices [2] are mostly intended for non-clinical, everyday home use. Simplicity of use and low price are essential for those devices. Some clinical application such as recording on babies [3] or Alzheimer's disease recognition [4] are often based on single channel EEG systems as well. In this paper we address the problem of eye blink artifacts detection in such single channel systems. We show how the new algebraic technique for numerical differentiation [5] can lead to the algorithm for joint detection and estimation of the locations of the eye blink artifacts in a single channel recording.

S. Supek and A. Sušac (Eds.): Advances in Biomagnetism – BIOMAG2010, IFMBE Proceedings 28, pp. 175–178, 2010.
www.springerlink.com

In the following section we model the artifact-free part of the signal as a piecewise polynomial and we interpret the occurrence of a spike shaped artifact as a discontinuity point in that model. Next, we describe an algorithm which allows us to detect and locate such an irregularity point. The main steps of the algorithm, presented in section II, are based on the very classical operational calculus [6] (which is widely used in electrical circuit). The solution admits an easy implementation, as a combination of finite impulse response (FIR) filters.

In section III we compare the proposed approach with three other approaches: (1) the very basic threshold approach, (2) an approach that combines the use of median filter, matched filter and nonlinear energy operator (NEO) [7] and (3) the wavelet based spike detection algorithm [8]. Since the projection of the levator muscle activity depends on the brain region where EEG is recorded we did the comparison on single channel recordings from 32 different brain locations. To be able to have a complete control over the eye blink locations and the signal to noise ratio we also did a comparison of the four approaches on artificially created signal where we used eye blink templates obtained from the real EEG recordings. The results are presented with receiver operating characteristic (ROC) for each of the 32 locations. Representing results using ROC curves is standard in detection problems. They indicate that the proposed approach compares favorably with the threshold and NEO based approach, while it performs similarly as the wavelet approach, but it has significantly lower computational complexity and can be implemented for online detection.

II. Description of our approach

The association *differential algebra-operational calculus* leads to simple and efficient algebraic approach for parameter estimation, initiated in [6]. This approach is applied here in order to obtain an eye blink artifacts detection method. This can be seen as a change-point detection [9] problem: Given $y(t)$, a noisy observation of a piecewise regular signal $x(t)$, we want to detect the abrupt changes and estimate their locations. The problem is widely studied in the literature where the statistical approach naturally takes the dominant role (see [9], [10] and their references). Here, we apply a purely algebraic approach (see [6], [11] and [12] for the details).

We represent our signal with a piecewise polynomial model. The rapid variation induced by an eye blink artifact in interpreted as an irregular point in the signal. Consider an interval I_τ^T, with the origin τ and length T. For $t \in I_0^T$, let set $x_\tau(t) = x(t+\tau)$. Next, choose T such that there is at most one discontinuity point t_τ in I_τ^T. Now, if t_τ is such an artifact location (an irregular point) then we know that the nth order derivative of $x(t)$ has the form:

$$\frac{d^n x_\tau}{dt^n} = [x_\tau^{(n)}](t) + \sum_{k=0}^{n-1} \left(\mu_k^0\, \delta(t)^{(k)} + \mu_k\, \delta(t-t_\tau)^{(k)} \right), \quad (1)$$

where the exponent (k) stands for the derivative of order k, μ_k (resp. μ_k^0) is the jump of the kth order derivative at the point t_τ (resp. 0) and $[f]$ represents the regular part of f.

In a sequel we assume a piecewise polynomial model of order not exceeding 1. Then, by taking $n = 2$ in equation (1) the regular part, $[x_\tau^{(2)}](t)$, vanishes and we obtain:

$$\frac{d^2 x_\tau}{dt^2} = \mu_0^0\, \delta(t) + \mu_1^0\, \dot{\delta}(t) + \mu_0\, \delta(t-t_\tau) + \mu_1\, \dot{\delta}(t-t_\tau). \quad (2)$$

To handle more easier expressions like $\delta(t)$, we rewrite equation (2) in the operational domain:

$$s^2 \hat{x}_\tau(s) = \mu_0^0 + s\mu_1^0 + (\mu_0 + s\mu_1)e^{-t_\tau s}. \quad (3)$$

As the jump parameters μ_0^0, μ_1^0, μ_0 and μ_1 are unknown we use the algebraic framework of [6] to eliminate them and obtain:

$$t_\tau^2 (s^2\hat{x}_\tau)^{(2)} + 2t_\tau (s^2\hat{x}_\tau)^{(3)} + (s^2\hat{x}_\tau)^{(4)} = 0 \quad (4)$$

We apply the algorithm described in [13] (see also [14] and [15]) to obtain back in the time domain a second order equation in t_τ where the coefficients are iterated integrals of the signal. In the discrete time, we set $\tau = kT_s$ and $T = MT_s$ where T_s is sampling time. Using classical numerical integration with sliding windows $I_k^M = [kT_s\ (k+M)T_s]$ we end up with a second order equation

$$a_k t_k^2 + b_k t_k + c_k = 0, \quad (5)$$

where each coefficient is the output of a FIR filter, with the noisy observation as the input. In this equation we use the notation t_k for the discrete counterpart of t_τ. The detection will be based on the following observation [13]: If there is no discontinuity point in I_k^M then all three coefficients of equation (5) are equal to 0. Else these coefficients are all different from 0. Therefore, as a detection criteria we use the product of the three coefficients: $Q_k = |a_k b_k c_k|$. Due to the noise, an eye blink artefact will be detected when Q_k is greater than a given threshold σ, representing the noise level.

III. Results and conclusion

In order to asses the efficiency of the proposed algebraic approach we compare it with three other approaches:

- The basic threshold approach, where it is assumed that the eye blink artifacts have the largest amplitudes in the recorded signal and only a threshold is applied to detect them.
- The approach that combines median filtering, matched filtering and NEO. In the literature this approach is known to emphasize spike shape waveforms as the eye blinks [7].
- The approach that is based on wavelet decomposition, where scales are calculated depending on the input parameters: minimum and maximum spike width [8].

The comparison has been done on 7.5 minutes of 32 channel EEG recording, at the sampling frequency of 1kHz. The electrodes were placed according to the international 10-20 system. The recording was unipolar with reference calculated as the averaged sum of all 32 electrodes. The ground electrode was placed on the forehead. All electrode impedances were less then $5k\Omega$.

The parameter T is set to 0.5s, what gives us satisfying resolution since the duration of eye blink artifacts is typically longer than 0.5s.

Since we address the problem of the eye blink detection and localization in single channel EEG, all four approaches were applied on each of the 32 channels separately. Since the recorded signals are from different brain regions, we are able to evaluate how different electrodes positions affect the results of the different approaches.

In order to have reference information on when the examinee actually blinks, apart from the EEG, two additional electrodes were used. One was placed above and the other below the examinee's right eye. In 7.5 minutes of recording we detected 65 eye blinks. The results for each of the four approaches are presented with 32 ROC curves, one for each channel (fig. 1). Every curve is made from 50 points that are calculated for different threshold levels. The detection is taken as correct if the eye blink occurred in 1s window around the point that is found to be above the threshold level. For each of the 50 points, the probability of correct detection is defined as the ratio between the number of correct detections for a given threshold and the total number of spikes. The probability of false alarm is defined as the ratio between the number of false alarms and the maximal possible number of false alarms for a given threshold. From the results shown on fig. 1 we conclude the following:

- Since electrodes FP1 and FP2 are located very close to the eyes all four approaches have almost perfect detection rate (all eye blinks are detected with almost no false alarms).

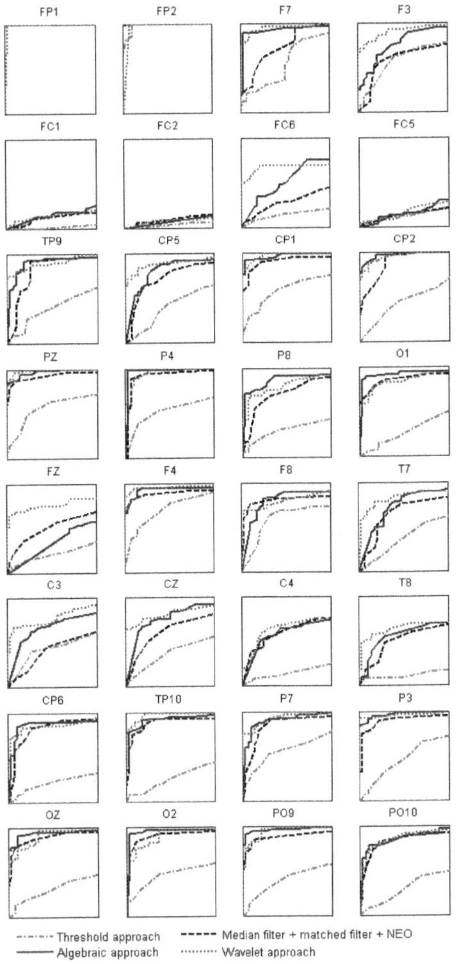

Fig. 1 32 ROC curves obtained when real EEG signal is used. The vertical axis denotes probability of correct detection and goes from 0 to 1, while the horizontal axis denotes probability of false alarm and goes from 0 to 0.03

- Poor detection results for channels FC5, FC1, FC2 and FC6 suggest that the projection of the eye muscle activity is very low in this region, what leads to conclusion that eye blink detection is not necessary in some cases.
- Advantages of using either algebraic approach, wavelet approach or combination of median filter, matched filter and NEO are the greatest when the algorithms are applied on signals from the electrodes that are placed further from the eyes (from TP9 to PO10 on fig. 1). Here we notice the property of the algorithms to emphasis spike shape waveforms in the noisy environment (since threshold approach gives poor results the recorded signal is obviously very noisy in these brain regions). For the electrodes F7, F3, FZ, F4 and F8, where the projection of the eye activity is larger (since they are located closer to the eyes), threshold approach performs similar to the other three approaches.

Simple to implement, the proposed algebraic approach allows extremely fast computation in comparison with the other approaches. The processing time (averaged from 50 iterations) of 7.5 minutes long input signal was 77.30s for the NEO based approach, 3.22s for the algebraic approach (FIR filters with 500 samples long impulse responses were used) and 27.31s for the wavelet based approach (minimum spike width was set to 300ms and maximum to 700ms, number of scales was set to 4 and biorthogonal mother wavelet was used). The implementation is done in MATLAB and run under Microsoft XP OS on the computer with 1.73GHz Intel Core2 processor and 2GB of RAM.

Simulation done on the real EEG recording gives a good overview of the algorithms performance on signals from different brain regions. However, this simulation has a drawback since we can not be sure that the eye artifacts are the only spike shaped artifacts in the signal. Any other muscle movement by the examinee or some electrodes shifting may cause such artifacts as well. Since we are not aware of these additional artifacts their detection by the algorithms will be considered as a false alarm.

In order to have a complete control over the artifacts locations we made an artificial signal by using white noise and randomly distributed eye blink template created by averaging eye blinks from channel T8 from the real EEG recording. The result of the comparison is again presented with ROC curve (fig.2). Now the algebraic and the wavelet approach performs similarly (while algebraic approach has much lower computational complexity) and better than approach based on combination of median filter, matched filter and NEO.

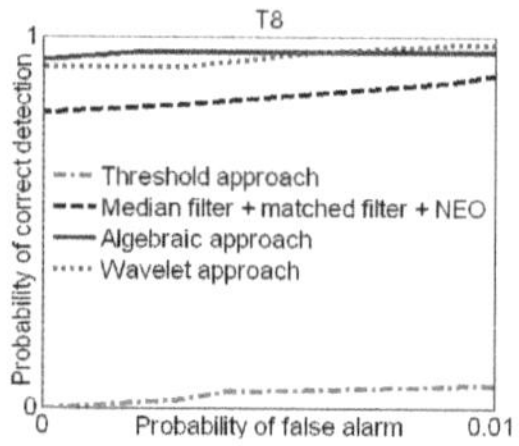

Fig. 2 ROC curve obtained when synthesized EEG signal is used

The algebraic approach performs better with respect to the other approaches when simulation is done on the artificial signal (fig. 2) than when it is done on the real signal (fig. 1). This suggests that, as we mentioned earlier, the real signal had some other spike shape artifacts that we were not aware of and their detections were counted as false alarms.

Easy implementation (FIR filters), high robustness and the real time processing make the proposed approach quite eligible for eye artifact detection in single channel EEG systems. This is particularly interesting knowing that, due to low costs and simplicity of use, single channel devices become more and more common, especially in home use consumer devices, as well as in some clinical applications. Moreover this approach can also be applied in multichannel EEG and MEG systems.

Acknowledgements

The authors thank the University of Zagreb for providing them with the EEG.

References

1. Thulasidas M, Guan C, Ranganatha S, Wu J K, Zhu X, Xu W. Effect of Ocular Artifact Removal in Brain Computer Interface Accuracy *26th Annual Inter Conference of the Engineering in Medicine and Biology Society.* 2004;6:4385–4388.
2. Othmer S, Kaiser D. Implementation of Virtual Reality in EEG Biofeedback *Cyberpsychology and Behavior.* 2000;3:415-420.
3. Walls-Esquivel E, Vecchierini M F, Héberlé C, Wallois F. Electroencephalography (eeg) recording techniques and artefact detection in early premature babies *Clinical Neurophysiology.* 2007;37:299–309.
4. Cho S Y, Kim B Y, Park E H, et al. Automatic recognition of Alzheimer's disease with single channel EEG recording *International Conference of the IEEE Engineering in Medicine and Biology.* 2003;3:2655-2658.
5. Mboup M, Join C, Fliess M. Numerical differentiation with annihilators in noisy environment *Numerical Algorithms.* 2009;50:439–467.
6. Fliess M, Sira-Ramírez H. An algebraic framework for linear identification *ESAIM: COCV.* 2003;9:151–168.
7. Land B, Spence A. Spike (action potential) detection and sorting *Journal Club 25,* *http://www.nbb.cornell.edu/neurobio/land/PROJECTS/spikeSort/.* 2002.
8. Nenadic Z, Burdick J W. Spike detection using the continious wavelet transform *IEEE Trans Biomed Eng.* 2005;52:74-87.
9. Basseville M, Nikiforov I V. Detection of Abrupt Changes: Theory and Application *Prentice-Hall.* 1993.
10. Lavielle M. Using penalized contrasts for change-point problem *Signal Processing.* 2005;85:1501–1510.
11. Fliess M, Mboup M, Mounier H, Sira-Ramírez H. *Questioning some paradigms of signal processing via concrete example.* México: Editiorial Lagares 2003.
12. Mboup M. Parameter estimation for signals described by differential equations *Applicable Analysis.* 2009;88:29 - 52.
13. Mboup M. A Volterra filter for neuronal spike detection *Research report, INRIA, http://hal inria fr/inria-00347048/fr/.* 2008.
14. Fliess M, Join C, Mboup M. Algebraic change-point detection *in press in AAECC, DOI:10.1007/s00200-010-0119z.* 2010.
15. Tiganj Z, Mboup M. Spike Detection and Sorting: Combining Algebraic Differentiations with ICA in *Independent Component Analysis and Signal Separation, 8th International Conference*(Paraty, Brazil):475-482 2009.

Author: Zoran Tiganj
Institute: INRIA Lille - Nord Europe
Street: Parc Scientifique de la Haute Borne 40
City: Villeneuve d'Ascq
Country: France
Email: zoran.tiganj@inria.fr

Relevant Observations for Averaging Stimulus Evoked Magnetic Fields across Trials and across Subjects

Norman Zacharias[1], Cezary Sielużycki[1], Artur Matysiak[2], Reinhard König[1], and Peter Heil[1]

[1] Leibniz Institute for Neurobiology, Magdeburg, Germany
[2] Department of Biomedical Physics, Institute of Experimental Physics, University of Warsaw, Warsaw, Poland

Abstract— **The analysis of electroencephalographic (EEG) and magnetoencephalographic (MEG) data commonly involves arithmetic averaging of the evoked magnetic fields across trials, but also across subjects ("grand average"), with subsequent comparisons of grand averages across stimuli or conditions. The classical, and most frequently used, model in the analysis of stimulus evoked brain responses in EEG and MEG is based on the assumptions that the arithmetic mean is a good estimator of the evoked responses, that the evoked responses are independent of background activity, and that the latter averages out. The comparison of evoked responses ideally requires homogeneity of the variance (homoscedasticity). Here, we explore the issue of homoscedasticity in a typical MEG experiment. Fifteen subjects were stimulated with tones of different onset intervals, while magnetic fields were measured with a whole-head MEG device. Our analyses revealed an approximately homogeneous standard deviation (SD) of the responses across repeated trials of the same stimulus, SD_{trials}, i.e. SD_{trials} was independent of the corresponding mean response amplitude. In contrast, the $SD_{subjects}$ derived from averaging the trial-averaged responses across subjects increased approximately linearly with the corresponding mean amplitude. This result shows that the evoked responses, including the M100, from different subjects and stimulus conditions differed largely by scaling factors. The latter findings question the usefulness of the common practice of comparing arithmetic averages of trial-averaged responses across subjects to different stimuli or conditions, e.g. by subtraction of responses to different conditions, and thus have widespread implications.**

Keywords—**auditory cortex, arithmetic averaging, standard deviation, magnetoencephalography, homoscedasticity.**

I. Introduction

In search for a good estimator of event-related potentials or magnetic fields, the continuously recorded EEG or MEG signals are commonly divided into trials, or epochs, with fixed reference to the feature of interest in the repetitive stimulation (e.g. stimulus onset), followed by arithmetic averaging of signals across trials [1], [2]. This arithmetic averaging across trials is based on the assumptions of the classical signal-plus-noise (SPN) model, i.e., that the stimulus-evoked response is deterministic and independent of ongoing activity and environmental noise, which are considered zero-mean stochastic processes and thus expected to average out, provided the number of trials is large enough. Hence, the arithmetic mean is thought to be a good estimator of the evoked response. The resulting trial-averaged evoked responses often still display small signal-to-noise ratio and are therefore frequently averaged further, across subjects ("grand average"). The signals averaged in this way are then compared across stimuli or conditions, e.g. by subtraction. Such comparisons, however, tacitly rely on the assumption of homoscedasticity associated with the mean. Recently, we have shown [3] that the $SD_{subjects}$ associated with the peak amplitude of the largest component of auditory evoked magnetic fields (AEFs), the M100, is not homogeneous. Instead, $SD_{subjects}$ increased approximately linearly with the mean, revealing that the M100-peak amplitudes of different subjects obtained in response to a set of stimuli differ mainly by scaling factors. However, geometric averaging of M100-peak amplitudes across subjects, which is equivalent to arithmetic averaging of log-transformed amplitudes, resulted in an almost homogeneous variance across different stimulus conditions [3]. Here, we examine whether this might also apply to averaging across trials. We exploit the variation in the amplitude of AEFs with time relative to stimulus onset, as well as the variation in the amplitude of AEFs as a function of the stimulus onset interval (SOI; [4], [5]) for a given subject and the variation across subjects to address these issues.

II. Materials and Methods

Fifteen subjects (8 males, 7 females; age range 23–52 years, mean 29 years) participated in three experiments each, which were performed on separate days. The subjects were seated under a 248-channel whole-head MEG system (Magnes 3600 WH) and listened to tones presented to their left ears. Tones had carrier frequencies of 1.5 kHz and 1.6 kHz and were shaped into bursts of 100-ms duration with 5-ms linear onset and offset ramps. In each experiment, the mean stimulus onset interval (SOI) was 2.5 s, but the actual SOIs varied from tone to tone in an unpredictable sequence. SOIs were selected from a set of fixed values, ranging from 0.25 s to 12.5 s, ensuring that all SOIs

S. Supek and A. Sušac (Eds.): Advances in Biomagnetism – BIOMAG2010, IFMBE Proceedings 28, pp. 179–182, 2010.
www.springerlink.com

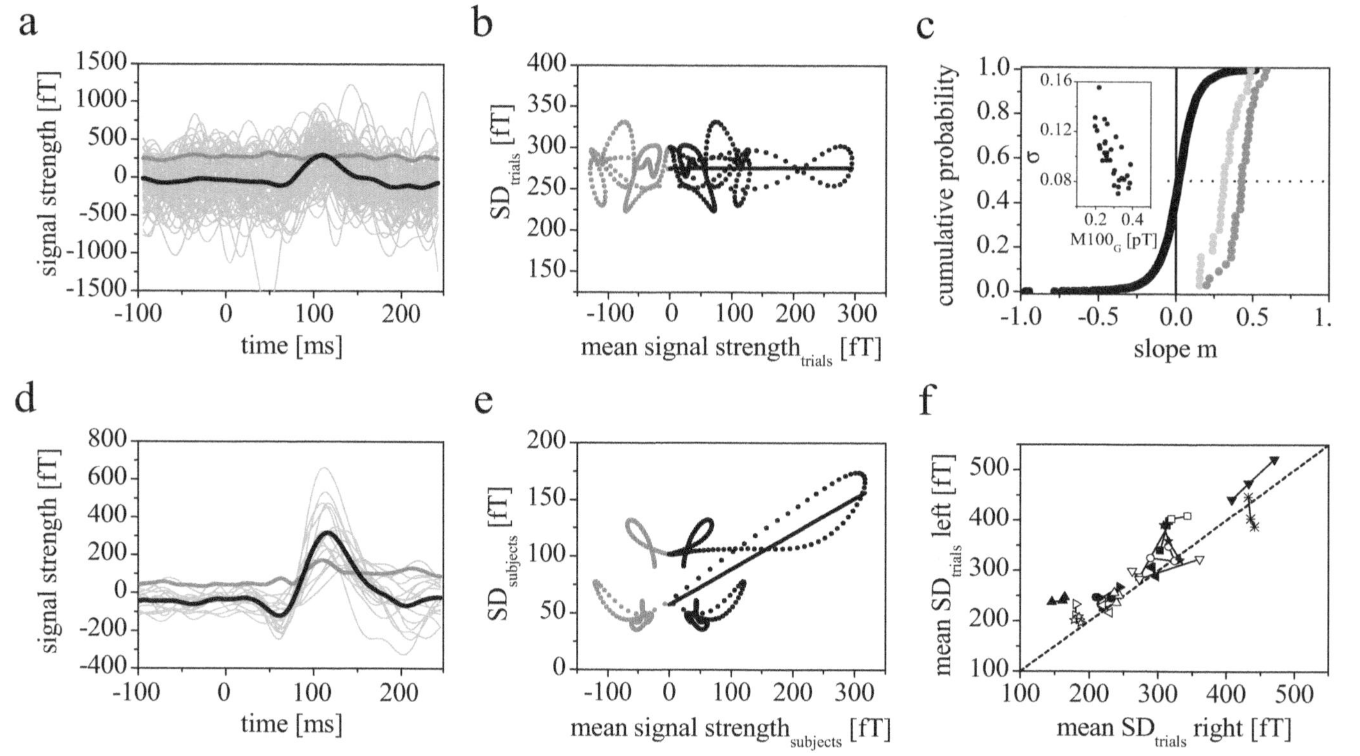

Fig. 1(a) Artifact-free epochs, from 100 ms before to 245 ms after tone onset, of one subject in response to 1.5-kHz tones of a SOI of 2 s, recorded at one sensor (thin grey lines), their arithmetic average (thick black line) and the corresponding standard deviation, SD_{trials} (thick grey line). (b) Plot of SD_{trials} against the arithmetic mean of the data shown in (a). Negative means (grey symbols) were inverted prior to calculation of a line-fit (black, nearly horizontal, line) through the data. (c) Cumulative probability of slopes derived from line-fits as in (b) (black symbols). The 4800 slopes have a mean and SD of 0.013±0.129. The inset plots the SD of slope distributions, calculated for different SOIs separately, as function of the geometric mean of the M100-peak amplitude. Note the significant negative correlation. Light and dark grey symbols show the distributions of slopes derived from the relationships between $SD_{subjects}$ and the corresponding grand mean for the left and right hemisphere, respectively. (d) Trial-averaged evoked magnetic fields of all 15 subjects to tones of one SOI in one experiment (grey lines) recorded at the maximum sensor over the left hemisphere, their arithmetic mean ("grand average"; thick black line) and corresponding $SD_{subjects}$ (thick grey line). (e) Plot of $SD_{subjects}$ against the corresponding grand mean for the data of (d). Negative mean values (grey symbols) were sign-inverted prior to the line fit (black line). The resulting slopes are shown in (c). (f) Plot of the mean SD_{trials}, computed by averaging SD_{trials} from 5 sensors and for all SOIs in a given experiment, derived for the left hemisphere against the corresponding value for the right hemisphere. Different symbols represent different subjects. Symbols from a given subject are connected by lines. Note their clustering

occurred the same number of times (n=110) in an experiment. The sound level was adjusted to about 80 dB SL with reference to the individual subject's threshold for a 1.5-kHz test tone. Data were acquired continuously at a sampling rate of 678.17 Hz and with an online band-pass filter of 0.1 Hz to 100 Hz. For analysis, data were filtered offline with a band-pass filter of 0.5 Hz to 30 Hz. Epochs contaminated by eye-blinks were rejected, leaving on average 93±8 uncontaminated epochs for each stimulus. The analysis focused on the cluster of five sensors over each hemisphere showing the largest amplitude of the M100. For those sensors we calculated the standard deviation of signal strengths across trials, SD_{trials}, at each of the 235 sampling points over the time interval from 100 ms before to 245 ms after tone onset.

III. Results

Figure 1a shows, for one representative subject and experiment, all artifact-free signals (n=96; thin grey lines) recorded at one sensor in response to 1.5-kHz tones and a SOI of 2 s, as well as their arithmetic mean (thick black line) and the corresponding SD_{trials} (thick grey line). SD_{trials} is large, between about 250 and 300 fT, but appears independent of time relative to stimulus onset and of the mean signal amplitude. This is confirmed in Figure 1b, which plots the 235 values of SD_{trials} obtained at all sampling points against the corresponding mean amplitudes. A line-fit through these data yielded a slope of 0.002. The distribution of the 4 800 slopes obtained from equivalent plots for all 15 subjects, the 2×5 sensors, and the 32 SOIs tested in the three experiments, is shown in Figure 1c (black symbols). The distribution has a mean of 0.013 and a SD of 0.129. Hence, the mean is not different from zero. The sizable SD is due to the low signal-to-noise ratios for some SOIs. When we calculated separate distributions of slopes for the different SOIs, we observed that the SD of the slope distribution was the smaller, the larger the geometric mean amplitude of the M100 ($M100_G$; see inset in Fig. 1c). Large mean amplitudes were evoked by tones with long SOIs. The correlation was highly significant (r^2=0.531; $p<0.001$; n=32). In contrast, the means of these distributions for separate SOIs were independent of $M100_G$ (r^2=0.056; $p>0.05$), yielding a mean and SD of 0.016±0.02. Thus, overall, SD_{trials} is independent of the mean amplitude across trials.

SD_{trials} not only reflects measurement noise, but also ongoing brain activity, as evident from the differences between subjects. For each subject and for a given experiment, we computed the mean of the SD_{trials} by averaging the individual values of SD_{trials} across sample points, SOIs, and the five sensors of a cluster over each hemisphere. Figure 1f plots the mean SD_{trials} from the left against those from the right hemisphere, with different subjects identified by different symbols, and data from the same subjects connected by lines. Two observations are worth emphasizing: (1) While the data points fall reasonably close to the diagonal, the majority of data points (37 of 45) fall above the diagonal (Wilcoxon signed paired rank test; $p<0.0001$), indicating significantly higher mean SD_{trials} values in the left hemisphere. Since no heartbeat artifact correction procedure was applied to the data, we attribute this difference to the effects of the heartbeat, consistent with the common location of the heart in the left half of the body and its greater effect on magnetic signals recorded over the left hemisphere [3]; (2) There are large differences in mean SD_{trials} between subjects, while the values of a given subject derived from the different experiments performed on separate days are remarkably similar. Hence, the ongoing activity recorded by the magnetometers appears rather subject-specific in our experiments.

In contrast to SD_{trials}, the SD computed across subjects, $SD_{subjects}$, is not independent of the corresponding mean. Figure 1d shows, for one experiment and SOI (2.5 s), the trial-averaged signals recorded at the sensor with the largest signal over the left hemisphere of all 15 subjects (thin grey lines). The grand average (thick black line) and corresponding standard deviation, $SD_{subjects}$ (thick grey line), are also shown. Here, $SD_{subjects}$ appears to scale with the mean amplitude. $SD_{subjects}$ is largest approximately at the time point when the grand mean is the largest. The match would have been perfect had we first aligned the AEFs of the individual subjects with respect to a common reference point, such as the peak of the M100, prior to averaging. The scaling of $SD_{subjects}$ with the mean is confirmed in Figure 1e, which plots the $SD_{subjects}$ against the corresponding mean. Because of the sign reversal of the mean amplitude around the P50 to M100 deflections (see Fig. 1d), slope estimates were derived using the absolute values of the means (grey symbols in Fig. 1e show the signed, negative means). For the example illustrated in Figure 1e, the resulting slope characterizing the relationship between $SD_{subjects}$ and corresponding mean was 0.412. The distributions of the 32 slopes obtained from equivalent plots for all SOIs and experiments for both hemispheres are also shown in Figure 1c (light and dark grey symbols). The distributions have means and SDs of 0.32±0.10 for the left and 0.43±0.09 for the right hemisphere, clearly larger than zero. The positive slopes reveal that the trial-averaged signals of different subjects are largely scaled with respect to each other, as observed previously by us for the peak amplitude of the M100 [3]. The on average steeper slopes between $SD_{subjects}$ and mean amplitudes from the right compared to the left hemisphere are likely due to the somewhat greater heterogeneity of the shapes of the average signals in the right compared to the left hemisphere in our cohort of subjects.

IV. Discussion

Our observation that, on average, SD_{trials} and corresponding means are independent of each other (Fig. 1a–c) is consistent with the classical notion that stimulus evoked responses are independent of ongoing activity (and of noise). This notion has been challenged by the proposition that such event-related responses may instead be due to phase-synchronization or phase-resetting of ongoing oscillatory brain activity induced by the stimulation [6], [7], but recently it has been demonstrated by Yeung et al. [8] that the arguments provided for this scenario cannot resolve the issue. Mäkinen et al. [9] argued that in the case of phase-synchronization or resetting, the SD_{trials} should be minimal at the point of phase reset (e.g. near the peak of the M100). However, they did not find such a minimum in their data. On the contrary, they observed slightly larger SD_{trials} around the peak of the tone-evoked M100 than during the remainder of the epochs. We also found no such minimum, but also no evidence for enhanced variability around the M100, which would have shown up as a bias of the SD_{trials} versus mean slope distribution in Fig. 1c towards positive values. In fact, in our data, SD_{trials} was homoscedastic. The reason for these different observations in the study of Mäkinen et al. [9] and ours may lie in the degree of response habituation in the course of the acoustic stimulation [10]. Habituation will induce additional variability of the evoked response. In the study of Mäkinen et al. [9], the tones were repeated at a constant SOI of 0.8 s, while in our experiments SOIs varied from tone to tone between 0.25 s and 12.5 s. This variation may have diminished any habituation effects.

SD_{trials} predominantly reflects ongoing brain activity rather than measurement noise. This conclusion can be drawn from the variability of mean SD_{trials} between subjects. The slight, but significant, inter-hemispheric differences are likely caused by heartbeat artifacts (see [3]). Interestingly, mean SD_{trials} from measurements performed on the same subjects on different days varied little compared to the inter-subject variability (Fig. 1f). It thus appears that, in our experiments, the measurable ongoing activity can also be viewed as an individual characteristic, much like the shape of the M100 waveform. We found no correlation between the mean SD_{trials} in a given subject, hemisphere, and experiment and the amplitude of the M100 (not shown), suggesting that ongoing and evoked activity are independent, at least between subjects (but see [11], [12]). In any event, our findings provide justification for the common practice of arithmetic averaging across trials or epochs to obtain estimates of event-related magnetic fields.

When it comes to averaging evoked responses across subjects ("grand averages"), the issue is more complicated. Our analyses here provide further support for our previous conclusion, which was based on the M100-peak amplitude and its variability [3], that the response amplitudes of different subjects or hemispheres and in response to different stimuli differ by scaling factors, and extend this conclusion to a wider time range around the peak of the M100. As predicted by scaled responses, the $SD_{subjects}$ increased roughly linearly with increasing arithmetic mean (Figs. 1d and e). The deviations from linearity are most pronounced at small mean amplitudes and can be attributed to the additional ongoing activity.

V. Conclusions

The commonly employed arithmetic average of evoked magnetic fields across trials, but also across subjects ("grand average"), is thought to provide reliable estimators of the evoked responses and assumes that the evoked responses are independent of background activity. However, in our data evoked responses from different subjects appear to differ largely by scaling factors, since the SD across subjects grows almost linearly with the mean. This heterogeneity of the inter-subject variance of the amplitudes as a function of time relative to stimulus onset (Fig. 1e), or as a function of stimulus condition for a given point in time (e.g. the M100-peak latency [3]), questions the usefulness of the common practice of subtracting arithmetic means of responses evoked by different stimuli or under different conditions. Instead, it may be better to average evoked responses geometrically, which leads to a more homogeneous SD, and to compute ratios rather than differences. Ultimately, it would be best to first transform the data in such a way that the variance is stabilized, and average the transformed data arithmetically.

Acknowledgments

This work was supported by the Deutsche Akademische Austauschdienst (DAAD) within its project based Personnel Exchange Programme (PPP) between Germany and Poland and the Deutsche Forschungsgemeinschaft (SFB-TRR 31).

References

1. Dawson G.D. (1954) A summation technique for the detection of small evoked potentials. Electroencephalogr Clin Neurophysiol. 6:65–84
2. Hämäläinen M. et al. (1993) Magnetoencephalography – theory, instrumentation, and applications to noninvasive studies of the working human brain. Rev Mod Phys 65:413–497
3. Zacharias N. et al. (2010) Evoked magnetic fields recorded in different subjects, to different stimuli and at different sensors differ by scaling factors: implications for signal averaging. Submitted
4. Hari R. et al. (1982) Interstimulus interval dependence of the auditory vertex reponse and its magnetic counterpart: implications for their neural generation. Electroencephalogr Clin Neurophysiol 54:561–569
5. Sams M. et al. (1993) The human auditory sensory memory trace persists about 10 sec: neuromagnetic evidence. J Cog Neurosci 5.3:363–370
6. Sayers B. et al. (1974) The mechanism of auditory evoked EEG responses. Nature 247:481–483
7. Makeig S. et al. (2002) Dynamic brain sources of visual evoked responses. Science 295:690–694
8. Yeung N. et al. (2004) Detection of synchronized oscillations in the electroencephalogram: an evaluation of methods. Psychophysiology 41:822–832
9. Mäkinen V. et al. (2005) Auditory event-related responses are generated independently of ongoing brain activity. NeuroImage 24:961–968
10. Callaway III E. (1973) "Habituation of averaged evoked potentials in man." in Habituation vol. 2, H. V. S. Peeke and M. J. Herz, Eds. New York: Academic: pp.153–174
11. Brandt M.E. (1991) Pre-stimulus spectral EEG patterns and the visual evoked response. Electroencephalogr Clin Neurophysiol 80:16–20
12. Rahn E. and Başar E. (1993a) Prestimulus EEG-activity strongly influences the auditory evoked vertex response: a new method for selective averaging. Int J Neurosci 69:207–220

Norman Zacharias

Leibniz Institute for Neurobiology
Brenneckestraße 6
39118 Magdeburg
Germany
norman.zacharias@ifn-magdeburg.de

Quantification of the Time and Frequency Signatures of Visual Cortical Activation in the Developing Brain: A Study with MEG and Wave-Cross Spectrogram

Xinyao Guo, Jing Xiang, Yangmei Chen, Lu Meng, Xiaoshan Wang, and Yingying Wang

MEG Center, Department of Neurology, Cincinnati Children's Hospital Medical Center, Cincinnati, USA

Abstract— **Our previous studies have demonstrated that Morlet-wavelet transform with an extra large sigma value could precisely determine the frequency signatures of neuromagnetic signals. Unfortunately, the increase of frequency sensitivity is associated with a decrease of temporal resolution. To solve this problem, we have developed wave-cross analysis to quantify neuromagnetic signals with both high temporal and frequency resolutions. The objective of the present study is to measure the time and frequency signatures of visual cortical activation in children using this new method. Twelve healthy children were studied with a whole-head magnetoencephalography (MEG) system. Visual evoked magnetic fields (VEFs) were evoked with full-field pattern-reversal checks. MEG data were transformed from time-domain to frequency domain using wave-cross spectrogram. Neuromagnetic sources were volumetrically localized with a wavelet-based beamformer. Three response peaks were identified at 73±7 ms (M75), 111±8 ms (M100) and 149±12 ms (M145). The latency of M75 and M100 decreased with age. The amplitude of M75 decreased with age while the amplitude of M100 increased with age. The amplitude ratio of M100/M75 increased significantly with age. The frequency bands up to 300 Hz have been identified. The frequency in the M73 appeared higher than M145, the later responses. In addition, the frequency signatures of the neuromagnetic signals also changed with age. The temporal and frequency signatures of the development of visual function in childhood are noninvasively quantifiable with MEG and Wave-cross. Our preliminary data have confirmed that wave-cross, a new time-frequency analysis method, could precisely determine the frequency and temporal signatures of brain activation. The results lay a foundation for quantitative identification of developmental delay and/or abnormalities of visual function in children with brain disorders.**

Keywords— **Magnetoencephalography, visual cortex, children, Wave-cross spectrogram, visual evoked magnetic fields.**

I. Introduction

Recent advances have found that noninvasive biomarkers are critical to the study of functional brain maturation and developmental delay in children. Magnetoencephalography (MEG) is a new technology for detecting neuromagnetic fields associated with electric brain activities. It is generally accepted that MEG has both a high spatial and temporal resolution. Visual evoked magnetic fields (VEFs) in the visual cortex in response to a pattern reversal visual stimulus have revealed several detectable components. Since visual evoked responses correlate to myelination and synaptic transmission, VEFs are potential biomarkers for evaluation of visual function. The latency and amplitude of VEFs are commonly used to analyze the VEFs. However, the frequency signatures of VEFs are rarely used for evaluating visual function. Since brain activation changes with time, the conventional Fourier transform may not suitable for the study of the frequency signatures. Recent years, wavelet transform has been used in the study of brain function. Our previous studies have demonstrated that Morlet-wavelet transform with an extra large sigma value could precisely determine the frequency signatures of neuromagnetic signals. Unfortunately, the increase of frequency sensitivity is associated with a decrease of temporal resolution. To solve this problem, we have developed wave-cross analysis to quantify neuromagnetic signals with both high temporal and frequency resolutions. The objective of the present study is to measure the time, frequency as well as amplitude signatures of visual cortical activation in children using the new method.

II. Subjects and Methods

A. Subjects

Twelve healthy normal children (age: 6-17 years, mean age: 11; 6 female and 6 male) were studied with MEG. These children were recruited from the surrounding Cincinnati area. Since these children were developmentally normal and had no irregular findings reported from MRI, we considered that they were representative of the normal population. A written informed consent, at Cincinnati Children's Hospital Medical Center (CCHMC), was obtained from the parent/legal guardian of each child. Inclusion criteria for participation were: (1) healthy, without history of neurological disorder, psychiatric disease, or brain injury; (2) normal hearing, vision, and hand movement; (3) current age between 6 years – 17 years old. Exclusion criteria for participation: (1) a child could not keep still; (2) a child could not look at a fixing point or did too much eye blink; (2)

S. Supek and A. Sušac (Eds.): Advances in Biomagnetism – BIOMAG2010, IFMBE Proceedings 28, pp. 183–186, 2010.
www.springerlink.com

subjects with learning and /or speech/language disability; (3) subjects with unidentifiable magnetic noise.

B. Methods

Stimuli: Visual stimuli were generated by BrainX (Xiang et al., 2001), a program based on DirectX (version 9; Microsoft Inc., Redmond, WA, USA). The visual stimulus was a black and white pattern reversal checkerboard with a yellow dot fixated in the center. The visual stimuli were projected through a small hole in the magnetically shielded room (MSR) via an XGA portable multimedia projector (model PLC-XP41, Sanyo Electronics Ltd., Chatsworth, CA). Stimuli were projected onto a translucent screen located approximately 32 cm from the subject's nasion. The visual angle of the checkerboard (the visual field diameter) was 30 degree; the visual angle of each square (the check) within the checkerboard was 1 degree and 15 min. The luminance of the white areas was about 80 cd·m−2. Reversal rate between the white and black checker was 1 Hz.

MEG Recording: The supine position, his or her arms resting on either side, during the entire procedure. The sampling rate of the MEG recording was 6000 Hz per channel. The MEG signals were recorded in a magnetically shielded room (MSR) using a whole head CTF 275-Channel MEG system (VSM MedTech Systems Inc., Coquitlam, BC, Canada) in the MEG Center at CCHMC. Before data acquisition commenced, three electromagnetic coils were attached to the nasion, left and right pre-auricular points of each subject. These three coils were subsequently activated at different frequencies for measuring each subject's head position relative to the MEG sensors in real-time. The data was recorded with a noise cancellation of third order gradients and without on-line filtering. One hundred trials of MEG data were recorded for each subject. Subjects were asked to keep still. If the head movement during one recording was beyond 5 mm, the dataset would be indicated as bad and an additional trial would be recorded. The entire MEG recording was monitored with an audio-video system.

MRI Scan: Three-dimensional Magnetization-Prepared Rapid Acquisition Gradient Echo (MP_RAGE) sequences were obtained for all subjects with a 3T scanner (Siemens Medical Solutions, Malvern, PA). Three fiduciary points were placed in identical locations to the positions of the three coils used in the MEG recordings, with the aid of digital photographs, to allow for an accurate co-registration of the two data sets. All anatomical landmarks digitized in the MEG study were made identifiable in the magnetic resonance images (MRI).

Data Analyses: VEFs were investigated via DataEditor (CTF Systems Inc.), a program that enables visualization of MEG data. MEG data were averaged and DC-offset corrected with respect to the pre-stimulus baseline. All channels were overlapped together for identification of evoked responses. The waveform was then visually inspected for the VEF components and eye blinks and/or artifacts. If a trial had noticeable eye blinks and/or artifacts, it was marked as bad and excluded from the averaging or source localization. The latency and amplitude of each component were then measured with DataEditor. We took the peak latency and amplitude of each component using overlapped waveforms instead of a single channel because the relative position between the subject's head and the MEG sensor arrays might vary across subjects. Frequency analysis was done with Morlet continuous wavelet transform. The frequency window for wavelet transform was 1-2000 Hz. The time window for wavelet transform was 0-600 ms. Our analysis focused on neuromagnetic activation in 40-400 ms following visual stimulation. Since wavelet transform is sensitive to time but not frequency at high-frequency signals, a large sigma value (the number of waves) of Morlet continuous wavelet transform could enhance its sensitivity to frequency at high-frequency signals. On the contrary, wavelet transform is sensitive to frequency but not time in a low-frequency range. Therefore, a small sigma value of Morlet continuous wavelet transform could enhance its sensitivity to frequency at high-frequency signals. Building on this principle, we developed wave-cross method (see the following formula 1 and 2), which allowed us to set two sigma values.

$$\mathrm{Wc(t)} = \mathrm{W(t,s1)} \times \mathrm{W(t,s2)} \tag{1}$$

$$\mathrm{W(t,s)} = c_\sigma \pi^{-\frac{1}{4}} e^{-\frac{1}{2}t^2} \left(e^{i\sigma t} - \kappa_\sigma\right) \tag{2}$$

In the above formula, Wc(t) indicates the final results of wave-cross analysis based on two Morlet wavelet transforms: W(t,s1) and W(t,s2). Noteworthy, each wavelet transform has its own sigma value (s1 and s2). To produce one spectrogram for each time data set, a cross-product was computed for each time- frequency point with the coefficients corresponding to the two sigma values with the mother wavelet (2). If signals appeared in the giving sensitive time (a small sigma value) and sensitive frequency (a big sigma value) ranges, they would be enhanced. In the spectrogram, it looked like crossing one time-frequency point. Thus, the method was named as wave-cross analysis. Each trial of MEG data was transformed from time-domain waveform to frequency domain spectrogram individually with MEG Processor. Once all trials were transformed to spectral data; all trials were then accumulated one spectrogram.

III. Results

Three peaks were identified at 73±7 ms (M75), 111±8 ms (M100) and 149±12 ms (M145) (see Figure 1). The latency of M75 and M100 decreased with age (r1 = -0.62, r2 = -0.59, $p < 0.01$, respectively). The amplitude of M75 decreased with age ($r = -0.49$, $p < 0.05$) while the amplitude of M100 increased with age ($r = 0.63$, $p < 0.01$). The amplitude ratio of M100/M75 increased significantly with age ($r=0.75$, $p <0.005$). The spectrograms revealed focal increases of spectral power in 80-140 Hz in the visual cortex across all participants (12/12, 100%). Neuromagnetic signals in 150-300 Hz were identified in the visual cortex in the left and right hemispheres for 8 participants (8/12, 67%), however, it had inter-individual variation. Neuromagnetic signals up to 600 Hz were identifiable (see Figure 2). In comparison to low-frequency signals, high-frequency neuromagnetic signals were highly localized (Figure 3). The frequency in the M73 appeared higher than M145, the later responses. The data in Figure 1-3 are based on a single representative participant.

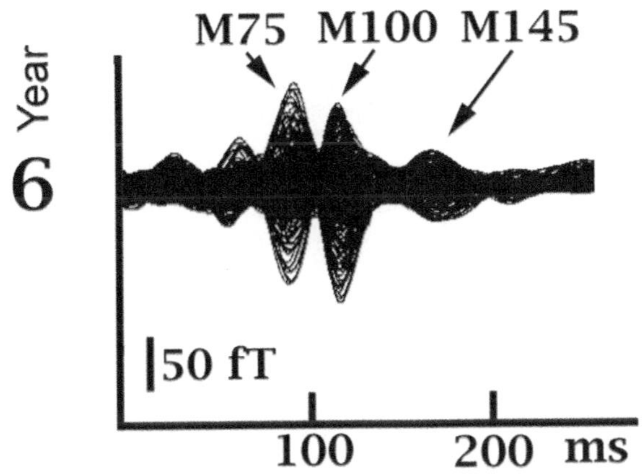

Fig. 1 Waveform from a representative participant showing the main neuromagnetic responses (deflections) of visual evoked magnetic fields in children. Note that three components, M75, M100 and M145 are clearly identifiable

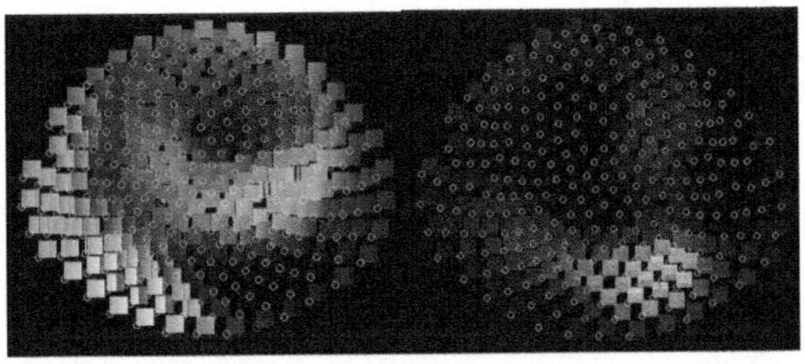

Fig. 2 Spectral contour maps based on wave-cross showing the neuromagnetic distribution of visual evoked brain activation. The left image indicate activation in a low-frequency range (1-40 Hz) revealed with a small sigma value (1.0) in wavelet transform. The right image in indicates activation in a high-frequency range (100-200 Hz) with a big sigma value (48.0). Noteworthy, low-frequency brain signals are diffused while high-frequency brain signals are highly localized

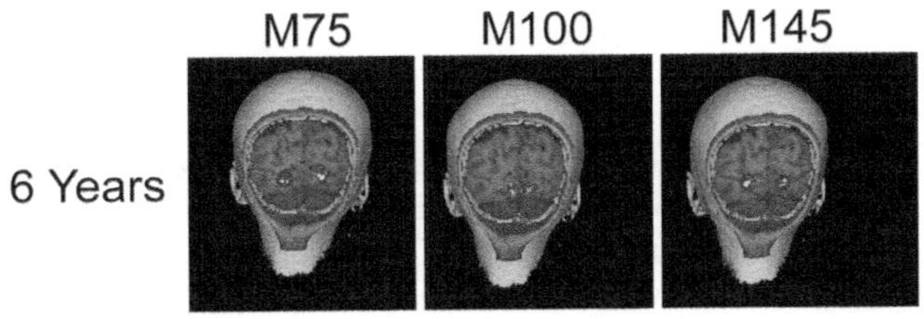

Fig. 3 Magnetic source imaging based on wave-cross analysis from a representative participant showing the spatial location and extent of the three main responses. The analysis frequency is in 150-300 Hz. The M75, M100 and M145 components are localized to subareas of the visual cortex. The locations of M75 and M145 appear far from the location of M100

IV. Discussion

Several advanced MEG data analysis methods including wavelet transform and accumulated spectrograms have been reported previously (1,2). The present study moved one step further by slightly modifying the Morlet wavelet transform to compute a new spectrogram with two sigma values. Those methods have been integrated into a systematic approach to analyze VEFs in the child's brain. We adopted accumulated spectrograms because the MEG data recorded from the brain commonly have very strong low-frequency signals. Since high-frequency neuromagnetic signals are much weaker than low-frequency signals, high-frequency signals are commonly obscured by low-frequency signals. In this study, we evaluated high frequency neuromagnetic signals in VEFs using Morlet wavelet transform based wave-cross analysis. According to our preliminary data, the temporal and frequency signatures of the development of visual function in childhood are noninvasively quantifiable with MEG and Wave-cross. Our preliminary data have confirmed that wave-cross, a new time-frequency analysis method, could precisely determine the frequency and temporal signatures of brain activation. The results lay a foundation for quantitative identification of developmental delay and/or abnormalities of visual function in children with brain disorders.

High-frequency brain signals were preliminary found in animal experiments [3] and invasive recordings in clinical patients [4]. As the development of high-sampling rate MEG/EEG systems, there is a rapid increasing list of publications on high-frequency signals (60 -890 Hz) in the somatosensory cortex in the healthy human brain (2). However, high-frequency brain signals in the visual cortex have rarely been studied. Our results indicate that the visual cortex generate high-frequency signals (> 60 Hz). This observation is strongly supported by a previous report (5). Funke and colleagues (5) have found that the cat geniculate geneartes signals up to 800 Hz. Bakhtazad and colleagues have noted that the perception of Kanizsa square is altered by lateral displacements of the lower inducers degrees. In parallel, high frequency components of cortical responses (20-100 Hz) gain in strength with misalignments. Gamma oscillations emerge at or about the psychometric threshold. Gamma oscillations appear in short bursts (approx. 50 ms) in the time window between 200 and 500 ms after stimulus onset. It seems that gamma oscillations are associated with image structures and may be induced by local properties of the target. Building on previous reports (5,6) and our data, we consider that high-frequency brain signals are a new window in the electrophysiological analysis of the visual cortex.

V. Conclusion

The results have demonstrated that the spatial, temporal and frequency signatures of function activation in occipital cortex can be quantitatively and noninvasively determined with of MEG and wave-cross analysis. The development of visual function is associated with neuromagnetic changes in terms of frequency and latency of VEFs in children.

Acknowledgment

This study was supported by a Trustee Grant from Cincinnati Children's Hospital Medical Center, Cincinnati, OH, USA.

References

1. Xiang J, Xiao Z (2008) Spatiotemporal and frequency signatures of noun and verb processing: A wavelet-based beamformer study. J Clin Exp Neuropsychol 31: 648-657.
2. Kotecha R, Xiang J, Wang Y, Huo X, Hemasilpin N, Fujiwara H et al. (2008) Time, frequency and volumetric differences of high-frequency neuromagnetic oscillation between left and right somatosensory cortices. Int J Psychophysiol 72(2):102-110.
3. Engel J Jr, Bragin A, Staba R, Mody I. (2009). High-frequency oscillations: what is normal and what is not? Epilepsia. 50(4):598-604.
4. Xiang J, Wang Y, Chen Y, Liu Y, Kotecha R, Huo X, Rose DF, et al. (2010). Noninvasive localization of epileptogenic zones with ictal high-frequency neuromagnetic signals. J Neurosurg Pediatr. 5(1):113-22.
5. Funke K, Kerscher N. (2000) High-frequency (300-800 Hz) components in cat geniculate (dLGN) early visual responses. J Physiol Paris. 94(5-6):411-25.
6. Bakhtazad L, Shumikhina S, Molotchnikoff S. (2003). Analysis of frequency components of cortical potentials evoked by progressive misalignment of Kanizsa square. Int J Psychophysiol. 50(3):189-203.

Author: Xinyao Guo
Institute: MEG Center, Department of Neurology, Cincinnati Children's Hospital Medical Center
Street: 3333 Burnet Avenue
City: Cincinnati
Country: USA
Email: xinyaoguo@gmail.com

Cluster-Based Algorithm for ROI Analysis and Cognitive State Decoding Using Single-Trial Source MEG Data

Gustavo Sudre[1], Yang Xu[2], Rob Kass[2], Doug J. Weber[3,4], and Wei Wang[3,4]

[1] Center for the Neural Basis of Cognition, Carnegie Mellon University, Pittsburgh, US
[2] Machine Learning Department, Carnegie Mellon University, Pittsburgh, US
[3] Department of Physical Medicine and Rehabilitation, University of Pittsburgh, Pittsburgh, US
[4] Department of Bioengineering, University of Pittsburgh, Pittsburgh, US

Abstract— **We present a cluster-based decoding algorithm for discovering regions of interest (ROIs) from EEG/MEG data in source space (or optimal cluster of sources) and predicting multiple conditions in a single experimental trial. Our algorithm automatically identifies contiguous brain regions that yield maximum mean test statistics from hypothesis tests over individual brain sources. We show that by utilizing these sources of interest it is possible to predict the experimental conditions via Bayesian decoding using single trials. Furthermore, we examine how these sources evolve spatially and temporally over the brain surface by running our algorithm at sequential time intervals. We demonstrate our method in a visually-guided wrist movement task where the goal is to infer the direction of wrist movement. MEG data were recorded using an Elekta Neuromag® device while subjects performed center-out wrist movements in 4 directions, and sources were estimated from raw data using the Minimum Norm Estimate method [1]. Our results demonstrate that the algorithm is a potential alternative to traditional ROI analysis using p-values, successfully discovers optimal clusters of sources relevant to the task, and achieves significant decoding accuracies in leave-one-out cross validation.**

Keywords— **clustering, single trial, ROI, magnetoencephalography, source-space decoding.**

I. Introduction

Identifying regions of interest (ROI) in neuroimaging data is of crucial importance in neuroscience [2]. Although one can use prior knowledge to narrow down the regions in the brain responsible for signals modulated by a given task, obtaining statistical significance from the data adds robustness and credibility to experimental results. The use of magnetoencephalography (MEG) offers superior temporal and spatial resolution for studying brain functions, and therefore it is important to develop a method to select areas in time and space that are relevant to the task.

The cluster-based decoding algorithm presented here discovers spatially contiguous brain sources of interest and uses these to infer information about multiple experimental conditions via Bayesian decoding. The algorithm was used to automatically find significant sources under a spatial contiguity constraint, avoiding the inclusion of spurious sources. In addition, a decoding scheme was developed which utilizes the discovered cluster of sources to predict different conditions from single trials. Improving MEG decoding from single-trials can have significant impact in multiple scenarios, such as using MEG for training to interact with brain-computer interfaces, and for providing neurofeedback to promote neuroplasticity [3]. The algorithm used in this study offers an alternative to traditional p-value methods for finding ROI in the context of decoding. Moreover, this algorithm can be applied repeatedly through time to track spatial and temporal evolution of sources of interest and be used to determine the optimal cluster for decoding.

II. Methods

A. Experimental Paradigm

All procedures were approved by the Institutional Review Board at the University of Pittsburgh, and all experiments were performed in accordance with the approved protocol. Two subjects performed a 4-target wrist center-out task while holding a MEG-compatible joystick with their right (dominant) hand. A trial was considered successful when the subject held the cursor position in the center for 1 second and then moved to one of the 4 targets (up, down, left, or right) within 1.5sec without overshooting, holding the cursor position at the target for another second (center and peripheral targets disappeared after the second had elapsed, indicating the start of the next movement). Subjects were instructed to keep eyes focused on a cross-hair in the center of the screen, and to perform movements solely with the wrist. MEG data were recorded at 1KHz for 120 successful repetitions per target for each subject using an Elekta Neuromag® scanner with 102 magnetometers and 204 planar gradiometers. Electromyography (EMG) of wrist flexor and extensor muscles was used to determine movement onset and electrooculography (EOG) was used to detect eye-movements. Both subjects also had structural MRI scans to aid in the source localization process.

S. Supek and A. Sušac (Eds.): Advances in Biomagnetism – BIOMAG2010, IFMBE Proceedings 28, pp. 187–190, 2010.
www.springerlink.com

B. Processing MEG Signals

The temporal extension of the Signal Space Separation method [4] was applied to the raw MEG signals after visual inspection for bad channels. The signal was then band-pass filtered between 1-40Hz and down-sampled to 125Hz. Freesurfer [5] was used to segment and create a 3-D model of the cortical surface from the MRI scans. MNE suite [6] was then used to co-register MRI and MEG data, and to later apply the Minimum Norm Estimate method [1] to the MEG signals. For classification purposes, MEG source-localized data were then parsed based on the onset of the peripheral targets. Trials containing eye-movements or eye-blinks (i.e. whenever the EOG channels passed a given threshold during a successful repetition) were discarded from further analysis.

C. Cluster-Based Decoding Algorithm

The idea behind a cluster-based approach is to search for a single "optimal" brain region that shows significantly differential responses under different conditions. For simplicity, we initially assume that there is a single ROI, namely that there is only one contiguous brain region that gives most distinguishable response to different conditions. We demonstrate in Section III, however, that our method can be easily generalized to discover multiple clusters.

Like traditional ROI-discovery methods [7], our algorithm starts with source-based hypothesis tests (e.g. ANOVA). The resulting test statistics are then thresholded such that the thresholding p-value is defined adaptively according to a predefined desired number of sources. The candidate sources that exceed the threshold are then grouped via a clustering algorithm. Specifically, the 3-D topology of the candidate sources are input to the clustering algorithm, which induces a constraint on spatial contiguity of sources in the resulting clusters (in other words, sources that are close to each other are grouped together). Bayesian hierarchical clustering (BHC) [8,9] was used for grouping the sources. BHC is an agglomerative clustering algorithm which greedily determines pairs of clusters to merge via a statistical hypothesis test and builds a binary tree bottom-up. It is shown to be superior than traditional clustering methods [8]. More importantly, BHC automatically infers the number of clusters, which is desirable towards the goal of autonomously separating spatially contiguous clusters. This also means that the algorithm can easily incorporate multiple clusters for the subsequent decoding analysis. Once the sources are clustered, the optimal cluster having the maximal test statistic averaged over its sources is chosen for decoding.

The decoding algorithm applies linear discriminant analysis (LDA) to reduce the dimensions of source signals in the discovered optimal cluster and then uses Bayesian decoding to infer different experimental conditions. Once an optimal cluster is chosen from the clustering method, the sources within that cluster are used for decoding. A linear discriminant analysis can be used to reduce the dimension if the number of sources is large. In this case, signals from these sources under multiple conditions are pooled together and input to the LDA algorithm. Finally, Bayesian decoding is used to predict conditions. Specifically, the dimension-reduced signals in repeated trials within the same condition are modeled via a multivariate Gaussian distribution, and the repeated trials constitute the training data which are used to compute the probability density. If C_i denotes a condition, where $i=1,\dots,M$, and the dimension-reduced source signals from the optimal cluster is x (e.g. a low-dimensional vector), then via Bayes rule it follows that

$$p(C_k|x) = \frac{p(x|C_k)p(C_k)}{\sum_{i=1}^{M} p(x|C_i)p(C_i)} \propto p(x|C_k)p(C_k) \qquad (1)$$

Here the likelihood $p(x|C_k)$ is a multivariate Gaussian and a prior distribution $p(C_k)$ can be defined over the conditions (e.g. discrete uniform distribution). To predict the most likely condition given an observed single-trial MEG source signal, it is similar to find condition k which yields the highest posterior, i.e. maximizing $p(C_k|x)$ over $k=1,\dots,M$.

III. Results

A. Synthetic Data

The algorithm was initially tested with a synthetic dataset to verify that it could select clusters in the correct brain location and time. A few representative repetitions of the synthetic signal are shown in Figure 1. Specifically, 4 different categories of signals were generated representing 4 distinct conditions, with 50 trials per condition, each of 600ms duration in 5ms steps. Posterior sources on the cortical surface were chosen first to show maximal modulation at approximately 150ms. Subsequently, medial sources were modulated at around 450ms, roughly simulating the onset of activity in the cortex during a visuomotor task. The source locations were limited to the left hemisphere. Modulation was characterized by a Gaussian distribution with mean at the time of maximal modulation, and standard deviation of 20ms. Then, the signal was multiplied by a different scalar to create the 4 different conditions. Finally, uniformly distributed random noise in the interval [0,1] was added to each time point of the signal in every trial.

Figure 1 shows that the algorithm successfully identifies clusters of sources modulated with respect to their condition. Before 100ms there is only random noise in the signal, so the accuracy is at chance level (~25%) as expected.

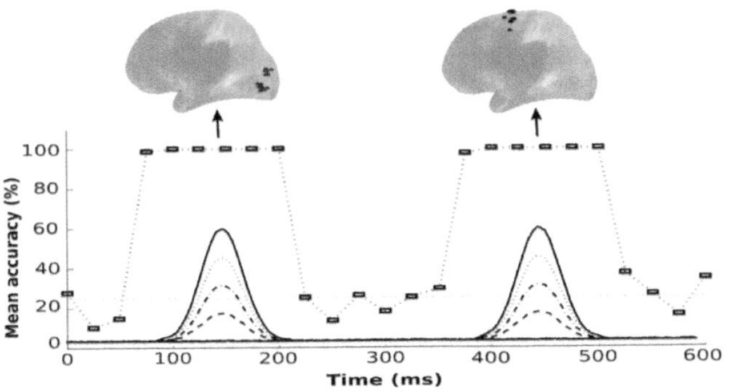

Fig. 1 Classification accuracies using synthetic data and the clusters detected and used by the algorithm. Percent accuracies measured out of 50 repetitions per condition. Gaussian-shaped traces (not in Y-scale with accuracy plot) show the synthetic signal for all sources (one trace per condition). Thin dotted line in accuracy plot shows chance level (25%), and markers represent the time windows when the predictions were made (approximately every 60ms). Note that the algorithm selects a cluster for each of the markers above, but only a couple representative snapshots were chosen for display

Then, sources in the posterior region start to be modulated until shortly after 200ms, and decoding results are perfect because of the high degree of modulation per target in each source. A similar event happens between 350 and 500ms, but this time the more rostral sources are modulated. In both time periods, the algorithm also identifies the correct cluster of sources responsible for the modulation. Note that the time specificity of the decoding results is slightly blurred by the filters used in the data, as well as by the size of the decoding window.

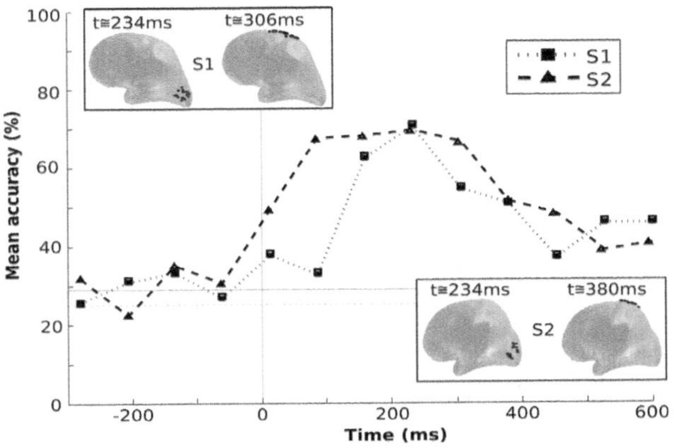

Fig. 2 Task decoding over time and the clusters the algorithm identified and used for decoding. Boxed insets correspond to brain surface plots for subject 1 (S1) and subject 2 (S2) at their peak decoding accuracy and at the time point when the cluster migrated to a motor region. Mean movement onset was 440ms for S1 and 430ms for S2. Vertical solid line corresponds to peripheral target onset, thin dotted line is chance accuracy, and solid horizontal line is the upper limit of its 95% confidence interval

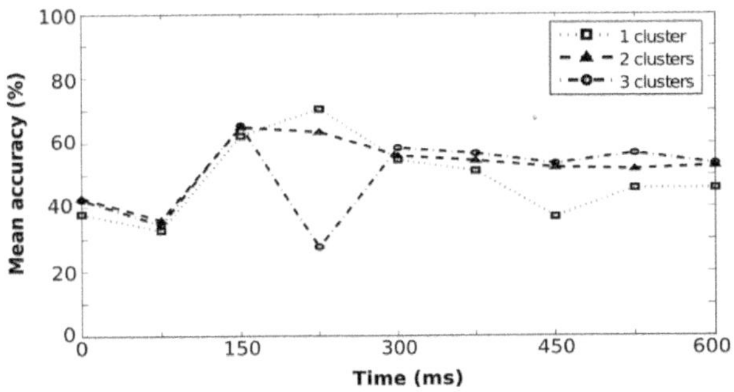

Fig. 3 Decoding accuracy results for subject 1 obtained by varying the maximal number of clusters selected by the algorithm

B. *Experimental Data*

We further applied our algorithm to the whole head MEG data obtained in the experiment described in Section IIa. After discarding trials contaminated by eye-movements, each subject had at least 100 successful repetitions per target. Figure 2 shows the results of classifying each repetition in a leave-one-out cross-validation framework. More specifically, at each time point one trial for an arbitrary direction is left-out for testing, and the rest of the trials are used for training. This process is iterated for each trial and target direction, such that the final prediction rate corresponds to the number of trials correctly predicted divided by the total number of trials.

Decoding accuracies start rising well above chance level (25%) at about 100ms for both subjects, reaching a peak of approximately 70% at around 230ms after peripheral target onset. The clustering algorithm identified sources in the visual cortex for this first peak in both subjects. The next spatially different cluster used for decoding occurred around 350ms, showing sources near motor cortex. All sources used for decoding significantly above chance level had p-values smaller than 10^{-6}. In addition, it is important to point out that the algorithm selected sources in the left hemisphere for all the points showing significant decoding accuracies. This is expected because both subjects performed the task with their right (dominant) hand.

The results in Figure 2 were obtained by limiting the number of clusters the algorithm identified for decoding to one. We then wanted to find whether increasing the number of clusters would also increase decoding accuracy. Figure 3 shows the decoding results from varying the number of clusters. It is clear that increasing the number of allowed clusters did not cause a significant increase in the decoding accuracy. In fact, the accuracies decreased for a few time points with the inflated number of clusters, which demonstrates that the algorithm originally selects the optimal cluster for decoding. Subject 2 showed similar results to the ones shown in Figure 3.

IV. Discussion

Although many studies have explored the possibility of decoding sensor signals in EEG/MEG [10,11], few have attempted to perform single-trial decoding in MEG source space. We have presented an algorithm that discovers ROI based on task-specific modulations. Our algorithm successfully identifies task-related sources at multiple instances in time, showing the evolution of the modulation from visual to motor cortical regions in a visuomotor task. Moreover, it infers the condition of single-trial MEG data with accuracy well-above chance level in a 4-target center-out paradigm.

Although the decoding result obtained here is similar to those in previous studies of decoding motor parameters in MEG sensor space [11], it is important to note the following differences. In addition to the different techniques used to filter the signals in these previous works, which have a direct effect on decoding results, it is still unclear if performing decoding in source space instead of sensor space would affect the decoding accuracy. Other parameters such as the sampling frequency used and the number of repetitions recorded per condition also have a direct impact in such results. Still, working in source space provides better spatial resolution for researchers interested in mapping brain function to cortical locations, which is more difficult when only analyzing sensor-space data. In the future, it would be desirable to compare the results presented here with those from sensor-space decoding. It is not trivial, however, to draw direct comparisons between source and sensor based approaches.

It is not surprising that increasing the number of clusters did not improve the decoding accuracies in the center-out task presented here. If the newly added clusters do not contain much further information about the task, it would be the same as adding noise to the classifier, which would affect the decoding results. However, whether a single cluster suffices to decode should be highly dependent on the task, hence another goal for future work is to consider how to automatically select the optimal number of clusters that would be suitable for decoding and ROI analysis.

Finally, while the algorithm has been presented here in the context of decoding, it can easily be generalized as a multivariate technique for neural data analysis by considering the different brain states being investigated (e.g. baseline and responses to stimuli) as classes to be decoded.

Acknowledgment

We thank the Center for Advanced Brain Magnetic Source Imaging (CABMSI, specifically Mrs. Anna Haridis and Dr. Anto Bagic) and Magnetic Resonance Research Center (MRRC), both at the University of Pittsburgh Medical Center, for providing the scanning time for MEG and MRI data collection. This work was partially supported by NSF Cooperative Agreement EEC-0540865, U.S. Army (TATRC) Agreement W81XWH-07-1-0716, and Grant Number 5 UL1 RR024153 from NCRR of NIH, and a special grant from the Office of the Senior Vice Chancellor for the Health Sciences at University of Pittsburgh, as well as a student travel fund from Center for Neural Basis of Cogni-tion. Additional funding support was provided by NIH grants from the NIBIB (1R01EB007749) and NINDS (1R21NS056136) for Douglas Weber. This project is also supported by NIH grants R01 MH064537 and R01EB005847 for Rob Kass and Yang Xu.

References

1. Hämäläinen, M.S., and Ilmoniemi, R.J. (1994). Interpreting magnetic fields of the brain: Minimum norm estimates. Medical & Biological Engineering & Computing, 32, p. 35-42.
2. Brett M, Johnsrude I, Owen A. (1996) The problem with functional localization in the human brain. Nature Reviews Neuroscience. 383(March):243-249.
3. Birbaumer N, Cohen LG. (2007) Brain-computer interfaces: communication and restoration of movement in paralysis. The Journal of physiology. 579:621-36.
4. Taulu S, Simola J. (2006) Spatiotemporal signal space separation method for rejecting nearby interference in MEG measurements. Physics in Medicine and Biology. 51(7):1759-68.
5. Freesurfer at http://surfer.nmr.mgh.harvard.edu/.
6. MNE at www.nmr.mgh.harvard.edu/martinos/userInfo/data/sofMNE.php
7. Friston KJ, Holmes AP, Worsley KJ, et al. (1994) Statistical parametric maps in functional imaging: A general linear approach. Human Brain Mapping. 2(4):189-210.
8. Heller, K. and Ghahramani, Z. (2005) Bayesian hierarchical clustering. ICML '05: Proceedings of the 22nd International Conference on Machine Learning, pp. 297-304.
9. Xu, Y., Heller, K. and Ghahramani, Z. (2009) Tree-based inference for Dirichlet process mixtures. Proceedings of the 12th International Conference on Artificial Intelligence and Statistics (AISTATS) 2009, 16-18 April 2009, Florida USA.
10. F Lotte, M Congedo, A Lécuyer, F Lamarche and B Arnaldi. (2007) A review of classification algorithms for EEG-based brain–computer interfaces. J. Neural Eng. 4. R1-R13.
11. Waldert S, Preissl H, Demandt E, Braun C, N. (2008) Hand movement direction decoded from MEG and EEG. Journal of Neuroscience. 28:1000-1008.

Author: Wei Wang

Institute: Department of Physical Medicine and Rehabilitation
University of Pittsburgh
Street: Keystone Building
3520 Fifth Av. Room 318
City: Pittsburgh, PA
Country: USA
Email: wangw4@upmc.edu

Decoding Subjective Simultaneity from Neuromagnetic Signals

Kohske Takahashi[1,2], Shohei Hidaka[3], and Katsumi Watanabe[1,4,5]

[1] Research Center for Advanced Science and Technology, The University of Tokyo, Tokyo, Japan
[2] Japan Society for the Promotion of Science, Tokyo, Japan
[3] Department of Psychological and Brain Sciences, Indiana University, USA
[4] Japan Science and Technology Agency, Saitama, Japan
[5] National Institute of Advanced Industrial Science and Technology, Ibaraki, Japan

Abstract— **The present study examined neural correlates of subjective simultaneity by using magnetoencephalography. Observers were asked to judge whether the visual and auditory stimuli occurred simultaneously. The subjective judgment for 90-ms-asynchronous stimuli showed trial-by-trial variation, and we successfully classified subjective simultaneity using neuromagnetic signals. We submitted raw MEG signals, a wavelet transform, and nonlinear dynamics to a naive Bayes classifier. In the case of raw signals and nonlinear dynamics, the classifier trained with the VA (where the visual stimulus was given first) or AV (where the visual stimulus was given second) data could predict the subjective simultaneity of the other VA (or AV) data at a rate better than chance. The classification rate using nonlinear dynamics was comparable to that using raw signals, despite the fact that the dimension was considerably low (101 vs. 88,000 dimensions). In the case of the wavelet transform, the classifier trained with the VA data was able to decode the AV data, and vice versa. These results suggest that (1) we can decode subjective simultaneity using MEG signals, (2) nonlinear dynamics may encode simultaneity specific to the order of the audiovisual inputs, (3) the time-frequency characteristics of neural activity may predict subjective simultaneity independently of the physical order of the audiovisual inputs, and (4) the neural activity (time-frequency characteristics) reflecting subjective simultaneity may share a common mechanism among different sensory modalities.**

Keywords— **magnetoencephalography (MEG), decoding, time perception, subjective simultaneity.**

I. Introduction

The temporal interval between two events is a key aspect of integrating multisensory inputs. A smaller interval leads to a feeling of subjective simultaneity. Several studies have found the neural activity relevant to subjective time perception [1] through the use of electroencephalograms (EEG) and magnetoencephalograms (MEG). Most of these studies examined neural correlation of biased simultaneity (or temporal order) by presenting additional stimuli.

However, time perception in humans is inherently variable; even if the temporal intervals of two events are kept constant, observers show trial-by-trial variation with regard to subjective simultaneity. The goal of this study is to examine whether neuromagnetic signals can represent the trial-by-trial variation of subjective simultaneity, and if so, what aspect of neuromagnetic signals (e.g., time-frequency or nonlinear dynamics) is responsible for subjective simultaneity. For this purpose, we applied a decoding approach [2]. If the classifier successfully predicts trial-by-trial subjective simultaneity in cases where the stimuli are completely identical, this implies that subjective simultaneity is encoded in the neural activity, independent of the stimulus.

Observers performed an audiovisual simultaneity judgment task in an MEG scanner. We presented visual and auditory stimuli with small temporal asynchrony. We first tested whether the event-related magnetic fields (ERF) in "simultaneous" and "non-simultaneous" responses would differ in peak amplitudes and latencies [3, 4]. Further, we performed a classification analysis using a wavelet transform and the nonlinear dynamics of neuromagnetic signals to examine whether subjective simultaneity can be decoded.

II. Methods

Neuromagnetic responses were measured using a 160-channel whole-scalp MEG system (Yokogawa PQ1160C) in a magnetically shielded room. The magnetic signals were low-pass filtered at 500 Hz, digitized at 1000 Hz, and stored for off-line analysis. Visual stimuli were presented on a screen in the MEG scanner using a high luminance LCD projector, and auditory stimuli were presented to the right ear using a magnetic-free earphone. The visual stimulus was a white square presented in the center of the screen for 50 ms, and the auditory stimulus was a 50 ms beep sound.

Eight observers performed an audiovisual simultaneity judgment task in the MEG scanner. In each trial, one visual stimulus and one auditory stimulus were presented in temporal asynchrony. The asynchrony was randomly chosen from five stimulus onset asynchronies (SOA; 0, ±90, ±180 ms). The observers were asked to report whether the auditory and visual stimuli occurred simultaneously, by pressing the left or right button. The 0-ms- and ±180-ms SOAs were repeated 60 times and the ±90-ms-SOAs were repeated 120 times.

S. Supek and A. Sušac (Eds.): Advances in Biomagnetism – BIOMAG2010, IFMBE Proceedings 28, pp. 191–194, 2010.
www.springerlink.com

III. Results and Discussions

A. Behavioral Data

Table 1 Percentages of "simultaneous" judgments for each SOA

SOA (ms)	–180	–90	0	+90	+180
% Simultaneous	26.9	58.8	85.6	65.6	26.5

Table 1 shows the percentage of "simultaneous" responses that were prompted for each SOA. The observers judged that most of the ±180-ms-SOA trials were "non-simultaneous" and most of the 0-ms-SOA trials were "simultaneous." The subjective simultaneity judgments for the ±90-ms-SOA were highly variable. In the following MEG analyses, we used a VA condition (the visual stimulus was presented 90 ms before the auditory stimulus) and an AV condition (the visual stimulus was presented 90 ms after the auditory stimulus). We submitted all trails to the analyses without artifact rejection or correction. We confirmed that the temporal interval of visual and auditory stimuli did not differ between "simultaneous" and "non-simultaneous" trials by directly measuring the change of luminance and sound pressure (90.5 ms vs. 89.9 ms, $p = .15$), which indicates that the variability of subjective response is not due to the temporal jitter of visual and auditory stimulus presentation.

B. Event Related Magnetic Fields (ERF)

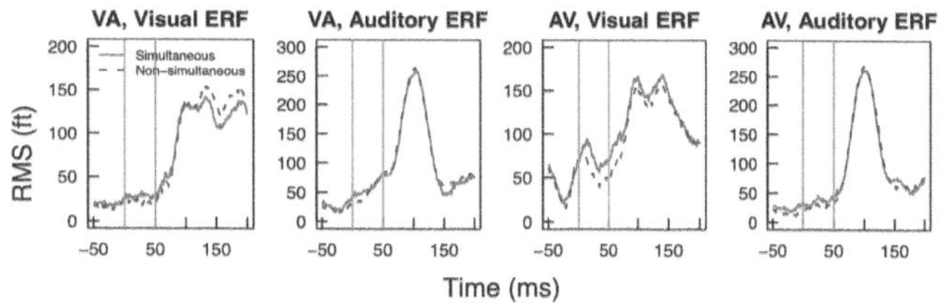

Fig. 1 RMS of ERFs in ±90-ms-SOA conditions. The red and blue lines indicate the ERF in "simultaneous" and "non-simultaneous" responses, respectively. The green lines indicate the onset and offset of the stimuli (a flash for the visual ERF and a sound for the auditory ERF)

The latency or amplitude of ERFs may represent subjective simultaneity ([3, 4]). To test the difference between ERFs, we picked six visual and six auditory channels that were strongly responsive to the stimuli on the basis of RMS value of each channel. Figure 1 shows the ERFs at visual and auditory channels (the mean of the RMS of the six channels). Visual inspection showed that both visual and auditory ERFs appeared approximately 90 ms after stimulus onset. In the quantitative comparison between the ERF in the trials of "simultaneous" responses and those of "non-simultaneous" responses, we did not find any significant difference in the latencies and amplitudes (Table 2, all $ps > .10$). Thus, the ERF did not represent subjective simultaneity in the present task. The peak amplitudes [3] and the latencies [4] of ERF are known to correlate with subjective temporal order of two events. The absence of neural correlation here may imply that the neural mechanisms underlying simultaneity and temporal order judgments are different.

Table 2 Latencies and amplitudes of the peaks of ERF. The S and NS indicate "simultaneous" and "non-simultaneous" responses, respectively

Condition		VA		AV	
Channel		Visual	Auditory	Visual	Auditory
Latency (ms)	S	122	100	112	102
	NS	113	101	104	99
Amplitude (fT)	S	170.6	278.6	192.1	269.5
	NS	171.2	282.5	203.9	275.9

C. Classification Analyses

Treating neuromagnetic responses as high-dimensional signals, we tried to "decode" subjective simultaneity. As part of the decoding approach [2], we trained a naive Bayes classifier [5] using the neuromagnetic signals from some of the trials (the training dataset), and performed a validation test using the other trials (the test dataset), to assess whether the classifier successfully discriminates between "simultaneous" and "non-simultaneous" classes. If the discrimination rate was above chance (0.5 in the case of two classes), this would imply that the classifier could "decode" subjective simultaneity from neuromagnetic signals. We performed a separate classification analysis for each observer.

We conducted four types of cross-validation test: first, where both training and test datasets were a VA condition (VA→VA); second, where both datasets were an AV condition (AV→AV); third, where the training dataset was a VA condition and the test dataset was an AV condition (VA→AV); and fourth, where the training dataset was an AV condition and the test dataset was a VA condition (AV→VA). Note that VA→VA and AV→AV were within-condition validation, and VA→AV and AV→VA were between-condition validation. We randomly chose the training and test trials without overlap, and the numbers of "simultaneous" and "non-simultaneous" trials were made equal in order to avoid the bias caused by the different prior probabilities. We determined the number of training and test trials on the basis of the number of "simultaneous" responses and "non-simultaneous" responses of each observer. We repeated the training and validation test more than 100 times with different random sampling.

a) Decoding Using Raw Neuromagnetic Signals

Table 3 Percentages of correct classification of subjective simultaneity. * significantly ($p < 0.05$) higher than chance (0.5) in the validation test

	VA→VA	AV→AV	VA→AV	AV→VA
Raw signal				
Training	90.4	91.9	90.0	91.7
Test	54.4 *	58.7 *	52.2	50.2
Wavelet transform				
Training	80.2	86.4	79.1	85.6
Test	51.1	52.9	54.6 *	52.7 *
Nonlinear dynamics				
Training	87.3	89.5	87.3	89.5
Test	52.2 *	56.7 *	51.8	51.8

We submitted the raw neuromagnetic signals of all channels and samples (ranging from –50 to 499 ms after the onset of the stimulus at 1 ms intervals) to the classifier (i.e., 160 channels × 550 samples = 88,000 dimensions). Table 3 shows the correct classification rates. The classifier worked well for the training dataset; the correct rates were significantly high but not perfect, indicating that over-fitting did not take place. For the test dataset of within-condition validation (VA→VA, AV→AV), the classifier was able to predict subjective simultaneity at a rate better than chance. In contrast, the same classifier failed to predict subjective simultaneity in between-condition validation; that is, the classifier that was trained with the VA data could not decode subjective simultaneity of the AV data, and vice versa.

b) Decoding Using Time-Frequency Characteristics

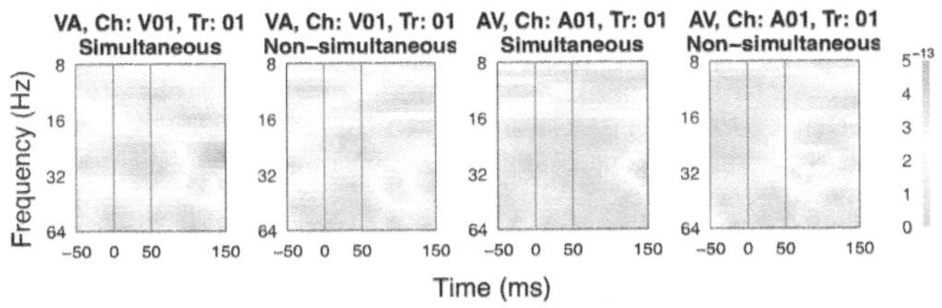

Fig. 2 Representative scalograms of the wavelet transform. The green lines indicate the onset and offset of the stimulus (a flash for the VA and a sound for the AV condition)

Neural oscillations are known to be relevant to a wide range of cognitive functions [6], including time perception (e.g., gamma-band oscillation [1]). Therefore, we tried to decode subjective simultaneity using the time-frequency characteristics of neuromagnetic signals. We performed the Morlet wavelet transform on the raw neuromagnetic signals. We chose a different set of six channels for the VA and AV conditions, on the basis of the strength of the ERF. The sampled time range was –50-149 ms with a 1 ms interval, after the onset of the stimulus. The range of frequency was 24 bins of 8-64 Hz. Figure 2 shows the representative scalogram of the wavelet transform. We then submitted the power of time-frequency (the modulus of the wavelet transform) to the classifier (6 channels × 200 samples × 24 frequencies = 28,800 dimensions). The classifier worked well for the training dataset, without over-fitting (Table 3). In the cross-validation test, in contrast to the raw neuromagnetic signals, the classifier could predict subjective simultaneity in between-condition validation (VA→AV and AV→VA), Thus, the time-frequency characteristics of neuromagnetic signals predict subjective simultaneity independently of the physical order of the audiovisual inputs.

The fact that the classifier worked in between-condition validation was surprising, since the order of audiovisual inputs was reversed between conditions and we picked a different set of channels for the VA and AV conditions. Although the reason why the classifier failed to pass the within-condition validation is unclear, these results still imply that subjective simultaneity can be decoded, independent of the physical order of audiovisual inputs. The classifier could predict subjective simultaneity using the signals of sensory-specific channels and only the early response (<150 ms) to the first stimulus. Hence, the time-frequency characteristics of the sensory area related to the first stimulus, independent of the sensory modality, may be relevant to subjective simultaneity that is stochastically variable. Thus, the neural activity reflecting subjective simultaneity may share a common mechanism among different sensory modalities.

c) Decoding Using Nonlinear Dynamics

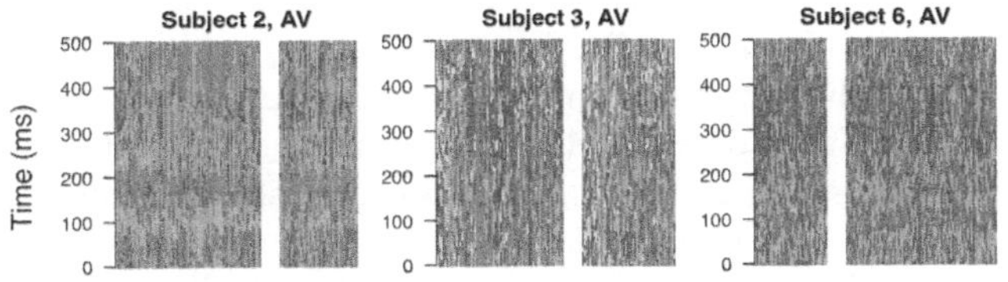

Fig. 3 Representative symbol sequence estimated based on the nonlinear dynamics of neuromagnetic signals. The vertical axis indicates the time from the onset of the stimulus. Each column shows a symbol sequence for each trial. Each color indicates the estimated symbol. The trials of "non-simultaneity" response are shown on the left panel and those of "simultaneous" response are on the right panel

The nonlinear dynamics of neuromagnetic signals may include crucial information that can be used to discriminate observers' subjective state [7]. To test this possibility, we estimated a generating partition on neuromagnetic signals, wherein the continuous time series was discretized into a given number of symbols. We sampled 101 points (ranging from 0 to 500 ms after the stimulus onset at 5 ms intervals) from the raw signals of 160 channels. The 160 dimensions of the channels were reduced to 70-90 dimensions (different for each observer) using a principal component analysis (PCA). We then estimated the generating partition (10 symbols or less) on the reduced matrix (101 × 70-90 dimensions) using the Symbol False Nearest Neighbor (SFNN) method [8, 9]. Thus, we obtained the estimated symbol sequence of 101 dimensions for each trial (Figure 3) Finally, we submitted the symbol sequence (101 dimensions) to the classifier. The results showed that the classification rates were comparable to those using the raw signals (Table 3): in within-condition test, the classifier could predict subjective simultaneity at a rate better than chance. Note that the dimensions of data submitted to the classifier were 88,000 in the raw signal and only 101 in the nonlinear dynamics.

d) Comparison among Decoding Methods

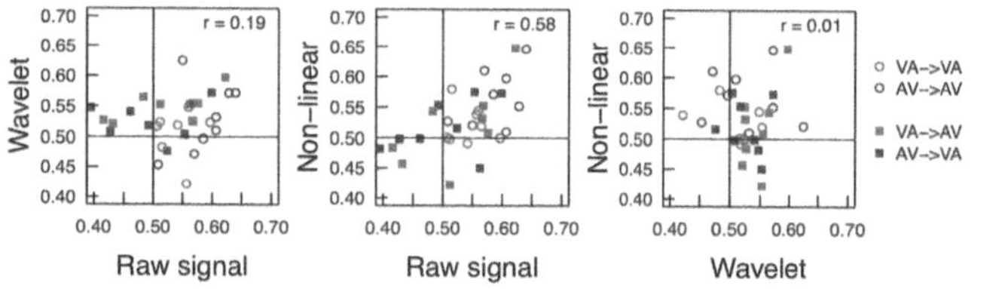

Fig. 4 Within-subject correlation among the three types of decoding methods. Each point represents the individuals' data. Red and blue symbols indicate that the training dataset were VA and AV conditions, respectively. Open circles and filled squares indicate within-condition and between-condition classification, respectively

As we submitted different features to the classifier, we analyzed within-subject correlation among three decoding methods (Figure 4). We found significant positive correlation in the classification rate between the raw signals and the nonlinear decoding (Figure 4, middle panel). Together with the comparable classification rate and the failure of between-condition validation, these two methods focus on the similar features of neuromagnetic signals. We concluded that the low-dimensional nonlinear dynamics would encode simultaneity specific to the order of the audiovisual inputs.

On the other hand, the classification rate in the wavelet transform correlated to neither the raw signals nor the nonlinear dynamics. In addition, the classifier with the wavelet transform passed the between-condition validation, i.e., it was insensitive to the order of the visual and auditory stimuli. Therefore, it seems clear that the wavelet method emphasizes different features from the other methods. Although the sample in wavelet method included only a small number of channels showing ERF, and also small time range (<150 ms) in comparison with the other two methods, we suggest that extracting time-frequency characteristics (e.g., gamma-band oscillation [1, 6]) enables to pass the between-condition validation. Further research that applies the wavelet transform to all channels and that uses a long time range will clarify those features that are crucial for decoding subjective simultaneity insensitive to the stimulus order.

IV. Conclusion

The naive Bayes classifier successfully decoded subjective simultaneity independent of the stimulus condition, using high-dimensional neuromagnetic signals. The dissociation of the results of decoding using nonlinear dynamics and time-frequency characteristics may imply that different types of neural dynamics are relevant to subjective simultaneity. This study will provide a useful foundation for further research aimed at revealing the underling neural mechanisms of subjective time perception.

Acknowledgment

This research was supported by the Hokuriku Innovation Cluster for Health Science (MEXT, Japan), Japan Society for the Promotion of Science, and Japan Science and Technology Agency.

References

1. Senkowski D, Talsma D, Grigutsch M et al. (2007) Good times for multisensory integration: Effects of the precision of temporal synchrony as revealed by gamma-band oscillations. Neuropsychologia 45:561-571
2. Haynes J, Rees G (2006) Decoding mental states from brain activity in humans. Nat Rev Neurosci 7:523-534
3. McDonald JJ, Teder-Sälejärvi WA, Di Russo F et al. (2005) Neural basis of auditory-induced shifts in visual time-order perception. Nat Neurosci 8:1197-1202
4. Vibell J, Klinge C, Zampini M et al. (2007) Temporal order is coded temporally in the brain: early event-related potential latency shifts underlying prior entry in a cross-modal temporal order judgment task. J Cogn Neurosci 19:109-120
5. Zhang H (2004) The optimality of naive bayes. In Barr V, Markov Z (eds.) FLAIRS Conference, AAAI Press
6. Senkowski D, Talsma D, Herrmann CS et al. (2005) Multisensory processing and oscillatory gamma responses: effects of spatial selective attention. Exp Brain Res 166:411-426
7. Pereda E, Quiroga RQ, Bhattacharya J (2005) Nonlinear multivariate analysis of neurophysiological signals. Prog Neurobiol 77:1-37
8. Kennel MB, Buhl M (2003) Estimating good discrete partitions from observed data: symbolic false nearest neighbors. Phys Rev Lett 91: 084102
9. Buhl M, Kennel MB (2005) Statistically relaxing to generating partitions for observed time-series data. Phys Rev E Stat Nonlin Soft Matter Phys 71:046213

Methods for Determination of Functional Connectivity in Brain

Katarzyna J. Blinowska

University of Warsaw/Department of Biomedical Physics, Warsaw, Poland

Abstract— **The methods applied for estimation of functional connectivity are presented with special emphasis on the estimators of directedness such as Directed Transfer Function and Partial directed Coherence. Time-varying estimator of directedness Short-time Directed Transfer Function is described.**

Keywords— **functional connectivity, multivariate autoregressive model, Directed Transfer Function, Partial Directed Coherence, Granger causality.**

I. INTRODUCTION

The problem of the determination of the functional connectivity in brain have been for a long time in the center of attention of neurophysiologists, since it is crucial for understanding of information processing mechanisms in brain. In the last years the increasing interest in the determination of directional connectivity is observed. It may be connected with the tremendous increase of the experimental data gathered by the electrophysiological and fMRI techniques. The understanding of mental processes requires not only the information on localization but also on mutual relations between the activated structures. In this context the directed measures of connectivity showing the direction of information transfer are of special interest and the impact of this paper will be concentrated on them.

Here, the development of the methods of estimation of functional connectivity will be presented. First, the bivariate methods will be mentioned and their limitation will be pointed out. Next, the multivariate methods, in particular the methods based on the extension of Granger causality principle to the multichannel systems will be described. Finally, the methods for determination of dynamical fast changing transmission of brain activity will be presented.

II. BIVARIATE MEASURES OF CONNECTIVITY

A. Correlation and Coherence

For a long time the standard methods of establishing relations between signals have been cross-correlation and coherence (the formulas can be found in any textbook on signal analysis e.g. [1]).

The amplitude of the correlation describes the similarity of the signals and, if for certain delay τ function R_{xy} has maximum, we can assume that signal Y is delayed by τ in respect to signal X.

Cross-spectrum is a counterpart of cross-correlation in the frequency domain; it can be expressed as a Fourier transform of cross-correlation. Both functions are therefore statistically equivalent, however they emphasize different features of signals cross-correlation—in time, cross-spectrum in frequency domain. Coherence is a normalized form of cross-spectrum.

From the phase of coherence the delay between frequency components of the signals may be calculated, hence we may judge about the direction of propagation. However we must remember that the phase is determined modulo 2π, therefore the direction may be sometimes misjudged.

Another problem concerning correlation and coherence between two signals is connected with the fact that we don't know if the dependence between the signals does not come from the influence of the third channel which drives them. Bivariate coherence patterns are usually very dense, showing multiple connections and are therefore not very informative. In case of bivariate estimators of direction the situation is even worse, since the propagation is detected always when there is a delay between the signals. Obtained patterns of connectivity are very complicated and often misleading, even the reversal of propagation may be obtained [2], [3].

B. Non-linear Measures

Several non-linear measures of connectivity such as: non-linear correlation, mutual information, transfer entropy, general synchronization, phase synchronization have been introduced (review may be found in [4]). These measures are bivariate, so they suffer from the properties common to all bivariate estimators. The non-linear estimators are very sensitive to noise, and additionally these which rely on the embedding theorem require long stationary data. On the other hand, it has been demonstrated that linear estimators of connectivity such as e.g. Directed Transfer Function (DTF) are very robust to noise and perform quite well even for non-linear signals [5], therefore the non-linear approaches are hardly recommended, especially that e.g. in case of EEG signals non-linear character is rather an exception.

S. Supek and A. Sušac (Eds.): Advances in Biomagnetism – BIOMAG2010, IFMBE Proceedings 28, pp. 195–198, 2010.
www.springerlink.com

III. Granger Causality and Granger Causality Index

The testable definition of causality was introduced in the field of economics by Granger [6]. Granger causality principle states that if some series $Y(t)$ contains information in past terms that helps in the prediction of series $X(t)$, then $Y(t)$ is said to cause $X(t)$. More specifically, if we try to predict a value of $X(t)$ using p previous values of the series X only, we get a prediction error e:

$$X(t) = \sum_{i=1}^{p} A'_{11}(j)X(t-j) + e(t) \tag{1}$$

If we try to predict a value of $X(t)$ using p previous values of the series X and p previous values of Y we get another prediction error e_1:

$$X(t) = \sum_{j=1}^{p} A_{11}(j)X(t-j) + \sum_{j=1}^{p} A_{12}(j)Y(t-j) + e_1(t) \tag{2}$$

If the variance of e_1 (after including series Y to the prediction) is lower than the variance of e we say that Y causes X in the sense of Granger causality. Similarly we can say that X causes Y in the sense of Granger causality when the variance of e_2 is reduced after including series X in the prediction of series Y:

$$Y(t) = \sum_{j=1}^{p} A_{22}(j)Y(t-j) + \sum_{j=1}^{p} A_{21}(j)X(t-j) + e_2 \tag{3}$$

Granger causality index is based directly on the definition of causality, namely it shows, if the information contributed by second channel improves the prediction of first channel. The logarithm ratio of residual variances for one and two-channel models is computed:

$$\mathrm{GCI}_{1\to 2} = \ln(e/e_1) \tag{4}$$

This definition may be extended to the multichannel case by considering how the inclusion of given channel changes the residual variance ratios. GCI is an estimator working in the time domain. In many applications the estimators dependent on frequency are more appropriate. Such estimators derived from Granger causality concept, but operating on multichannel data are: Directed Transfer Function and Partial Directed coherence.

IV. Multivariate Estimators of Directedness

A. Multivariate Autoregressive Model

Granger causality was defined for two channels, however Granger in his later work [7] pointed out that the causality principle holds only, if there are no other channels influencing the process. To account for the whole multivariate structure of a process of k channels the multichannel autoregressive model (MVAR) has to be considered.

For a multivariate k-channel process $\mathbf{X}(t)$:

$$\mathbf{X}(t) = (X_1(t), X_2(t), \ldots, X_k(t))^{\mathrm{T}} \tag{5}$$

The model takes the form

$$\mathbf{X}(t) = \sum_{j=1}^{p} \mathbf{A}(j)\mathbf{X}(t-j) + \mathbf{E}(t) \tag{6}$$

where $\mathbf{E}(t)$ are vectors of size k and the coefficients $\mathbf{A}$ are $k{\times}k$-sized matrices.

Equation (4) can be easily transformed to describe relations in the frequency domain. After changing the sign of $\mathbf{A}$ and application of Z transform we get:

$$\begin{aligned} \mathbf{E}(f) &= \mathbf{A}(f)\mathbf{X}(f) \\ \mathbf{X}(f) &= \mathbf{A}^{-1}(f)\mathbf{E}(f) = \mathbf{H}(f)\mathbf{E}(f) \\ \mathbf{H}(f) &= \left(\sum_{m=0}^{p} \mathbf{A}(m)\exp(-2\pi i m f \Delta t)\right)^{-1} \end{aligned} \tag{7}$$

From the form of the above equations we see that the model can be considered as a linear filter with white noises $\mathbf{E}(f)$ on its input and the signals $\mathbf{X}(f)$ on its output. The matrix of filter coefficients $\mathbf{H}(f)$ is called the transfer matrix of the system. It contains information about all relations between data channels in the given set including the phase relations between signals. From transfer matrix cross-spectra and partial coherences can be found. Partial coherence is given by the formula:

$$C_{ij}(f) = \frac{\mathbf{M}_{ij}(f)}{\sqrt{\mathbf{M}_{ii}(f)\mathbf{M}_{jj}(f)}}, \tag{8}$$

where $\mathbf{M}_{ij}$ is a minor of spectral matrix (matrix of spectra and cross-spectra) with the i-th row and j-th column removed. Partial coherence is different from zero only when the given relation between channels is direct. If a signal in a given channel can be explained by a linear combination of some other signals of the set, the partial coherence between them will be low.

B. Directed Transfer Function

Basing on the properties of transfer function of MVAR Directed Transfer Function (DTF) was introduced in 1991 by Kaminski and Blinowska [8] in the form:

$$\mathrm{DTF}^2_{j\to i}(f) = \frac{|H_{ij}(f)|^2}{\sum_{m=1}^{k} |H_{im}(f)|^2} \tag{9}$$

DTF describes causal influence of channel j on channel i at frequency f. The above equation defines a normalized version of DTF, which takes values from 0 to 1 producing a ratio between the inflow from channel j to channel i to all the inflows to channel i.

The non-normalized DTF which is directly related to the coupling strength [9] is defined as:

$$\theta_{ij}^2(f) = \left|H_{ij}(f)\right|^2 \tag{10}$$

DTF is a very robust estimator in respect to noise. It was demonstrated [8] that it is able to detect correct propagation in case when noise was several times higher than the signal. Directed Transfer Function found many applications e.g.: for localization of epileptic foci [10], for estimation of EEG propagation in different sleep stages and wakefulness [11], for determination of transmission between brain structures of an animal during a behavioral test [12], for estimation of cortical connectivity [13], [14] and many others.

DTF shows not only direct, but also cascade flows, namely in case of propagation 1→2→3 it shows also propagation 1→3. In order to distinguish direct from indirect flows direct Directed Transfer Function (dDTF) was introduced [15].

The dDTF is defined as a multiplication of a modified DTF by partial coherence. The modification of DTF concerned normalization of the function in such a way as to make the denominator independent of frequency. The dDTF ($\chi_{ij}(f)$) showing direct propagation from channel j to i is defined as:

$$\chi_{ij}^2(f) = F_{ij}^2(f) C_{ij}^2(f)$$
$$F_{ij}^2(f) = \frac{\left|H_{ij}(f)\right|^2}{\sum_f \sum_{m=1}^{k} \left|H_{im}(f)\right|^2} \tag{11}$$

Where $C_{ij}(f)$ is partial coherence. $\chi_{ij}(f)$ has a nonzero value when both functions $F_{ij}^2(f)$ and $C_{ij}^2(f)$ are nonzero, in that case there exists a direct causal relation between channels $j \rightarrow i$.

Distinguishing direct from indirect transmission is essential in case of signals from implanted electrodes, for EEG signals recorded by scalp electrodes it is not really important.

C. Partial Directed Coherence

The partial directed coherence (PDC) was defined by Baccala and Sameshima in 2001 [16] in the following form:

$$P_{ij}(f) = \frac{A_{ij}(f)}{\sqrt{\mathbf{a}_j^*(f)\mathbf{a}_j(f)}} \tag{12}$$

In the above equation $A_{ij}(f)$ is an element of $\mathbf{A}(f)$ – a Fourier transform of MVAR model coefficients $\mathbf{A}(t)$, where $\mathbf{a}_j(f)$ is j-th column of $\mathbf{A}(f)$ and the asterisk denotes the transpose and complex conjugate operation. Although it is a function operating in the frequency domain, the dependence of $\mathbf{A}(f)$ on the frequency has not a direct correspondence to the power spectrum. From normalization condition it follows that PDC takes values from the interval [0,1]. PDC shows only direct flows between channels. Unlike DTF PDC is normalized to show a ratio between the outflow from channel j to channel i to all the outflows from the source channel j so it emphasizes rather the sinks, not the sources.

In neurophysiological applications rather sources, not the sinks are of primary interest, therefore later on the estimator called Generalized Partial Directed Coherence (GPDC) was proposed [17], where normalization factor in the denominator similar to the one applied in DTF was introduced. GPDC is given by the formula:

$$\text{GPDC}_{j \rightarrow i}(f) = \frac{A_{ij}(f)}{\sum_{i=1}^{k} \left|A_{ij}(f)\right|^2} \tag{13}$$

DTF and PDC first found the application in analysis of EEG or ECoG. More recently MVAR starts to be applied also to the fMRI signals [18], [19]. However in case fMRI signals, because of poor time resolution, only main connections acting during whole task may be identified. The dynamical fast transmissions involved in the information processing may be identified by means of Short-time Directed Transfer Function (SDTF).

V. Estimators of Dynamical Propagation

In order to grasp the dynamical changes of propagation we have to apply very short measurement window, however it deteriorates strongly the statistical properties of the estimate. In case when multiple repetitions of experiment are available, it is possible to apply ensemble averaging over realizations. We divide a nonstationary recording into shorter time windows, short enough to treat the data within a window as quasi-stationary. The estimation of MVAR coefficients is based on calculation of the correlation matrix R_{ij} of k signals X_i from multivariate set [9]. We calculate the correlation matrix between channels for each trial separately. The resulting model coefficients are based on the correlation matrix averaged over trials. The correlation matrix has a form:

$$\widetilde{R}_{ij}(s) = \frac{1}{N_T}\sum_{r=1}^{N_T} R_{ij}^{(r)}(s) = \frac{1}{N_T}\sum_{r=1}^{N_T}\frac{1}{N_S}\sum_{t=1}^{N_S} X_i^{(r)}(t) X_j^{(r)}(t+s) \tag{14}$$

The averaging concerns correlation matrices for short data windows; the data are not averaged in the process. The choice of the window size is always a compromise between quality of the fit and time resolution. In the case of parametric model the number of data points kN_T has to be bigger (preferably by order of magnitude) than the number of parameters, which in case of MVAR is equal to k^2p (p model order).

SDTF was used to find the dynamic propagation in case of performance of movement task and its imagination [19], [20] and for evaluation of transmissions during Continuous Attention Test [21], [22]. The results in the form of movies are presented at http://brain.fuw.edu.pl/~kjbli [23], [24]. SDTF may be applied as well for the spike trains [25].

Short-time direct Directed Transfer Function was used in the analysis of ECoG activity during word repetition [25]. In the quoted paper the method of the evaluation of the significance of results by means of semiparametric method is given, which solves the problem of multiple repetition effect present in case of time-frequency distributions. For testing the hypothesis of directedness surrogate data tests are usually applied. Good approach is the use of surrogates, which preserve the spectral properties of signals randomizing only the phases.

Acknowledgment

The author acknowledges the support of the Polish Ministry of Science and Higher Education (Decision no.119/N-COST/2008/0) and NEUROMATH Action.

References

1. Marple SL Jr. (1987) Digital Spectral Analysis with Applications, Prentice-Hall, San Diego
2. Blinowska KJ, Kuś R, Kamiński M, (2004) Granger causality and information flow in multivariate processes, Phys Rev E, 70:050902 (2004); Virt J Biol Phys Res, 8(11)
3. Kuś R, Kamiński M, Blinowska KJ, (2004) Determination of EEG activity propagation: pair-wise versus multichannel estimate, IEEE Trans Biomed Eng, 51, 1501-10
4. Pereda E, Quiroga R Q ,Bhattacharya J (2005) J.,Nonlinear multivariate analysis of neurophysical signals. Progress in Neurobiol.77:1-37
5. Winterhalder M, Schelter B, Hesse W, et al. (2005) Comparison of linear signal procesing techniques to infer directed interactions in multivariate neural systems. Signal Processing 85:2137-2160
6. Granger CWJ, (1969) Investigating causal relations in by econometric models and cross-spectral methods, Econometrica, 37: 424-38
7. Granger CWJ , (1980) Testing for causality: a personal viewpoint, J Econ Dyn Control 2: 329-352
8. Kamiński M, Blinowska K.J (1991) A new method of the description of the information flow in brain structures, Biol Cybern, 65: 203-210
9. Kaminski M, Ding M, Truccolo W, Bressler S (2001) Evaluating causal relations in neural systems: Granger causality, directed transfer function and statistical assessment of significance. Biol Cybern 85: 145-157
10. Franaszczuk PJ, Bergey GJ, Kamiński M (1994) Analysis of mesial temporal seizure onset and propagation using the directed transfer function method, Electroenceph Clin Neurophys, 91:413-27
11. Kamiński M, Blinowska KJ, Szelenberger W (1997) Topographic analysis of coherence and propagation of EEG activity during sleep and wakefulness, Electroenceph Clin Neurophys, 102: 216-27
12. Korzeniewska A, Kasicki S, Kamiński M, Blinowska KJ (1997) Information flow between hippocampus and related structures during various types of rat's behavior, J Neurosci Meth, 73: 49-60
13. Astolfi L, Cincotti F, Mattia D, Babiloni C, et al. (2005) Assessing cortical functional connectivity by linear inverse estimation and directed transfer function: simulations and application to real data. Clin Neurophys, 116(4): 920-932
14. Babiloni F, Cincotti F, Babiloni C, et al. (2005) Estimation of the cortical functional connectivity with the multimodal integration of high-resolution EEG and fMRI data by directed transfer function. Neuroimage, 24(1): 118-131
15. Korzeniewska A, Mańczak M, Kamiński M, et al., (2003) Determination of information flow direction between brain structures by a modified Directed Transfer Function method (dDTF), J Neurosci Meth, 125: 195-207
16. Baccala LA, Sameshima K (2001) Partial directed coherence:A new conceptin neural structure determination, Biol. Cybern. 84:463-474
17. Baccala LA, Takahashi DY, Sameshima K (2007) Generalized Partial Directed Coherence. Proc.of 15th Conf DSP2007, IEEE1:162-166
18. Sato JR, Takahashi DY, Arcuri SM, et al. (2009) Frequency domain connectivity identification:an application of partial directed coherence to fMRI 30:452-461
19. Ginter J Jr, Blinowska KJ, Kaminski M, Durka PJ, (2001) Phase and amplitude analysis in time-frequency space - application to voluntary finger movement. J Neurosci Meth 110: 113-124
20. Kus R, Ginter J. Jr, Blinowska KJ (2006) Propagation of EEG activity during finger movement and its imagination. Acta Neurobiolog Exp 66(3): 195-206
21. Kus R., Blinowska K. J., Kaminski M., Basińska-Starzycka A. (2008) Transmission of information during Continuous Attention Test, Acta Neurobiologiae Experimentalis, 68, 103-112
22. Blinowska KJ, Kuś R, Kaminski M, Janiszewska J (2010) Brain Topohraphy, in print
23. http:// brain.fuw.edu.pl/%7Ekjbli/DTF_MOV.html
24. http://brain.fuw.edu.pl/~kjbli/CAT_mov.html
25. Kocsis B, Kaminski M (2006) Dynamic changes in the direction of the theta rhythmic drive between Supramammillary Nucleus and the Septohippocampal system. Hippocampus16:531-540
26. Korzeniewska A, Crainiceanu C, Kus R, Franaszczuk PJ, Crone NE (2008) Dynamics of event-related causality (ERC) in brain electrical activity. Human Brain Mapping 29:1170-1192

Author: K.J. Blinowska
Institute: Department of Biomedical Physics
Street: Hoza 69
City: Warszawa
Country: Poland
Email: kjbli@fuw.edu.pl

Estimating Directions of Information Flow between Cortical Activities Using Phase-Slope Index

Kensuke Sekihara[1], Julia Owen[2], Hagai Attias[3], David Wipf[2], and Srikantan S. Nagarajan[2]

[1] Department of Systems Design and Engineering, Tokyo Metropolitan University, Tokyo, Japan
[2] Biomagnetic Imaging Laboratory, University of California, San Francisco, San Francisco, USA
[3] Golden Metallic Inc., San Francisco, USA

Abstract— **This paper reports our preliminary experiments on analyzing the direction of information flow between cortical activities. Our analysis method combines the Champagne source reconstruction algorithm[1], with the phase slope index (PSI) analysis, which was proposed by Nolte *et al.*[2]. We applied the source PSI analysis to MEG data measured with a hand grasping. The Champagne algorithm reconstructed source time courses for the VOI, and the PSI was then computed between all voxel locations and a seed point set at ipsilateral primary motor area. Our PSI analysis shows that the contralateral primary motor area has a large negative PSI value, indicating a good agreement with our physiological knowledge that the contralateral primary motor area is first activated and the ipsilateral primary motor area is activated later. Our analysis also shows that several areas in cerebellum have fairly large positive PSI values, suggesting that the information flows from the contralateral primary motor area to cerebellum via the ipsilateral primary motor area. Consistent results were obtained from two data sets: the one measured with subject's left-hand grasping and the other with subject's right-hand grasping.**

Keywords— **functional connectivity, information flow, flow direction, cross-spectral matrix, phase-slope index.**

I. Introduction

This paper presents results of our preliminary experiments on estimating the directions of information flow between cortical activities. Our analysis combines recently proposed source reconstruction method, referred to as the Champagne algorithm[1], with the analysis of the phase slope index (PSI) proposed by Nolte *et al.*[2]. The idea behind the PSI is that the phase of the cross spectrum between two different source activities depends on the time needed for the information to flow between those areas. Thus the derivative of the phase with respect to the frequency, called the "phase slope", is proportional to the time needed for the information flow. Therefore, the sign of the phase slope indicates the direction of the information flow.

We applied the source PSI analysis to MEG data recorded during a hand grasping experiment. We computed the PSI between all voxel locations and a seed point set in ipsilateral primary motor area. The results of our source PSI analysis show that the contralateral primary motor area has a large negative PSI value, indicating a good agreement with our physiological knowledge that the contralateral primary motor area is first activated and the ipsilateral primary motor area is activated later. Our analysis also shows that several areas in the cerebellum have fairly large positive PSI values, suggesting that the information flows from the contralateral primary motor area to cerebellum via the ipsilateral primary motor area. In our preliminary experiments, consistent results were obtained from two data sets: the one measured with the subject's left-hand grasping and the other with the subject's right-hand grasping.

II. Source PSI analysis

A. Champagne Source Reconstruction Algorithm

Let us denote the number of sensors M and the number of voxels N. We denote an $M \times 3N$ composite lead field matrix over all voxel locations as L_V and a $3N \times 1$ augmented source activity vector as $s(t)$. The champagne algorithm assumes dual-condition measurements in which the task (target) measurement $b(t)$ and the control (baseline) measurement $b_c(t)$ are available. Here, $b(t)$ and $b_c(t)$ can be modeled as,

$$b(t) = L_V s(t) + b_I(t) + v(t), \tag{1}$$

$$b_c(t) = b_I(t) + v(t), \tag{2}$$

where $b_I(t)$ is interference magnetic fields and $v(t)$ is the sensor noise. The algorithm assumes that the interference $b_I(t)$ is represented by a small number of independent factor activities, so the control measurement is expressed as

$$b_c(t) = Bu(t) + v(t), \tag{3}$$

where $u(t)$ is a $d \times 1$ factor vector and d is the number of factors. B is an $M \times d$ factor mixing matrix.

S. Supek and A. Sušac (Eds.): Advances in Biomagnetism – BIOMAG2010, IFMBE Proceedings 28, pp. 199–202, 2010.
www.springerlink.com

We assume that $u(t)$ and $v(t)$ are Gaussian,.i.e., omitting the explicit time notation, we have

$$p(u) = \mathcal{N}(u|0,I), \quad (4)$$
$$p(v) = \mathcal{N}(v|0,\Lambda), \quad (5)$$

where Λ is a diagonal precision matrix of v, and the precision matrix of the factor is set equal to the identity matrix I. Under these assumptions, the Champagne algorithm first estimates B and Λ from the control period using a Variational Bayesian Factor Analysis method[3]. Then, using B and Λ, the algorithm estimates the source activity $s(t)$ from the task measurement $b(t)$. Here, we again apply the Gaussian prior on $s(t)$, i.e.,

$$p(s) = \mathcal{N}(s|0,\Phi), \quad (6)$$

where Φ is the block-diagonal precision matrix in which the diagonal elements are 3×3 matrices that represent the prior assumption of the source activity at all voxel locations. To estimate s, we combine Eqs (1) and (3), and derive,

$$b = [L_V, B] \begin{bmatrix} s \\ u \end{bmatrix} + v = Fx + v, \quad (7)$$

where $F = [L_V, B]$ and $x = [s, u]^T$.

Thus, the estimation of x is formulated using the following linear Gaussian model,

$$\text{prior}: \quad p(x) = \mathcal{N}(x|0,\widetilde{\Phi}), \quad (8)$$
$$\text{likelihood}: \quad p(b|x) = \mathcal{N}(b|Fx,\Lambda), \quad (9)$$

where

$$\widetilde{\Phi} = \begin{bmatrix} \Phi & 0 \\ 0 & I \end{bmatrix}. \quad (10)$$

It is well-known that, under this Gaussian model, the minimum-mean squared error estimate of x, $\widehat{x}$, is obtained as

$$\widehat{x} = \Gamma^{-1} F^T \Lambda b, \quad (11)$$

where

$$\Gamma = F^T \Lambda F + \widetilde{\Phi}. \quad (12)$$

The update rule for $\widetilde{\Phi}$ is derived by maximizing the data likelihood:

$$\log p(b|\widetilde{\Phi}) = \frac{1}{2}\left(\log|\widetilde{\Phi}| - \log|\Gamma| - \widehat{x}\widetilde{\Phi}\widehat{x}^T\right) + \mathscr{C}, \quad (13)$$

where $\mathscr{C}$ represents all variables that do not contain $\widetilde{\Phi}$. By computing $\partial \log p(b|\widetilde{\Phi})/\partial\widetilde{\Phi}$ and setting this derivative to zero, we have

$$\widetilde{\Phi}^{-1} - \Gamma^{-1} + Q = 0, \quad (14)$$

where $Q = \widehat{x}\widehat{x}^T$. We can derive $\widetilde{\Phi}^{\text{new}}$ that satisfies the above equation, such that

$$\left(\widetilde{\Phi}^{\text{new}}\right)^{-1/2} = S^{-1/2}\left(S^{1/2} Q S^{1/2}\right)^{1/2} S^{-1/2}, \quad (15)$$

where

$$S = \widetilde{\Phi}^{1/2} \Gamma^{-1} F^T \Lambda F \widetilde{\Phi}^{-1/2}. \quad (16)$$

Further details of the Champagne algorithm is presented in [1].

B. Phase-Slope Index

Let the Fourier transform of the reconstructed time course at the jth voxel be $\sigma_j(f)$. The cross spectrum between $\sigma_j(f)$ and $\sigma_k(f)$ is defined as $C_{j,k}(f)$, such that

$$C_{j,k}(f) = \langle \sigma_j(f) \sigma_k^*(f) \rangle. \quad (17)$$

When the information propagation from the ith voxel to the jth voxel takes time τ, the phase of $C_{j,k}(f)$ is proportional to $2\pi\tau f$. That is, we have the relationship,

$$\frac{\partial}{\partial f} \mathscr{P}\{C_{j,k}(f)\} \sim \tau, \quad (18)$$

where $\mathscr{P}\{A\}$ indicates the phase of the complex quantity A. The derivative of the phase of $C_{j,k}(f)$ with respect to f is referred to as the phase slope. According to Eq. (18), if the phase slope has a positive value, the information flows from the kth voxel to the jth voxel locations, and if the phase slope has a negative value, the information flows from the jth voxel to the kth voxel. The phase slope can be assessed by using the phase-slope index (PSI), which is defined as

$$\Psi_{j,k} = \Im\left[\sum_f C_{j,k}(f + \delta f) C_{j,k}^*(f)\right], \quad (19)$$

where $\Im[A]$ indicates the imaginary part of the complex quantity A.

III. Application to hand-motor MEG data

The effectiveness of the source PSI metric described in this paper was tested using two motor-evoked MEG data sets. The measurement was performed using the 160-channel MEGVISION ™ whole-cortex biomagnetometer (Yokogawa Electric Inc., Tokyo, Japan). We collected two sets of data: the first data-set with subject's self-paced right-hand grasping and the second data-set with the same subject's self-paced left-hand grasping. The average interval between the grasps was approximately 15 sec. Total 72 trials were acquired for

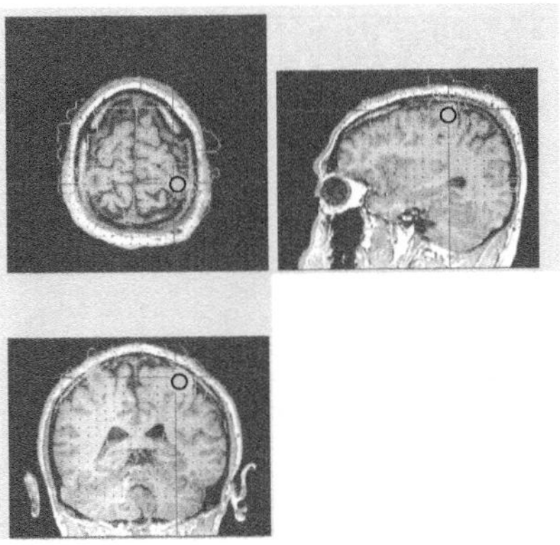

Fig. 1 The results of the source reconstruction for the left-hand grasping data. A small circle shows the seed point at ipsilateral primary motor area. In the axial and coronal slices of subject's MRI, the right-hand side shows the left hemisphere and the left-hand side shows the right hemisphere

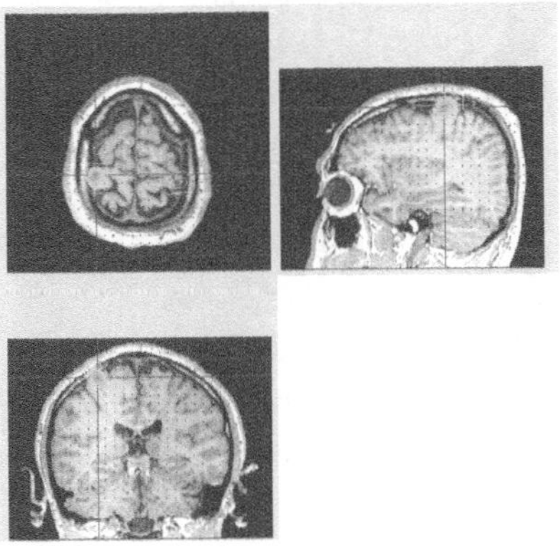

Fig. 2 Results of mapping negative PSI values for the left-hand grasping data. Here, the positive PSI values were set to zero, and contours represent relative magnitude of negative PSI values

the right-hand grasping and 67 trials for the left-hand grasping. All the analysis described below was applied to the non-averaged trials. The onset of the hand movement was measured using an optical device, and it is defined as the time-origin. The time-frequency representations of the sensor data showed that significantly strong beta band (18–30 Hz) activity exists around 2–3 seconds after the onset of the movements for the both data sets. Thus, we set a time window for the task data to a period between 2.1 sec and 3.1 sec, and a time window for the control data to a period between -2.9 sec and -1.9 sec. The (non-averaged) sensor data was band-pass filtered with the bandwidth of the beta band, i.e., the bandwidth between 18 and 30 Hz.

The Champagne algorithm was first applied to the left-hand grasping data. Here, the band-pass filtered data in each (task or control) window was concatenated over trials to create an extended (task or control) data set. The results of the source reconstruction are shown in Fig. 1. Here, the source power distribution $\langle \|s(t)\|^2 \rangle$ was overlaid onto the subject MRI. Throughout this paper we use the radiology convention, i.e., in the axial and coronal slices of subject's MRI, the right-hand side shows the left hemisphere and the left-hand side shows the right hemisphere. The results in Fig. 1 show fairly strong bilateral activation of primary motor cortices (M1), By setting the seed point at the ipsi-M1 as shown in this figure, we computed PSI values between the seed and all voxel locations.

We first set the positive PSI values to be zero, and map the magnitude of negative PSI values. The negative PSI mapping results are shown in Fig. 2. In these results, the contralateral motor area is clearly detected, and the results suggest that there is an information flow from the contra-M1 to the ipsi-M1. These results agree with our physiological knowledge.

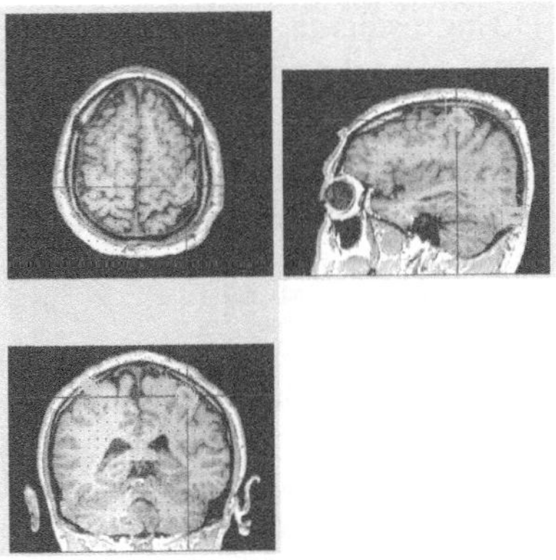

(a)

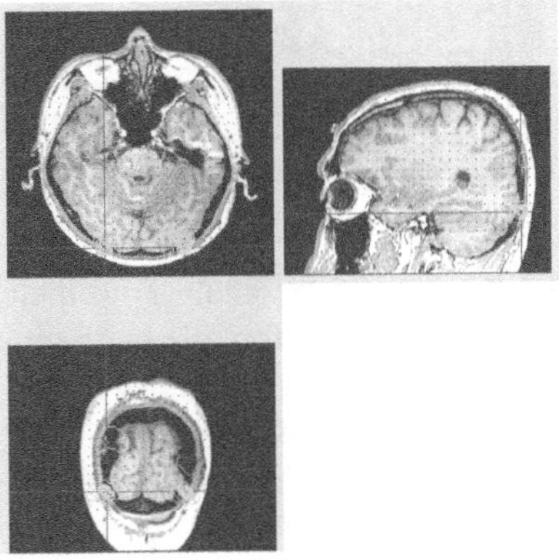

(b)

Fig. 3 Results of mapping positive PSI values for the left-hand grasping data. Here, the negative PSI values were set to zero, and contours represent relative magnitude of positive PSI values

The results of mapping positive PSI, in which the negative PSI values were set to zero, are shown in Fig. 3. The results

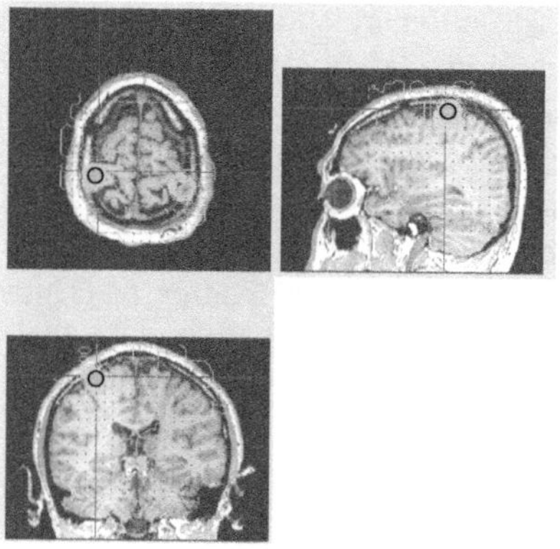

Fig. 4 The results of the source reconstruction for the right-hand grasping data. A small circle shows the seed point at ipsilateral primary motor area

in Fig. 3(a) show the activation at the seed point. The positive PSI mapping also detects bilateral activation near cerebellum in Fig. 3(b). The results in Figs. 2 and 3 suggest that the information flows from the contra-M1 to cerebellum*.

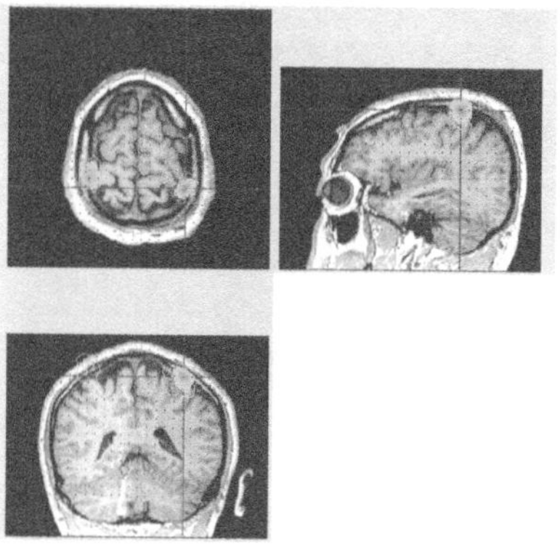

Fig. 5 Results of mapping negative PSI values for the right-hand grasping data. Here, the positive PSI values were set to zero, and contours represent relative magnitude of negative PSI values

We applied the source PSI analysis to the data collected with the right-hand grasping from the same subject. We again set a seed point at the ipsi-M1 (as shown in Fig. 4), and computed the PSI values at all voxel locations with respect to the seed location. The negative PSI image is shown in Fig. 5, and the positive PSI image is shown in Fig. 6. The negative PSI mapping again shows the contra-M1, suggesting that the information flows from the contra-M1 to ipsi-M1. The positive PSI mapping shows the activation at the seed location, as well as unilateral activations near cerebellum. Thus, the results in Figs. 5 and 6 again suggest that the information flows from the contra-M1 to cerebellum.

*Note that the PSI cannot tell that this information flow is via ipsi-M1.

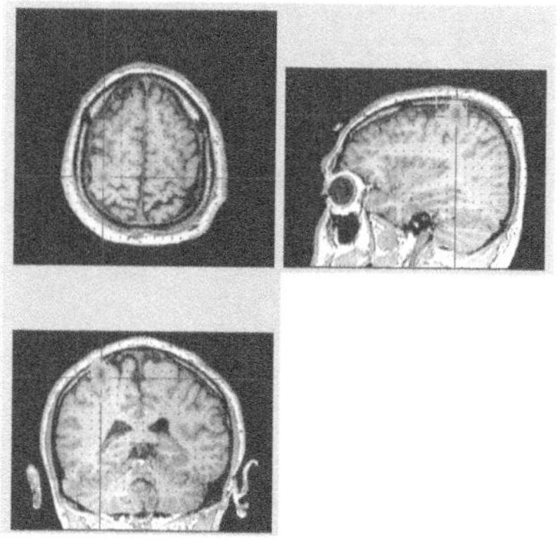

(a)

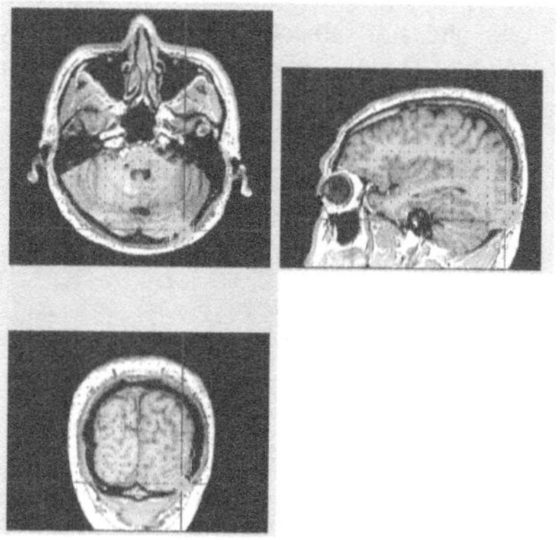

(b)

Fig. 6 Results of mapping positive PSI values for the right-hand grasping data. Here, the negative PSI values were set to zero, and contours represent relative magnitude of positive PSI values

Acknowledgement

We would like to thank Hiroaki Tanaka, and Eiichi Okumura for providing hand-grasping MEG data.

References

1. Wipf D. P., Owen J. P., Attias H. T., Sekihara K., Nagarajan S. S.. Robust Bayesian estimation of the location, orientation, and time course of multiple correlated neural sources using MEG *NeuroImage.* 2010;49:641-655.
2. Nolte G., Ziehe A., Nikulin V. V., et al. Robustly estimating the flow direction of information in complex physical systems *Phys. Rev. Lett.*. 2008;100:234101-4.
3. Nagarajan S. S., Attias H., 2nd K. E. Hild, Sekihara K.. A probabilistic algorithm for robust interference suppression in bioelectromagnetic sensor data *Stat. Med.*. 2007;20:3886-3910.

Estimating Causality Measures from Reconstructed Source Time Courses When Large Background Activities Exist

Kensuke Sekihara, Julia Owen, Hagai Attias, and Srikantan S. Nagarajan

[1] Department of Systems Design and Engineering, Tokyo Metropolitan University, Tokyo, Japan
[2] Biomagnetic Imaging Laboratory, University of California, San Francisco, San Francisco, USA
[3] Golden Metallic Inc., San Francisco, USA

Abstract— **This paper proposes to use state-space modeling to incorporate the observational noise in the MVAR formulation. The difference of the proposed method from recently-reported investigations in the same direction[1, 2] is that the proposed method allows the noise to have a temporal structure, and thus the method can handle the background interference (brain noise) as the observational noise. The proposed method first estimates the MVAR coefficients of the brain noise from the control period, and then constructs the state-space model using these coefficients. The expectation maximization (EM) algorithm estimates the MVAR coefficients of the brain signals. The estimated MVAR coefficients are then used to compute Granger-causality-based measures such as the partial directed coherence and the directed transfer function. The results of numerical experiments that validate the method's effectiveness are presented.**

Keywords— **causality measure, Granger causality, autoregressive process, partial directed coherence, state-space modeling.**

I. Introduction

Analyzing a causal relationship among cortical source activities using Granger causality has gained great interest. Such analyses require computation of the multivariate vector autoregressive (MVAR) coefficients of the source time courses. One major source of errors here is the noise contained in the reconstructed source time courses. This paper proposes to use the state-space modeling to incorporate the observational noise in the MVAR formulation. The difference of the proposed method from recently-reported investigations in the same direction[1, 2] is that the method allows the noise to have a temporal structure, and the method can handle the background interference (brain noise) as the observational noise. A pre-requisite of the proposed method is that a control (baseline) period is available, and the method first estimates the MVAR coefficients of the brain noise from the control period, and then constructs the state-space model using these coefficients to derive the MVAR coefficients of the target source time courses. Using the proposed method, we can compute Granger-causality-based measures even when large background interference exists. The results of numerical experiments that validate the method's effectiveness are presented.

II. Method

A. Assumptions

It is assumed that total q signal sources of interest exist, and we would like to analyze the causal relationship among these sources. Here, q is less than the number of sensors, M. The time courses of these sources are denoted $s_j(t)$, where $j = 1,\ldots,q$. We define the source time course vector $s(t)$ such that $s(t) = [s_1(t),\ldots,s_q(t)]^T$ where the superscript T indicates the matrix transpose. We assume that $s(t)$ is represented by the MVAR process, such that

$$s(t) = \sum_{j=1}^{p} A(j)s(t-j) + e, \tag{1}$$

where time t is expressed using a unit-less value, $A(j)$ is the coefficient matrix, p is the model order, and e is the residual vector that obeys the white Gaussian process.

The composite lead-field matrix for the q sources are denoted L_S:

$$L_S = [l_1,\ldots,l_q], \tag{2}$$

where l_j is the lead field vector for the jth-source location and orientation. Then, the measured biomagnetic field $b(t)$ is expressed as

$$b(t) = L_S s(t) + b_I(t) + n, \tag{3}$$

where $b_I(t)$ represents the magnetic field generated from interference sources such as the background activity. In the above equation, n represents the sensor noise. In this paper, we assume that the control (baseline) measurement, $b_c(t)$:

$$b_c(t) = b_I(t) + n, \tag{4}$$

is available.

S. Supek and A. Sušac (Eds.): Advances in Biomagnetism – BIOMAG2010, IFMBE Proceedings 28, pp. 203–206, 2010.
www.springerlink.com

We also assume that the locations and the orientations of the q sources of interest are known, i.e., the composite lead field matrix L_S is known. Then, using the pseudo inverse of L_S, we can try to estimate $s(t)$ such that

$$y(t) = L_S^+ b(t) = L_S^+ L_S s(t) + L_S^+ b_I(t) + L_S^+ n, \tag{5}$$

where $y(t)$ is the source time course estimate and $L_S^+ = (L_S^T L_S)^{-1} L_S^T$. Let us define $u(t)$ such that $u(t) = L_S^+ b_I(t)$, which represents the influence of the interference on the estimate of the source time courses. We also denote the sensor noise influence w: $w = L_S^+ n$. We then have the relationship,

$$y(t) = s(t) + u(t) + w. \tag{6}$$

B. State-Space Modeling

In Eq. (6), w is white Gaussian, because the sensor noise n is assumed to be white Gaussian. However, the interference term $u(t)$ should have a temporal structure, because it is generated from brain source activities that are not of interest. Thus, we model the temporal property of $u(t)$ using the MVAR process, such that

$$u(t) = \sum_{j=1}^{\widetilde{p}} B(j) u(t-j) + \widetilde{e}, \tag{7}$$

where $\widetilde{p}$ is the model order, $B(j)$ is the coefficient matrix, and $\widetilde{e}$ is the residual vector. Therefore, according to Eq. (6), the estimated source time courses $y(t)$ consists of two independent MVAR processes, $s(t)$ and $u(t)$, and the white Gaussian process w. Using a state space vector x_t, which is defined as

$$x_t = [s(t), \ldots, s(t-p), u(t), \ldots, u(t-\widetilde{p})]^T, \tag{8}$$

Eq. (6) is converted into

$$x_t = F x_{t-1} + \varepsilon, \tag{9}$$
$$y(t) = H x_t + w. \tag{10}$$

In Eq. (9), the state transition matrix F is expressed as[3],

$$F = \begin{bmatrix} F_s & 0 \\ 0 & F_u \end{bmatrix}, \tag{11}$$

where F_s is a $qp \times qp$ matrix defined as

$$F_s = \begin{bmatrix} A(1) & A(2) & \cdots & A(p-1) & A(p) \\ I & 0 & \cdots & 0 & 0 \\ 0 & I & \cdots & 0 & 0 \\ \vdots & \vdots & \ddots & \vdots & \vdots \\ 0 & 0 & \cdots & I & 0 \end{bmatrix}, \tag{12}$$

and F_u is a $q\widetilde{p} \times q\widetilde{p}$ matrix defined as

$$F_u = \begin{bmatrix} B(1) & B(2) & \cdots & B(\widetilde{p}-1) & B(\widetilde{p}) \\ I & 0 & \cdots & 0 & 0 \\ 0 & I & \cdots & 0 & 0 \\ \vdots & \vdots & \ddots & \vdots & \vdots \\ 0 & 0 & \cdots & I & 0 \end{bmatrix}. \tag{13}$$

In the above equations, I is the identity matrix and the notation 0 expresses a matrix whose elements are all zeros.

In Eq. (9), ε is a $q(p+\widetilde{p}) \times q$ matrix defined as

$$\varepsilon = [\underbrace{e, 0, \ldots, 0}_{q \times p}, \underbrace{\widetilde{e}, 0, \ldots, 0}_{q \times \widetilde{p}}]^T. \tag{14}$$

In Eq. (10), H is a $q \times q(p+\widetilde{p})$ matrix defined as

$$H = [\underbrace{I, 0, \ldots, 0}_{q \times p}, \underbrace{I, 0, \ldots, 0}_{q \times \widetilde{p}}]. \tag{15}$$

C. Kalman Smoother Equations and EM Update Rules

Based on the state space representations in Eqs. (9) and (10), the posterior probability of the state-space vector x_t is Gaussian, and its distribution is assumed to be:

$$f(x_t | y(t_0), \ldots, y(t)) = \mathcal{N}(x_t | \bar{x}_t, V_t), \tag{16}$$

where $\bar{x}_t$ and V_t are the mean and covariance of x_t based on the observations from $y(t_0)$ to $y(t)$. Then, we can derive the following Kalman filter equations:

$$P_t = F V_{t-1} F^T + R_\varepsilon, \tag{17}$$
$$K = P_t H^T (H P_t H^T + R_w)^{-1}, \tag{18}$$
$$\bar{x}_t = F \bar{x}_{t-1} + K [y(t) - H F \bar{x}_{t-1}], \tag{19}$$
$$V_t = P_t - K H P_t, \tag{20}$$

where R_w indicates the covariance matrices of the noise vector w, and R_ε indicates the covariance matrix of the residual vector ε; R_ε is expressed as

$$R_\varepsilon = \langle \varepsilon \varepsilon^T \rangle = \begin{bmatrix} R_e & 0 & 0 & 0 \\ 0 & 0 & 0 & 0 \\ 0 & 0 & R_{\widetilde{e}} & 0 \\ 0 & 0 & 0 & 0 \end{bmatrix}, \tag{21}$$

where R_e and $R_{\widetilde{e}}$ are the covariance matrices of the residual vectors e and $\widetilde{e}$.

Kalman smoother equations are:

$$J_t = V_t F^T P_t^{-1}, \tag{22}$$
$$\widehat{x}_t = \bar{x}_t + J_t (\widehat{x}_{t+1} - F \bar{x}_t), \tag{23}$$
$$\widehat{V}_t = V_t + J_t (\widehat{V}_{t+1} - P_t) J_t^T. \tag{24}$$

Here, $\widehat{x}_t$ is the minimum mean squared error estimate of the state vector x_t based on all of the observed data. $\widehat{V}_t$ is the corresponding covariance matrix of x_t.

We assume that we can predetermine R_w using the sensor noise estimate. The parameters, F and R_ε, can be estimated using the EM algorithm. The update rules for F and R_ε are[4]:

$$F^{\text{new}} = \Phi_B \Phi_A^{-1}, \quad (25)$$

and

$$R_\varepsilon^{\text{new}} = \Phi_A - F^{\text{new}} \Phi_B, \quad (26)$$

where

$$\Phi_A = \frac{1}{K-1}\left[\sum_{t=2}^{K} J_{t-1}\widehat{V}_t + \widehat{x}_t(\widehat{x}_{t-1})^T\right], \quad (27)$$

$$\Phi_B = \frac{1}{K-1}\left[\sum_{t=1}^{K-1} \widehat{V}_t + \widehat{x}_t(\widehat{x}_t)^T\right]. \quad (28)$$

Our proposed procedure is summarized as follows. In a preparation step, using the control measurement $b_c(t)$ and source lead field L_S, the VAR coefficient matrices $B(j)$ and $R_{\tilde{e}}$ are determined. In the E step of the EM algorithm, the Kalman filter and smoother equations in Eq. (17)–(24) are computed to obtain $\widehat{x}_t$ and $\widehat{V}_t$. In the M step, F and R_ε are updated according to Eqs. (25) – (28). Here, the values of $B(j)$ are fixed to the predetermined ones, so only the first $q \times qp$ components of F are updated. The covariance matrix, R_ε is updated according to Eq. (26). Here, $R_{\tilde{e}}$ is fixed to the predetermined value, so only the part of R_e, the first $q \times q$ components of R_ε, are updated.

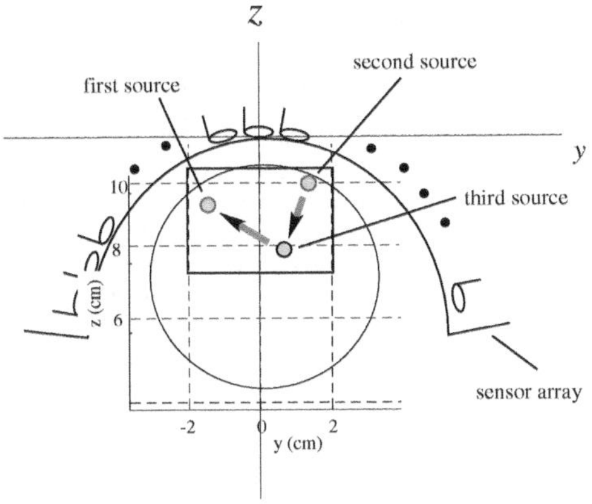

Fig. 1 Source-sensor configuration. The three small circles represent the sources, and bold arrows represent their causal relationship

III. Numerical experiments

Numerical experiments were conducted to check the effectiveness of the proposed method. We used a sensor alignment of the 275 whole-head sensor array. Three sources were assumed to exist on the vertical single plane: ($x = 0$ cm). The source-sensor configuration and the coordinate system are depicted in Fig. 1. Denoting the source time courses as $s_1(t)$, $s_2(t)$, and $s_3(t)$, they were generated using the MVAR process[5],

$$\begin{bmatrix} s_1(t) \\ s_2(t) \\ s_3(t) \end{bmatrix} = \begin{bmatrix} 0.8 & 0 & 0.4 \\ 0 & 0.9 & 0 \\ 0 & 0.5 & 0.5 \end{bmatrix} \begin{bmatrix} s_1(t-1) \\ s_2(t-1) \\ s_3(t-1) \end{bmatrix} + \begin{bmatrix} -0.5 & 0 & 0 \\ 0 & -0.8 & 0 \\ 0 & 0 & -0.2 \end{bmatrix} \begin{bmatrix} s_1(t-2) \\ s_2(t-2) \\ s_3(t-2) \end{bmatrix} + e. \quad (29)$$

This MVAR process simulates the causal relationship depicted in Fig. 1, in which the second source has a directional causal influence on the third source and the third source has that on the first source. Spontaneous MEG was added as the interference, and the signal-to-interference ratio (SIR) was controlled to 4. The recently-proposed Champagne algorithm[6] was applied to reconstruct the source distribution. The results on the plane of $x = 0$ cm are shown in Fig. 2. The information on the source locations and orientations were extracted from these results, and the composite lead field matrix L_S was constructed.

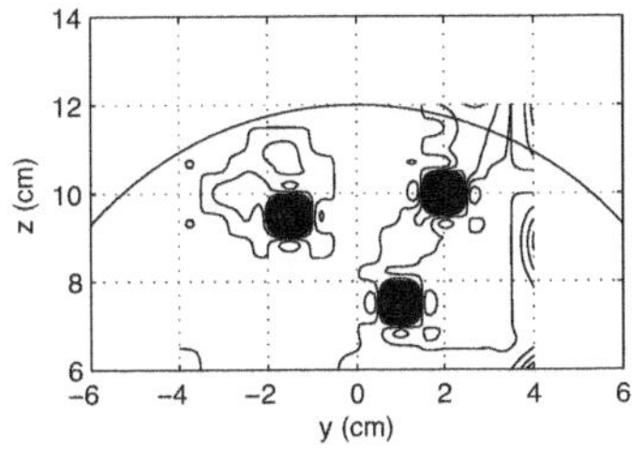

Fig. 2 Reconstructed source image on the plane of $x = 0$ cm

We first applied the conventional method to estimate the MVAR coefficient. That is, without using the state-space model, we simply assume the relationship,

$$y(t) = \sum_{j=1}^{p} A(j) y(t-j) + e. \quad (30)$$

To estimate the MVAR coefficients, we used the algorithm described in [7], through the software package called `arfit`, and computed the partial directed coherence (PDC)[8] and the directed transfer function (DTF)[9]. The results are shown

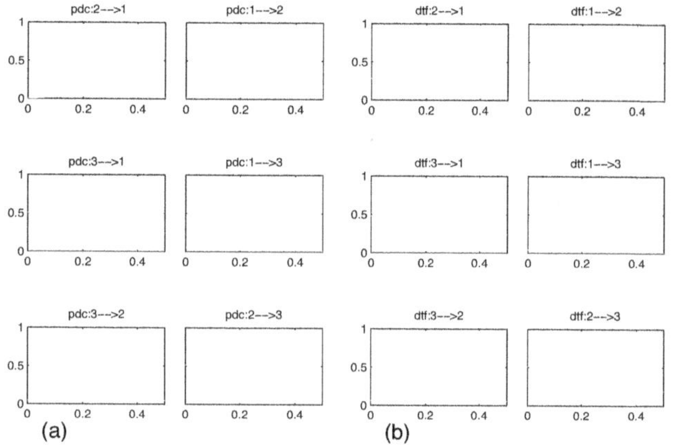

Fig. 3 (a)Results of computing partial directed coherence (PDC) and (b)directed transfer function (DTF) by using the conventional method for estimating MVAR coefficients for the case of SIR set to 4

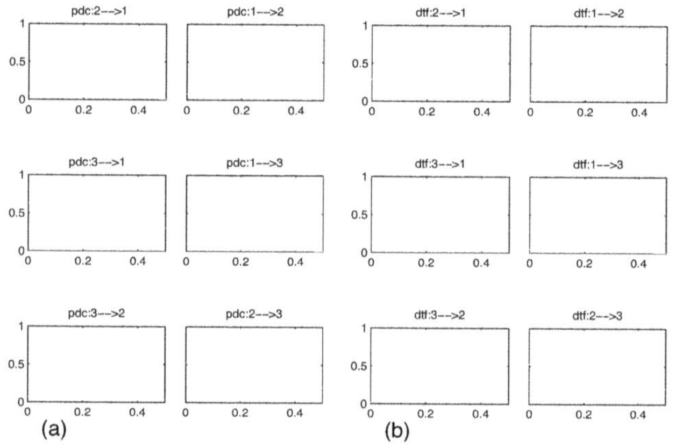

Fig. 5 (a)Results of computing partial directed coherence (PDC) and (b)directed transfer function (DTF) by using the proposed method for estimating MVAR coefficients for the case of SIR set to 0.7

in Fig. 3. Here, we can see that the influence of the interference is small and the causality measures can detect the assumed causal relationships. We next generated the sensor data with SIR set equal to 0.5, and repeated the same procedure to obtain PDC and DTF. The results are shown in Fig. 4. These results contain significant influence of the interference, and the causality relationship becomes unclear, particularly between the second and the third sources. We then applied the proposed method to the same generated data, and the results are shown in Fig. 5. Here, the results are very similar to those of the high SIR case in Fig. 3.

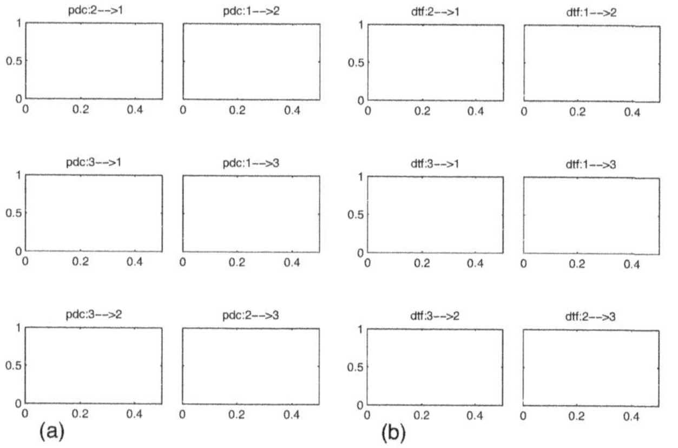

Fig. 4 (a)Results of computing partial directed coherence (PDC) and (b)directed transfer function (DTF) by using the conventional method for estimating MVAR coefficients for the case of SIR set to 0.7

REFERENCES

1. Cheung B. L. P., Riedner B., Tononi G., Van Veen B.. State-space multivariate autoregressive models for estimation of cortical connectivity from EEG in *Proceedings of 31st Annual International Conference of the IEEE EMBS*(Minneapolis):61-64 2009.
2. Nalatore H., Ding M., Rangarajan G.. Denoising neural data with state-space smoothing: Method and application *Journal of Neuroscience Methods.* 2009;179:131-141.
3. Doblinger G.. Adaptive Kalman smoothing of AR signals disturbed by impulses and colored noise in *Proceedings of IEEE symposium on advances in digital filtering and signal processing*(Victoria, Canada):72-76 1998.
4. Shumway R. H., Stoffer D. S.. An approach to time series smoothing and forecasting using the EM algorithm *Journal of Time Series Analysis.* 1982;3:253-263.
5. Ding M., Chen Y., Bressler S. L.. Granger causality: Basic theory and application to neuroscience in *Handbook of Time Series Analysis* (Schelter B., others . , eds.):500-600Wiley-VCH 2006.
6. Wipf D. P., Owen J. P., Attias H. T., Sekihara K., Nagarajan S. S.. Robust Bayesian estimation of the location, orientation, and time course of multiple correlated neural sources using MEG *NeuroImage.* 2010;49:641-655.
7. Schneider T., Neumaier A.. A Matlab package for the estimation of parameters and eigenmodes of multivariate autoregressive models *ACM transactions on Mathematical Software.* 2001;27:58-65.
8. Sameshima K., Baccala L. A.. Using partial directed coherence to describe neuronal ensemble interactions *Journal of Neuroscience Methods.* 1999;94:93-103.
9. Kaminski M., Ding M., Truccolo W. A., Bressler S. L.. Evaluating causal relations in neural systems: Granger causality, directed transfer function and statistical assessment of significance *Biol. Cybern..* 2001;85:145-157.

Constructing Surrogate Data to Control for Artifacts of Volume Conduction for Functional Connectivity Measures

Forooz Shahbazi, Arne Ewald, Andreas Ziehe, and Guido Nolte

Fraunhofer Institute FIRST, IDA Group, Berlin, Germany

Abstract— **The major problem in examining interactions between sources of brain activity from MEG and EEG data are the 'artifacts of volume conduction' denoting the fact that the sensors measure a largely unknown superposition of brain activities. In this work we suggest a method to test for artifacts of volume conductions. The test is absolutely general: it is applicable to both linear and nonlinear methods and both on the sensor and source level.**

The idea of the method is to construct surrogate data which are statistically as close as possible to the original data but which are a superposition of independent sources. As a first step we decompose the data via Independent Component Analysis (ICA) resulting in signals which are as independent as possible. In a second step we destroy any remaining dependencies by shifting the n.th signal by a time (n-1)*T where T must be substantially larger than any autocorrelation time. Finally, these shifted signals are mixed again into sensors using the mixing matrix found by the ICA algorithm. Any interaction measure can now be calculated both for the original and the surrogate data. If a specific effect can be seen in both data sets we consider this effect as insufficient evidence for an observed brain interaction.

We applied this method to real and imaginary parts of coherency (real EEG data), 1:2 phase locking (real EEG data) and Granger causality (simulated data). We found that the real part of coherency is almost perfectly consistent with mixtures of independent sources while the imaginary part cannot be explained with the surrogate data at all. Results for phase locking can only be explained partly by volume conductions. For true directional interactions Granger causality is attenuated but not removed in the surrogate data as compared to the original data.

Keywords— **Functional connectivity measures, Volume conduction, Surrogate Data, Coherency method, Phase Locking, Granger causality, ICA.**

I. INTRODUCTION

One major problem in examining interactions between sources of brain activity from noninvasive electrophysiological measurements like MEG or EEG are the artifacts of volume conduction. Volume conduction is the term used to describe the effects of recording neuronal activity at a distance from their source generator. When the recording sensors are not connected directly to the nerve or muscle due to a medium of some sort which separates the two, e.g., a correlation between several electrodes can be detected even if the single sources are independent. This effect frequently occurs because the activity of a single source inside the brain is mapped simultaneously to many EEG channels. In fact, volume conduction plays a significant role in almost all noninvasive electrophysiological recordings, since the sensors are never in direct contact with the nerve cells generating the signals [1]. Therefore, volume conduction substantially affects the results of brain connectivity measures leading in general to highly significant dependencies between channels even if all sources are independent. Although this effect is not new, it becomes increasingly acknowledged both on the sensor level [2] and on the source level [3].

In this paper, a new method to control for artifacts of volume conduction for functional connectivity measures is proposed. It is based on producing surrogate data from the original data which are statistically as close as possible to the original data but are at the same time a superposition of independent sources. Different methods of connectivity measures are applied to both original and surrogate data and the results are compared to each other to evaluate the reliability of each method.

This report is divided into three main sections. The first introduces the construction of the surrogate data. The second section represents the results of applying different connectivity measures to both the surrogate and original data. In the third section the limitations of this method are discussed.

II. METHOD

To evaluate the reliability of different methods of connectivity, surrogate data are constructed as superposition of independent generated with statistical properties as the original data but based on a superposition of independent sources. This is achieved via Independent Component Analysis (ICA) [4] following the shifting of the time series of each independent component relative to each other. As the output components in ICA might not be perfectly independent, shifting each component in time space would result in completely independent components. Later, these components are multiplied by the inverse of the coefficient matrix calculated by ICA algorithm. And the result is

S. Supek and A. Sušac (Eds.): Advances in Biomagnetism – BIOMAG2010, IFMBE Proceedings 28, pp. 207–210, 2010.
www.springerlink.com

referred as surrogate data. This will be explained in detail in the following.

A. Construction of Surrogate Data

The general idea behind the construction of surrogate data is to destroy a specific property to be tested while keeping all other properties unchanged as good as possible. E.g., in univariate nonlinear data analysis it is common to destroy nonlinear effects by phase randomization which leaves all spectral, i.e. linear, properties unchanged. The property to be destroyed here is interaction between sources, such that the surrogate data are merely a mixture of independent sources. This corresponds exactly to the assumption behind Independent Component Analysis (ICA) and we will use ICA techniques to construct such data in three steps:

First, we decompose the data with a demixing matrix W found by an ICA algorithm. The respective components are as independent as possible with respect to the dependence measure of the ICA method. Specifically, let $x_i(t)$ be the measured signal at time t in the *i.th* channel and $\vec{x}(t) = (x_1(t),\ldots,x_N(t))^T$ for N channels. Then, the demixing matrix W, found by an ICA algorithm, transforms the data into ICA channels:

$$\vec{S}(t) \equiv W\vec{x}(t) \qquad (1)$$

Second, the temporal signals of the ICA components, $s_i(t)$, are now as independent as possible. Note, that if they were truly independent we could shift these signals relative to each other in time without affecting the statistical properties. E.g., if $s_1(t)$ and $s_2(t)$ are statistically independent, then the pair $(s_1(t), s_2(t))$ has the same statistical properties as $(s_1(t), s_2(t+T))$ for any value of T. Independence is here only approximately the case, and hence such shifting does affect the statistical properties but only very moderately. We will use this to construct 'perfectly' independent signals $v_i(t)$ while at the same time only minimally affecting the statistical properties:

$$v_i(t) \equiv s_i(t+(i-1)T) \qquad (2)$$

Where T is chosen as large as possible and should be at least as large as one trial. The shifting is understood to be cyclic, e.g. for K trials the *K+1.th* trial is understood to be identical with the first trial.

For large T, the $v_i(t)$ become 'quasi-independent', i.e. independent with respect to all measures which quantify relations between channels (or sources) within each trial and which are then averaged across all trials, as opposed to relations between different trials. This is, formally speaking, only a subgroup of measures. In practice, however, it covers all measures which are commonly used. In other words, the dependence between different $s_i(t)$ didn't disappear but was put out of reach of all dependence measures used in practice. This property holds if T is large compared to any autocorrelation time. If T is set to the trial length (for epoched data) the implicit assumption is that different trials are independent of each other before the shifting.

Third, from the quasi-independent signals we reconstruct surrogate data by mixing them back into channels as

$$\vec{x}_{surr}(t) \equiv W^{-1}\vec{v}(t) \qquad (3)$$

These surrogate data are now a mixture of independent signals - omitting here and in the following the term 'quasi'. Any connectivity measure found from these surrogate data must be an artifact of volume conduction. Any connectivity result, to be tested for artifacts of volume conduction, will be recalculated using the surrogate data. If a specific effect is found both for the original and surrogate data we conclude that the evidence for true brain connectivity is insufficient. This does not mean that the observed effect *is* an artifact of volume conduction but rather that this alternative explanation cannot be rejected and that hence the respective claim about brain connectivity is unsubstantiated, but not necessarily wrong. The limitations of this interpretation will be explained in detail in the discussion section.

III. Applications

A. Coherency

We constructed the surrogate data based on the real electroencephalography (EEG) data in rest condition [5] for a single subject for illustration purposes. The coherence function gives the linear correlation between two signals x and y as a function of the frequency. We divide the data into segments and apply a Hanning window with subsequent

Fourier transformation on the data for each channel and segment resulting in complex Fourier coefficients $\hat{x}_{ik}(f)$ at frequency f for the *i.th* channel and *k.th* segment. The cross-spectrum is now defined as

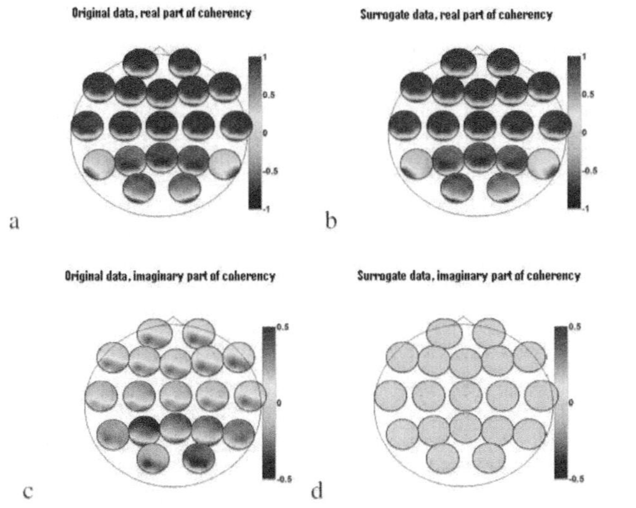

Fig. 1 a) Real part of coherency for original data, b) real part of coherency for surrogate data, c) imaginary part of coherency for original data, d) imaginary part of coherency for surrogate data

$$G_{ij}(f) = \left\langle \hat{x}_{ik}(f)\hat{x}^*_{jk}(f) \right\rangle_k \tag{4}$$

Where $\langle \cdot \rangle_k$ denotes average over all segments. Complex coherency is defined as

$$\gamma_{ij}(f) = \frac{G_{ij}(f)}{\sqrt{G_{ii}(f)G_{jj}(f)}} \tag{5}$$

We show both real and imaginary parts of coherency in one subject at 10Hz in Figure 1. The real part can almost perfectly be explained by the surrogates while the imaginary part cannot be explained at all. This is consistent with the finding that non-interacting sources cannot explain a non-vanishing imaginary coherency [6].

B. *Phase Locking*

Phase locking looks for latencies at which the phase difference between the signals varies little across trials [7]. The procedure ignores amplitudes and analyzes only phase relationships. We are here interested in 1:2 phase locking which is defined as

$$P_{ij}(f) = \left\langle \exp\left(i(2\Phi_{ik}(f) - \Phi_{jk}(2f)\right)\right\rangle_k \tag{6}$$

where $\Phi_{ik}(f)$ are the phases of the signals according to

$$r_{ik}(f)\exp\left(i\Phi_{ik}(f)\right) \equiv \hat{x}_{ik}(f) \tag{7}$$

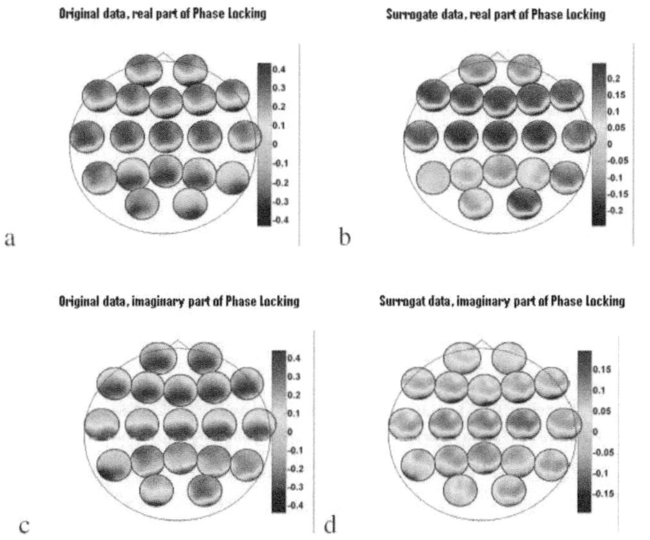

Fig. 2 a) Real part of phase locking for original data, b) real part of phase locking for surrogate data, c) imaginary part of phase locking for original data, d) imaginary part of phase locking for surrogate data

In Figure 2 we show both real and imaginary parts of 1:2 phase locking in one subject at f=10Hz, which corresponds to a coupling between 10 and 20Hz. In contrast to coherency also a non-vanishing imaginary part can in general be generated by mixtures of independent sources. We observed that both the real and imaginary parts of phase locking can only partially be explained by mixtures of independent sources.

C. *Granger Causality*

Granger causality is a measure to estimate the causal effect of one signal to another. We estimated the net causality between two simulated activities and normalized the result by an estimate of the standard deviation [5]. Granger Causality was calculated a) for truly unidirectional systems where the first channel drives the second, which should always result in a positive net causality, and b) for mixed noise consisting either of 2 or 3 independent sources mixed into 2 channels. Results are given in Fig.3 for original and surrogate data. We found that for true interacting systems the surrogates can partially but not fully explain the result. For mixed noise, the method almost perfectly identifies the effect as false if there are not more sources than sensors. For more sources the identification is largely but not perfectly possible.

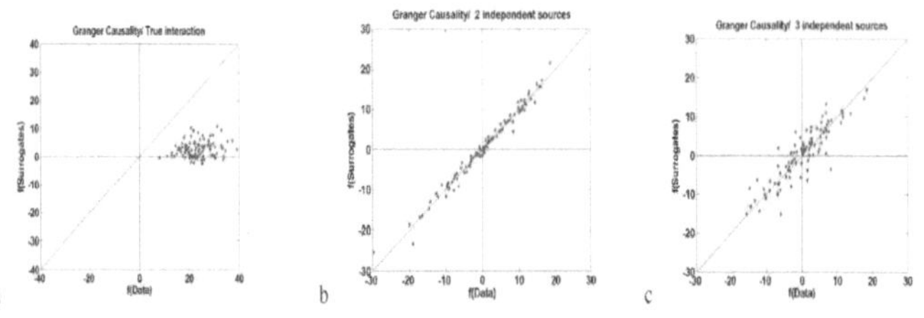

a b c

Fig. 3 a) Granger causality for true interactions with positive causality b) Granger causality for two independent sources and two channels, c) Granger causality for three independent sources and two channels

IV. Discussion

We proposed a method to test for artifacts of volume conduction by recalculating a connectivity measure using surrogate data which were constructed as a superposition of independent sources with statistical properties as close as possible to the original data. We argue that if a specific effect can be seen both in the original and surrogate data the trivial hypothesis that the effect is a mixing artifact cannot be ruled out. We suggest using this method as a sanity check in the interpretation of connectivity measures as indicators of brain interaction. We here would like to also elaborate on the limitations of this method.

First, if a specific result does not pass this test, that doesn't mean that the observed effect is necessarily an artifact. True interactions may be undistinguishable from mixtures. However, given that mixing is the dominant cause of relations between channels and estimated sources, we consider it as necessary to give in that case additional evidence to support a claim about brain interactions.

Second, we expressed the data as a superposition of as many independent sources as there are channels. It is conceivable that a substantial difference to surrogate data is found because the true data are a mixture of more sources which are all relevant for the statistical properties. We believe that this problem becomes decreasingly relevant for many channels, and also almost perfect recovery of the real part of coherency in real EEG data indicates that this problem might be negligible in practice. However, the detailed implications are not clear to us, yet.

Third, in many situations the question is not whether there is true brain interaction but whether this interaction is different between two conditions or populations. If, e.g., imaginary coherency is different in two groups a possible explanation could be that the brain interaction is identical in both group which just differ by noise level. The proposed method is not capable of generally distinguishing a change of interaction from a change of other factors like noise level. We would really like to say that this will be dealt with in future work, but, to say the truth, we have no idea how to do that.

References

[1] Andrew F.Westdorp, Volume conduction effects On Correlation analysis Of EEG Data, IEE 1993.

[2] Srinivasan R, Winter WR, Ding J, Nunez PL (2007) EEG and MEG coherence: measures of functional connectivity at distinct spatial scales of neocortical dynamics. J Neurosci Methods. 166:41-52

[3] Schoffelen JM, Gross J (2009) Source connectivity analysis with MEG and EEG. Hum Brain Mapp. 30:1857-65.

[4] Andreas Ziehe, Klaus- Robert Mueller,(1998),TDSEP- an Efficient Algorithm for Blind Separation Using Time Structure, ICANN, 675-680.

[5] Nolte G, Ziehe A, Nikulin VV, Schlögl A, Krämer N, Brismar T, Müller KR (2008) Robustly estimating the flow direction of information in complex physical systems.Phys Rev Lett. 100:234101.

[6] Nolte G., Bai O., Wheaton L.,Mari Z., Vorbach S. and Hallett M. (2004) Identifying true brain interaction from EEG data using the imaginary part of coherency, Clinical Neurophysiology, Vol. 115, No. 10, 2292-307.

[7] Ernesto Pereda, Rodrigo Quian Quiroga, Johdeep Bhattacharya,(2005), Nonlinear Multivariate analysis of Neurophysiological Signals, Neurobiography 77, 1-37.

Author: Forooz Shahbazi
Institute: Fraunhofer FIRST, IDA group
Street: Kekulestr. 7, 12489
City: Berlin
Country: Germany
Email: Forooz.shahbazi@first.fraunhofer.de

MEG Detects Alpha-Power Modulations in Pulvinar

Yohan Attal[1,2], Jerome Yelnik[1,3], Eric Bardinet[1,2], Marie Chupin[1,2], and Sylvain Baillet[4]

[1] Université Pierre et Marie Curie Paris 6, Centre de Recherche de l'Institut du Cerveau et de la Moelle épinière, UMR-S975, Paris, France
[2] CNRS, UMR 7225, Paris France
[3] Inserm, U975, Paris, France
[4] MEG Program, Department of Neurology, Medical College of Wisconsin/Froedtert Hospital, Milwaukee, USA

Abstract— **Deep brain structures such as the thalamus are crucial to multiple brain processes and related disorders. Most MEG imaging inverse models considered that electromagnetic surface signals are predominantly produced by cortical, hence relatively superficial neural currents. Using a realistic model of cortical and sub-cortical MEG generators, we sought to detect modulations of ongoing brain activity in deeper brain regions using a resting eyes closed / eyes opened paradigm. Results confirmed that MEG can access sub-cortical brain structures: task-related amplitude modulations in the alpha band were found in parieto-occipital brain areas and a sub-territory of the thalamus: the pulvinar.**

Keywords— **MEG, source analysis, resting state, pulvinar, α activity.**

I. Introduction

Neuroimaging evidence regarding the functional organization of sub-cortical nuclei [1-6] is rather scarce compared to cortical structures.

However, it is well established that sub-cortical structures play a key role in multiple processes in the healthy or diseased brain. To access these deeper structures using magnetoencephalography (MEG) source imaging, we have recently developed a global-brain MEG source model, which integrates a number of structural and electrophysiological properties of cortical and sub-cortical structures as priors to source inverse modeling [7].

In this paper, we sought to use this model to detect power modulations of ongoing brain activity in deeper brain structures. Because detecting deeper brain activity with MEG has been both considered as controversial and challenging, we have focused our investigations on well-established power modulations in the alpha frequency range using minimal paradigm design [8,9]. Hence, alpha ongoing oscillations are well-known to be dominant during an eyes-closed resting condition. Of primary interest to the evaluation of the global source model, alpha modulations have been discussed as a primary mediator of thalamo-cortical communications [10,11].

Further, former studies of the brain resting-state combining functional Magnetic Resonance imaging (fMRI) and EEG evidenced engagement of thalamus in the modulation of resting cerebral fluctuations [12].

This paper details the global source model used in this study and how source localization was performed on ongoing MEG data.

II. Materials and Methods

A. Modeling Cortical and Sub-cortical Structures

Individual T1-weighted MRI volume data (1x1x1.5mm voxel size, axial scans) from 6 healthy subjects was processed to extract cortical and sub-cortical tessellations. The neocortex was segmented by the automatic pipeline available in the BrainVISA/Anatomist (http://brainvisa.info) software package using [13]. Structural boundaries of hippocampus and amygdala are ill-defined on T1-weigthed MRI images. A specific detection and segmentation technique was used to obtain surface tessellations of these two structures [14]. This technique is fully automatic and also implemented in the BrainVISA environment.

The basal ganglia and related structures were segmented from the individual MR volumes using a deformable atlas approach. This latter combines detailed histological slices registered to the post-mortem MRI data of a specimen anatomy. The methodology of the atlas warping procedure to the individual anatomy of the subjects in this study is detailed in [15].

To study cortical and sub-cortical MEG activations at the group level, individual brain anatomies were registered into the standard Montreal Neurological Institute geometric referential using affine transformation (see Section B).

Central grey nuclei and related structures in the model consisted of the putamen, thalamus, reticular perithalamic nucleus (RPN), lateral geniculate nucleus (LGN) and external pallidum (EGP).

S. Supek and A. Sušac (Eds.): Advances in Biomagnetism – BIOMAG2010, IFMBE Proceedings 28, pp. 211–214, 2010.
www.springerlink.com

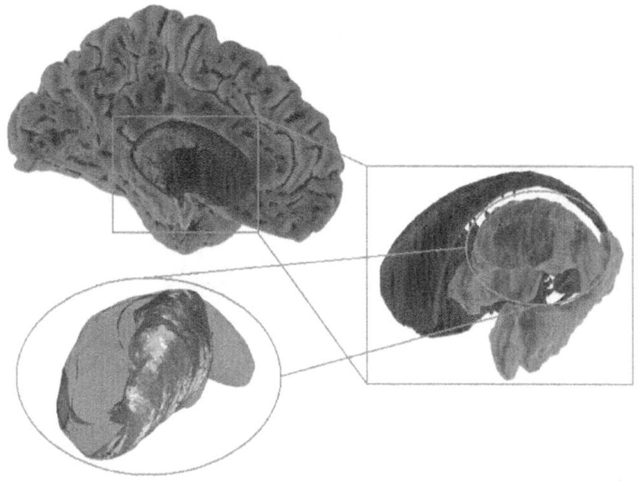

Fig. 1 The global brain model used for MEG source imaging. The cortical mesh is shown in grey (only left hemisphere is shown) and the sub-cortical and deeper structures are detailed: hippocampus (pink), amygdala (yellow), putamen (red), caudate nucleus (magenta), thalamus (orange), reticular perithalamic nucleus (RPN, green), lateral geniculate nucleus (LGN, dark blue) and external pallidum (EGP, light blue). Two subdivisions of thalamic sub-territories are further detailed: ventral lateral nucleus (yellow), ventral intermediate nucleus (light blue)

The global anatomical model is displayed in Fig. 1. The electrophysiological properties in the model were defined in the context of an imaging MEG source model consisting of equivalent current dipoles (ECD) distributed orthogonally to the cortical mantle at each vertex location. This model requires extensions adapted extensions to the mass electrophysiology deeper and/or sub-cortical structures.

ECDs are convenient compact models of the current flow resulting from postsynaptic potentials (PSP). An ECD is typically oriented along the direction originating from the cell soma and pointing toward the principal axis of the dendritic tree. Consequently, the expected strength and direction of elementary neural currents depend on the density and types of neural cells that make the detailed architecture of each anatomical structure. As a first approximation, only pyramidal and stellate cells were considered as possible elementary MEG generators. Pyramidal cells are typically considered as 'open-field' neurons because of their global longitudinal morphology resulting in a strong vector sum of the PSPs along their dendritic trees. Conversely, dendrites of stellate cells form a radial arborescence about the cell body, and the resulting net vector sum of PSPs is much weaker than that of pyramidal cells. For hippocampus, ECD were placed as for the neocortex at each vertex of the tessellations, and oriented normally to the local surface. This model would ideally require further sophistication by accounting for the finer involuted grey-matter structure of the hippocampus. This will be the objective of future developments.

Central grey nuclei contain multiple types of neural cells with a large variety of cell densities, hence of expected currents strengths. Therefore their large-scale electrophysiology cannot be approximated with a surface distribution of current dipoles. Regular volumic grids were adjusted to the surface envelope of each nucleus. These grids were designed to support a volumic distribution of elementary current dipoles. Some of the nuclei are expected to sustain current flows with a preferred orientation though. Dipoles in the pallidum, perithalamic nucleus and LGN were oriented along the principal axis of their respective surface envelope. The thalamus and putamen essentially consist of closed-field cells – i.e. with no preferred source orientation – hence ECD were placed at each node of the inner volumic grid as described above, with a random distribution of ECD orientation across the grid. More details on the electrophysiological model can be found in [7]. The characteristics of the global source model are summarized in Table 1.

The MEG gain matrices for each structure were computed on the 151-channel sensor array of the CTF/VSM Omega whole-head system. The analytical MEG forward model using a spherical geometry was computed as implemented in BrainStorm (http://neuroimage.usc.edu/brainstorm).

The neural current density time series at each location of the global source model were estimated from the weighted minimum norm image model using BrainStorm.

Table 1 Global characteristics of the source model (P stands for pyramidal cells and S stands for stellate cells)

Structures (Left hemisphere)	Number of sources	Surface (cm^2) / Volume (cm^3)	Cell type
Cortex	12,125	745.9	P
Hippocampus	2,196	15.3	P
Thalamus	715	5.8	S
LGN	154	0.16	P
EGP	308	1.03	P
Putamen	770	6.18	S
RPN	382	1.3	P

B. Experimental Evaluation

Six healthy volunteers (5 men and 1 woman, 33 year-old on average) participated in this study. The experimental procedures were approved by the local ethical committee and all participants gave written informed consent.

The MEG session consisted of a block-designed "eyes opened - eyes closed" paradigm, each block lasting for 30 s. MEG signals were recorded on a CTF 151 channels whole-head system (CTF Systems Inc., Port Coquitlam, Canada), at a 1250 Hz sampling rate.

The peak of alpha frequency varies across subjects [9]. Therefore, data were band-pass filtered within the individual alpha band that was determined after inspection of the MEG RMS power spectrum across channels, of each subject (Table 2). For each subject, distributed source imaging was performed using BrainStorm as explained in Section A.

Table 2 Peak of the alpha frequency across subjects

Subjects	S1	S2	S3	S4	S5	S6
Alpha band (Hz)	8-13	10-13	9-13	8-12	9-12	8-13

Non-parametric permutation tests [16] were used to detect modulations of source amplitudes by contrasting the eyes opened vs. eyes closed conditions. All source maps were registered and normalized to the MNI reference frame for group statistical analysis. Surface and volume grid source amplitudes were interpolated in the volume of the 3D MNI template using a procedure based on 3D nearest-neighbor interpolation with Shepard's weighting. Group average source maps were thresholded at $p < 0.0001$.

III. Results

Strong modulation of alpha power was detected at the sensor level between the two experimental conditions, illustrated Fig. 2. Source analysis of these modulations using the global brain source model revealed significant bilateral amplitude modulations originating in the posterior parietal and occipital cortex. More specifically, significant source amplitude modulations were captured in the pulvinar thalamic region as shown on Fig. 2.

This experimental evidence complements our early results with simulations detailed in [7] and emphasizes that activity of deeper brain structures may be detected and localized with MEG, if adequate modelling is provided.

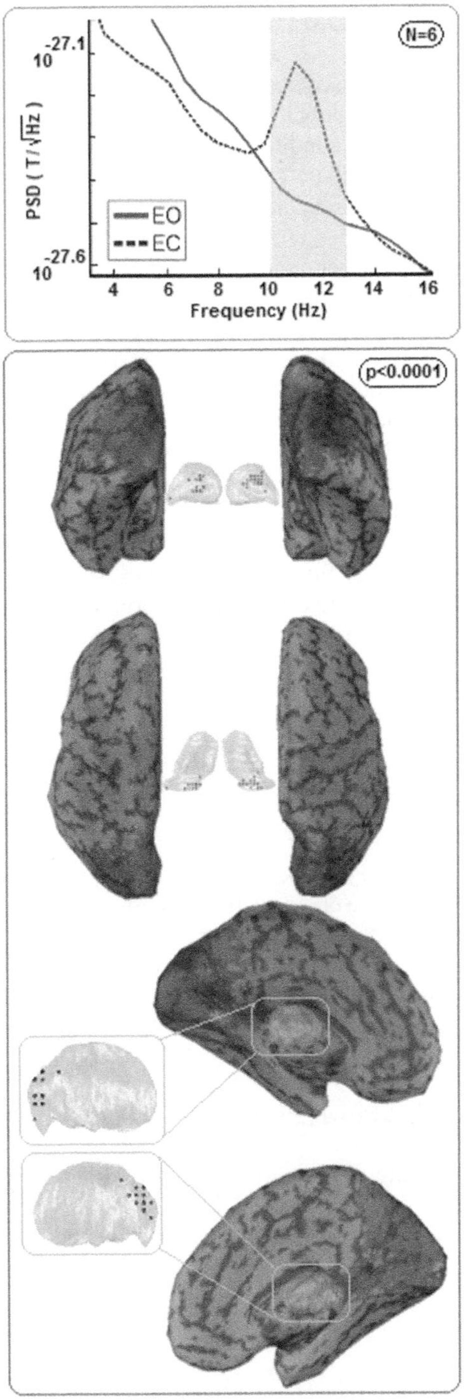

Fig. 2 Upper inset: group average RMS power spectral density of eyes opened (red line) vs. eyes closed (dotted line) across all MEG sensor time series; N=6 subjects. Lower inset: Significant variations of source amplitudes were identified in the pulvinar by contrasting the two experimental conditions. Results are shown at the group level with data registered on the MNI Colin27 brain template ($p < 0.0001$). Results are shown on the cortical envelope and through a glass view of the thalamus where pulvinar activations are displayed by blue dotted regions

IV. Discussion

The thalamus is a key participant in many synchronized neuronal oscillations. The global source model used in this study contributes to a better conditioning of the MEG inverse problem, by combining realistic anatomical constraints with priors on the respective current density expected in distinct regions of the brain.

This study used the larger number of data samples obtained from a steady-state paradigm. Accessing the fine transient dynamics of cortico-sub-cortical interactions remains to be demonstrated with a more specific paradigm.

The global source model requires further sophistication. In the case of the hippocampus, we have chosen to consider only the Ammon's fields CA1, CA2 and CA3 where pyramidal neurons are organized in a similar way as in the neocortex. However, the hippocampus is a folded structure that should also be taken into account for enhanced realism. This will be the objective of future developments.

In addition, structures without expected organized current flows might benefit from recent developments in the modelling of elementary current sources using current multipoles [17].

V. Conclusions

The global source model was evaluated on experimental MEG data on a classical eyes-opened/eyes-closed paradigm. Modulations of activities were detected in parieto-occipital cortical regions and in the pulvinar part of the thalamus. These results suggest that complete, realistic modeling of MEG source distribution can help detect activation in deeper brain structures that were thought to be invisible to MEG.

Acknowledgment

The authors are grateful to the MEG center of la Pitié-Salpêtrière hospital, especially D. Schwartz, A. Ducorps and F. Bergame for their precious assistance during data acquisition.

References

[1] Tesche C.D. (1996) *Non-invasive imaging of neuronal population dynamics in human thalamus* Brain Res.

[2] Gross, et al. (2001) Dynamic *imaging of coherent sources: Studying neural interactions in the human brain*, Proc Natl Acad Sci USA

[3] Jerbi, et al. (2007) *Coherent Neural Representation of Hand Speed in Humans Revealed by MEG Imaging*, Proc Natl Acad Sci USA

[4] Llinas et al. (1998) T*he neuronal basis for consciousness*, Phil.Trans. R. Soc. Lond. B

[5] Llinas et al. (1999) *Thalamocortical dysrhytmia: a neurological and neuropsychiatric syndrome characterized by magnetoencephalography. Proc Natl Acad Sci USA 96:15222-15227.*

[6] Rosanova et al. (2009) *Natural frequencies of human corticothalmic-circuits,* J. Neurosci., June 17,29(24):7679-7685

[7] Attal et al. (2009) *Modeling and detecting deep brain activity with MEG and EEG* IRBM--Biomed. Eng. & Res. *30*, 133-3

[8] Berger (1933) H. *Uber das Elektrenkephalogramm des Menschen.* Arch F Psychiat; 98:231–54.

[9] Miller (2000) *Time and the Brain*, CRC Press.

[10] Fisch and Spehlmann's *EEG primer : basic principles of digital and analog EEG*. Amsterdam: Elsevier; 1999.

[11] Barry (2007) *EEG differences between eyes-closed and eyes-open resting conditions*, Clinical Neurophysiology 118 (2007) 2765–2773

[12] Mantini et al, (2007) *Electrophysiological signatures of resting state networks in the human brain,* PNAS vol. 104 no.32, 13170-13175

[13] Mangin, et al. (2004) *Object based morphometry of the cerebral cortex*, IEEE Trans Med Imaging 23, no. 8, 968-82

[14] Chupin, et al. (2007) *Anatomically-constrained region deformation for the automated segmentation of the hippocampus and the amygdala: Method and validation on controls and patients with Alzheimer's disease*, Neuroimage, Volume 34, page 996—1019.

[15] Yelnik, et al. (2007) *A three-dimensional, histological and deformable atlas of the human basal ganglia. I. Atlas construction based on immunohistochemical and MRI data*, Neuroimage.

[16] Pantazis, et al. (2005) *A comparison of Random Field Theory and Permutation Methods for the statistical Analysis of MEG data*, NeuroImage.

[17] Jerbi, et al. (2004) *Localization of realistic cortical activity in MEG using current multipoles*, NeuroImage 22

Author: Yohan Attal
Institute: CNRS, UMR 7225
Street: 47 Bd de l'hopital
City: 75651 Paris Cedex 13
Country: France
Email: yohan.attal@upmc.fr

Characterization of Spontaneous Neuromagnetic Brain Rhythms Using Independent Component Analysis of Short-Time Fourier Transforms

Pavan Ramkumar[1], Aapo Hyvärinen[2], Lauri Parkkonen[1,3], and Riitta Hari[1]

[1] Brain Research Unit, Low Temperature Laboratory, Helsinki University of Technology, Finland
[2] Department of Mathematics and Statistics, University of Helsinki and Helsinki Institute of Information Technology, Helsinki, Finland
[3] Cognitive Neuroimaging Unit, INSERM-CEA/NeuroSpin, Gif-sur-Yvette, France

Abstract— **Unsupervised spatiotemporal and spectral characterization of spontaneous neuromagnetic brain rhythms over long time scales (minutes) would be useful for basic and clinical neuroscience. We recently showed that after applying a sparsifying transform (the short-time Fourier transform or STFT) to MEG data, independent component analysis (ICA) identified sources of oscillatory activity [1]. STFT on MEG data results in a 3-way data structure with temporal (time points), spatial (channels or source signals) and spectral (frequency bins) dimensions. Here, we propose to treat the 3-way data by using ICA to impose sparseness in the space–frequency dimension, resulting in a "spatial Fourier-ICA" (sFICA). Results of sFICA applied to STFTs of source-level MEG data from subjects who received natural stimulation or were resting suggest that sFICA is an efficient technique to identify both stimulus-related and intrinsic dynamics of neuromagnetic brain rhythms.**

Keywords— **Magnetoencephalography, cortical oscillations, spatiotemporal source localization, inverse problem, independent component analysis.**

I. Introduction

Spontaneous neuromagnetic brain rhythms have been characterized by their spatial and spectral features [2]. The most prominent of these oscillations are the ~10 Hz posterior alpha rhythm and the mu rhythm with ~10- and ~20-Hz components. These rhythms have been studied, e.g., by following their envelope dynamics as a function of external stimulation or tasks, and by modelling their cortical generators with local current dipoles. The envelopes of these brain rhythms are modulated in a region- and frequency-specific manner under a variety of complex conditions. Hence, the value of brain rhythms as markers of normal brain function, and as indicators of the subject's perceptual or cognitive state is apparent. Meanwhile, there is a growing interest to understand the behavior and functional significance of spontaneous brain rhythms during less controlled conditions (resting state or natural stimulation) where conventional averaging of the stimulus over several trials is not possible. Future studies of brain signalling during resting state or natural stimulation would benefit from a characterization of the dynamics of brain rhythms over time scales of minutes.

Here we propose a method to automatically identify statistically independent brain rhythms from raw MEG data. Our method takes into account prior knowledge about certain properties of brain rhythms. Specifically, since many brain rhythms are spatially rather locally generated and spectrally narrowband, but not necessarily temporally sparse, we sought independent generative processes from the data by maximizing spatial and spectral sparseness of these processes. We thus applied independent component analysis (ICA), a well-known sparseness-maximizing algorithm. However, ICA applied on the raw MEG data may not find sources of rhythmic activity [1]. Instead, applying a suitable sparsifying transform, such as the short-time Fourier transform, prior to ICA greatly improves the likelihood of identifying sources of oscillatory processes.

Short-time Fourier transforms (STFTs) of time series result in a 3-way data structure with temporal (time points), spatial (channels or source signals) and spectral (frequency bins) dimensions. The 3-way data can be treated as a 2-dimensional matrix in several ways. We have earlier [1] imposed sparseness on the time-frequency dimension with no constraint on the spatial dimension. In this work, we propose to apply ICA to STFTs of cortically-constrained minimum norm estimates of MEG data along the spatio-spectral dimension. We call this method spatial Fourier-ICA (sFICA). Each statistically independent component identified by sFICA is characterized by its cortical distribution, its spectral profile, and its envelope dynamics over the entire duration of the recording.

II. Materials and Methods

A. *Subjects*

Eleven healthy adults (6 females, age range 23–41 years) participated in the study after written informed consent. The

S. Supek and A. Sušac (Eds.): Advances in Biomagnetism – BIOMAG2010, IFMBE Proceedings 28, pp. 215–218, 2010.
www.springerlink.com

recordings were approved by the Ethics Committee of the Helsinki and Uusimaa Hospital District. Data from eight of the eleven subjects were used in this analysis.

B. *Experimental Setup*

The experiment consisted of three parts: (i) *natstim*: the subjects received a sequence of natural stimuli lasting about 8 min, comprising auditory, visual and tactile stimulus blocks of durations ranging from 6 to 33 s presented in a randomized order so that one block succeeded the next without any rest period in between, (ii) *restfix*: the subjects were resting quietly for about 10 min with their eyes open, fixating on a crosshair on the screen, and (iii) *nat&rest*: the subjects received a natural stimulus sequence similar to *natstim* but alternating with 15-s rest blocks, altogether lasting about 12 min. The stimuli were adapted from those used in an earlier study [3]. Auditory blocks consisted of tone beeps and speech. Visual blocks consisted of silent home-made video clips of faces, hands or buildings. Tactile stimuli were delivered at 4 Hz, in 15-s blocks, using pneumatic diaphragms attached to the index, middle and ring fingers of both hands; the order was random but homologous left and right fingers were always stimulated simultaneously.

C. *Measurements*

Measurements for each condition (*natstim*, *restfix* and *nat&rest*) were repeated once, resulting in a total of 6 measurements. In addition, after the experiment, a 2-min recording was conducted without a subject in the scanner (*emptyroom*) to estimate noise statistics. MEG signals were acquired with a 306-channel Elekta-Neuromag VectorView MEG system (Helsinki, Finland), lowpass-filtered at 200 Hz and digitized at 600 Hz. Anatomical MRIs were obtained using a 3-T General Electric Signa MRI scanner (Milwaukee, USA). During the MEG recording, four small head position indicator (HPI) coils, whose locations had been digitized with respect to anatomical landmarks, were continuously energized to track the subject's head position with respect to the MEG sensors. Electro-oculogram was monitored in all 11 subjects and electrocardiogram in 7.

D. *Preprocessing*

Raw MEG data were manually examined for 'bad channels' that were omitted from Signal Space Separation (SSS) [4] ("MaxFilter" version 2.0, Elekta Neuromag, Helsinki, Finland) employed to suppress external interference and to correct for head movements. Finally, the data were downsampled by a factor of 4.

E. *Minimum-Norm Inverse Solution*

For each subject, the cortical surface of the brain was reconstructed from an anatomical MRI using FreeSurfer (http://surfer.nmr.mgh.harvard.edu, Martinos Center for Biomedical Imaging, Massachusetts General Hospital, USA). Cortically constrained source points were defined at a spacing of 15 mm, such that each source point contained one current dipole oriented along the surface normal, and two transverse dipoles perpendicular to it. A single-compartment boundary element model (BEM) was used to model the conductivity of the cranium. A forward solution was computed for each source point. Next, using the MNE-suite [5] an empirical noise covariance matrix was computed from the 2-min *emptyroom* recordings, separately for each subject's data. A linear, depth-weighted [6], minimum-norm inverse operator **G** of dimensions (N_s x N_c) (where N_s is the number of source points and N_c is the number of channels; $N_s >> N_c$) was computed with a loose orientation constraint favoring source currents perpendicular to the local cortical surface by a factor of 2.5 with respect to the currents along the surface [7].

F. *Spectral Estimation in Source Space*

For each preprocessed data matrix **Y** (N_t time points x N_c channels), we computed the Hamming-windowed, complex-valued short-time Fourier transform over N_w non-overlapping windows, each one second in duration. The discrete Fourier transform of each time window $\mathbf{Y}_w$ may be interpreted as a matrix multiplication from the right by the Vandermonde matrix of the N_f-th roots of unity, where N_f is the number of frequency bins. Let us denote this matrix by **H** such that

$$\hat{\mathbf{Y}}_w = \mathbf{Y}_w \mathbf{H} \tag{1}$$

Here, $\mathbf{Y}_w$ is (N_c channels x N_t/N_w time points) and and **H** is (N_t/N_w time points x N_f frequency bins). The (N_c x N_wN_f) STFT across all time windows is then obtained by

$$\hat{\mathbf{Y}} = [\hat{\mathbf{Y}}_1 \ \ \hat{\mathbf{Y}}_2 \ \ldots \hat{\mathbf{Y}}_{N_w}] \tag{2}$$

The STFT in source space can then be obtained by left-multiplying the STFT by the linear inverse operator **G**, *viz*.

$$\hat{\mathbf{X}} = \mathbf{G}\hat{\mathbf{Y}} \tag{3}$$

$\hat{\mathbf{X}}$ (N_s x N_wN_f) can be recognized as a 3-way (N_s x N_w x N_f) tensor written out as a 2-way matrix. We then rearrange $\hat{\mathbf{X}}$ as a N_w x N_sN_f matrix such that each row is a collection of Fourier coefficients of all source points concatenated together horizontally for its respective time window *w*. Our 15-mm spacing between the source points resulted in N_s ~3000, including both hemispheres. A 128-point FFT used to compute the STFT led to 64 bins for each 1-s window. The 64 bins spanned a range of 0–75 Hz but we used only 21 bins spanning the range of 5–30 Hz for the analysis.

G. Spatial Fourier-ICA

For each subject's data, we temporally concatenated the rearranged STFT $\hat{\mathbf{X}}_k$; k = 1, 2, …, 6 runs (two runs each of *natstim*, *restfix*, and *nat&rest*) resulting in $\hat{\mathbf{X}}^{\mathrm{T}}_{\mathrm{concat}} = [\hat{\mathbf{X}}_1^{\mathrm{T}}\ \hat{\mathbf{X}}_2^{\mathrm{T}}\ \ldots\ \hat{\mathbf{X}}_6^{\mathrm{T}}]$. For the *restfix* runs, we discarded the first 1-min of the data before concatenation. To $\hat{\mathbf{X}}_{\mathrm{concat}}$, we applied a complex-valued independent component analysis, treating each row as an observed signal which is assumed to be a linear mixture of sources: we reduced the dimensionality of the data using principal component analysis (PCA) to 25, whitened the reduced data and then applied ICA using the complex-valued FastICA algorithm [8]. The estimation was symmetric and the nonlinearity was of the form $\log(1+|.|^2)$ [9].

H. Visualization and Statistics

Each complex-valued independent component (IC) resulted in a set of Fourier coefficients for each cortical source point. We visualized the spatial map of the IC by taking the mean of the squared Fourier amplitude (squared magnitude of the complex Fourier coefficients) across all frequency bands. Since the distribution of mean squared Fourier amplitude over the whole brain is highly non-Gaussian, we applied a threshold to display only those source points with top 5-percent squared Fourier amplitude for each component map. Since the source points represent only a sparse subset of the vertices in the triangular tessellation of the cortical surface, iterative smoothing was applied using the MNE suite: a blurred image was produced from a sparse image by applying a blurring matrix at each iteration step [9]; the smoothing factor was 15. To visualize the spectral profile, we plotted the mean and standard deviation of the Fourier power spectrum over all source points exceeding the 95-percentile threshold. The absolute values of the column of the estimated mixing matrix corresponding to the row of the estimated IC matrix give the time course of the Fourier amplitude envelope. We separately plotted the time courses of each of the 6 runs.

III. Results

Each individual IC is associated with a spatial map, a spectral profile and a time course. The time course of the IC represents the envelope of the oscillatory current source.

The method identified several ~10-Hz components in the calcarine and parieto-occipital sulci and several ~10-Hz and ~20-Hz components around the central sulcus; these components were clearly suppressed during visual and tactile stimulation, respectively. A few components were located in the auditory cortex, the lateral and medial temporal regions, the precuneus, and the posterior cingulate cortex; the spectral characteristics of these components were varied and, based on visual inspection, they did not clearly react to any particular stimulus type. The spatial distributions of the components were similar during stimulation vs. rest conditions. Fig. 1 shows the spatial map of one component for a single subject. The map is overlaid on reconstructed inflated surfaces of the left and right hemispheres. The component has local maxima in the calcarine sulcus of both left and right hemispheres. Fig. 2 shows the corresponding spectral profile and envelope time course for one of the *nat&rest* recordings. The spectrum, computed for source points with amplitude larger than the statistical threshold, indicates strongest amplitude at ~13.5 Hz. The time course, overlaid against the stimulus blocks color coded by type of stimulus, indicated clear suppression of the envelope during visual stimuli (indicated by red arrows).

IV. Discussion and Conclusion

Independent component analysis is an effective algorithm to separate sparse, statistically independent sources from a linear mixture. To separate statistically independent MEG oscillations, we applied ICA to a novel ordering of the spatio–temporo–spectral 3-way data to maximize spatial and spectral sparseness. Sparseness in both these dimensions is desirable to render the statistically independent components as physiologically independent as possible. The method provides an automatic way to identify different underlying frequency bands in the data and the spatial generators of different oscillations. This approach could advance quantitative studies of brain rhythms over time scales of minutes.

At present, our method is able to characterize oscillatory sources in single-subject data. It would also be beneficial to make inferences at the level of a group of subjects. However, since component maps are non-Gaussian, non-parametric tests could be applied for this purpose [10]. Moreover, the method needs to be extended to relate the time courses of independent components to the stimuli in a statistically sound manner. Ultimately, if the method is able to identify robust sources of rhythmic activity, it can be used to address neuroscientific questions concerning the properties of brain rhythms while the subject is resting performing tasks, and processing long natural stimulus sequences.

V. References

1. Hyvärinen A, Ramkumar P, Parkkonen L, Hari R. 2010. Neuroimage, 49:257–271.
2. Hari R, Salmelin R. 1997. Trends Neurosci 20:44–49.
3. Malinen S, Hlushchuk Y, Hari R. 2007. Neuroimage 35:131–139.
4. Taulu S, Kajola M. 2005. J App Phys 97:124905–124910.
5. http://www.nmr.mgh.harvard.edu/martinos/userInfo/data/sofMNE.php
6. Dale AM, Liu AK, Fischl BR, Buckner RL, Belliveau JW, Lewine JD, Halgren E. 2000. Neuron 26:55–67.
7. Lin F-H, Belliveau JW, Dale AM, Hämälainen MS. 2006. Hum Brain Mapp 27:1–13.
8. Hyvärinen A. 1999. IEEE Trans Neural Netw 10:626–634.
9. Bingham E, Hyvärinen A. 2000. Int J Neural Syst 10:1–8.
10. Nichols TE, Holmes AP. 2001. Hum Braim Mapp 15: 1–25.

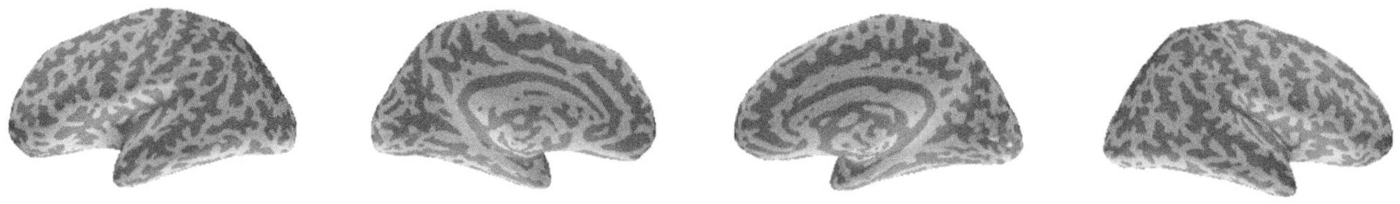

Fig. 1 Spatial map of an IC from a single subject overlaid on her inflated brain surface. Dark regions represent sulci and bright regions gyri. The surfaces from left to right are the lateral and medial views of the left hemisphere, and the medial and lateral views of the right hemisphere

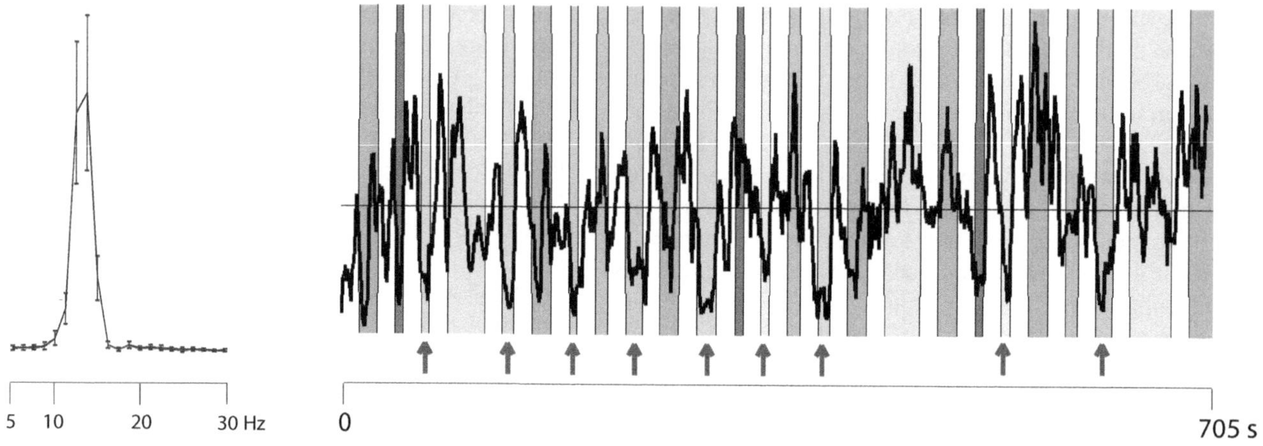

Fig. 2 The spectral and temporal characteristics of the component shown in Fig. 1. Left: Mean ± standard deviation of spectral amplitude over the 5–30 Hz range for source points exceeding the 95-percentile threshold. Right: Envelope time course of one of the *nat&rest* runs. Visual stimulus blocks are colored in green and cyan and indicated by red arrows, auditory stimuli in red and pink, and tactile stimuli in violet. The white regions are the interspersed rest periods

Prestimulus Oscillatory Brain Activity Influences the Perception of the McGurk Effect

Julian Keil[1], Niklas Ihssen[2], and Nathan Weisz[1]

[1] Department of Psychology, University of Konstanz and Center for Psychiatry Reichenau, D-78457 Konstanz, Germany
[2] School of Psychology, Bangor University, Gwynedd LL57 2AS, United Kingdom

Abstract— **The McGurk effect is a perceptual illusion, which demonstrates an interaction between auditory and visual sensory systems in speech perception. If a mismatch between the perception of a sound and the accompanying visual input occurs, it has been observed that the unified perception of both modalities fuses to a novel percept that neither matches the sound nor the visual stimulus. This effect only appears in 60-80% of trials containing mismatching information. The present study was designed to clarify the conditions under which this effect occurs and to identify cortical sources associated with it. We were interested in the potential influence of induced ongoing brain oscillations associated with varying perception. Our results indicate that the perception of the McGurk effect is associated with higher pre-stimulus beta activity in parietal, frontal, and temporal cortical areas as well as with higher post-stimulus beta activity in parietal cortical areas.**

Keywords— **MEG, Beta, McGurk, Induced Oscillation, Audiovisual Integration.**

I. Introduction

The integration of multimodal information is crucial in our day-to-day lives. While there is a substantial body of literature on the neural basis of unimodal information processing, less is known about the neural basis of multimodal information integration. Even less is known about the neural basis of processing mismatching multimodal information.

The McGurk effect is a perceptual illusion that was first described by McGurk and MacDonald in 1976 [1]. It occurs when auditory and visual information mismatch, i.e., if the video of an actor pronouncing the syllable "ga" is presented together with the audio stream of the syllable "ba". In this case, the information of the two modalities fuses to a novel percept (e.g. "da").

It has been demonstrated that when the perception of incongruent stimuli is compared with the perception of congruent stimuli, gamma band activity increases over left posterior parietal, occipital, and left inferior frontal cortices [2]. Animal studies [3] have provided evidence for an influence of visual information on higher auditory areas.

Most research on audiovisual information integration has focused on the comparison between the perception of congruent and incongruent stimuli. In contrast, in the present study we were specifically interested in the varying perception elicited by identical incongruent stimuli. As the McGurk effect is not present in all incongruent trials, this paradigm allows varying behavioral outcomes (e.g. fusion of information vs. response to only one modality) to be differentiated for identical stimuli.

The present study sought to elucidate (i) the conditions under which either a fusion or unimodal response occurs, and (ii) to identify the cortical sources associated with this differential outcome. Specifically, we were interested in the potential influence of ongoing ("background") brain oscillations associated with varying perception.

II. Methods

A. *Subjects*

12 (4 m/8f, mean age 25.2) paid volunteers participated in this study. All participants gave their written informed consent. All participants were right-handed, had normal hearing, and normal or corrected-to-normal vision.

B. *Experimental Design and Apparatus*

Participants were informed about the procedure and familiarized with the facilities. They were then prepared for the recording session and seated in the magnetically shielded room.

The experiment consisted of 390 trials in which we showed videos of an actor pronouncing the syllables "aba", "ada", or "aga". The stimuli were presented via Psyscope [4] on a MiniMac. Two thirds of the trials contained a mismatching audio stream (e.g. visual "ada" combined with auditory "aba"). Using a forced choice task, participants had to indicate by button press whether they had perceived "aba", "ada", "aga" or something else ("other"). Videos were stopped after the first frame for a randomized duration in order to make the onset of the audio stream and mouth movement unpredictable (See Figure 1 for timing). The Videos Stimuli were presented on a screen inside the magnetically shielded MEG acquisition room via a video

S. Supek and A. Sušac (Eds.): Advances in Biomagnetism – BIOMAG2010, IFMBE Proceedings 28, pp. 219–222, 2010.
www.springerlink.com

projector (DLA-G11E, JVC, Friedberg, Germany) and a set of mirrors positioned outside the room. The audio streams were presented with an analogue-to-digital converter (Motu 2408) via amplifiers (Servo 200, Samson) and a 6.1 m length, 4 mm diameter tube system (Etymotic Research, ER30). Sounds were corrected for the filters inherent in the tube system.

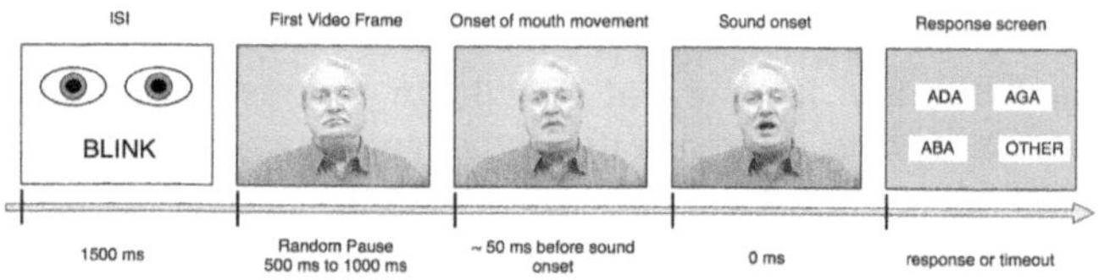

Fig. 1 Timing of the stimulation. Videos were paused after the first screen for a randomized duration. Mouth movement occurred ~50 ms before the sound onset. Congruent or incongruent sounds were presented at time 0. After the video, participants had to indicate their perception

C. Data Acquisition and Analysis

Magnetoencephalographic (MEG) recording was done with a 148-channel magnetometer (MAGNES 2500 WH, 4D Neuroimaging, San Diego, USA). A subject-specific headframe coordinate reference was defined by means of five anatomical landmarks. These head fiducials, five coils, and the subject's head shape were digitized with a Polhemus 3Space Fasttrack at the start of each session. The subject's head position relative to the pickup coils and the MEG sensors were estimated before and after each session to ensure that no large movements occurred during data acquisition.

Subjects were lying supine in a comfortable position. They were instructed to lie still during the stimulation and to avoid eye movements and blinks as best as possible. Continuous data sets were recorded with a sampling rate of 678.17 Hz (bandwidth 0.1 - 200 Hz). A video camera installed inside the MEG chamber allowed subjects' behaviour and compliance to be monitored throughout the experiment.

After data acquisition, epochs of four seconds (+/- 2s) around each sound onset were extracted from the raw data. Epochs were visually inspected for EOG, ECG, or movement artefacts. Trials were categorized according to the combination of type of video, type of audio, and type of response into four categories: (i) congruent (match between auditory and visual stimulus and correct response), (ii) fusion (mismatch between auditory and visual stimulus and response that neither matched the auditory nor the visual information), (iii) auditory (mismatch between auditory and visual stimulus and response that matched the auditory information), (iv) and visual (mismatch between auditory and visual stimulus and response that matched the visual information). The numbers of trials for the four different categories were equalized for each subject by random omission. This was done to ensure an identical improvement in signal-to-noise ratios for all trials. Resulting epochs were filtered with a 1 Hz high-pass filter (zero-phase, Butterworth) and then averaged and used for the analysis of oscillatory activity. As the pre-response activation was also of interest, no baseline was defined in the analysis of the oscillatory activity; and time-frequency and source analysis outputs for the conditions were compared directly.

A wavelet time-frequency transformation based on Morlet wavelets was used to assess the changes of induced oscillatory activity in the time-frequency domain. Average event-related activity was subtracted from single trials before computing the time-frequency transformation, in order to remove event-related oscillatory activity. This procedure resulted in single-trial estimates of oscillatory power between 2 Hz and 80 Hz in 2 Hz steps.

Source activity was projected onto individual anatomical MRI-images and subsequently normalized onto a standard MNI brain [5] using SPM2 [6], in order to calculate group statistics and for illustrative purposes.

Dynamic imaging of coherent sources (DICS, [7]) – a frequency-domain adaptive spatial filtering algorithm – was used to identify the sources of the effects found in the time-frequency domain. This algorithm has proven to be particularly powerful when localizing oscillatory sources [8].

In order to define relevant time and frequency windows, a cluster-based (at least 2 sensors per cluster) dependent-samples t-test with Monte-Carlo randomization was performed on the sensor data. This method allows the identification of clusters of significant differences in 2D and 3D (time, frequency and space). It effectively controls for multiple comparisons [9]. Clusters were defined as significant if the probability of observing larger effects from shuffled data was below 5%. The cluster-level test statistic is defined as the sum of the t statistics in 2D or 3D space in the respective cluster.

In order to scrutinize the probable neuronal generators of the observed sensor effects, statistical comparisons on the source level were computed using dependent-samples t-tests.

All aspects of offline treatment of the MEG signals were accomplished using fieldtrip, an open source signal processing toolbox for Matlab [10]. Anatomical structures corresponding to the statistical effects were found using the MNI brain and Talairach atlas [11].

III. Results

When comparing fusion trials versus trials with visual responses, we found two effects, one before sound onset and one thereafter. Using wavelet time-frequency analyses on the sensor level, we identified an increase in induced beta (15 Hz – 33 Hz) band power between -250 ms and -150 ms before onset of the auditory stimulus (Figure 2).

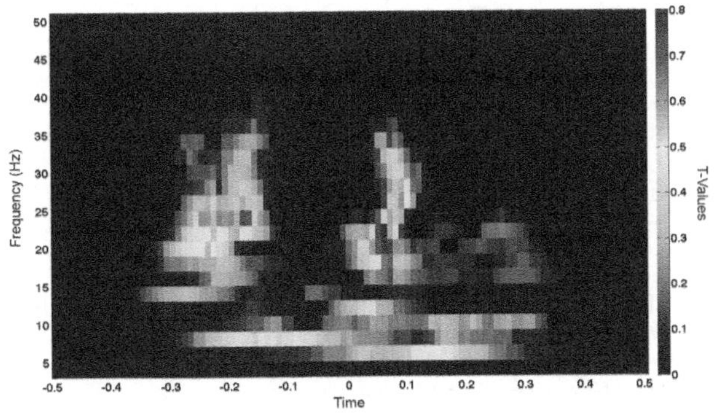

Fig. 2 Time-Frequency representation (t-values) of the two beta-band effects found on sensor level. Time point 0 marks the onset of the sound. (Masked for significant effects)

The DICS beamformer indicated the left Inferior Frontal Gyrus, Precuneus and Medial Frontal Gyrus as possible sources of this effect (Figure 3).

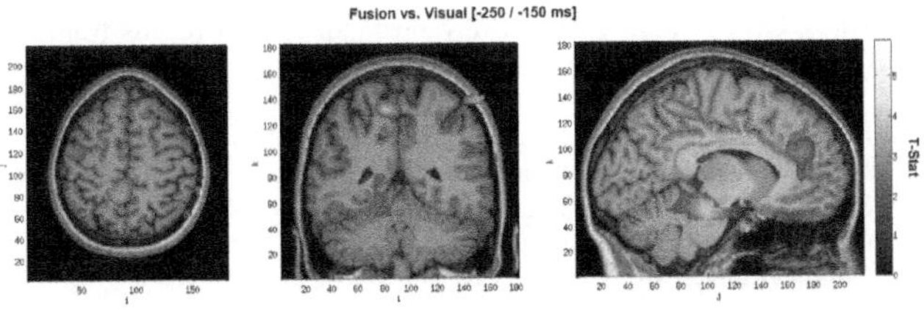

Fig. 3 DICS source projection of the pre-stimulus beta band effect (15 Hz – 33 Hz; -250 ms - -150 ms). (Masked for significant effects)

We also found an increase in beta (14-32 Hz) band power between 30 ms and 140 ms after the onset of the auditory stimulus (Figure 1). With the DICS beamformer we identified the Precuneus and Middle Frontal Gyrus as possible sources of this effect (Figure 4).

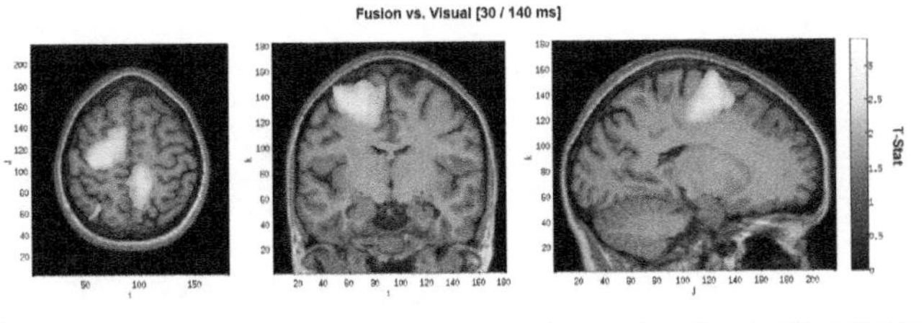

Fig. 4 DICS source projection of the post-stimulus beta band effect (14 Hz – 32 Hz; 30 ms – 140 ms). (Masked for significant effects)

IV. Discussion

In the present study, we used the McGurk paradigm to examine cortical processes related to varying perceptions elicited by identical incongruent stimuli. Earlier studies focused on cortical processes related to the processing of congruent versus incongruent stimuli. For instance, animal studies [3] identified direct projections from visual to auditory areas. ERP studies [12, 13] found that visual information influences auditory processing at various stages throughout the perceptual process. Common to all these studies is the finding that higher-order processing stages contribute to the perception of the McGurk effect. Corroborating the role of oscillatory cortical activity in audiovisual speech perception, Kaiser [2] found an increase in gamma band activity over left posterior parietal, occipital, and left inferior frontal cortex.

In contrast to the studies mentioned above, we analyzed varying perception elicited by identical incongruent stimuli. We found an increase in induced beta band activity before the actual perception of the McGurk effect in the left Inferior Frontal Gyrus, Precuneus, and Medial Frontal Gyrus (comprising the Frontal Eye Field). This pattern of co-activation indicates a top-down modulation of both visual and auditory cortex. It is possible that the likelihood to perceive an unified percept – in this case the McGurk effect – depends on the state of this network as reflected in ongoing oscillatory cortical activity. Further, certain expectations may shape the functional architecture of this network to enhance the processing of perceptual features.

We also found an increase in induced beta band activity during the perception of the McGurk effect in the Precuneus and Middle Frontal Gyrus (comprising the Frontal Eye Field). This replicates earlier findings on the role of higher-order processing stages involved in McGurk effect. The present results add to these findings by showing oscillatory activity is not merely elicited by mismatching auditory and visual information but related to the actual illusory fusion of both modalities.

A crucial question for further studies is to investigate the direction and flow of information between the identified cortical sources and also between primary sensory areas and higher perceptual areas. Another promising approach is the use of a simpler sensory integration paradigm, such as the ventriloquist effect [14], which may help to disentangle crossmodal interactions at primary sensory stages from interactions occurring at higher-order stages during audiovisual speech perception.

V. Conclusions

When the presentation of incongruent stimuli leads to a fusion of visual and auditory information rather than to a focus on the visual modality alone, both pre- and poststimulus oscillations in the beta band are modulated. Higher-order processing areas are involved before the perception of the McGurk effect as well as during the actual fusion of information. We conclude that specific ongoing brain oscillations facilitate the perception the McGurk effect, which may also influence the perception of natural multimodal stimuli.

Acknowledgment

This study was supported by DFG (Deutsche Forschungsgemeinschaft), TRI (Tinnitus Research Initiative) and Zukunftskolleg of the University of Konstanz. We thank Prof. Thomas Elbert and Dr. Sabine Heim for support and input.

References

1. McGurk, H., & MacDonald, J. (1976). Hearing lips and seeing voices. Nature , 264 (5588), 746-8.
2. Kaiser, J. (2004). Hearing Lips: Gamma-band Activity During Audiovisual Speech Perception. Cerebral Cortex, 15 (5), 646-653.
3. Bizley, J., Nodal, F., Bajo, V., Nelken, I., & King, A. (2007). Physiological and anatomical evidence for multisensory interactions in auditory cortex. Cerebral cortex, 17 (9), 2172-89.
4. Cohen JD, MacWhinney B, Flatt M, Provost J (1993) PsyScope: A new graphic interactive environment for designing psychology experiments. Behavioral Research Methods, Instruments, and Computers 25: 257–271.
5. Montreal Neurological Institute (MNI), Montreal, Canada; http://www.bic.mni.mcgill.ca/brainweb
6. SPM2 http://www.fil.ion.ucl.ac.uk/spm
7. Gross, J., Kujala, J., Hamalainen, M., Timmermann, L., Schnitzler, A., & Salmelin, R. (2001). Dynamic imaging of coherent sources: Studying neural interactions in the human brain. Proceedings of the National Academy of Sciences of the United States of America, 98 (2), 694-9.
8. Liljeström, M., Kujala, J., Jensen, O., & Salmelin, R. (2005). Neuromagnetic localization of rhythmic activity in the human brain: a comparison of three methods. NeuroImage , 25 (3), 734-45.
9. Maris, E., & Oostenveld, R. (2007). Nonparametric statistical testing of EEG- and MEG-data. Journal of Neuroscience Methods, 164 (1), 177-90.
10. Donders Institute for Brain, Cognition and Behaviour, Radboud University Nijmegen, the Netherlands http://www.ru.nl/neuroimaging/fieldtrip
11. Talairach atlas MRC Cognition and Brain Sciences Unit, Cambridge, England. http://imaging.mrc-cbu.cam.ac.uk/ imaging/ MniTalairach
12. Saint-Amour, D., De Sanctis, P., Molholm, S., Ritter, W., & Foxe, J. (2007). Seeing voices: High-density electrical mapping and source-analysis of the multisensory mismatch negativity evoked during the McGurk illusion. Neuropsychologia , 45 (3), 587-97.
13. Joassin, F., Maurage, P., Bruyer, R., Crommelinck, M., & Campanella, S. (2004). When audition alters vision: an event-related potential study of the cross-modal interactions between faces and voices. Neuroscience letters, 369 (2), 132-137.
14. Alais, D., Burr, D. (2004). The ventriloquist effect results from near-optimal bimodal integration. Current Biology, 14, 257-262.

Author: Julian Keil
Institute: University of Konstanz
Street: Universitaetsstrsse 10, P.O. Box D25
City: 78457 Konstanz
Country: Germany
Email: julian.keil@uni-konstanz.de

Relating Motor Cortical Oscillations to Motor γ-Aminobutyric Acid (GABA)

William Gaetz, James C. Edgar, D.-J. Wang, and Timothy P.L. Roberts

Department of Radiology, The Children's Hospital of Philadelphia and
the University of Pennsylvania, Philadelphia, PA, USA

Abstract— **The human motor cortex exhibits characteristic beta (15 – 30 Hz) and gamma oscillations (60-90 Hz) in the context of transient finger movement tasks. The functional significance of these oscillations, such as post-movement beta rebound (PMBR) and movement-related gamma synchrony (MRGS), remain unclear. Considerable animal and human non-invasive studies suggest that the networks supporting these motor cortex oscillations depend critically on the inhibitory neurotransmitter γ-Aminobutyric acid (GABA). However, a direct relation between MEG measured motor cortex oscillation power and frequency with resting GABA concentrations in adult human participants has not been demonstrated.**

In the present study, motor cortical responses were measured from 8 healthy adults participating in a cued button-press task using the right index finger. PMBR and MRGS measures were obtained using beamformer differential source localization. In particular, PMBR and MRGS measures were obtained from the time-frequency plots obtained from beamformer localized primary motor (MI) sources. For each participant, complimentary magnetic resonance spectroscopy (MRS) GABA measures from the motor hand knob of the left central sulcus were also obtained. GABA concentration was estimated as the ratio of the motor cortex GABA integral to a reference GABA phantom.

A significant linear relation was observed between MI GABA concentration and PMBR power ($R^2 = 0.42$, p=0.04), with no such relation observed for GABA and PMBR frequency. Conversely, a significant correlation was observed between MI GABA concentration and MRGS frequency ($R^2 = 0.67$, p=0.01), with no correlation observed between GABA concentration and MRGS power.

Present findings support a role for GABA in the generation and modulation of endogenous motor cortex rhythmic activity.

Keywords— **Movement-related gamma synchrony (MRGS), Post-movement beta rebound (PMBR), Magnetic Resonance Spectroscopy (MRS), γ-Aminobutyric acid (GABA), Event-related desynchrony / synchrony (ERD/ERS).**

I. Introduction

Animal [1, 2] and non-invasive human studies [3-6] have demonstrated a strong relationship between cortical oscillation power and frequency and the inhibitory neurotransmitter γ-Aminobutyric acid (GABA). Non-invasive magnetoencephalographic (MEG) observations have recently shown that visual cortex gamma (~50Hz) oscillation frequency is directly related to MRS measured GABA concentrations in adults [7]. Adult *motor cortex* is also known to exhibit several characteristic oscillations in both the beta (15-30 Hz) and gamma bands (60-90 Hz). In particular, beta-band power is known to decrease at movement onset and then "rebound" approximately 0.5 s following movement termination [8, 9]. In addition, gamma-band oscillations have been observed to increase in power at movement initiation (~300 ms in duration) [10], termed movement-related gamma synchrony or *MRGS*, and resting beta-band motor cortical oscillations have been shown to be augmented by administration of GABAergic compounds [3-6]. Moreover, the motor cortical PMBR period has been associated with an increased state of transient motor cortical inhibition, likely due to intracellular GABA currents [11]. Although the evidence for an association between beta and gamma motor oscillations and GABA is intriguing, a direct relation between MEG measures of synchronous motor cortical oscillation power and frequency with resting GABA concentrations in adult participants has not been demonstrated. The present study examined the relationship between MEG measured PMBR and MRGS and complimentary measures of motor cortex GABA. We hypothesized that PMBR power would vary directly with GABA concentration. In addition, we hypothesize that MI gamma frequency would vary directly with GABA concentration, as reported recently in an MEG study comparing visual cortex (VI) gamma with resting GABA concentrations [7].

II. Methods

A. MEG Methods

Eight healthy adult subjects (3F) participated (mean age = 33, range 22.9 to 45.1 yrs). Subjects pressed a keypad using the right index finger (100 trials) in response to a color change in a visually-presented fixation cross (4 s ISI on average). MEG data was continuously recorded using a 275-channel system (VSM Medtech, Canada). Contralateral motor cortex was localized using differential SAM beamformer methods. To localize post-movement beta rebound (PMBR), beta band power was localized for the 15-30 Hz frequency band using an active time window of 0.5 to 1.0 s, and a control window of -1.8 to -1.3 s, with the button press equal to time zero. Time-frequency percent-change plots obtained from the peak MI PMBR source waveforms were

S. Supek and A. Sušac (Eds.): Advances in Biomagnetism – BIOMAG2010, IFMBE Proceedings 28, pp. 223–226, 2010.
www.springerlink.com

then used to calculate PMBR frequency and power measures in units of percent change. PMBR frequency and power was reported as the frequency corresponding to the largest mean power observed within a 0.5 to 1.5 s time window for the 15-30 Hz frequency range.

Movement-related gamma-band peak locations are known to occur in close temporal relation to movement onset [8]. Differential beamformer images of the 60-90 Hz frequency band were created using an active time window of -0.1 to 0.2 s, and a control window of -1.8 to -1.5 s. Time-frequency percent-change plots from the individually-determined MI gamma-band peak locations were then analyzed for frequency and power. PMGS peak frequency and power was reported as the frequency corresponding to the largest mean power observed within a 0.1 to 0.2 s time window for the 60-90 Hz frequency range.

B. *MRS Methods*

Single voxel (30x30x30mm) GABA MRS was obtained using the MEGAPRESS [12] spectral editing sequence, via subtraction, with TE=68ms at 3T. Voxels were centered on the motor "hand-knob" of the left central sulcus. High-order shimming allowed FWHM line-widths < 15Hz. After Fourier transformation, phase correction was applied to the unsubtracted spectra and propagated to the subtracted spectrum. The integral under the GABA resonance (3 ppm) was obtained by spectral peak-fitting using a Gaussian resonance. A reference phantom of 50mM GABA placed in the head coil with the subject was also interrogated (using a 14x14x30mm voxel). GABA levels were estimated as the ratio of the motor cortex GABA integral to the phantom.

III. Results

MEG Results: Beamformer differential PMBR peak locations were observed from bilateral MI in all participants (typically stronger contralaterally). Time-frequency plots obtained from contralateral PMBR MI peak locations showed the expected beta-band ERD typically arising ~0.3 s prior to the button press and continuing ~0.4 s after button press. Characteristic PMBR was observed as a rapid increase in beta-band power starting ~0.5 s after button press. PMBR typically reached maximum power within 0.5 s, and remained elevated (relative to pre-movement levels) throughout the 2 s post-movement epoch (see Figure 1).

Beamformer differential images of MRGS 60-90 Hz activity consistently localized to the contralateral MI (see Figure 1). MRGS sources were strongly lateralized to contralateral MI, and unlike PMBR sources, did not include ipsilateral MI activity. Time-frequency plots of source waveforms from peak locations showed that MRGS duration was approximately 300 ms, and observed to arise coincidentally with the button press at time zero. Comparison of the PMBR and MRGS motor locations showed that PMBR localized to a more superior aspect of primary motor cortex than MRGS peak locations (paired t-test (corrected); $p<0.01$).

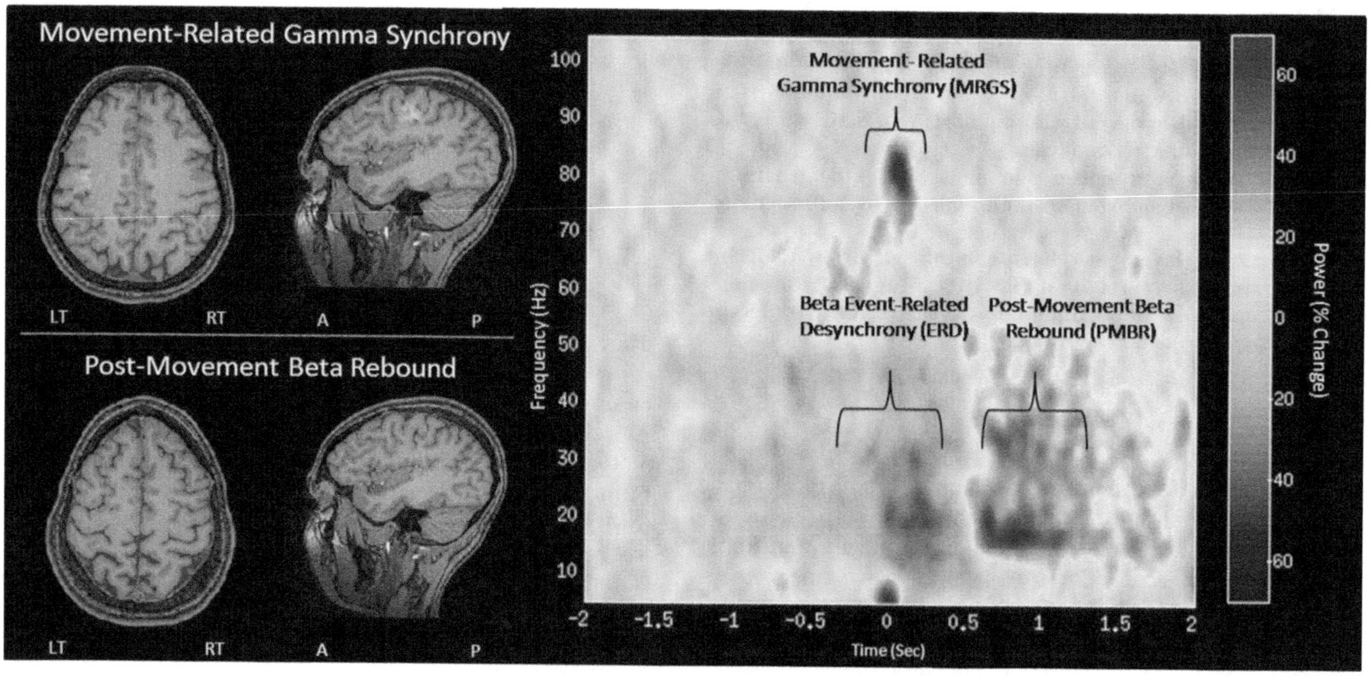

Fig. 1 Movement-related gamma synchrony (MRGS) and post-movement beta rebound (PMBR) locations for a representative subject. The time-frequency total power plot from the PMBR peak location (lower left) is shown on the right. Time-frequency colors indicate increases (red) and decreases (blue) in source power with respect to a pre-stimulus baseline (-2 to -1.5 sec), plotted in units of percent change

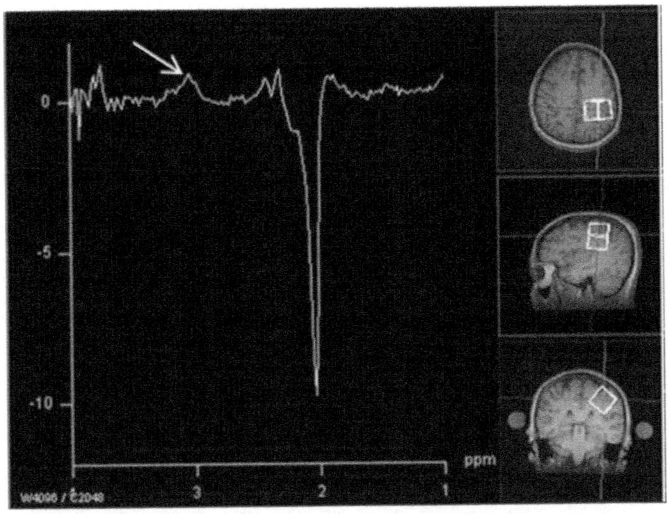

Fig. 2 Spectrum showing GABA resonance (arrow) obtained from radiological left-hemisphere motor cortex. MRI images to the right show the MRS voxel in relation to the motor "hand knob". GABA reference phantoms can be observed in the coronal MRI (bottom right)

MRS Results: A clear motor cortex GABA peak (measured relative to the 50mM phantom) was observed in all subjects (e.g. Figure 2 left). As anticipated, a significant association was observed between MI GABA concentration and PMBR power (R^2 = 0.47, p=0.03; see Figure 3). No relation was observed for GABA and PMBR frequency.

As anticipated, MI GABA was positively associated with MRGS frequency (R^2 = 0.67, p=0.01; see Figure 4). No such relation was observed for GABA and MRGS power.

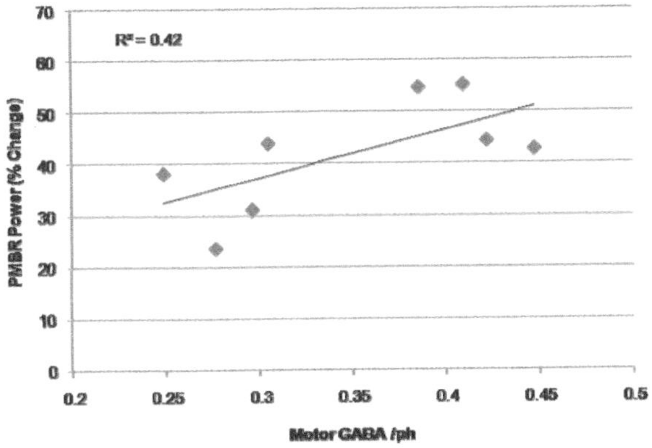

Fig. 3 The linear association between MI GABA concentration (x axis) and peak PMBR power (y axis) is shown. Increases in GABA concentration were associated with increased PMBR power (R^2 =0.42, p 0.04)

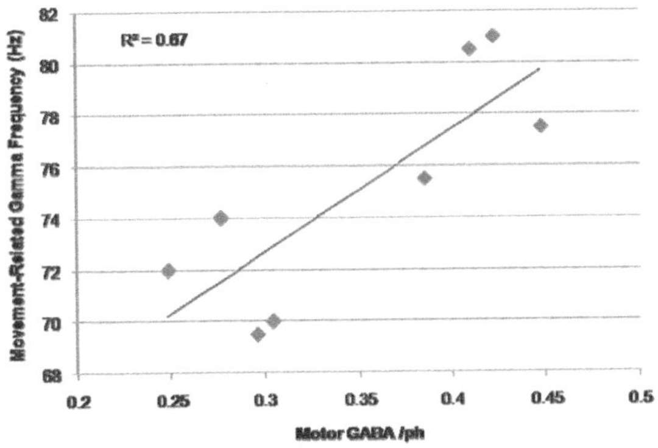

Fig. 4 The linear association between MI GABA concentration (x axis) and peak MRGS (y axis) is shown. Relatively large GABA levels are associated with higher MRGS frequencies (R^2 =0.67, p 0.01)

IV. CONCLUSIONS

GABA levels and motor cortex electrophysiological oscillatory activity were successfully measured in all subjects. As hypothesized, GABA levels correlated positively with MI beta and gamma activity. To our knowledge, these data are first to demonstrate a direct association between MEG measures of MI cortical oscillation activity and the inhibitory neurotransmitter GABA.

Previous observations have implicated a transient increase in MI beta-band power, such as PMBR, with a state of reduced motor cortical excitability. For example, Chen et al. [10] reported decreased Motor Evoked Potential (MEPs) amplitudes from an intrinsic hand muscle (FDI) when stimulating the contralateral MI with transcranial magnetic stimulation (TMS) pulses during a period of post-stimulus beta synchrony. A series of pharmaceutical studies have since demonstrated that motor cortex beta rhythms are augmented by benzodiazepine administration [4-6], thus supporting the hypothesis that motor cortex excitability is regulated through GABAergic inhibition. Biophysically designed computer networks which model motor cortex as an interconnected network of excitatory and inhibitory neurons have shown that an increase in beta-band power is directly related to the conductance of GABAa mediated inhibition [5]. In the present study, whereas PMBR power may reflect the level of GABA conductance, it is not yet clear how this rhythmic *post-movement* inhibition relates to motor function. Future studies relating PMBR measures to motor performance, motor learning (i.e., plasticity), and development may help explain the functional relevance of PMBR. Present results do, however, help account for the considerable inter-subject variability typically observed when measuring motor cortical oscillations from adult human participants using MEG.

Although an association between MI oscillatory activity and GABA was observed, the specific mechanisms and underlying networks relating motor cortical oscillation power (PMBR) and frequency (MRGS) to GABA concentration remain unclear. Recent MEG measures of VI gamma activity induced by visual gratings showed that decreased GABA concentration in VI predicted a lower visual cortex gamma oscillation frequency [7]. A similar finding was observed in the present study for the MI gamma-band activity only. This suggests a common GABA mechanism governing VI and MI cortical gamma networks. Neural network models based on hippocampus CA1 field-potential recordings suggest that the synaptic release of GABA generates inhibitory postsynaptic potentials (IPSPs) that briefly and rhythmically attenuate random axonal activity. This attenuation modulates excitatory inputs to interneurons and thus promotes the gamma-frequency field oscillation [13]. In the case of MRGS, these modulating signals may arise briefly in relation to movement initiated proprioceptive signals, or possibly from sub-cortical structures such as the sub-thalamic nucleus, also known to exhibit coherent gamma bursts in relation to transient movements [14]. Future investigations are needed to further explore MRGS in relation to specific movement tasks, and in participants with relatively high or low GABA MI levels.

In conclusion, our results demonstrate that multimodal imaging can non-invasively assess associations between brain oscillations and brain chemistry. It is hoped that continued work in this area will led to a better understanding of the mechanisms influencing the generation and modulation of endogenous rhythmic brain activity. Such insights may help explain the underlying functional mechanisms as they relate to normal development of the motor system. Finally, it is likely that work in this area will provide insight to the neural mechanisms underlying the abnormal oscillatory activity frequently observed patient populations (e.g., schizophrenia and autism), with work in this area helping to create treatments that target very specific neurotransmitters and, by extension, specific cortical network abnormalities in these patient populations.

Acknowledgment

The authors would like to thank Drs. Suresh Muthukumaraswamy and Krish Singh for developing and sharing Matlab based beamformer source analysis software, and for helpful methodological and theoretical discussions.

References

1. Roopun, A.K., Middleton, S.J., Cunningham, M.O., LeBeau, F.E., Bibbig, A., Whittington, M.A., Traub, R.D., 2006. A beta2-frequency (20-30 Hz) oscillation in nonsynaptic networks of somatosensory cortex. Proc Natl Acad Sci U S A 103, 15646-15650.
2. Yamawaki, N., Stanford, I.M., Hall, S.D., Woodhall, G.L., 2008. Pharmacologically induced and stimulus evoked rhythmic neuronal oscillatory activity in the primary motor cortex in vitro. Neuroscience 151, 386-395.
3. Wanquier, A., 1998. EEG and neuropharmachology, in electroencephalography. Basic Principles, Clinical Applications, and Related Fields, 660– 670.
4. Hall, S., Barnes, G., Furlong, P., Seri, S., Hillebrand, A., Hum Brain Mapp (In Press). Neuronal network pharmacodynamics of GABAergic modulation in the human cortex determined using pharmacomagnetoencephalography.
5. Jensen, O., Goel, P., Kopell, N., Pohja, M., Hari, R., Ermentrout, B., 2005. On the human sensorimotor-cortex beta rhythm: sources and modeling. Neuroimage 26, 347-355.
6. Baker, M.R., Baker, S.N., 2003. The effect of diazepam on motor cortical oscillations and corticomuscular coherence studied in man. J Physiol 546, 931-942.
7. Muthukumaraswamy SD, Edden RA, Jones DK, Swettenham JB, Singh KD. (2009). Resting GABA concentration predicts peak gamma frequency and fMRI amplitude in response to visual stimulation in humans. PNAS, (2009).
8. Pfurtscheller G, Neuper C. Motor imagery activates primary sensorimotor area in humans. Neurosci Lett. 1997 Dec 19;239(2-3):65-8.
9. Pfurtscheller G, Stancák A Jr, Neuper C. (1996). Post-movement beta synchronization. A correlate of an idling motor area? Electroencephalogr Clin Neurophysiol. 1996 Apr;98(4):281-93.
10. Cheyne, D., Bells, S., Ferrari, P., Gaetz, W., Bostan, A.C., 2008. Self-paced movements induce high-frequency gamma oscillations in primary motor cortex. Neuroimage 42, 332-342.
11. Chen, R., Corwell, B., Hallett, M., 1999. Modulation of motor cortex excitability by median nerve and digit stimulation. Exp Brain Res 129, 77-86.
12. Waddell KW, Avison MJ, James M. Joers JM, and Gore JC. A practical guide to robust detection of GABA in human brain by J-difference spectroscopy at 3 T using a standard volume coil. Magnetic Resonance Imaging, Volume 25, Issue 7, September 2007, Pages 1032-1038.
13. Traub RD, Cunningham MO, Gloveli T, LeBeau FE, Bibbig A, Buhl EH, Whittington MA. GABA-enhanced collective behavior in neuronal axons underlies persistent gamma-frequency oscillations. Proc Natl Acad Sci U S A. 2003 Sep 16;100(19):11047-52.
14. Lalo, E., Thobois, S., Sharott, A., Polo, G., Mertens, P., Pogosyan, A., Brown, P., 2008. Patterns of bidirectional communication between cortex and basal ganglia during movement in patients with Parkinson disease. J. Neurosci. 28, 3008–3016.

Author: William Gaetz
Institute: Children's Hospital of Philadelphia
Street: 34th and Civic Center Boulevard
City:Philadelphia PA
Country: USA
Email: bill.gaetz@gmail.com

Connectivity in the Human Fetal Brain

I. Kostovic and M. Pletikos

Croatian Institute for Brain Research, School Of Medicine, University of Zagreb, Zagreb, Croatia

Abstract— **It is generally accepted that complexity of neuronal connections in the human brain underlies complexity of cognitive and behavioral functions. In this paper we explain how developmental cellular events and connectivity patterns form special framework for early functions and establishment of adult circuitry. At the beginning of fetal period distributions of circuitry elements (axons, dendrites and synapses) is strictly bilaminar outside of proliferative zones. The first quantifiable number of synapses occurs after 15 postconceptional weeks (PCW) in the transient subplate zone. This early synapses are substrate of endogenous, sensory non-driven activity. Distribution within the connectivity compartments is random and no modular units were observed. Important phenomenon in development of afferent input is waiting phenomenon: axons wait for several months until they gradually invade target cortical area. These waiting synaptic compartments may be visualized in both in vivo and postmortem images. The bipolar orientation of synapses, with predominance of deep circuitry is basis for cortical surface-positive electric response. The majority of described circuitry is transient. It is only after 23 PCW that sensory driven afferents establish connections in the corresponding cortical target area. At the end of gestation another phenomenon take place and that is a dramatic reorganization of fetal circuitry due to the retraction of axons and gradual disappearance of transient fetal layers. In conclusion, early fetal circuitry shows transient laminar distribution and transient endogenous activity. Sensory driven activity is late phenomenon and further development of connectivity proceeds after dramatic spatial and cellular reorganization of cortex.**

Keywords— **development, connectivity, human, brain.**

I. Introduction

Human cerebral cortex is characterized by increased number of cortical processing units, appearance of new cortical areas, enlargement of association areas, expanded connectivity of layer three pyramidal neurons and extremely long period of structural maturation. The expanded connectivity is related to increased number and size of cortical areas articles [1]. The increased number of cortico-cortical connections forms distributed networks articles [2]. This complex system of connections forms the connection matrix called connectome [3]. It is generally accepted that complexity of human connectome underlies complexity of cognitive and behavioral functions.

The developmental events leading to the development of the earliest cortical circuitry (axons dendrites synapses), early networks cannot be studied directly in the human brain. These events were described experimentally in primates [4], [5]. However development of the connections in the human fetal cortex was successfully analyzed using histochemical, electromicroscopical and MR diffusion tensor imaging [6], [7], [8].

In this paper we explain basic spatial and temporal parameters of developing early fetal circuitry.

II. Materials and Metods

For the reconstruction of the circuitry elements we have used sections from the Zagreb Neuroembriological Collection and in vivo MRI by the procedure for the human tissue samples that was approved by Internal Review Board of the Ethical Committee at the school of Medicine, University of Zagreb.

III. Results

A. When It Begins and How It Is Distributed?

The first synapses appear around 8 PCW while the first quantifiable synapses occur after 15 weeks. Distribution of the first circuitry elements is strictly bilaminar outside of proliferative zones. Two laminas contain circuitry elements, the subplate zone below cortical plate and the marginal zone above cortical plate. The early circuitry elements are present across the cerebral wall and are present in both neocortex and limbic cortex. The external afferents to the early circuitry originate in the brainstem [6], [9], [10].

B. What Is Functional Meaning of the Early Synapses?

It is very likely that the early circuitry is endogenous and that it displays spontaneous oscillation [11], [12].

S. Supek and A. Sušac (Eds.): Advances in Biomagnetism – BIOMAG2010, IFMBE Proceedings 28, pp. 227–229, 2010.
www.springerlink.com

C. Spatial and Temporal Parameters of the Midfetal Transient Circuitry

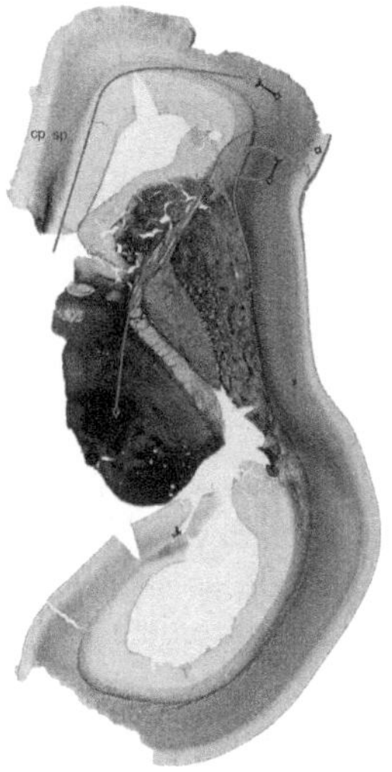

Fig. 1 Transient fetal pathways

The midfetal period is characterized fully developed transient fetal pattern (Fig 1.). The most significant compartment for the development of circuitry is the transient subplate zone [10], [13]. Transient nature of the circuitry within the subplate and the marginal zone may be defined on the basis of the following parameters: (1) growing afferent fibers which are not in the final position (2) presence of the extracellular space for navigation and guidance of the axons transient nature of the circuitry elements (3) transient arrangement of the axons in the axonal strata (4) absence of sensory driven and default circuitry

1. Growing afferent fibers use prominent subplate zone as a waiting compartment. Cerebral pathways grow over months during their path finding, tangential growth and target selection in the subplate [6], [10], [14]. Waiting is important phenomenon in development of afferent input: axons wait for several months until they gradually invade target cortical area.

2. The waiting synaptic compartments may be visualized in both in vivo and postmortem images due to the high content of the extracellular matrix (ECM). The subplate zone shows very low MR signal intensity due unique mixture of three major tissue components: numerous cells dispersed within the loose plexiform network of the fibers and embedded in an abundant and hydrophilic [14].

3. Many neurons of the transient subplate zone are transformed into the gyral white matter or disappear. Accordingly their synapses disappear [10].

4. General pattern of growth of pathway is through axonal strata of intermediate zone and graduate ingrowth of the deep anlage of the subplate zone. The axonal strata are fibrilar compartments which provide the connectivity within frontal and posterior occipital and temporal lobe. This basic anteroposterioral organization becomes obscure during perinatal development and sagittal organizations remains only in the frontal an occipital polar regions [14]. We believe that basic tangential anisotropic organization of the fiber systems and random disperse network in the subplate provides essential framework for the later networks development. (See later.)

5. According to most investigators transient thalamocortical circuitry is not sufficient for successful sensory (visual, auditory, tactile and other) transmission. This transient circuitry was described as non-sensory driven [7], [11]. However this transient circuitry shows the most advanced neuronal differentiation and numerous synapses in both neocortical and limbic areas. In the limbic cortex of the hippocampus the majority of elements are located in superficially located marginal zone. The bipolar orientation of synapses, with predominance of deep circuitry is basis for cortical surface-positive electric response.

D. Late Fetal and Perinatal Development

With appearance of synapses in the deep cortical plate around 25 weeks of gestation begins the new period of synapse development. Rapid development of synapses within the cortical plate is enhanced by relocation of the thalamocortical afferents from subplate to cortical plate [7], [15], [16], [17]. Due to the maturation of the thalamocortical circuitry there is possibility to evoke cortical responses after auditory, visual and sensory stimulation in the preterm infants [18].

Late fetal period is characterized by the growth of the cortico-cortical connection. During this period long associative bundles begin to appear in the human cerebrum and by the birth seem to be in the position. This period is characterized with simultaneous coexistence of endogenous and sensory driven circuitry [7]. The transient subplate zone is still present in newborn cortex. Endogenous and sensory-driven activity form basis for transient cortical function and transient behavioral phenomena over the prolonged period of the development. The prolonged coexistence of endogenous transient subplate zone circuitry seems to be a salient feature of the developing human cerebral cortex [19], [20].

In the period of prenatal development we noticed a rapid development of the limbic cortex (hippocampus, cingular and precingulate cortex).

Cortical executive systems can be studied by different approaches using transmagnetic stimulation [21]. It can be concluded by their studies that executive (efferent) system develop parallel to the afferent system and elaboration of endogenous system.

E. Early Postnatal Development

Early postnatal development is characterized by the growth of the short cortico-cortical connections, rapid synaptogenesis and dendritic development [22] and development of local circuitry neuron [23]. Development of cortical networks in this period can be studied by resting state functional MR [24]. It was noticed that resting state network related to primary sensory motor functions develops earlier and faster than default resting state networks. This is consistent with gradual emergence of the consciousness [25].

IV. Conclusion

Endogenous circuitry of the human cortex develops during the earliest fetal life. Whiten the transient cellular zones and is characterized by strictly laminar distribution. The peak of the transient circuitry occurs during the midfetal development in transient cellular waiting compartments which are developed more in human then in any other mammalian brain. This transient circuitry is gradually replaced by permanent circuitry of the cortical plate. However the coexistence of transient and permanent circuitry is prolonged throughout the first year of life. The change of the diffuse circuitry with focused networks seems to be characteristic of the developing networks and occurs primarily in primary somatosensory networks. It is surprising associative cortical areas related to emotions and memory seems to have preconscious development and may be minor but developmentally significant initial default network. It is encouraging that the development of cortical circuitry can be analyzed by modern imaging methods and will help to understand the developmental origin of mental and cognitive disorders.

References

1. Rakic P. Specification of cerebral cortical areas *Science.* 1988;241:170–176.
2. Selemon LD, Goldman-Rakic PS. Common cortical and subcortical targets of the dorsolateral prefrontal and posterior parietal cortices in the rhesus monkey: evidence for a distributed neural network subserving spatially guided behavior *J Neurosci.* 1988;241:170–176.
3. Sporns I, Tonioni G, K?tter R. The Human Connectome: A Structural Description of the Human Brain *PLoS Comp Biol.* 2005;4:e42.
4. Rakic P. Neurons in rhesus monkey visual cortex: systematic rela-tion between time of origin and eventual disposition *Science.* .
5. Rakic P. Evolution of the neocortex: a perspective from developmental biology *Nat Rev Neurosci.* 2009;10:724–735.
6. Kostovic I, Jovanov-Milosevic N. The development of cerebral connections during the first 20-45 weeks' gestation *Semin Fetal Neonatal Med.* 2006;11:415–422.
7. Kostovic I, Judas M. Transient patterns of cortical lamination during prenatal life: do they have implications for treatment? *Neurosci Biobehav Rev.* 2007;31:1157–1168.
8. Huang H, Xue R, Zhang J et al. Anatomical characterization of human fetal brain development with diffusion tensor magnetic resonance imaging *J Neurosci.* 2009;29:4263–4273.
9. Nobin A, Bj?rklund A. Topography of the monoamine neuron systems in the human brain as revealed in fetuses *Acta Physiol Scand Suppl.* 1973;388:1–40.
10. Kostovic I, Rakic P. Developmental history of the transient subplate zone in the visual and somatosensory cortex of the macaque monkey and human brain *J Comp Neurol.* 1990;297:441–470.
11. Penn AA, Shatz CJ. Brain waves and brain wiring: the role of endogenous and sensory-driven neural activity in development *Ped Res.* 1999;45:447–458.
12. Hanganu IL, Kilb W, Luhmann HJ. Functional synaptic projections onto subplate neurons in neonatal rat somatosensory cortex *J Neurosci.* 2002;22:7156–7176.
13. Kostovic I, Molliver ME. New Interpretation Of Laminar Development Of Cerebral-Cortex - Synaptogenesis In Different Layers Of Neopallium In Human Fetus *Anat Rec.* 1974;178:395–395.
14. Kostovic I, Judas M, Rados M et al. Laminar organization of the human fetal cerebrum revealed by histochemical markers and magnetic resonance imaging *Cereb Cortex.* 2002;12:536–544.
15. Kostovic I, Goldman-Rakic PS. Transient cholinesterase staining in the mediodorsal nucleus of the thalamus and its connections in the developing human and monkey brain *J Comp Neurol.* 1983;219:432–447.
16. Kostovic I, Rakic P. Development of prestriate visual projections in the monkey and human fetal cerebrum revealed by transient cholinesterase staining *J Neurosci.* 1984;4:24–42.
17. Kostovic I, Judas M. Correlation between the sequential ingrowth of afferents and transient patterns of cortical lamination in preterm infants *Anat Rec.* 2002;267:1–6.
18. Vanhatalo S, Kaila K. Development of neonatal EEG activity: From phenomenology to physiology *Semin Fetal Neonatal Med.* 2006;11:471–478.
19. Kostovic I, Judas M. Prolonged coexistence of transient and permanent circuitry elements in the developing cerebral cortex of fetuses and preterm infants *Dev Med Child Neurol.* 2006;48:288–293.
20. Hadders-Algra M. Putative neural substrate of normal and abnormal general movements *Neurosci Biobehav Rev.* 2007;31:1181–1190.
21. Eyre JA, Miller S, Clowry (GJ. Functional corticospinal projections are established prenatally in the human foetus permitting involvement in the development of spinal motor centres *Brain.* 2000;123:51–64.
22. Petanjek Z, Judas M, Kostovic I et al. Lifespan Alterations of Basal Dendritic Trees of Pyramidal Neurons in the Human Prefrontal Cortex: A Layer-Speci?c Pattern *Cereb Cortex.* 2008;18:915–929.
23. Mrzljak L, Uylings HBM, Kostovic (I. Prenatal Development of Neurons in the Human Prefrontal Cortex: II. A Quantitative Golgi Study *J Comp Neurol.* 1992;316:485–496.
24. Fransson P. Spontaneous low-frequency BOLD signal fluctuations: an fMRI investigation of the resting-state default mode of brain function hypothesis *Hum Brain Mapp.* 2005;26:15–29.
25. Lagercrantz H, Changeux JP. The Emergence of Human Consciousness: From Fetal to Neonatal Life *Ped Res.* 2009;65:255.

Author: Ivica Kostovic, MD, Dsc
Institute: Croatian Institute for Brain Research, School of Medicine
Street: Salata 12
City: Zagreb
Country: Croatia
Email: ikostov@hiim.hr

Electrical Brain Function of Prematurely Born Babies

Sampsa Vanhatalo

Helsinki University Central Hospital, Department of Children's Clinical Neurophysiology, Helsinki, Finland

Abstract— **Routine recordings of preterm human electroencephalography (EEG) have been successfully performed for over four decades. Over these years, the EEG recording techniques, as well as the ways of EEG interpretation have remained essentially unchanged. Recent studies have demonstrated that major components of the preterm EEG signal are lost with the conventional EEG recording. At the same time, basic studies have suggested that brain development *per se* relies on activity that is readily ignored in the existing EEG practise. This paper summarizes a new framework of EEG interpretation that combines the knowledge regarding functional and structural brain development. An approach of this kind may ultimately open novel avenues to conduct rational, evidence based studies on brain pathophysiologies and their treatments.**

Keywords— **preterm EEG, spontaneous activity transient, SAT, FbEEG.**

I. Introduction

Recording of electrical brain activity by scalp electroencephalography (EEG) from preterm babies has been part of clinical routines for four decades [1,2]. Basic features of EEG activity during the early phases of development were characterized in 1960s in series of elegant studies on both human babies and other mammal species, such as sheep, dogs, and guinea pigs (for a review, see ref [3] and references therein). It was noted by then already, that main aspects of the early EEG trace are surprisingly similar between different species, and hence explorations of physiological and pathological mechanisms of brain activity may be studied using rather straightforward extrapolations from animal models to human babies.

The subsequent development (from the late 1960s onwards) of this field, however, led the human and animal research onto separate tracks. Animal studies became dominated by an interest in microelectrodes and subcellular level questions. At the same time, studies on human babies became steered towards identification of features in the EEG signal that have empiric, statistical correlates with babies' clinical status or outcome. This has given rise to a literature with rich repertoire of phenomenological ('graphological') EEG features, which comprise the core of prevailing neonatal EEG classification, and is centered around two main aspects:

1. *Continuity, ie. intermittency in the activity.* The most remarkable feature in the early preterm (scalp) EEG is the absence of continuous oscillations/fluctuations. The EEG signal of an early preterm baby consists of short (up to a few seconds) episodes of robust activity (see Table 1) followed by periods of relative silence, often visually perceived as 'inactive' trace.
2. *Change in frequency content.* There is a common perception that the activity content of the preterm EEG changes from the initial abundance of low frequencies to the later (relative) abundance of faster frequencies.

EEG interpretation of this kind has dominated the literature and clinical practise for more than three decades. It is not considered to involve any direct or implicit framework that could bind together the underlying brain mechanisms and/or structures with the recorded EEG signal. Moreover, akin to most EEG classifications, neonatal EEG classification is heavily biased by 'frequency thinking': EEG phenomena are mostly classified according to their visually perceived frequency content. In the case of preterm EEG, this has led to a myriad of descriptive terms that mainly refer to the variations of the slow (*'delta'*) EEG waves (cf. Table 1) recorded by different authors, with different device configurations, or possibly from different brain areas.

In addition to the stagnancy in EEG classification, it is also notable that the specifications of recording methods have been practically unchanged since its first implementation in 1950s. Most importantly, neonatal EEG is still recorded with a few channel configuration by using amplifiers that effectively cut out the largest EEG components of the preterm EEG, the infraslow activity (cf. refs [3,4]).

II. Merger of Human and Animal Studies

Not surprisingly, an EEG tradition with complex nomenclatures of graphemes and strict conventions of recording setups has been clinically successfull but scientifically even counterproductive: Until recently, it has quite effectively precluded comparative, translational

S. Supek and A. Sušac (Eds.): Advances in Biomagnetism – BIOMAG2010, IFMBE Proceedings 28, pp. 230–233, 2010.
www.springerlink.com

studies between experimental models and human babies. This situation has been recently changed by advances in both EEG techniques and developmental neuroscience.

Recent introduction of EEG methods with an ability to record undistorted, *i.e.* completely unfiltered EEG signal (so called Full-band EEG, *i.e.* DC coupled amplification and DC stable electrode-skin interface enabling recording of the full physiological range of signals [4]) have disclosed an abundance of brain activity at infraslow range. Interestingly, these activities are not oscillations like the faster frequencies, but they rather comprise complex events with slow 'carrier wave' and nested oscillations at several higher frequency bands in a 'multiband fashion' [5]. Moreover, recordings with higher density of EEG electrodes on the scalp (using EEG caps) have shown that much of neonatal brain activity goes unnoticed with the current sparse spacing of electrodes [3,6,7].

New observations in the basic neurobiology have established the view that growth of neuronal networks depends on endogenous activation within the networks. This activation is generated spontaneously in *all* structures in the central nervous system. These are intermittent bouts self-organizing and self-terminating activity [8] that may 'spread' by triggering activation in other brain areas connected to it. Such activity depends on immature mechanisms of neuronal transmission, especially on the immature form of GABAergic transmission. Endogenous activity of this kind is believed to act as an 'epigenetic link'

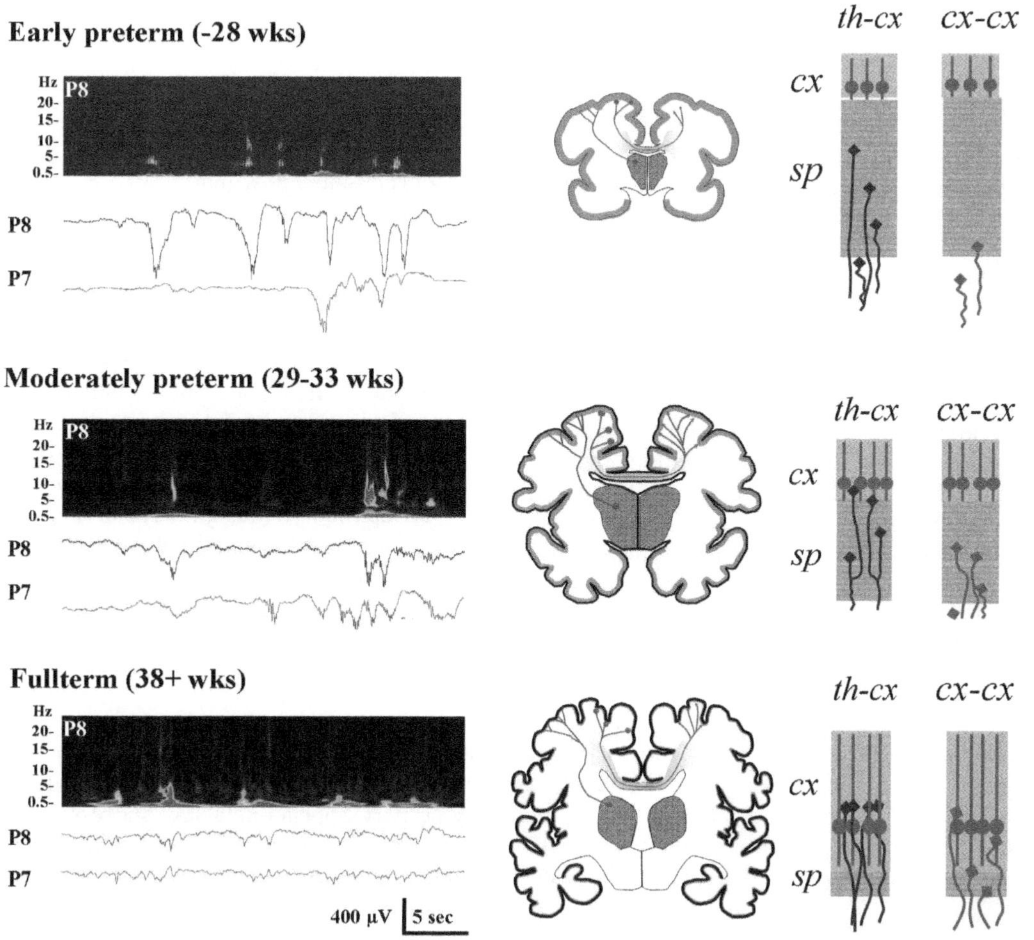

Fig. 1 *Major developmental milestones of cortical function (left side) and structure (right side) of the human preterm baby.* Left side shows one EEG trace from each parietal area lobe (P8 is right and P7 is left; Fz reference). In addition, a time-frequency representation (by wavelet) of the P8 trace is shown to demonstrate the multiband nature of the EEG events (see main text). Reprinted with permission from [10]. For details, see the reference

that enables up to few thousand genes construct the immensely complex brain networks: Genes control brain development by regulating the molecular mechanisms underlying this network activity. Importantly, an activity of this kind precedes the sensory inputs that will later arrive from sensory organs to contribute to the finer tuning of brain networks.

Recent studies in animal models have suggested that these glutamate-driven spontaneous activity transients (SATs [5]) take place only during the early development when they are permitted by the 'immature' GABAergic transmission [9,10]. This, in turn, is due to an early developmental lack of potassium chloride co-transporter (KCC2) expression, which leads to a situation where GABA-A channel activation may depolarize neuronal membranes [10].

III. Human histology and EEG

Our studies using *in situ* hybridization in histological specimens of human fetal brain tissue showed that KCC2 gene expression begins as late as during the last trimester of pregnancy in human babies [5]. It shows a marked variation between brain areas. The onset of expression occurs over a prolonged period of time, which is typical of all human brain development [11].

Using FbEEG technique, we showed that the main component in the EEG of an early premature baby is the intermittently occurring *Spontaneous Activity Transient* (SAT [5]), which consists of a particularily slow (1-5 seconds) and high amplitude (upto $800\mu V$) carrier wave and several faster frequencies (1-30Hz) nested within it. Later (Vanhatalo, unpublished) studies on in several tens of babies have disclosed further details: Spatial analysis by comparison of 20-64 channels of EEG have shown that, during the last three months of pregnancy, SAT events reduce in amplitude and they transform into a polyphasic and more globally synchronous activation pattern (exemplified in Fig. 1). Interestingly, high pass filtering, akin to the conventional EEG recording settings, distorts these events so that they become variants of the delta-related graphoelements commonly described in neonatal EEG literature (see also Table 1).

Qualitative and temporal comparison between brain histological maturation and the maturational change in the SAT events suggests that they are interrelated. We have recently proposed a framework for EEG analysis that attempts to combine the structural and functional changes (see Fig. 1). This framework proposes that the change from focal to global SAT occurrence reflects the growth of longer range cortico-cortical connections. Increase in SAT area on the cortex, together with the evolving gyration will lead to transformation of SAT events from uniphasic to multiphasic.

IV. Conclusions

It seems possible now that

i. The early brain activities can be recorded in a routine manner in the neonatal intensive care units, so that no significant temporal and spatial information is lost.
ii. Analysis of the EEG signal can be approached using a framework that is built upon knowledge of cortical events, their structural underpinnings, and even molecular level mechanisms.
iii. By recognizing these brain events (SATs) from human EEG signal, a genuinely evidence based approach can be applied to design translational studies on various pathophysiologies, their mechanisms and possible therapeutic strategies.

Table 1 Frequently used terms in the neonatal EEG literature, with comments from a neurophysiological point of view. See also main text. Adapted from ref [10]

EEG terms	*Common feature*	*Comment*
Delta waves from 0.3 to 2 Hz Delta-brush Delta crest Spindle-shaped bursts of fast activity Rapid rhythm Rapid bursts Spindle-like fast Fast activity at 14-24 Hz Ripples of prematurity Positive slow wave	Cerebral (cortical) EEG event which has both low and high frequency component at the same time	The non-distorted EEG correlate of this cortical activity has no or barely any oscillatory component at delta frequencies. The higher frequency activity appears as a few successive cycles at about 5-30Hz. These events are clearly not genuine spindles (akin to the sleep spindles later in life) or any other frequency-wise defined oscillations, neither are they "fast" as defined in current EEG literature (see [4]). Cortical activity of this type does probably form a physiological entity (or spectrum) with some variations depending on e.g. the context (endogenous vs. evoked), vigilance state, level of brain maturation, and cortical area.

References

1. Lamblin MD, Andre M, Challamel MJ et al. (1999) Electroencephalography of the premature and term newborn. Maturational aspects and glossary. Neurophysiol Clin, 29: 123-219
2. Tharp BR. (1990) Electrophysiological brain maturation in premature infants: an historical perspective. J Clin Neurophysiol. 7:302-14
3. Vanhatalo S, Kaila K. (2006) Ontogenesis of EEG activity: from phenomenology to physiology. Seminars in Fetal and Neonatal Medicine 11:471-478
4. Vanhatalo S, Voipio J, Kaila K (2005) Full-Band EEG (FbEEG): an emerging standard in electroencephalography. Clin. Neurophys, 116:1-8
5. Vanhatalo S, Palva M, Andersson S et al. (2005) Slow endogenous activity transients and developmental expression of K-Cl cotransporter 2 in the immature human cortex. Eur. J. Neurosci 22:2799-2804
6. Grieve PG, Emerson RG, Isler JR, Stark RI (2004) Quantitative analysis of spatial sampling error in the infant and adult electroencephalogram. Neuroimage 21:1260-1274.
7. Roche-Labarbe N, Aarabi A, Kongolo G, Gondry-Jouet C, Dümpelmann M, Grebe R, Wallois F (2008) High-resolution electroencephalography and source localization in neonates. Hum Brain Mapp. 29:167-176.
8. de la Prida LM, Huberfeld G, Cohen I, Miles R (2006) Threshold behavior in the initiation of hippocampal population bursts. Neuron, 49: 131-142
9. Sipilä ST, Kaila K (2008) GABAergic Control of CA3-driven network events in the developing hippocampus. In Inhibitory Regulation of Excitatory Neurotransmission, ed. M. Darlison (Springer-Verlag, Heidelberg. 2008), Results Probl Cell Differ, 44: 99-121
10. Vanhatalo S, Kaila K (2009) Spontaneous and evoked activity in the early human brain. In. The Newborn Brain: Neuroscience & Clinical Applications 2nd edition (Eds. H. Lagercrantz, Hanson, Peebles, Ment & P. Evrard) Cambridge University Press, 2009
11. Kostovic I, Judas M (2006) Prolonged coexistence of transient and permanent circuitry elements in the developing cerebral cortex of fetuses and preterm infants. Dev Med Child Neurol, 48: 388-393

Author: Sampsa Vanhatalo
Institute: Children's Clinical Neurophysiology
Street: Helsinki University Central Hospital, P.O. Box 280
City: 00029 HUS (Helsinki)
Country: Finland
Email: sampsa.vanhatalo@helsinki.fi

Early Development of Sensory Systems in Prematurely Born Babies

Sampsa Vanhatalo

Helsinki University Central Hospital, Department of Children's Clinical Neurophysiology, Helsinki, Finland

Abstract— **There has been an increasing clinical demand to assess the function and integrity of thalamocortical (th-cx) circuitries very early in life in preterm babies. The conventional somatosensory evoked potential (SEP) paradigm has failed, mainly due to its inability to adapt to the specific properties of the early developing th-cx circuitry. Based on the anatomical and physiological knowledge from experimental studies, a paradigm has been introduced to enable bedside testing of somatosensory mechanisms as early as before 30th week after conception. This may open a novel way of early identification of patients at developmental risks.**

Keywords— **somatosensory evoked response, SER, SEP, subplate.**

I. Introduction

Despite major advances in the survival rates through the stay in the neonatal intensive care unit (NICU), a major proportion of the extremely low birth weight infants still develop with debilitating neurocognitive dysfunctions [1]. While an early detection of these future problems is often difficult because of a lack of neurological capabilities to be clinically tested, it has become evident that many of them may be associated with compromised thalamo-cortical (th-cx) connectivity. Th-cx connections can be damaged by the in preterms rather frequent intracranial hemorrhages, which are routinely captured by brain ultrasound examination. Due to the immaturity of nervous system, however, clinical examinations can not detect whether a possible brain lesion is associated with a damage to the th-cx connections responsible for conveying (motor and) sensory information. For an early diagnosis, and even more so for potential early interventions, it would be of high interest to have a possibility to assess somatosensory pathways immediately after possible vascular lesions. This would mean a test of sensory circuitry as early as during the first postnatal weeks of an early preterm baby, ie. at around the 28-32th conceptional weeks.

II. Clinical status presence

It is unfortunate in this context, that the conventional SEP responses are obtained in only a variable proportion of preterm babies [2], discouraging their use as a routine monitoring tool. The reported failure is likely due to the conventional SEP paradigm that is not optimal for early sensory system. It is now known, that the responses in the early brain are variable, they can be markedly long and low in frequency, and they show fatigue after few stimuli (for a more detailed account, see ref [3]).

In this context, it is helpful to note that several studies in animal models, both *in vivo* and *in vitro*, have disclosed the mechanisms by which cortical tissue reacts to stimulation by th-cx afferents. It has been shown that the early afferent th-cx inputs terminate only in subplate (SP), which is a prominent structure (upto four times thicker than somatosensory cortex in humans) laying underneath the 'cortex proper' during the second half of pregnancy [4]. Their stimulation leads to a long-lasting, but easily fatiguing activation of SP networks, which in turn can trigger rapid oscillatory reaction in the overlying cortex. The ensuing cortical reaction to sensory stimulus resembles, but is not identical with, the endogenous, spontaneous activity transients (SAT) shown to take place in the cortex during the same developmental period [5].

III. Structural development

Development of afferent innervation from thalamus to cortex (th-cx afferents) begins in mid-gestation. The first fibers arrive during the 24-26th week of pregnancy [6], ie. at the time when the first babies will survive in the NICU, and there are sufficiently good connections to see cortical somatosensory reactions as early as during 26-28th week after conception. A schematic drawing of this development relative to seomarosensory responses is shown in Fig. 1.

IV. Recent studies

In our recent studies [7,8], we have recruited very preterm babies to record their electroencephalography (EEG) activity during tactile stimulation of peripheral body parts. Tactile stimulation can be readily performed in the incubators where the babies are treated for their prematurity.

S. Supek and A. Sušac (Eds.): Advances in Biomagnetism – BIOMAG2010, IFMBE Proceedings 28, pp. 234–237, 2010.
www.springerlink.com

(A video of such study during routine EEG recording, see http://www.helsinki.fi/science/eeg/tSEP_video/) We found that individual stimuli will result in robust EEG responses (somatosensory evoked responses, SER) that have a duration longer than one second, and amplitudes up to hundreds of microvolts [8] (Fig. 2).

Importantly, these responses have a complex character: They involve both a slow 'carrier wave' as well as a faster oscillation riding on the top of it (see also [9,10]). The conventional SEP response, major N (or N1 response) is also seen by mere visual inspection in many responses [8] (Fig. 2). It is, however, intriguingly variable, explaining the reported difficulties in its identification from the typical average responses.

We showed recently that SERs can be recorded also from one EEG derivation, but only if that is properly placed over the primary sensory area (contralateral centro-temporal for hand stimuli, and midline centroparietal for foot stimuli) (see Fig. 3). Since brain monitors with one or two recording are widely used in the NICUs, our observation opens a

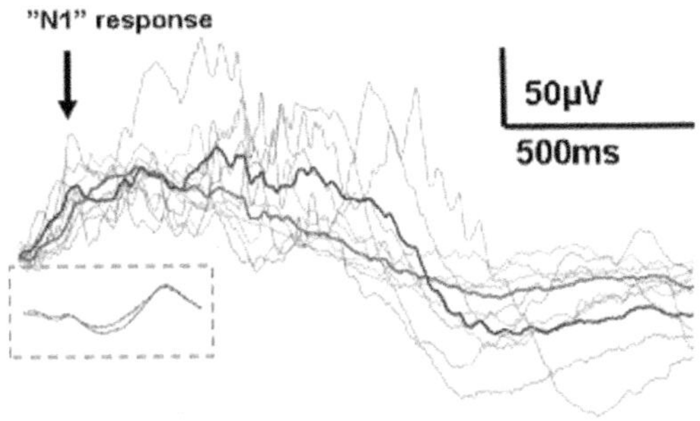

Fig. 2 Comparison of the conventional SEPs (inset) and the SERs obtained by tactile stimulation in a preterm baby. Note the dramatically different time and amplitude scales in the responses really seen in the EEG vs. the responses typically recorded in the clinical practise. Figure is adapted from ref [8]. Note that time and amplitude scales are the same in all traces

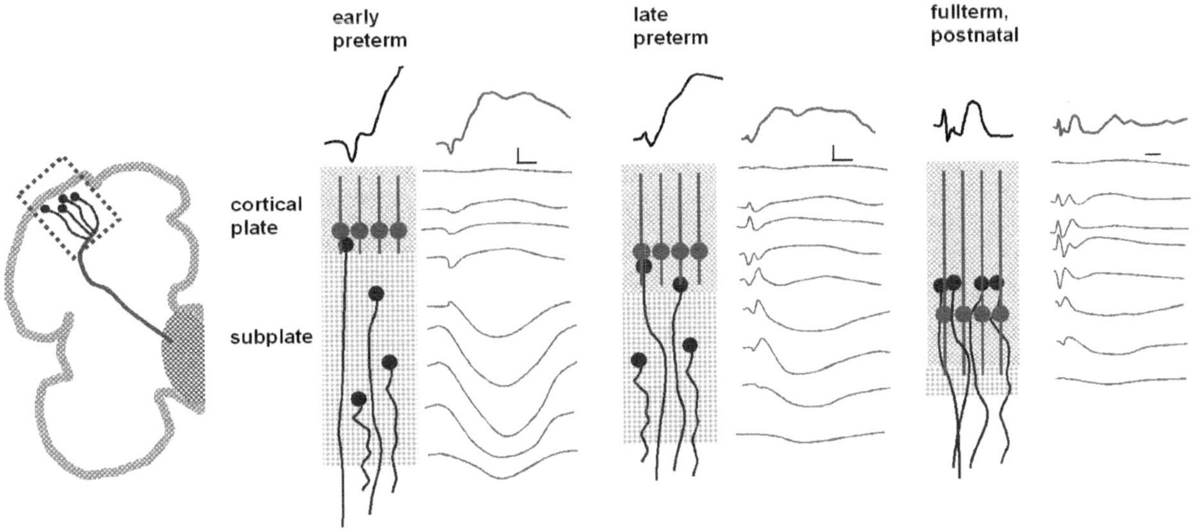

Fig. 1 Correlation between the maturation of sensory thalamo-cortical connections and the cortical responses (SEP, SER). Note the changing depth profile from subplate to cortex, as well as the changing temporal structure from a very slow to a rapid responses. Figure adapted from ref [3]

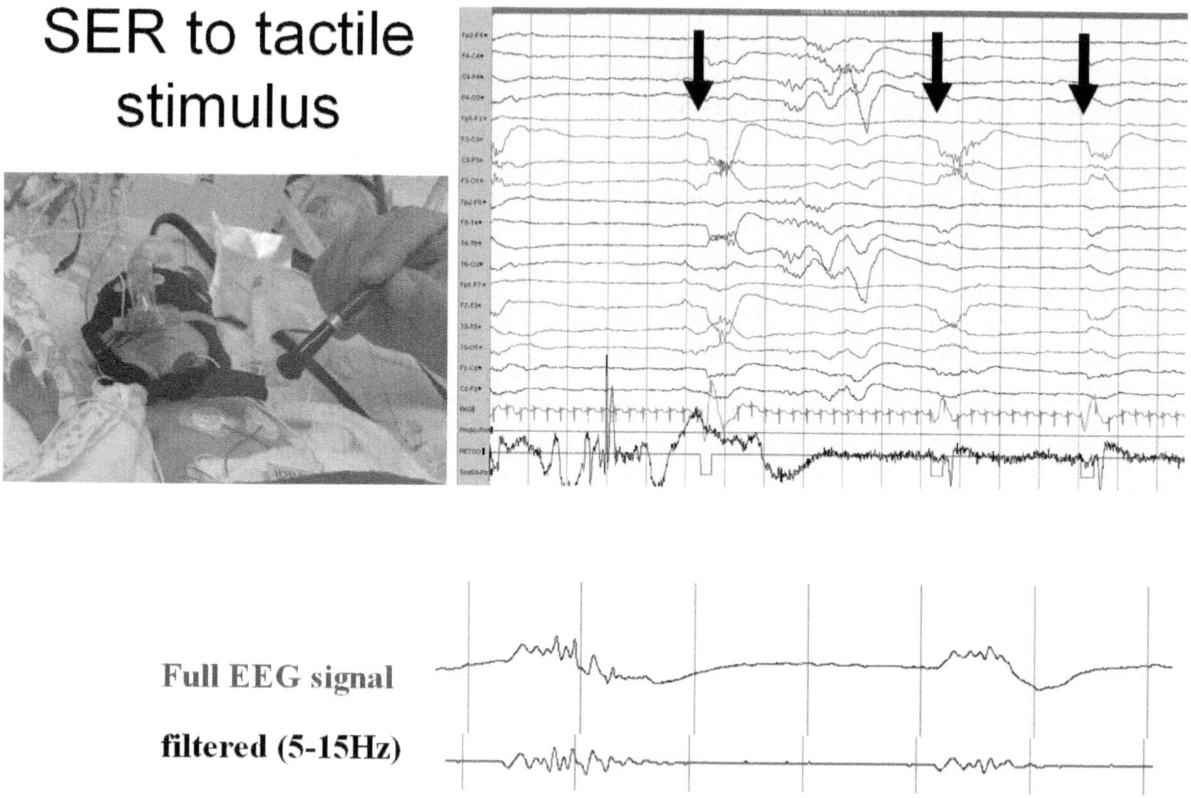

Fig. 3 Performance of the tactile stimulus on a preterm baby (32 weeks after conception) in the incubator, using a brush stimulator [8]. Responses are shown in the raw EEG recording (arrows), and one channel is extracted to show the appearance of the signal (current source density montage) without filtering and after filtering at 5-15Hz frequency band

possibility that assessment of the integrity of somatosensory pathways may become a routine part of the clinical neurological assessment of the youngest premies.

Our preliminary analysis of SERs also suggests, that they are surprisingly resistant as a whole to parenchymal brain damage by bleeding, but their internal structure (frequency content) may be changed by brain lesions (Kontio et al, unpublished).

V. Conclusion

A routine assessment of SERs during EEG recording makes EEG method a genuinely multimodalic tool. Coupled with the FbEEG recording (temporal dimension [13]) and high density EEG caps (spatial dimension), SERs can add the subcortical (system dimension [3]) component to a comprehensive, High Fidelity (HiFi [11]) assessment of preterm babies.

References

1. Marlow N, Wolke D, Bracewell MA, Samara M (2005) Neurologic and developmental disability at six years after extremely preterm birth. N Engl J Med 352:9–19
2. Taylor MJ, Boor R, Ekert PG (1996) Preterm maturation of the somatosensory evoked potential. Electroenceph Clin Neurophysiol 100: 448–452.
3. Vanhatalo S, Lauronen L (2006) Neonatal SEP - back to bedside with basic science. Semin Fetal Neonatal Med, 11: 464-470
4. Kostović I, Judas M (2007) Transient patterns of cortical lamination during prenatal life: Do they have implications for treatment? Neurosci Biobehav Rev, 31:1157-1168.
5. Vanhatalo S, Palva M, Andersson S et al. (2005) Slow endogenous activity transients and developmental expression of K-Cl cotransporter 2 in the immature human cortex. Eur. J. Neurosci 22:2799-2804

6. Kostović I, Jovanov-Milosević N (2006) The development of cerebral connections during the first 20-45 weeks' gestation. Semin Fetal Neonatal Med. 11:415-22
7. Vanhatalo S, Kaila K (2009) Spontaneous and evoked activity in the early human brain. In. The Newborn Brain: Neuroscience & Clinical Applications 2nd edition (Eds. H. Lagercrantz, Hanson, Peebles, Ment & P. Evrard) Cambridge University Press
8. Vanhatalo S, Jousmäki V, Andersson S, Metsäranta M (2009) An easy and practical method for routine, bedside testing of somatosensory systems in extremely low birth weight infants (ELBW) . Pediatric Research 66:710-3
9. Hrbek A, Karlberg P, Olsson T (1973) Development of visual and somatosensory evoked responses in pre-term newborn infants. Electroencephalogr Clin Neurophysiol, 34: 225-232.
10. Milh M, Kaminska A, Huon C, et al. (2007) Rapid cortical oscillations and early motor activity in premature human neonate. Cereb Cortex, 17:1582-1594.
11. Vanhatalo S, Metsäranta M, Andersson S (2008) High-fidelity recording of brain activity in the extremely preterm babies: Feasibility study in the incubator. Clin Neurophysiol, 119:439-445.

Sampsa Vanhatalo
Children's Clinical Neurophysiology
Helsinki University Central Hospital
P.O. Box 280
00029 HUS (Helsinki)
Finland
sampsa.vanhatalo@helsinki.fi

Maturation of the Cortical Somatosensory Processing as Studied by MEG

Päivi Nevalainen[1], Leena Lauronen[2], and Elina Pihko[3]

[1] BioMag laboratory, Hospital district of Helsinki and Uusimaa, HUSLAB, Helsinki University Central Hospital (HUCH), P.O. Box 340, FIN-00029 HUS, Finland

[2] Department of Clinical Neurophysiology, Hospital for Children and Adolescents, HUCH, Helsinki, Finland

[3] Brain Research Unit, Low Temperature Laboratory, Helsinki University of Technology, P.O.Box 5100, FIN-02015 TKK, Espoo, Finland

Abstract— **We present here a review of our studies on maturation of the somatosensory system during development. The data of most subjects have previously been published in several articles. Altogether 79 healthy newborns, 25 older infants (aged 6 months to 6 years), 13 adolescents (12 to 18 years), and 12 adults have participated in our studies. MEG was measured after electric stimulation of median nerve (MN) (some newborns and adults and all adolescents) or tactile stimulation of index finger (all age groups).**

In newborns, the first response after MN stimulation peaks at 30 ms with anteriorly flowing underlying current, similar to the well-known adult N20m. The immaturity is seen in the subsequent prominent cortical responses from the primary somatosensory cortex (SI) both for tactile and electric stimulation: the direction of the equivalent current dipole (ECD) of the neonatal M60 continues to point anteriorly contrasting the posteriorly oriented ECD of P35m (MN stimulation) or M50 (tactile stimulation) in adults. In newborns, a later response, M200, localized more inferiorly and laterally than M60, most likely represents activity from the secondary somatosensory cortex (SII). M200 ECD is stronger in quiet than active sleep and with longer interstimulus intervals.

During the first 6 months of life, M60 develops into a W-shaped response with a notch separating two distinct peaks. In later infancy, this notch gradually increases in amplitude crossing the baseline and revealing an underlying ECD pointing posteriorly. By the age of 2 years it has developed into an adult-like M50 response, which then persists through childhood and adolescence.

MEG is a useful method to study the function and development of the somatosensory cortices from the newborn age, when SEFs are still characterized by several immature features. By 6 years of age the SEFs attain the adult-like pattern reflecting functional maturation of the somatosensory cortex.

Keywords— **Development, infant, newborn, somatosensory system, plasticity.**

I. Introduction

Until recently, objective investigation of the developing central nervous system has been hampered because of a lack in noninvasive research methods. Consequently, little is known about processing of sensory information (*e.g.* somatosensory) and development of these processes in infants and children. At the same time the very low birth weight (VLBW) infants, whose survival has tremendously increased in the last decades, form a new group of patients susceptible to neurological impairments, such as cerebral palsy (CP). Thus, a need for new prognostic tools planning of rehabilitation and treatment resources to those in need exists.

Normal and abnormal functional development of the somatosensory system has been thus far investigated mainly with somatosensory evoked potentials (SEPs). Although SEPs provide a millisecond timescale temporal resolution, the conventional recordings with only a few electrodes have not allowed source localization. Furthermore, source localization using electroencephalography-based methods in infants is challenging because of the confounding effects caused by changes in anatomical structures, such as fontanels and skull thickness, on conductivity. Magnetoencephalography (MEG) is less sensitive to these conductivity differences and offers thus a unique tool for precise temporal as well as spatial investigation of the cortical functions, particularly in infants.

II. Materials and Methods

Our aim was to determine the utility of magnetoencephalography (MEG) and somatosensory evoked magnetic fields (SEFs) in investigating the somatosensory system of infants and children. We studied 79 healthy newborns, 25 infants and children, 13 adolescents, and 12 adults. In addition, we aimed at characterizing the normal developmental pattern of somatosensory information processing.

In most subjects we applied tactile stimulation of the contralateral index finger with an interstimulus interval (ISI) of 2 s. In 11 newborns, additional sessions with ISIs of 0.5 and 4 s were performed, and the ipsilateral index finger was stimulated in a separate session in 10 other newborns to detect possible ipsilateral responses. In addition, in 12 newborns we applied electrical stimulation of the median nerve at the wrist. The ISI was 2 s and the stimulation intensity was set just above the motor threshold (Table 1).

Table 1 Numbers of healthy newborns recorded with different stimulation parameters

Stimulation	Number of newborns
Tactile stimulation (ISI 2 s) of the index finger	71
Ipsilateral tactile stimulation	10
Additional interstimulus intervals	11
Electrical median nerve stimulation	12

The MEG recordings were performed in the BioMag Laboratory at the Helsinki University Central Hospital with a whole head helmet-shaped sensor array consisting of 306 independent channels: 204 planar gradiometers and 102 magnetometers (Vectorview, Elekta Neuromag Oy, Helsinki, Finland). For sleep stage monitoring, electroencephalography (EEG) was recorded with one to three electrodes placed at F4, P4, Cz, or P3 and electro-oculography (EOG) from two electrodes, one above the left and the other below the right eye canthi. Both EEG and MEG were bandpass filtered at 0.03 – 257 Hz. The sampling rate varied in different studies between 987 and 1002 Hz.

Four position indicator coils were attached on a cloth cap in infants and children. In older subjects the coils were attached on the skin. With a three-dimensional digitizer, we determined the coil positions relative to anatomical landmarks to construct an individual Cartesian coordinate system. In this coordinate system the preauricular points determined the x-axis, which pointed to the right. The y-axis was perpendicular to the x-axis pointing towards the nasion and the z-axis, perpendicular to the x-y plane, pointed upwards.

When necessary, we fed the infant before placing him/her on a bed next to the MEG measuring helmet. In infant studies the MEG device was in supine position. Newborns and 6-month-olds lay with one hemisphere downwards over the "occipital" part of the helmet. Older children and adults lay on their back. One or two researchers were in the room with the infants and children holding the stimulator on the index finger and observing the subject's behavior. The researcher(s) coded the behavioral state onto special trigger channels linked to the raw data file. EEG and behavioral coding served for off-line sleep stage determination. No sedation was used during the measurements. In the measurements of adults the sleep stage was determined based on electrophysiological data only. For the adolescents, the MEG device was in upright position and the subject was sitting comfortably watching a self chosen film without audio.

All newborns, infants and adults were measured while they were asleep. The sleep stage was determined from the electrophysiological data and the behavioral coding. In newborns the sleep stage was characterized as quiet sleep (QS) or active sleep (AS). For older infants and adults, we classified the sleep stages according to the guidelines from the classical EEG criteria [1]. Periods when the sleep stage could not be unambiguously specified were excluded from further analysis. Periods with movement artifacts were discarded from the averages. For further information of artifact removal see original articles [2–4].

The equivalent current dipole (ECD) model was calculated to estimate the location, strength and orientation of the sources in a spherical conductor model. The time 100 ms before the stimulus was used as a baseline. For tactile stimulation trials of all studies the signals were digitally low-pass filtered at 90 Hz prior to analysis. For MN stimulation we applied no further filtering. The peak of each deflection was determined by modeling single dipoles with 1-ms intervals around the visually determined peaks. The ECD with the greatest dipole was then selected for further analysis. Please see the original articles for more detailed methods [2–4].

III. Results

The SEFs of healthy newborns markedly differ from those of adults. As expected, the response latencies are delayed and the deflections wider. The first response of the newborns peaks at 30 ms (M30) after median nerve (MN) stimulation. It is accounted for by an equivalent current dipole (ECD) with an anterior direction of current flow, similar to the well-known adult N20m. However, after this initial M30 the ECD continues to point anteriorly sharply contrasting the posteriorly oriented ECD of P35m in adults. [2] Tactile stimulation does generally not elicit a clear N20m/M30 in neither adults nor infants. The opposite ECD directions are still demonstrated in the main early cortical responses: posteriorly oriented ECD of M50 in adults and anteriorly oriented ECD of M60 in newborns. [2–4]

Transformation from this M60 to the adult form begins already within the first 6 months of life, during which the neonatal M60 shifts into a W-shaped response with a notch separating two distinct peaks (Fig.1). [2,4] During later infancy, this notch gradually increases in amplitude crossing the baseline. Finally, by the age of 2-3 years it has developed into an adult-like response, the M50 with an underlying ECD pointing posteriorly, which then persists through childhood (6 y) [4] and adolescence (our own unpublished observation).

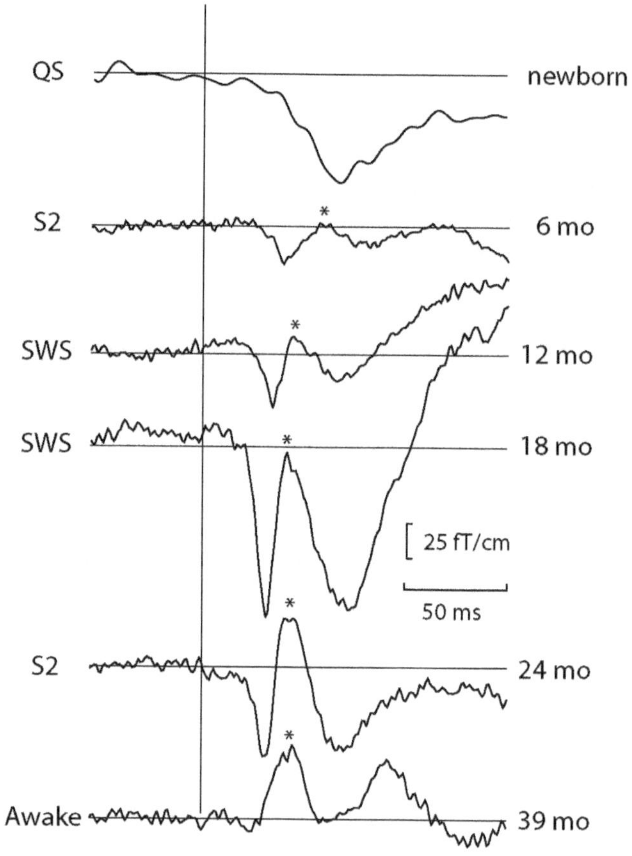

Fig. 1 Age-related changes in the somatosensory evoked responses to tactile stimulation of the index finger. Responses from one representative gradiometer channel are shown. The measurements from 6 to 39 months are from the same infant. The star denotes the emerging M50 response. QS = quiet sleep, S2= sleep stage 2, SWS = slow wave sleep

In healthy newborns, besides the M60, tactile stimulation of the index finger elicits a long latency response, M200. Whereas the M60 presumably originates from the primary somatosensory cortex (SI), the generator of the M200 is localized more inferiorly and laterally, most likely in the secondary somatosensory cortex (SII). M200 source strength is affected by both the sleep stage (active vs. quiet sleep) and the interstimulus interval (ISI). It is stronger in quiet sleep (QS) and with longer ISIs. An M200-like response is also often present after stimulation of the ipsilateral hand. However, in 2 newborns also a late response with similar location and orientation with M60 was present after ipsilateral stimulation. [3]

IV. Discussion

In these studies we have been able to demonstrate that MEG measurements can be reliable conducted in newborns and young infants. We showed previously unknown, major differences in the first cortical responses from the primary somatosensory area between newborns and adults. We further demonstrated how the newborn activation pattern slowly matures to the adult pattern over the first years of life.

One of the greatest challenges faced in these studies was establishing a reliable method for infant MEG measurements and data analysis.

An MEG measurement session involving infants requires time and patience. All our recordings were performed when the infants were in natural sleep. Most of the failures in infant measurements have resulted from the infant not falling asleep within the time limits. When the infant did fall asleep, our success rate in the recordings has been extremely high with a few infants being excluded because of problems with the head digitization. SEFs could be identified and modeled in all infants in whom the measurement was successfully completed.

V. Conclusions

We have shown in a relatively large number of infants that SEFs can be reliably recorded from both the SI and SII already a few hours after full-term birth. At this age the opposite ECD orientations of the early SEFs of newborns compared with adults demonstrate immaturity of cortical processing. Similarly, the increase in M200 in quiet sleep is in contrast with disappearance of SII responses in SWS sleep in adults reflecting differences in cortical function and functional connectivity. The orderly systematic change in SEFs parallels behavioral development of the sensorimotor functions.

Acknowledgements

We wish to express our gratitude to our subjects and their families for preparing the way for this study. This work was supported by the Sigrid Jusélius Foundation, Yrjö Jahnsson Foundation, Finnish Cultural Foundation, the Pediatric Research Foundation, the Biomedicum Helsinki Foundation, the Jenny and Antti Wihuri Foundation, the Academy of Finland, the Helsinki Pediatric Graduate School, and The Finnish Medical Foundation.

References

1. Rechtshaffen A, Kales A, (1968) A Manual of Standardized Terminology: Techniques and Scoring System for Sleep Stages of Human Subjects. Washington, DC: US Government Printing Office.

2. Lauronen L, Nevalainen P, Wikström H et al., (2006) Immaturity of somatosensory cortical processing in human newborns. NeuroImage 33:195-203
3. Nevalainen P, Lauronen L, Sambeth A et al., (2008) Somatosensory evoked magnetic fields from the primary and secondary somatosensory cortices in healthy newborns. NeuroImage 40:738-745
4. Pihko E, Nevalainen P, Stephen J et al., (2009) Maturation of somatosensory cortical processing from birth to adulthood revealed by magnetoencephalography. Clin Neurophysiol 120:1552-1661

Author: Päivi Nevalainen
Institute: BioMag Laboratory, Hospital District of Helsinki and Uusimaa, HUSLAB, Helsinki University Central Hospital
Street: P.O.Box 340
City: FIN-00029 HUS
Country: Finland
Email: paivi.nevalainen@helsinki.fi

Pediatric CNS Pathophysiology

Julia M. Stephen[1], Sven Braeutigam[2], Paul L. Furlong[3], Urs Ribary[4,6], Timothy P.L. Roberts[5], and Nazin Virji-Babul[6]

[1] Mind Research Network, Albuquerque, NM USA
[2] University of Oxford, Oxford UK
[3] Aston University, Birmingham, UK
[4] Simon Fraser University, Vancouver, BC Canada
[5] Children's Hospital of Philadelphia, Philadelphia, PA USA
[6] Down Syndrome Research Foundation, Burnaby, BC Canada

Abstract— **Research in pediatric central nervous system pathophysiology is focused around three primary goals: identification of neurodevelopmental disorders, understanding the differences in brain development which underlie these disorders, and improving treatment for these young children. Autism spectrum disorders (ASDs) are a complex set of disorders which are characterized by difficulties in language and social interactions. These behavioral measures are highly variable and a number of underlying causes can generate similar behavioral effects. Therefore, it is important to identify neurophysiological markers to better identify and characterize these disorders. Recent ASD findings using MEG show atypical latency and amplitude responses and poor cortical connectivity in children with ASDs across the cognitive spectrum from basic auditory processing, multisensory integration, to face and semantic processing. These results further support the view that ASDs are a complex neurologically-based disorder. On the other hand, the cause of Down syndrome is well understood as originating from a partial or full replication of chromosome 21. However, the cognitive and neurological consequences of this chromosomal abnormality are not yet well understood. Using a simple observation and motor execution task, poor functional connectivity in sensory-motor areas, particularly in the gamma band range, has been identified in children with Down syndrome and is consistent with behavioral deficits in the sensory-motor realm. Additional studies are needed to better understand whether targeted identification of these abnormalities can facilitate treatment in this disorder. Finally, while epilepsy can be reliably diagnosed, seizure control is still limited in many cases where the seizure onset zone is not readily apparent. Advances in pre-surgical evaluation and intra-operative co-registration will be described. These studies describing pediatric CNS pathophysiology will be discussed.**

Keywords— **Autism Spectrum Disorders, Down Syndrome, Epilepsy, Children, MEG.**

I. AUTISM SPECTRUM DISORDERS

A. Introduction

Autism Spectrum Disorders (ASDs) are a set of neurodevelopmental disorders characterized by impaired verbal and nonverbal communication, abnormal reciprocal social interaction and restricted and repetitive patterns of interests and behaviors. Previous studies have identified a broad spectrum of atypical behavioral responses in children with an ASD, including hyper- and hypo-sensory sensitivities in addition to the cognitive deficits described above. Over the last decade, the neuroanatomical and neurophysiological understanding of autism has increased substantially, but the diagnosis of ASDs is based exclusively on behavioral criteria, and calls have been made for interdisciplinary approaches to combine biological and psychological theory. A better understanding of the underlying physiological differences in children with ASD may lead to earlier identification of ASDs and better characterization of this broad spectrum of disorders.

B. Auditory Processing (Roberts and colleagues)

Roberts and colleagues have focused on understanding auditory processing deficits in children with ASDs in conjunction with language impairment [1] (Roberts et al. 2010). Previous results from their group identified a restricted dynamic range in response to 200 and 1000 Hz tones in children with an ASD over the right hemisphere suggesting abnormal auditory processing in the right hemisphere [2] (Gage et al. 2003). The most recent study characterized the latencies and amplitudes of the 50 and 100 ms peaks (M50 and M100) in the bilateral auditory source estimates in response to 200, 300, 500 and 1000 Hz tones in both hemispheres. There were no group differences identified in the M50 latencies or amplitudes. For the M100, there was an overall group effect for latencies in the right hemisphere with the ASD children demonstrating delayed latencies relative to typically developing children (11 ms difference). In addition, there was no effect of latency based on language impairment, suggesting that the latency difference is specific to the diagnosis of ASD. The right hemisphere 500 Hz condition produced a sensitivity of 75%, a specificity of 81% and a positive predictive value of 86% for correct identification of diagnostic group based on latency. These results suggest that analysis of basic auditory processing

S. Supek and A. Sušac (Eds.): Advances in Biomagnetism – BIOMAG2010, IFMBE Proceedings 28, pp. 242–245, 2010.
www.springerlink.com

may provide an important electrophysiological endophenotype for ASD.

Finally, the typically developing children in this study demonstrated a decrease in M100 latency with increasing age, whereas the children with ASD did not. In a related diffusion tensor imaging/MEG study focusing on typically developing children Roberts et al. [3] (2010) identified the same decrease in M100 latency with age, an increase in fractional anisotropy in the acoustic white matter radiations, and a direct correlation between M100 latency and FA measures. While further studies are needed, this suggests that decreased FA in white matter may be a possible underlying mechanism for the delayed M100 latencies identified in ASD children.

C. *Multisensory Processing (Stephen and Colleagues)*

The work of Stephen and colleagues has focused on probing the disconnection theory of autism using a basic multisensory integration paradigm [4] (Stephen et al. submitted,Okada and Stephen 2009). In an effort to identify early markers of atypical brain development, the babySQUID pediatric MEG system [5] (Okada et al. 2006) was used to focus specifically on auditory-somatosensory integration in children 18 months to 5 years of age. This paradigm consisted of a simple unisensory auditory 800 Hz tone stimulus, unisensory tactile index finger stimulus (air-balloon pressure stimulus) and synchronously presented auditory and somatosensory conditions. The right hemisphere was studied based on previous results suggesting larger differences in this hemisphere between typically developing children and children with an ASD [2] (2 Gage et al. 2003). It was hypothesized that the children with an ASD would exhibit delays in multisensory processing relative to typically developing children due to poor cortical connectivity.

Consistent with the multisensory integration literature, the onset of the multisensory integration response was identified as the time when the synchronously presented auditory stimuli were larger than the sum of the unisensory responses. In all but one ASD child, the synchronous response was enhanced relative to the sum of the unisensory responses. A significant difference in the onset latency of the multisensory integration response was identified in typically developing children between ages 6 months to 5 years, with the youngest children showing the longest latency consistent with the known changes in myelination across this age range. In addition, a significantly longer multisensory integration latency difference was identified in typically developing children relative to ASD children. There were no significant differences in unisensory peak latencies for the auditory or somatosensory responses individually.

Recent results also suggest that local coherence measures over right hemisphere increase with age in neurotypical children. This may provide a complimentary measure of connectivity to help increase the specificity for identification of young children with an ASD.

D. *Face and Semantic Processing (Braeutigam and Colleagues)*

The broad objective of Braeutigam and colleagues' efforts is to investigate dynamic brain activity in autism focusing on two aspects of neural processing which are affected to varying degrees amongst individuals with ASD, namely face processing, and semantic processing. Using MEG as the main investigative tool, they have been utilising both developmental and longitudinal study designs together with identity recognition and semantic incongruity detection tasks. The MEG data have been analysed using a variety of approaches including evoked fields, induced and phase-locked gamma band responses, and source estimation.

Our results suggest that during the course of development in individuals with ASD, the cortical activity associated with the processing of human faces assumes a different-from-normal localisation in extrastriate brain regions [6,7] (Bailey 2005, Kylliäinen 2006b). Nevertheless, face and object processing in children with autism appears to follow a trajectory partly similar to that seen in typically developing children [8] (Kylliäinen 2006a) implying qualitative changes in the neural processing between childhood and adult life. Specifically, in middle childhood the neural mechanisms underlying face and object processing are possibly less specialised than in adults, with greater early activation of posterior occipital cortices and less specific activation of ventral occipito-temporal cortex.

The data obtained in individuals with ASD reading sentences ending with a semantically incongruous word suggest an atypical lateralisation of the N400 response as well as an anomalous long latency response LPC [9] (Braeutigam 2008). In contrast, meaningful sentences elicited long-latency gamma-oscillation not observed in the control group. These findings might indicate that appreciation of sentence context in individuals with ASD involves both more and different neural stages. It is currently unresolved, however, whether such neuronal stages reflect genuinely anomalous pathways, i.e. pathways that are never utilised in typically developing individuals.

Overall, these research findings support, to some extent, models describing autism in terms of anomalous cognitive strategies rather than cognitive impairments or deficits per se. We believe that MEG will contribute to autism research for many years to come, however, important challenges need to be met as to whether the observed electrophysiological anomalies are indeed specific to autism, or merely indicative of a wider class of abnormal conditions. Time will tell how the current research can contribute to a better understanding of heterogeneity of the autistic spectrum, its diagnosis and long-term treatment.

II. Down syndrome (Ribary and Virji-Babul and colleagues)

Down syndrome (DS) is a genetic condition resulting from the presence of an additional whole or part copy of chromosome 21, which disrupts many aspects of cognitive and sensory-motor development. Structural brain imaging studies have shown that there are significant differences in the structure of the brain in DS, however there are considerable gaps in our understanding of the functional significance of these structural differences and in particular how these structural differences relate to the dynamics of basic sensory-motor processes. The objective of Ribary and Virji-Babul's efforts is to study the underlying functional connectivity dynamics during sensory-motor integration, especially relating to action execution/observation or during voluntary finger movements in adults with [10] (Virji-Babul 2008) and without DS [11] (Virji-Babul 2010). Cortical activations were analyzed in 3D space, across frequency and time using a 151 channel MEG system (VSM MedTech). Subjects were engaged in lifting a cup approximately 2-3 inches off a pad and return it to the same position and finally return to their initial hand position (action execution) or observing the experimenter performing the same action (action observation) or simply pressing a button on the keypad using their right dominant index finger. SAM beamformers were used to obtain spatial distributions and synchronizations between pairs of significant sources were determined by using phase locking values (PLV). PLV analysis was performed in multiple 4Hz bands from 3-46 Hz.

During execution, the motor responses were strongly lateralized to the ipsilateral rather than the contralateral side in the DS group. Observation of movement activated a network of cortical regions that was similar to the control group, however there was no significant peak activity in the motor areas. In addition, the overall pattern of neural activation was more scattered and less organized in the DS group. These results further support the hypothesis of a dysfunction in the execution/observation matching system in adults with DS. Relating to voluntary finger movements, findings indicated significant differences in the PLV locations between the controls and the DS. Compared to high PLVs in the gamma band for distinct locations within the left hemisphere and distinct cross hemisphere connectivity in the lower frequency bands (theta and alpha) in control subjects, the DS group was dominated by high PLVs in the lower frequency bands with very few connectivity observed in the gamma band. These data suggest that individuals with DS have altered functional connectivity dynamics particularly in the gamma band during sensory-motor processing.

III. Epilepsy (Furlong and colleagues)

Epilepsy is one of the most prevalent neurological disorders in paediatric age. It affects 42,000 children under 16 years of age in the UK (1 in 279) or 58,000 (1 in 242) children and young people under 18 years of age. In approximately 10-20% of these young patients, pharmacological treatment is unable to secure seizure control. Children with epilepsy achieve less than would be predicted from their IQ, suggesting an additive psychological and social burden of repeated seizures. For quality of life and health economics considerations, surgical treatment of drug-resistant epilepsies in pediatric age is being increasingly advocated and offered worldwide. Patients of pediatric age benefit most from surgical intervention, enhancing the potential for more normal neural development. These patients are the most challenging to assess for their epilepsy, so development of non-invasive measures have an important translational component. Repeated measures in these patients provide important neurodevelopmental data relating to pre- and post-lesionectomy and monitoring of subsequent functional and developmental progression.

Our current research efforts include: the identification of epileptogenic cortex utilising novel beamformer measures employing a signal kurtosis measure to localise the source; combining simultaneous EEG/fMRI studies with MEG; non-invasive localisation and progression monitoring of hearing, speech and language cortex; assessment of pre- and post- surgical effects of hearing, speech, language and memory processing; the study of the relationship between epileptogenic cortical activity and cognitive impairment.

Studies utilising Single Pulse Electrical Stimulation (SPES) through intracranial electrodes have provided robust localisation of hyperexcitable cortex in patients of pediatric age [12] (Flanagan et al 2009). In a recent pilot study (Agirre-Arrizubieta et al, in preparation) we have compared pre-surgical inter-ictal MEG source localisations of epileptiform activity, using both kurtosis and frequency measures, together with SPES assessments of ictal zones. MEG-SPES were highly concordant (90%) and thus may provide an enhanced pre-surgical assessment tool.

A critical issue is to ensure that surgical removal of the epileptogenic cortex does not result in functional deficits. This is particularly relevant for language and memory functions in candidates for resection in the temporal lobe. Whilst cerebral plasticity allows for functional rearrangement of language dominance before 6 years of age, resection of the temporal lobe in the language-dominant hemisphere at an older age can result in significant functional impairment. At the present time, the intracarotid amobarbital procedure (Wada test) is the gold-standard procedure to determine the hemisphere of speech representation, but an associated morbidity in up to 10% of the patients has been reported.

We recently reported on the feasibility of language lateralisation paradigms using MEG in pediatric age, together with pilot data in a pediatric patient assessed for epilepsy surgery [13] (Fisher et al 2009). MEG data was analyzed using Synthetic Aperture Magnetometry (SAM), a beamforming technique that allows the localisation of frequency specific spectral power changes associated with the task [14] (Hillebrand et al., 2005). Our findings suggest that spectral power decrease recorded with MEG during a covert verb generation task could provide a robust and non-invasive indicator of language lateralization in children and serve as a positive non-invasive criterion for the determination of lateralisation.

IV. Conclusions

Studies using MEG to understand pediatric pathophysiology have increased rapidly in the last 10 years. The non-invasive nature of the methodology along with the high temporal resolution to study functional connectivity suggests that MEG may contribute not only to a better understanding of normal neurophysiological development in young children but also to better identification and intervention in young children with CNS pathophysiology. Early intervention has the potential to greatly improve the long-term outcome in pediatric populations with neurodevelopmental disorders and more studies are needed to identify and understand normal and abnormal development.

References

1. Roberts TPL, Khan S, Rey M et al. (2010) MEG detection of delayed auditory evoked responses in autism spectrum disorders: Towards an imaging biomarker for autism. Autism Research, In press.
2. Gage NM, Callen B, Roberts TP (2003) Cortical sound processing in children with autism disorder: an MEG investigation. Neuroreport 14(16):2047-2051.
3. Roberts TPL, Khan SY, Blaskey L et al. (2009) Developmental correlation of diffusion anisotropy with auditory-evoked response. NeuroReport 20(18):1586-91.
4. Okada Y, Stephen JM (2009) MEG studies of human brain development. Proc. Japanese Biomagnetism & Bioelectromagnetic Society, Kanazawa City, Japan, 2009.
5. Okada Y, Pratt K, Atwood C et al. (2006) BabySQUID: A mobile, high-resolution multichannel magnetoencephalography system for neonatal brain assessment. Rev Sci Instrum 77:24301-24309.
6. Bailey AJ, Braeutigam S, Jousmäki V et al. (2005) Abnormal activation of face processing systems at early and intermediate latency in individuals with autism spectrum disorder: a magnetoencephalographic study. Eur J Neurosci 21:2575-2585.
7. Kylliäinen A, Braeutigam S, Hietanen JK et al. (2006) Face- and gaze-sensitive neural responses in children with autism: a magnetoencephalographic study. Eur J Neurosci 24:2679-2690.
8. Kylliäinen A, Braeutigam S, Hietanen JK et al. (2006) Face and gaze processing in normally developing children: a magnetoencephalographic study. Eur J Neurosci 23:801-810.
9. Braeutigam S, Swithenby SJ, Bailey AJ (2008) Contextual integration the unusual way: a magnetoencephalographic study of responses to semantic violation in individuals with autism spectrum disorders. Eur J Neurosci 27:1026-1036.
10. Virji-Babul N, Moiseev A, Cheung T et al. (2008) Changes in mu rhuthm during action observation and execution in adults with Down syndrome: implications for action representation. Neurosci Lett 436:177-180.
11. Virji-Babul N, Moiseev A, Cheung T et al. (2010) Spatial-temporal dynamics of cortical activity underlying reaching and grasping. Hum Brain Mapp 31:160-171.
12. Flanagan D, Valentín A, García Seoane JJ et al. (2009) Single-pulse electrical stimulation helps to identify epileptogenic cortex in children. Epilepsia 50:1793-1803.
13. Fisher AE, Furlong PL, Seri S et al. (2008) Interhemispheric differences of spectral power in expressive language: a MEG study with clinical applications. Int J Psychophysiol 68:111-122.
14. Hillebrand A, Singh KD, Holliday IE et al. (2005) A new approach to neuroimaging with magnetoencephalography. Hum Brain Mapp 25:199-211.

Corresponding author:

Author: Julia M. Stephen, Ph.D.
Institute: Mind Research Network
Street: 1101 Yale Blvd NE
City: Albuquerque, NM
Country: USA
Email: jstephen@mrn.org

Fetal Development Assessed by Self-Organization Indices

Dirk Hoyer[1] and Uwe Schneider[2]

[1] University Hospital, Biomagnetic Center, Department of Neurology, Friedrich Schiller University of Jena, Germany
[2] University Hospital, Department of Obstetrics, Friedrich Schiller University of Jena, Germany

Abstract— **Background: The fetal precursors of mental and cardiovascular disease in later age are an important issue of developmental medicine. We investigate the question whether key parameters of fetal development reflect self-organization principles. From the systems theoretical point of view self-organization means the evolution of a previously less organized system into a higher organized form in the absence of external control. This process is characterized by a move from a large region of state space to a persistent smaller one, under the control of the system itself, and the development of long term correlations over time or space of previously independent locally operating variables.**

Aim: The object of the present work is to describe associations between the development of fetal heart rate patterns and characteristics of self-organization.

Study design: From 78 normal fetuses, gestational age 23-40 weeks heart beat interval series, 30 min duration, 1 kHz sampling rate were recorded by magnetocardiography. As indices of fetal development several indices which are typically associated with self-organization were analyzed.

Results: With development the variability of heart rate increased, the number of acceleration increased, and autonomic information flow increased over large time scales. These features are associated with essential self-organization characteristics, namely necessary variability, loss of degrees of freedom, and emergence of global order.

Conclusion: Heart rate characteristics interpreted following the systems theoretic concept of self-organization may provide new insights into the fetal development with implications for more appropriate diagnosis and prenatal treatment.

Keywords— **Prenatal diagnosis, self-organization, autonomic information flow, heart rate variability, fetal development.**

I. Introduction

Developmental processes in nature and physiology are based on self-organization mechanisms of complex interconnected networks, which can be found on different scales such as between molecules, cells, tissues, and organs leading to a global adjustment of the organism [1-3]. Consequently, those mechanisms may also apply in the fetal developmental process and its adjustment to pathological disturbances.

Developmental disorders during the fetal period can become permanently programmed and impact later children, adolescent, and adult health [4]. Respective mal-adjustments of the hypothalamic-pituitary-adrenal axis and sympathoadrenal system are reflected in heart rate rhythms modulated by the autonomic nervous system (ANS), but so far they have not sufficiently been understood during the fetal developmental process. Heart rate recordings, obtained by magnetocardiographic (MCG) measurements, allow the monitoring of the development of the fetal autonomic control over the third trimester.

Nonlinear heart rate dynamics over different time scales seem to provide information about self-organization related aspects of fetal development [5, 6]. Previously, it was found that nonlinear heart rate dynamics over different time scales, assessed by autonomic information flow (AIF) functions, reflect re-organization and risk in adult patients with cardiovascular and multiple organ diseases [7-9].

The objective of the present work is to highlight associations between the fetal development of heart rate patterns and essential systems theoretical features of self-organization during the normal fetal development.

II. Fetal development and self-organization

From the systems theoretical point of view self-organization means the evolution of a previously less organized system into a higher organized form in the absence of external control. This kind of process is characterized by a move from a large region of state space to a persistent smaller one, under the control of the system itself, and the development of correlations over time or space for previously independent locally operating variables (e.g. [10-12]).

Essential aspects of cardiovascular behavior are associated as follows:

- Variability: A certain level of non-predictable variability, such as of heart rate fluctuations, reflects sufficient capabilities of the organism in the search for optimal behaviour. Reduced heart rate variability (HRV) in adult heart failure patients is significantly correlated with lower survival rate [13]. Also in fetuses with developmental disorders decreased HRV indicates their restricted capabilities (e.g.[14]).

S. Supek and A. Sušac (Eds.): Advances in Biomagnetism – BIOMAG2010, IFMBE Proceedings 28, pp. 246–249, 2010.
www.springerlink.com

- Loss of degrees of freedom: The loss of random like behaviour in favour of structured pattern can be seen in the development of fetal behavioural states. Active and quiet behaviour already appear at the beginning of the third trimester, but completely developed behavioural states according to [15] do not appear before 32 weeks gestational age (GA). The same holds for movement related heart rate accelerations (AC), which reflect synchronized autonomic, cardiovascular, and motor activity.
- Emergence of global order: Global adjustments require information transfer over longer distances and longer time scales. In adult heart failure patients the survival rate is significantly higher in those with increased long scale autonomic information flow (AIF), which may reflect a better self-stabilization in the autonomic-cardiovascular system [16].

In the present work these aspects will be considered by respective HRV characteristics, namely by variability (SDNN), loss of degrees of freedom (AC), and emergence of global order (long term AIF).

III. Methods

A. Subjects

Seventy eight normal fetuses (gestational age 23-40) in baseline condition without any developmental disorders were included in the study (details in ([5]). The study was approved by the local ethics committee of the Friedrich Schiller University. All women signed a written, informed consent form.

B. Data Aquisition

All measurements were done by a vector-magnetograph ARGOS 200 in a magnetically shielded room in the Biomagnetic Center, Friedrich Schiller University, Jena.

The pregnant women were positioned supine or with a slight twist to either side to prevent compression of the inferior vena cava by the pregnant uterus. The Dewar was positioned as close as possible to the fetal heart without contact over to the maternal abdomen.

The MCG signal was recorded over a period of thirty minutes with a sampling rate of 1024 Hz. The fetal heart beats were detected like previously reported [21].

Two examples of typical heart rate pattern are demonstrated in Fig. 1. The early (28 weeks GA) is characterized by short lasting heart rate deceleration (DC) like patterns. In contrast, the maturated case (34 weeks GA) shows longer lasting heart rate accelerations (AC).

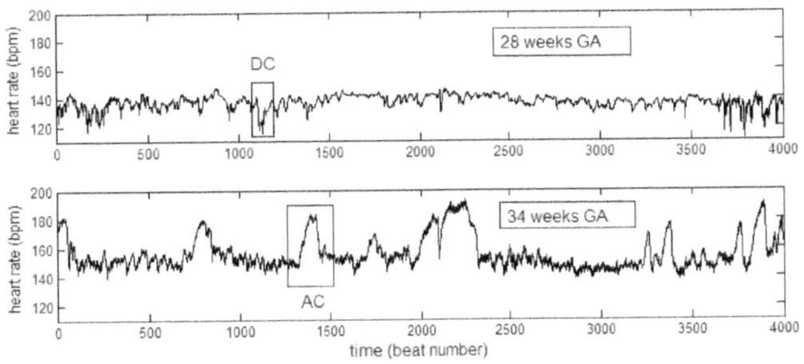

Fig. 1 Typical 30 min recordings of different gestational age. Example AC and DC are indicated by boxes, adapted from [5]

C. Heart Rate Variability

SDNN (standard deviation of normal-to-normal beat intervals, ms): overall variability reflects sympathetic and vagal heart rate modulations (calculated according to [18]).

D. Accelerations/Deceleration

Heart rate accelerations (AC) and decelerations (DC) were evaluated for prominent durations according to results of [5]. Peaks of 50-89 consecutively increased heart rate values were considered as AC, peaks of 10-49 decreased values as DC. The increase/decrease had to exceed a threshold of 10 bpm above/below a baseline (moving average over 401 beats).

E. Autonomic Information Flow

Autonomic Information Flow (AIF) describes the communication over temporal or spatial scales based on the transfer of information which can advantageously be assessed by Shannon entropy. In the present work complex correlations over different time scales were assesses as AIF indices based on results from previous studies [7,19,20]. Computational details are reported elsewhere [20].

AIF_NN: information flow over one heart beat (normal-to-normal) interval (short term information transfer).

AIF_fVLF: mean information flow over of the periods of the fetal very low frequency band fVLF (0.02-0.08 Hz), frequency boundaries according to [21], (long term information transfer).

F. Statistical Analysis

All indices were plotted versus weeks GA. The dependencies of the indices from GA were evaluated by regression lines. $p < 0.05$ suggest a significant linear relationship between GA and index.

IV. Results

Heart Rate Variability: SDNN increased and dispersed with GA (Fig. 2).

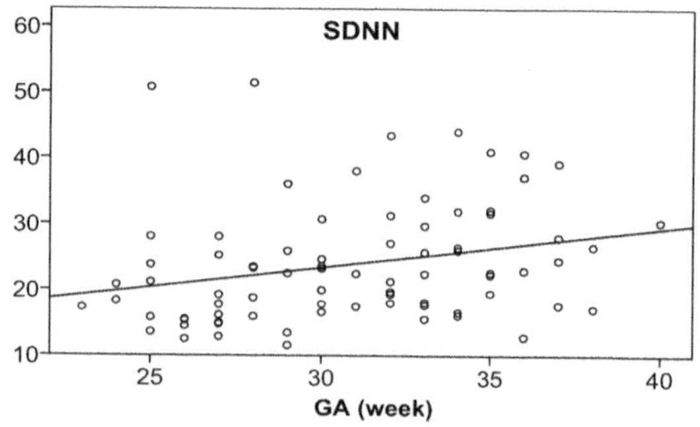

Fig. 2 SDNN vs. gestational age (GA) (linear increase, p=0.021), adapted from [5]

Acceleration and Deceleration: The number of AC increased starting from about 30 weeks GA. In contrast, the number of DC disappeared with increasing GA (Fig. 3a,b).

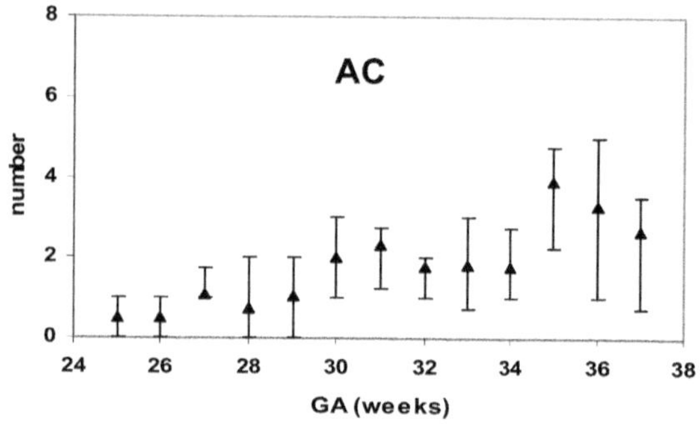

Fig. 3a Number of Accelerations (length 50-89 successive interbeat intervals) per 30 min vs GA

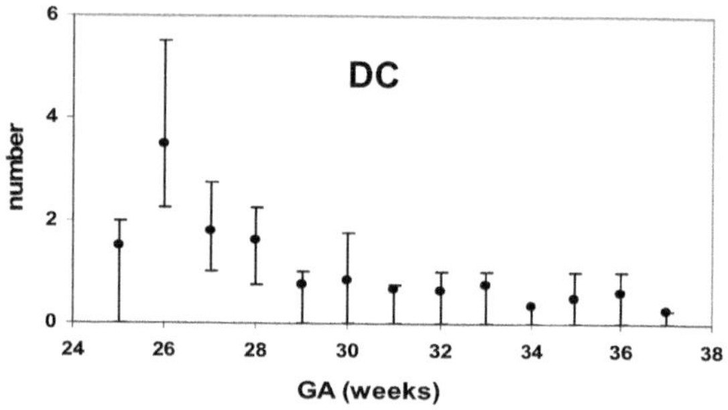

Fig. 3b Number of Decelerations (length 10-49 successive interbeat intervals) per 30 min vs. GA

Autonomic Information Flow: Only long term AIF significantly increased with GA, but short term AIF remained on the same level (Fig. 4 a,b).

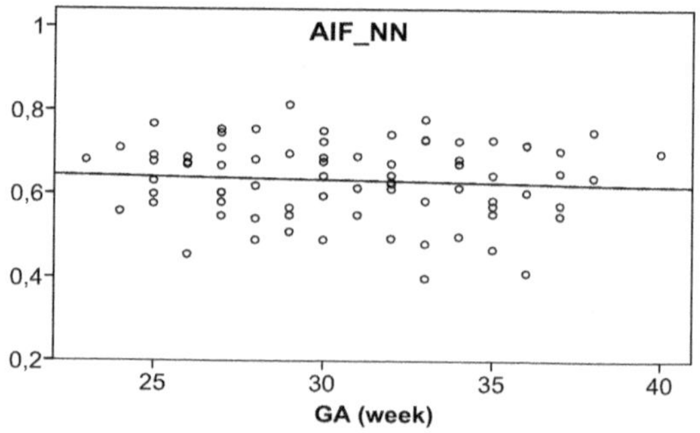

Fig. 4a Nonlinear short term correlations (AIF_fNN) vs. GA, linear regression line, p=0.586), adapted from [5]

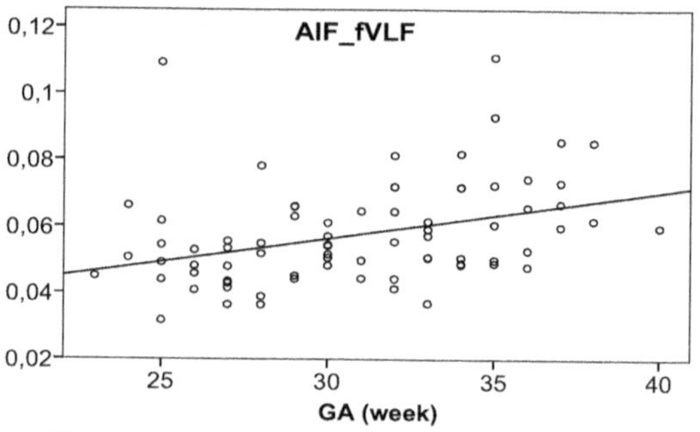

Fig. 4b Nonlinear long term correlations (AIF_fVLF) vs. GA, significant linear increase, p<0.001), adapted from [5]

V. Conclusions

The present work aimed to evaluate associations between typical characteristics of self-organization and the fetal development of autonomic rhythms.

The relationships between different scales such as the presentation of highly dimensional ANS networks by the low dimensional order parameter "autonomic heart rate modulations" is a generally valid characteristic of nonlinear dynamics and complex systems theory [11]. In the present work we found typical characteristics of fetal development related to self-organization, namely increasing variability (SDNN), increasing number of synchronized activations (AC), and increasing long scale communication (AIF), represented in heart rate patterns.

HRV: The increasing amount and dispersion of the slower heart rate fluctuations reflect increasing capability of global systemic regulation. The positive prognostic value of SDNN is known from clinical survival studies of adult heart failure patients [13]. Its interpretation in the context of synchronization and information flow may give new insights into both, adults cardiovascular reorganization, and fetal development.

AC/DC: About 2-3 AC of more than 15 bpm over at least 15 sec within 30 minutes attribute the normal frequency of this type of synchronisation between motor and cardiovascular activity in a healthy term fetus. Heart rate DC indicate premature, non-stable or pathological behaviour [22]. In the present work AC/DC like patterns were investigated. The development from mainly short-term DC in the youngest fetuses of about 25 weeks GA to mainly adequate long-term AC in the later period of more than 30 weeks GA clearly reflects both, the reduction of degrees of freedom, and the emergence of globally coordinated behavior.

AIF: The time scale dependent AIF characteristics reflect the emergence of those connections which are beneficial for the global complex communication and stability. Accordingly, the long term AIF increases. Maximization of mutual information over longer time scales is a fundamental criterion of self-organization in physiological systems [23]. These results are in line with [17, 24]. The stabilizing effect of increasing long term AIF was previously demonstrated in a simulation study of a nonlinear dissipative system with two feedback loops [8]. In a survival study of adult patients with chronic heart failure, those patients with stronger long term AIF showed a better outcome [7].

These results, obtained from fetal heart rate characteristics, confirm associations between the development of fetal autonomic control system and typical characteristics of self-organization. This knowledge about possible mechanisms of self-organization reflected in nonlinear heart rate dynamics may not only improve prenatal diagnostic but also substantiate early therapeutic implications.

Acknowledgment

This work has in part been supported by the Deutsche Forschungsgemeinschaft – DFG: HO 1634/9 -1,2, HO 1634 12-2, KA 1726/2, Schn 775/2-3).

References

1. Buchman TG (2002) The community of the self. *Nature* 420: 246-51.
2. Glass L (2001) Synchronization and rhythmic processes in physiology. *Nature* 410: 277-84.
3. Strogatz SH (2001) Exploring complex networks. *Nature* 410: 268-76.
4. Van den Bergh MR, Mulder EJ, Mennes M, et al. (2005) Antenatal maternal anxiety and stress and the neurobehavioural development of the fetus and child: links and possible mechanisms. a review. *Neurosci Biobehav Rev* 29: 237-58.
5. Hoyer D, Heinicke E, Jaekel S, et al. (2009) Indices of fetal development derived from heart rate patterns. Early Hum Dev 85: 379-386.
6. Van Leeuwen P, Cysarz D, Lange S, et al. (2007) Quantification of fetal heart rate regularity using symbolic dynamics Chaos 17: 015119.
7. Hoyer D, Maestri R, La Rovere MT, et al. (2008) Autonomic response to cardiac dysfunction in chronic heart failure: a risk predictor based on autonomic information flow. PACE 31: 214-220.
8. Hoyer D, Frank B, Gotze C, et al. Interactions between short-term and long-term cardiovascular control mechanisms. Chaos 17: 015110.
9. Hoyer D, Friedrich H, Zwiener U, et al. (2006) Prognostic impact of autonomic information flow in multiple organ dysfunction syndrome patients. Int J Cardiol 108: 359-369.
10. Eigen M (1972) Selforganization of matter and the evolution of biological macromolecules. Naturwissenschaften 58: 465-523.
11. Haken H (1983) Synergetics: An Introduction. Nonequilibrium Phase Transition and Self-Organization in Physics, Chemistry, and Biology. Springer Verlag, Berlin, New York.
12. Kauffman SA (1993) The Origins of Order: Self-Organization and Selection in Evolution. Oxford University Press, Oxford, New York.
13. Kleiger RE, Stein PK, Bigger JT Jr. (2005) Heart rate variability: measurement and clinical utility. Ann Noninvasive Electrocardiol 10: 88-101.
14. Rotmensch S, Liberati M, Vishne TH, et al. (1999) The effect of betamethasone and dexamethasone on fetal heart rate patterns and biophysical activities. A prospective randomized trial. Acta Obstet Gynecol Scand 78: 493-500.
15. Nijhuis JG, Prechtl HF, Martin CB, et al. (1982) Are there behavioural states in the human fetus? Early Hum Dev 6: 177-195.
16. Aoyagi N, Struzik ZR, Kiyono K et al. (2007) Autonomic imbalance induced breakdown of long-range dependence in healthy heart rate. Methods Inf Med 46: 174-178.
17. Schneider U, Frank B, Fiedler A, et al. (2008) Human fetal heart rate variability-characteristics of autonomic regulation in the third trimester of gestation. J Perinat Med 36: pp. 433-441.
18. TaskForce (1996) Heart rate variability: standards of measurement, physiological interpretation and clinical use. Circulation 93: 1043-1065.
19. Hoyer D, Friedrich H, Stein PK, et al. (2008) Autonomic information flow improves prognostic value of heart rate patterns after abdominal aortic surgery. J Crit Care 23: 255-262.
20. Hoyer D, Pompe B, Chon KH, et al. (2005) Mutual information function assesses autonomic information flow of heart rate dynamics at different time scales. IEEE Trans Biomed Eng 52: 584-592.
21. David M, Hirsch M, Karin J, et al. (2007) An estimate of fetal autonomic state by time-frequency analysis of fetal heart rate variability. J Appl Physiol 102: 1057-1064.
22. Dawes GS, Moulden M, Redman CW (1992) Short-term fetal heart rate variation, decelerations, and umbilical flow velocity waveforms before labor. Obstet Gynecol 80: 673-678.
23. Becker S (1996) Mutual information maximization: models of cortical self-organization. Network: Computation in Neural Systems 7: 7-31.
24. Van Leeuwen P, Lange S, Bettermann H, et al. (1999) Fetal heart rate variability and complexity in the course of pregnancy. Early Hum Dev 54: 59-69.

Author: Dirk Hoyer, Ph.D., Prof.
Institute: Biomagnetic Center, Dept. Neurology, University Hospital
Street: Erlanger Allee 101
City: 07747 Jena,
Country: Germany
Email: Dirk.Hoyer@med.uni-jena.de

Altered Long-Range Phase Synchronization and Cortical Activation in Children Born Very Preterm

Sam M. Doesburg[1,2], Urs Ribary[2,3,4], Anthony T. Herdman[3], Teresa Cheung[4,5], Alexander Moiseev[4], Hal Weinberg[3], Michael F. Whitfield[1,2], Anne Synnes[1], Mario Liotti[3], Daniel Weeks[6], and Ruth E. Grunau[1,2]

[1] Pediatrics, University of British Columbia, Vancouver, Canada
[2] Child and Family Research Institute, Vancouver, Canada
[3] Psychology, Simon Fraser University, Burnaby, Canada
[4] Down Syndrome Research Foundation, Burnaby, Canada
[5] Physics, Simon Fraser University, Burnaby, Canada
[6] Psychology, University of Lethbridge, Lethbridge, Canada

Abstract— **Children born very preterm, even with broadly normal IQ, commonly show selective difficulties in visuospatial processing and executive functioning. Very little, however, is known what alterations in cortical processing underlie these deficits. We recorded MEG while eight children born very preterm (≤32 weeks gestational age) and eight full-term controls performed a visual short-term memory task at mean age 7.5 years (range 6.4 – 8.4). Previously, we demonstrated increased long-range alpha and beta band phase synchronization between MEG sensors during STM retention in a group of 17 full-term children age 6-10 years. Here we present preliminary evidence that long-range phase synchronization in very preterm children, relative to controls, is reduced in the alpha-band but increased in the theta-band. In addition, we investigated cortical activation during STM retention employing synthetic aperture magnetometry (SAM) beamformer to localize changes in gamma-band power. Preliminary results indicate sequential activation of occipital, parietal and frontal cortex in control children, as well as reduced activation in very preterm children relative to controls. These preliminary results suggest that children born very preterm exhibit altered inter-regional functional connectivity and cortical activation during cognitive processing.**

Keywords— **preterm birth, beamformer, short-term memory, neural synchrony, functional connectivity.**

I. Introduction

Children born very preterm commonly exhibit selective deficits in visual processing and executive function, even when general intelligence is within the normal range [1]. Although considerable research has been conducted investigating neuroanatomical alterations associated with very preterm birth [see 2 for review], little is known about what differences in cortical processing underlie the selective cognitive difficulties common in this population.

A wealth of recent experimental evidence indicates that synchronization of neural oscillations is relevant for a variety of perceptual and cognitive processes. Active processing within a cortical region, for example, has been reliably related to local synchronization of gamma-band oscillations in a variety of contexts [3]. Synchronization of oscillations *between* cortical regions in a number of frequency ranges has also been associated with the formation of transient, functionally integrated networks for the performance of perceptual and cognitive tasks [3]. Previously, we demonstrated in a group of seventeen full-term children that visual short-term memory (STM) retention yields increased alpha and beta band synchronization between MEG sensors, indicative of the formation of a transient large-scale network supporting maintenance of the memory trace [4].

To investigate alterations in cortical processing in school age children who were born very preterm, we recorded MEG while very preterm children and age-matched controls performed a visual STM task. We present multiple preliminary results regarding phase synchronization between MEG sensors, indicative of task-dependent functional coupling between cortical regions, and from beamformer localization of gamma-band activation, reflecting the engagement of cortical regions, during task performance.

II. Methods

Subjects: We recorded MEG while 8 very preterm children born ≤ 32 weeks gestation and 14 controls born at full-term performed a visual STM task. To examine group differences in long-range synchronization and gamma-band activation 8 age-matched pairs were selected (mean age 7.5 years, range 6.4 - 8.4 years). All 14 controls (age 6-10 years) were used in the beamformer analysis of gamma-band activation in full-term children. None of the very preterm children had significant brain injury on neonatal ultrasound, and all were born appropriate birth-weight for gestational age. All subjects had normal or corrected-to-normal vision.

Task: Subjects performed a visual STM task (Fig. 1) with stimuli adapted with permission [5]. On each of 180 trials

S. Supek and A. Sušac (Eds.): Advances in Biomagnetism – BIOMAG2010, IFMBE Proceedings 28, pp. 250–253, 2010.
www.springerlink.com

subjects viewed an initial stimulus (S1) for 1000 ms, followed by a 900 ms retention interval during which they were required to maintain S1 in STM, followed by the presentation of a second stimulus (S2) for 1000 ms. Subjects were required to press a button on a response box to indicate if the shapes were the same or different. Additional task details are available in [4].

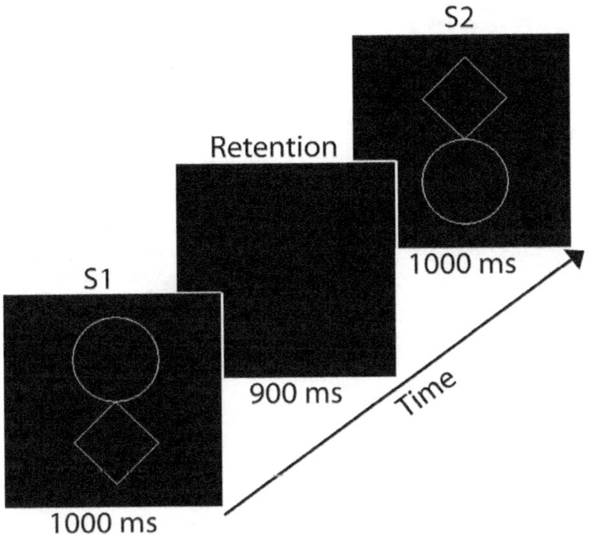

Fig. 1 Time course of the stimulus display on a single trial

MEG Recording and Processing: MEG was digitized continuously at 1200 Hz using a 151 channel CTF system (Burnaby, Canada) and stored on disk for off-line analysis. Data were then standardized relative to sensor locations within and between subjects using a continuous head localization and correction procedure [6]. Epochs were extracted surrounding each trial and the record of ocular and nonocular artifacts was removed using a principal components analysis based procedure [7].

Phase Synchronization Analysis: Due to computational limitations imposed by pairwise comparisons, and to avoid spurious synchronization arising from volume conducted signals propagating to nearby sensors, we extracted data from a montage of 19 sensors distributed roughly evenly over the surface of the head to image the global pattern of long-range phase synchronization (sensors LF11, RF11, LF32, RF32, LC14, RC14, LO22, RO22, LO41, RO41, LT21, RT21, LT42, RT42, LO43, RO43, ZF03, ZCO3, ZP02). Data from each sensor were then digitally filtered at 1 Hz intervals between 6 and 60 Hz (passband $= f \pm 0.05f$, where f is the filtered frequency). We then calculated the analytic signal

$$\varsigma(t) = f(t) + i\tilde{f}(t) = A(t)e^{i\phi(t)} \quad (1)$$

of the filtered waveform for each epoch, $f(t)$, where $\tilde{f}(t)$ is the Hilbert transform of $f(t)$ and $i = \sqrt{-1}$, to obtain the instantaneous phase, $\phi(t)$, and amplitude, $A(t)$, at each sample point. Phase locking values (PLVs) were computed from the differences of the instantaneous phases for analyzed sensor pairs for each analyzed frequency. For example, sensors j and k, at each point in time, t, across N available epochs [8]:

$$PLV_{j,k,t} = N^{-1}\left|\sum_N e^{i[\phi_j(t)-\phi_k(t)]}\right| \quad (2)$$

Phase locking values were then standardized relative to a baseline interval occurring before the onset of S1 in order to index changes in synchrony relevant to task demands by subtracting the mean baseline PLV at that frequency from the PLV at each data point and dividing the difference by the standard deviation of the PLV for the baseline interval at that frequency. Group differences were assessed by averaging data across subjects for each frequency and time point, and subsequently subtracting averaged data between groups. See [4] for additional details on synchronization analysis.

Beamformer Analysis: We employed synthetic aperture magnetometry (SAM) to localize gamma-band activation during STM retention in controls, and well as group differences between very preterm and control children. Beamformer analysis implements a spatial filter, estimating the contribution of each voxel in brain space to magnetic signals recorded at MEG sensors, by minimizing correlations with all other analyzed voxels. We imaged 30-50 Hz activity for each subject in four 225 ms time windows during STM retention (0-225, 225-450, 450-675, 675-900 ms) relative to a 225 ms baseline interval preceding the onset of S1. We averaged across the 14 controls to image gamma-band network dynamics in full-term controls. Group differences were assessed by averaging across the eight subjects in each group and subtracting averaged images across groups. Additional details about SAM can be found in [7]. Covariance matricies were regularized if noise spectral density exceeded an acceptable range.

III. RESULTS

Long-Range Phase Synchronization: Preliminary results indicate that global patterns of long-range phase synchronization, calculated by averaging PLV across all 171 analyzed sensor pairs, was reduced in the alpha-band but increased in the theta-band in very preterm children, relative to full-term controls. These sustained group differences were centered at ~10 Hz and ~6 Hz, respectively.

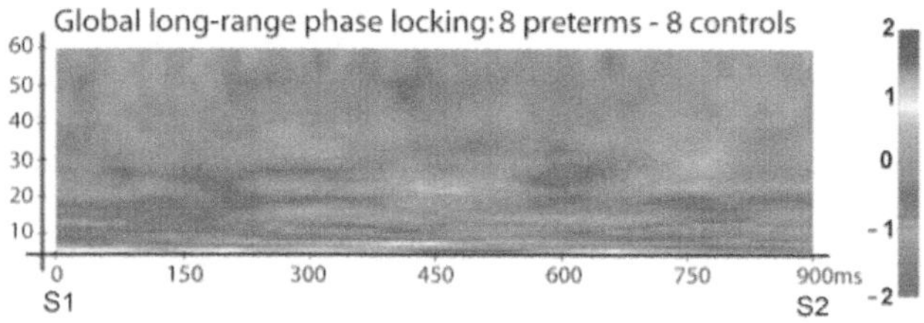

Fig. 2 Alterations in global long-range phase locking during STM retention associated with very preterm birth. Data were averaged across all 171 analyzed sensor pairs, and differences are expressed in units of standard deviation from the pre-S1 baseline. Blue regions represent decreases in long-range synchronization; red and yellow regions denote increases

Beamformer Analysis of Gamma-Band Activation: Our preliminary analysis of gamma-band activation in 14 controls indicates sequential stages of cortical activation during STM retention (Fig 3A). During the 0-225 ms interval activation was observed in early visual cortex and higher, posterior parietal, visual cortex. The subsequent 225-450 ms interval was characterized by activation in posterior parietal cortex and prefrontal cortex. During the 450-675 ms and 675-900 ms time windows residual activation was seen throughout the network of activated occipital, parietal and prefrontal cortical regions, with the addition of motor cortex activation during the 675-900 ms time window.

Analysis of group differences yielded preliminary results indicating that very preterm children generally show less activation in within the identified network of task relevant cortical regions, although greater activation is observed for very preterm children in specific instances (Fig 3B). In the 0-225 ms interval very preterm children exhibited reduced activation in early visual cortex and in parietal cortex. In the 225-450 ms time window pronounced reductions were observed in posterior parietal cortex and prefrontal cortex. During the subsequent 450-675 ms interval reduced activation was found in early visual cortex and posterior parietal cortex, whereas *increased* activation was observed in prefrontal cortex for very preterm children. Reduced activation was seen for very preterm children throughout the task-relevant network of early visual, posterior parietal and prefrontal cortical regions during the final 675-900 ms interval.

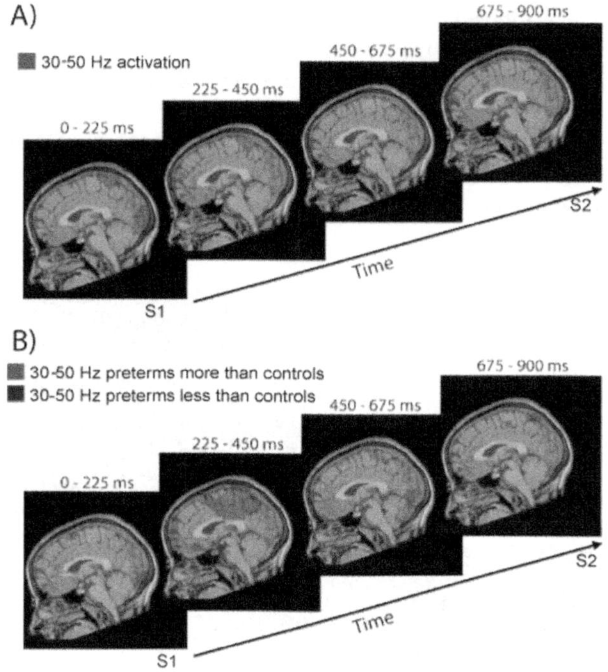

Fig. 3A Gamma-band activation in four time intervals during STM retention in full-term controls. 3B Altered gamma-band activation in children born very preterm in four time intervals during STM retention; blue regions indicate reduced activation for very preterm children, orange regions indicate increased activation for very preterm children

IV. Discussion

Long-Range Phase Synchronization: Synchronization between cortical regions has been implicated in the formation of transient, functionally integrated networks for the performance of particular tasks and to encode the features of percepts [3]. We previously reported long-range MEG synchronization in the alpha and beta bands during STM retention in children in this age range [4]. Such coherent large-scale network activity likely corresponds to functional coupling within a distributed network of cortical regions responsible for memory trace maintenance. Results of the present study indicate that this pattern of synchronization is altered in children born very preterm, with a pronounced and sustained loss of synchronization in the alpha-band as well as a sustained increase of synchronization in the theta-band. These preliminary results suggest that neural mechanisms underlying task-dependent functional coupling may be altered in children born very preterm. Previous research has associated slowing of peak oscillatory frequency in the MEG from the alpha-band toward the theta-band with dysrhythmic thalamocortical interactions [9]. Although such previous findings typically pertain to resting spectral power in MEG, the similarity of these alterations to those observed in the present data set suggest that dysrhythmic thalamocortical interactions may be relevant for altered long-range synchronization in very preterm children.

Gamma-Band Cortical Activation: Gamma-band activation within cortical regions is understood as reflecting active processing [3]. Preliminary results of gamma-band activation during STM retention in 14 control subjects indicate sequential activation of task-relevant cortical regions. Activation of early and higher visual cortex in the 0-225 ms time window presumably reflects processing of S1. Activation of prefrontal and parietal cortex during the subsequent 225-450 ms interval likely reflects recruitment of a network of executive and higher visual cortical areas responsible for maintaining the memory trace in STM. Activation of motor cortex in the 675-900 ms interval presumably reflects response preparation in anticipation of S2, and residual prefrontal, parietal and occipital activation in the 450-675 ms and 675-900 ms intervals likely reflect memory maintenance. This approach highlights the advantage of combined spatial and temporal resolution of MEG which permits the imaging of oscillations in specific frequency ranges in short time windows to reveal sequential processing in distributed cortical networks relevant to cognition.

This study provides preliminary evidence of altered gamma-band activation in children born very preterm, indicating that they have difficulty recruiting cortical regions relevant to STM retention. Reduced activation in very preterm children is seen in early visual cortex and parietal cortex 0-225 ms into the retention interval, suggesting problems in visual processing, consistent with previous behavioral research indicating selective visual processing difficulties [1]. Pronounced reductions in gamma-band activation in prefrontal and parietal cortex during the 225-450 ms interval, during which control children appeared to establish a network of prefrontal and posterior parietal cortical regions to maintain the memory trace in STM, highlights the apparent difficulty very preterm children exhibit in sufficiently activating cortical regions critical for task performance, consistent with previous work showing selective behavioral deficits in executive function in preterm children [1]. Interestingly, very preterm children showed increased gamma-band activation in prefrontal cortex predominantly during the 450-675 ms time window. This could be interpreted either as delayed executive processing (which occurred in the 225-450 ms interval for the full-terms) or as compensatory processing in very preterm children. If this indeed reflects compensatory processing, it is noteworthy that despite increased prefrontal activation in the 450-675 ms interval, reduced activation in parietal cortex is observed. This could be interpreted as increased executive effort leading to nonetheless diminished recruitment of higher visual cortex. Very preterm children also showed reduced activation in prefrontal, parietal and occipital cortex during the 675-900 ms interval, consistent with the overall pattern of reduced activation of task-relevant cortical areas.

V. Conclusions

Our preliminary findings indicate that cortical processing during STM retention is altered in children born very preterm. Altered long-range phase synchronization suggests that neural mechanisms mediating transient functional connectivity between brain regions may be abnormal in this group. Reduced and/or delayed gamma-band activation in task-relevant cortical regions suggests that children born very preterm may have difficulty recruiting networks of brain regions to support cognitive processing.

Acknowledgments

This study was supported by the Kennedy Shriver National Institute of Child Health and Human Development grant R01 HD039783 to R.E.G. and the BC Leading Edge Endowment Fund to U.R. We thank Ivan Cepeda, Gisela Gosse, Katia Jitlina, Julie Unterman and John Gaspar for their help collecting data.

References

1 Anderson PJ, Doyle LW (2008) Cognitive and educational deficits in children born very preterm. Semin Perinatol 32: 51–58
2 Hart AR, Whitby EW, Griffiths PD et al., (2008) Magnetic resonance imaging and developmental outcome following preterm birth: review of current evidence. Dev Med & Child Neurol 50: 655-663
3 Varela F, Lachaux JP, Rodriguez E et al., (2001) The brainweb: phase synchronization and large-scale integration. Nat Rev Neurosci 2): 229–239
4 Doesburg SM, Herdman AT, Ribary U et al., (in press) Long-range synchronization and local desynchronization of alpha oscillations during short-term memory retention in children. Exp Brain Res DOI 10.1007/s00221-009-2086-9
5 Beery KE, Buktenica NA, Beery NA. (2004) Beery-Buktenica Developmental Test of Visual-Motor Integration, 5th Ed. Psychological Corporation
6 Wilson H, Moiseev A, Podin S et al., (2007) Continuous head localization and data correction in MEG, Int Congr Ser 1300, pp 623–626.
7 Herdman AT, Cheyne D (2009) A practical guide to MEG and beamforming, In Handy (ed) Brain signal analysis: advances in neuroelectric and neuromagnetic methods. MIT Press, Cambridge, pp 99-140.
8 Lachaux JP, Rodriguez E, Martinerie J, Varela F (1999) Measuring phase synchrony in brain signals. Hum Brain Mapp 8: 194–208
9 Llinás RR, Ribary U, Jeanmonod D et al., (1999) Thalamocortical dysrhythmia: a neurological and neuropsychiatric syndrome characterized by magnetoencephalography. Proc Natl Acad Sci USA 96(26): 15222–15227

The Frequency Profile of Somatosensory Evoked Magnetic Fields in the Developing Brain

Y. Wang[1, 2], J. Xiang[2], D.F. Rose[2], T. Holroyd[3], E. Harris[4], and T.J. deGrauw[2]

[1] Department of Biomedical Engineering, University of Cincinnati, Cincinnati, OH, 45229, USA
[2] Division of Neurology, Cincinnati Children's Hospital Medical Center, Cincinnati, OH, 45229, USA
[3] MEG Core Facility, National Institute of Mental Health, Bethesda, MD, 20892, USA
[4] Division of Child Psychiatry, Cincinnati Children's Hospital Medical Center, Cincinnati, OH, 45229, USA

Abstract— **The objective of this paper is to study the variations of magnetic fields evoked by index finger stimulation in four age groups (6-9, 10-13, 14-17 and 18-49 years old) of normal population. We recorded magnetic fields during unilateral electrical stimulation of the index finger in 60 healthy children (7±1.2, 12±1.3 and 15±1.1) and 20 healthy adults (30±5.7) using a 275-channel whole head magnetoencephalography (MEG) system. Each group consisted of 10 females and males. Electrical stimulation was applied in randomized order to the right or left index finger with two digital rings. One hundred trials were recorded for both the right and left index finger stimulations. MEG data were analyzed in both the time and frequency domains. Brain regions associated with finger stimulations were localized with the equivalent current dipoles (ECDs). Time-frequency analysis was at the source level with a vector beamformer and Morlet wavelets. Our data have revealed evoked neuromagnetic signals in low-frequency ranges (10-40 Hz). In addition, the spectrograms revealed consistent focal increases of spectral power around 120-200 Hz in the left and right hemispheres in children. Furthermore, the neuromagnetic activation above 200 Hz was also observed, which were in the frequency range of 600-890 Hz. The preliminary data suggest that neuromagnetic signals above 120 Hz in the developing brain are noninvasive detectable.**

Keywords— **Magnetoencephalography, virtual sensor, beamformer, index finger stimulation.**

I. Introduction

Neurons in human brain generate oscillatory activity in a wide frequency range up to 1000 Hz [1, 2]. It was reported that up to 1/3 of children with epilepsy present with medically intractable seizures [3]. Some patients with two or three foci can benefit from surgery. However, removing the epileptogenic zones and preserving normal functional areas can be very challenging. Thurs, pre-surgery functional areas need to be carefully localized so that any functional deficits can be avoided. Somatosensory cortex is very important for pre-surgery mapping.

Magnetoencephalography (MEG) is a non-invasive imaging technique with high temporal and good spatial resolution [4], which measures magnetic fields from the cortical pyramidal neurons [5]. Neuroelectric source imaging is limited by the variations of the electrical conductivity related to skull and cerebrospinal fluid, while neuromagnetic fields are not distorted. Thus, MEG can detect higher frequency signals up to 1000 Hz, which might open a new window to help researchers understand the mechanism of human brain.

The present study aimed to investigate the frequency components of magnetic fields evoked by index finger stimulations in the developing brain.

II. Material and Methods

A. Participants

Four age groups (6-9, 10-13, 14-17, and 18-49 years old) of healthy participants volunteered after having given informed consent according to the Institutional Review Board at Cincinnati Children's Hospital Medical Center (CCHMC). None of the participants had any neurological impairment or trauma. They all had normal or corrected-to-normal vision and normal hearing. Group A had ten boys and ten girls (7±1.2, mean age±sd). Group B had ten boys and ten girls (12±1.3, mean age±sd). Group C had ten boys and girls (15±1.1, mean age±sd). Group D had ten females and ten males (30±5.7, mean age±sd).

B. Stimulus

Two Digitimer Constant Current Stimulator model DS7A electrical stimulation systems (Digitimer Ltd., Welwyn Garden City, England) were used to stimulate independently the participant's right and left index fingers via Digital Ring Electrodes (Oxford Instruments Medical, Hawthorne, NY). The electrical stimulus was randomized by varying interstimuli intervals (+/− 10 ms) using BrainX [6, 7], and adjusted to an intensity that delivered stimulation large enough to be detected by the participant yet not cause pain.

S. Supek and A. Sušac (Eds.): Advances in Biomagnetism – BIOMAG2010, IFMBE Proceedings 28, pp. 254–257, 2010.
www.springerlink.com

C. Data Acquisition

MEG signals were recorded from a 275-channel whole head MEG system (VSM MedTech Ltd., Port Coquitlam, BC, Canada) in a magnetic shielded room (Vacuum-Schmelze, Hanau, Germany). During the recording session, participant needs to lie as still as possible on a bed with their head inside Dewar. They were instructed to avoid eye blinks and head movements during the five-minute recording session. Before data acquisition commenced, three electromagnetic coils were placed on the nasion, at the left and right pre-auricular points of each participant in order to superimpose MEG responses upon 3D MRI later. MEG data were recorded at a sampling rate of 6,000 Hz with a noise cancellation of third order gradients and without any online bandpass filters. The data were online averaged over trials.

To convert the source of MEG responses into functional brain images, individual MRI scans (1.5T Sigma scanner, GE Medical Systems, Milwaukee, WI) were obtained. T1-weighted axial anatomical images with in-plane resolution of 256*256 and 128 slices (1.4 mm thickness) were recorded using spoiled gradient echo imaging. Three fiduciary points were placed in identical locations to the positions of the three coils used in the MEG recordings. Using MRIViewer (VSM MedTech Ltd., Port Coquitlam, BC, Canada), the Digital Imaging and Communications in Medicine (DICOM) format was converted to the CTF MEG System's own format for reading and viewing volumetric MR images. These MRI data files retained information regarding the fiducial landmarks such that any three-dimensional location with respect to the MEG head coordinate system can be automatically co-registered with the corresponding voxel location in the MRI.

D. Data Analyses

During the MEG recordings, 100 trials for each hand were averaged online by DataEditor (VSM MedTech Ltd., Port Coquitlam, BC, Canada). MEG data were preprocessed by using 3rd Gradient for environmental noise reduction and removing DC offset based on pre-trigger.

Frequency analysis was done with Morlet continuous wavelet transform. This study focused on high-frequency neuromagnetic signals (30-1000 Hz); low frequency signals were analyzed in waveform and therefore were filtered out before wavelet transform. The frequency window for wave-let transform was 10-2000 Hz. The time window for wave-let transform was 0-60 ms. Our analysis focused on neuro-magnetic activation in 10-40 ms following finger stimulation. Since wavelet transform is sensitive to time but not frequency at high-frequency signals, a large sigma value (the number of waves) of Morlet continuous wavelet transform could enhance its sensitivity to frequency at high-frequency signals. On the contrary, wavelet transform is sensitive to frequency but not time in a low-frequency range. Therefore, a small sigma value of Morlet continuous wavelet transform could enhance its sensitivity to frequency at high-frequency signals. Building on this principle, we developed wave-cross method (see the following formula 1 and 2), which allowed us to set two sigma values. To produce one spectrogram for each time data set, a cross-product was computed for each time- frequency point with the coefficients corresponding to the two sigma values

$$Wc(t) = W(t,s1) \times W(t,s2) \quad (1)$$

$$W(t,s) = c_\sigma \pi^{-\frac{1}{4}} e^{-\frac{1}{2}t^2} (e^{i\sigma t} - \kappa_\sigma) \quad (2)$$

In the above formula, Wc(t) indicates the final results of wave-cross analysis based on two Morlet wavelet transforms: W(t,s1) and W(t,s2). Noteworthy, each wavelet transform has its own sigma value (s1 and s2). To produce one spectrogram for each time data set, a cross-product was computed for each time- frequency point with the coefficients corresponding to the two sigma values with the mother wavelet (2). If signals appeared in the giving sensitive time (a small sigma value) and sensitive frequency (a big sigma value) ranges, they would be enhanced. In the spectrogram, it looked like crossing one time-frequency point.

Each trial of MEG data was transformed from time-domain waveform to frequency domain spectrogram individually with MEG Processor. Once all trials were transformed to spectral data; all trials were then accumulated one spectrogram. Magnetic sources were estimated with wavelet-based beamformers. To visualize magnetic sources, source imaging was combined with individual MRI to form magnetic source imaging (MSI).

III. Results

In the left index finger stimulation condition, there were three reproducible neuromagnetic responses (deflections) in the averaged raw signal data. In the right index finger stimulation condition, there were three reproducible responses recognized in all participants. In addition to brain signals below 120 Hz, the spectrograms revealed consistent focal increases of spectral power around 120-200 Hz in the left and right hemispheres. In addition, the neuromagnetic activation above 200 Hz was also observed, which were in the frequency range of 600-890 Hz. Neuromagnetic signals up to 2000 Hz were analyzed. Those high-frequency signals were probably from the deep brain areas. The neural sources for neuromagnetic activation in 120-200 Hz were localized in the finger areas for both left and right stimulation for most of the children (see Figure 1 and 2).

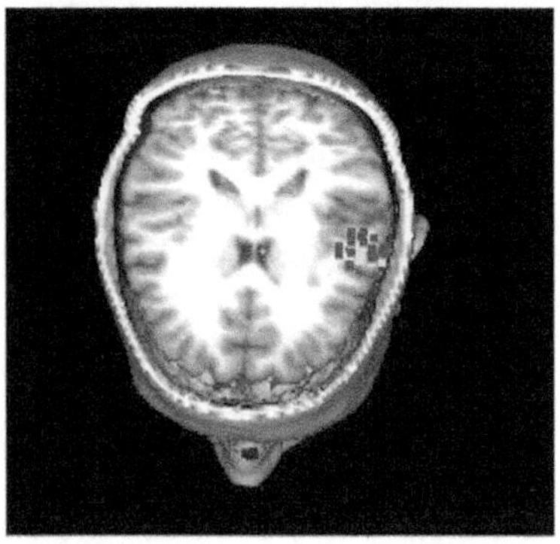

Fig. 1 Magnetic source imaging from a representative subject following electrical stimulation to the right index finger showing the location of high-frequency neuromagnetic signals. Red indicates regions of activation; blue indicates regions of suppression as determined using pseudo-Student-t values. Activity is shown in the 120-200 Hz frequency band

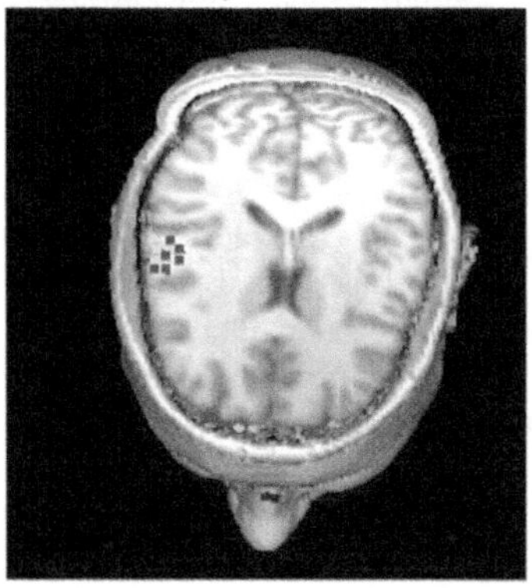

Fig. 2 Magnetic source imaging from a representative subject following electrical stimulation to the left index finger showing the source location of high-frequency brain signals. Red areas indicate regions of activation; blue areas indicate regions of suppression as determined using pseudo-Student-t values. Activity is shown in the 120-200 Hz frequency band

IV. Discussion

High-frequency neural signals are a new window in the electrophysiological analysis of the somatosensory cortex. High-frequency signals (> 100 Hz) have been found in MEG recently in the brain. According to our preliminary data, the temporal and frequency signatures of the development of somatosensory function in childhood are noninvasively quantifiable with MEG and Wave-cross. Several advanced MEG data analysis methods including wavelet transform and accumulated spectrograms have been reported previously [6, 8]. The present study moved one step further by slightly modifying the Morlet wavelet transform to compute a new spectrogram with two sigma values. Those methods have been integrated into a systematic approach to analyze SEFs in the child's brain. We adopted accumulated spectrograms because the MEG data recorded from the brain commonly have very strong low-frequency signals. Since high-frequency neuromagnetic signals are much weaker than low-frequency signals, high-frequency signals are commonly obscured by low-frequency signals. In this study, we evaluated high frequency neuromagnetic signals in SEFs using Morlet wavelet transform based wave-cross analysis. Our preliminary data have confirmed that wave-cross, a new time-frequency analysis method, could precisely determine the frequency and temporal signatures of brain activation. The results lay a foundation for quantitative identification of developmental delay and/or abnormalities of the somatosensory function in children with brain disorders.

Wavelet-based beamformer has been used in the localization of somatosensory evoked magnetic fields. There are a few differences between the standard SAM beamformer and our wavelet-based beamformer: (1) SAM beamformer assumes that the magnetic source is "static" in a time window, but wavelet-base beamformer does not; (2) SAM beamformer analyzes source with a band-pass filter selected by users and assumes brain activation in one frequency band, but wavelet-based beamformer analyze signals in all possible frequency ranges. Though further investigation and confirmation are necessary, we consider that wavelet based beamformer is superior to the standard SAM beamformer because functional brain activation could be in multiple frequency ranges and changing rapidly with time.

V. Conclusions

In conclusion, in an effort to characterize the maturational changes in cortical somatosensory processing, the present investigation analyzed evoked somatosensory responses in children with beamformer and wavelet. Our results demonstrate that the spatial and frequency signatures of SEFs in both low- and high-frequency ranges can be quantitatively determined. The frequency signature of SEFs could be also useful for improving the accuracy of functional mapping of the somatosensory cortex for children.

Acknowledgment

This study was supported by a Trustee Grant from Cincinnati Children's Hospital Medical Center. We would like to thank our research technologist, Kendall O'Brien for performing the MRI scan.

References

1. Xiang J, Liu Y, Wang Y, Kotecha R, Kirtman EG, Chen Y, Huo X, Fujiwara H, Hemasilpin N, DeGrauw T, Rose D. Neuromagnetic correlates of developmental changes in endogenous high-frequency brain oscillations in children: A wavelet-based beamformer study. Brain Res. 2009; 1274:28-39.
2. Xiang J, Liu Y, Wang Y, Kirtman EG, Kotecha R, Chen Y, Huo X, Fujiwara H, Hemasilpin N, Lee K, Mangano FT, Leach J, Jones B et al. Frequency and spatial characteristics of high-frequency neuromagnetic signals in childhood epilepsy. Epileptic Disord. 2009; 11:113-125.
3. Cascino GD. Surgical treatment for epilepsy. Epilepsy Res. 2004; 60:179-186.
4. Xiang J, Holowka S, Sharma R, Hunjan A, Otsubo H, Chuang S. Volumetric localization of somatosensory cortex in children using synthetic aperture magnetometry. Pediatr Radiol. 2003; 33:321-327.
5. Hari R. On brain's magnetic responses to sensory stimuli. J Clin Neurophysiol. 1991; 8:157-169.
6. Kotecha R, Xiang J, Wang Y, Huo X, Hemasilpin N, Fujiwara H, Rose D, Degrauw T. Time, frequency and volumetric differences of high-frequency neuromagnetic oscillation between left and right somatosensory cortices. Int J Psychophysiol. 2008.
7. Wang Y, Xiang J, Kotecha R, Vannest J, Liu Y, Rose D, Schapiro M, Degrauw T. Spatial and frequency differences of neuromagnetic activities between the perception of open- and closed-class words. Brain Topogr. 2008; 21:75-85.
8. Xiang J, Xiao Z. Spatiotemporal and frequency signatures of noun and verb processing: A wavelet-based beamformer study. J Clin Exp Neuropsychol. 2008:1-10.

Author: Yingying Wang
Institute: Cincinnati Children's Hospital Medical Center
Street: 3333 Burnet Avenue, MLC 2015
City: Cincinnati, OH 45229
Country: U.S.A
Email: yingying.wang@cchmc.org

Detecting Gross Fetal Movements Using Fetal Magnetocardiography

Peter Van Leeuwen[1], Daniel Geue[2], Silke lange[3], and Dietrich H.W. Grönemeyer[1]

[1] Dept. Radiology and Microtherapy, University Witten/Herdecke, Bochum, Germany
[2] VISUS Technology Transfer, Bochum, Germany
[3] p-Wert.de, Witten, Germany

Abstract— **Fetal development is associated with the emergence of fetal movements (FM). Gross FM involve the limbs and/or the trunk and are accompanied by changes in heart rate and heart rate variability (HRV). To detect such FM, we developed a semi-automatic technique which analyzes changes in signal morphology of multichannel fetal magnetocardiogram (FMCG) signals over time. The technique exploits the changes QRS signal shape and amplitude in specific channels resulting from gross FM. From our database of FMCG records, we selected 26 five-minute FMCGs in 14 healthy pregnant women (31st-41st week of gestation) in which FM were suspected on the basis of fetal heart rate accelerations. In each recording, after the removal of the maternal artifact, the fetal QRS complexes were identified using template matching. In each channel, the minimum and maximum amplitudes of each QRS complex were determined. In the actogram, the beat-to-beat changes in maximum and minimum values, respectively, were quantified using the variance over a 20 beat window. Automatically determined, rapid increases in variance were considered reflecting changes in fetal position. The timings of these increases were examined together with changes in fetal heart rate (determined by the RR intervals derived from the QRS complexes). We found between 1-4 episodes of rapid increases in variance in each of the 26 actograms. These episodes were associated with an immediate decrease of 31.7±18.8 ms from overall RR interval (heart rate increase: 11.3±6.5 bpm). Beat-to-beat RR interval variability was reduced from overall 7.4±3.2 ms to 3.8±1.7 ms immediately following the begin of the actographically determined episodes. The results suggest that the majority of the episodes reflected FM and that this actographic approach may be used to automatically scan FMCG recordings for such movement. This will be helpful in categorizing FMCG recordings with respect to fetal behavioral state and improve the assessment of fetal HRV.**

Keywords— **biomagnetic signals, fetal actography, fetal magnetocardiography, heart rate variability, fetal movements.**

I. Introduction

In a variety of studies it has been shown that fetal magnetocardiography (FMCG) is well-suited for high quality evaluation of fetal heart rate variability (FHRV) [1] [2] [3]. However, in the interpretation of FHRV, a number of confounding factors will affect the calculated values. Among others, these include the heart rate, gestational age and the state of the autonomous nervous system [1] [4]. Of particular importance, however, is fetal behavioral state. FHRV is strongly and systematically affected by the wake or sleep, the passive or active state of the fetus [5] [6] [7]. The emergence of these states differentiates in increasingly complex behavioural patterns as pregnancy progresses. Thus, when assessing FHRV in a clinical setting, information on fetal behavioral state is essential. Normally, this is done on the basis of the evaluation of heart rate patterns in conjunction with ultrasound imaging reflecting gross and fine fetal movements (FM). As the registration of FMCG signals is strongly disturbed by interference from ultrasound equipment, combining these registration modes is not an option. Therefore, an alternative approach enabling the identification of fetal movement would be useful.

Recently, we developed a method for the automatic detection of gross fetal movements using multichannel FMCG signals [8]. This approach, expanding on a method previously reported [9], exploits the effect that fetal gross movement has on fetal QRS morphology. Due to the change in position of the fetus, the position and direction of the heart vector will be translated and rotated with respect to the sensing system. In a multichannel biomagnetometer system this will result in changes in waveform shape and amplitude in specific channels. The algorithm we developed automatically detects the channels with the greatest changes and uses these changes to identify the begin and termination of FM. The aim of the present work was to assess this approach in a group of subjects through the third trimester of pregnancy. In particular, we were interested in the concurrence of gross FM and heart rate accelerations with the accompanying changes in FHRV.

II. Methods

A. Subjects

The subjects were selected from a pool of data studied previously [4] [10] according to the following criteria: healthy pregnancy, gestational age >30 weeks, registration of spontaneous data (i.e. without intervention), presence of at least one clear heart rate acceleration from baseline lasting at least 5 s. This resulted in 26 FMCG registrations from 14 pregnant women (31st-41st week of gestation).

S. Supek and A. Sušac (Eds.): Advances in Biomagnetism – BIOMAG2010, IFMBE Proceedings 28, pp. 258–261, 2010.
www.springerlink.com

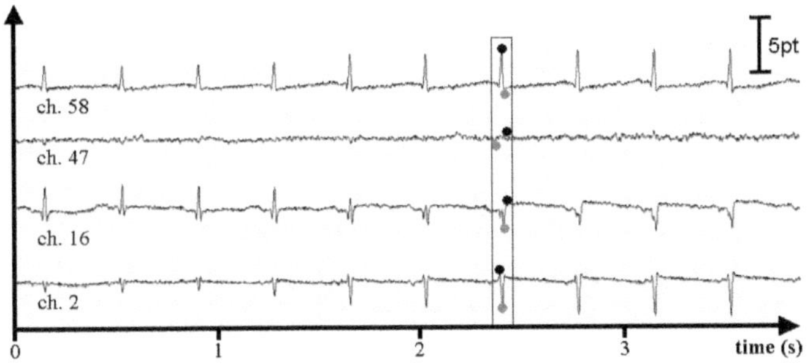

Fig. 1 Changes in QRS amplitude resulting from fetal movement in selected channels: unidirectional in channel 2 (negative increase in the minimum) and channel 58 (positive increase in the maximum); reversal in amplitude in channel 16; no change in channel 47, (●=min, ●=max)

B. Data Acquisition

The FMCG were recorded using a 61 channel biomagnetometer (Magnes 1300C, 4D Neuroimaging, San Diego) at a sampling rate of 1 kHz with a bandpass of 1-200 Hz. During the 5 min. registration, the mother was in a quiet, supine position with the sensor placed directly over abdomen. Sensor diameter (32.4 cm) and area of coverage (825 cm^2) were large enough to cover the entire abdomen. To reduce the effects of external noise, all measurements were performed in a standard shielded room (AK3b, Vacuumschmelze, Hanau).

C. Data Processing

In each recording, maternal heart beats were automatically identified by correlation to a maternal QRS signal template, constructed on the basis of an average of 10-20 manually identified QRS signal traces. The maternal PQRST signal components were then digitally averaged and this average was subsequently subtracted from the signal traces. In the resulting traces, fetal beats were similarly identified on the basis of a representative fetal QRS signal template constructed using 20-30 manually selected fetal QRS traces of similar morphology. Visual inspection of the identified fetal R peaks was used to remove falsely marked artifacts and identify and mark missed beats. The resulting R peaks were then used 1) to construct the fetal actogram, and 2) to produce a complete fetal RR interval time series for the data set. The latter formed the basis for the analysis of fetal heart rate and HRV.

D. Constructing the Fetal Actogram

In each of the 61 channels, a window of 100 ms was constructed around all of the identified R-peaks (Fig.1). The signals in these windows were detrended and offset to compensate for low frequency drifts. The absolute minimum and

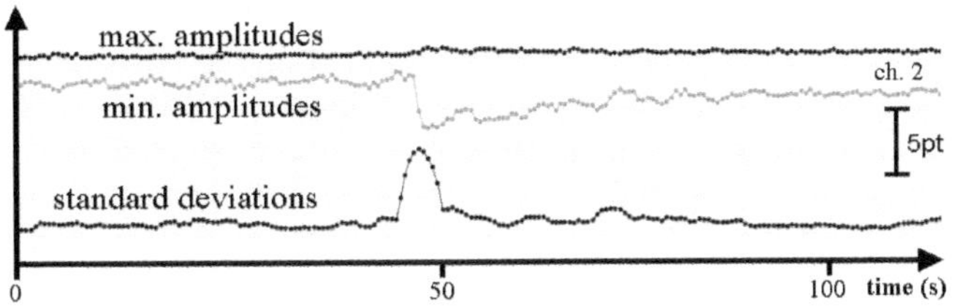

Fig. 2 Plot of the maximum (black) and minimum (grey) amplitudes of channel 2 (from Fig. 1) as well as the sum of their standard deviations calculated over 20 samples

maximum amplitudes in each QRS-complex window were determined in each channel (see box in Fig. 1).

These amplitudes were plotted (Fig. 2, top) and the standard deviation (SD) was calculated for both minimum and maximum amplitudes over a 20 sample window. The sum of the two SDs is higher when there are unidirectional increases or decreases in amplitude as well in the case of amplitude reversals (Fig. 2, bottom). In any one channel, periods of higher summed-SD-values represent substantial changes in signal amplitude – i.e. changes in QRS shape – of consecutive QRS-complexes. Such a channel thus registers the change in the position of the fetal heart vector resulting from FM. Such periods were detected by automatic peak detection using an adaptive threshold. The 61 channels were then sorted by magnitude of summed-SD and the 3 channels with the highest values were selected and their values averaged. Timing of the suspected begin and end of FM was identified using slope detection which takes the variance of the series into account.

E. Calculated Fetal Heart Rate Measures

The fetal actograms were visually compared to the RR interval time series (Fig. 2). At each of the detected time points representing the begin a suspected FM, we examined

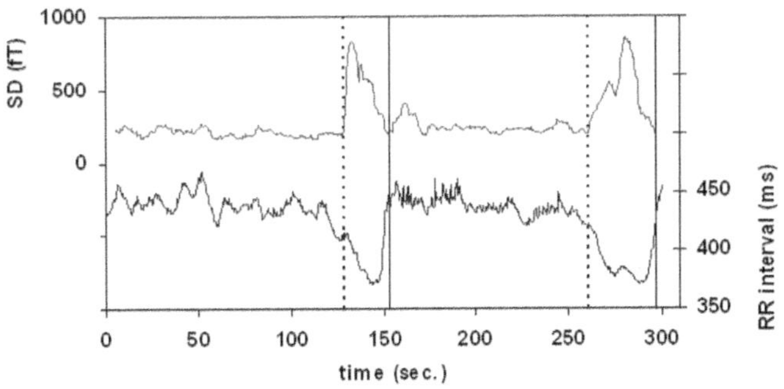

Fig. 3 Actogram based on the standard deviation of the summed amplitudes (SD in femto Tesla, top) and the corresponding RR time series (bottom) of an FMCG in the 35th week. Detected movement begin at the dashed vertical lines and end at the continuous vertical lines

heart rate, determined on the basis of the RR interval durations, and short term FHRV, determined on the basis of the root mean square of successive differences (RMSSD). In order to quantify the cardiac effect of the suspected movements, we compared 1) the minimum RR interval found within the first 10 s after onset detection to the mean RR interval of the complete data set and 2) RMSSD over the first 10 s after onset detection to the RMSSD of the complete data set.

F. Timing of Actogram FM Episodes

We also examined the timing of the actogram episode begin to the timing of the begin of heart rate changes. We scanned for fetal heart rate acceleration on the basis of windows of 15 pairs of consecutive beats including at least 13 beat pairs displaying a shortening of RR interval duration and an RMSSD < 3 ms. The precise begin of acceleration was determined visually as the first beat initiating a continuous decrease in RR interval.

III. RESULTS

A. Identified FM Episodes in the Actograms

In all the constructed actograms, the algorithm identified at least 1 and up to 4 episodes of substantial changes in signal morphology per data set (example in Fig. 3, upper trace). A total of 59 episodes were found (median per data set = 2) with a mean duration of 31 ± 20 s (± SD).

B. Changes in Heart Rate

These episodes were associated with higher momentary heart rates: within 10 s of the begin of an episode the mean RR interval was 398 ± 20 ms whereas mean RR interval over the complete series was 430 ± 21 ms (Fig. 4, left). This corresponded to a mean difference of 31.7 ± 18.8 ms (heart rate increase from 140 ± 6 to 151 ± 7 bpm). Associated with the increase in heart rate was a decrease in short-term variability: RMSSD was reduced from overall 7.4 ± 3.2 ms to 3.8 ± 1.7 ms immediately following the begin of the actographically determined episodes (Fig. 4, right). For both measures, the difference was statistically significant at the level of 0.001 (Wilcoxon test).

C. Timing of Actogram FM Episodes and Heart Rate Changes

As can be seen in Fig. 4, not all identified actographic episodes were associated with a heart rate acceleration. The exceptions correspond to data pairs with higher values at

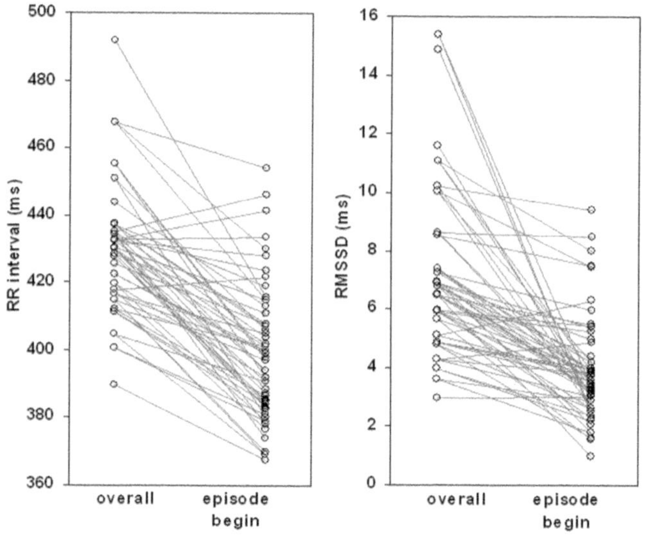

Fig. 4 Differences between overall values of heart rate (RR interval duration, left) and short term variability (RMSSD, right) to values after the begin of an identified episode in the actogram. Single value pairs are connected by lines

episode begin. A clear acceleration could not be identified in 7 episodes. In the remaining 52 episodes (88%), the changes in heart rate occurred after the timing of the sharp increases in the actogram. The mean time lag was 3.3 ± 3.6 s. In the majority (85%) the actogram detection preceded the change in heart rate. In 9% the timings concurred and in 6% of the episodes the acceleration onset occurred before the onset detected in the actogram.

IV. DISCUSSION

In this work, we applied an algorithm for the automatic detection of fetal movement in the fetal magnetocardiogram to a group of fetuses in the 3rd trimester of healthy pregnancy. The FMCG had been selected on the basis of the presence of heart rate accelerations which may often be associated with FM. We reasoned that, if the algorithm correctly identifies changes in fetal position, then the timing of these changes in the actogram should concur with increase in fetal heart rate. The results show that the majority (88%) of the episodes found in the actogram occurred together with heart rate acceleration. Furthermore these episodes were associated with lower FHRV, commonly found in movement-induced heart rate increases. This is a clear indication that the changes in QRS wave morphology registered by the FMCG denote movement behavior in the fetus.

The time lag of about 3 s between initiation of movement and heart rate increase correspond well to other published results in which FM was determined using Doppler actography [11]. Using an FMCG actography approach similar to that used here, Zhao and Wakai [9] noted that heart rate increases more often than not preceded detected movements. However their method, based on a single channel with the greatest amplitude deflection of QRS amplitude, incorporated less information compared to our approach. This may have led to a delay in the detection of movement. Another possible source of these discrepancies may lie in the difficulties involved in determination of the begin of heart rate acceleration.

Not all accelerations were associated with automatically detected episodes in the actogram. In some cases, visual inspection of the actogram showed amplitude increases which were not identified by the detection procedure, suggesting that the algorithm may be improved. In other cases, no actographic evidence for movement was present, indicating other physiological causes. As only gross trunk movement will result in a reorientation of the cardiac vector, these accelerations may result from smaller movements such as those involving solely limb displacement. Furthermore, changes in autonomic state due to stress may also be involved. Taking this into consideration, the method might aid in the discrimination between fetal behavioral states.

We also found a number of actogram episodes which did not concur with accelerations (Fig. 4). These may have resulted from artifacts resulting from maternal movement. On the other hand, periods of repeated movements may lead to variable heart rates making it difficult to identify baseline heart rate.

V. Conclusions

The results presented here indicate that signal amplitude changes in FMCG may be used for the identification of fetal movements and the associated changes in fetal heart rate. On the basis of a larger data set, it should be possible to optimize the algorithm, offering a tool with which FMCG traces can be automatically scanned for gross fetal movement. This information, combined with the heart rate and HRV will permit a more dependable determination of fetal behavioral state. This, in turn, may enable a reliable and clinically relevant assessment of heart rate parameters associated with prenatal health.

References

1. Wakai RT (2004) Assessment of fetal neurodevelopment via fetal magnetocardiography. Exp Neurol 190 (Suppl 1):65-71. DOI 10.1016/j.expneurol.2004.04.019
2. Van Leeuwen P, Lange S, Geue D, Grönemeyer D (2007) Heart rate variability in the fetus: a comparison of measures. Biomed Tech 52:61-65. DOI 10.1515/BMT.2007.012
3. Cuneo BF, Strasburger JF, Wakai RT (2008) Magnetocardiography in the evaluation of fetuses at risk for sudden cardiac death before birth. J Electrocardiol 41:116.e1-6. DOI 10.1016/j.jelectrocard.2007.12.010
4. Lange S, Van Leeuwen P, Geue D, Hatzmann W, Grönemeyer D Influence of gestational age, heart rate, gender and time of day on fetal heart rate variability. Med Biol Eng Comput 2005; 43(3): 481-486. DOI 10.1007/BF02344729
5. Nijhuis JG, Prechtl HF, Martin CB, Bots RS (1982) Are there behavioural states in the human fetus? Early Hum Dev 6:177-195. DOI 10.1016/0378-3782(82)90106-2
6. Frank B, Pompe B, Schneider U, Hoyer D (2006) Permutation entropy improves fetal behavioural state classification based on heart rate analysis from biomagnetic recordings in near term fetuses. Med Biol Eng Comput 44:179-187. DOI 10.1007/s11517-005-0015-z
7. Lange S, Van Leeuwen P, Schneider U, Frank B; Hoyer D, Geue D, Grönemeyer D (2008) Heart rate features in fetal behavioral states. Early Human Dev, in press. DOI 10.1016/j.earlhumdev.2008.07.004
8. Geue D, Van Leeuwen P, Lange S, Grönemeyer D (2007) Automatic detection of gross fetal movements using actography based on magnetocardiography. Biomed Tech 52 (suppl):CD-ROM.
9. Zhao H, Wakai RT Simultaneity of foetal heart rate acceleration and foetal trunk movement determined by foetal magnetocardiogram actocardiography. (2002) Phys Med Biol 47:839-846. DOI 10.1088/0031-9155/47/5/310
10. Van Leeuwen P, Geue D, Lange S, Hatzmann W and Gronemeyer D (2003) Changes in the frequency power spectrum of fetal heart rate in the course of pregnancy. Prenat Diagn 23:909-916,. DOI 10.1002/pd.723
11. Dipietro JA, Caulfield LE, Irizarry RA, Chen P, Merialdi M, Zavaleta N (2006) Prenatal development of intrafetal and maternal-fetal synchrony. Behav Neurosc 120:687-701. DOI 10.1037/0735-7044.120.3.687

Author: Peter Van Leeuwen
Institute: Dept. Biomagnetism, Chair of Radiology and Microtherapy, University of Witten/Herdecke
Street: Universitätsstr. 142
City: 44799 Bochum
Country: Germany
Email: petervl@microtherapy.de

Fetal Maternal Heart Rate Entrainment under Controlled Maternal Breathing

Peter Van Leeuwen[1], Daniel Geue[2], Marko Thiel[3], Silke Lange[4], D. Cysarz[5], Mamen C. Romano[3], Jürgen Kurths[6], and Dietrich H.W. Grönemeyer[1]

[1] Dept. Radiology and Microtherapy, University Witten/Herdecke, Bochum, Germany
[2] VISUS Technology Transfer, Bochum, Germany
[3] Dept. Physics, University of Aberdeen, United Kingdom
[4] p-Wert.de, Witten, Germany
[5] Integrated Curriculum for Anthroposophic Medicine, University Witten/Herdecke, Herdecke, Germany
[6] Potsdam Institute for Climate Impact Research and Humboldt University, Berlin, Germany

Abstract— **There is evidence that, during pregnancy, fetal and maternal cardiac activity may coordinate over short periods of time. Aim of this work was to investigate whether controlled paced maternal respiration has an effect on the occurrence of fetal-maternal heart rate synchronization. In 6 healthy pregnant women (34th - 40th week of gestation) we obtained simultaneous 5 min. fetal and maternal magnetocardiograms (MCG) at maternal respiration rates of 10, 12, 15 and 20 cpm as well as under spontaneous breathing. Fetal and maternal RR interval time series were constructed for each MCG data set and synchrograms were obtained using the stroboscopic technique. Synchronization epochs (SE) >10 s were identified in these original data. Furthermore, "twin surrogate" data sets of the maternal MCG were constructed, combined with the fetal MCG data and SE were identified in these surrogate data. In the original data, there was a higher number of SE found at 20 cpm respiratory rate compared to other rates. This was not evident in the surrogate data. Fewer SE were found at lower rates (10 cpm) both in the original and surrogate data. Examination of the phase of fetal R peaks relative to maternal RR cycles showed that in the original data there was a clear phase preference in the 20 cpm data. This preference was not found in the surrogate data. These results were reproduced on the basis of a mathematical model incorporating the essential elements of fetal and maternal heart rates and their variability. We conclude that fetal-maternal heart rate entrainment may be induced by high maternal respiratory rates and that chance fetal-maternal heart rate coordination may be inhibited by low maternal respiratory rates.**

Keywords— **magnetocardiography, fetal maternal heart rate synchronization, respiration, surrogate techniques, modeling.**

I. Introduction

Prenatal interaction between mother and fetus has been postulated and various studies confirm a relationship between maternal and fetal physiological conditions via fetal heart rate. These include, among others, changes in fetal heart rate and heart rate variability associated with altered maternal arterial oxygen content [1] [2], maternal hypothermia [3] and maternal exercise [4]. Furthermore, links between maternal and fetal heart rate have been examined. A positive correlation between these rates has been found over periods of one and 24 hours [5] and the entrainment of the fetal heart rate rhythm to the maternal diurnal rhythm has been observed [6].

On the other hand, short term interaction between fetal and maternal heart rate is elusive. In previous work, we have shown that fetal and maternal cardiac activity may coordinate over short periods of time [7]. However, surrogate analysis suggested that there is also a high probability of epochs with spurious fetal-maternal heart rate coupling resulting simply from the naturally changing frequencies of two independent cardiac oscillators.

Other work using cross spectral analysis has shown that maternal respiratory sinus arrhythmia may also be correlated to changes in fetal heart rate in the high frequency band related to maternal breathing [8]. We thus postulated that conditions conducive to physiological cardiac interaction between mother and fetus might be achieved by controlling maternal heart rate via regulation of her respiratory sinus arrhythmia. Aim of this work was to assess the occurrence of epochs of fetal-maternal heart rate synchronization under conditions of spontaneous and controlled breathing of the mother.

II. Methods

We examined six healthy pregnant women between the 34th and 40th week of gestation. The cardiac activity of both the mother and her fetus was recorded simultaneously using a 61 channel biomagnetometer (Magnes 1300C, 4D Neuroimaging, San Diego). In each subject, six 5 minute recordings were acquired under the following conditions: spontaneous breathing, paced respiration at 15 cpm, 10 cpm, 20 cpm and 12 cpm and finally spontaneous breathing. The data were recorded at a sampling rate of 1 kHz with a bandpass of 1-200 Hz.

In each of the resulting 5 minute MCG data sets, fetal and maternal R peaks were identified to an accuracy of 1 ms and

S. Supek and A. Sušac (Eds.): Advances in Biomagnetism – BIOMAG2010, IFMBE Proceedings 28, pp. 262–265, 2010.
www.springerlink.com

the corresponding RR interval time series were constructed., Synchrograms were generated on the basis of the relative timing of fetal and maternal R peaks using the stroboscopic technique described by Schäfer et al [9]. The phases of the fetal R peaks were plotted on a time axis with respect to m maternal RR interval cycles whereby $1 \le m \le 4$. Horizontal groupings of data points indicate phase synchronization and the number (n) of parallel lines in an epoch of such horizontal groupings embedded in m maternal RR intervals indicates the coupling ratio n:m. We set $2 \le n \le 7$ and examined six non-redundant n:m ratios: 2:1, 3:2, 4:3, 5:3, 5:4, 7:4. These grouping were identified as synchronization epochs (SE) according to an algorithm developed by Toledo et al [10]. For an SE to be considered as representing possible fetal-maternal interaction, its duration needed to be 10 s or longer (for details, see [7]).

For each data set in the original data (= each pair of fetal-maternal RR interval time series), the number of SEs, their duration and n:m combination were determined. These results were examined with respect to the six respiratory conditions. Furthermore in order to investigate whether there was a preference in the timing of the fetal R peaks relative to the maternal RR cycle, the phase of each SE found in each data set was determined in all data sets. The distributions of these phases over all subjects were examined for the n:m combinations of which more than 10% of the SE were present: this was true for the combinations 3:2, 4:3 and 5:3 (see results). For each n:m combination, the number of histogram bins was set to $8 \times m$, corresponding to a resolution of 8 classes per maternal beat. The distributions were visually inspected for a systematic bias for specific phases under each of the six respiratory conditions.

In order to distinguish epochs resulting from physiological interaction and those due to transient combinations of heart rates, so-called "twin surrogate" data sets of the maternal MCG were constructed. The surrogates are generated by a mathematical algorithm which is based on the recurrences of a system. The generated time series corresponds to a trajectory of the system starting at different (random) initial conditions. Not only do the surrogate series mimic both the linear and nonlinear dynamical properties of the system but also, for any original series, an arbitrary number of surrogate series can be produced (for details, see [11]). For each data set, the fetal MCG signals were combined with 20 surrogate maternal MCG signals. These 20 fetal maternal pairs were then processed and analyzed using the same procedure described above for the original data. The results of the analyses of the original and surrogate data were then compared.

We also compared the heart rate properties of each pair of original and surrogate data sets with respect to the number of beats, the mean RR interval, their standard deviation and their root mean square of successive differences, their low frequency and high frequency power as well as their ratio and finally the approximate entropy (ApEn). In each data set, the mean value of each heart rate measure was calculated in both the original and the corresponding 20 surrogate RR interval time series. The 20 surrogate values were averaged and the difference (average surrogate value – original value) was determined.

Finally, we constructed mathematical models of the fetal and maternal cardiac cycles in order to simulate and control the primary factors involved in these cardiac oscillators. The models took the respective heart rates, aspects of short-term and long-term heart rate variability (HRV) determined in the real data, as well as maternal respiratory arrhythmia and noise into account [12]. We compared the number of SEs found in the model data to those in the original and surrogate data.

III. Results

We compared the heart rate and its variability of the original maternal and surrogate maternal RR interval time series. The linear properties of the time series were generally well retained in the surrogate data, apart from a lower variance as expressed by the standard deviation in the surrogate data. Also, the complexity was reduced (Table 1). Inspection of the surrogate maternal RR interval time series showed that in some of the reconstructed data a reduced RR interval range and/or a repetition of specific subseries of RR interval was present.

Table 1 Differences (means ± standard deviation) between the heart rate properties of the original and the surrogate maternal RR interval time series

Heart rate property	Difference (surrogate – original)	p-value
number of beats	4.2 ± 12.8	0.004
mean RR interval (ms)	-1.4 ± 16.3	0.109
SDRR (ms)	-5.5 ± 6.4	0.000
RMSSD (ms)	0.3 ± 2.4	0.329
low frequency (LF) power (ms^2)	-1.3 ± 11.2	0.679
high frequency (HF) power (ms^2)	-1.0 ± 12.2	0.931
LF / HF	-0.01 ± 0.20	0.891
approximate entropy	-0.042 ± 0.020	0.000

SDRR = standard deviation of the RR intervals; RMSSD = root mean square of successive differences.

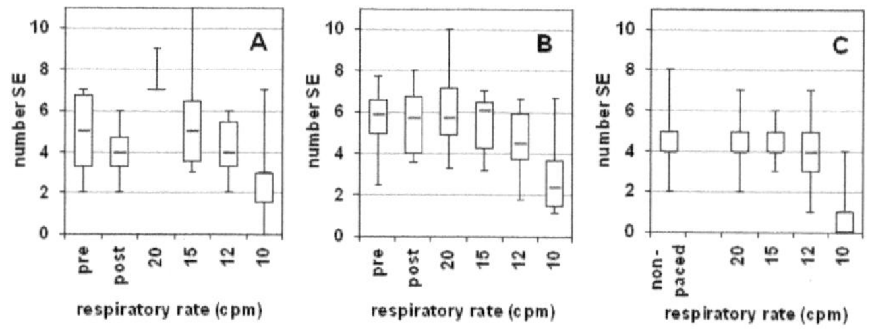

Fig. 1 Boxplots showing the number of synchronization epochs (SE) at different respiratory rates in the original (A), surrogate (B) and model (C) data

In the original data the number of SEs found in the data sets of spontaneous breathing was 4.8±2.2 prior to the controlled breathing and 4.0±1.5 after. The numbers were similar at the respiratory rate of 15 and 12 cpm (5.7±1.6 and 4.2±1.6) whereas at the fast rate (20 cpm) more SEs were present (7.3±0.9) and at the slow rate (10 cpm) there were fewer (2.8±2.6) (see Fig. 1A). In the surrogate data, the number of SEs lay in the mid-range for all conditions (pre and post pacing: 5.6±2.0, 5.6±1.9; 20 cpm: 6.2±2.6, 15 cpm: 5.4±1.6, 12 cpm: 4.6±1.9, 10 cpm: 3.0±2.3). Only the data at the slow rate (10 cpm) tended to low values (see Fig. 1B). For the model data, the results were similar to those of the surrogate data (Fig. 1C).

The durations of the SEs were similar for all six respiratory conditions in the original (between 15±6 s and 18±10 s) and in the surrogate data (between 15±6 s and 21±17 s). In the model data the duration was longer at higher frequencies (22±11 s) and shorter at low frequencies (13±1 s). In the original data the *n*:*m* combination 3:2 occurred most frequently (46%), followed by 5:3 (23%) and 4:3 (17%). The other combinations (2:1, 5:4 and 7:4) each had less than 10% of the SEs. The result for the surrogate data were similar.

Examination of the phase of the fetal R peaks relative to the maternal RR cycles in the three *n*:*m* combinations which contained at least 10% of the number of SE showed that, in the original data, phase preferences were evident in the 20 cpm data (Fig. 2). This was particularly clear for the combinations 4:3 and 5:3. Furthermore, a phase preference for the occurrence of SE was also visible at the respiratory rate of 12 cpm in the 3:2 combination. None of the surrogate data showed a similar preference.

IV. Discussion

The main finding in this study is that the presence or absence of interaction between fetal and maternal cardiac activity may be influenced by maternal respiration. In particular, higher paced breathing rates led to a greater number of SE when compared to normal or slower paced rates or to varying rates found in spontaneous breathing. Furthermore, at these higher respiratory rates we found a tendency for the SE to occur with a higher frequency at specific times of the maternal cycles. The fact that neither the surrogate nor the model data displayed an increased number of SEs at 20 cpm respiratory rate suggests that a physiological mechanism leading to coupling of the heart rate frequencies was present in the original data. Regularization of maternal heart rate via respiratory sinus arrhythmia induced by higher respiratory rates may thus create conditions in which the fetal cardiac system may be led to couple its timing to that of the maternal system.

Previous work in our group has shown that, in uncontrolled fetal-maternal heart rate acquisitions, most of the SEs found may be attributed to the naturally changing frequencies of two independent cardiac oscillators and do not necessarily represent true physiological interaction. Nonetheless, examination of a large number of subjects and acquisitions showed evidence of occasional physiological synchronization on the basis of non-uniform distribution of the phase preferences of the fetal beats found in the original but not the surrogate data. In the results of the present work, the phase preference found in the data recorded at a relatively normal respiratory rate of 12 cpm (Fig. 2) may be due to this kind of spontaneous interaction. It is clear, however, that synchronization occurred more consistently at fast maternal breathing rates.

We also found a lower probability for SE at the respiratory rates of 10 cpm (Fig. 1). The fact that this occurred not only in the original but also in the surrogate and model data

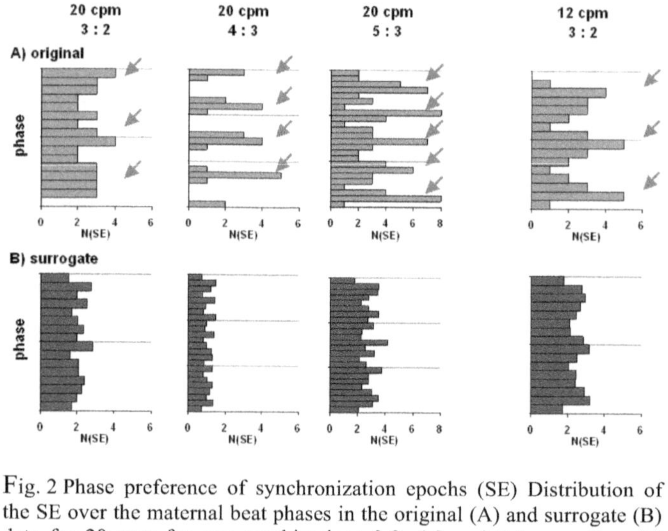

Fig. 2 Phase preference of synchronization epochs (SE) Distribution of the SE over the maternal beat phases in the original (A) and surrogate (B) data for 20 cpm for *n*:*m* combinations 3:2, 4:3 and 5:3 as well as 12 cpm with respect to the *n*:*m* combination 3:2. Arrows indicate phase preferences

suggests that breathing at low rates may create conditions in which the spurious occurrence of fetal-maternal heart rate coupling is hindered. We reasoned that chance coordination between the rates could be inhibited by high short-term beat-to-beat variability associated with the higher respiratory modulation amplitude found at such slow breathing. We did indeed find that the RMSSD and HF power values were higher at a respiratory rate of 10 cpm when compared to those of the other rates.

With respect to the reliability of the approach used to generate surrogate data, the results indicate that most of the descriptors of heart rate variability were similar between the original and the surrogate data sets. The reduction in the variance and complexity in the surrogate data are most likely due to the repetition of specific subseries of RR intervals in the surrogate data, which is to be expected for this kind of block bootstrap. However, it has been argued that this should not invalidate the test for synchronization [11].

V. Conclusion

We conclude that the results suggest that fetal-maternal heart rate entrainment may be induced by high maternal respiratory rates and that chance fetal-maternal heart rate coordination may be inhibited by low maternal respiratory rates.

References

1. Jensen A (1996) in Comprehensive Human Physiology, eds Greger R, Windhorst U (Springer, Berlin) pp 2307-2337
2. Bekedam DJ, Mulder EJ, Snijders RJ, Visser GH (1991) The effects of maternal hyperoxia on fetal breathing movements, body movements and heart rate variation in growth retarded fetuses. Early Hum Dev 27:223-32 DOI:10.1016/0378-3782(91)90196-A
3. Hankins GD, Leicht T, Van Hook JW (1997) Prolonged fetal bradycardia secondary to maternal hypothermia in response to urosepsis. Am J Perinatol 14:217-219 DOI: 10.1055/s-2007-994130
4. Webb KA, Wolfe LA, Mcgrath MJ (1994) Effects of acute and chronic maternal exercise on fetal heart rate. J Appl Physiol 77:2207-2213
5. Patrick J, Campbell K, Carmichael L, Probert C (1982) Influence of maternal heart rate and gross fetal body movements on the daily pattern of fetal heart rate near term. Am J Obstet Gynecol 144:533-538
6. Lunshof S, Boer K, Wolf H, Vanhoffen G, Bayram N, Mirmiran M (1998) Fetal and maternal diurnal rhythms during the third trimester of normal pregnancy: Outcomes of computerized analysis of continuous twenty-four-hour fetal heart rate recordings. Am J Obstet Gynecol 178:247-254 DOI:10.1016/S0002-9378(98)80008-2
7. Van Leeuwen P, Geue D, Lange S, Cysarz D, Bettermann H, Grönemeyer D (2003) Is there evidence of fetal-maternal heart rate synchronization? BMC Physiology 3:2 DOI 10.1186/1472-6793-3-2
8. Ferrazzi E, Pardi G, Setti, PL, Rodolfi M, Civardi S, Cerutti S (1989) Power spectral analysis of the heart rate of the human fetus at 26 and 36 weeks of gestation. Clin Phys Physiol Meas 10B:57-60 DOI: 10.1088/0143-0815/10/4B/009
9. Schäfer C, Rosenblum MG, Kurths J, Abel HH (1998) Heartbeat synchronized with ventilation. Nature 392(6673):239-240 DOI: 10.1038/32567
10. Toledo E, Akselrod S, Pinhas I, Aravot D (2002) Does synchronization reflect a true interaction in the cardiorespiratory system? Med Eng Phys 24:45–52 DOI:10.1016/S1350-4533(01)00114-X
11. Thiel M, Romano MC, Kurths J, Rolfs M, Kliegl R (2006) Twin surrogates to test for complex synchronisation. Europhys Let 75:535–541 DOI: 10.1209/epl/i2006-10147-0
12. Van Leeuwen P, Geue D, Lange S, Grönemeyer D (2009) Modeling fetal-maternal heart rate interaction at different rates of maternal respiration. IEEE Eng Med Biol Mag in press.

Author: Peter Van Leeuwen
Institute: Dept. Biomagnetism, Chair of Radiology and Microtherapy, University of Witten/Herdecke
Street: Universitätsstr. 142
City: 44799 Bochum
Country: Germany
Email: petervl@microtherapy.de

Magnetomyography Combined with Fetal Magnetocardiography in the Registration of Uterine Contractions

Peter Van Leeuwen[1], Wolfgang Hatzmann[2], Sven Schiermeier[2], and Dietrich H.W. Grönemeyer[1]

[1] Dept. Radiology and Microtherapy, University Witten/Herdecke, Bochum, Germany
[2] Dept. of Obstetrics, University Witten/Herdecke; Witten, Germany

Abstract— **Fetal heart rate and its variability (HRV) reflect not only fetal development and behavioral state but may also be indicative of fetal stress. Particularly toward the end of pregnancy, contractions of the uterus can influence fetal heart rate and HRV. Conventionally, fetal heart rate and uterine contractions are measured and observed using cardiotocography. Multichannel biomagnetic measurements offer the possibility to record both smooth muscle and cardiac muscle activity simultaneously with higher spatial and temporal resolution. Here we describe magnetomyographic and magnetocardiographic measurements acquired in a pregnant woman while experiencing intermittent contractions. Using a 61 channel biomagnetometer which covered the complete maternal abdomen, we recorded MMG and FMCG for 15 min at a sampling rate of 1 kHz and a bandpass of 1-200 Hz. Appropriate signal processing such as filtering, template matching and digital subtraction / averaging of signal components permitted the identification of signals associated with both uterine contractions as well as with fetal cardiac activity. Data analysis showed that broad band high frequency signals related to the contractions were consistently associated with rapid sustained increases in fetal heart rate and decreases in short-term heart rate variability. We conclude that MMG may identify episodes of uterine contraction and that their effect on fetal heart rate can be assessed simultaneously on the basis of FMCG. This combination of MMG and FMCG may thus aid in fetal monitoring.**

Keywords— **biomagnetic signals, magnetomyography, fetal magnetocardiography, uterine contractions, heart rate variability.**

I. Introduction

Various aspects of fetal cardiac function, both physiological and pathological, have been examined, classified and quantified in prenatal studies using fetal magnetocardiography (FMCG) [1] [2]. The use of the fetal electrocardiogram (FECG) in this context has been limited due to its relatively poor signal characteristics. In contrast, FMCG permits the registration of an ECG-like signal as early as in the 2nd trimester of pregnancy thus giving unique access to fetal cardiac electrophysiological phenomena. In routine clinical examination, changes in fetal heart rate are usually assessed using cardiotocography (CTG). The method is based on the estimation of fetal heart beats registered by an pulsed Doppler ultrasound signal, quantified on the basis of autocorrelation. A second transducer allows the recording of uterine contractions. Changes in fetal heart rate can correlate with such contractions and inspection of fetal heart rate variations induced by labor is a central aspect of CTG evaluation.

Although CTG is ubiquitously used in the clinical surveillance of the fetal condition, its visual interpretation of the course of fetal heart rates has been associated with a high degree of intra- and interobserver variability [3] [4]. In recent work we compared fetal heart rate variability (HRV) measures as quantified by simultaneously acquired CTG and FMCG tracings [5]. We were able to show that there is a limit to the resolution of the CTG beat duration and that a further increase in the dynamic range of short-term variability may be achieved with FMCG based analysis. This suggests that fetal HRV as quantified by FMCG may be advantageous in fetal surveillance.

Although magnetomyography (MMG) has been used to register muscle activity in general [6] as well as during uterine contractions [7], the combination of MMG and FMCG has not been reported. Here we present a pregnancy in which contractions occurred during an FMCG acquisition, allowing the examination of their effect on fetal heart rate and HRV.

II. Methods

A. *Subject and Data Acquisition*

A 35 year old pregnant woman (gravida 3, para 1) was referred to our institute because of suspected fetal bradycardia in the 37th week of pregnancy. We performed a 15 min. FMCG/MMG using a 61 channel biomagnetometer (Magnes 1300C, 4D Neuroimaging, San Diego) at a sampling rate of 1 kHz with a bandpass of 1-200 Hz. The mother was in a supine position with the sensor placed directly over abdomen. Sensor diameter (32.4 cm) and area of coverage (825 cm^2) were large enough to cover the entire abdomen, the lower rim being over the pubis and the upper rim approximately over the xyphoid (Fig. 1, left). To reduce the

S. Supek and A. Sušac (Eds.): Advances in Biomagnetism – BIOMAG2010, IFMBE Proceedings 28, pp. 266–269, 2010.
www.springerlink.com

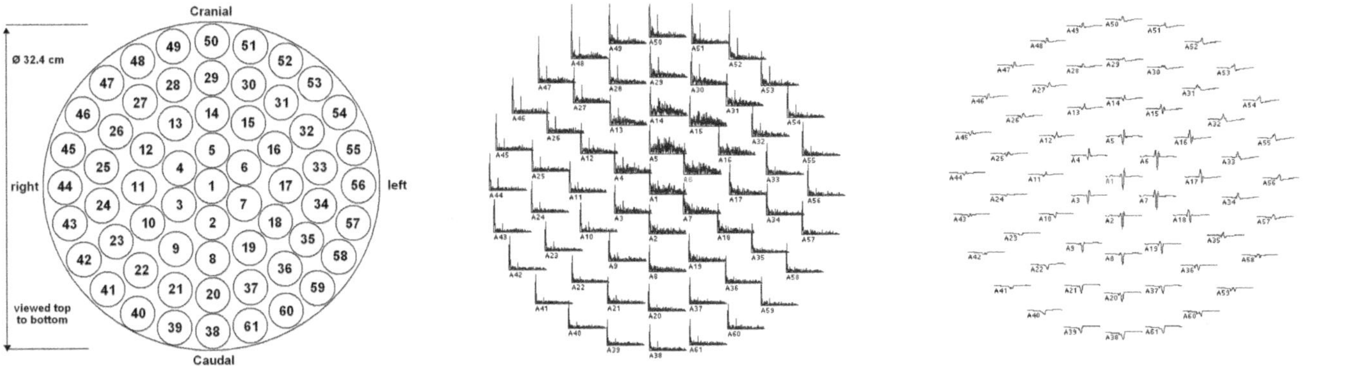

Fig. 1 Sensing system and biomagnetic signals Left: 61 channel sensor configuration. Center, MMG activity: spectra of the MMG signals between 0-200 Hz with increased power centered around channels 5, 6, 14, 15. Right, FMCG activity: fetal PQRST traces with high amplitude signals centered around channels 1, 2, 6, 7

effects of external noise, all measurements were performed in a standard shielded room (AK3b, Vacuumschmelze, Hanau).

B. Data Processing

Using weighting factors obtained from a set of reference coils which are located approximately 19cm above the detection coils, the gradients of the ambient field were estimated. These factors were used to perform a further noise suppression, correcting the signals for the effect of more distant disturbing signal sources.

In the noise reduced data, maternal heart beats were automatically identified by correlation to a maternal QRS signal template, constructed on the basis of an average of 10-20 manually identified QRS signal traces. The maternal PQRST signal components were then digitally averaged and this average was subsequently subtracted from the signal traces. In the resulting traces, fetal beats were similarly identified on the basis of a representative fetal QRS signal template constructed using 20-30 manually selected fetal QRS traces of similar morphology. Visual inspection of the identified fetal R peaks was used to remove falsely marked artifacts and identify and mark missed beats. The resulting R peaks were then used to produce a complete fetal RR interval time series for the data set and this formed the basis for the analysis of fetal heart rate and HRV.

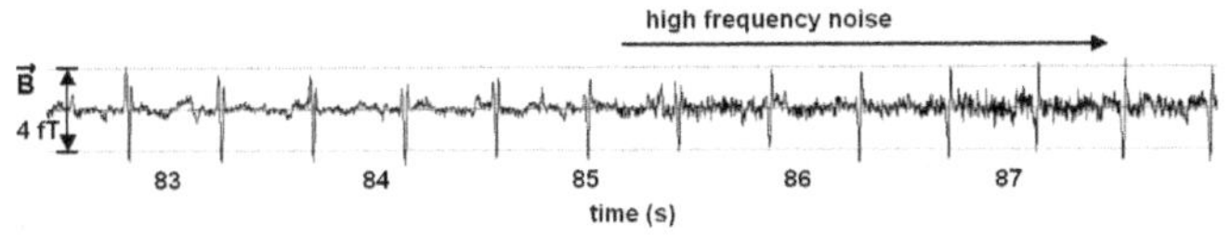

Fig. 2 FMCG channel 6 showing the sudden occurrence of high frequency signals after sec. 85

As the mother reported experiencing uterine contractions during the data acquisition, the signals in all 61 channels were inspected for corresponding evidence of magnetomyographic deflections. We noted occasional episodes of high frequency noise confounding the FMCG signal (Fig. 2) and presumed that muscle contractions were producing these MMG signals. In order to examine these MMG signal components in more detail, the fetal cardiac signals were also digitally removed, as described above for the maternal signals. Contraction episodes (CE), characterized by higher frequency signals in a band roughly between 20 - 200 Hz, with a duration > 1 s were marked (see Fig. 3).

C. Data Analysis

The location of the high frequency signals associated with the CE as well as the location of fetal signals were noted with respect to the sensor configuration. Fetal momentary heart rate was examined with respect to CE timing. Short-term fetal HRV was quantified using the root mean square of successive differences (RMSSD) calculated over windows of 9 fetal beats. The course of RMSSD was then also examined with respect to the occurrence of CE.

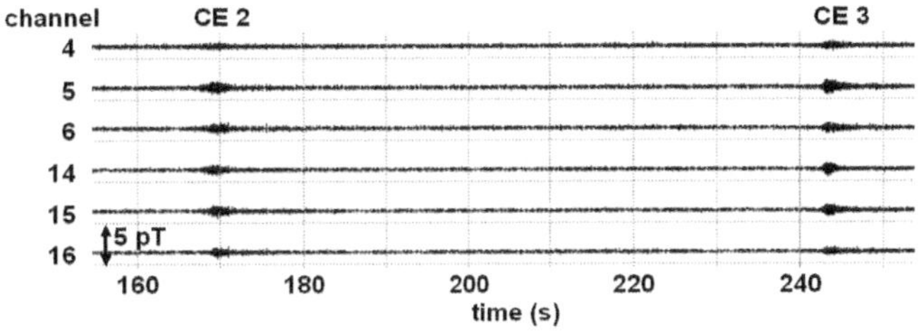

Fig. 3 Occurrence of contraction episodes (CE). High pass filtered (60 Hz) signal traces in selected channels

III. RESULTS

A. Overall Heart Rate

No clinically significant episodes of bradycardia – defined as heart rates < 100 bpm for > 10 s – were found in the 15 min recording. The mean momentary heart rate was 135 ± 11 bpm (± standard deviation) within a range of 99-156 bpm.

B. Contraction Episodes (CE)

We found 7 CE ranging from 1.4 – 19.2 s (Table 1). Examination of both time and frequency domain characteristics showed that this high frequency activity was centered in the mid-upper portion of the sensor, slightly to the left (Fig. 1, center). The fetus was in a vertex left position and, accordingly, the main fetal signal components were found in the mid-lower left of the sensor (Fig. 1, right).

C. Changes in Heart Rate and Heart Rate Variability

The fetal heart rate had a baseline of 120-130 bpm. From this baseline, the heart rate showed repeated rapid increases. As can be seen in Fig. 4 (top), these accelerations concurred in the presence of CE. In six of the seven CEs, the increment was 16 bpm or more, indicating that the CE were associated with sustained increases in heart rate (Table 1).

With respect to HRV, RMSSD values lay between 0.5 and 46.2 ms (mean ± standard deviation: 9.4 ± 7.1) over the complete 15 min acquisition. A drop to low RMSSD values consistently occurred at the initiation of the CE and remained low for several seconds thereafter (Fig. 4, bottom). The beat-to-beat differences were shorter than 6 ms, at five of the seven CE, shorter than 3 ms (Table 2). Thus, concurrent with the heart rate accelerations, short-term HRV of the fetus was reduced considerably.

Table 1 Timing and duration of contraction episodes (CE) as well as their effect on fetal heart rate and short term variability (root mean square of successive differences: RMSSD)

CE	begin at s	duration (s)	heart rate (bpm) before	heart rate (bpm) maximum after	RMSSD (ms) before	RMSSD (ms) minimum after
1	85	3.0	136	153	10.2	1.1
2	168	3.1	131	153	4.9	1.6
3	225	1.4	120	138	16.8	4.4
4	243	4.5	127	155	7.8	1.6
5	379	5.3	122	153	9.4	2.5
6	569	19.2	117	141	18.4	5.1
7	705	6.2	142	149	4.5	2.7

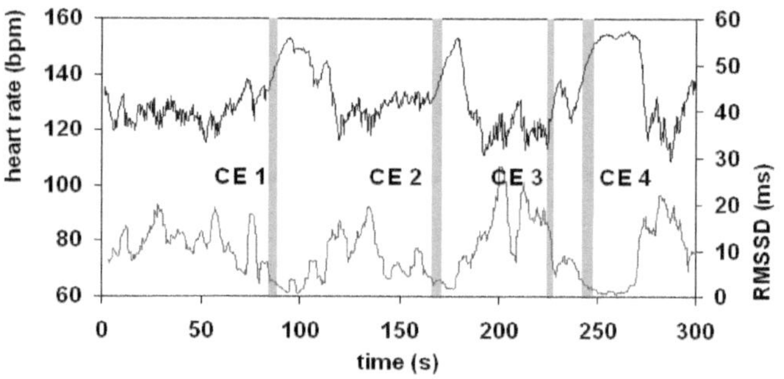

Fig. 4 Changes in fetal heart rate (top) and heart rate variability (bottom) in the first 300 s of the data. Grey bars indicate the timing of the contraction episodes (CE). Note the consistent rapid heart rate increase and drop in RMSSD at and following the occurrence of CE

D. Birth

Six days following the MMG / FMCG recordings, a healthy female was born spontaneously in the 39th week of gestation. There were no signs of cardiac arrhythmias.

IV. DISCUSSION

In previous studies, it has been shown that, using multichannel biomagnetometers, the registration of uterine contractile activity and the localization and synchronization of its propagation is feasible [7][8]. In that work, the authors investigated the low frequency band of the uterine contraction < 1 Hz which they presume represents the plateau and repolarization phase of the action potentials.

The results presented here show that it is possible to simultaneously register uterine contractile activity and fetal cardiac activity in late pregnancy on the basis of magnetomyography and fetal magnetocardiography. The acquired data permitted the analysis of fetal heart rate behavior in the context of contractions of the maternal abdomen.

We examined the spatial distribution of the signals in both the time and frequency domain and were able to identify signals of different origin. These could be clearly associated with muscle activity on the one hand and fetal cardiac conduction on the other. The former was primarily localized in the upper abdomen and characterized by an increase in spectral power in a broad band at frequencies mostly above 20 Hz. On the other hand, the fetal signals characterized by their PQRST morphology were found in the mid lower abdomen. These corresponded to the lie and orientation of the fetus.

Not only were the spatial localizations of the signals consistent with their presumed physiological sources, their behavior over time corresponded to the expected interaction between uterine contraction and fetal heart rate response. The uterine activity most likely resulted from uncoordinated local contractions which are known as Alvarez waves or Braxton-Hicks contractions. These occur in the second half of pregnancy with a frequency of 4-10 within 10 min. If the contractions are associated with a nominal increase in intrauterine pressure, they may be associated with a corresponding temporary increase in heart rate. The data presented in this work clearly reflects the fetal cardiac response to the contractile activity of the uterus.

Fetal heart rate behaviour is central to prenatal surveillance, the level of the baseline heart rate and its variation over different time scales being indicators for fetal well being or stress. In the clinical routine this is usually assessed on the basis of CTG. As mentioned in the introduction, the low specificity of this method may be due to its lack of high temporal resolution. FMCG, which permits acquisition of the fetal heart period at high sampling rates, overcomes this weakness. Because of this, it is particularly suited to quantify short-term heart rate variability. In the data presented here, the sensitivity of RMSSD to the timing of contractile activity is evident.

Depending on their intensity and duration as well as their influence on the fetus, the umbilical chord and the placenta, uterine contractions may result in differing effects on fetal heart rate. Generally speaking, short episodes of acceleration are considered a normal physiological reaction to the effects of increased pressure. On the other hand decelerations are often associated with pathological conditions and require increased vigilance. If the results of the single case presented here can be duplicated in a reasonable number of subjects, then combined MMG and FMCG may aid in fetal monitoring.

V. Conclusions

We conclude that large array biomagnetometers characterized by good abdominal coverage and a number of channels sufficient to separately identify signals from different electrical sources may be used to perform magnetomyography and fetal magnetocardiography simultaneously. Such systems may help improve prenatal surveillance through a more precise analysis of acute events relating to fetal well being.

References

1. Peters M, Crowe J, Pieri JF, Quartero H, Hayes-Gill B, James D, Stinstra J, Shakespeare S (2001) Monitoring the fetal heart non-invasively: a review of methods. J Perinat Med 29:408-416.
2. Van Leeuwen P (2004) Fetal magnetocardiography: time intervals and heart rate variability. Neurol Clin Neurophysiol 46:1-7.
3. Donker DK, van Geijn HP, Hasman A (1993) Interobserver variation in the assessment of fetal heart rate recordings. Eur J Obstet Gynecol Reprod Biol 52:21-28. DOI 10.1016/0028-2243(93)90220-7
4. Bernardes J, Costa-Pereira A, Ayres-de-Campos D, van Geijn HP, Pereira-Leite L (1997) Evaluation of interobserver agreement of cardiotocograms. Int J Gynaecol Obstet 57:33-37. DOI 10.1016/S0020-7292(97)02846-4
5. Van Leeuwen P, Schiermeier S, Geue D, Lange S, Hatzmann W, Grönemeyer D (2007) Simultaneous cardiotocographic and magnetocardiographic registration of fetal cardiac activity. Biomed Tech; 52(Suppl): CD-ROM.
6. Reincke M. Magnetomyography with the SQUID (1993) Biomed Tech 38:276-81.
7. Eswaran H, Preissl H, Wilson JD, Murphy P, Robinson SE, Lowery CL (2002) First magnetomyographic recordings of uterine activity with spatial-temporal information with a 151-channel sensor array. Am J Obstet Gynecol 187:145-151. DOI 10.1067/mob.2002.123031
8. Ramon C, Preissl H, Murphy P, Wilson D, Lowery C, Eswaran H (2005) Synchronization analysis of the uterine magnetic activity during contractions. Biomed Eng Online 4: 55. DOI 10.1186/1475-925X-4-55

Author: Peter Van Leeuwen
Institute: Dept. Biomagnetism, Chair of Radiology and Microtherapy, University of Witten/Herdecke
Street: Universitätsstr. 142
City: 44799 Bochum
Country: Germany
Email: petervl@microtherapy.de

Characterizing Fetal Sympatho-Vagal Balance through Multivariate Time-Varying Autoregressive Modeling of Magnetocardiografic Data

D. Gutiérrez[1], F.M. García-Guevara[1], I. Kiefer[2], R. Draganova[2], and H. Preissl[2]

[1] Centro de Investigación y de Estudios Avanzados (Cinvestav), Monterrey, Mexico
[2] MEG-Center, Tübingen, Germany

Abstract— **We propose a method to analyze fetal beat-to-beat heart rate variability (HRV) obtained from magnetocardiographic (MCG) measurements at different times of gestation using a multivariate time-varying autoregressive (MTVAR) model. Our approach is based on treating a group of HRV signals from a single fetus and measured at different gestational periods as a dynamical system, from which a suitable MTVAR model is obtained. Then, changes in the values of the coefficients of the MTVAR model are associated to changes in fetal sympatho-vagal balance. Such association is made by comparing the dynamics of the MTVAR coefficients to the SDNN/RMSSD ratio, which is known to be a potential marker for sympatho-vagal balance. Furthermore, an overall behavior of the sympatho-vagal balance during the gestation is estimated from the average values of the MTVAR coefficients and compared to the tendencies previously reported in the literature. In order to demonstrate the applicability of the proposed modeling approach, real magnetocardiographic data from a healthy fetus is analyzed. Preliminary results show that the MTVAR model provides good insight into the sympatho-vagal dynamics with greater specificity compared to traditional measures.**

Keywords— **Fetal magnetocardiography, sympatho-vagal balance, heart rate variability, time-varying autoregressive modeling.**

I. Introduction

Fetal heart rate (fHR) monitoring is one of the most useful techniques for clinical fetal evaluation and neurodevelopmental research [1]. Doppler ultrasound-based cardiotocography (CTG) systems are widely used for assessing fHR and many other parameters of fetal development. However, CTG systems lack the precision required to accurately assess beat-to-beat variations, especially in early stages of gestation. In comparison to CTG, fetal magnetocardiography (fMCG) provides a precise signal that can be detected throughout the last half of pregnancy and often earlier [2]. Furthermore, fMCG allows for the temporal resolution of the QRS complex detection to be greatly enhanced [3], then enabling precise fetal beat-to-beat heart rate variability (HRV) analysis.

Although heart rate is affected by many factors, the utility of fHR monitoring is based on the increasing influence of the autonomic nervous system (ANS) during gestation. The separate rhythmic contributions from the sympathetic and parasympathetic (or vagal) branches of the ANS are mainly responsible of modulating fHR. Vagal tone causes fHR to fall, while sympathetic tone causes it to rise. Thus, fHR is regulated primarily by balancing sympathetic and vagal tone [4]. For this reason, much effort has been placed in the development of different methods to accurately quantify the sympatho-vagal balance by means of HRV analysis. In previous studies, linear HRV analyses, both in the time and frequency domains, have proved to be viable in the fetus [5,6]. Classical parameters of the time domain such as the standard deviation of the normal to normal beat intervals (SDNN) and the root mean square of successive differences of the normal beats (RMSSD) have been commonly used in earlier studies on fetal HRV [7], but they are not accurate enough and require a more extensive physiological validation to become an standard in the quantification of the sympatho-vagal balance. On the other hand, the representation of the frequency spectrum in the human fetus is currently a matter of debate [8].

In this paper, we propose a modeling approach for the characterization of the sympatho-vagal balance which is based on a multivariate time-varying autoregressive (MTVAR) model of the fetal HRV. The MTVAR modeling approach has been previously found useful in characterizing the dynamics of electroencephalographic (EEG) data from epileptic seizures [9]. In our case, the goal is to correlate the dynamics of the coefficients associated to the MTVAR model to (i) changes in the vagal components of the HRV when analyzing measurements corresponding to a single gestational period, and (ii) overall changes of the sympahto-vagal balance during different gestational periods when analyzing a group of measurements. In the first case, the correlation is investigated by comparing the dynamics of the MTVAR model's coefficients to the mean heart rate (mHR), the SDNN/RMSSD ratio, and to the fetal HR pattern (fHRP) index [10]. For the case when a group of fetal HRV measurements acquired at different gestational periods is analyzed as a dynamic system, we propose to use the mean value of the time-varying coefficients as an indicator of the overall development of the sympatho-vagal balance through gestation. The trend of such indicator is expected to

S. Supek and A. Sušac (Eds.): Advances in Biomagnetism – BIOMAG2010, IFMBE Proceedings 28, pp. 270–273, 2010.
www.springerlink.com

follow the tendency reported in the literature, where the sympatho-vagal balance has been reported to decrease towards term due to an increased vagal influence [11].

Therefore, in Section II we describe the proposed MTVAR modeling approach. In Section III we show the applicability of our method through numerical examples using real fMCG data, and in Section IV we discuss the results and future work.

II. Methods

Let us denote the HRV signal at a given gestational period as vector $\boldsymbol{x}_k = [x_{k,1}\ x_{k,2}\ \cdots\ x_{k,N}]^T$, where k corresponds to the gestational variable such that $k = 1, 2, \ldots, K$ equally spaced gestational periods, and N is the number of samples. Furthermore, let us define the following autoregressive model for the measurements [12]:

$$\boldsymbol{x}_\nu = \boldsymbol{\omega} + \sum_{l=1}^{p} A_l \boldsymbol{x}_{\nu-l} + \boldsymbol{\varepsilon}_\nu \qquad (1)$$

where ν is the dynamic variable, p is the order of the model, A_l are the coefficients of the model, $\boldsymbol{\omega}$ is a vector of intercept terms that allows to model nonzero mean time series, and $\boldsymbol{\varepsilon}_\nu$ is a vector of uncorrelated random noise.

For the case when ν corresponds to equally spaced gestational periods, i.e. $\nu = k$, then (1) becomes a multivariate model for a time series of K-dimensional state vectors, where $A_1, A_2, \ldots, A_p \in \mathcal{R}^{K\times K}$ are the coefficient matrices of the model. Hence, the problem of computing a model from the measured data corresponds to find the values of the p coefficients such that (1) best fits the data according to a selection criterion [13]. Finally, the MTVAR modeling problem corresponds to the case when the values of the coefficient matrices are updated every τ samples, which results in a group of time-varying coefficients.

The MTVAR models have been used to study the dynamics of different types of multivariate time series. For example, this modeling approach was used in [9] to study EEG data of generalized seizures for the identification of changes in frequency structure over time, which may suggest that seizures are product of multiple processes. In [14], the coefficients of a MTVAR model are used as unique identifiers of EEG data from different mental tasks, so they can be used in brain-computer interface as control commands. In our case, a similar approach is proposed: the dynamics of the MTVAR coefficients will be associated to the dynamics of the sympatho-vagal balance. In order to validate such association, we compare the dynamics of the computed MTVAR coefficients to the mHR, the SDNN/RMSSD ratio, and the fHRP index, as all them are currently used as reference parameters in clinical applications and have shown good potential as markers to reflect sympatho-vagal balance [15].

III. Numerical Examples

We conducted a series of experiments in which we applied the MTVAR model described in Section II to real fMCG data from a healthy patient at the 24, 26 and 28 week of gestation. The data were collected at the MEG-Center in Tübingen, Germany. The sample rate was 1250 Hz and recordings were performed during 30 minutes intervals. Then, the algorithm described in [3] was applied for the detection of the R peaks, and from them the HRV signal was constructed. An example of the resulting signal for the recordings at week 26 is shown in Figure 1. The HRV signals corresponding to weeks 24 and 28 are shown as supplementary material in [16].

We performed two different kind of analysis to the HRV signals. Firstly, we analyzed each gestational period individually to evaluate changes in the vagal components of the HRV during the time of the recordings. Secondly, we took all available measurements and obtained the MTVAR model. From the dynamics of the corresponding coefficients, we characterized the overall tendency of the sympatho-vagal balance relative to the measurement period. In both cases, the autoregressive models were obtained using the computer implementation proposed in [17], and the optimal order for the models was selected according to Schwarz's Bayesian Criterion [18] in every case.

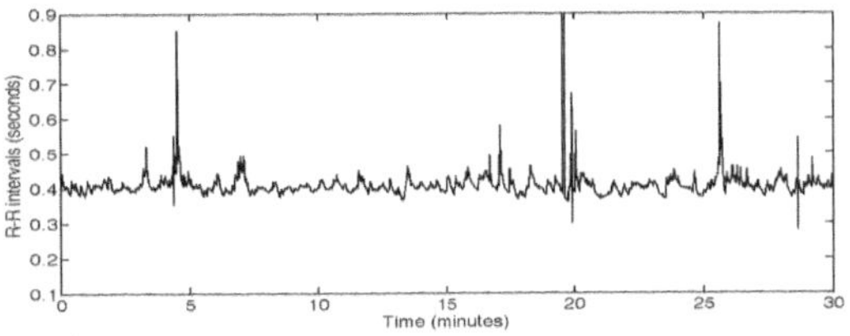

Fig. 1 HRV signal corresponding to the recordings at the 26 week of gestation

A. *Univariate Analysis*

In this case, we computed the autoregressive model for the gestational periods individually. In each case, the model was computed using a window with a fixed size τ = 256 heart beats [19]. The window was shifted every sample to cover the complete HRV signal, which allowed to update the coefficients of the model. Given that the modeling process not necessarily chooses the same order for each

HRV signal, the number of coefficients might differ. Since we are interested in the dynamics of the coefficients and not in their values themselves, we computed the compound variability of the coefficients by means of the multivariate Length Transform [20] given by

$$\lambda_i = \sum_{j=i}^{\tau+i-1} \sum_{l=1}^{p} |A_{l,j} - A_{l,j-1}|, \; i = 1, 2, \ldots \quad (2)$$

where the index i stands for the coefficients' update time. Figure 2 shows the result of this process for the case of the HRV data previously shown in Figure 1. The results corresponding to the data of weeks 24 and 28 are shown in [16].

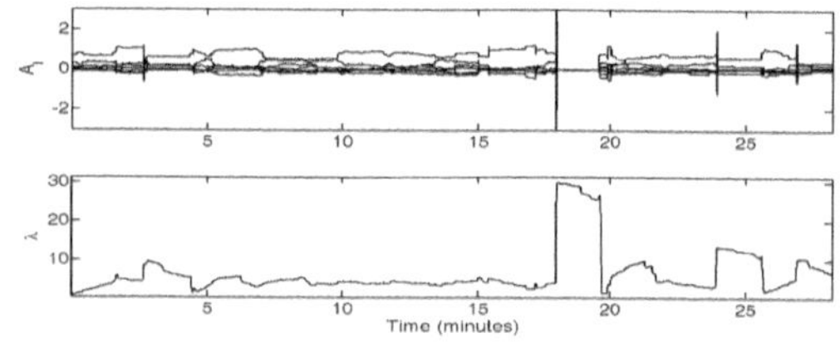

Fig. 2 Dynamics of the autoregressive model's coefficients for the HRV signal corresponding to the recordings in Figure 1 (top) and their multivariate Length Transform (bottom)

The variability information of the autoregressive coefficients given by λ is then compared to the mHR, the SDNN/RMSSD ratio, and the fHRP index, which are computed as described in [7, 10]. In order to make a more accurate comparison, two additional measures are proposed:

- SDNN/λ ratio, which is motivated by the similirarities between the RMSSD and λ. However, we expect λ to provide an enhanced quantification of the short-term variability;
- State detection index (SDI) which is based in the SDNN/λ ratio and uses an statistical criterion to classify fHR rate accelerations within the lenght of the study in three levels: (1) above SDNN/λ ratio's mean, (2) above the mean plus one standard deviation, and (3) above the mean plus two times the standard deviation.

Therefore, we computed all the previously described measures for the HRV signals. The results are shown in Figure 3 for the data corresponding to week 26, and in [16] for weeks 24, and 28. Note that the mHR is given in heart beats per minute, the SDNN/RMSSD and SDNN/λ ratios are adimensional, and the fHRP and SDI indexes indicate different discrete levels related to the sympatho-vagal balance of the fetus [15]. Our results show a strong correlation between sustained accelerations in fHR and the SDNN/λ ratio, while providing greater specificity in comparison to the SDNN/RMSSD ratio. This fact is more clearly shown by comparing the fHRP against the SDI: they both correspond to discrete versions of the SDNN/RMSSD and SDNN/λ ratios, respectively. However, the fHRP is not capable to determine the activity state of the fetus most of the time (which is indicated by a zero-value in the fHRP graph in Figure 3). On the other hand, the SDI index is directly linked to fHR accelerations, specially in terms of the duration and magnitude of the events within the processing window, then it provides a better insight on the activity state of the fetus.

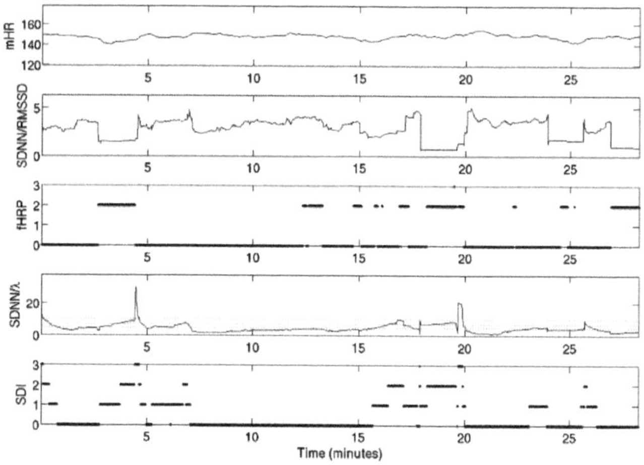

Fig. 3 HRV univariate analysis of the data corresponding to the 26 week of gestation. The dotted-lines in the SDNN/λ ratio (fourth row) indicate, from lower to higher magnitude, the mean value, the mean plus one standard deviation, and the mean plus two standard deviations, respectively. Then, the SDI (fifth row) corresponds to a three-level categorization of the SDNN/λ ratio

B. *MTVAR Modeling*

We computed the MTVAR model of the measurements for the case when they are analyzed together as a dynamical system. Since the coefficients A_l are now matrices, we studied the dynamics of the diagonal components only, as it is assumed that the measurements at each gestational period are independent between each other. Therefore, we calculated λ of the diagonal elements at each gestational period and compared their dynamics. Figure 4 shows the variations in time of the diagonal elements in matrix A_1 (figures corresponding to matrices A_2, ..., A_p are shown in [16]), as well as their length transforms. In comparison to the univariate analysis, the MVTAR modeling approach also shows the relative dynamics of the system as function of the gestational period. However, note that the short-term characteristics are kept, as shown around the 18 minutes of the recording of week 26, where an increase in the variability is clearly noted as probable result of increased fetal activity.

Finally, we computed the mean value of λ over time for each gestational period and normalized it to the total energy for the dynamical system. As result, we obtained the mean energy content (in percent) at each gestational period, which turned out to be of 28%, 55%, and 17%, for week 24, 26, and 28 respectively. These percentages could be used as an overall indicator of the relative changes of the sympatho-vagal balance at the analyzed gestational periods.

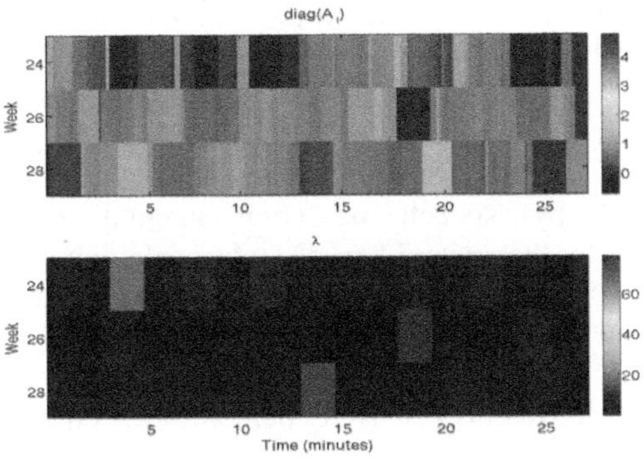

Fig. 4 MTVAR analysis. Figures show the dynamics of the diagonal elements of A_1 (top) and the corresponding length transform (bottom)

IV. Conclusions

We presented a method to characterize the sympatho-vagal balance by using the dynamics of the coefficients of an MTVAR model as indicator, where the variability content of the coefficients was enhanced through the length transform. Our characterization is in agreement with traditional measures, but seems to provide a higher level of specificity. Still, further physiological validation of the proposed measures is required.

Acknowledgment

This work was supported by the International Bilateral Collaboration Program between the National Council of Science and Technology (CONACyT-Mexico) and the German Research Foundation (DFG). D. Gutiérrez was also supported by CONACyT-Mexico under Grant 101374. R. Draganova was supported by the DFG (BI 195/50).

References

[1] Martin CB (1982) Physiology and clinical use of fetal heart rate variability. Clin Perinatol 9:339-352
[2] Peters M, Crowe J, Pieri JF, Quartero H, Hayes-Hill B, James D, et al. (2001) Monitoring te fetal heart non-invasively: a review of methods. J Perinat Med 29:408-416
[3] Wilson JD, Govindan RB, Hatton JO, Lowery CL, Preissl H (2008) Integrated approach for fetal QRS detection. IEEE Trans Biomed Eng 9:2190-2197
[4] Malliani A, Pagani M, Montano N, Mela GS (1998) Sympathovagal balance: a reappraisal. Circulation 23:2640-2643
[5] Van Leeuwen P, Geue D, Lange S, Hatzmann W, Groenemeyer D (2003) Changes in the frequency power spectrum of fetal heart rate in the course of pregnancy. Prenat Diagn 23:909-916
[6] Schneider U, Frank B, Fiedler A, Kaehler C, Hoyer D, Liehr M, Haueisen J, Schleussner E (2008) Human fetal rate variability-characteristics of autonomic regulation in the third trimester of gestation. J Perinat Med 36.433-441
[7] Grimm B, Haueisen J, Huotilainen M, Lange S, Van Leeuwen P, Menendez T, et al. (1997) Recommended standards for fetal magnetocardiography. Pacing Clin Electrophysiol 26.2121-2126
[8] David M, Hirsch M, Karin J, Toledo E, Akselrod S (2007) An estimate of fetal autonomic state by time-frequency analysis of fetal heart rate variability. J Appl Physiol 102:1057-1069
[9] Krystal AD, Prado R, West M (1999) New methods of time series analysis of non-stationary EEG data: eigenstructure decompositions oftime varying autoregressions. Clin Neurophysiol 110:2197-2206
[10] Nijhuis JG, Prechtl HFR, Martin CB, Bots RSG (1982) Are there behavioral states in the human fetus? Early Hum Dev 6:177-195
[11] Zhuravlev Y, Rassi D, Mishoin AA, Emery SJ (2002) Dynamic analysis of beat-to-beat fetal heart rate variability recorded by SQUID magnetometer: quantification of sympatho-vagal balance. Early Hum Dev 66:1-10
[12] Neumaier A, Schneider T (2001) Estimation of parameters and eigenmodes of multivariate autoregressive models. ACM Trans Math Softw 27:27-57
[13] Lütkepohl H (1993) Introduction to multiple time series analysis. Springer-Verlag, Berlin
[14] Anderson CW, Stolz EA, Shamsunder S (1998) Multivariate autoregressive models for classification of spontaneous electroencephalographic signals during mental tasks. IEEE Trans Biomed Eng 45(3):277-286
[15] Task Force (1996) Heart rate variability. Standards of measurement, physiological interpretation, and clinical use. Task force of the European society of cardiology and the north American society of pacing end electrophysiology. Eur Heart J 17:354-381
[16] Available at http://biomag2010.gutierrezruiz.com
[17] Schneider T, Neumaier (2001) Algorithm 808: ARfit – A Matlab package for the estimation of parameters and eigenmodes of multivariate autoregressive models. ACM Trans Math Softw 27:58-65
[18] Schwarz G (1978) Estimating the dimension of a model. Ann Statist 6:461-464
[19] Ronneburger A, Kähler C, Hoyer D, Schleussner E, Schneider U, Hauisen J, et al. (2004) Influence of fetal state of activity and duration of recording time on linear and non-linear parameters of fetal heart rate variability investigated by fetal magnetocardiography. Biomed Eng 48(Suppl):143
[20] Gutiérrez D, Preissl H, Eswaran H, Lowery CL (2007) A study of fetal sympatho-vagal balance at various gestational periods using the length transform on magnetocardiographic data. Comp Cardiol 34:685-688

Corresponding author:

Name: D. Gutiérrez, Ph.D.
Institute: Cinvestav Unidad Monterrey
Email: dgtz@ieee.org

Objective Assessments of Bone-Conducted Ultrasonic (BCU) Hearing by Auditory Evoked Fields

Seiji Nakagawa

Institute for Human Science and Biomedical Engineering (HSBE),
National Institute of Advanced Industrial Science and Technology (AIST), Ikeda, Osaka, Japan

Abstract— **Bone-conducted ultrasound (BCU) is perceived even by the profoundly sensorineural deaf. We have been developing a novel hearing-aid using BCU perception, which transmits a 30-kHz bone-conducted carrier that is amplitude-modulated by speech. The fundamental capabilities of the BCU hearing-aid (BCUHA) were objectively assessed by magneto-encephalography. Frequency- and temporal-resolution were assessed by measuring mismatch fields (MMFs). As well, the capability to discriminate two-channel BCUs presented to the left and right mastoids was verified by the laterality of auditory evoked magnetic field N1m. In the frequency-discrimination tests, amplitude-modulated BCU (AM-BCU) and air-conducted sound (AC) showed almost the same MMF-magnitudes when frequency/modulation-frequency of the standard-stimuli were 1.0 kHz, even though MMF-magnitudes for BCU were smaller than those of AC at 0.125 kHz. In the temporal-discrimination tests, MMFs for changes of stimulus-duration were measured. MMF-magnitudes for AM-BCUs reached more than 80% of those for air-conducted sounds (ACs). In addition, in the laterality-discrimination tests, N1m evoked by contralateral stimuli were larger in amplitude and shorter in latency than those evoked by ipsilateral stimuli for both AM-BCUs and ACs. This result agrees with earlier reports of AC auditions, and suggests that two-BCU channels presented to the left and right mastoid were separately localized. These findings objectively indicate that the BCUHA has practical frequency- and temporal resolution, and give a rationale for the development of a multi-channel (at least two) BCUHA.**

Keywords— **Bone-conduction, Ultrasound, Mismatch field, N1m, Discrimination capability.**

I. Introduction

Several studies have reported that bone-conducted ultrasounds (BCUs) are perceived by the profoundly deaf, i.e., those who can hardly sense sounds even with conventional hearing-aids [1-3], as well as the normal-hearing. In 1991, Lenhardt et al. reported that BCUs modulated by speech sounds were intelligible to some extent [1], suggesting the possibility of the development of a novel hearing-aid based on BCU perception. However, Dobie disputed Lenhardt's results, which had been obtained from subjective psychological experiments [4]; and ever since the subject has been controversial.

We previously objectively supported Lenhardt's argument by using magnetoencephalography (MEG) [2, 3] and we developed a novel hearing-aid using BCU perception (BCU hearing-aid: BCUHA) for the profoundly deaf [5]. In the BCUHA, ultrasounds are amplitude-modulated by speech or environmental sounds and presented to the mastoid or the sternocleidomastoid by a vibrator [5]. With the BCUHA, both the speech and pitch of the carrier signal, equals to ten-odd kHz [3, 6], are perceived simultaneously. To assess and optimize the BCUHA, the features of BCU perception need to be better specified.

In this study, the fundamental capabilities of our BCUHA were objectively assessed by MEG in normal-hearing and profoundly deaf subjects. Frequency- and temporal-resolution were assessed by measuring mismatch fields (MMFs), which generally reflect the perceptual properties of sound discrimination. Furthermore, the capability to discriminate between two-channel BCUs presented to the left and right mastoids was verified by the laterality of the auditory evoked magnetic field, N1m.

II. Experiment I: MMF Measurements to Evaluate Frequency-Discrimination Capability

We have already reported that difference limens for frequency (DLFs) for BCU are larger than that of air-conducted sound (AC) below 0.125 kHz and above 6.0 kHz: however, BCU and AC showed almost the same DLFs between 0.25 and 4.0 kHz [7]. To assess the frequency-discrimination capability of the BCUHA objectively, MMFs for changes of modulation-frequency of amplitude-modulated BCU (AM-BCU) were recorded.

A. Methods

Subjects: 16 normal-hearing subjects (12 males, 22• 39 years, right-handed) participated. Subjects were certified as being able to sufficiently sense BCU via both left and right mastoids. During recordings, subjects sat on a chair in a magnetically shielded room, and were requested to watch self-selected silent movies with subtitles.

Stimuli: AC/AM-BCU stimuli were presented in 4 different sessions. Each session consisted of one standard

S. Supek and A. Sušac (Eds.): Advances in Biomagnetism – BIOMAG2010, IFMBE Proceedings 28, pp. 274–277, 2010.
www.springerlink.com

(std, probability: 85%) and three types of deviant (dev, probability: 5% each) stimuli. Frequency/modulation-frequency of the stimuli in each session were as follows.

A. AC: 40 dB SL, std: 1000 Hz, dev: 1010, 1020, 1040 Hz
B. AC: 40 dB SL, std: 125 Hz, dev: 135, 145, 165 Hz
C. AM-BCU: 10 dB SL, std: 1000 Hz, dev: 1010, 1020, 1040 Hz
D. AM-BCU: 10 dBSL, std: 125 Hz, dev: 145, 165, 205 Hz

Carrier frequency of AM-BCU stimuli was set at 30 kHz. According to a previous report that the psychoacoustical dynamic range of BCU sensation is extremely narrow, less than 20 dB [8], the loudness of each stimulus was set to be almost constant through all the stimuli. Durations of all stimuli were 200 ms including linear rising and falling ramps of 50 ms each. Stimulus onset asynchrony (SOA) was set at 500 ms. In each session, stimuli were presented in a random order, and evoked responses were averaged more than 120 times for the deviants. The order of the sessions was counterbalanced between subjects. MMFs were delineated by separately subtracting standard responses from the deviant responses. Bone-conducted sounds were presented to the right/left mastoid using a newly devised ceramic vibrator.

MEG Recordings and Analyses: Recordings of event-related magnetic fields were carried out in a magnetically-shielded room using a 122-channel whole-head neuro-magnetometer (Neuromag-122^{TM}; Neuromag Ltd., Finland). The averaged data were digitally band-pass-filtered between 2 and 20 Hz. In each subject, we employed the MMF peak amplitude and latency with a channel that showed the maximum amplitude placed over the left temporal regions.

B. *Results and Discussions*

All subjects reported that they perceived two pitches for AM-BCU: one is of the carrier, identical to a pitch of 10~14-kHz AC, while another is the pitch of the modulator, i.e., 1 or 2 kHz.

MMFs were clearly observed for each deviant stimulus (Fig. 1). When frequency/modulation-frequency of the std. were 1 kHz, no significant differences of MMF amplitude and latency were observed between AC and BCU (Fig. 2). On the other hand, MMF magnitudes for BCU were smaller than those of AC at 0.125 kHz (p<0.01). The result agreed with previous psychophysical tests and indicated that the BCUHA has a practical frequency-discrimination capability at the speech frequency-band.

III. Experiment II: MMF Measurements to Evaluate Temporal-Resolution

In the previous psychophysical measurement, it was suggested that the auditory system had sufficient ability to process timing information in the envelopes of AM-BCUs [9]. To assess the temporal-resolution of the BCUHA objectively, MMFs for changes of stimulus-duration were measured in normal-hearing subjects.

A. *Methods*

Subjects: 11 normal-hearing subjects (8 males, 22• 52 years, right-handed) participated.

Stimuli: AC, BCU, AM-BCU, and complex air-conducted sound (Complex-AC), which has a pitch equal to that of AM-BCU, were presented in different sessions. Each session consisted of one standard stimulus (75 ms duration, probability 85%) and three types of deviant stimuli (52.5, 37.5, and 22.5 ms duration, probability 5 % each). Stimuli were presented in a random order, and evoked responses were averaged more than 120 times for each deviant. SOA

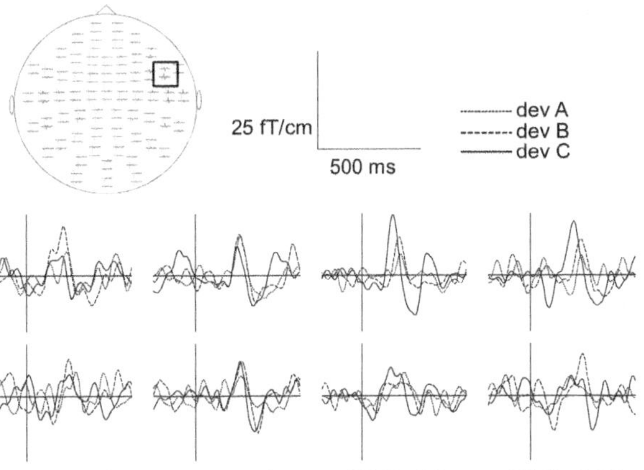

Fig. 1 MMF wave forms evoked by changes of the frequency observed in the right temporal region

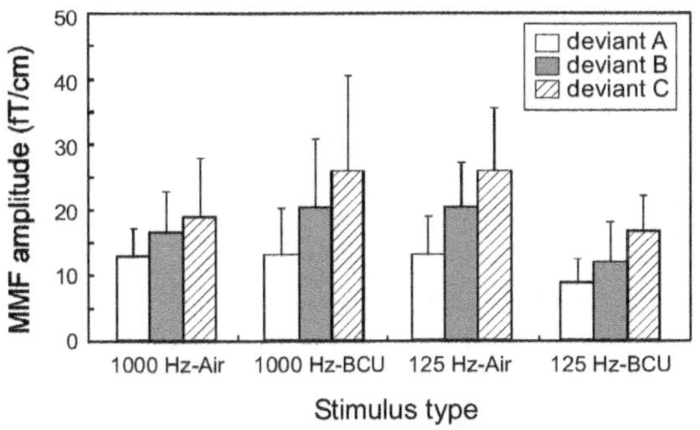

Fig. 2 MMF amplitudes evoked by changes of the frequency (mean+S.D.)

was set at 300 ms. The same stimulation-apparatus and MEG-recording conditions as in experiment I were applied. Latencies and ECD moments of MMF were compared among the stimuli.

B. Results and Discussion

In each session, MMFs were elicited for each deviant stimulus. For all kinds of stimuli, larger MMFs were observed for shorter deviants ($p<0.01$, Fig. 3). ECD moments were significantly different among the four kinds of stimuli ($p<0.01$, AC: BCU: AM-BCU: Complex-AC = 1.0: 0.53: 0.91: 0.92), whereas no differences were observed in MMF latency.

Almost the same amounts of MMFs were observed for AM-BCU and Complex-AC. Since MMF is generally more related to the perceptual properties of stimuli than the physical, the current results show the difference between BCU and AC in perceptual attributes rather than in their pathways. It is also indicated that the BCUHA has practical temporal resolution.

IV. Experiment III: N1m Measurements to Evaluate the Discrimination Capability of Multi-Channel Inputs

It is sometimes more difficult to localize bone-conducted sounds than ACs, because interaural time differences (ITDs) in BC hearing are small. However, even in BCU hearing, multi-channel inputs are lateralized properly [10]. In this study, the capability to discriminate two-channel BCUs presented to both left and right mastoids was verified objectively in both normal-hearing and profoundly deaf subjects. We evaluated the laterality of auditory brain magnetic fields. For AC, it has been reported that N1m responses evoked by the ipsilateral-ear stimuli are smaller in amplitude [11] and later in latency than those evoked by the contralateral-ear stimuli [11, 12]. If two-channel BCUs presented to the right/left mastoids can be separately localized, the same phenomena should be observed.

A. Methods

Subjects: 10 normal-hearing (7 males, 21•36 years, right-handed) and 4 profoundly deaf volunteers (2 males, 45• 62 years, right-handed) participated.

Stimuli: For the normal-hearing subjects, the following types of stimuli were presented in different sessions.

A. BCU: 30-kHz BCU tone-burst amplitude-modulated by a 1000-Hz tone. Intensity: 10 dB SL.
B. Bone-conducted audible sound (BC): 1000-Hz bone-conducted tone-burst. Intensity: 50 dB SL.
C. AC: 1000-Hz bone-conducted tone-burst. Intensity: 50 dB SL.

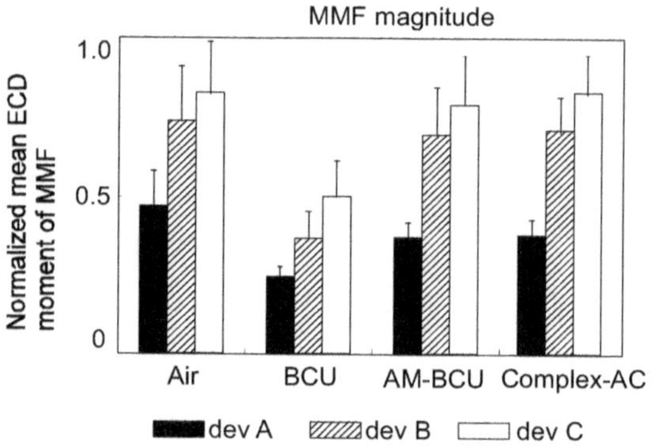

Fig. 3. MMF magnitudes evoked by changes of the stimulus-duration (mean+S.D.)

For the profoundly deaf subjects, only BCU was presented. Durations of all stimuli were 100 ms including linear rising and falling ramps of 10 ms each. Interstimulus intervals were varied randomly between 1900 and 2100 ms. The same stimulation-apparatus and conditions of MEG recordings and analyses as in experiment I were applied.

B. Results and Discussion

In both the normal-hearing and deaf subjects, N1m components were elicited in both temporal regions for each kind of stimuli (Fig. 4).

Fig. 5 shows N1m peak amplitude and latency, respectively, for each kind of stimuli. In the normal-hearing subjects, significant effects of the stimulus-type (AC/BC/BCU, $p<0.001$), the stimulation side (ipsilateral/contralateral, $p<0.001$), and the hemisphere (left/right, $p<0.01$) were observed for both amplitude and latency. N1m for BCU was significantly smaller in amplitude (vs AC: $p<0.01$, vs BC: $p<0.01$) and larger in latency (vs AC: $p<0.01$, vs BC: $p<0.01$) than AC and BC. In terms of the stimulation side, amplitudes were significantly larger (AC: $p<0.001$, BC: $p<0.001$, BCU: $p<0.005$) and latencies were shorter (AC: $p<0.01$, $\Delta=8.1\pm$ 4.1 ms, BC: $p<0.01$, $\Delta=6.0\pm3.4$ ms, BCU: $p<0.01$, $\Delta=7.0\pm3.8$ ms, Δ indicates a difference of the both stimuli) for contralateral than for ipsilateral stimuli. In profoundly deaf subjects, the effects of the stimulation side ($p<0.01$) were observed in both amplitude and latency, whereas the effects of the hemisphere were observed only in amplitude. Amplitudes were significantly larger (BCU: $p<0.05$) and latencies shorter (BCU: $p<0.05$, $\Delta=9.5\pm4.8$) for contralateral compared with ipsilateral stimuli.

Even in BCU, as well as in audible-frequency sounds, N1m responses evoked by the ipsilateral stimuli are smaller in amplitude and later in latency than those evoked by the contralateral stimuli. These results agree with earlier reports in AC [11, 12], and suggest that two-BCU channels presented to the left and right mastoids were separately localized, i.e, each BCU channel entered the ipsilateral auditory pathway before the superior olivary nucleus. This finding also provides a rationale for the development of a multi-channel (at least two-channels) BCUHA.

V. Conclusions

The results objectively indicated that the BCUHA has practical frequency- and temporal-resolution. In addition, two-BCU channels presented to the left and right mastoid, as for both ears, were separately localized. This finding also provides a rationale for the development of a two-channel BCUHA. The present results will also be useful for the further development of this BCUHA.

In experiments III, the same phenomena were shown even in the profoundly deaf subjects, as well as the normal-hearing subjects. Among several hypotheses about the mechanisms of BCU perception, one proposes a transformation into low-frequency audible sound due to certain bio-mechanical non-linearity [4]. However, it was thought that detection of audible low-frequency sounds would be impossible for the profoundly deaf subjects who participated in the experiment, because they had little sound sensitivity below 20 kHz. As well, no significant audible-frequency components were recorded by the physio-acoustical measurements on/around the human head during BCU presentation [13]. The current results suggested that BCU is perceived by a special mechanism which is different from that of audible-frequency sound perception even though BCU enters the auditory pathway before the superior olivary nucleus.

Acknowledgment

This work was supported by the Industrial Technology Research Grant Program of the New Energy and Industrial Technology Development Organization (NEDO), Japan.

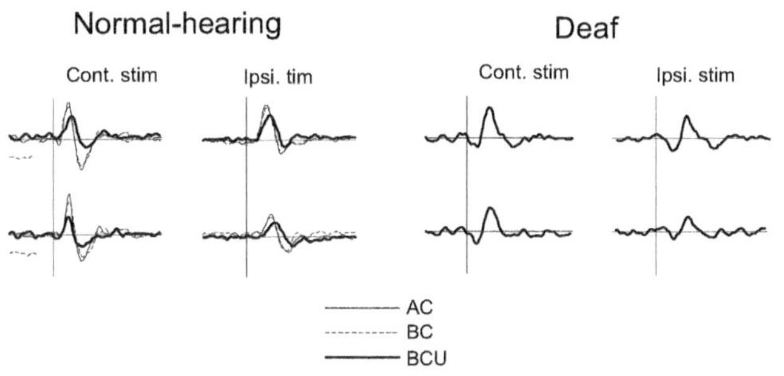

Fig. 4. MMF wave forms evoked by changes of stimulus-laterality observed in the right temporal region

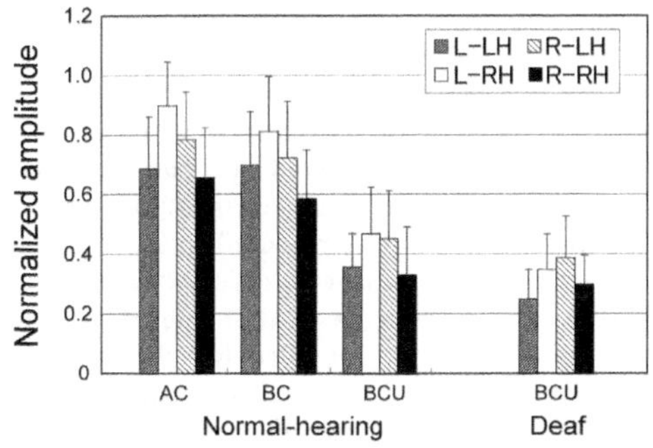

Fig. 5. MMF amplitudes evoked by changes of stimulus-laterality (mean+S.D.)

References

1. Lenhardt ML, Skellett R, Wang P et al. (1991) Human ultrasonic speech perception. Science 253:82-85
2. Hosoi H, Imaizumi S, Sakaguchi T et al. (1998) Activation of the auditory cortex by ultrasound. The Lancet 351:496-497
3. Nakagawa S, Sakaguchi T, Yamaguchi Yet al. (1999) Characteristics of auditory perception of bone-conducted ultrasound. Tech Rep IEICE. WIT99-15:89-94
4. Dobie RA, Wiederhold WL (1992) Ultrasonic hearing. Science 255:1584-1585.
5. Nakagawa S, Okamoto Y, Fujisaka Y (2006) Development of a bone-conducted ultrasonic hearing aid for the profoundly sensorineural deaf. Trans. Jpn. Soc. Med. Biol. Eng. 44:184-189
6. Nakagawa S, Tonoike M (2005) Measurement of brain magnetic fields evoked by bone-conducted ultrasounds: Effect of frequencies. Unveiling the Mystery of the Brain: ICS1278:333-336
7. Fujimoto K, Nakagawa S (2005) Non-linear ultrasonic perception. Hear. Res. **204**: 210-215.
8. Nishimura T, Nakagawa S, Sakaguchi T et al. (2003) Ultrasonic masker clarifies ultrasonic perception in man. Hear. Res. 175:171-177
9. Hotehama T, Nakagawa S (2006) Lateralization of amplitude-modulated bone-conducted ultrasounds: discrimination of interaural time and intensity differences. J. Acoust. Soc. Am., 120:3082-3082.
10. Hotehama T, Nakagawa S (in press) Modulation detection for amplitude-modulated bone-conducted sounds with sinusoidal carriers in the high- and ultrasonic-frequency range. J. Acoust. Soc. Am.
11. Hari R, Mäkälä JP (1988) Modification of neuromagnetic responses of the human auditory cortex by masking sound. Exp. Brain Res. 71:87-92
12. Pantev C, Lütkenhöner B, Hoke M et al (1986) Comparison between simultaneously measured auditory-evoked magnetic fields and potential elicited by ipsilateral, contralateral, and binaural tone burst stimulation. Audiology 25:54-61
13. Ito K, Nakagawa S (in press) Perception mechanisms of bone-conducted ultrasound assessed by acoustic characteristics in the external auditory meatus. Jpn J. Aplli. Phys.

Author: Seiji Nakagawa
Institute: Nat'l Inst. of Advanced Industrial Sci. and Tech. (AIST)
Street: 1-8-31 Midorigaoka
City: Ikeda, Osaka
Country: Japan
Email: s-nakagawa@aist.go.jp

Observing Mismatch Fields from Human Auditory Discrimination of Heavily Damped Frequency Component Adjustments

S. Gardeen[1], A. Leuthold[2], and J.H. Broadhurst[1]

[1] University of Minnesota/Department of Physics, Minneapolis, MN, 55455
[2] University of Minnesota/Department of Neuroscience, VA Medical Center, Minneapolis, MN, 55417

Abstract— **The aim of this study was to test the brain's ability to discriminate frequency shifts of heavily damped components of a complex sound. Since the auditory system is tonotopically organized, heavily damped transients should generate more activity than lightly damped partials. The increased activity increases the 'categorical resolution'. This work attempts to test this by detecting the 'Mismatch Field' from human subjects during a defined hearing task.**

Since the hearing system evolved in a complex environment, complex synthetic audio samples are constructed similar to the sound of a guitar. The sounds presented to the subject consisted of repetitive 'standard' tones, with occasional 'deviant' epochs. Standards differ from deviants in that one of the anharmonic frequency components of the heavily damped partials is transposed. The partials representing the more sustained vibrations of the string are identical for both standard and deviant. The randomized samples were presented to the subject's right ear with an average random inter-stimulus interval of 1 second over a duration of 17 minutes. The subject was then submitted to an active discrimination task where they actively attempted to count the deviants.

The magnetic field responses from the brain were then detected using whole head magnetometer. The grand averaged samples were then band-pass and localized. Although the subjects could not actively discriminate the sound samples, mismatches were observed originating from the auditory cortex.

These results suggest the auditory system is able to distinguish at a sensory level, shifts in the heavily damped spectral structure of complex sounds even though some cannot do so attentively. This could be due to the greater analysis given by the human brain to determining the pitch center rather then the sound timbre, yet some trained musicians have the ability to distinguish these subtle differences.

Keywords— **MEG, mismatch, sensory memory, auditory discrimination.**

I. Introduction

Identifying a sound source solely by its acoustical energy is remarkable human skill, especially when considering the different number of voices and the quality of transmissions (i.e. cell phones, etc.). Studying this aspect of the auditory system can effectively be done with the sub-second resolution of Magnetoencephalography (MEG) or its electrical counterpart Electroencephalography (EEG). [1] The mainstream interpretation of the Mismatch Field is an event related field elicited when, in a sequence of standard sounds, an occasional deviant sound is discriminated from a standard. The standard forms a well-represented sensory trace that is maintained in sensory memory for at least of few seconds. When a deviant is encountered the incoming stimulus is compared to the sensory memory trace and if a difference is discriminated, a mismatch response is elicited, all at a pre-attentive level [2][3] (might not be the only mechanism eliciting the mismatch but is the one which is of interest).

The auditory system processes sound waves by first reconstructing them in the frequency domain; from the cochlea up to the cortex, the tonotopicity is maintained. One property of this transformation into the frequency domain is that more frequency components are required to reconstruct a heavily damped or transient vibration, another formulation of the Uncertainty Principle. This work postulates that more transient sounds should have more firing neurons than more sustained sounds. Due to the wider dispersion of frequencies associated with the more rapidly damped transient response, more cochlear neurons would be involved in the preliminary tonotopic analysis of the sound. The larger the numbers of active neurons, greater are the number of possible categorizations for a particular network. This gives an explanation for why one can classify a familiar voice, out of many known voices, after just the first syllable.

Since voice differences are very subtle, the sound produced in this work is modeled after a musical instrument consisting of two basic sound production mechanisms: a string and a resonator box. This work will test how sensitive the human brain is to subtle acoustical differences of the resonator box.

II. Methods

Mismatch field responses were detected from subjects who were submitted at a loudness of 80 dB SPL to a randomized sequence of two different complex sound samples: a standard and a deviant. Both samples were constructed by a superposition of a heavily damped sound source with a lightly sound source. Both standard and

S. Supek and A. Sušac (Eds.): Advances in Biomagnetism – BIOMAG2010, IFMBE Proceedings 28, pp. 278–281, 2010.
www.springerlink.com

deviant were constructed with the same lightly damped subsample but different heavily damped sub samples, the difference being a shift in the frequency component at 370 Hz (standard) to 410 Hz (deviant). The frequency component structure was modeled after that of an acoustic guitar, a common real-world sound. The heavily damped sounds represent vibrations of the sound box and the sustained sounds that of the more familiar string vibrations.

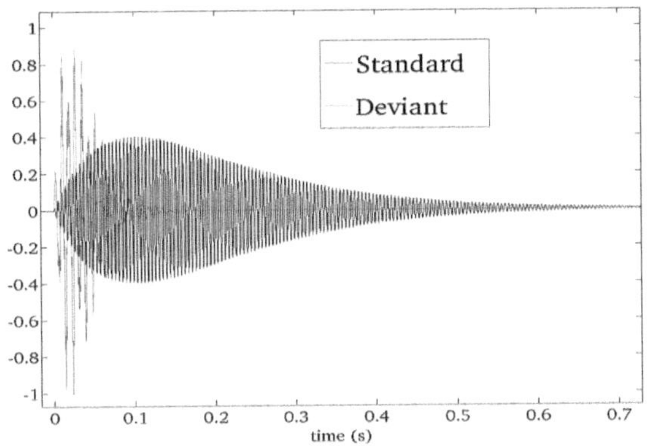

Fig. 1 The audio samples in the time domain where the green and red indicate the heavily damped portions of the complex sample

The digital audio files were constructed by enveloping the different sound source frequency component superposition sets with solutions to the critically damped oscillator for a non-zero initial velocity.

$$\psi_{ij}(t) = C_j t e^{-\frac{t}{\tau_j}} A_i Sin(2\pi f_i t) \qquad (1)$$

The parameters $\{C_j, \tau_j\}$ indicate the relative amplitude and rise time of the sets of Fourier components $\{A_i, f_i\}$. The rise time is considered to be a cue for source identification [4].

The sound sample sequence was randomized except for at least 3 standards preceding each deviant. The inter-stimulus interval (ISI) was also random with an average of about 1 s between epochs. This was done to prevent any sort of anticipation processing before the stimulus arrived. The brain responses were measured using a whole-head 248 channel axial gradiometer system (Magnes 3600WH; 4D-Neuroimaging, San Diego) The sound was presented to the subjects' right ear for 17 minutes, a total of 682 standards and 134 deviants. As monaural listeners achieve sound discrimination, no attempt was made to use binaural signals, with the possible left-right confusion in the magnetic field information. After the magnetic measurements, the subjects then listened to a similarly constructed 5-minute audio file and asked to actively attempt to discriminate the sounds in the sequence, by counting when a deviant was encountered.

Data was sampled at ~1kHz and the grand averaging was done on all standards and deviants but also on only the standards immediately preceding the deviants. This was done to get a balanced statistical comparison of the standard with the deviant and compare the best-established standard memory trace. Channels were selected if the difference between the maximum and minimum field exceeded a certain threshold per hemisphere. About 15-30 channels were used for the localization per subject. The averages were then filtered, first with a 4th order high pass Butterworth filter at a .5 Hz cutoff, and then with a 5th order low pass Butterworth filter with a 30 Hz cutoff.

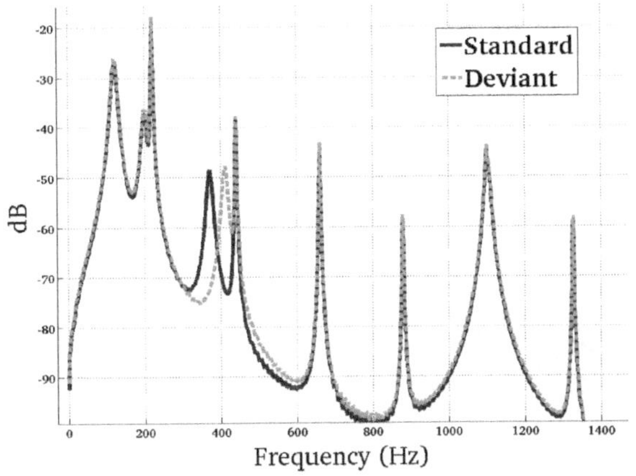

Fig. 2 The audio samples in the frequency domain. The wider peaks are the Fourier components of the heavily damped portion of the sound

Localizations of both the standard and deviant responses were done using the RAP-MUSIC algorithm implemented in Brainstorm (http://neuroimage.usc.edu/brainstorm) for a single sphere head model approximated from T1-weighted anatomical MRI data. [5] A mismatch field was determined by performing a Welch's t-test on each average at every time sample. Due to possible morphological differences in cell arrangement in the brains of different subjects, the comparison of standard to deviant mismatch was made between signals applied to the same subject and no global fitting was considered.

III. RESULTS

Since the work is probing the pre-attentive ability to discriminate subtle differences in sound, not all subjects exhibited mismatch fields but the majority do to some degree. No subject was able to actively discriminate the standard from the deviant.

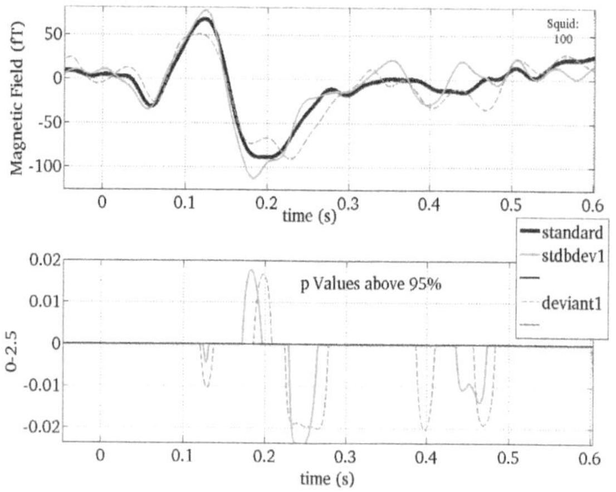

Fig. 3 The thick lines represent the averages of the standards where the thinner line is just the average before the deviant. The deviant is the dashed line. The lower plot is a Welch's t-test comparing only standard to deviant. The dotted line is the comparison between the deviant and the general standard. Only values that would indicate a 95% confidence in distinct means are plotted. Note that in the range from 150 ms to 300 ms that both standards differ noticeably from the deviant

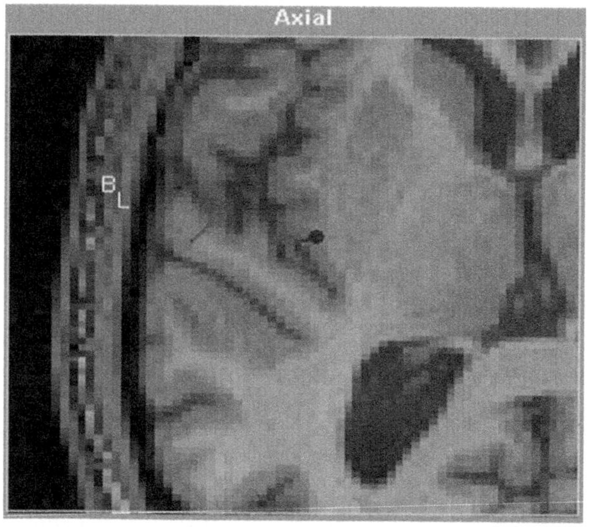

Fig. 4 The standard localization in this figure is green where the deviant localization is blue. It could be that the mismatch field is due to the novel processing done by neurons down and around the fissure. Both localizations (150ms – 350ms) had over 99% correlations with the signal subspace

IV. Discussion

The trace and the localization information above suggest that although the sounds are not consciously discriminable, the brain shows the pre-attentive ability to discriminate the sounds. Since the sounds are modeled after a guitar, most subjects seem to be conditioned to cue on the pitch representation which is the same for both sounds. Since context doesn't require listeners (except for some professional musicians) to cue on the intricacies of the sound source (of the same instrument), most aren't able to do so without some form of training. This explains the subtleness of the mismatches; with training a more robust mismatch could be elicited.

Speech consists of syllables where consonants are partitioned by vowel sounds. Vowels sounds tend to be more sustained and consonants less so and more damped. One explanation for there being fewer vowels and more consonants is due to the frequency component richness of the heavily damped consonant sounds. This allows for more categories to discriminate different heavily damped sounds. The vowels serve to encapsulate and chunk the consonant sounds.

Most studies don't address the temporal information that can be discerned from the spectral information. This work suggests that the human brain is sensitive to shifts in these heavily damped frequency components. Further work should then focus on the distinction between what is lightly damped and heavily damped. This can be done by changing the single damping parameter, τ_i, using the sound production model discussed in the methods section.

Other timbre or sound identification studies [4] address spectral envelope and spectral centroid. This work attempts to complement those studies by approaching the question of source identification from a fundamental sound production viewpoint.

V. Conclusion

This work shows the resilience of the human brain to discriminate subtle differences in complex sounds at a pre-cogitative level. The idea that sound source identification is accomplished by cueing on heavily damped sounds is supported by the results here.

References

1. Hämäläinen. (1993). Magnetoencephalography—theory, instrumentation, and applications to noninvasive studies of the working human brain. Reviews of Modern Physics, 65(2), 413.
2. Cowan, N. (1984). On short and long auditory stores. Psychological Bulletin, 96(2), 341-370.
3. Naatanen, R. (2001). The perception of speech sounds by the human brain as reflected by the mismatch negativity (MMN) and its magnetic equivalent (MMNm). Psychophysiology, 38(1), 1.

4. Caclin, A., Brattico, E., Tervaniemi, M., Naatanen, R., Morlet, D., Giard, M. H., et al. (2006). Separate neural processing of timbre dimensions in auditory sensory memory. Journal of Cognitive Neuroscience, 18(12), 1959-1972.
5. Leahy, R. M., Mosher, J. C., Spencer, M. E., Huang, M. X., & Lewine, J. D. (1998). A study of dipole localization accuracy for MEG and EEG using a human skull phantom. Electroencephalography and Clinical Neurophysiology, 107(2), 159-173.

Cortical Dynamics Underlying the Visual Perception of Stationary and Moving Stimuli

Wenqi Sun[1], Alexander Moiseev[2], Faisal Beg[1], Urs Ribary[1,2], and Naznin Virji-Babul[1,2]

[1] Simon Fraser University, Burnaby, B.C. Canada
[2] Down Syndrome Research Foundation, MEG Laboratory, Burnaby, B.C. Canada

Abstract— **In this study, we investigated the neural responses during the perception of stationary dots, randomly moving point light dots and point light dots representing human motion. We recorded cortical responses from ten right handed subjects with a 151 channel whole head MEG system (VSM MedTech, Coquitlam, BC). For each subject a three-dimensional spatial distribution of source activity was computed using differential SAM beamforming method. The data were spatially normalized to a common brain template and group-averaged for each condition. Locations showing statistically significant differences between conditions were chosen for time frequency analysis. We compared the random motion and human motion with the stationary display and then compared the random and human motion. Perception of random motion was associated with greater desynchronization in the alpha band in the left lingual and (L) superior parietal region and (R) precuneus and mid-temporal regions as compared with stationary dots. Perception of human motion led to greater desynchronization in the (L) precentral region and the (R) mid-temporal and post-central areas in comparison with random motion. Late activation in the gamma band activity was observed in both motion conditions in the visual areas in comparison to the static displays. These data suggest that (a) the perception of motion involves both the dorsal and ventral pathways and the interaction between these subsystems may be important for visual discrimination and (b) that action cues depicting human motion are sufficient to selectively activate premotor and sensory-motor regions.**

Keywords— **Perception, biological motion, point-light displays.**

I. Introduction

The recognition of action is a fundamental prerequisite for the development of imitation and motor learning [1]. In humans, mirror neurons, which respond to both the observation and execution of an action, have been found in the ventral premotor cortex, inferior parietal lobe [2] and the superior temporal sulcus (STS) [3] suggesting a common coding between perception and action. These regions form a complex network in which the visual representation of motion activates an appropriate motor representation.

One approach used to study the perception of action involves point-light stimuli [4] in which complex dynamic events are reduced to the movements of a few points strategically positioned on the joints or corners of moving people or objects. Point-light stimuli are unrecognizable when presented statically but are rapidly recognized once set in motion. Thus, point-light displays clearly depict action cues but only depict very sparse and largely unrecognizable cues to static form. We have previously shown that the perception of human and meaningful object motion share overlapping cortical networks involving the occipital, parietal and sensory-motor regions [5]. Here we contrast brain activation induced during the observation of static dots (SD), randomly moving point light dots (RM) and point light dots representing human motion (HM).

II. Methods

A. Study Participants

10 healthy right-handed adult subjects (aged 20-40 years) participated in this study. All participants provided informed, written consent and had normal or corrected to normal vision. This study was approved both by the Simon Fraser University Research Ethics Board and the Down Syndrome Research Foundation Research Ethics Committee and conducted in accordance with the Declaration of Helsinki.

B. Stimuli

The stimuli consisted of 2-second movies depicting point light displays of HM, RM and SD. A 3-D VICON motion capture system was used to measure the motion of human actors and the resulting motion capture data was converted into point-light movies. The human motion consisted of fifteen dots representing the head and limbs. The human figure was seen walking, jumping, running, twirling or hopping. Random motion videos and the static displays were created from the trajectories and displays of the human motion videos.

C. Procedure

A blocked design was used consisting of 3 sets of 150 movies from each condition. Twenty different movies were

S. Supek and A. Sušac (Eds.): Advances in Biomagnetism – BIOMAG2010, IFMBE Proceedings 28, pp. 282–285, 2010.
www.springerlink.com

created for each condition. The movies were presented in a random order and repeated 2-3 times during each run. Participants were asked to visually fixate a point in the center of the screen. To time lock the onset of the movement, a photo sensitive diode was inserted on the screen to detect changes in luminance and triggered at onset of the video presentation. This trigger was simultaneously recorded with the MEG signals.

D. MEG Data Acquisition

The MEG data was acquired using a 151-channel whole head MEG system (VSM Medtech, Coquitlam, Canada) installed within a magnetically shielded room. Three sets of approximately 7 min of continuous data were initially collected at a rate of 600 samples per second with a 150-Hz low pass filter, using third-order gradiometer noise cancellation. The data was manually inspected for artifacts and corresponding segments were marked as bad.

E. Data Processing

The dual state SAM beamformer spatial filtering algorithm [6] was used to identify areas of action or stimuli-related activity. For forward modeling of the source fields, a multiple-sphere approximation of the scalp surface of each participant was obtained based on their digitized head shape. Coregistration of the functional and anatomical data was performed by matching the digitized head shape to the subject's MRI image. If the latter was missing, the best matching MRI from a database of about one hundred images was used. The MRI that had the best fit with the subject's head shape was used for co-registration.

A three-dimensional distribution of pseudo-T statistic reflecting differences in the source power in [8 - 18] Hz band for active [0, .5s] and control [-.5s, 0] intervals with respect to the stimulus were initially obtained for each subject. Individual distributions were spatially normalized using SPM99 [7] and group-averaged. These images were very similar for all conditions (see fig. 2, A, B, C). Therefore three differential pseudo-T images: human motion (active)-static dots (control) (HM-SD), random motion (active)-static dots (control) (RM-SD) and human motion (active)-random motion (control) (HM-RM) were constructed using the stimulus interval [0, 0.5s] (fig. 2, bottom). Statistically significant maxima and minima of these differential images (using a 95% significance threshold obtained by bootstrap) were selected as SAM virtual sensor locations. Locations obtained this way from all differential images were pooled together for further analyses.

The time-frequency distributions (TFDs) of the virtual sensor signals were analyzed by computing spectrograms. A 0.4-second long Hamming window was moved in 5 msec steps in the interval [-1, 1 s] relative to the event marker. For every channel, we first calculated trial-averaged TFDs for each subject, and then computed a group average. The power for each frequency line was normalized on its mean in the interval [-0.6s, 0] to obtain the time-frequency changes of activity for each condition.

Although these TFDs are interesting by themselves, to visualize the differences between conditions we calculated the differential TFDs. The results (in decibels) were plotted using color encoding. Figure 1 shows an example of how these differences were obtained from one virtual sensor.

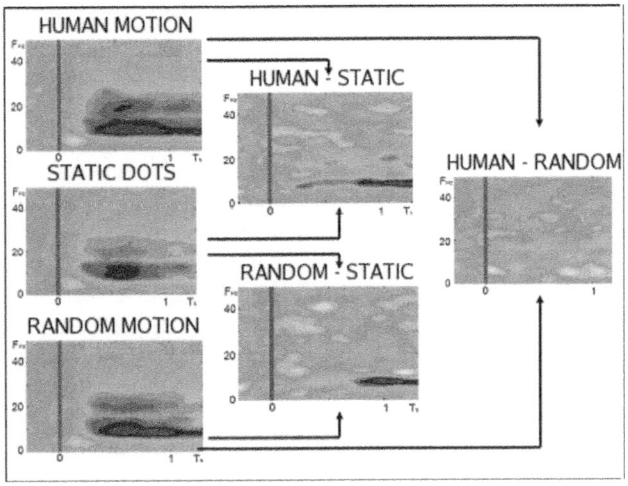

Fig. 1 Visual description of differential TFDs of one virtual sensor

III. RESULTS

The top panel of Fig. 2 shows the group averaged SAM pseudo-T distributions for the 8-18 Hz frequency for the individual conditions. The bottom panel displays the differential pseudo-T image from one condition (HM-RM). Figures 3-5 shows the differential time frequency plots for the following conditions: RM-SD, HM-SD, and HM-RM. In each differential image, areas of blue indicate stronger desynchronization for the primary condition and red indicates less desychronization for the primary condition.

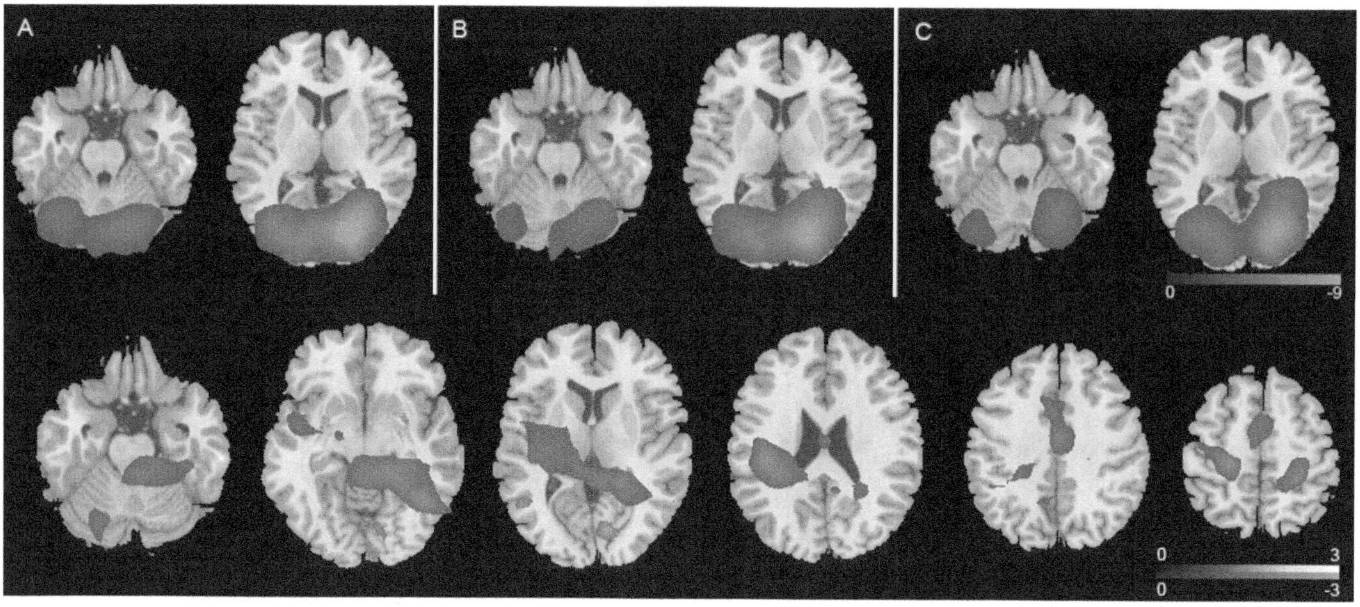

Fig. 2 Group-averaged SAM pseudo-T distributions for [8 – 18] Hz frequency band. Panels A, B, C – horizontal slices showing cerebella and visual activity for static, random and dynamic conditions; [0, 0.5] s active and [-0.5, 0] s control intervals were used. Bottom panel – pseudo-T image using possible as active and random as control; [0, 0.5s] window was used for both conditions

For RM-SD (Fig. 3) we observed greater desynchronization in the (L) lingual gyrus for the RM motion condition, early during the visual presentation at approximately 500 msec following stimulus onset. Between 750 ms to 1 second we observed stronger desychronization in the (R) precuneus and (R) mid-temporal regions during the RM display. Desynchronization was typically observed in the alpha band.

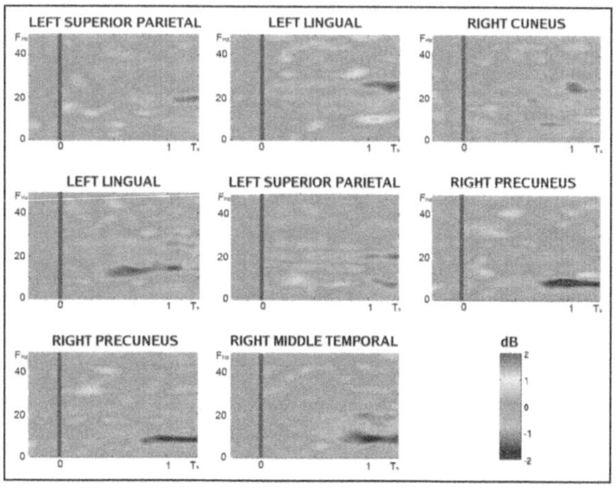

Fig. 3 TFD of the Random motion-Static dots condition

Further differences between the two conditions were evident at the end of the presentation with increased desynchronization observed in the (L) lingual gyrus and (R) cuneus for RM. Interestingly, this later stronger desynchronization was observed in the beta band.

Figure 4 shows the differential images from the TFD of HM-SD. Greater desynchronization was observed early following stimulus presentation in the (R) precuneus and (R) supramarginal gyrus combined with less desychronization in the (L) superior parietal region for the HM condition. Both were observed in the alpha band. Further increased synchronization was observed at approximately 750 ms to 1 second in the (R) precuneus, mid-temporal and cuneus and (L) lingual gyrus for the HM condition, again in the alpha band. Less desynchronization was observed in the gamma band in bilateral occipital and parietal regions for the human motion condition. In addition, the perception of HM led to increased desynchronization in the (L) precentral gyrus in the beta band. For the HM-RM condition (Figure 5), only three areas showed differences in activation between the two motion conditions: the (R) mid-temporal region, (R) post-central region and the (L) precentral region both showed greater desynchronization for the HM condition.

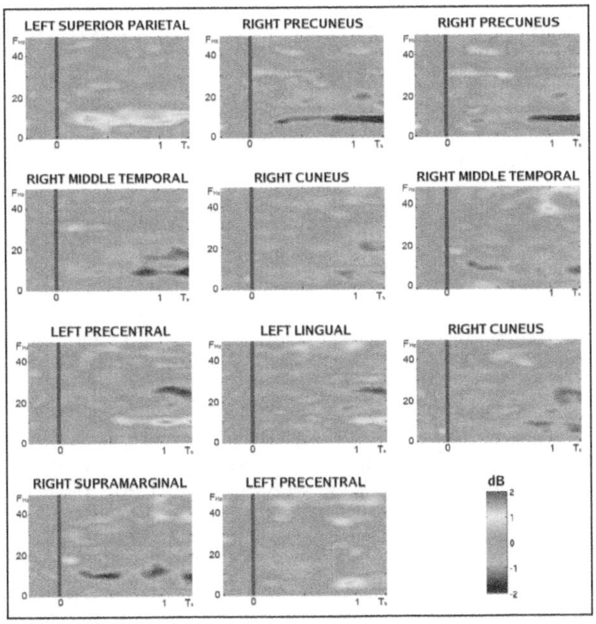

Fig. 4 TFD of the Human motion-Static dots condition

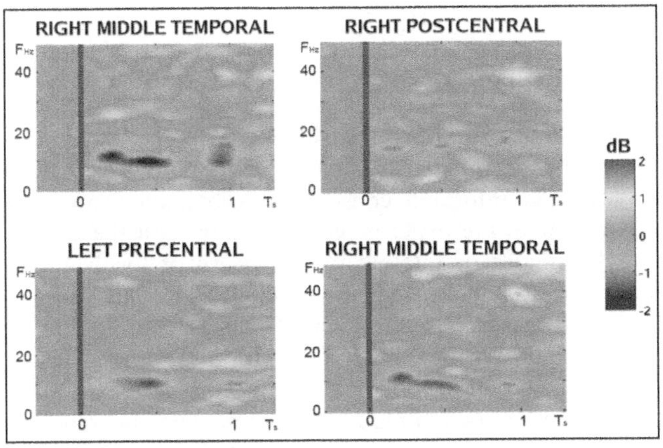

Fig. 5 TFD of the Human motion-Random motion condition

IV. Discussion

Our data suggest that the perception of motion, whether human or random, selectively activates motion processing areas within the visual cortex as well as areas within the parietal and temporal regions. Both random and human motion displays activated the (L) occipital and parietal regions and the (R) occipital, parietal and mid-temporal regions. While the perception of motion has traditionally been considered to involve separate dorsal and ventral pathways, our data is consistent with recent work [8] showing that there may be much more overlap and interaction between these systems than previously thought.

The major difference that we observed between the two types of motion displays was the selective activation of the (L) premotor and (R) sensory-motor regions during the perception of human motion. These data are consistent with recent studies showing that action cues depicting human motion are sufficient to selectively activate premotor and sensory-motor regions [9]. Further work is currently underway to examine the significance of this hemispheric specialization within the context of visual recognition and motion perception.

Acknowledgments

This work is supported by the Natural Sciences and Engineering Research Foundation (NSERC) Canada and the BC Leading Edge Endowment Fund (UR). We thank all the participants who took part in this study.

References

1. Rizzolatii G, Arbib MA. (1998). Language within our grasp. Trends in Neuroscience, 21, 188-194
2. Rizzolatti G and Craighero L. (2004). The mirror-neuron system. Annual Review of Neuroscience 27:169-192.
3. Iacoboni M, Koski LM, Brass M, Bekkering H, Woods RP, Dubeau MC, Mazziotta JC, Rizzolatti G. (2001). Reafferent copies of imitated actions in the right superior temporal cortex. Proceedings of the National Academy of Sciences of the United States of America 98(24):13995-13999.
4. Johansson G (1973) Visual perception of biological motion and a model for its analysis. Percept Psychophys 14:201–211.
5. Virji-Babul N, Cheung T, Weeks D, Kerns K, Shiffrar M. 2007. Neural activity involved in the perception of human and meaningful object motion. Neuroreport 18(11):1125-1128.
6. Robinson, S. & Vrba, J. (1999) Functional neuroimaging by synthetic aperture magnetometry (SAM). *Recent Advances in Biomagnetism,* Sendai, Japan, Tohoku Univ. Press, pp. 302–305
7. SPM software at http://www.fil.ion.ucl.ac.uk/spm/
8. Farivar, R., Blanke, O., & Chaudhuri, A. (2009). Dorsal-ventral integration in the recognition of motion-defined unfamiliar faces
9. Saygin AP, Wilson SM, Hagler DJ. Bates, E and Sereno MI.(2004). Point light Biological motion perception activates human premotorcortex. The Journal of Neuroscience, 24(27):6181– 6188

Author: N. Virji-Babul
Institute: Down Syndrome Research Foundation
Street: 1409 Sperling Avenue
City: Burnaby, B.C.
Country: Canada
Email: naznin@dsrf.org

Neural Interactions between Dorsal and Ventral Visual Subsystems While Perceiving 3-D Structure from 2-D Motion

Sunao Iwaki[1,2] and John W. Belliveau[2]

[1] Institute for Human Science and Biomedical Engineering, National Institute of Advanced Industrial Science and Technology (AIST), Ikeda, Japan

[2] Athinoula A. Martinos Center for Biomedical Engineering, Massachusetts General Hospital (MGH), Boston, USA

Abstract— **Two-dimensional optic flow is an important cue to perceive 3D structure of objects. Recent neuroimaging studies suggest the involvement of both the ventral and the dorsal visual pathways in the perception of 3D structure from motion (3D-SFM), though the neural dynamics underlying the 3D-SFM is not fully understood. Here, we combine magnetoencephalography (MEG) and fMRI to detect the dynamic brain responses to 3D-SFM. We manipulated the coherence of randomly moving dots to create different levels of 3D perception and investigated the associated changes in brain activity. Results of the fMRI analysis were used to impose plausible constraints on the MEG inverse calculation to improve spatial resolution of the spatiotemporal activity estimates. Time-frequency analysis was also employed to elucidate spatio-temporal dynamic changes in the spontaneous brain activities.**

MEG-fMRI combined analysis showed that the activities the posterior infero-temporal (pIT), parieto-occipital (PO), and intra-parietal (IP) regions were increased at different latencies during highly coherent motion conditions in which subjects perceived a robust 3D object. Results of the time-frequency analysis indicated the suppression of alpha- and beta-band activities in these regions which reflect the commitment of these areas in the perception of 3D-SFM. Current results suggest that the interactions between the dorsal and ventral visual subsystems are crucial for the perception of 3D object from 2D optic flow.

Keywords— **3D structure perception from motion, MEG, fMRI, combined analysis, event-related desynchronization.**

I. Introduction

Recognizing three-dimensional structure of the object from two-dimensional retinal motion requires the visual motion to be integrated into the internal representation of the 3-D object. Many psychophysical studies have been made to investigate how the visual system extracts the three-dimensional structure of objects from the two-dimensional motion of random dots (structure-from-motion: 3D-SFM) [1]. Electrophysiological studies of non-human primates suggested the involvement of middle-temporal (MT) and medial-superior-temporal (MST) neurons in the perception of 3-D structure from the optical flow [2,3]. Recent neuroimaging studies using functional magnetic resonance imaging (fMRI) techniques suggested the involvement of the parieto-occipital junction (POJ), the superior-occipital gyrus (SOG), and the ventral occipito-temporal junction (OTJ) in the perception of 3-D structure from motion [4,5] though the neural dynamics underlying the reconstruction of a 3-D structure from optic flow is not fully understood.

In this study, we used neuromagnetic measurement (magnetoencephalography: MEG) and fMRI to detect dynamic brain activities during the perception of 3D-SFM in human.

II. Methods

A. Subjects and Visual Stimuli

Nine normal right-handed subjects with no history of neurological disorders participated in the study as volunteers. Written informed consent was obtained from each subject prior to the experiment complying with the human research policies of internal review board of the National Institute of Advanced Industrial Science and Technology (AIST) and Massachusetts General Hospital.

The visual stimuli consisted of 1000 randomly-placed dots, which started to move 500 ms after the onset of presentation with various motion coherences. The coherence of the motion was controlled from 0 to 100 % (Fig. 1). A stimulus that is fully coherent had all the dots moving as if they belonged to a rotating spherical surface with a radius of

Fig. 1 Examples of the random-dot motion stimuli used in this study. Coherence of the motion was changed from 100 % to 0 % to generate different 3-D object perception

S. Supek and A. Sušac (Eds.): Advances in Biomagnetism – BIOMAG2010, IFMBE Proceedings 28, pp. 286–289, 2010.
www.springerlink.com

10 degree in visual angle. On the other hand, the 80, 60, 40, 20, and 0 % coherence stimuli contain dots having the same speed as the fully coherent stimuli but the directions of the 20, 40, 60, 80, and 100 % of the dots were randomized, respectively.

B. Data Acquisition

a) MEG

Neuromagnetic signals were measured while the subjects viewed the visual stimuli using a whole-cortex type SQUID system with either 122 or 306 planar gradiometers. The stimulus-related epochs of 2000 ms, including a 1000 ms pre-stimulus baseline, were recorded with a sampling rate of 600 Hz. More than 80 epochs were averaged for each condition. Electro-oculogram (EOG) was simultaneously recorded and the epochs with EOG signal change exceeded 150 mV were discarded to eliminate artifacts. Subjects were asked to react as soon as the motion stimuli disappeared by lifting their index finger when they perceive rotating 3-D sphere or by middle finger when they saw random-motion.

b) fMRI

The scanning was conducted using a 3 Tesla scanner. For functional imaging, the single shot echo-planer imaging sequence was used with the imaging parameters TR 3000 ms, TE 40 ms, FA 90 deg, 40 axial slices, 3 mm thickness with 0 mm gap, 64x64 matrix, and FOV 220 mm, which covered the entire brain. Three 14-min functional scans were divided into 12-second phases, randomly alternating between different stimulus (coherency) conditions and resting (fixation) periods. Within each phase, motion stimuli were presented every 4 s. Subjects' task was the same as that in the MEG experiment

C. Data Processing

3-D reconstruction of the cortical surface was performed using Freesurfer software [6,7] for each subject. Statistical Parametric Mapping (SPM) was used to infer statistical significance of the signal changes in fMRI data [8]. The BOLD images obtained during fMRI scanning were re-aligned using the first image as a reference and spatially smoothed using Gaussian kernels of 6 mm. A boxcar wave function was applied as a reference function, and a statistical parametric map was generated for each voxel using general linear model (GLM). The results of the fMRI analysis were used to impose plausible constraints on the MEG inverse calculation using 'weighted' minimum-norm approach [9,10] to improve spatial resolution of the spatio-temporal activity estimates. In this study, we introduced fMRI weighting, which was determined by thresholding the fMRI statistical parametric map for each condition vs. fixation condition at $p<0.001$ (FDR (False Detection Rate) corrected), on the minimum L2-norm linear inverse operator used to map measured MEG signal into estimated neural source distributions [10].

The event-related time-frequency wavelet decomposition (using Morlet's wavelet basis) [11] of the recorded MEG raw signals between 5 and 60 Hz was used to assess event-related changes in spontaneous oscillatory brain activity, and the time-frequency energy of each single trial was averaged separately for each stimulus condition in the time-frequency domain.

III. Results

a) Behavioral Results

Figure 2 shows the behavioral results obtained during MEG and fMRI acquisitions. In the 100 and 80 % coherent conditions, the subjects perceive rotating 3-D spherical surface in the most of the trials, whereas 3-D object was rarely perceived in the 0 and 20 % conditions.

b) Spatio-Temporal Imaging of the Event-Related Responses

Figure 3 shows the results of the fMRI group statistics, in which fMRI contrasts between the 3-D perception condition (100 and 80 % coherent conditions) vs. the random motion condition (0 and 20 % coherent conditions) were depicted. Increased brain activities were observed in the parieto-occipital (PO), the intra-parietal (IP), and the posterior inferior temporal (pIT) regions as well as the motion sensitive MT area in the 3-D perception condition.

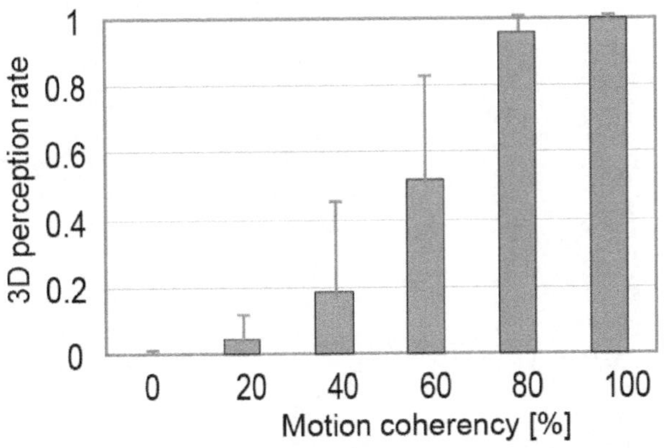

Fig. 2 Subjects' behavioral performance during the MEG and fMRI experiments. In the 100 and 80 % coherent conditions, rotating 3-D sphere was perceived in more than 95 % of the trials, while 3-D object was not perceived in the most of the trials in the 0 and 20 % coherent conditions

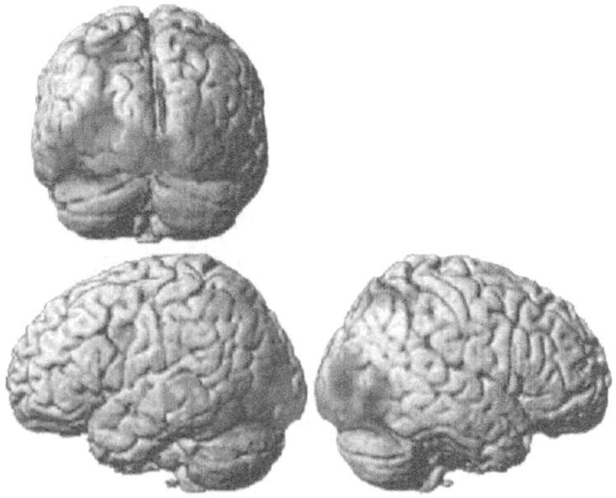

Fig. 3 Results of the fMRI group statistics depicting the contrasts between the 3-D perception (100 and 80 % coherent) conditions and the random motion (0 and 20 % coherent) conditions

The spatiotemporal activity estimates obtained from the fMRI-weighted MEG inverse analysis were shown in Fig. 4, which were the results of the inter-subject averaging of the activity estimates by using spherical morphing technique. Increased neural activity in the PO area was observed both in the 3-D perception (100 and 80 % coherence) conditions and the intermediate (60 % coherence) condition in the latency range between 220 and 300 ms, while the activities in the IP and the pIT regions were observed only in the 3-D perception conditions in the latencies between 340 and 450 ms after the onset of motion. Also, the activity in the primary visual areas during earlier latencies were decreased in the 3-D perception conditions compared to the random motion and intermediate (0, 20, 40, and 60 % coherent) conditions.

c) Event-Related Changes in the Spontaneous Activities

Statistical significance of the event-related suppression of the spontaneous activities from the pre-stimulus baseline was assessed by inter-subject t-test, in which spectral power decrease at each time and frequency in the 3-D perception condition was tested (Fig. 5).

Suppression of alpha- and beta-band activities started around 300 ms after the onset of random dot motion in SFM condition at occipital, bilateral occipito-temporal and occipito-parietal recording channels.

IV. Discussion

The results shown here indicate that the PO, IP, and pIT regions play an important role in the perception of 3-D object structure from random-dot motion. These results are in agreement with those from the previous studies of 3-D SFM using fMRI [4,5] in terms of their locations, and add further insight into the temporal characteristics of the neural activities in these regions.

At the same time, the event-related suppression (event-related desynchronization: ERD) of alpha- and beta-band spontaneous brain activities was detected in the occipital and parieto-occipital regions. The suppression of spontaneous activities was observed about 300 ms after the onset of random-dot motion which was used as a cue for 3-D shape perception. Local suppression of alpha- and beta activity in occipital and parieto-occipital area was thought to indicate the engagement of these regions to the relevant sensory or cognitive functions, which is in agreement with the results obtained

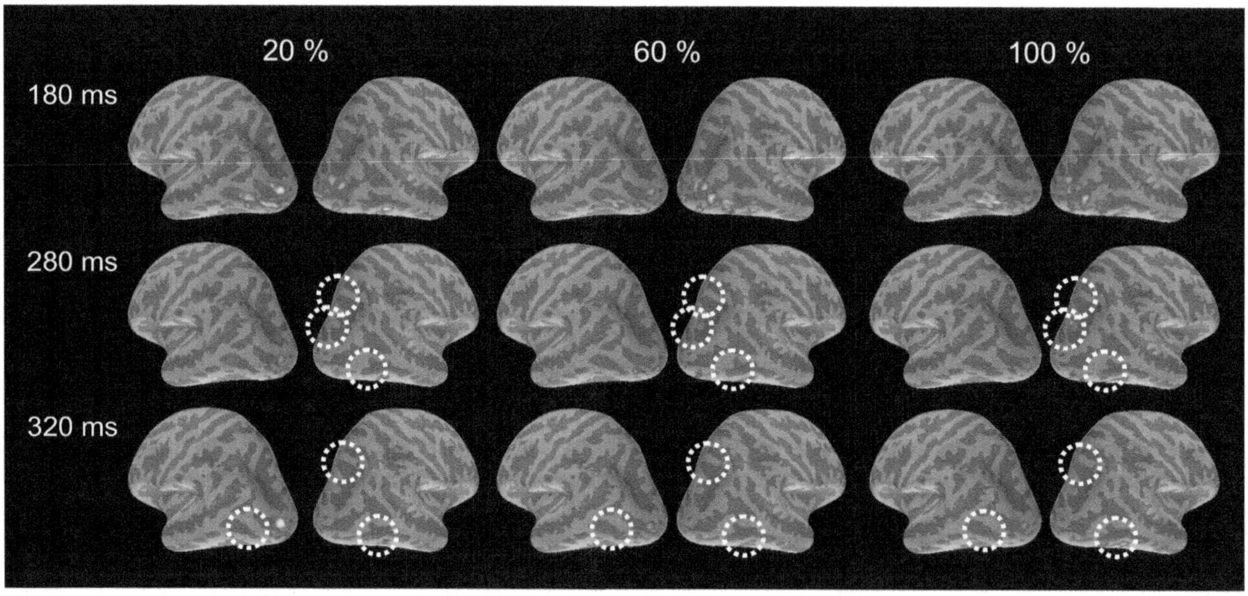

Fig. 4 The results of the inter-subject averaging of the activity estimates over all seven subjects. The blue circles denote the sites where we observed the major changes between the conditions

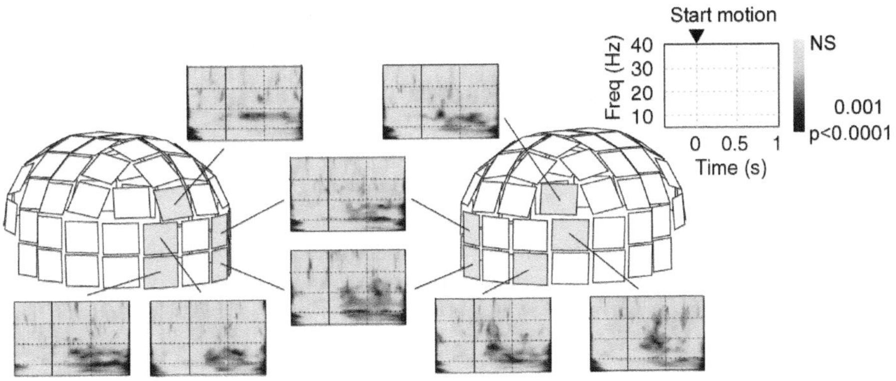

Fig. 5 Results of inter-subject t-test to detect the statistically significant suppression of the time-frequency energy from the pre-stimulus baseline during SFM condition

by MEG-fMRI combined analysis of event-related responses to 3D-SFM stimuli.

The change in the neural activities in the MT area in conjunction with the perception of global motion is reported in the previous fMRI studies [12], the PO areas were known to be involved in the integration of spatial aspects of the visual motion [13], the IP region processes the mental imagery [14,15], and the pIT regions are responsible for the object recognition [16]. Together with these previous findings, the current results suggest that the perception of moving 3-D objects from the 2-D retinal motion depends on the higher visual processing in the dorsal visual pathway, i.e., (a) the global motion processing in MT, (b) the spatial integration of the global motion in the PO region, and (c) the mental imagery processing in the IP region, as well as the ventral object recognition system in the pIT.

V. Conclusion

We used MEG-fMRI combined analysis and time-frequency analysis of MEG data to visualize dynamic brain activities during the perception of 3D-SFM. The results suggest that the perception of moving 3-D object from 2-D random-dot motion includes both perception and integration of global motion and 3-D mental image processing as well as the object recognition, that are accomplished by the co-operative engagement of both the ventral and the dorsal visual information processing streams.

References

1. Wallach H, O'Connell DN (1953) The kinetic depth effect. J. Exp. Psychol., 45: 205-217.
2. Bradley DC, Chang GC, Andersen RA (1998) Encoding of three-dimensional structure-from-motion by primate area MT neurons. Nature, 392: 714-717.
3. Sugihara H, Murakami I, Shenoy KY et al. (2000) Response of MSTd neurons to simulated 3D orientation of rotating planes. J. Neurophysiol., 87: 273-285.
4. Orban GA, Sunaert S, Todd JT et al. (1999) Human cortical regions involved in extracting depth from motion. Neuron, 24: 929-940.
5. Paradis AL, Cornilleau-Peres V, Droulez J et al. (2000) Visual perception of motion and 3-D structure from motion: an fMRI study. Cereb. Cortex, 10: 772-783.
6. Dale AM, Fischl B, Sereno MI. (1999) Cortical Surface-Based Analysis I: Segmentation and Surface Reconstruction. Neuroimage, 9: 179-194.
7. Fischl B, Sereno MI, Dale AM. (1999) Cortical Surface-Based Analysis II: Inflation, Flattening, and a Surface-Based Coordinate System. (1999) Neuroimage, 9: 195-207.
8. Friston KJ (1998) Imaging neuroscience: principles or maps?, Proc. Natl. Acad. Sci. USA, 95: 796-802.
9. Hamalainen M, Hari R, Ilmoniemi RJ.et al. (1993) Magnetoencephalography - theory, instrumentation, and application to noninvasive studies of the working human brain. Rev. Mod. Phys., 65: 413-497.
10. Dale AM, Liu AK, Fischl BR et al. (2000) Dynamic statistical parametric mapping: combining fMRI and MEG for high-resolution imaging of cortical activity. Neuron, 26: 55-67.
11. Tallon-Baudry C, Bertrand O, Delpuech C et al (1996) Stimulus specificity of phase-locked and non-phase-locked 40 Hz visual responses in human. J. Neurosci., 16: 4240-4249.
12. Vaina LM, Belliveau JW, des Roziers EB et al. (1998) Neural systems under-lying learning and representation of global motion. PNAS, 95: 12657-12662.
13. Hotson JR, Anand S (1999) The selectivity and timing of motion processing in human temporo-parieto-occipital and occipital cortex: a transcranial magnetic stimulation study. Neuropsychologia, 37: 169-179.
14. Iwaki S, Ueno S, Imada T et al (1999) Dynamic cortical activation in mental image processing revealed by biomagnetic measurement. Neuroreport, 10: 1793-1797.
15. Jordan K, Heinze HJ, Lutz K et al. (2001) Cortical activations during the mental rotation of different visual objects. Neuroimage, 13: 143-152.
16. Haxby JV, Grady CL, Horwitz B et al (1991) Dissociation of object and spatial visual processing pathways in human extrastriate cortex. PNAS, 88: 1621-1625.

Author: Sunao Iwaki
Institute: National Institute of Advanced Industrial Science and Technology (AIST)
Street: 1-8-31 Midorigaoka
City: Ikeda
Country: Japan
Email: s.iwaki@aist.go.jp

Objective Detection of the Flicker Fusion Threshold with Pupil Diameter and Visual Evoked Magnetic Field

Nobuyoshi Harada[1], Sunao Iwaki[1], and Mitsuo Tonoike[2]

[1] Institute for Human Science and Biomedical Engineering,
National Institute of Advanced Industrial Science and Technology, Ikeda, Japan
[2] Department of Medical System Engineering, Faculty of Engineering, Chiba University, Chiba, Japan

Abstract— **Objective detection of a flicker fusion threshold was investigated by examining the differences in responses of pupil diameter and the visual evoked magnetic field (VEF) for flickering visual stimulation with a frequency over or under the flicker fusion threshold. The pupil diameter increment was larger and the decrement was smaller in frequencies over 7.5 Hz. The amplitude of OFF responses increased with an increase in the frequency of flickering. Differences in pupil diameter and OFF responses for VEF demonstrated the frequency of the flicker fusion threshold in a non-arbitrary way.**

Keywords— **MEG, VEF, Flicker Fusion Threshold, OFF responses.**

I. Introduction

The progressive diminution of frequency of the flicker fusion threshold (FF) was observed during the development of the fatigue state [1]. The FF was defined as the frequency of intermittent light stimulus that provoked a flickering sensation. The variation in FF was employed as an index in the development of the fatigue state of the visual system encompassing effects regarding the excitation level of the sensory area of the cortex, and the arousal level of consciousness [2]. This index was employed for monitoring the stages of fatigue in the areas of public health, occupational health and ergonomics.

The detection of FF is normally decided by a subjective judgment of the flickering state of a visual stimulus co-coordinated with a response by pressing a button [2],[3]. Obviously, this process is not free from the arbitrariness of human response; therefore, a solution for this is important in order to accurately measure FF. Non-arbitrary and objective detection of FF was investigated by examining the differences in responses of pupil diameter and the visual evoked magnetic field (VEF) for flickering visual stimulation with a frequency over or under the FF.

II. Methods

Experiment 1

The variation in pupil diameter response induced by a visual stimulation sequence with higher and lower frequencies than FF was observed with six males (mean age ± standard deviation, 35.5±9.3 years). The FF was decided from second maximum score of frequency from five time measurements of threshold frequencies before the experimental session. The average FF of the six subjects was 41.5±5.0 Hz. Two flickering visual stimuli of luminance 310 cd/m^2, with a frequency 7.5 Hz under or over the determined FF were presented to subjects with a light-emitting unit. The duration of the flickering light stimulus was 10 s with a 1.5 s gap, and this was repeated 10 times at both frequencies. The source of intermittent light stimulus was generated with a halogen light unit (ELI-100N40G; Mitsubishi Rayon Co. Ltd.) connected to a 10 m fiber optic cable (1 cm diameter). A fiber optic cable on the other side of the halogen light unit was located 1 m in front of the subject. The experiments were carried out in complete darkness.

Pupil diameter was observed between periods of light stimulus with an imaging unit that contained an infrared CCD camera (Free View Telescope Type; Takei Scientific Instruments Co. Ltd.), located 1.5 m in front of the subject. The exposure of the infrared CCD camera was automatically controlled. Numbers of pixels of pupil diameter was defined with image of infrared CCD with square flame in image which was 100 pixels square.

A rapid enlargement of the pupil due to the turning off of the optical stimulation and a rapid contraction of pupil due to the turning on of the optical stimulation were observed. A zero base pupil diameter was defined based on the data taken 1.5 s before a gap in optical stimulation, for detecting the difference in pupil response.

S. Supek and A. Sušac (Eds.): Advances in Biomagnetism – BIOMAG2010, IFMBE Proceedings 28, pp. 290–293, 2010.
www.springerlink.com

Experiment 2

The subjects were eight males (28.9±4.0 years) and the average FF of the subjects was 44.1±5.1 Hz. Four flickering stimuli of 1500 ms duration, 2.5 or 7.5 Hz under or over the threshold were randomly presented to subjects. The flickering sensation was clearly provoked at 7.5 Hz under the threshold and was not apparent when the frequency was 7.5 Hz over the threshold. Flickering sensation was not so unstable, and possible to change in -2.5 Hz and +2.5 Hz condition. Subjects were stimulated 100 times for each condition, and the presentation of visual stimuli were randomized on four conditions. The flickering of light was presented in the following sequence: a flicker signal was turned on and maintained for 1500 ms then darkness was maintained for an average of 1500 ms (between 1000–2000 ms); thereafter, another flicker signal at a different frequency was turned on and maintained for 1500 ms. The ON and OFF response of VEF for the optical stimulation sequence was averaged 100 times with four flickering stimuli, 2.5 or 7.5 Hz over or under the threshold. The VEF ON response was observed 100 ms after the onset of visual stimulus, and the OFF response was observed 1600 ms after the onset of visual stimulus. The system of visual stimulation was employed same system of experiment 1.

The 122 channel whole-cortex SQUID system (Neuromag Inc.) was used to measure MEG responses. Neuromagnetic responses were sampled continuously at 400 Hz (0.03–100 Hz analog bandwidth). Trigger signals defined the onset of the visual stimulus of the flickering stimulus duration. MEGs of 100 responses were averaged for 2500 ms, beginning 500 ms before the onset of stimulus onset and corrected for a zero baseline from -200–0 ms. Offline digital band pass filtering reduced the frequency range to 0.5–30 Hz. A 0.5 Hz high pass filter was chosen as a stable baseline for ease of comparison with the VEF OFF response. A 30 Hz low pass filter was adequate enough to detect a P100m component. Noises in MEG responses were rejected on the period of over ranged 3000 fT of MEG response. Averaged waveforms were observed on channels at a maximum amplitude of the P100m component in the MEG sensor of occipital area.

III. RESULTS AND DISCUSSION

Experiment1

Increases and decreases in pupil diameter from the baseline were observed in the period between stimulations, due to the process of adapting to dark or light environments.

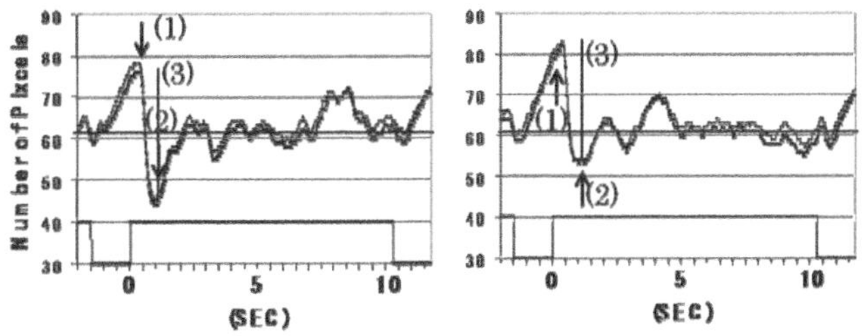

Fig. 1a Two figures show waveforms of pupil diameter variations during the 1.5 s rest time between 10 s visual stimuli with a frequency 7.5 Hz under (left) or 7.5 Hz over (right) the flicker fusion threshold, with a luminance of 310 cd/m^2. The numbers in the figure indicate (1) increase, (2) decrease, (3) variation (peak to peak), of pupil diameters

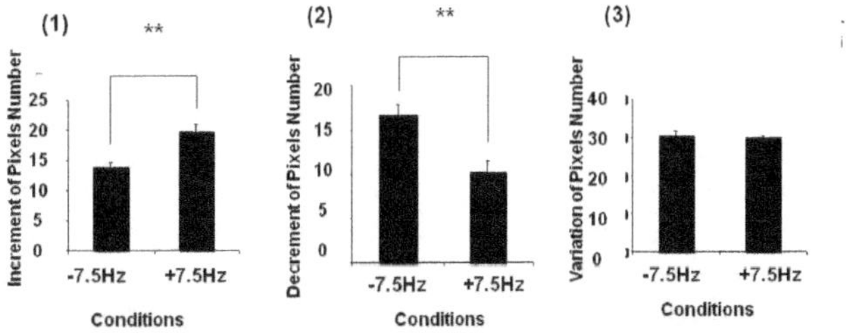

Fig. 1b Three graphs show number of pixels, (1) increments of pupil diameter on the adaptation to the dark environment, (2) decrements of pupil diameter on the adaptation to the light environment, (3) variation of pupil diameter from dark to light environment. The graphs show the mean + SEM for 6 subjects. **: p<0.01

Fig. 1a shows the differences of the pupil responses between the conditions in which flickering was perceived (-7.5 Hz) and flickering could not be perceived (+7.5 Hz). The increase in pupil diameter was 13.83±0.91 pixels when the frequency was 7.5 Hz under the threshold and 19.83±1.25 pixels when the frequency was 7.5 Hz over the threshold. A significant difference in variation was observed at +7.5 Hz when compared to -7.5 Hz ($F(1/11) = 10.59$, $P = 0.022$).

The pupil contraction response induced by turning on the optical stimulation (Fig. 1a2). A decrease in pupil diameter was observed at -7.5 and +7.5Hz. This decrease was 16.67±1.15 pixels at -7.5 Hz, and 10.17±1.30 pixels at +7.5 Hz). A significant decrease in variation was observed between the two conditions ($F(1/11) = 11.86$, $P = 0.018$) (Fig. 1b2). There was no difference in pupil diameter change from large to small, or from a dark to light environment for the two flickering conditions ($F(1/11) = 0.21$, $P = 0.665$) (Fig. 1a,b3).

The degree of pupil enlargement and contraction was greater when the frequency was 7.5 Hz above the threshold, as the subjects did not perceive the flicker. The variations in pupil diameter in those conditions are such that the pupil variation has a contraction tendency in the condition in which the subject perceives the flicker, and the pupil variation has an enlargement tendency in the condition in which the subject does not perceive the flicker. Adaptation for a light environment was inhibited under the condition of flicker recognition.

Experiment 2

VEF was observed in the V1 area on the MEG channel with a maximum amplitude of the P100m component. The ON response is a response to a change in the light environment from the dark state. The OFF response is a response to a change in the light environment to the darkened state. Averaged waveforms of ON and OFF responses were examined for four flickering conditions. There was no significant difference in the ON response between the four frequencies. However, a large OFF response was observed at frequencies 2.5 and 7.5 Hz higher than the threshold frequency, and a small OFF response was observed 2.5 and 7.5 Hz lower than the threshold frequency. The amplitude of OFF responses increased with the increment of the frequency of flicker ($F(3/39) = 3.05$, $P = 0.045$). As shown in Fig. 2, the order of the frequencies from the threshold frequency

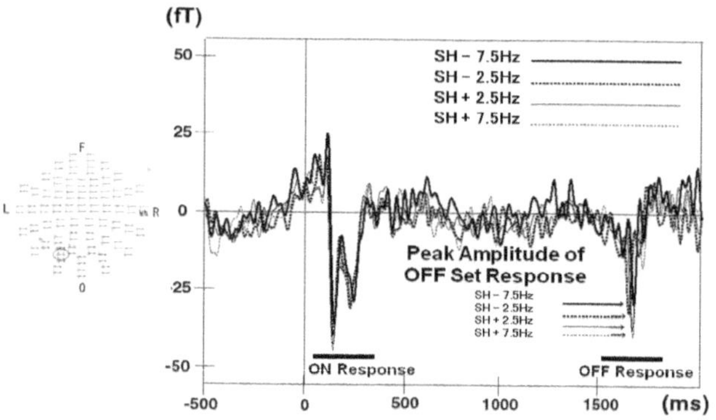

Fig. 2 Averaged waveforms of visual evoked fields (VEFs) of magnetoencephalographs, elicited by four flickering stimuli of 1500 ms durations, 2.5 or 7.5 Hz over or under the flicker fusion threshold. The line conditions indicate the frequency of the flicker fusion threshold, with the thick continuous line representing a frequency 7.5 Hz under the threshold, the thick dotted line was 2.5 Hz under the threshold, the thin continuous line was 2.5 Hz over the threshold and the thin dotted line was 7.5 Hz over the threshold

in terms of the amplitude variation during the OFF response was +7.5Hz, +2.5Hz, -2.5Hz, and -7.5Hz. The four arrows in the lower right of the graph denote the positions of the maximum amplitudes of the OFF responses with respect to the four frequencies. It was reported, OFF effect for blocking time of α wave enchased with increment of intensity or

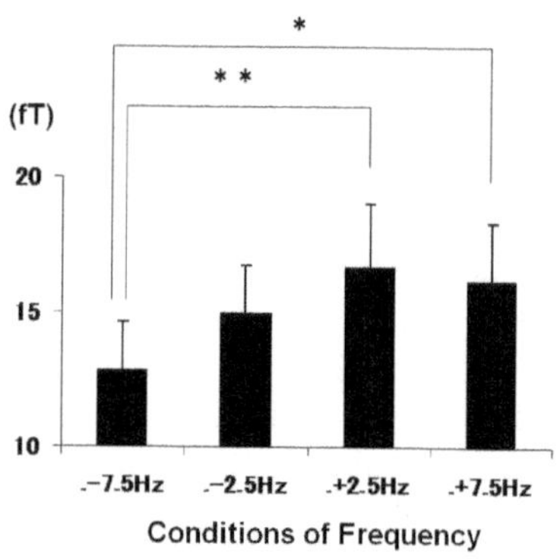

Fig. 3 Relationship between the amplitude of the OFF response of VEF at different frequencies above and below the flicker fusion threshold. The graphs show the mean + SEM for eight subjects. * indicates $p < 0.05$; ** indicates $p < 0.01$

duration of visual stimulus [3]. Adaptation to a light environment was enhanced sensitivity to change in contrast around [4]. Increment of intensity or duration of visual stimulus, was possible to enhance to adaptation for light environment to a dark environment.

At a frequency higher than the threshold frequency, adaptation to the optical steady state advances more, which is apparently why the greater OFF response was detected in response to turning off the light. At a frequency lower than the threshold frequency, adaptation to the optical steady state is insufficient, and the OFF response was reduced. As is evident from the results, in the visually evoked magnetoencephalographic OFF response to a flicker light stimulation sequence, the inhibition of adaption to the optically steady state corresponding to flicker perception was detected as a decrease in the amplitude of the OFF response. The increment of frequency enhanced adaptation to the light environment, and increased the amplitude of VEF of OFF response.

The results show that the response, which is the perception of a flickering light, directly influences the activity of the cerebral cortex primary visual area; in other words, the response itself is an immediate reflection of the activity of the cerebral cortex. Accordingly, by observing not only the variation in pupil diameter, but also the visually evoked magnetoencephalographic OFF response or brain potential diagram OFF response to a flicker light stimulation sequence, it is possible to monitor the state where the subject perceives the flickering of light. The change of FF with development of the fatigue state possibly reflects the variation of differences of OFF response of VEF. Changed FF was detected with differentiation of OFF response with new visual stimulation at a low frequency.

IV. Conclusions

The frequencies over or under the FF and duration of stimulation affected the processes of adapting to a flickering light. The increment of frequency for flicker stimulus changed the "OFF" and "ON" response of pupil diameter from "closing" sift to "opening" sift, and changed responses for OFF responses for VEF from "inhibition" to "enhancement". Differences in pupil diameter and OFF responses for VEF demonstrated the frequency of the FF in a non-arbitrary way.

References

1. Simon E. and Enzer N. (1941) Measurements of fusion frequency of flicker as a test of fatigue of the central nervous system, J. Industr. Hyg. & Toxicol. 23:83-89,
2. Hashimoto K., (1963) Physiological meaning of the critical flicker frequency (CFF) and some problem in their measurement: theory and practice of the flicker test, Industrial medicine ("Sangyou Igaku"), 5:9 :3-18
3. Ruth M.C.(1937) Human occipital brain potential as affected by intensity-duration variables of visual stimulation, J. Exp. Psychol. 21: 6: 625-641
4. Ross J. and Speed D.H (1991) Contrast adaptation and contrast masking in human, Proceedings: Biological Sciences, 21:1315:61-70

Use macro [author address] to enter the address of the corresponding author:

Author: Nobuyoshi Harada
Institute: National Institute of Advanced Industrial Science and Technology
Street: 1-8-31, MIdorigaoka
City: Ikeda
Country: Japan
Email: n-harada@aist.go.jp

Effects of Adaptation on Visual Sensitivity and MEG Responses to the Envelope of AM Flicker

Yosuke Okamoto[1,2] and Seiji Nakagawa[2]

[1] Japan Society for the Promotion of Science (JSPS), Tokyo, Japan
[2] Institute for Human Science and Biomedical Engineering, National Institute of Advanced Industrial Science and Technology (AIST), Osaka, Japan

Abstract— **For a sinusoidal flickering light amplitude-modulated (AM) sinusoidally, flicker with a periodicity of a temporal envelope (modulation-frequency) of the light is perceived even though there is no such frequency component. However, the mechanisms underlying the perception are still unclear. In human visual system, it is known that there are two temporal channels with a lower and a higher peak frequency. Therefore, we investigated whether the two temporal channels affected the perception of the envelope of the AM flickering light. Visual sensitivity and magnetoencephalographic (MEG) responses to the envelope periodicity were measured when the sensitivity of each of the temporal channel was reduced by adaptation. Sinusoidal flickering lights with a low (2 Hz) and a high (16 Hz) temporal frequency were used as adapting stimuli. The sinusoidal flickering lights of 2 and 16 Hz and an AM flickering light with a modulation-frequency of 2 Hz and a carrier-frequency of 16 Hz were presented as test stimuli. The results showed that the sensitivity and MEG responses to the sinusoidal flickering light decreased more when the sensitivity of the temporal channel tuned for the test stimulus frequency was reduced than when that of the other temporal channel was reduced. On the other hand, the sensitivity and MEG responses to the envelope periodicity decreased more when the sensitivity to the frequency corresponding to the envelope periodicity was reduced than when that to the carrier frequency was reduced, even though a frequency component corresponding to the envelope periodicity did not exist in the stimulus. The results indicated that the envelope component of AM flickering light was affected by the low-frequency channel and suggest the possibility that the envelope component arises before or while the AM flickering signal passes the low-frequency channel.**

Keywords— **Flicker, Amplitude modulation, Adaptation, Detection threshold, MEG.**

I. Introduction

When two or more sinusoidal flickering lights with different temporal frequencies are added, other flicker at the difference frequency is produced. This phenomenon is referred to as temporal beat, and some studies have used such a flicker for investigating temporal visual processing [1-3]. In addition, flicker at a modulation frequency is perceived when a temporal white noise or a sinusoidal flicker is amplitude-modulated (AM) [4-6]. Flicker perception for these flickering lights indicates that flicker with the periodicity of the temporal envelope of luminance oscillation is perceived even though no such frequency component is present.

The mechanism underlying the perception of the envelope periodicity is still unclear. Studies measured electroretinograms (ERG) for temporal beats suggested that the temporal beat were detected on the based on a distortion product due to a luminance non-linearity generated at an early stage (within retina) of visual processing [2, 3]. However, psychophysical measurements of masking and adaptation effects of the spatial and temporal beats suggested that beat detection in the spatial and temporal domains is not based purely on distortion products produced by a peripheral luminance non-linearity [1]. They also implied that the detection of temporal beats could rely on a cortical mechanism because it is shown that temporal beats are produced under dichoptic stimulation [7].

It is known that the contrast threshold for a high-contrast sinusoidal grating is elevated at the adapting and neighboring spatial frequency. However, the threshold elevation is small at the other nearby spatial frequencies [8]. This result indicates that, in the visual system, there are a relatively large number of channels which tuned to narrow spatial frequency ranges. On the other hand, for the temporal frequency, threshold elevation is observed over a broad frequency range [9, 10]. After adapting to a low temporal frequency, the detection threshold is elevated broadly over the low frequency range. Also, after adapting to a high temporal frequency, the threshold elevation is observed broadly over the high frequency range. This result is explained by the suggestion that there are only two temporal channels in the visual system: one with lower peak frequency and the other with higher peak frequency [9, 11-13].

The purpose of this study was to discuss where the envelope component arises in the visual system by investigating whether the two temporal channels affected the envelope perception. An AM flickering light with a low modulation frequency (envelope periodicity) and a high carrier frequency was used as the test stimulus. The visual sensitivity and magnetoencephalographic (MEG) responses to the

S. Supek and A. Sušac (Eds.): Advances in Biomagnetism – BIOMAG2010, IFMBE Proceedings 28, pp. 294–297, 2010.
www.springerlink.com

envelope periodicity were measured when the sensitivity of the temporal channel tuned for low (or high) temporal frequency was reduced by adapting to low (or high) temporal frequency. Under the same adaptation condition, the sensitivity and MEG responses to sinusoidal flickering lights were also measured to reconfirm whether the temporal channels were affected by the adaptation stimuli.

II. Detection Threshold

A. Adaptation Effects on Sinusoidal Flicker

A steady light and sinusoidal flickering lights of 2 and 16 Hz were used as adaptation stimuli. The test stimuli were the sinusoidal flickering lights of 2 and 16 Hz. The detection thresholds for the test stimuli were estimated using the method of limits. In each experiment, there were an initial adaptation period of 90 s and 15 s "top-up" adaptation periods. After the adaptation period, the test stimulus was presented for 2 s. Each period was signaled by an auditory tone. Two ascending and 2 descending runs were conducted for each subject. The mean luminance of each stimulus was 20 cd/m^2. The stimuli were presented via a 5x5 array of green light emitting diodes (LEDs) that subtended 2.2° of the visual angle. Five healthy right-handed adults with normal or corrected-to-normal vision participated in the study.

Fig. 1 shows the averaged detection thresholds for test stimuli under the different adaptation conditions. Results showed that the detection threshold for the flickering light of 2 Hz increased more after adapting to 2 Hz than after adapting to 16 Hz. Also, the detection threshold for the flickering light of 16 Hz increased more after adapting to 16 Hz than after adapting to 2 Hz.

B. Adaptation Effects on AM Flicker

An AM flickering light with a carrier frequency of 16 Hz and a modulation frequency of 2 Hz was used as a test stimulus. The luminance waveform and its frequency spectrum were shown in Fig. 2. Frequency components were observed at the carrier frequency and its sidebands, but not at the frequency corresponding to the envelope periodicity. The adaptation stimuli and the procedures of threshold measurements were identical to those used in the detection threshold measurements of the sinusoidal flickering lights. The thresholds for detecting the flicker with the envelope periodicity after adapting to 2, 16 Hz and a steady light were estimated.

Fig. 3 shows the averaged threshold. The results showed that the detection threshold for the envelope periodicity increased more after adapting to the 2 Hz (envelope periodicity) than after adapting to 16 Hz (carrier frequency).

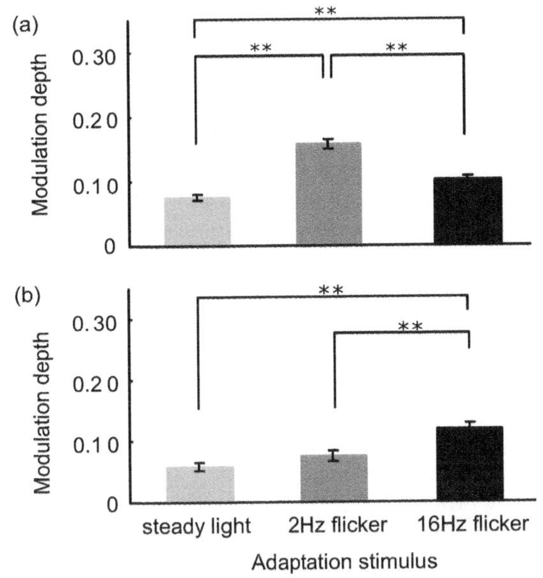

Fig. 1 Detection thresholds for the sinusoidal flickering lights of (a) 2 Hz and (b) 16 Hz. Error bars indicate SEM ($^{**}p$<0.01)

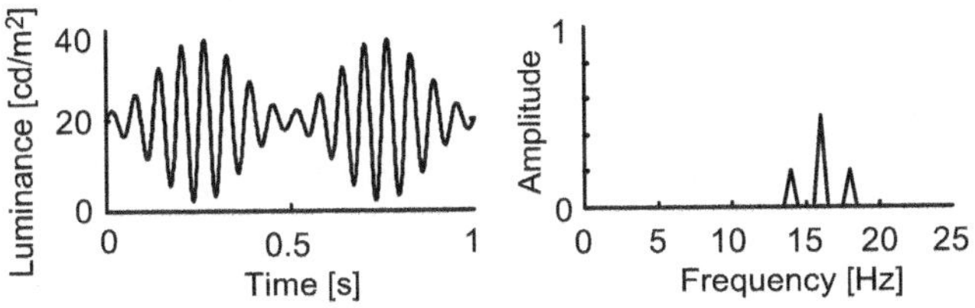

Fig. 2 A waveform and a frequency spectrum of the AM flickering light

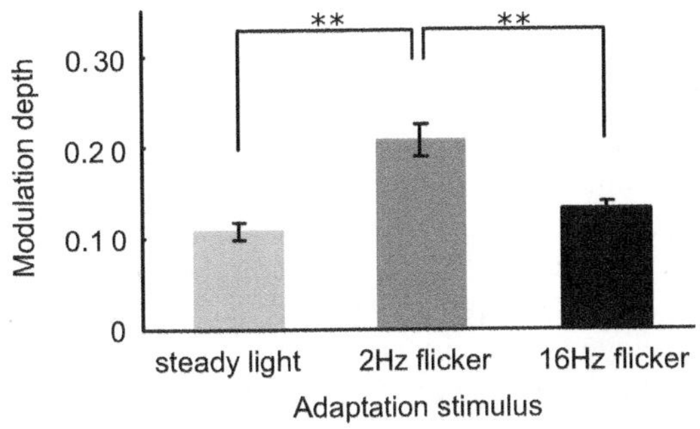

Fig. 3 Detection thresholds for flicker with the envelope periodicity of the AM flickering light. Error bars indicate SEM ($^{**}p$<0.01)

III. MEG RESPONSES

A. Adaptation Effects on Sinusoidal Flicker

The adaptation stimuli used for the detection threshold measurements were also used for the MEG measurements. The test stimuli were sinusoidal flickering lights of 2 and 16 Hz with the modulation depth of 0.8. The adaptation stimulus was presented initially for 120 s, then for 30 s as a top-up. Following the adaptation stimulus, subjects viewed the test stimulus for 4 s. Each test stimulus was presented 5 times. The subjects were requested to fixate on the stimulus when it appeared. The cortical responses were measured by using a whole-head 122-channel MEG system (Neuromag-122™, Neuromag Ltd.). The recordings were performed in a dimly lit, magnetically shielded room. Five adults participated in the MEG measurement. MEG responses obtained from 16 sensors located over the occipital area (which included the primary visual cortex) were used for the analysis. Power spectra were calculated by fast Fourier transform (FFT) for the data obtained from each sensor during the test periods. A prominent peak component was observed at the test stimulus frequency. Therefore, the strengths of MEG responses at the frequency of the test stimulus (2 or 16 Hz) were analyzed.

Fig. 4 shows the power of MEG signals for the test stimuli under the different adaptation conditions. The results showed that the strength of the MEG responses for the flickering light of 2 Hz decreased more after adapting to the 2 Hz than after adapting to 16 Hz. Also, the strength of the MEG responses for the flickering light of 16 Hz decreased more after adapting to 16 Hz than after adapting to 2 Hz.

B. Adaptation Effects on AM Flicker

An AM flickering light with a modulation frequency of 2 Hz and a carrier frequency of 16 Hz was used as the test stimuli. The modulation depth was set at 0.8. The adaptation stimuli and the MEG recording procedures were identical to those for the sinusoidal flickering lights. Power spectra of the data during the test periods were calculated. Prominent peak components were observed at the frequency corresponding to the envelope periodicity as well as at the carrier frequency of the AM flickering light. Therefore, the strengths of MEG responses at the frequency corresponding to the envelope periodicity were analyzed.

Fig. 5 shows the power of MEG signals for the envelope periodicity under the different adaptation conditions. The results showed that the strength of the MEG responses decreased more after adapting to 2 Hz (envelope periodicity) than after adapting to 16 Hz (carrier frequency).

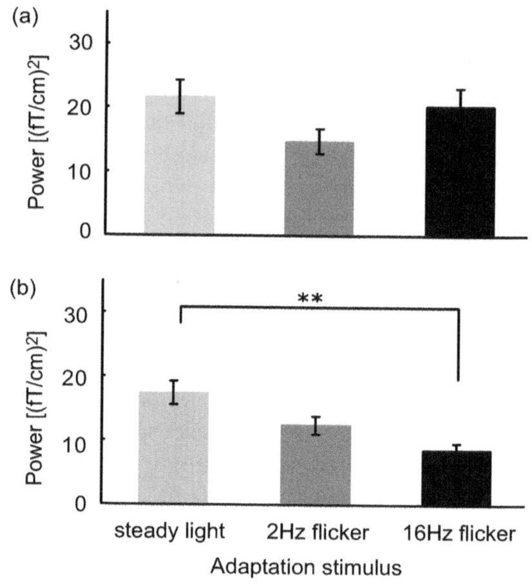

Fig. 4 Strength of MEG responses to sinusoidal flickering lights of (a) 2 Hz and (b) 16 Hz. Error bars indicate SEM (**p<0.01)

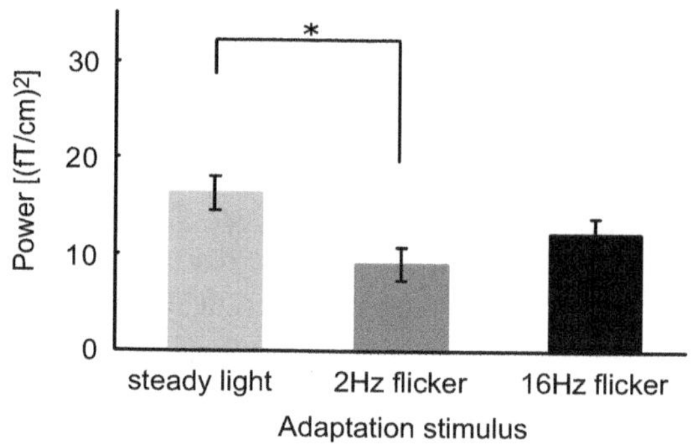

Fig. 5 Strength of MEG responses to flicker with the envelope periodicity of the AM flickering light. Error bars indicate SEM (*p<0.05)

IV. DISCUSSION

The detection threshold to the sinusoidal flicker at the low (or high) temporal frequency increased more when the subjects adapted to the low (or high) temporal frequency than when they adapted to the high (or low) temporal frequency, as the previous studies suggested [10, 11]. Also, the MEG responses to the sinusoidal flicker at the low (or high) temporal frequency decreased more after adapting to the low (or high) temporal frequency than after adapting to the high (or low) temporal frequency. The result that the adaptation effect to the low (or high) temporal frequency was greater after adapting to the low (or high) temporal frequency was due to the reduction of sensitivity of the low- (or high-) frequency channel by adapting to the low (or high) frequency.

For the flicker with the envelope periodicity of the AM flickering light, detection threshold increased and MEG responses decreased more when the subjects adapted to the frequency corresponding to the envelope periodicity (the low frequency) than when they adapted to the carrier frequency (the high frequency). These results indicated that the adaptation effect on the flicker with the envelope periodicity was greater when the visual sensitivity to the frequency corresponding to the envelope periodicity was reduced than when that to the carrier frequency was reduced, even though the AM flickering light has a frequency component at the carrier frequency but not at the frequency corresponding to the envelope periodicity. The present results showed that the envelope component was affected by the low-frequency channel. Therefore, this interpretation suggests the possibility that the envelope component of AM flickering light arises before or while the AM flickering signal passes the low-frequency channel.

Regarding to the two temporal channels in the human visual system, it was suggested that the first stage involves low-pass filtering by photoreceptors, while the second involves high-pass filtering associated with neural transmission, mainly due to lateral antagonism in the retina but also potentially in the brain [14]. If we follow their suggestion, the present results suggest the possibility that the envelope component arises before the signal passes both of the two temporal channels. However, the previous result that the sensitivity to flicker with the envelope periodicity decreased when the carrier frequency of AM flickering light was high [6] suggests that the envelope component arises after the process of low-pass filtering. This is not consistent with the suggested possibility that the envelope component is generated in the visual pathway before both of the temporal channels. However, properties and a configuration of the two temporal channels are not yet clear. For example, a recent study suggested that there is a high band-pass channel before a low band-pass channel [13]. Therefore, we need to take some proposals for the two temporal channels into consideration to estimate the site where the envelope extraction is careered out.

Acknowledgment

This study was supported by a Grant-in-Aid for JSPS Fellows from the Japan Society for the Promotion of Science.

References

1. Hammett, S.T., and Smith, A.T. (1994). Temporal beats in the human visual system. Vision Research 34, 2833-2840
2. Burns, S.A., Elsner, A.E., and Kreitz, M.R. (1992). Analysis of nonlinearities in the flicker ERG. Optometry and Visual Science 69, 95-105.
3. Burns, S.A., and Elsner, A.E. (1996). Response of retina at low temporal frequencies. Journal of the Optical Society of America A 13, 667-672.
4. Gorea, A., Wardak, C., and Lorenzi, C. (2000). Visual sensitivity to temporal modulations of temporal noise. Vision Research 40, 3817-3822.
5. Stockman, A., and Plummer, J. (1998). Color from invisible flicker: a failure of the Talbot-Plateau law caused by an early 'hard' saturating nonlinearity used to partition the human short-wave cone pathway. Vision Research 38, 3703-3728.
6. Okamoto, Y., Nakagawa, S., Fujii, K., and Yano, T. (2009). Visual sensitivity and cortical response to the temporal envelope of amplitude-modulated flicker. Journal of the Optical Society of America A. 26, 2346-2352.
7. Baitch, L.W., and Levi, D.M. (1988). Evidence for nonlinear binocular interactions in human visual cortex, Vision Research 28, 1139-1143
8. Campbell, F.W., and Robson, J.G. (1968). Application of Fourier analysis to the visibility of gratings. Journal of Physiology 197, 551-566.
9. Moulden, B., Renshaw, J., and Mather, G. (1984). Two channels for flicker in the human visual system. Perception 13, 387-400.
10. Hammett, S.T., and Snowden, R.J. (1995). The effect of contrast adaptation on briefly presented stimuli. Vision Research 35, 1721-1725.
11. Anderson, S.J., and Burr, D.C. (1985). Spatial and temporal selectivity of the human motion detection system. Vision Research 25, 1147-1154.
12. Snowden, R.J., Hess, R.F., and Waugh, S.J. (1995). The processing of temporal modulation at different levels of retinal illuminance. Vision Research 35, 775-789.
13. Cass, J., and Alais, D. (2006). Evidence for two interacting temporal channnels in human visual processing. Vision Research 46, 2859-2868.
14. Rovamo, J., Raninen, A., and Donner, K. (1999). The effects of temporal noise and retinal illuminance on foveal flicker sensitivity. Vision Research 39, 533-550.

Author: Yosuke Okamoto
Institute: National Institute of Advanced Industrial Science and Technology (AIST)
Street: 1-8-31 Midorigaoka
City: Ikeda, Osaka
Country: Japan
Email: yos-okamoto@aist.go.jp

Long-Range Coupling of Prefrontal Cortex and Visual (MT) or Polysensory (STP) Cortical Areas in Motion Perception

Lucia Maria Vaina[1,2], Finnegan Calabro[2], Fa-Hsuan Lin[1], and Matti S. Hämäläinen[1]

[1] Athinoula A. Martinos Center for Biomedical Imaging, Massachusetts General Hospital, Charlestown, U.S.A.
[2] Brain and Vision Research laboratory/Biomedical Engineering, Boston University, Boston, U.S.A.

Abstract— **To investigate how, where and when moving auditory cues interact with the perception of object-motion during self-motion, we conducted psychophysical, MEG, and fMRI experiments in which the subjects viewed nine textured objects during simulated forward self-motion. On each trial, one object was randomly assigned its own looming motion within the scene. Subjects reported which of four labeled objects had independent motion within the scene in two conditions: (1) visual information only and (2) with additional moving-auditory cue.**

In MEG, comparison of the two conditions showed: (i) MT activity is similar across conditions, (ii) late after the stimulus presentation there is additional activity in the auditory cue condition ventral to MT, (iii) with the auditory cue, the right auditory cortex (AC) shows early activity together with STS, (iv) these two activities have different time courses and the STS signals occur later in the epoch together with frontal activity in the right hemisphere, (v) for the visual-only condition activity in PPC (posterior parietal cortex) is stronger than in the auditory-cue condition. fMRI conducted for visual-only condition reveals activations in a network of parietal and frontal areas and in MT.

In addition, Dynamic Granger Causality analysis showed for auditory cues a strong connection of the AC with STP but not with MT suggesting binding of visual and auditory information at STP. Also, while in the visual-only condition PFC is connected with MT, in the auditory-cue condition PFC is con-nected to STP (superior temporal polysensory) area.

These results indicate that PFC allocates attention to the "object" as a whole, in STP to a moving visual-auditory object, and in MT to a moving visual object.

Keywords— **visual search, auditory facilitation, MEG, dynamic Granger causality.**

I. Introduction

Current models of brain organization suggest a hierarchical progression of processing from simple to complex aspects of unimodal stimuli within visual and auditory systems [1, 2]. However, recent studies of connection patterns and physiological data suggest that only a small fraction of the cortex is truly unimodal. Convergence between sensory systems occurs even at the early putative unisensory processing stages and at extremely short post-stimulus latencies, arising through either feedback or feed-forward circuits. In multi-sensory processing [3, 4] recruitment of the early sensory areas into the cognitive-perceptual processes can be partly attributed to feedback mechanisms. This calls for investigating the auditory and visual processing hierarchies in both temporal and spatial dimensions. This is what we have addressed in this study through MEG in healthy human observers performing search tasks in a 3Dmotion field.

Motion is pervasive in our everyday environments, and perceiving the direction and speed of our own and object motion is essential for functioning effectively, such as walking, coping with traffic, or even detecting objects on our path. These are fundamental everyday tasks which are often helped by additional cues, especially auditory. The combined use of psychophysics and brain imaging techniques is essential to discover the underlying principles governing cross-modal integration during the perception of motion. A grasp of these fundamental principles is not only critical for understanding the function of the healthy brain but, importantly, brings the promise of designing effective strategies to help patients recover and deal with their challenging everyday life environments.

Employing unimodal (visual) and cross-modal (visual and auditory) tasks we investigated psychophysically whether the auditory cues enhance perception of object motion in a search task simulating a forward moving observer searching among a set of moving objects an object moving in depth. Our fundamental hypothesis is that simultaneous congruent information from another sensory modality may facilitate the perception of object motion. Using anatomically constrained MEG in conjunction with dynamic time and frequency domain Granger causality we investigated the direction and dynamics of connectivity between the functional areas involved in selective, unimodal visual attention and cross-modal attention and facilitation for visual perception [5].

II. Materials and Methods

A. Subjects

Three experimental conditions were used: vision only (in both psychophysics and MEG), vision and non-spatially localized auditory cue (psychophysics and MEG) and visual

S. Supek and A. Sušac (Eds.): Advances in Biomagnetism – BIOMAG2010, IFMBE Proceedings 28, pp. 298–301, 2010.
www.springerlink.com

with spatially localized auditory cues (psychophysics only). Subjects participating in MEG also had a high resolution structural MRI and fMRI for localizing the MT+ region.

Eleven subjects, recruited from graduate and undergraduate college students whose age ranged between 19-29 years (7 women, 4 men) participated in the first two experimental conditions and 9 subjects (ages ranging between 20-28 years; 5women, 4 men) participated in the third experimental condition. All subjects gave Informed Consent according to the Boston University and Martinos Center for Biomedical Imaging Human Subjects Committees.

None of the participants had any history of neurological, psychiatric or ophthalmological/neuro-ophthalmological diseases, or were on any medication, they were all right handed and had normal to corrected vision. All the subjects who participated in the imaging experiments satisfied the inclusion criteria for participation in imaging, according to the Martinos Center for Biomedical Imaging requirements.

B. Psychophysics

To address whether moving auditory cues enhance the perception of object-motion during self-motion, subjects viewed nine computer generated textured spheres (1.5° in diameter) during 1sec simulated forward self-motion. On each trial, one of the spheres was randomly assigned its own looming motion (forward or backward) within the scene. In the last frame of the motion four of the spheres were labeled from 1-4. Subjects were asked to report which of four labelled spheres had independent motion within the scene (4AFC task), with (1) visual information only, (2) an auditory cue localized to the center of the screen, either moving in the same direction as the target ("Auditory moving") or stationary ("Auditory static"), or (3) an auditory stimulus co-localized to the target object (moving or stationary).

The visual stimuli were displayed on a calibrated 23" LCD screen. Auditory stimuli were presented via Bose QuietComfort noise canceling headphones. Subjects gave responses via a USB number pad and were instructed to maintain fixation throughout the testing block.

C. MEG and MRI Acquisition

MEG data were acquired at the Martinos Center using a 306-channel MEG (VectorView, Elekta-Neuromag, Helsinki, Finland) and in a magnetically shielded room. The MEG signals were band-pass filtered to the frequency range 0.03–200 Hz and digitized at 600 samples/s. Electro-oculogram (EOG) were recorded to monitor eye-movement artifacts. All epochs exceeding 150 μV or 3000 fT/cm at any EOG or MEG channel, respectively, were automatically discarded from the offline averages computed after filtering the data to 0.1…40 Hz.

For stimulus presentation, we employed a DLP projector and a back-projection screen, plastic tubes connected to frequency-compensated loudspeakers outside the magnetically shielded room (ADU1c, Unides Design, Helsinki). Responses were recorded with a fiber-optic response pad.

We also acquired a high-resolution 3D T1-weighted structural MRI using a 32 channel phase array head coil in a 3T scanner (Siemens-Trio, Erlagen, Germany). Using standard functional localizers in fMRI we localized the area hMT+. fMRI data was analyzed using Freesurfer.

D. MEG Source Estimates

Cortically-constrained depth-weighted l2 minimum-norm estimates were computed using the MNE software (http://www.nmr.mgh.harvard.edu/martinos/userInfo/data/sofMNE.php). The sources were restricted to the white matter surface extracted with FreeSurfer software (http://surfer.nmr.mgh.harvard.edu/). The dense triangulation of the cortex was decimated to about 9000 sources per hemisphere resulting in an average distance of 4 mm between adjacent sources. The orientations of the sources were approximately constrained to the cortical surface normal direction using the loose orientation constraint approach. For computation of the Dynamic Granger Causality measures of association between cortical regions, the source waveforms were extracted from each anatomical region of interest discussed in Section III.

E. Dynamic Granger Causality (DGC)

We estimated causal influences between the anatomical areas of interest delineated in MEG for each of the search experiment conditions. We obtained time series from regions-of-interest (ROIs) after MEG source localization as discussed above. For each pair of time series from two ROIs ("X" and "Y"), we calculated the Granger causality $G_{X\to Y}$ and $G_{Y\to X}$. We used the Schwarz's Bayesian Criterion to estimate the order of the AR model and stepwise least squares to estimate model parameters [6, 7]. Granger causality also has a frequency-domain representation [8]:

$$G_{X\to Y}(f) = -\ln(1-(\Sigma_{XX}-\Sigma_{XY}^2/\Sigma_{XY}^2)|H_{YX}(f)|^2/S_{YY}(f)),$$

where the subscripts are entries in the matrix, $S(f)=H(f)\Sigma H^H(f)$ is the spectral matrix, $H(f)=(\sum_{k=0}^{p} A_k \exp(j2\pi kf))^{-1}$ is the system transfer function computed from the coefficient matrix A_k, and Σ is the noise covariance matrix of $\left[\varepsilon_{y,x}(t)\,\varepsilon_{x,x}(t)\right]^T$ in the AR model.

To develop a DGC model, similar to the published study by Lin et al. [5], we used a time-window approach in both the time and frequency domains. Granger causality depends on the estimated optimal order of the AR model. This means that the DGC model must incorporate simultaneous estimates of both the AR model order and the window length. For this we will use the SURE metric [5] obtained by formulating the time series as a dynamic AR model. Thus if we use a univariate time series: $y(t)=\sum_{k=1}^{p}\alpha_k^w(t)y(t-k)+\varepsilon_y^w(t)$, where $\alpha_k^w(t)$ is the time-varying AR model estimated from the least squares criterion: $\alpha_k^w(t)=\arg\min\sum_{w}(y(t)-y(t-k))^2$ and w denotes the length of the temporal window, and p the order of the AR model. The residual variance of this dynamic AR model fit is estimated as $\varepsilon_{p,w}^2=\sum\left(\sum_{k=1}^{p}\alpha_k^w(t)y(t-k)-y(t)\right)^2 \big/ T$ and the SURE metric can be written as $SURE(p,w)=\log\varepsilon_{p,w}^2+\frac{2p\alpha}{w}$, where $\alpha=\left(\sum_{t=1}^{T}y_{t,p}^T M^{-1}(t)y_{t,p}\right)/T$ is a modification term and $M(t)=\left(\sum_{q=0}^{w-1}y_{t-q,p}y_{t-q,p}^T\right)/w$, with $y_{t-q,p}=[y(t-q-1),y(t-q-2),\ldots,y(t-q-p)]^T$. We designate the window length corresponding to minimal SURE metric as optimal. *Since our causality model takes into account both time and frequency domains it would help elucidate the dynamics of cortical spatiotemporal orchestration.*

We used this approach to detect modulation of effective connectivity during visual and visual-auditory search in order to reveal the spatiotemporal cortical network involved in these tasks. We were particularly interested to determine whether the correlations between auditory and visual cortices are mediated by he same connections.

III. Results

A. Psychophysics

In the "auditory off" condition, performance depended on the speed and direction of the object: in both experimental conditions subjects were significantly better when the target object moved towards them (positive speeds) than away, when retinal velocities were larger, suggesting that relative and retinal speed affected performance.

Non-spatial and spatially localized auditory cues were either stationary (25% of trials) or moving in the same direction as the target. With the non-spatial cue (centered at fixation), subjects showed no statistical difference between the auditory off and auditory static conditions (p=0.2, t=-1.38, df=8, Fig 1 left). For the moving auditory cue, there was a modest but statistically significant improvement of 3.3% compared to the static auditory cue (p=0.003, t=4.18, df=8, min 1.2% and max 8.1, Fig 1 right).

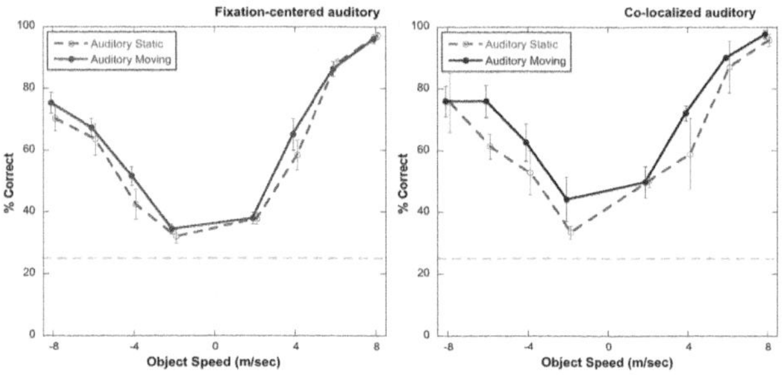

Fig. 1 Results from healthy controls (n=11) for the detection of object motion during forward self-motion. Performance significantly improved for an auditory cue moving in the same direction as the target compared to an uninformative, static auditory cue for both fixation-centered auditory cues (p=0.003, left), and spatially localized auditory cues (p=0.04, right). The co-localized cue was nearly twice as effective as fixation-centered at enhancing detection relative to a static auditory condition (p=0.06, t=2.1, df=10). Dashed line is chance performance (25%), and error bars are s.e.m

B. Dynamic Granger Causality (DGC)

Pair-wise DGC [5] was computed using the MEG source time courses. Nine regions-of-interest (ROIs) were identified: the left and right auditory cortices, left and right medial prefrontal cortices (MPFC), left and right MT, right posterior IPS complex, right post-central gyrus, and right superior temporal poly-sensory area (STP). In each condition, we calculated the average time courses across source locations in each ROI for each trial separately. Randomly selected 50% of trials were averaged to generate one evoked time course for each ROI. We used a 100 ms time window and 5-th order autoregressive model to calculate dynamic Granger causality using these evoked time courses. The random selection across trials was re-peated 30 times. Statistical significance of each causality estimate was evaluated by (i) averaging all DGC over 30 repetitions, (ii) subtracting the average baseline (-0.2 s to 0 s) DGC, and (iii) dividing each time point with the standard deviation of the DGC in the baseline to transform the DGC time courses calculated from each ROI pair to a t-statistic.

In Figs 2 and 3 we show differences in connectivity between the *Visual-auditory* and *Visual-only* tasks. In Fig 2 we illustrate that in the presence of the auditory cue, the auditory cortex and the polymodal area STP are strongly connected very early on, connection which is not present in the Visual-only task (Fig 2D).

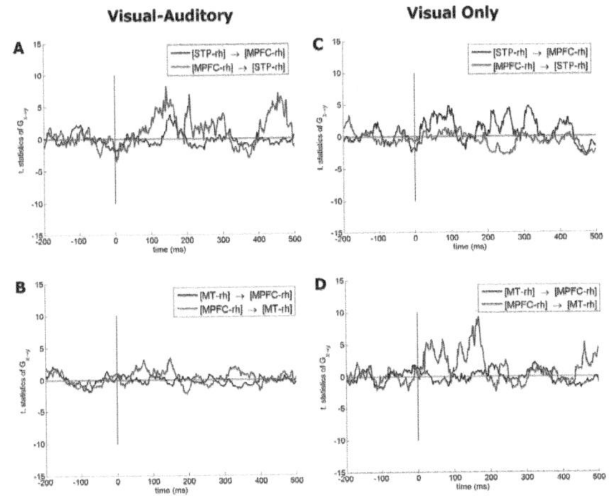

Fig. 2 Feedback connectivity from PFC is directed to MT+ in the visual-only condition (D), but moves to STP when the target is multi-sensory (A)

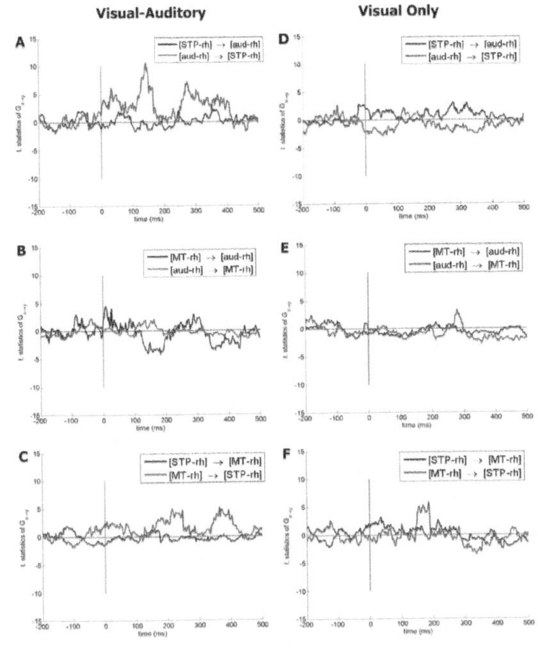

Fig. 3 Dynamic connectivity shows for an illustrative single subject. t-statistics show that both auditory cortex (top, visual- auditory condition only) and MT+ (bottom) input to STP, but there is no direct connectivity between auditory cortex and MT+ (middle)

Figure 3D shows that in the visual only task, the prefrontal cortex provides a very early input to MT+ and a 100 delay input to STP in the visual-auditory tasks, which suggests a top-down influence. It is interesting that the auditory cortex and MT (significantly involved in the motion task) are not connected (Fig. 3) even in the presence of the auditory cue.

IV. Conclusions

These results suggest that the binding of visual and auditory information may occur in STP, and this area is probably involved in the mutimodal task. The visual information from MT appears to arrive to STP later than the auditory information. Figure 2, addresses the question of attention modulation by the prefrontal cortex. Fig 2A shows that the PFC is strongly connected with STP in the Visual-auditory task, where the target is a multimodal object. PFC will modulate STP during the selection of the target. However, this connectivity is not established in the visual-only task where PFC is directly and strongly connected with area MT+ whose neurons are significantly involved in the perception of this stimulus and detecting the target-object. In the visual-auditory task MT+ is not connected with PFC, nor with the auditory cortex. These preliminary results show that the prefrontal cortex allocates attention to the "object" as a whole, in STP to a moving visual-auditory object, and in MT to a moving visual object.

Acknowledgments

This work was supported by NIH grant R01NS064100 to LMV, and the NCRR supported Center for Functional Neuroimaging Technologies P41 RR14075 (MSH, FHL).

References

1. Felleman, D.J. and D.C. Van Essen, Distributed hierarchical processing in the primate cerebral cortex. Cereb Cortex, 1991. 1(1): p. 1-47.
2. Rauschecker, J.P., et al., Serial and parallel processing in rhesus monkey auditory cortex. J Comp Neurol, 1997. 382(1): p. 89-103.
3. Schroeder, C.E., A.D. Mehta, and J.J. Foxe, Determinants and mechanisms of attentional modulation of neural processing. Front Biosci, 2001. 6: p. D672-84.
4. Bullier, J., Feedback connections and conscious vision. Trends Cogn Sci, 2001. 5(9): p. 369-370.
5. Lin, F.H., et al., Dynamic Granger-Geweke causality modeling with application to interictal spike propagation. Hum Brain Mapp, 2009. 30(6): p. 1877-86.
6. Neumaier, A. and T. Schneider, Estimation of parameters and eigenmodes of multivariate autoregressive models. ACM Trans. Math. Softw., 2001. 27: p. 27-57.
7. Schneider, T. and A. Neumaier, Algorithm 808: ARfit - A Matlab package for the estimation of parameters and eigenmodes of multivariate autoregressive models. ACM Trans. Math. Softw., 2001. 27: p. 58-65.
8. Brovelli, A., et al., Beta oscillations in a large-scale sensorimotor cortical network: directional influences revealed by Granger causal-ity. Proc Natl Acad Sci U S A, 2004. 101(26): p. 9849-54.

Author: Lucia Maria Vaina
Institute: Boston University-Biomedical Engineeering
Street: 44 Cummington St
City: Boston, Ma 02215
Country: USA
Email: vaina@bu.ede

In Search of Neural Signatures of Visual Binding : A MEG/SSVEF Study

Charles Aissani[1,2,3], Benoit Cottereau[1,2,3], Anee-Lise Paradis[1,2,3], and Jean Lorenceau[1,2,3]

[1] UPMC, Centre de Recherche de L'Institut du Cerveau et de la Moëlle épinière (Cricm), Paris, France
[2] CNRS, Cricm UMR 7225, Paris, France
[3] Inserm, Cricm UMR-S975, Paris, France

Abstract— **Visual processes are distributed in numerous specialized cortical areas. How neural responses are bound into assemblies to elicit a unified perception remains a central issue in cognitive neurosciences. We address this issue in an MEG experiment aiming at characterizing neural mechanisms of coherent motion integration.**

Stimuli consisted i2 vertical and 2 horizontal bars arranged in a square like shape and oscillating at 2.3Hz (f1) and 3Hz (f2), respectively. Such periodic stimulations elicit stereotypical oscillatory MEG activities phase-locked to visual stimulation at fundamental and superior harmonics in cortical areas processing the stimulus elements (Steady State Visual Evoked Field; SSVEF). Subjects' perception, either a rigidly or non-rigidly moving square, was modulated by subtle changes in luminance distribution along the bars, resulting in 4 experimental conditions. This design allows separating SSVEF relative to element computation (f, 2f, 4f…etc) and SSVEF linked to motion integration at intermodulation frequencies (nf1 + nf2).

MEG analyses reveal focal and percept independent activities at first harmonics on occipital sensors; sources reconstruction shows retinotopic segregation between 2.3Hz and 3Hz activities in V1 as well as widespread activities at second harmonics over occipital and parietal sensors. Contrasting coherent and incoherent percepts showed fourth-order intermodulation frequency enhancement localized in the right frontal cortex, likely the frontal eye field (FEF). No significant modulation in the gamma band was observed.

Overall, tagged stimulation and SSVEF analyses allow a precise localization of cortical activities related to the computation of focal visual elements, consistent with retinotopy. This study further highlights an electrophysiological signature of shape processing from ambiguous motion components at fourth-order intermodulation frequency, presumably in FEF.

Keywords— **SSVEF, form, motion, intermodulation, FEF.**

I. INTROCUCTION

One of the most challenging issues in visual cognitive neurosciences is to understand how a massively parallel and distributed visual system effortlessly binds neural responses elicited by different features into a unified percept. Amongst the different theoretical solutions to this «binding problem », the Binding-By-Synchrony hypothesis [1] proposes that synchronized activity in the gamma band is a temporal code for binding features into objects, each being characterized by a particular assembly of neurons firing in synchrony.

The aim of the present study was to explore new kinds of neurophysiological signatures of 2D form from motion integration using Steady State Visual Evoked Potential / Field (SSVEP/F) with MEG. Our approach builds upon the observation that a periodic stimulation elicits a phase locked response at the fundamental and superior harmonics of the stimulation frequency in relevant cortical areas. When two stimulations at distinct frequencies *f*1 and *f*2 are integrated to form a unified percept, the response in relevant areas may contain Fourier components at various combination frequencies $n1\ f1 + n2\ f2$ [2]. As a matter of fact, a recent work using two tagging frequencies within a binocular rivalry paradigm (2 frequencies by monocular stimulation), showed significant intermodulation frequencies associated with subject' integrate perceptions [3].

In this study, we recorded MEG activity elicited by oscillatory bar motion at two different frequencies that could, or not, be integrated into a coherently moving square-like shape in order to determine the extent to which cortical activity at intermodulation frequencies is an electrophysiological marker of sensory integration.

II. MATERIAL AND METHODS

A. Experimental Procedure

Participants: twelve naive subjects with normal vision participated in the study. All subjects provided informed written consent and were paid for their participation. The study was approved by the Local Ethical Committee.

Procedure and stimuli: Stimuli consisted of 2 horizontal and 2 vertical bars positioned at 3.0° of visual angle from a fixation point localized at the centre of the screen, arranged so as to form a square shape of 6.0° X 6.0° without visible corners. Four conditions, differing only by different luminance distributions along the bars, were used to

S. Supek and A. Sušac (Eds.): Advances in Biomagnetism – BIOMAG2010, IFMBE Proceedings 28, pp. 302–305, 2010.
www.springerlink.com

modulate perceptual integration: from coherent to incoherent motion [4]. The mean luminance of the display remained constant throughout the experiment. The screen was positioned at 85 cm from subjects' eyes; the stimuli were displayed on a calibrated video projector (resolution, 1024x768 pixels; refresh rate, 60 Hz) located outside the shielded recording room.

Each trial followed four steps (fig. 1): during the first 1.5 second (t=-1.5 to t=0), only the fixation point was presented on the screen. At t = 0, the four static bars were displayed for a random duration (between 450 and 550 ms). This static phase was directly followed by a 1.2 s. time window during which both pairs of bars moved sinusoidally along their orthogonal axis. Oscillation frequencies, f1=2.3 Hz and f2= 3 Hz for the horizontal and vertical bars respectively, were chosen so that: i. the phase locked response of neurons to vertical and horizontal motions could be discriminated, ii. the intermodulation frequency, 2f1x2f2=10.6 Hz, did not coincide with harmonics of either component. Under these conditions, a coherent motion corresponds to a rigid square moving along a Lissajou's trajectory, while incoherent motion appears as 2 pairs of bars moving independently along the vertical and horizontal axes. After motion ends, subjects coded their percept using three response buttons: coherent, incoherent or intermediate percept. In order to minimize the artifacts associated to motor preparation, the positions of the response codes varied randomly across trials.

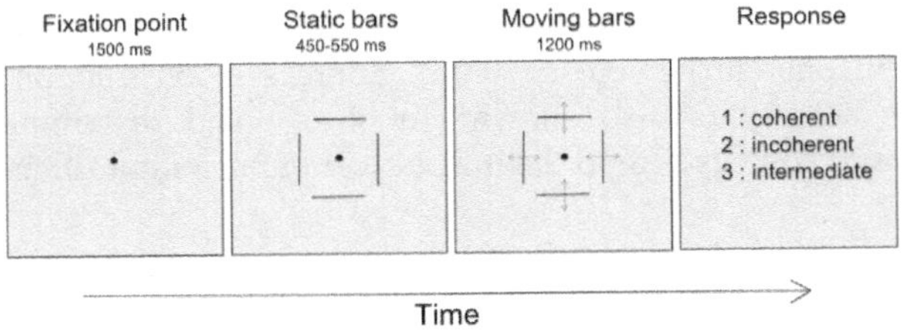

Fig. 1 Time flow of a trial

B. *MEG Analyses*

Recording: continuous magnetoencephalographic signals were collected using a whole-head MEG system with 151 axial gradiometers (CTF Systems, Port Coquitlam, British Columbia, Canada) at a sampling rate of 1250 Hz and were low-pass filtered on-line at 300 Hz. Head position was tracked during the experiment with respect to the MEG sensors using marker coils that were placed at the cardinal points of the head (nasion, left and right ear). Eye position was recorded with an ISCAN eye-tracking system. Recording also included the signal of a photodiode that detected the actual appearance of the stimuli on the screen. This allowed to average event-related magnetic fields (ERFs) precisely time-locked to the actual stimulus onset.

Trials contaminated with eye movements, blinks, or muscular artifacts were rejected off-line on visual inspection of their unfiltered ISCAN and MEG traces.

Time-frequency analyses: Time-frequency plots were produced using Morlet wavelet analysis. We used wavelet family with a m parameter of 10 (Wf = 0.235f and Wt = 3.74/f).

Sources reconstruction: The estimation of the cortical currents was performed according to a distributed approach where 10,000 sources localized on the cortical manifold of the Colin27 template from the Montreal Neurological Institute (http://www.mni.mcgill.ca/) were reconstructed. The forward model was developed with Brainstorm using a spherical head geometry. The inverse modeling was performed using a classical minimum-norm approach [5].

Spectral Analysis: Once the activities of all the elementary sources of the Colin cortical tessellation were obtained, we computed the power spectrum densities of the corresponding time-courses using the Welch periodogramm. The estimations of the active sources were approached by contrasting the power during the last 1030ms of the stimulation with the power during the baseline, defined as the 1030ms time window starting 1.7 seconds before the beginning of the motion. The contrast was performed using a base-2 log transformed increase of signal power relative to baseline.

III. Results

A. *Behavioral Results*

We first quantified how luminance distribution along the bars drove subjects' perception. Figure 1 shows percept response rate for the four conditions. Coherent and incoherent percepts were respectively increased ($p<0.01$) and decreased ($p<0.01$) with relative low contrast line-end. This confirms previous results [4]

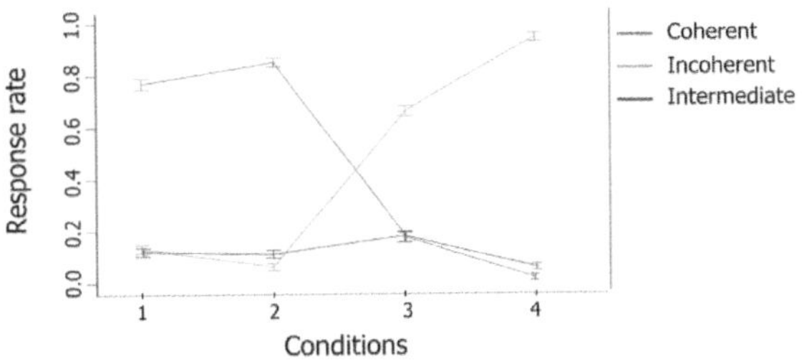

Fig. 2 Average response of 11 subjects for the four conditions

B. MEG Results

We first determined the overall cortical activity at harmonics of the stimulation frequency, independent of the experimental conditions. We used evoked fields calculated on all trials to finely study sensors topographies and cortical sources which produced fundamental and superior harmonic activities.

We found that fundamental power was restricted to several occipital sensors whereas 2nd harmonic power was distributed over numerous occipital and temporal sensors. We then reconstructed the cortical sources corresponding to the sensor topographies, while taking into account the position of observers' head for each run.

Notably, we observed a retinotopic segregation around the Calcarine scissure between fundamental activities (f1 and f2), corresponding to each frequency and therefore to each pair of bars (Fig 3) This latter result confirms that precise localization of the cortical structures encoding a stimulus is facilitated by the use of SSVEF and source reconstruction around the relevant frequencies.

In addition, we found widespread sources in parietal and occipital cortices with significant activities at 2nd harmonics ($p<10^{-6}$, not shown). Coherent and incoherent percepts did not yield significant differences for these frequencies.

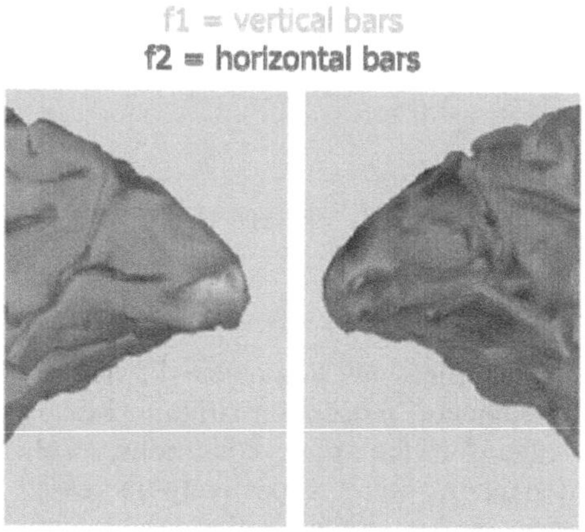

Fig. 3 cortical sources with significant activity (p<10-6) for f1 (vertical bar, yellow) and f2 (horizontal bars, red)

To test whether intermodulation frequencies represent an electrophysiological signature of 2D form from motion visual binding, we analyzed power modulation at these frequencies according to subjects' perception., Using Welch periodograms, we found increased 2f1+2f2 power during coherent perception in contrast with incoherent perception.

Figure 4a shows power differences between these two percept averaged around 10.6Hz +/- 0.2Hz for the last 1030 ms of stimulus motion. Notably, an increase of the 4th order intermodulation frequency is present on several right frontal sensors.

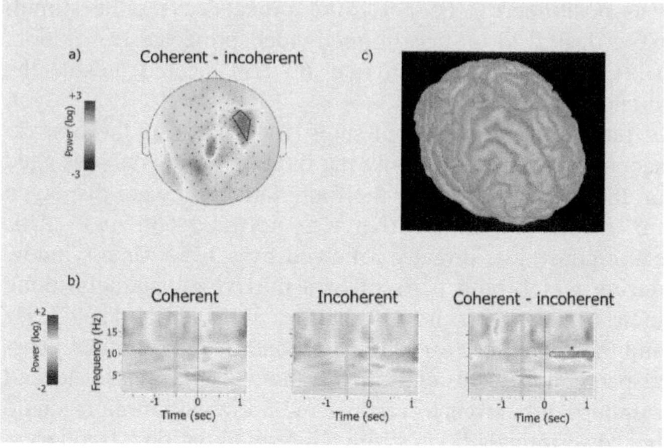

Fig. 4 a) Average topography of 2f1 + 2f2 +/- 0.2 Hz power-log activity for the last 1030 ms of moving stimulation (coherent – incoherent). b) time-frequency plots average across 5 sensors delineated in a) of power log activities for coherent, incoherent ERF and their differences. Significant cortical sources are shown in c) (paired t-test $p<10^{-4}$)

We further assessed the sustained character of this perception dependent activity by considering the time-frequency plots averaged across the five sensors delineated in Figure 4a. A paired t-test on the average power-log activity around 10.6Hz (+/-0.2Hz) for the last 1100 ms of stimulation confirms a significant increase of the 4th intermodulation frequency during a coherent as compared to an incoherent percept ($p<0.05$). Sources reconstruction of this acticity is also shown (fig 4.c)) and discriminate several sources behind the medial part of precentral sulcus.

C. Gamma Band Activity

To determine whether coherent motion perception is coupled to induced gamma band activity, we conducted a trial-by-trial frequency analysis according to subject perception. However, the observed power modulation in the gamma band was not significant.

D. Oculomotor Behaviour

Several studies suggest that FEF is involved in eye movement control [6]. Because the power of the 4th order intermodulation frequency that distinguish coherent and incoherent percepts is in the vicinity of FEF, we tested whether this difference could be related to the oculomotor behavior recorded simultaneously with the MEG recordings. However no spectral power modulation around the 1st and 2nd harmonics or around 10.6 Hz could be found. Although this might be due to an insufficient spatial precision of eye recordings, the 10.6Hz power modulation appears independent of macroscopic oculomotor behaviour.

IV. Discussion

Our results confirm the interest of SSVEF in localizing brain activities, as they nicely fit the expected retinotopic location of the stimulus components. The interpretation of f1 and f2 predominant power at the stimulation frequency in V1 remains however unclear. The back and forth periodic motion used herein should recruit different direction-selective neurons at each direction reversal and yield a response at harmonics of the stimulation frequency. One reason for the strong power found at the fundamental stimulation frequency may be weak MEG diffusion, which decreases spatial pooling summation on sensor recording. Alternately, f1 and f2 activity could be due to the particular characteristics of sinusoidal oscillating motion as compared to checkerboard counter phase stimulation.

We found 4th order intermodulation power increase on several right frontal sensors as a signature of 2D form from motion integration. Source reconstruction (fig 4.c)) indicates this activity corresponds to several sources anterior to the precentral sulcus, suggesting it might belong to FEF in regard of precedent localization of human FEF [6,7].

Although there is an agreement to consider that FEF is involved in eye movement control [8], several recent studies show FEF implication in perceptual processing, independently of any saccade preparation or production. In addition, FEF neurons have robust shape selectivity [9], as well as selectivity for direction and speed of motion [10]. Moreover, studies of connectivity suggest FEF is a "visual system hub"; in particular, reciprocal connections between FEF and many ventral and dorsal visual areas have been demonstrated [11]. Further, TMS stimulation of FEF has been found to improve subject ability to detect masked visual stimuli [12] and to improve MT excitability [13]. The present finding of enhanced power at intermodulation frequency and the lack of significant eye movements suggest that FEF could be involved in binding component bar motions into a global moving shape. Although one may wonders whether this finding nevertheless reflects attentional following during coherent motion, it remains unclear why this would entail activity at this particular frequency unless FEF integrates component motions across space and time.

V. Conclusions

The finding that activity at the intermodulation frequency (2f1+2f2) differentiates coherent and incoherent percept suggests that the use of multiple frequencies in SSVEF paradigm is a powerful method for identifying cortical structures involved in visual binding. Our results further point to unexpected functional processing roles of FEF that need additional exploration.

References

1. Gray, C.M., and Singer, W. (1989). Stimulus-specific neuronal oscillations in orientation columns of cat visual cortex. Proc. Natl. Acad. Sci. USA *86*, 1698–1702.
2. Sutoyo D and Srinivasan R, (2009). Nonlinear SSVEP responses are sensitive to the perceptual binding of visual hemifields during conventional 'eye' rivalry and interocular 'percept' rivalry .Brain Res.1251: 245-255
3. Regan, D., and Regan, M.P. (1988). Objective evidence for phaseindependent spatial frequency analysis in the human visual pathway.VisionRes.28,187–191.
4. J. Lorenceau and M. Shiffrar, (1992) The influence of terminators on motion integration across space, Vision Res. 32, 263–273.
5. B. Cottereau, K. Jerbi and S. Baillet (2007) Multiresolution imaging of MEG cortical sources using an explicit piecewise model, *NeuroImage* 38, pp. 439–451
6. Hagler D.J., Jr., and Sereno, M.I. (2006). Spatial maps in frontal and prefrontal cortex. Neuroimage 29, 567–577
7. Paus, T. Location and function of the human frontal eye-field: a selective review. Neuropsychologia, 1996, 34: 475–483.
8. J.D. Schall, (2002) The neural selection and control of saccades by the frontal eye field. *Philos. Trans. R. Soc. Lond. B Biol. Sci.* 357, pp. 1073–1082
9. Peng X, Sereno ME, Silva AK, Lehky SR, Sereno AB (2008) Shape selectivity in primate frontal eye field. J Neurophysiol 100:796–814
10. Xiao, Q., Barborica, A. & Ferrera, V.P. (2006) Radial motion bias in macaque frontal eye field. *Vis. Neurosci.* 23, 49–60.
11. Vezoli J, Falchier A, Jouve B, Knoblauch K, Young M, Kennedy H 2004 Quantitative analysis of connectivity in the visual cortex: extracting function from structure. Neuroscientist. 10, 476–482.
12. M.H. Grosbras and T. Paus, (2003) Transcranial magnetic stimulation of the human frontal eye field facilitates visual awareness, *Eur. J. Neurosci.* 18, pp. 3121–3126.
13. J. Silvanto, N. Lavie and V. Walsh (2006) Stimulation of human frontal eye fields modulates sensitivity of extrastriate visual cortex, *J. Neurophysiol.* 96, pp. 941–945.

Corresponding author:

Charles Aissani
Criem, équipe Cogimage - LENA
47, bd de l'hôpital
F-75013, Paris, France
Email: charles.aissani@upmc.fr

Repeatability of AEF and SEF from Static and Moving Head Positions

Jukka Nenonen, Lauri Parkkonen, Liisa Helle, Samu Taulu, and Antti Ahonen

Elekta Oy, P.O. Box 34, FIN-00531 Helsinki, Finland

Abstract— **We studied the degree of repeatability of auditory and somatosensory evoked fields (AEF and SEF). A single subject was studied repeatedly with a 306-channel Elekta Neuromag® system over a period of six years. In 21 recordings the subject kept his head as immobile as possible. Averaged AEF and SEF data were analyzed with equivalent current dipoles (ECDs). The mean 3D variations of the dipole localizations were 5.6 mm for SEF, 6.7 mm for ipsilateral AEF, and 5.9 mm for contralateral AEF.**

In addition, 17 recordings were performed with the subject moving his head continuously. In some of these recordings, additional interference was introduced by leaving the door of the magnetically shielded room open or by attaching magnetized pieces on the scalp. The ECDs obtained from the corresponding static head position served as the reference. After applying spatiotemporal signal space separation (tSSS) and head movement compensation (MC), the mean ECD localization differences to the reference localizations were 3.5 ± 2.6 mm for SEF, 4.0 ± 1.5 mm for ipsilateral AEF, and 4.0 ± 2.0 mm for contralateral AEF.

The combined tSSS and MC processing compensated effectively for the disturbances caused by head motion and additional interference sources. All ECD parameters were similar to the reference results for SEFs, but AEF dipole amplitudes decreased by 30% during head movement. Importantly, the dipole localization differences after tSSS and MC processing of data with motion and additional interference are smaller than the localization variability in the reference recordings.

Keywords— **Repeatability, Movement compensation, Signal space separation, AEF, SEF.**

I. Introduction

Magnetoencephalographic (MEG) responses, especially their latencies, are very replicable even after years [1]. The first studies of repeatability of MEG evoked responses were published in the 1990's using instruments with less than 20 channels (see, e.g, [2] and the references therein). Thereafter, relatively few reports have been published from repeated recordings with whole-head MEG devices (e.g., [2-6]). Auditory and somatosensory evoked fields (AEF and SEF, respectively) were analyzed with equivalent current dipoles in [2-5] and the reported 3D variances of the dipole localizations are summarized in Table 1 for AEF and Table 2 for SEF dipole results. In these studies, the subjects were instructed to keep the head still.

The Signal Space Separation (SSS) [7] is a novel signal processing method that, e.g., suppresses interference originating outside of the MEG sensor array. The spatiotemporal SSS (tSSS) method [8] improves the software shielding capability by removing artifacts caused by interference sources inside the sensor array, such as magnetized material on the head. In addition, the SSS method can be applied for transforming data between different head positions and for head movement compensation (MC) [9, 10].

Table 1 Variances of dipole localizations from repeated AEF measurements in other studies

Number of subjects and trials	Left hemisphere	Right hemisphere
Ref [2]: 5 subjects, 6 trials (monaural, left ear)	6.9 mm	3.5 mm
Ref [5]: 16 subjects, 2 trials (binaural)	8.5 mm	8.3 mm

Table 2 Variances of dipole localizations from repeated SEF measurements in other studies

Number of subjects and trials	
Ref [3]: 1 subject, 3 trials	4-7 mm
Ref [4]: 5 subjects, 2 trials	4-9 mm
Ref [5]: 16 subjects, 2 trials	8.4 mm

The aim of this study was to evaluate the repeatability of AEF and SEF from multiple MEG recordings of a single test subject. In addition, we evaluated the effectiveness of the combined tSSS and MC method, as suggested in [9], in recordings where the subject moved his head continuously.

II. Methods

A. Recordings

A healthy male volunteer was studied several times between the years 2003 and 2009 (42 yrs during the first recording) using the 306-channel Elekta Neuromag® systems at the BioMag Laboratory of Helsinki University Central Hospital and at the Low Temperature Laboratory of Helsinki University of Technology. The sampling rate and recording passband were 600 Hz and 0.1–172 Hz.

The subject was presented with a random sequence of visual (a flashing checkerboard pattern in one of the four quadrants), somatosensory (electrically to median nerve at the left or right wrist at the motor threshold, corresponding to 5–8 mA) and auditory (1-kHz 50-ms tone pips to the left or right ea) stimuli with an interstimulus interval of 600 ms. Presentation software (Neurobehavioral Systems, Inc., Albany, CA, USA) was used to produce the stimuli. In this study, we analyzed only the somatosensory evoked responses to the stimuli applied to the left median nerve, and the auditory evoked responses to right-ear stimulation.

In 21 sessions, the subject was first instructed to keep the head still during the recording. In 17 of those sessions, additional recordings were made when the subject moved the head continuously. Continuous head position tracking was applied, and the head positions were calculated off-line. In four recordings, additional interference was introduced by attaching magnetized pieces on the scalp or leaving the door of the magnetically shielded room open.

B. Interference Suppression

The reference recordings from static head positions were processed with the Signal Space Projection (SSP) method [11]. To improve the signal-to-noise ratio, evoked responses were averaged on-line in windows starting 100 ms before and ending 300 ms after stimulus. An epoch was rejected if the eye-movement EOG signal magnitude exceeded 200 µV, or a magnetometer or gradiometer signal magnitude exceeded, correspondingly, 5000 fT or 3000 fT/cm.

The SSS method [7] and its temporal extension, tSSS [8] were employed in the recordings with head movements. First, head positions were estimated from the continuous recordings [12]. The tSSS method was used to remove artifacts caused by head movements and additional interference sources. Finally, SSS was utilized to transform data from the estimated head positions into a static reference position. The tSSS and MC processing was performed with the MaxFilter^{TM} software (Elekta Oy). Thereafter, evoked responses were again selectively averaged.

C. Equivalent Current Dipole Analyses

SEF signals were modeled with a single equivalent current dipole (ECD) [13]. A homogeneous spherical head model was used, and the origin (0, 0, 60 mm) was optimized from the MR images of the subject. The baseline interval was set between -50 and 0 ms and a low-pass filter with a 100-Hz corner and a 10-Hz transition band was applied. The ECD position was fitted at the peak of the P30m response.

Two ECDs were used for AEF signals [13]. A low-pass filter with a 40-Hz corner and a 5-Hz transition band was applied. The peak response time of N100m was selected from the channel showing the largest signal amplitudes. Two channel selections consisting of about 150 channels were applied one at a time over each primary auditory response area. A single dipole at the maximum response amplitude was fitted for both hemispheres. After finding the dipole locations, all 306 channels were used to fit the dipole amplitudes as a function of time [13].

III. Results

A. Static Head Positions

The SEF and AEF dipole fitting results from 21 static head position recordings are collected in Table 3. The deviations of the mean XYZ-coordinates correspond to 3D variances of 5.6 mm for SEF, 6.7 mm for AEF right and 5.9 mm for AEF left hemisphere dipoles.

Table 3 ECD analysis results from 21 static head position recordings for SEF and left (L) and right (R) hemisphere AEF. The mean and standard deviations of the latency, XYZ coordinates, amplitude (Q), goodness of fit (G) and confidence volume (CV) are shown

	Lat [ms]	X [mm]	Y [mm]	Z [mm]	Q [nAm]	G [%]	CV [mm^3]
SEF							
mean	30.8	39.5	8.2	102.9	87.5	95.9	1.7
stdev	0.7	2.1	4.1	3.2	29.7	4.6	3.3
AEF R							
mean	96.2	61.5	4.3	61.6	49.0	85.7	2.3
stdev	2.8	3.6	4.2	3.9	9.7	4.5	3.8
AEF L							
mean	92.5	-58.5	-6.1	67.5	74.9	95.4	2.4
stdev	3.4	2.8	3.4	3.9	15.6	2.1	6.2

B. Interference Suppression and Movement Compensation

Examples of averaged responses before and after tSSS and MC are displayed in Figure 1. Corresponding contour maps from the same data are shown in Figure 2. In this case the subject had a magnetized piece attached on the scalp above the right somatosensory cortex and he was moving his head irregularly. The ECD results from the 17 recordings with head movements are summarized in Table 4.

Table 4 Mean and standard deviations of the dipole latency shift, localization difference and relative amplitude change. The ECD results were computed from the 17 averaged response sets after tSSS and MC and compared to the reference results from static head positions

	Shift [ms]	Diff [mm]	Q change [%]
SEF, mean	0.3	3.5	10.8
stdev	0.5	2.6	14.8
AEF R, mean	3.2	4.0	-28.1
stdev	2.0	1.5	16.1
AEF L, mean	2.6	4.0	-31.5
stdev	2.3	2.0	17.4

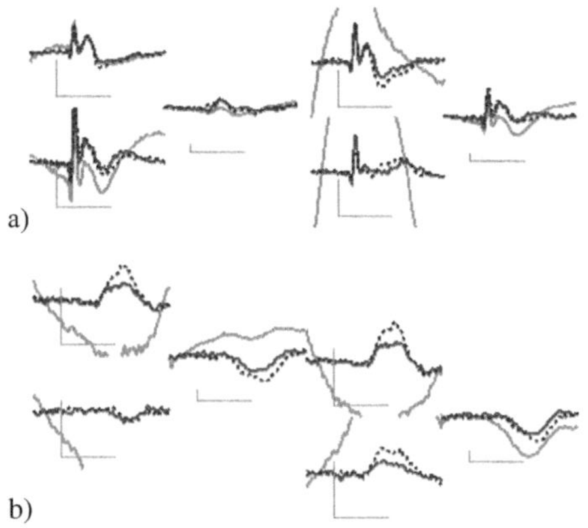

Fig. 1 Examples of averaged a) SEFs and b) AEFs in sensors over the primary response area on the right hemisphere. Four planar gradiometers and two magnetometers shown. Blue (dotted): Reference responses in a static head position (note that SSS was applied here instead of SSP). Gray: Head moving with attached magnetic particles below these sensors, spatial SSS, no MC. Red (solid): The previous responses after tSSS and MC. The AEF amplitudes decreased but the SEF amplitudes are similar to the reference responses. Scalebars correspond to 200 fT/cm or 200 fT and 100 ms

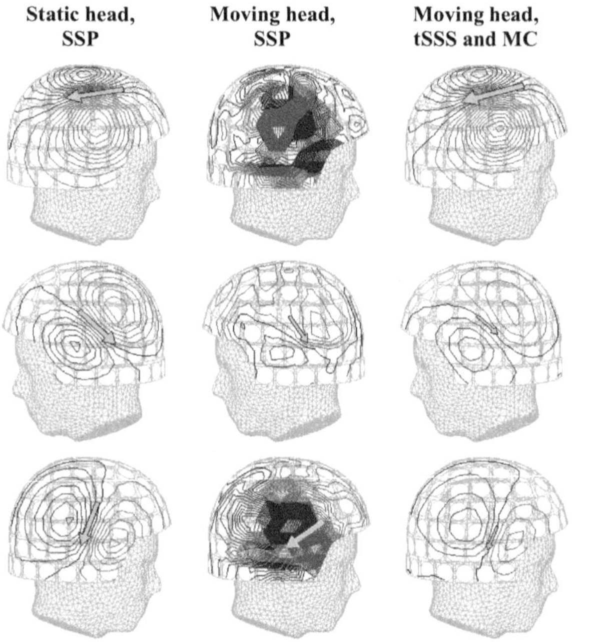

Fig. 2 Isocontour maps from the same data as in Figure 1. Top row: SEF patterns and dipoles at 30 ms. Middle row: AEF left hemisphere, bottom row: AEF right hemisphere patterns and dipoles at 90 ms after the stimulus. The step between adjacent contour lines is 100 fT

An example of the estimated head positions is given in Figure 3. The door of the magnetically shielded room was wide open and the subject was systematically drifting downward by about 5 cm in the course of the recording. Figure 4 and Table 5 present ECD localizations to P30m peaks of single non-averaged responses extracted from continuous raw data after tSSS without MC, and after combined tSSS and MC. Examples of the peak P30m field patterns in a single-trial response before and after movement compensation are shown in Figure 5.

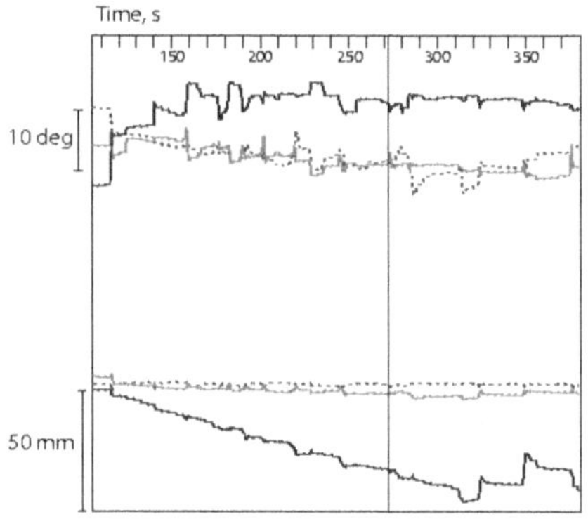

Fig. 3 Estimated head positions as a function of time. The lower curves represent the locations of the SSS expansion origin relative to the device coordinate origin: X (red dotted), Y (gray), Z (blue solid). The upper curves indicate rotation angles of the head around the coordinate axes: ax (red dotted), ay (gray), az (blue solid). The head was gradually moving downward (descending Z-coordinate) and occasionally turning

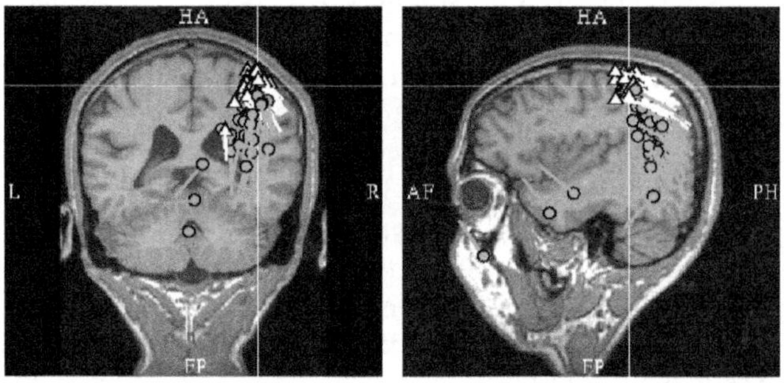

Fig. 4 Equivalent current dipoles fitted to the P30m peaks of single SEF responses are overlaid on coronal and sagittal MRI slices. The cross-hair cursor indicates the dipole localization from averaged responses in a static head position. Light blue circles point to dipole localizations obtained after tSSS but without MC, while white triangles indicate the corresponding localizations after combined tSSS and MC

Table 5 ECD localizations and moments from single responses. a) Dipole fitted to averaged SEF in a static head position. Averages and standard deviations from ECDs fitted to 72 responses are given b) after tSSS but without MC, and c) after combined tSSS and MC

	X [mm]	Y [mm]	Z [mm]	Q [nAm]
a) static	41.0	8.1	103.9	94.2
b) no MC	34.7	-8.7	76.2	139.8
stdev	10.8	17.2	28.8	70.6
c) after MC	38.8	6.4	102.7	112.7
stdev	6.6	7.5	10.3	56.1

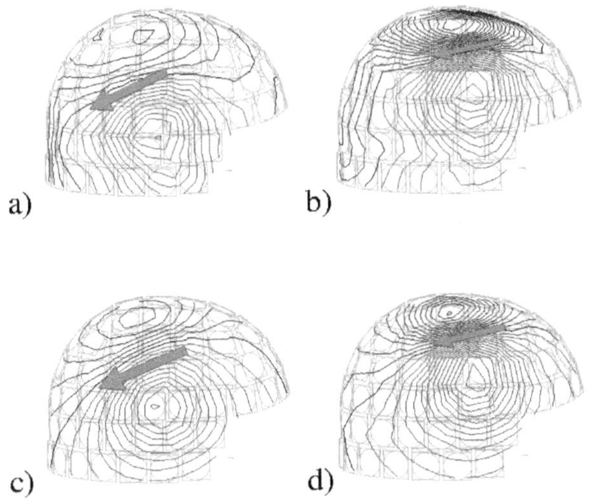

Fig. 5 Isocontour maps and dipoles of a single-trial SEF response at 271 s (the vertical line in Fig 3) when the head was 3 cm lower than the initial position, a) tSSS without MC, b) combined tSSS and MC. For comparison, contours and dipoles from averaged SEFs are also shown, c) tSSS without MC, d) tSSS and MC. The step between adjacent contour lines is 100 fT

IV. Conclusions

The dipole localizations and latencies from multiple repeated recordings were very reproducible. The ECD results from static head position recordings (Table 3) are in a good agreement with earlier reports [2-4].

Figures 2-5 demonstrate the importance of the interference elimination in cases with internal interference and head movements. As noted earlier [9], transformation of the head position by more than 3 cm increases noise in the results. This was also seen here; when the head position had drifted more than 3 cm from the initial position (Figure 3), the raw data after tSSS and MC showed increased noise and decreased signal to noise ratio. Still, single-trial SEF responses could be analyzed from the processed raw data.

The data in Table 4 show that the SEF dipole results are similar to the reference result. However, the head movements affected the AEFs: the response was slightly delayed and the mean amplitudes were on the average 30% weaker than in the reference data (seen also in Figure 1b). The AEF changes are probably due to additional acoustic noise and decreased salience of the auditory stimulus caused by the head motion. In all 17 cases the head movements were relatively slow. Faster and more regular movements tend to generate bursting background activation which decreases the AEF localization accuracy [5].

The mean dipole localization differences with respect to the reference localization in Table 4 were 3.5 ± 2.6 mm (SEF), 4.0 ± 2.0 mm (AEF left hemisphere) and 4.0 ± 1.5 mm (AEF right hemisphere), which are smaller than the total variances of the dipole fits from the reference recordings in Table 3.

In summary, the combined tSSS and MC method suppressed effectively the disturbances caused by head movements and magnetized impurities and successfully removed the localization bias due to such sources of error.

Acknowledgment

The authors thank Riitta Pietilä for help with the figures.

References

1. Hari R, Parkkonen L (2008) MEG in the study of higher cortical functions. Progress in Epileptic Disorders, Vol 5: Event-related Potentials in Patients with Epilepsy: from Current State to Future Prospects. Eurotext, Montrouge, France: Editions John Libbey, 2008, pp 103-112.
2. Virtanen J, Ahveninen J, Ilmoniemi R, et al (1998) Replicability of MEG and EEG measures of the auditory N1/N1m-response. Electroencephalogr Clin Neurophysiol 108: 291-298.
3. Castillo E, Simos P, Wheless J, et al (2004) Integrating sensory and motor mapping in a comprehensive MEG protocol: Clinical validity and replicability. NeuroImage 21: 973-983.
4. Schaefer M, Noenning N, Karl A, et al (2004) Reproducibility and stability of neuromagnetic source imaging in primary somatosensory cortex. Brain Topography 17: 47-53.
5. Nenonen J, Nurminen J, Kičić D, et al (2010) Correction of disturbances caused by temporary and continuous head movements. This volume.
6. Ou W, Golland P, Hämäläinen M (2007) Sources of variability in MEG. Med Image Comput Comput Assist Interv 10 (Pt 2): 751-759.
7. Taulu S, Kajola M. (2005) Presentation of electromagnetic multichannel data: The signal space separation method. J Appl Phys 97: 124905 (1-10).
8. Taulu S, Simola J (2006) Spatiotemporal signal space separation method for rejecting nearby interference in MEG measurements. Phys Med Biol 51: 1759-68.
9. Medvedovsky M, Taulu S, Bikmullina R, Paetau R (2007) Artifact and head movement compensation in MEG. Neurology, Neurophysiology, and Neuroscience 4: 1-10.
10. Wehner D, Hämäläinen M, Mody M, Ahlfors S (2008) Head movements of children in MEG: Quantification, effects on source estimation, and compensation. NeuroImage 40: 541–550.
11. Uusitalo M, Ilmoniemi R (1997) Signal-space projection method for separating MEG or EEG into components. Med Biol Eng Comput 35: 135-140.
12. Uutela K, Taulu S, Hämäläinen M (2001) Detecting and correcting for head movements in neuromagnetic measurements. NeuroImage 14: 1424-1431.
13. Hämäläinen M, Hari R, Ilmoniemi R, Knuutila J, Lounasmaa O (1993) Magnetoencephalography—theory, instrumentation, and applications to noninvasive studies of the working human brain. Rev Mod Phys 65: 413-497.

Author: Jukka Nenonen
Institute: Elekta Oy
Street: Siltasaarenkatu 18-20
City: Helsinki
Country: Finland
Email: jukka.nenonen@elekta.com

Functional Motor Mapping Using Corticokinetic Coherence

Mathieu Bourguignon[1,*], Xavier De Tiège[1], Marc Op de Beeck[1], Benoît Pirotte[2], Patrick Van Bogaert[1], Serge Goldman[1], Riitta Hari[3], and Veikko Jousmäki[1,3]

[1] Laboratoire de Cartographie Fonctionnelle du Cerveau, ULB-Hôpital Erasme, Brussels, Belgium
[2] Departement of Neurosurgery, ULB-Hôpital Erasme, Brussels, Belgium
[3] Brain Research Unit, Low Temperature Laboratory, Aalto University School of Science and Technology, Finland
*Corresponding author. mabourgu@ulb.ac.be

Abstract— **We present a novel method, corticokinetic coherence, for functional motor mapping by computing coherence between cortical magnetoencephalographic (MEG) signals and the kinetics of voluntary movements.**

Six subjects performed during an MEG recording self-paced flexion–extensions of the right-hand fingers at about 3 Hz, with a 3-axis accelerometer attached to the index finger. Cumulant density and coherence spectra were computed between the MEG and accelerometer signals.

In all subjects, accelerometer and coherence spectra showed peaks around 3–4 Hz and 6–8 Hz, corresponding to the movement frequencies. Coherence values (range 0.32–0.77; mean 0.55) exceeded the statistical threshold in all subjects.

The sources of the coherent MEG signals, found during peaks of the cumulant density, were located in the left rolandic sulcus, at the hand motor area.

Corticokinetic coherence appears to be a promising and robust tool for reliable and convenient functional mapping of the human motor cortex.

Keywords— **accelerometer, functional mapping, human, magnetoencephalography, motor cortex.**

I. Introduction

Functional mapping of the human motor cortex using magnetoencephalography (MEG) has until now relied on the use of cortico–muscular coherence (CMC) and premovement readiness fields [1, 2]. The CMC quantifies synchronization between MEG oscillations and surface electromyogram (EMG). The beta-band MEG oscillations of the primary motor (M1) cortex are coherent with the EMG of the isometrically contracting muscles [3]. In a fifth of subjects, however, CMC is too weak to allow proper M1 identification [4].

To search for a more reliable functional indicator of the human M1 cortex, we here describe a novel mapping method, the corticokinetic coherence (CKC). The CKC searches for synchronization between MEG oscillations and voluntary movement kinetics monitored by a 3-axis accelerometer. An accelerometer has been used earlier in MEG to provide trigger signals for averaging movement-related signals [5] and to demonstrate coherence between movement and M1 cortex during slow finger movements [6].

II. Materials and Methods

A. Subjects

Six healthy subjects (range 24–40 yrs; 4 males, 2 females) without any history of neuropsychiatric disease were studied. All were right-handed according to Edinburgh handedness inventory [7].

The subjects participated after informed consent, and the study had prior approval by the ULB-Erasme Hospital Ethics Committee.

B. Behavioural Task

Subjects were asked to make self-paced flexions/extensions of the right-hand fingers at about 3 Hz for 3 minutes with an accelerometer attached to the nail of the right index finger. During the flexion, the thumb touched the other fingers.

C. Recording

Cerebral activity was recorded using a whole-head neuromagnetometer (Vectorview & Maxshield; Elekta Neuromag Oy, Helsinki, Finland) at 0.1–330 Hz bandwidth and a sampling rate of 1 kHz. Head position inside the MEG helmet was continuously monitored using four head tracking coils. The locations of the coils with respect to anatomical fiducials were determined with an electromagnetic tracker (Fastrak, Polhemus, Colchester, VT, USA). The movements of the index finger were monitored using a 3-axis accelerometer (ADXL330 *i*MEMS Accelerometer, Analog Devices, Inc., Norwood, MA, USA) with 0–330 Hz passband and a sampling rate of 1 kHz.

High-resolution 3D-T1 magnetic resonance images (MRIs) were acquired on a 1.5 Tesla MRI scan (Intera, Philips, The Netherlands).

S. Supek and A. Sušac (Eds.): Advances in Biomagnetism – BIOMAG2010, IFMBE Proceedings 28, pp. 310–313, 2010.
www.springerlink.com

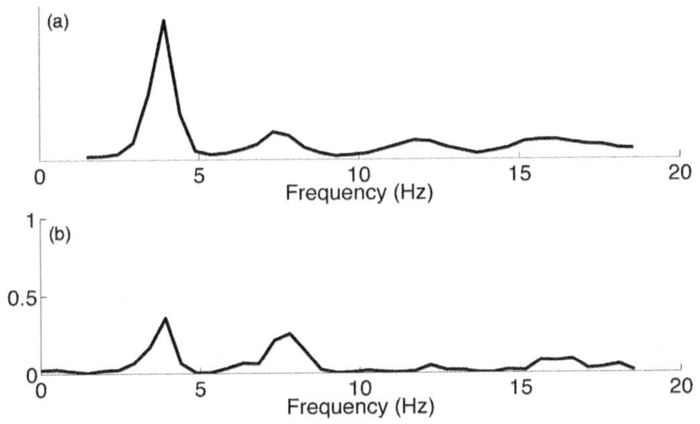

Fig. 1 (a) Spectrum of normalized acceleration and (b) coherence spectrum of a single MEG channel contralateral to hand movements of subject 6

D. *Data Preprocessing and Analysis*

Continuous MEG data were first preprocessed off-line using the signal-space-separation method to suppress external interferences and to correct for head movements [8]. To improve the signal-to-noise ratio of the coherence, MEG and accelerometer data were divided into overlapping 2048-ms epochs (frequency resolution 0.5 Hz) with 1638-ms epoch overlap [9]. Epochs contaminated by eye movements, muscle activity or artifacts in the MEG sensors were rejected. The acceleration was computed at every sample as the Euclidian norm of the three band-passed (1–200 Hz) accelerometer channels. Each epoch of acceleration was then normalized by its Euclidian norm.

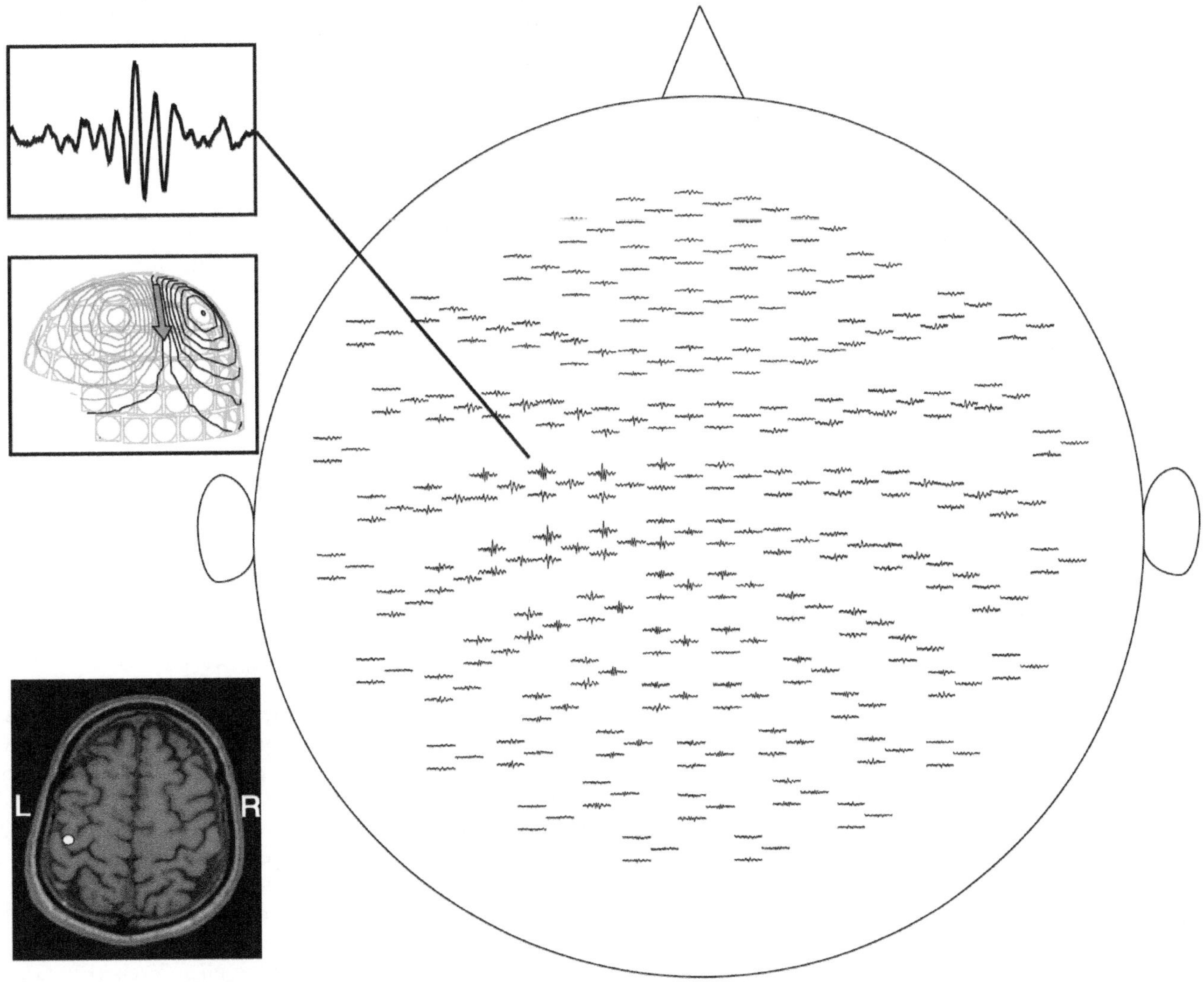

Fig. 2 Right: Spatial distribution of the cumulant density in Subject 6. The largest peaks of the cumulant density occur in the left rolandic sensors. Left: Enlarged channel showing the cumulant density (top), the magnetic field pattern obtained at the cumulant density global maxima (middle), and the corresponding source location at the posterior bank of the precentral gyrus (bottom)

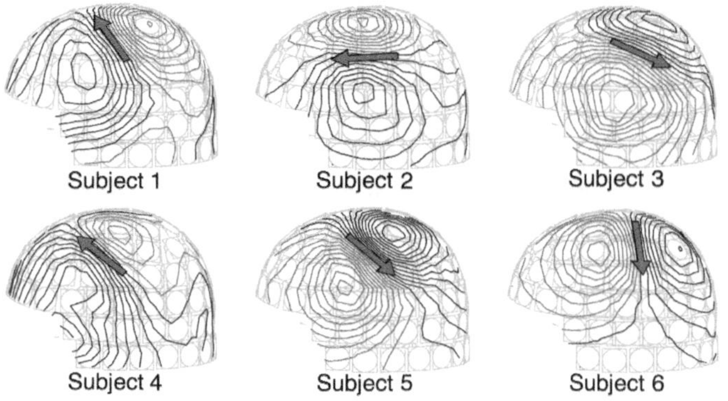

Fig. 3 Magnetic field patterns viewed from the left at the cumulant density global maxima in all subjects

Each epoch (indexed by k) of MEG (B^k) and normalized acceleration (A^k) were transformed using Fast Fourier Transform (FFT) algorithm. Power and cross spectra (P_{AA}, P_{BB} and P_{BA}) were first estimated

$$P_{XY} = \frac{1}{K}\sum_{k=1}^{K} FT(X_k)FT(Y_k)^*, \quad (1)$$

FT denoting the Fourier transform operator and * the complex conjugate; X and Y have to be replaced by A or B_i (i locating the MEG channel) as needed. The coherence spectra (Coh_{BA}) and cumulant densities (f_{BA}) were computed as previously suggested in [10].

$$\mathrm{Coh}_{XY} = \frac{|P_{XY}|^2}{P_{XX}\,P_{YY}} \quad (2)$$

$$f_{XY} = FT^{-1}(P_{XY}), \quad (3)$$

FT^{-1} being the inverse Fourier transform operator.

Cumulant density was band-pass filtered at 1–195 Hz, and notched at 50 Hz and its harmonics. Cortical sources explaining the main peak of the filtered cumulant density function were estimated by conventional equivalent current dipole modeling tools (Elekta Neuromag Oy, Helsinki, Finland) using spherical head models determined from individual MRIs. Sources were considered valid when the goodness-of-fit value exceeded 80% for a subselection of at least 100 sensors. Sources were then superimposed on the co-registered individual structural MRIs.

E. *Statistical Analysis*

The statistical significance of coherence was assessed using formula derived under the hypothesis of linear independence [10]

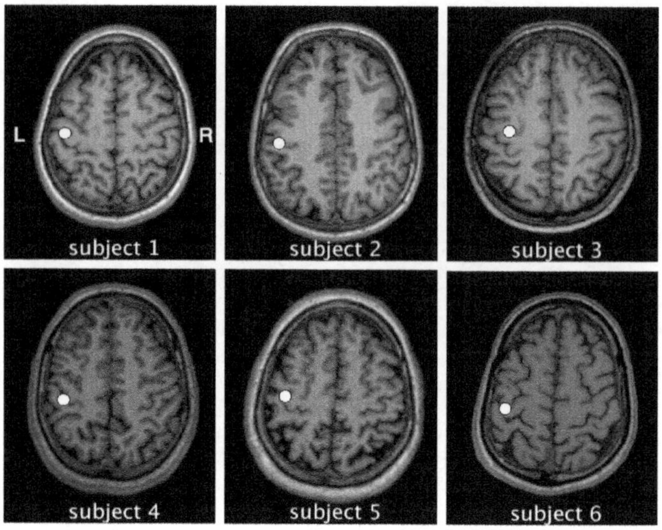

Fig. 4 Sources (white dots) corresponding to the cumulant density maxima in all subjects

$$Ct = 1 - p^{1/(L-1)} \quad (4)$$

where L is the number of disjoint epochs used, p is the chosen p-value for individual channels, and Ct is an estimation of the statistical threshold for the coherence. The p-value was computed to have a global false positive rate for all frequencies Nf and all channels Nc equal to P = 0.05 using Dunn-Šidák correction for multiple comparisons [11]:

$$p = 1 - (1 - P)^{1/N} \quad (5)$$

where N = Nf * Nc. As overlapping epochs were used, Ct is a conservative estimate of the threshold [9], and N is a conservative estimate of the number of degrees of freedom due to the dependence in the time courses of all channels. As a result, Ct is a valid statistical threshold for $P < 0.05$.

III. Results

The accelerometer and coherence spectra showed peaks at around 3–4 Hz and 6–8 Hz in all subjects, corresponding to the frequencies of hand movements (Fig. 1). Coherence values ranged from 0.32 to 0.77 (mean 0.55) and exceeded the statistical threshold in all subjects. The spatial distribution of the cumulant densities showed peaks at the rolandic sensors contralateral to the hand movement in all subjects (Fig. 2). At the times of the peaks of the cumulant density, clear dipolar field patterns, centered on the left rolandic sensors, were observed in all subjects (Fig. 3). The equivalent sources were located in the left precentral hand knob (Fig. 4) [12].

IV. Discussion

We found statistically significant CKC in all six subjects. The sources of coherent activity were located in the contralateral M1 cortex. These findings suggest that CKC method could be a robust tool for functional mapping of the M1 cortex.

The occurrence of the most significant phase locking between the M1 signals and finger movements at 3-4 Hz and 6-8 Hz corroborates a previous report on coherence around 3 Hz between hand velocity and MEG during a continuous visuomotor task [13]. These data suggest that slow oscillations in the M1 cortex may play an important role in the organization of voluntary movements.

From the technical point of view it is worth to note that the accelerometer did not produce any artifacts to the MEG signals, and that the use of the Euclidian norm of the accelerometer channels allowed us to measure the finger movement kinetics regardless of the hand position.

V. Conclusion

Corticokinetic coherence appears to be a promising and robust tool for reliable and convenient MEG mapping of human M1 cortex.

Acknowledgments

Brains Back to Brussels grant by the Institut d'Encouragement de la Recherche Scientifique et de l'Innovation de Bruxelles (Brussels, Belgium), ERC Advanced Grant #232946, and FRS-FNRS (Fond pour la Recherche Scientifique, research convention 3.4611.08 Belgium). Mathieu Bourguignon is Chercheur scientifique logistique F.R.S.-FNRS Belgium. Helge Kainulainen and Ronny Schreiber at the Brain Research Unit, Low Temperature Laboratory, Helsinki University of Technology for the technical support.

References

1. Baker SN (2007) Oscillatory interactions between sensorimotor cortex and the periphery. Curr Opin Neurobiol 17(6):649–655
2. Endo H, Kizuka T, Masuda T et al. (1999) Automatic activation in the human primary motor cortex synchronized with movement preparation. Brain Res Cogn Brain Res 8(3):229–239
3. Salenius S, Hari R (2003) Synchronous cortical oscillatory activity during motor action. Curr Opin Neurobiol 13(6):678–684
4. Pohja M, Salenius S, Hari R (2005) Reproducibility of cortex–muscle coherence. Neuroimage 26(3):764–770
5. Bowyer SM, Mason K, Weiland B et al. (2007) Localization of motor cortex by MEG using a tremorometer. Int Congr Ser 1300:321–324
6. Gross J, Timmermann L, Kujala J et al. (2002) The neural basis of intermittent motor control in humans. Proc Natl Acad Sci USA 99(4):2299-2302.
7. Oldfield RC (1971) The assessment and analysis of handedness: the Edinburgh inventory. Neuropsychologia 9(1):97–113
8. Taulu S, Simola J, Kajola M (2005) Applications of the Signal Space Separation Method. IEEE T Signal Proces 53(9):3359–3372
9. Bortel R, Sovka P (2007) Approximation of statistical distribution of magnitude squared coherence estimated with segment overlapping. Signal Process 87(5):1100–1117
10. Halliday DM, Rosenberg JR, Amjad AM et al. (1995) A framework for the analysis of mixed time series/point process data--theory and application to the study of physiological tremor, single motor unit discharges and electromyograms. Prog Biophys Mol Biol 64(2–3):237–278
11. Andrew C, Moonseong H (2005) A Comparison of Multiplicity Adjustment Strategies for Correlated Binary Endpoints. J Biopharm Stat15(5):839–855
12. Yousry TA, Schmid UD, Alkadhi H et al. (1997) Localization of the motor hand area to a knob on the precentral gyrus. A new landmark. Brain 120:141–157
13. Jerbi K, Lachaux JP, N'Diaye K et al. (2007) Coherent neural representation of hand speed in humans revealed by MEG imaging. Proc Natl Acad Sci USA 104(18):7676–7681

Face Processing in Children: Novel MEG Findings

Margot J. Taylor[1,2,3], Travis Mills[1], Linda Zhang[1], and Elizabeth W. Pang[2,3]

[1] Diagnostic Imaging and [2] Neurology, Hospital for Sick Children
[3] University of Toronto, Toronto, Canada

Abstract— **Face processing improves steadily across childhood, yet few studies have investigated the development of the neural sources underlying these processes. ERP studies have demonstrated marked developmental changes in early brain processes and these show systematic age-related changes. Using MEG, we investigated the developmental changes in brain sources in a face processing task.**

We studied 50 children (6-16yrs) and 22 young adults. Photographs of 240 neutral faces were used. Two blocks of 180 faces were presented. One block contained faces presented upright, and the other faces presented inverted. MEG activity was recorded on a 151 sensor CTF/MISL system. A structural MRI was acquired for all subjects. Source analyses were determined using "hands-off", objective, event-related beamforming protocols developed in-house.

The children's data were significantly different than what was seen in adults. As expected, M100 localised to bilateral occipital sources, decreased in latency and amplitude with age, and showed inversion effects. M170 was seen consistently in adults; however M170 was difficult to localise reliably in children. In contrast, a robust component at 140ms was present across the young children's age groups, but absent in adults.

These data showed qualitative differences between children and adults in the MEG components indexing the early stages of processing of faces. This suggests that measures of the neural responses to face processing in childhood depend in part on the neuroimaging technique used, with different components being more or less evident in ERPs and MEG.

Keywords— **face processing, development, M100, M140, M170.**

I. Introduction

The complexity of visual, emotional and semantic information conveyed by the human face makes face processing one of our most sophisticated cognitive abilities and, consistent with this complexity, distributed brain networks have been described for processing faces [e.g., 1,2]. Precise measures of the neuronal timing of face processing were first obtained in humans, with intracranial electrodes directly on the cortical surface [3]. These studies reported a negative component to faces, at about 200ms, at various sites on the fusiform and inferior temporal gyri that suggested the presence of discrete, specialized regions for processing faces. The scalp-recorded correlate of this response is the N170, an event-related potential (ERP) that has been extensively studied [e.g., 4] and thought to be generated in fusiform gyrus. The sensitivity of these cortical areas, particularly the fusiform gyrus, to faces was confirmed in fMRI [e.g., 5] and served as a catalyst for the large number of subsequent functional neuroimaging and neurophysiological studies on face processing.

MEG has also been applied to investigations of face processing skills in adults. The MEG component M170 has been shown to be sensitive to faces [6-11] with similar, but not identical effects as the N170 [10-11]. Face sensitive components occurring before 170ms, i.e., within the first 100ms, have also been described in MEG in adults [7,10-11].

Developmental changes examined in behavioural studies showed gradual increases in face processing task performance with age, with questions still remaining as to the reasons for this protracted maturation [e.g., 12-13]. Developmental ERP data demonstrate very rapid neural responses for face processing, even in young children—at the P1 component—with protracted age-related changes in both P1 and N170 [14-16]. The presence of N170 at 4 years of age has suggested that the neuronal substrate underlying face processing in adults is present early in childhood. The latency decreases of N170 from 4 years until adulthood supported a quantitative improvement with age; that is, face processing became faster and more efficient with development. Configural changes induced by face inversion have also been studied, as configural processing is a key aspect of face recognition [17-18]. In children, face inversion produced latency and amplitude effects on the P1 from four years of age, but effects on the N170 were seen only starting in mid childhood [14,15], suggesting these components reflect separable stages of face processing.

There have been relatively few investigations in MEG on developmental changes in face processing, and the studies that have been reported have not seen an M170 component comparable to the ERP literature [19-21]. Developmental improvements in face processing ability are widely documented, but the neural sources involved and their maturational timeline are still unclear. We used MEG to examine these processes with the hope of further elucidating the immature processing seen in children.

S. Supek and A. Sušac (Eds.): Advances in Biomagnetism – BIOMAG2010, IFMBE Proceedings 28, pp. 314–317, 2010.
www.springerlink.com

II. Methods

A. Participants

We studied 50 children divided into 5 age groups of 10 participants each (6-7, 8-9, 10-11, 12-13 and 14-16 years of age), and 22 young adults.

B. Stimuli

Greyscale photographs of 240 different faces (120 female) were used; faces were without earrings, glasses or facial hair. Two counter-balanced blocks of 180 faces were presented. To maintain attention, participants responded to occasionally repeating faces; only the MEG data to non-repeating faces are presented here. One block contained faces presented upright, and the other block of 180 presented the faces inverted.

C. Procedure

MEG data were recorded on a CTF Omega 151 channel system. Data were recorded continuously at 625Hz with an on-line bandpass of 0 to 200Hz. Subjects lay supine with their head in the dewar of the MEG, inside the dimly lit magnetically shielded room. Prior to data acquisition, three localization coils were placed at the nasion and pre-auricular points to localize the subject's head relative to the MEG sensors. The system allows head localisation to an accuracy of 1mm. Head localisations were completed before and after each block, to confirm that participants remained motionless.

Stimuli were back-projected onto a screen in front of the subjects, by a video projector situated outside the room, using the software Presentation (Neurobehavioral Systems, Albany CA). Faces were presented on a black background for 500ms with a random inter-stimulus interval of 1200-1500ms during which time a white fixation cross appeared. Subjects pressed a button as quickly and accurately as possible when they recognized that a face had repeated.

To ensure accurate MEG-MRI co-registration, immediately after completion of the MEG, the fiducial coils were replaced with MRI contrast markers in the same locations and an anatomic MRI was acquired for all subjects on a 1.5 Tesla Signa GE Advantage scanner, with a T1 3D-SPGR-pulse sequence.

D. MEG Data Analyses

Two consistent components were seen in the groups of children at the MEG mean sensor level corresponding to the M100 and M140 for both upright and inverted faces. Windows of interest were chosen post hoc from the groups' mean sensor data. The windows were 90-120ms to assess the M100 and 125-155ms to assess the second component. Constant windows were chosen to remain objective across all age groups yet sized wide enough to capture any age-related variability. Source images were created using an event-related adaptive beamformer [22, 23] to localize the principle source locations for each subject. After the location of the largest peaks was identified, estimated time courses were reconstructed for each subject so latency and amplitude comparisons could be made. All source images were thresholded to display the two largest peaks.

A realistic head model was computed from each subject's MRI to be used in the calculation of the beamformer and each 3D source image was overlaid onto each individual MRI. For group comparisons, individual MRIs were spatially normalized to the MNI (T1) template using SPM2.

E. Quantifying Latency and Amplitude Correlations in the Brain

To quantify latency and amplitude correlations as a function of age, post hoc time windows were chosen to bound the computer search to reasonable limits. A global maxima search was performed on each subject for each window to determine one peak latency and amplitude. Each subject's amplitude and latency for each window was plotted against age to assess possible developmental effects. Each plot was statistically validated by randomly permuting the index of each subject's value 1000 times to determine the null distribution, or the likelihood of a slope appearing by chance. The actual slope value was then compared to the null distribution to ensure it was distinct from the noise, thus rejecting the null hypothesis that the slope was a consequence of a random process.

III. Results

Grand averaged MEG results are shown in Figure 1 for the five age groups of children and the comparison group of adults. The over-plotted sensor tracings are on the left and the topographies of three peaks that were identified are shown on the right. As can be seen with the first major peak, and the first column of topographies, there is a very reliable M100 component that is seen across age groups. The source localisation of this component was in bilateral occipital regions and did not vary with age (Fig. 4, left panel). The amplitude of the M100 decreased with age ($p<.001$; Fig. 2) and showed a small latency decrease as well ($p<.01$). The M100 was also sensitive to inversion, being larger to upright then to inverted faces in the five groups of children (see Fig. 3).

After the M100, there was a very reliable component seen systematically in the four youngest groups of children that occurred at 140ms (M140). This component did not change with age across the four youngest age groups (see Fig. 1) and localised to the posterior occipital cortex (Fig. 4, middle panel). This localisation was very similar to what was seen for the M100. In the 14-16 year old group this component was much less evident, and in the adults it was not visible.

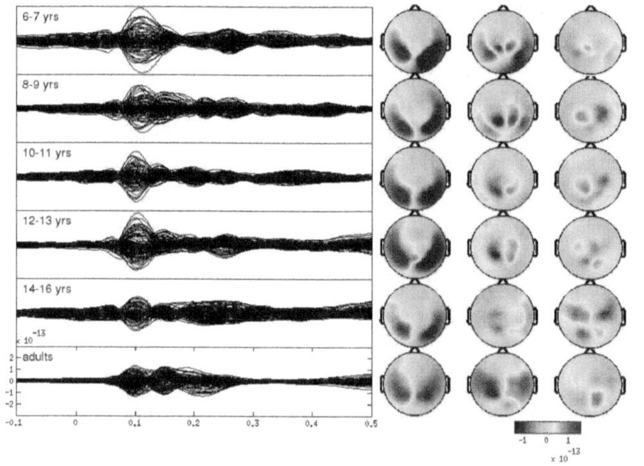

Fig. 1 Over-plotted MEG sensor data (left) for the 6 age groups with the topographic distributions (right) for the M100, M140 (M170 in adults) and a third component at around 200-220ms

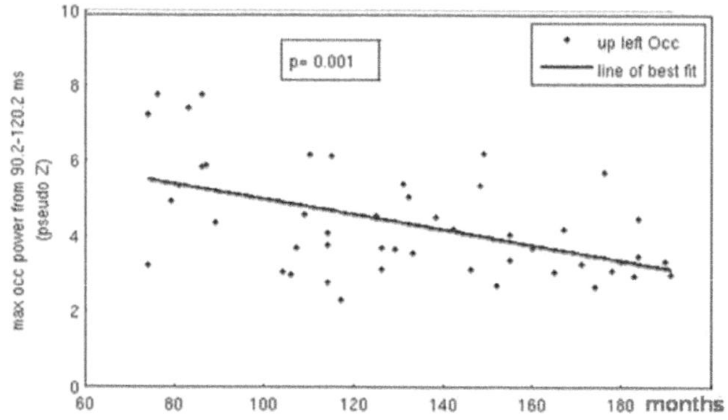

Fig. 2 M100 amplitudes for all of the children studied, showing the steady, significant decrease in amplitudes with age

In the adult group, a clear M170 was seen at 165ms latency which localized to the fusiform gyri bilaterally (see Fig. 4, right panel). The M140 seen in the youngest four groups of children was not the M170. The third component that was visible on the over-plotted sensor tracings and shown in the third column of the topographical plots in Figure 1 also did not localize to the fusiform or have the distribution typical of the M170 seen in adults. Fusiform activation in the latency range of the N170 taken from the ERP literature could be seen in the children's groups, but this was not an easily identifiable component and did not emerge from our hands-off analyses of these data.

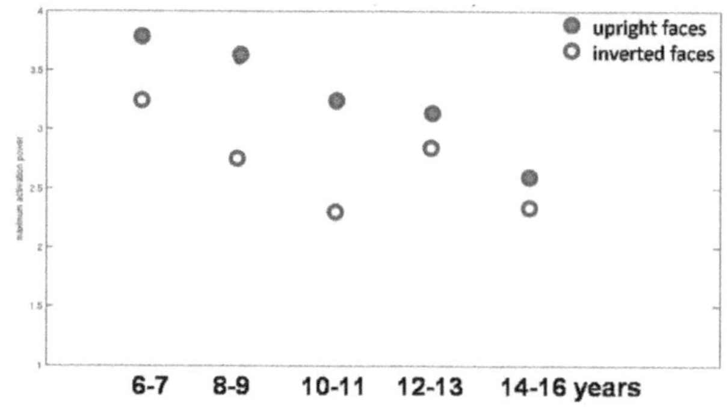

Fig. 3 Mean M100 amplitudes for the 5 groups of children, showing the consistently larger M100 for upright faces, as well as the decreasing amplitudes across childhood

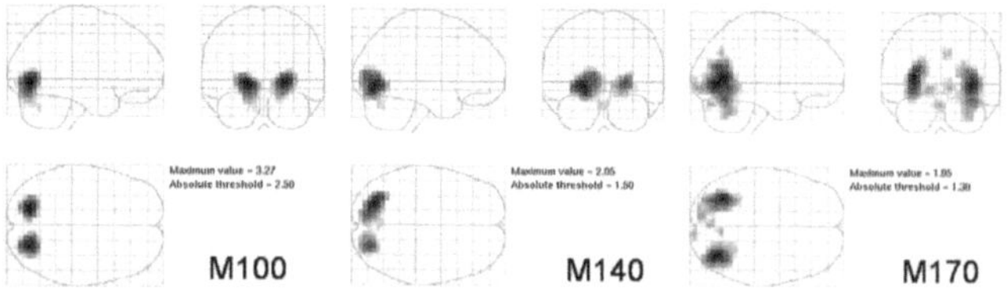

Fig. 4 Source localizations for the three components studied. The M100 was seen in all age groups, with the same bilateral occipital sources; the M140 was reliably seen in the four youngest age groups and also had a posterior, occipital localisation; the M170 was seen in the grand averages only in the adult group (see the bottom row, middle topographical map of Fig. 1) and showed the typical sources in the inferior temporal lobe, fusiform gyri

IV. Discussion

This is the largest developmental MEG data series with face stimuli. The data are reliable and obtained from children ranging in age from 6-16 years of age. The M100 shows similar effects as the P1 in ERP literature: age-related changes in latency and amplitude and effects of inversion, although these effects are much smaller than what is reported in the ERP literature. However, in an explicit face recognition task [24], P1 age-effects were smaller than in implicit face processing tasks [e.g., 14, 25].

Unlike the N170 seen with ERPs, in the present developmental data, there was not an easily identifiable comparable component with MEG. In the ERP literature the N170 is often seen as a bifid component in children [14, 15], with the earlier of the two components (N170a) showing a reduction in latency with age. This earlier component in ERPs localizes to a more posterior location in the brain then does the more lateral and inferior component, the N170b. The

M140 may be the neuromagnetic counterpart of the N170a in children. The N170a is also not seen in adults, disappearing in the early teen-aged years.

This still leaves a number of differences, however, between MEG and ERP data. The M100 showed less of an age-related change than did the P1. The M140 seen reliably in the present study and also in the few other reports of MEG to faces in children [19-21] was not the same as the M170, despite its interpretation as such in some studies. The M140 did not change in latency with age, while the N170a showed age-related decreases [14]. The M170 was not seen in the grand-averaged MEG data even in the 14-16yr olds, whereas the N170 is seen in the ERP grand averages from four years of age.

V. Conclusions

Although the M170 is readily recorded in adults and has similar sensitivity to faces as the N170, this was not the case in the developmental data. In this large series of children from 6-16 years, the topographical distribution of the M170 was not evident in the grand averages; instead, an earlier component, M140, was remarkable in its presence and lack of age-related changes in the younger four groups of children. This suggests that sources of the M170 are not mature even in mid adolescence, in contrast to N170. These MEG findings clarify some of the questions in the developmental ERP/MEG data and demonstrate the need to use complementary neuroimaging technologies to take advantage of their relative strengths and thus obtain a fuller picture of the developmental changes in the underling neural sources.

Acknowledgments

This work was supported in part by CIHR MOP-81161 to MJT, EWP and EJ Donner. The authors would like to thank Andrew Hamilton, Marion Malone, Laura Hopf and Sarah Bayless for their contributions.

References

1. Allison T, Puce A, McCarthy G (2000) Social perception from visual cues: role of the STS region. Trends Cogn Sci 4:267-278.
2. Haxby JV, Hoffman EA, Gobbini MI (2000) The distributed human neural system for face perception. Trends Cogn Sci 4:223-233.
3. Allison T, Ginter H, McCarthy G et al. (1994) Face recognition in human extrastriate cortex. J Neurophysiol 71: 821-825.
4. Bentin S, Allison T, Puce A et al. (1996) Electrophysiological studies of face perception in humans. J Cogn. Neurosci 8:551-565.
5. Puce A, Allison T, Asgari M et al. (1996) Differential sensitivity of human visual cortex to faces, letterstrings, and textures: A functional magnetic resonance imaging study. J Neurosci 16:5205–5215.
6. Halgren E, Raij T, Marinkovic K et al. (2000) Cognitive response profile of the human fusiform face area as determined by MEG. Cereb Cortex 10:69-81.
7. Liu J, Harris A, Kanwisher N (2002) Stages of processing in face perception: an MEG study. Nat Neurosci 5:910-916.
8. Swithenby SJ, Bailey AJ, Brautigam S et al. (1998) Neural processing of human faces: a magnetoencephalographic study. Exp Brain Res 118:501-510.
9. Taylor MJ, George N, Ducorps A (2001) Magnetoencephalographic evidence of early processing of direction of gaze in humans. Neurosci Lett 316(3):173-177.
10. Itier RJ, Herdman AT, George N et al. (2006) Inversion and contrast-reversal effects on face encoding and recognition assessed by MEG. Brain Res 1115: 108-120.
11. Bayle DJ, Taylor MJ (2009) Attention inhibition of early cortical activation to fearful faces. Brain Res *in press*.
12. Pimperton H, Pellicano E, Jeffery L et al. (2009) The role of higher level adaptive coding mechanisms in the development of face recognition. J Exp Child Psychol 104(2):229-238.
13. Chung MS, Thomson DM (1995) Development of face recognition. Br J Psych 86:55-87.
14. Taylor MJ, Batty M, Itier RJ (2004) The faces of development: A review of early face processing over childhood. J Cogn Neurosci 16: 1426-1442.
15. Taylor MJ, Edmonds GE, McCarthy G et al. (2001) Eyes First! Eye processing develops before face processing in children. NeuroReport 12(8):1671-1676.
16. Taylor MJ, McCarthy G, Saliba E et al. (1999) ERP evidence of developmental changes in encoding faces. Clin Neurophysiol 110: 910-915.
17. Rhodes G, Brake S, Atkinson AP (1993). What's lost in inverted faces? Cognition 47:25-57.
18. Maurer D, Le Grand R, Mondloch CJ (2002) The many faces of configural processing. Trends Cogn Sci 6:255-260.
19. Kylliäinen A, Braeutigam S, Hietanen JK et al. (2006) Face and gaze processing in normally developing children: a magnetoencephalographic study. Eur J Neurosci. 23(3):801-10.
20. Kylliäinen A, Braeutigam S, Hietanen JK et al. (2006) Face- and gaze-sensitive neural responses in children with autism: a magnetoencephalographic study. Eur J Neurosci 24(9):2679-90.
21. Kimura I, Kubota M, Hirose H et al. (2004) Children are sensitive to averted eyes at the earliest stage of gaze processing. Neuroreport 7;15(8):1345-8.
22. Cheyne D, Bakhtazad L, Gaetz W (2006) Spatiotemporal mapping of cortical activity accompanying voluntary movements using an event-related beamforming approach. Hum Brain Map 27:213-29.
23. Robinson SE, Vrba J (1999) Functional neuroimaging by synthetic aperture magnetometry (SAM). Recent Advances in Biomagnetism. In: Yoshimoto T, Kotani M, Kuriki S et al. (Eds.), Tohoku University Press, Sendai, Japan, pp. 302-305.
24. Itier RJ, Taylor MJ (2004) Face recognition memory and configural processing: a developmental ERP study using upright, inverted and contrast-reversed faces. J Cogn Neurosci 16(3):487-502.
25. Batty M, Taylor MJ (2006) The development of emotional face processing during childhood. Dev Sci 9:2,207-220.

Bottom-Up and Top-Down Driven Attentional Effects on Auditory Evoked Fields

Hidehiko Okamoto, Henning Stracke, Lothar Lagemann, and Christo Pantev

Institute for Biomagnetism & Biosignalanalysis, University of Muenster, Muenster, Germany

Abstract— **Under natural environments, picking up relevant sound signals, which are embedded in other irrelevant sound signals, is an important ability for animals including humans. Bottom-up and top-down driven attention seems to play an important role in this process. In the present studies, we investigated magnetic fields in humans that were evoked by pure tones embedded in band-eliminated noises during two different stimulus sequencing conditions (constant vs. random) and under two different auditory attentional conditions (focused vs. distracted) by means of magnetoencephalography.**

Constant stimulus sequencing was composed of identical pure tones overlaid with band-eliminated noises, whereas random sequencing was composed of pure tones of random frequencies overlaid with band-eliminated noises. We demonstrated that under focused auditory attention auditory evoked neural responses were sharper and larger in the constant sequencing condition than the random sequencing condition. On the contrary, the auditory evoked fields under distracted auditory attention were larger, but not sharper in the constant sequencing condition compared to the random sequencing condition.

Therefore, our results indicate that bottom-up driven auditory attention may mainly amplify excitatory neural networks, whereas the top-down auditory attention may both amplify and sharpen the neural activity via excitatory and inhibitory neural networks. The combination of bottom-up and top-down driven auditory attention would improve auditory performance in noisy environments.

Keywords— **attention, sharpening, gain, N1m, auditory cortex.**

I. Introduction

Humans can process task-relevant sound signals despite the usual presence of task-irrelevant noises. Bottom-up and top-down auditory attention eases this perception process. Recent magnetoencephalography (MEG) [1, 2] studies revealed that auditory focused attention not only enhances task-relevant neural activity, but also suppresses task-irrelevant neural activity in human auditory cortex.

Although top-down driven auditory attention can sharpen the population-level frequency tuning in human auditory cortex [1], it is still unsettled whether these attentional effects depend on the specific frequency region within the tonotopic maps. A functional magnetic resonance imaging (fMRI) study [3] showed that attentional modulation takes place mainly on the cortical level. However, not only top-down, but also bottom-up driven attention seems to play an important role for neural processing of sound signals in noisy environments. In the noisy environments, we can automatically segregate a specific auditory signal from other noises based on bottom-up auditory inputs (auditory scene analysis [4-10]) without focusing auditory attention on it. However, it still remains elusive whether bottom-up driven involuntary neural processing could amplify and sharpen population-level neural activity as 'salient' processing.

The goal of the present study was to investigate by means of MEG whether population-level frequency tuning can be modulated by differential stimulus sequencing under auditory focused attention and distracted auditory attention conditions.

II. Materials and Methods

14 (active listening session) and 15 (distracted listening session) healthy subjects participated in the present study. Subjects were different between active and distracted listening conditions. All subjects were strongly right-handed, and their hearing thresholds were within normal hearing range. Subjects gave written informed consent for their participation in the study in accordance with procedures approved by the Ethics Commission of the Medical Faculty, University of Muenster.

We presented pure tones as test stimuli (TS) simultaneously with band-eliminated noises (BENs). The TS had a duration of 600 msec (10 msec rise and fall times), and a frequency of either 250, 350, 450, 570, 700, 840, 1000, 1170, 1370, 1600, 1850, 2150, 2500, 2900, 3400, or 4000 Hz. In the active listening session, the 50 % TS contained a silent gap of 10 msec duration starting at latency 285 msec. In the distracted listening session, the TS contained no silent gap. The TS with temporal gaps were targets for behavioral responses during the active listening session and ensured the subjects' compliance regarding the focus of attention. In the distracted listening session, subjects were instructed to pay attention on a silent movie of their choice. The sound onset

S. Supek and A. Sušac (Eds.): Advances in Biomagnetism – BIOMAG2010, IFMBE Proceedings 28, pp. 318–321, 2010.
www.springerlink.com

asynchrony between two subsequent TS was fixed to 3000 msec (Fig. 1).

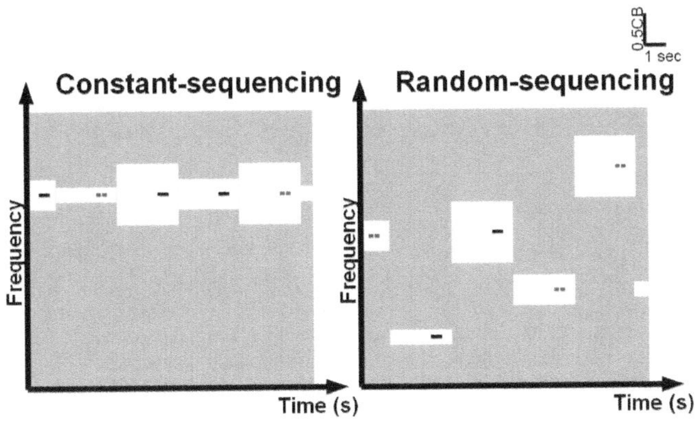

Fig. 1 Concept and time course of the auditory stimulation with respect to the constant sequencing (Left) and random sequencing (Right) conditions in the active listening session. Pass-bands and stop-bands of the band-eliminated noises (BENs) are represented by the light gray and white areas, respectively. The notch-bandwidth of a BEN (white area) is either 1/4, 1/2, or 1 critical band (CB). Target and non-target test stimuli (TS) are represented as red lines with gap (target TS, requiring a button press) and black lines without gap (non-target TS), respectively. In the distracted listening condition, only the non-target TS are presented

The simultaneously presented BENs were prepared as follows: Spectral frequency bands with widths of either 1/4 critical band (1/4 CB), 1/2 critical band (1/2 CB), or 1 critical band (1 CB) centred at the frequency of the simultaneously presented TS were eliminated from the white noise. All sound stimuli were prepared as sound files and presented under control of Presentation (Neurobehavioral Systems, Albany, CA, United States). SRM-212 electrostatic earphones (Stax, Saitama, Japan) transduced air-conducted sounds which were delivered through silicon tubes and earpieces fitted to each subject's ears. The hearing threshold for the 1000 Hz TS was determined for each ear of each individual at the beginning of the MEG session. The 1000 Hz TS was presented binaurally at intensity of 35 dB above individual sensation level. The power of the other TS, which were also presented binaurally, was adjusted to the power of the 1000 Hz TS. The total power of the binaurally presented BENs was 15 dB larger than TS power.

In order to investigate the effects of stimulus sequencing during auditory focused attention, we contrasted two different conditions within subjects: 'constant sequencing' and 'random sequencing' (c.f., Fig. 1). In the constant sequencing session, 30 TS with identical frequency were successively presented simultaneously with either the 1/4, 1/2, or the 1 CB BEN. In the random sequencing session, 30 TS with different frequencies, which were pseudo-randomly chosen from the frequencies used in the constant sequencing blocks, were presented. As in the constant sequencing condition, BENs with notches of either 1/4, 1/2, or 1 CB were presented simultaneously and pseudo-randomly (Fig. 1). In total, 160 trials (10 trials for 16 frequencies) for each BEN condition in each sequencing condition were presented and subjected to data analysis.

Auditory evoked fields were measured with a helmet-shaped 275 channel whole head magneto-gradiometer (Omega; CTF Systems, Coquitlam, British Columbia, Canada) in a silent magnetically shielded room (Fig. 2). Epochs of data elicited by TS (- 300 msec pre-TS-onset interval to a 300 msec post-TS-onset) were averaged selectively for each BEN and attentional condition (irrespective of frequency) after rejection of artifact epochs.

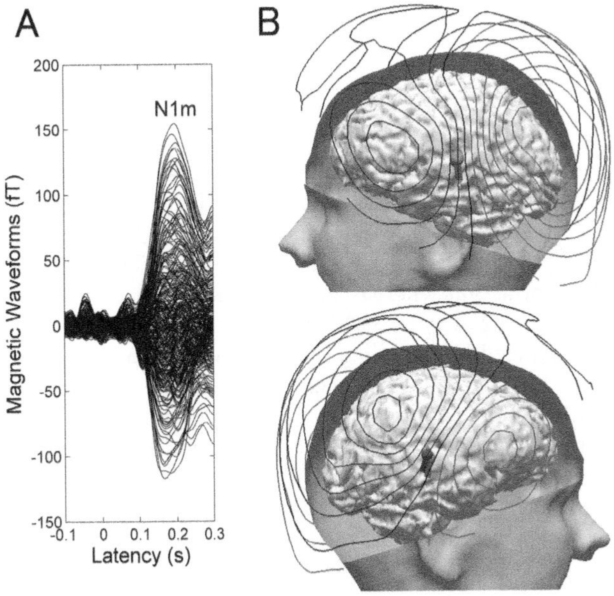

Fig. 2 A: MEG waveforms of one representative subject. Clear N1m-responses elicited by the sound onset are visible. B: Magnetic contour maps and estimated single dipoles at the latency of the maximal N1m response are illustrated together with skin and brain modeled from the individual MRI. Red and blue contour lines represent outbound and inbound flows of magnetic fields from and to the brain. The spheres and barrels indicate the locations and orientations of single dipoles in left (green) and right (red) hemispheres

For the N1m analysis, the averaged magnetic field signals were 30 Hz low-pass filtered and were baseline-corrected relative to the 300 msec pre-stimulus interval. The source locations and orientations were estimated by means of two single equivalent current dipoles based on the grand-averaged MEG waveforms for each subject. Finally, the source space projection method [11] was applied to estimate source strengths for each BEN condition (BEN_1/4CB, BEN_1/2CB, and BEN_1CB) and each stimulus sequencing condition (constant sequencing and random sequencing).

Because the subject groups differed between the active and distracted listening conditions, the maximum N1m source strengths and latencies elicited by the TS for each condition were analyzed separately via repeated-measures analyses of variance (ANOVAs) using the factors BEN-TYPE (BEN_1/4CB, BEN_1/2CB, and BEN_1CB) and SEQUENCING (Constant and Random).

III. Results

Clearly identifiable averaged auditory evoked fields were obtained from all subjects. Fig. 2 shows an example of individual magnetic field waveforms and equivalent current dipoles elicited by TS with a clear N1m. The averages of the N1m source strengths of each condition with their 95 % confidence intervals are presented in Fig. 3. In the active listening session, the repeated-measures ANOVA evaluating N1m source strengths resulted in significant main effects for SEQUENCING ($F(1, 27) = 14.1$, $p < 0.001$) and BEN-TYPE ($F(2, 54) = 45.4$, $p < 0.001$). Moreover, there were significant interactions between SEQUENCING and BEN-TYPE ($F(2, 54) = 65.8$, $p < 0.001$). The latter interactions show that whereas there was a very small sequencing effect in the wide BEN condition, the narrow BENs show an enhanced N1m source strength for the constant sequencing as compared to the random sequencing condition. In the distracted listening session, the repeated-measures ANOVA evaluating N1m source strengths resulted in significant main effects for SEQUENCING ($F(1, 29) = 34.3$, $p < 0.001$) and BEN-TYPE ($F(2, 58) = 36.6$, $p < 0.001$). However, there was no significant interaction between SEQUENCING and BEN-TYPE ($F(2, 58) = 0.33$, $p = 0.72$), which was observed in the active listening session.

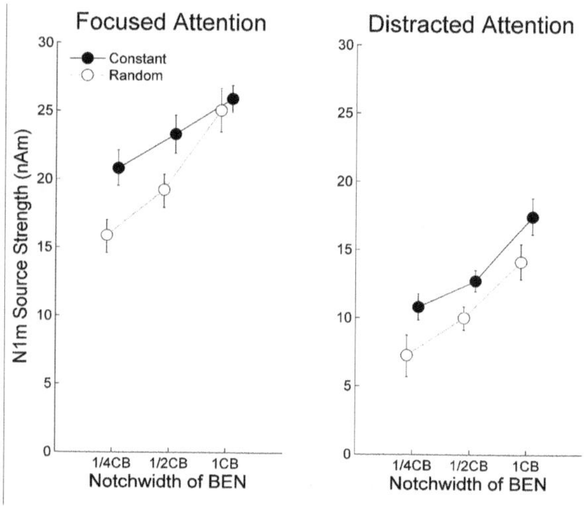

Fig. 3 Group means of the N1m source strengths for each experimental condition. Filled circles denote the N1m source strengths elicited by the test stimulus (TS) during the constant sequencing (Constant TS), and open circles denote the random sequencing conditions (Random TS)

IV. Discussion

In the present study, we used the combination of TS and simultaneously presented BEN (cf. Fig. 1) to investigate population-level gain and sharpening effects on the auditory evoked N1m response [12]. The neural activity evoked by the TS partially overlapped with the neural activity evoked by the BENs. However, in the present study the N1m response does not originate from this overlapping neural activity, since the neurons that could be activated by both TS and BEN had already been activated by the continuously presented BEN which preceded the TS onset. Thus, the neural activity corresponding to both TS and BEN did not contribute to the N1m auditory evoked fields elicited by the TS onset. The neural activity overlap would be rather large in the narrow BEN condition, and therefore a comparably small N1m response would be elicited by the TS. As described in our studies [1, 2, 13, 14], the neural gain effect driven by constant sequencing is reflected in an additive increment of N1m source strength across BEN conditions, whereas sharpening of population-level frequency tuning results in larger N1m source strength gain in the narrow BEN compared to the wider BEN conditions. The significant main effect of SEQUENCING found in the present study confirmed that constant sequencing could enhance the corresponding neural activity in both active and distracted listening conditions. However, a significant interaction between SEQUENCING and BEN-TYPE was observed only in the active listening condition (Fig. 3). Therefore, our findings suggest that constant stimulus sequencing during auditory focused attention can enhance task-relevant neural activity and suppress task-irrelevant neural activity in a frequency-specific manner, resulting in sharper population-level frequency tuning in human auditory cortex, whereas the bottom-up driven attention can only enhance the corresponding neural groups, but has merely little influence on the other neural groups. Interactions between excitatory and inhibitory neural networks, intensified by constant stimulus sequencing in auditory focused attention, sharpens population-level frequency tuning in a frequency-specific manner, leading to enhanced auditory performance in noisy environments. Bottom-up driven involuntary neural modulation presumably helps to automatically track a constant auditory stream in a noisy environment, without reducing the other potentially salient neural activity.

V. Conclusions

Constant stimulus sequencing during auditory focused attention can improve population-level frequency tuning in humans in a frequency-specific manner. In contrast, bottom-up driven attention seems to amplify the corresponding neural activity without sharpening the population-level frequency tuning in the auditory cortex. The combination of top-down and bottom-up driven attention will lead to better auditory performance in noisy environments.

Acknowledgements

We are grateful to Andreas Wollbrink and Karin Berning for technical assistance. This work has been supported by the "Deutsche Forschungsgemeinschaft" (Pa 392/10-3) and Tinnitus Research Initiative (BD169968).

References

1. Okamoto, H., et al., Attention improves population-level frequency tuning in human auditory cortex. J Neurosci, 2007. 27: p. 10383-90.
2. Stracke, H., H. Okamoto, and C. Pantev, Interhemispheric support during demanding auditory signal-in-noise processing. Cereb Cortex, 2009. 19: p. 1440-7.
3. Petkov, C.I., et al., Attentional modulation of human auditory cortex. Nat Neurosci, 2004. 7: p. 658-63.
4. Alain, C., Breaking the wave: effects of attention and learning on concurrent sound perception. Hear Res, 2007. 229: p. 225-36.
5. Alain, C., S.R. Arnott, and T.W. Picton, Bottom-up and top-down influences on auditory scene analysis: evidence from event-related brain potentials. J Exp Psychol Hum Percept Perform, 2001. 27: p. 1072-89.
6. Alain, C. and L.J. Bernstein, From sounds to meaning: the role of attention during auditory scene analysis. Curr Opin Otolaryngol Head Neck Surg, 2008. 16: p. 485-489.
7. Micheyl, C., et al., The role of auditory cortex in the formation of auditory streams. Hear Res, 2007. 229: p. 116-31.
8. Snyder, J.S. and C. Alain, Toward a neurophysiological theory of auditory stream segregation. Psychol Bull, 2007. 133: p. 780-99.
9. Bregman, A.S., Auditory scene analysis : the perceptual organization of sound. 1990, Cambridge, Mass.: MIT Press. xiii, 773 p.
10. Denham, S.L. and I. Winkler, The role of predictive models in the formation of auditory streams. J Physiol Paris, 2006. 100: p. 154-70.
11. Tesche, C.D., et al., Signal-space projections of MEG data characterize both distributed and well-localized neuronal sources. Electroencephalogr Clin Neurophysiol, 1995. 95: p. 189-200.
12. Sams, M. and R. Salmelin, Evidence of sharp frequency tuning in the human auditory cortex. Hear Res, 1994. 75: p. 67-74.
13. Okamoto, H., et al., Bottom-up Driven Involuntary Auditory Evoked Field Change: Constant Sound Sequencing Amplifies, but not Sharpens Neural Activity. J Neurophysiol, in press, DOI 10.1152/jn.00530.2009
14. Okamoto, H., et al., Frequency-specific modulation of population-level frequency tuning in human auditory cortex. BMC Neurosci, 2009. 10: p. 1.

Attention and Processing of Relevant Visual Information While Simulated Driving: a MEG Study

A. Fort[1], R. Martin[2], S. Daligault[3], A. Jacquet-Andrieu[4], and C. Delpuech[3]

[1] INRETS-LESCOT, Bron, France
[2] ISH, Lyon, France
[3] CERMEP/INSERM U821, Bron, France
[4] Institut Gaspard Monge / LabInfo – UMR 8049, Paris, France

Abstract— **It is a well-known fact that attention is crucial for the driver. In this innovative study, we aim to assess the impact of attentional demand modulation on driver's cerebral activity during simulated driving using magnetoencephalography (MEG). A car simulator equipped with a steering wheel, turning indicators, an accelerator and a brake pedal has been specifically designed to allow the recording of brain activity linked to the processing of relevant visual information for car driving.**

A radio broadcast is presented to the subjects while they drove in the simulator. During half of the driving scenarios subjects can ignore the broadcast (simple task, ST), and during the other half, they have to actively listen to it in order to answer 3 questions (dual task, DT). Evoked magnetic responses are computed for the traffic lights turning from green to amber. A modeling technique (Minimum-Norm Current Estimates) is used to estimate the cortical sources of these activities and their temporal time-course.

Preliminary results show the activation of a large distributed network similar in both ST and DT conditions. This network mainly involves sensory visual areas as well as parietal and frontal regions known to play a role in selective attention and motor areas. The activity within these areas is differentially modulated by the attentional state of the subject. For instance, several visual areas (primary and temporal visual areas) are less active in the DT condition than in the ST condition suggesting a modulation of the perceptual visual processing due to the secondary task. Frontal and parietal regions linked to attentional processing are also modulated. Moreover, this study demonstrates the feasibility to record MEG activity during an interactive simulated driving task.

Keywords— **MEG, attention, decision-making. driving, car simulator.**

I. Introduction

Driving is a complex task that requires sustained and selective attention. Drivers continuously have to adapt their behavior to the changing environment and to anticipate future situations. Because of our attentional capacities, only a limited amount of information can be processed in depth at any one time. Therefore, to have a safe and efficient behavior, drivers have to extract and process the most relevant pieces of information according to their objectives and previously acquired knowledge of such a situation and to ignore irrelevant source of information.

In this driving context and particularly when a critical situation occurs, even minor diversions of attention from the driving task can prove to be disastrous. Nevertheless, attentional defaults are a common aspect of everyday driving [1]. Hence, some epidemiological and naturalistic driving studies estimate that attentional defaults are a major contributing factor in at least a quarter of all car collisions [2, 3].

Until now, the impact of attentional defaults on driving has mainly been assessed by studying behavioral performances, naturalistic observations or self-reports [4]. These approaches can be very informative but they only provide information about the observable impact of attentional defaults. Neuroimaging techniques can bring new information in the investigation of specific attentional defaults on driving performance by examining the different stages of information processing.

Electro- (EEG) and magneto-encephalography (MEG) are particularly adapted for this purpose. Indeed, the analyses of the event related potentials (ERP) or evoked magnetic fields coming from these techniques allow characterizing the dynamics and topography of brain activity induced by the perception and the processing of a particular event with an excellent time resolution (fractions of a millisecond) [5]. Thus, EEG and MEG make it possible to examine different stages of information processing and allocation of attention.

Currently, only a few researchers in the transport domain have begun to use EEG to assess the impact of a conversation (with a passenger or via mobile phone) or the task (free driving or following a vehicle) on the processing of a relevant [6, 7] or not relevant information (arbitrary sounds in the environment) [8-10] for the driving task. These studies analyzed the characteristics of the ERP component P3 and found a deterioration of the cognitive processing by increasing attentional demand. However, these analyses concern only 3 or 4 sensors and does not allow for a deep examination of the spatial distribution of this activity. If EEG is easier to use in a driving simulator than the more cumbersome MEG, MEG is more efficient in estimating spatial

S. Supek and A. Sušac (Eds.): Advances in Biomagnetism – BIOMAG2010, IFMBE Proceedings 28, pp. 322–325, 2010.
www.springerlink.com

distributions. However, MEG implies to set up a car simulator adapted to the MEG environment which may explain why it has only been used once in this context [11]. In this experiment the authors investigated the impact of conversation on a visual detection task (red light) while watching a driving video. This experiment showed differential activation according to attention conditions in areas involved in visual perception and multitasking. However, the subjects did not interact with the driving scene, and the red light was un-related to the driving scene.

In the current study, we report an innovative MEG experiment using an interactive car simulator to examine the impact of an increase of attentional demand on the neuronal processes involved in the processing of relevant visual cue during driving.

II. Materials and methods

A. *Subjects*

Twelve right-handed male volunteers (mean 25.4 ± 2.1) with no history of neurological disorders and normal or corrected to normal vision (contact lenses) participated in this study. All participants have held a driving license for at least 3 years and drove at least 3 000 kilometers per year. Participants gave their informed consent in accordance with the Code of Ethics of the World Medical Association (Declaration of Helsinki) and were compensated for their participation.

B. *Stimuli and Procedure*

Subjects drove in a car simulator adapted to account of the MEG constraints to allow the recording of brain activity during the driving task. The simulator was equipped with a steering wheel, turning indicators, an accelerator and a brake pedal. The visual scene of an urban driving environment was projected onto a screen.

The subjects were seated at a distance of 87 cm in front of the screen (with a visual angle of 34°). Head motions were restricted by using a head stabilizer bladder. Instructions were given to drive as they would normally do, to respect all traffic signs, and to follow direction signs. In particular subjects were instructed to follow traffic light rules whereby they have to stop at an amber light.

Beforehand, participants completed either one or two practice trials (5mins) in order to familiarize themselves with the simulator controls. Following this, participants drove in 18 different, randomly controlled, driving scenarios each 5 min in length. During half of the scenarios (9) subjects were presented with radio broadcast (comfortable sound level) which they previously instructed to ignore – simple task. In the other trails subjects had to actively listen to the broadcast in order to answer three questions previously asked – dual task.

C. *Data Acquisition and Analysis*

The experiment was conducted at the MEG Centre of CERMEP (Bron, France). MEG recording was performed using a whole-head MEG system (Omega 275, CTF, VSM MedTech Ltd.) with 275 radial gradiometers over the scalp and 33 reference channels for ambient field correction. Signals were digitized at a sampling rate of 300 Hz and were recorded continuously applying band-pass filtering from 0 to 75 Hz.

Vertical and horizontal eye movements (electrooculogram, EOG) were recorded for artifact control. Head coils were placed on the nasion and on left and right pre-auricular points prior to scanning for continuous head localization recording. Trials were rejected offline for excessive eye blinks, eye movements, head movement, muscular, or electromagnetic artifacts.

A trigger marked the onset of each stimulus of interest (traffic light turning from green to amber) for trial extraction from the continuous MEG data. For trials with correct responses, the averaging epoch was defined from 200 ms before stimulus onset to 1000 ms, and data were baseline-corrected based on a 200 ms pre-stimulus interval.

To evaluate the underlying neuronal activity, source-space activity was estimated for each time point in each condition and subject for the traffic light turning from green to amber, using a distributed model: the minimum-norm-estimation method (MNE). The MNE is an inverse modeling technique, which estimates cortical generator structures without any a priori assumptions regarding the location and/or number of current sources. This inverse source modeling and the consecutive statistical analysis thereof were conducted with the BrainStorm® Toolbox (freely available at: http://neuroimage.usc.edu/brainstorm/). The differences between ST and DT were estimated in a group analysis using a Student t-test with permutations which provides a correction for multiple comparisons.

III. Results

A. *Neuronal Network Activated by Traffic Light Processing*

The minimum norm method allows us to explore neuronal generator sources of MEG activity induced by the change of traffic light from green to amber during simulated driving. The results show the activation of a widely distributed neuronal network in both simple and dual tasks. For the purpose of simplicity, only the major activities observed in this context will be described.

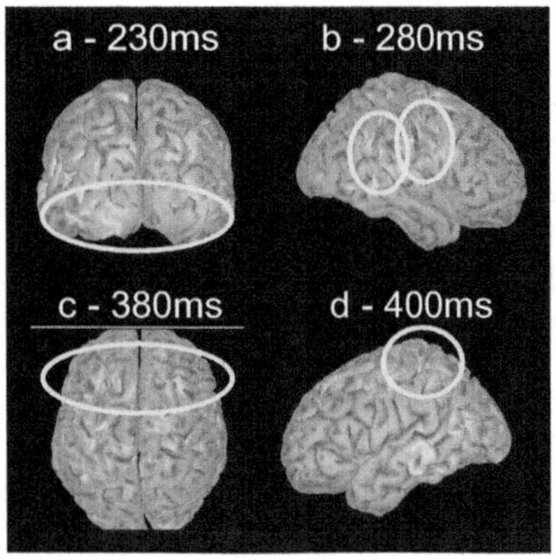

Fig. 1 MNE results showing the main neuronal network activated by traffic lights in ST according to the time

A first wide source of activity was observed within the primary and secondary visual areas (Brodmann areas BA 17, 18) from 140 ms to the end of the analysis window (Fig. 1a). This activity was maximal around 230 ms and larger in the left hemisphere than in the right. Then, a strong activity was observed around the right temporo-parietal junction (BA 37, 39) from 200 ms with a maximum around 280 ms in both ST (35 pAm) (Fig. 1b) and DT (29 pAm). Then, it spreads out to the posterior areas of the right frontal lobe (BA 6, 9, 44). This activity started around 200 ms and reached its maximum around 300 ms in ST (24 pAm) (Fig. 1b) and 325 ms in DT (20 pAm). The superior parietal areas (BA 5, 7) were also activated (Fig. 1d). In ST, this activity first occurred on the left hemisphere from 170 ms with two peaks: one at 215 ms in ST (17.5 pAm) as in DT (15pAm) and one at 400 ms (30 pAm) in ST and at 430 ms (30 pAm) in DT. Finally, the bilateral medial frontal areas (supplementary motor areas, BA 6) were activated with a maximum around 380 ms in both ST (Fig. 1c) and DT (24 vs 22.5 pAm).

B. Impact of the Dual Task on Traffic Light Processing

In order to determine the impact of a change of driver's attentional demand, MEG activity recorded in ST and DT conditions are compared. This comparison revealed a wide neuronal network including the areas cited above as well as other brain areas involved in attentional processing.

Hence, the activity related to the perception of traffic light turning from green to amber is significantly modulated in the primary visual areas (BA 17; around 280 ms; Fig. 2a), in the temporo-parietal junction (BA 37, 39; around 400 ms; Fig. 2b) and in the left superior parietal area (BA 5, 7; around 530 ms; Fig. 2c). Within these regions, only the last showed a stronger activation in DT than in ST. In addition, two other regions were activated to a lesser extent in both ST and DT conditions separately and in which a decreased activity following the increase of attentional demand is observed: the left dorsolateral prefrontal cortex (BA 10-46; around 530 ms; Fig. 2c) and the middle and inferior temporal cortex (BA 20, 21; around 550 ms; Fig. 2d).

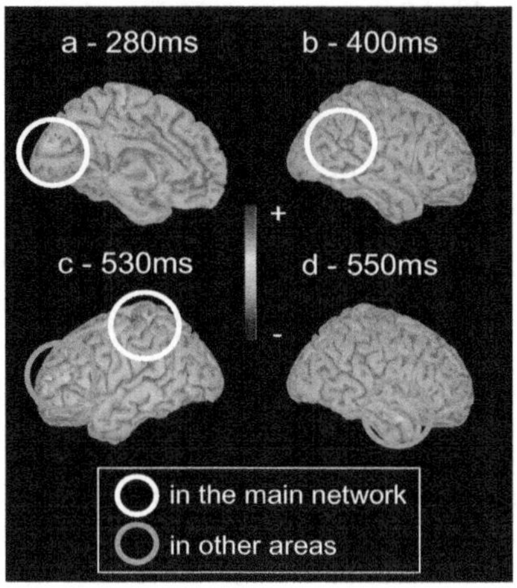

Fig. 2 Statistically significant differences (t-test corrected for multiple comparisons) between ST and DT according to the time

IV. Discussion

Using MEG, we investigated brain activity induced by the processing of visual stimuli relevant to the driving task: traffic light turning from green to amber. Then our analysis focused on the impact of an increase in attentional demand by comparing this activity according to the attentional state of subjects.

Concerning the main network involved in the processing of relevant visual cues for driving, the results revealed that, regardless of the attentional state of the subject, it implies visual areas and a right fronto-parietal network. Primary and secondary visual areas showed a stronger activity in the left hemisphere than in the right. This difference could be due to the crossed architecture of the visual streams inside the brain. As all traffic lights appear in the right field of view, they are firstly processed in the left cerebral hemisphere. The fronto-parietal network observed here is consistent with the one described in the literature and involved during tasks requiring attention orientation to relevant cues [For a review see 12]. Thus, the temporo-parietal junction has been observed repeatedly in a variety of tasks requiring the redirection of attention to task-relevant information [13, 14] and the superior parietal lobule would be engaged in tasks involving the shift of attention at a frequency basis, as it is the case in our study [15]. All activities recorded are consistent with the driving task, which strongly engages attention and implies to shift attention between a great numbers of potential relevant cues for safe driving. An activation of the bilateral medial frontal areas which seem to be linked to the motor responses (brake response) has also been observed.

Concerning the impact of an increase in attentional demand from simple to dual tasks, the results demonstrated a modulation of brain activity in the main network described above as well as in two other brain regions. Hence, a decrease in brain activity in primary visual areas is observed. These results suggest modification of the early visual processes by the increase of attentional demand. Modulations of activity in attention neural network have also been observed. The dorsolateral prefrontal cortex shows lower activations in high attentional demanding conditions. This region may be involved in action selection and could play a specific role in top-down control of neuronal activity in regions processing task-relevant representations [16, 17]. According to this, our results reflect a diminution of top-down attention for the selection of relevant information (traffic lights) and action selection (to brake or not) in the dual task situation. A diminution of activity is also observed around the right temporo-parietal junction known to play a role in redirection of attention [13, 14], which suggests a modulation of attention orientation in high attentional demand condition. Finally, the posterior parietal cortex showed stronger activity in high attentional demand condition. As this region has been shown to be involved in the shift of attention, this increase in the dual task condition is not surprising, since subjects had to shift their attention between the radio broadcast and the driving scene [15].

Further analyses are currently being conducted to better understand the impact of an increase of attentional demand on neuronal process involved while driving. These analyses concern anticipation and motor preparation which are critical while driving.

By demonstrating the feasibility of recording MEG activity during an interactive simulated driving task, this study opens new possibilities to investigate issues relative to drivers' activity.

Acknowledgment

This work is part of the project "DACOTA" (ADEME 0566C0176, ANR-05-PDIT-004-02; PREDIT-GO 4 Technology for Security) which was granted by the French National Agency for Research. We would like to thank all the partners of this study: C. Pangaud, S. Gatt, G. Foliot, C. Collet, A. Dittmar.

References

1. Stutts J, Feaganes J, Reinfurt D et al. (2005) Driver's exposure to distractions in their natural driving environment. Accid Anal Prev 37:1093-1101
2. Klauer S G, Dingus T A, Neale V L et al. (2006) The Impact of Driver Inattention On Near-Crash/Crash Risk: An Analysis Using the 100-Car Naturalistic Driving Study Data.Virginia Tech Transportation Institute, DOT HS 810 594
3. Stutts J C, Reinfurt D W and Rodgman E A (2001) The role of driver distraction in crashes: an analysis of 1995-1999 Crashworthiness Data System Data. Annu Proc Assoc Adv Automot Med 45:287-301
4. Lee J D (2008) Fifty years of driving safety research. Hum Factors 50:521-528
5. Luck S J (2005) An introduction to the event-related potential technique. MIT Press, Cambridge
6. Strayer D L and Drews F A (2007) Multi-tasking in the automobile. In Attention: From theory to practice, Kramer A F, Wiegmann D A and Kirlik A, Oxford University Press,
7. Bruyas M-P, Chapon A, Lelekov-Boissard T et al. (2006) Évaluation de l'impact de communications vocales sur la conduite automobile. Recherche Transports Sécurité 91:99-119
8. Raabe M, Rutschmann R M, Schrauf M et al. (2005) Neural correlates of simulated driving: auditory oddball responses dependent on workload, HCI International, Las Vegas, Nevada (USA)
9. Rakauskas M E, Ward N J, Bernat E et al. (2005) Psychophysiological measures of driver distraction and workload while intoxicated, Third International Driving Symposium on Human Factors in Driver Assessment, Training and Vehicle Design
10. Wester A E, Böcker K B E, Volkerts J C et al. (2008) Event-related potentials and secondary task performance during simulated driving. Accident Analysis & Prevention 40:1-7
11. Bowyer S M, Hsieh L, Moran J E et al. (2009) Conversation effects on neural mechanisms underlying reaction time to visual events while viewing a driving scene using MEG. Brain Research 1251:151-161
12. Nobre A C (2001) The attentive homunculus: Now you see it, now you don't. Neuroscience & Biobehavioral Reviews 25:477-496
13. Corbetta M and Shulman G L (2002) Control of goal-directed and stimulus-driven attention in the brain. Nat Rev Neurosci 3:201-215
14. Mitchell J P (2008) Activity in right temporo-parietal junction is not selective for theory-of-mind. Cereb Cortex 18:262-271
15. Vandenberghe R, Gitelman D R, Parrish T B et al. (2001) Location- or feature-based targeting of peripheral attention. Neuroimage 14:37-47
16. Rowe J B, Stephan K E, Friston K et al. (2005) The prefrontal cortex shows context-specific changes in effective connectivity to motor or visual cortex during the selection of action or colour. Cereb Cortex 15:85-95
17. Rowe J B, Toni I, Josephs O et al. (2000) The prefrontal cortex: response selection or maintenance within working memory? Science 288:1656-1660

Corresponding author:
Author: Alexandra Fort
Institute: INRETS-LESCOT
Street: 25 avenue François Mitterrand
Case 24
City: Bron
Country: France
Email: alexandra.fort@inrets.fr

Temporal Dissection of Stimulus-Driven and Task-Driven Processes during Perceptual Decision about 3D SFM Stimuli

Agnieszka Miskiewicz[1,2,3], Stéphane Buffat[4], Jean Lorenceau[1,2,3], and Anne-Lise Paradis[1,2,3]

[1] UPMC, Centre de Recherche de L'Institut du Cerveau et de la Moëlle épinière (Cricm), Paris, France
[2] CNRS, Cricm UMR 7225, Paris, France
[3] Inserm, Cricm UMR-S975, Paris, France
[4] IRBA, Brétigny-sur-Orge, France

Abstract— **Analyzing the visual properties of a stimulus, such as its shape or its motion, involves a temporally constrained cascade of processes including sensory integration, attentional selection and perceptual decision. In natural conditions, these processing stages may temporally overlap. In this study, we used a time constrained paradigm designed to segregate these stages, and recorded MEG activity to characterize their cortical correlates.**

The experimental sequence comprised a structure-from-motion (SFM) target flanked by pre and post masks, which limited the stimulus-driven processes in time.

The MEG responses were recorded in three perceptual tasks bearing on the presence or absence of the target SFM stimulus, on its direction of motion or on its 3D shape.

Subjects' responses were delayed to the end of the trials using a random stimulus-response mapping.

The analysis of the evoked magnetic fields reveals temporally non-overlapping responses for stimulus-driven and task-related processes.

Source reconstruction reveals that the occipital, dorsal and ventral stimulus-related areas follow different temporal profiles, The response in IPS slowly rising after the beginning of the activity in the lateral occipital cortex and falling back to the baseline prior the delayed motor response, may reflect accumulating evidence on the presented stimulus leading to a motor response, in line with the results from single cell studies in monkey LIP [1].

Keywords— **visual motion, 3-D shape, perceptual decision, MEG, source reconstruction.**

I. Introduction

Analyzing the visual properties of a stimulus, such as its shape or its motion in the 3-D environment, involves a temporally constrained cascade of processes including sensory integration, attentional selection and perceptual decision. In natural conditions, visually dynamic events require the integration of information about form and motion. Recent studies have challenged the traditional dichotomy between form- and motion-related processes in the ventral and dorsal visual pathways, and there is now evidence that form and motion processing interact in several ways. Both figural and low-level form cues guides motion perception, whereas motion cues can induce figural percept (see [2] for a review). However, more work is needed to better understand the perceptual decision stages and the attentional modulation required by the task. The experiments described in this paper aims at investigating these issues.

To segregate the different processing stages, we used structure from motion stimuli which yield both motion and shape percepts from motion cues only, and designed a time constrained paradigm with dynamical masks. We recorded MEG activity to characterize the cortical correlates of such perception in three perceptual tasks bearing on the presence or absence of a target SFM stimulus, its direction of motion or its 3D shape. We reasoned that the masks would isolate a time window for measuring the integration of form and motion cues by interrupting the early visual processing. By carefully choosing supra-threshold stimuli, we ensured that this integration yield to a perceptual decision, the MEG correlate of which should be detectable.

We specifically tested when the attentional modulation induced by the task takes place in the cortical activity, and if this modulation could be found in the occipital and intraparietal areas that are involved in the stimulus integration. We also searched a magnetic correlate of the perceptual decision processes in IPS, a region known to be involved in the accumulation of evidence in visual task implying motion stimuli [1].

II. Materials and Methods

A. Experimental Procedure

Participants. Sixteen people (age 28.5±8.1, right-handed, normal or corrected-to-normal vision, 7 men) took part in the study after giving their informed consent. The study was approved by the Local Ethical Committee.

SFM stimuli and motion masks. Structure-from-motion (SFM) stimuli were made of 140 white dots (0.19° visual angle) randomly distributed on a black background and back-projected on a 3D virtual surface. The surface had three possible shapes and oscillated in depth with three possible motion directions. Stimuli were viewed through a

S. Supek and A. Sušac (Eds.): Advances in Biomagnetism – BIOMAG2010, IFMBE Proceedings 28, pp. 326–329, 2010.
www.springerlink.com

20° circular mask hiding the boundaries of the 3D surface. Overall, nothing but motion cues could be used to identify the shape.

For the masks, dots followed a random walk with instant speeds comprised between 0°/s and 7°/s to match that of the SFM stimuli. More details about the display can be found in [3], where identical stimuli were used.

Experimental design. During the MEG recording, participants underwent six runs of 90 trials (see Figure 1) each.

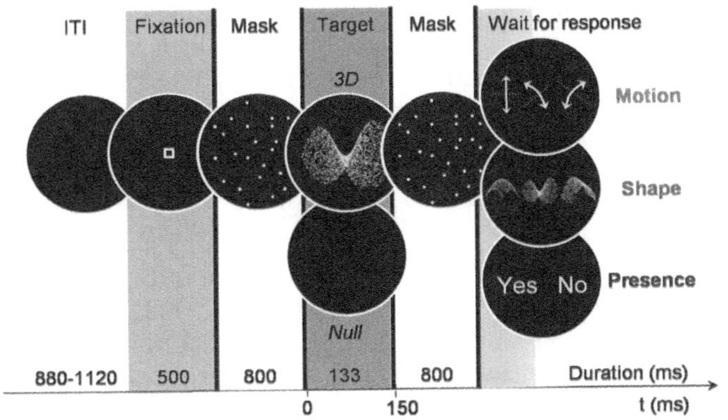

Fig. 1 Schematic time flow of a trial

Depending on the experimental run, participants had to:

- identify the target shape among three (Shape task);
- identify its motion direction among three (Motion task);
- answer whether a 3D stimulus appeared or not among the mask (Presence task). Run order was randomized across subjects.

The target, either a 3D stimulus (56 %) or a black screen (44 %) was flanked by dynamical masks to limit the duration of its visual integration. Because the stimulus-response mapping varied across trials, participants had to wait for the response screen to give their answer, so that motor activity was delayed after the end of the second mask.

Participants fixated the center of the screen but were advised to broaden their attention on the whole viewing window to ease identification. When a null target occurred in the shape and motion tasks, they had to choose a default response, which was specified at the beginning of the experiment and was randomized across subjects.

Stimulation was projected at 90 cm from the participants, using a calibrated LCD video device (60 Hz refresh rate) located outside the shielded room.

Data acquisition. MEG was recorded continuously during the runs (sampling rate: 1250 Hz), using a 151-channel whole-head system (axial gradiometer, CTF System, Canada). We ensured that the head position relative to the sensors did not change by more than 5 mm along the whole experiment. Electro-oculogram was also measured for offline artefact rejection.

B. MEG Analyses

Trials with signal artefacts (blinks, saccades, muscle signal…) were rejected using semiautomatic procedures. A 300-ms baseline was defined in the fixation period.

Event-related magnetic field (ERF). Trials were averaged by task and type of target, resulting in six ERF by subject.

Source estimation. The cortical current density was estimated using overlapping spheres for the head model and a minimum-norm constraint for the solution of the inverse problem (http://neuroimage.usc.edu/brainstorm). Individual anatomy was available for half subjects; scalp tessellation was then computed using BrainVISA (http//brainvisa.info). For the other half, we used a default head model based on the same template as that used to define the cortical surface on which results were projected (MNI "Colin 27" brain).

Sensors and regions of interest (ROI). The response to the 3D target condition in the Presence task was used to delineate groups of sensors and brain areas activated by structure-from-motion perception. We selected four regions of interest (see Figure 2), the activity of which was representative of the profiles encountered in all activated regions.

ANOVA. ERF were averaged within sensor groups and analyzed time by time in a repeated-measure ANOVA with factors Task (Shape, Motion); Target (3D, Null) and Sensors (Occip, Temp, Pariet, Centr, Front). We also performed a 2 x 2 (Task, Target) ANOVA in the selected areas.

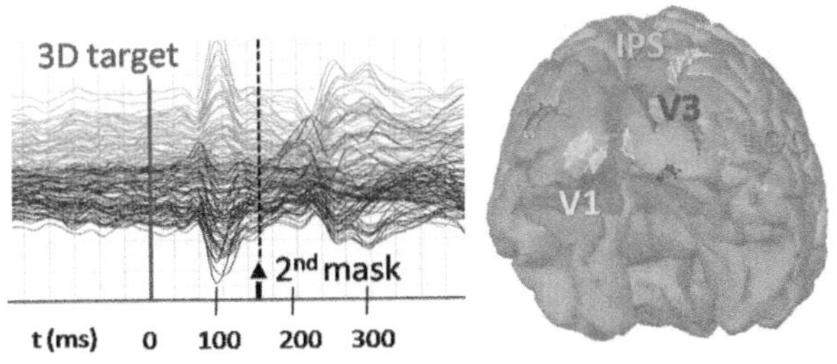

Fig. 2 Activity related to the 3D target in the Presence task (left) and selected regions of interest (right). V3: superior occipital gyrus; LOC: lateral occipital cortex; IPS: intraparietal sulcus

III. Results

A. Temporal Dissociation of Task and Target Effects

The ANOVA performed on the sensors revealed a temporal dissociation of the task and target effects, with significant ($p < 0.001$ uncorrected) target x sensors and task x sensors crossed effects occurring respectively between 100 and 200 ms, and between 300 to 320 ms. These crossed effects are explained by the topographies depending on the target in the first time window, and on the task in the second time window. More precisely, the target-related effect corresponds to modulations in the occipital sensors, whereas the task-related effect better corresponds to modulations in the parietal and temporal sensors.

B. Cascade of Activity from V1 to IPS

At the source level, both the time window of the target effects and the activity profiles (see Figure 3) reveal the successive involvement of the regions of interest in processing the 3D stimulus.

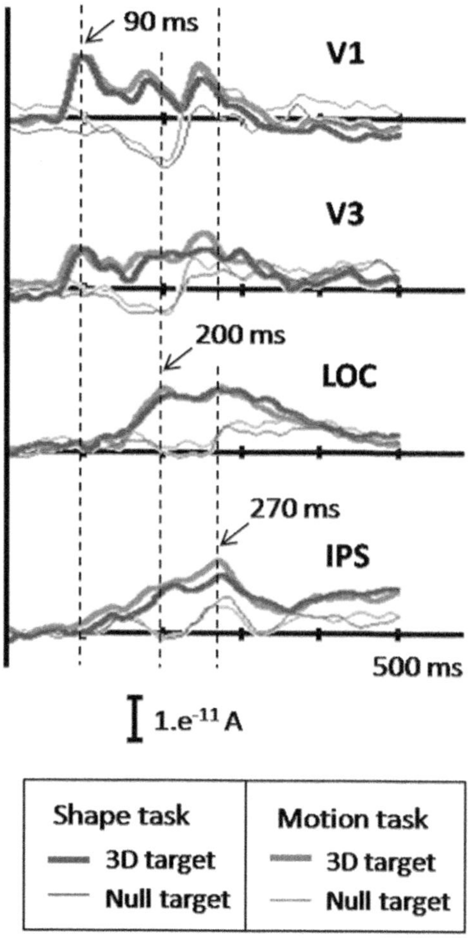

Fig. 3 Reconstructed activity in the regions of interest

The four areas indeed show very different time course after the 3D target, with tiered peak latencies. In V1, we observe a peak at 90 ms (M100 for target apparition) and another one at 240 ms, well accounted for by the M100 of the 2nd mask (onset at 150 ms). In line with this, we find a target effect (with $p < 0.01$ uncorrected during more than 20 samples) during the first peak (85 to 190 ms). We also find a late task effect during the M100 of the 2nd mask (230 to 250 ms). V3 also shows an early peak at 90 ms and corresponding target effects between 90 and 210 ms. LOC, with a later peak at 200ms, shows target effects between 200 and 350 ms. Neither V3 nor LOC reach significant task effect. Eventually, IPS activity rises slowly after the beginning of the activity in LOC, peaks at 170 ms, and falls back to the baseline prior the delayed motor response. It does not show any significant effect of target or task.

The activity related to the 2nd mask, which is visible in the null target condition, reveals the same succession of activity in the different areas.

IV. Discussion

We recorded activity related to structure from motion integration in a mere detection task (Presence), as well as in identification tasks (Shape and Motion) requiring selective attention to the shape or 3D motion of the stimulus.

A. Task-Related Modulation

The regions of interest highlighted by the processing of SFM stimuli in the detection condition did not appear to be significantly modulated by the identification tasks, which is consistent with previous fMRI results [4] showing that the attentional selection of 3D motion and shape involve different areas from stimulus-driven activity. Rather, the task effect seems to involve more anterior visual brain regions as well as right fronto-polar areas. Further study is required to clarify whether task difficulty [5] or prospective memory due to the delayed response [6] might account for such frontal activity.

B. From V1 to IPS

In the regions of interest, the time window of the target effect, as well as the activity profiles, revealed the successive involvement of these areas in processing the 3D stimulus. This cascade being consistent with what has been described recently for structure from motion stimuli [7].

Note that we find the implication of the superior occipital gyrus (putative V3) at earlier latencies than LOC activation, consistent with our previous proposal that this area could be involved in an intermediate stage of gross shape extraction.

Although one can wonder to what extent spiking activity can be related to MEG signal, we find in IPS a late climbing activity similar to that described in cell studies [1]. Also, the particular timing of IPS activity relative to MT/LOC is consistent with the idea that IPS might accumulate evidence about the moving stimulus before reaching a perceptual decision, as long as information about the target is required, and available from earlier areas. Further studies modulating both the amount of information necessary to reach a decision and the possible duration of stimulus integration (between masks) would be needed to confirm the causal relation of the activity in these areas.

C. Target-Related Activity and Masking

In this study, we used a time constrained paradigm with masks immediately preceding and following the target in order to segregate stimulus integration stages from task related-activity and hopefully perceptual decision. Our hypothesis however, was that the dynamical masks would successfully interfere with the integration of the visual cues leading to the perception of the 3D target stimulus. It is noteworthy, that contrary to [7], we neither found massive and sustained activity in the infero-temporal areas nor a rebound of activity in the early visual areas (V1 and V3) around 500 ms. This is a first hint that target visual integration is interrupted by the masks.

Another indication is given by the response to the 2nd mask after a Null target (Figure 4.A).

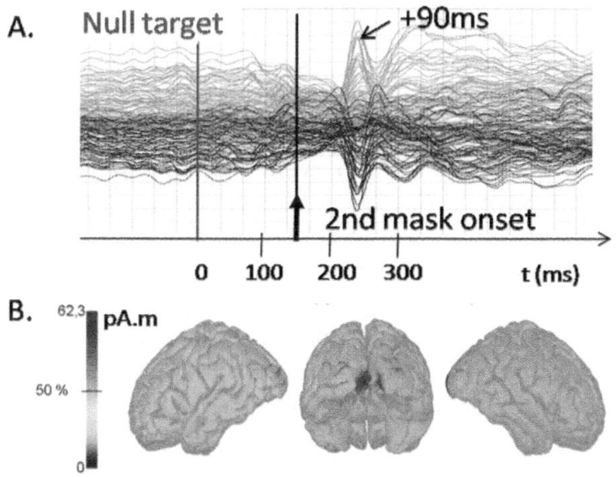

Fig. 4 Temporal window of target integration (A) and brain areas (B) defined by the response to the masks

In this condition, the mask-related M100 clearly appears in the ERF, delimiting the moment when target-related signal may not be observable anymore. Such response to the mask can also be found for the 3D target condition in V1, and to a less extent in V3 (see figure 3). Eventually, the sources of the response to the mask (Figure 4.B) appear to overlap largely with the areas involved in the 3D target processing, suggesting that the mask effectively compete with the 3D target for access to cascade of visual processing.

V. Conclusions

ERF revealed temporally non-overlapping effects of stimulus-driven and task-related processes in the perception of structure-from-motion stimuli.

Source reconstruction showed that the occipital, dorsal and ventral stimulus-related areas follow different temporal profiles and confirms a cascade of activity from V1 to IPS passing through MT/LOC for moving object perception.

The rising activity found in IPS may reflect accumulating evidence about the motion stimulus presented, in line with the results from single cell studies in monkey LIP [1].

Acknowledgment

This work was supported by a doctoral grant from DGA to AM and a MRIS-DGA grant (2006.34.059) to JL.

References

1. Huk AC, Shadlen MN (2005) Neural activity in macaque parietal cortex reflects temporal integration of visual motion signals during perceptual decision making. J Neurosci 25 (45): 10420-36.
2. Kourtzi Z, Krekelberg B, van Wezel RJA (2008) Linking form and motion in the primate brain. Trends Cogn Sci 12: 230-236
3. Miskiewicz A, Buffat S, Paradis AL, Lorenceau J (2008) Shape and motion interactions at perceptual and attentional levels during processing of structure from motion stimuli. J Vis 8 (16): 17 1-14.
4. Paradis AL, Droulez J, Cornilleau-Peres V, Poline JB (2008) Processing 3D form and 3D motion: respective contributions of attention-based and stimulus-driven activity. Neuroimage 43: 736-47.
5. Mangina CA, Beuzeron-Mangina H, Ricciardi E, Pietrini P, Chiarenza GA, Casarotto S (2009) Neural correlates of "analytical-specific visual perception" and degree of task difficulty as investigated by the Mangina-Test: a functional magnetic resonance imaging (fMRI) study in young healthy adults. Int J Psychophysiol 73: 150-6.
6. Burgess PW, Scott SK, Frith CD (2003) The role of the rostral frontal cortex (area 10) in prospective memory: a lateral versus medial dissociation. Neuropsychologia 41: 906-18.
7. Jiang Y, Boehler CN, Nonnig N, Duzel E, Hopf JM, Heinze HJ, Schoenfeld MA (2008) Binding 3-D object perception in the human visual cortex. J Cogn Neurosci 20: 553-62.

Corresponding author:
Anne-Lise Paradis
Cricm, équipe Cogimage - LENA
47, bd de l'hôpital
F-75013, Paris, France
Email: anne-lise.paradis@upmc.fr

MEG Study of Amygdala Responses during the Perception of Emotional Faces and Gaze

Thibaud Dumas[1,2,3,4], Yohan Attal[2,3,4], Marie Chupin[2,3,4], Roland Jouvent[1], Stéphanie Dubal[1], and Nathalie George[2,3,4]

[1] CNRS, USR 3246, Centre Emotion, Paris, France
[2] CNRS, UMR 7225, Centre de Recherche de l'Institut du Cerveau et de la Moelle épinière (CRICM), Paris, France
[3] Université Pierre et Marie Curie-Paris 6, UMR-S975, Paris, France
[4] Inserm, U975, Paris, France

Abstract— **The uncovering of the time course of amygdala activation is a challenge for non invasive electromagnetic brain imaging methods. Here, we propose a new method to detect the activation of this deep brain structure with magnetoencephalography (MEG). The originality of our model is to use a realistic anatomical and electrophysiological model of the neocortex and amygdala, together with distributed source imaging performed simultaneously throughout the cortical surface and the amygdala volume. In a first step, we analyzed the extent to which simulated activation of the amygdala may result in detectable MEG signals. This simulation study demonstrated that MEG is sensitive to amygdala sources. Then, our method was applied during an experimental protocol where fearful and neutral faces with direct and averted gaze were presented. Such stimuli are known to activate the amygdala from fMRI and intracranial EEG studies. Early peaks of amygdala activity were detected at 100 and 160ms post-stimulus onset. This shows an amygdala contribution to the first stages of face processing.**

Keywords— **MEG, amygdala, ERF, emotional faces, gaze.**

I. Introduction

Numerous fMRI and PET scan studies have shown that amygdala is preferentially activated during emotional face presentation [1], and may also be responsive to gaze direction [2] with a preference for direct gaze [3]. However, there are still few studies exploring the time course of amygdala responses. In this domain, the most precise data come from depth electrode studies in epileptic patients [4]. However, such studies cannot be performed routinely in healthy subjects. Here, we propose a new non-invasive method that allows studying amygdala responses with the excellent temporal resolution of MEG.

It has long been held that sub-cortical sources were not detectable with MEG, due to their distance to the scalp as well as to their lack of pyramidal cell. However, an increasing number of studies have recently brought evidence that MEG could detect the activation of deep brain structures including the amygdala [5,6]. In the present study we used a new method specifically dedicated to the detection of amygdala activation with MEG and EEG [7]. The originality of this method is, first, to be based on individual subject's anatomy obtained from segmented structural MRI [8] and, second, to be based on a distributed source localization method that includes the segmented neocortex, hippocampus and amygdala altogether.

In classical MEG and EEG source modeling, the cortical grey matter is supposed to be the origin of the signal measured, because of its composition of pyramidal neurons and the disposition of these neurons in macro-columns, which are considered to be the elementary MEG/EEG generators. However, anatomical data show that the basolateral nucleus of the amygdala has a neuronal composition very similar to that of the neocortex [9]. An histological study shows that 93% of this nucleus cells are pyramidal cells [10] and that this nucleus represents half of the number of amygdala neurons. Moreover with an average volume of 44,5 mm^3 and a composition of 12,2 millions of neurons, the amygdala has a mean density of 272 millions of neurons/cm^3 [11]. By contrast, the neocortex has a density of 44 millions of neurons/cm^3 [12]. Thus, the amygdala is six times as dense as the neocortex. This suggests that the histological properties of the amygdala may compensate for its distance to the scalp and make amygdala activity detectable with MEG.

II. Materials and Methods

A. *Anatomical Modelling (Forward Problem)*

For each subject, a tessellated envelope of the neocortex was obtained with The Anatomist/brainVISA free software solutions (http://brainvisa.info). Moreover, the hippocampus and amygdala were segmented from each individual MRI with the recently developed method of Chupin et al. [8]. This method is based on a competitive region-growing approach and was operated in the brainVISA environment. Thus, we obtained three tessellated surfaces that were

S. Supek and A. Sušac (Eds.): Advances in Biomagnetism – BIOMAG2010, IFMBE Proceedings 28, pp. 330–333, 2010.
www.springerlink.com

merged into a global head model (Fig. 1). Equivalent current dipoles (ECD) were evenly distributed perpendicularly to the surface of the neocortex and hippocampus in order to model the macrocolumns of pyramidal cells in these structures. As for the amygdala, because of its heterogeneous composition in terms of nuclei and histological aspect, we chose to transform the segmented amygdala surface into a volumic grid, where randomly oriented ECD were disposed in each intersection for each subject (see [7] for more details). The average numbers of source disposed in the model for each structure are summarized in Table 1.

Table 1 Average number of sources distributed in the model

Structures (Left hemisphere)	Number of sources	Standard deviation
Neocortex	12,986	1231.8
Hippocampus	2,549	144.3
Amygdala	396	72.8

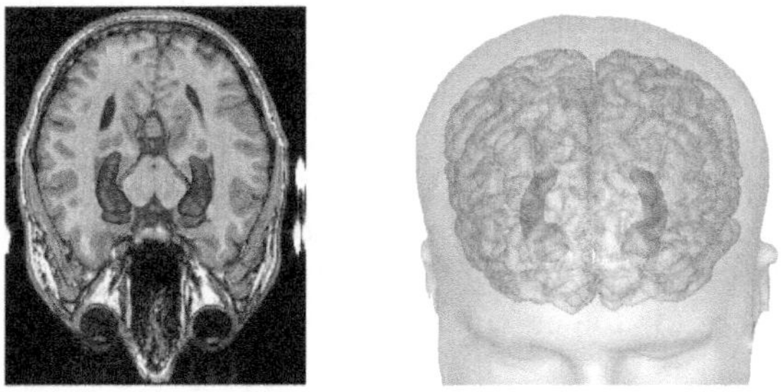

Fig. 1 Anatomical modeling used for MEG source imaging. Obtention of the tessellated surfaces from segmented individual MRI scans, cortical mesh (grey), amygdala (green) and hippocampus (red)

B. Source Localization (Inverse Problem)

The gain matrix for every sources of the global head model (including the neocortex, amygdala and hippocampus) was obtained for each subject. Then, the amplitude of activation of each source, at each time point, was estimated by distributed source imaging based on weighted Minimum Norm Estimation (L2-MCE) computed with Brainstorm free Matlab Toolbox (http://neuroimage.usc.edu/brainstorm).

C. Subjects

Six healthy, right-handed volunteers (4 women; 21-30 years; mean=25 years) provided written informed consent to take part in the study, which was approved by the local ethics committee (CPP IdF VI, protocol nb. 07024). All subjects had normal vision and no history of neurological or psychiatric disorders.

D. Stimuli and Task

The protocol consisted in presenting fearful and neutral faces with direct and averted gaze. Thus there were four conditions combining emotion (fearful or neutral) and gaze direction (direct or averted). Stimuli were 16 different individual's faces selected from the Karolinska Directed Emotional Face database [13], under their fearful and neutral frontal view versions. The eye positions of theses faces were modified manually to obtain the averted gaze version of the stimuli. All pictures were put in gray level, resized and cropped to an oval shape, and standardized in. Bbrightness and contrast of each stimulus was modified with Adobe Photoshop 7.0.1 in order to ensure that there was no significant difference between the experimental conditions. Faces subtended a visual angle of 6° (vertically) and 3.5° (horizontally).

Each trial started with a central fixation point displayed for 700 to 900 ms. Then a face stimulus was displayed for 500 ms. It was followed by a blank screen presented for 1 to 2 seconds before the next trial started. There were six blocks of 64 trials (total=384 trials). In each block, the 16 different individual's faces were seen once under each of the 4 experimental conditions of gaze (direct/averted) and emotion (fearful/neutral). Eight to twelve target stimuli consisting in a blue circle were added to each block. The subjects' task was to press a button whenever they detected the target stimulus.

E. Data Acquisition

The study took place in the MEG Center of the CENIR imaging platform of Pitie-Salpetriere Hospital, Paris (France). MEG signals were recorded on a CTF 151 channels whole head system (CTF Systems Inc., Port Coquitlam, Canada), at 1250 Hz sampling frequency with a 200-Hz low-pass filter. Three fiducials coils were placed over the nasion, the left and the right pre-auricular points on each subject's face in order to determine the head position within the MEG helmet, and to provide co-registration with the anatomical MRI. Head position was recorded and controlled before each block. An Electrooculogram (EOG) was included to detect eye movements as well as blinks.

F. Preparation for Source Analysis

MEG signals contaminated by eye movements, blinks, muscular activity or other artifacts were rejected manually. Then, MEG data were averaged between -200 ms and +600 ms (0 ms=face onset) in order to get the event-related field under each experimental condition (fearful or neutral face with direct or averted gaze). For each average, the MEG signal was low-pass filtered at 40 Hz, and the 200 ms pre-stimulus period was used for baseline correction.

III. Results

A. Evaluation of the Amygdala Model

Using the forward problem, it is possible to model the magnetic field measured by MEG sensors which is induced by the activation of any subset of sources. This however implies introducing some electrophysiological indices in the model, which are related to the current density of the dipole moments in the region of interest. For the amygdala, these indices were estimated from micro-MEG recordings in the hippocampus of animal models (amygdala data not available) [14]. Then, we calculated the magnetic field produced by a growing patch of active sources in the amygdala volume. The patch was started in each node of the volumic grid covering the amygdala, and it included a various number of adjacent nodes depending on its size. Figure 2 represents the strength of the magnetic field generated by this virtual amygdala activation. It indicates that 0.2-0.3 cm^3 of activation were sufficient to result in MEG signal above noise level detection threshold. Note that, if anything, the MEG signal calculated here may have been underestimated because it was based on the current density in the hippocampus where the neuronal density is inferior to that in the amygdala.

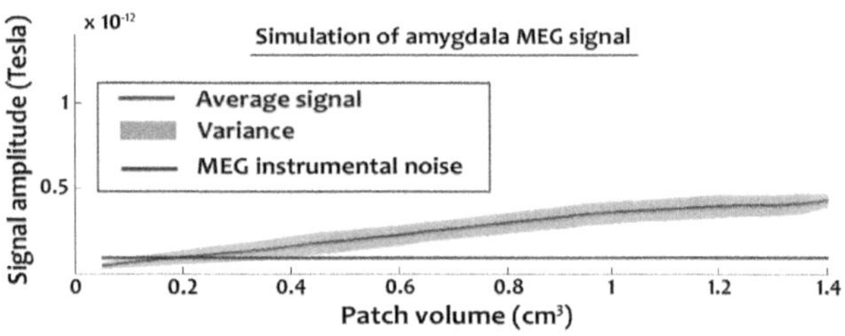

Fig. 2 Simulation of amygdala MEG signal with forward problem

B. Preliminary Experimental Validation

For every subject and at every time point, after averaging event-related fields in each experimental condition, we estimated source amplitudes over the global head model including the neocortex, the amygdala and the hippocampus. We then extracted the time course of activity for the sources located in the amygdala. At each time point, we computed the mean of the absolute amplitude values across all sources in the left and right amygdala respectively. Figure 3 represents the grand average of these activation time courses across the six subjects. A prominent peak of activity was observed at about 160 ms after stimulus onset, preceded by a smaller peak at 100 ms. This pattern was consistently observed across our six subjects.

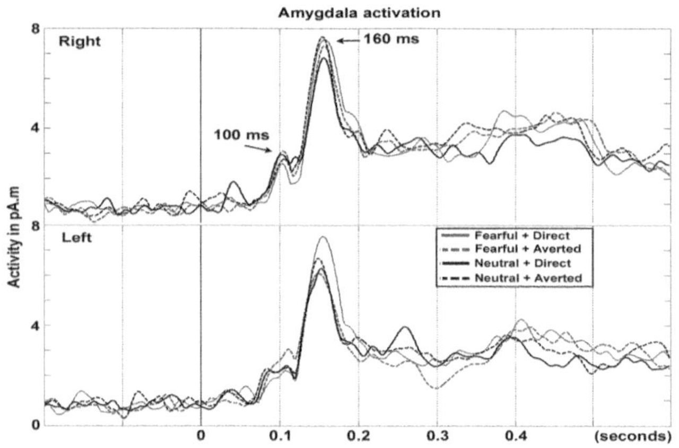

Fig. 3 Average amygdala time course obtained after Minimum Norm Estimation with our method. Mean on six subjects

Prominent activations were of course also observed in extrastriate cortex, particularly in the fusiform regions, in the early time range (Figure 4). The mean amplitude of fusiform sources was about 4 times as large as that of the amygdala sources. Yet, our model allowed localizing a temporally structured pattern of activity in the amygdala.

Finally, the limited number of subjects included in this preliminary study did not allow revealing any significant effect of the emotion or gaze conditions.

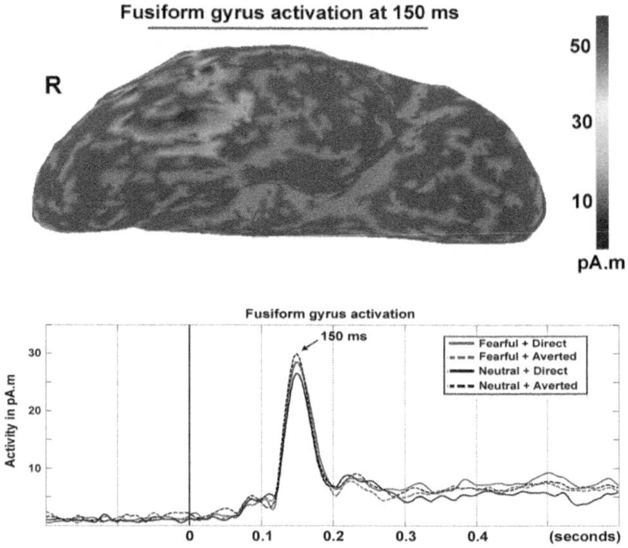

Fig. 4 Average fusiform gyrus time course obtained after Minimum Norm Estimation with our method. Mean on six subjects. On top left, an anatomical view of fusiform gyrus activation at 150 ms post-stimulus is presented for a typical subject on a ventral view of the right brain hemisphere

IV. Discussion

The aim of the present study was to propose a new method for estimating amygdala contribution to MEG signal. This method is based on a head model that is the closest possible to individual structural information. This information can be highly variable between subjects, which can contribute to the variability of MEG signal topography at the scalp surface. Therefore, taking into account individual anatomy is important in order to estimate brain sources optimally. While recent studies that have shown amygdala responses with MEG used individual structural MRI data, none included a specific model of the medial temporal lobe structures [5, 6, see also 15]. The model that we used here was based on an automated segmentation procedure of the whole amygdala volume [8], where sources were then evenly distributed. As the amygdala is in fact heterogeneous, comprising of several anatomically and functionally differentiated nuclei, with only the basolateral nucleus containing a high proportion of pyramidal cells, it is possible that our method would be improved by taking into account more refined structural information from the amygdala. This might improve the quality of amygdala source estimation.

Still, our method allowed showing clear peaks of amygdala activity at 100 ms and 160 ms following face onset. These peaks are coherent with the results from intracranial measurements in epileptic patients [16,17]. Note that these peaks of activity were concomitant with peaks of activity in fusiform, regions. It is likely that these latter regions contributed predominantly to the MEG signal in the early time window of face processing. Yet, amygdala seems to be also active in this early time range.

We did not find any significant effect of emotion or gaze on the amygdala activity. While this may appear contradictory with previous studies [1-4], several explanations may be proposed. First, we have included only 6 subjects so far. More subjects will be needed to assess the effects of gaze and emotion statistically. Second, it is interesting to note that with intracranial EEG, an effect of emotion was obtained only during an explicit emotion recognition task [4].

V. Conclusions

A new method to estimate sources in the amygdala from MEG recordings is presented. It allowed showing the contribution of amygdala to early magnetic responses to faces. Future study will determine whether these early amygdala activities and/or later ones are modulated by the emotion and gaze direction of the face.

References

1. Vuilleumier & Pourtois. (2007), 'Distributed and interactive brain mechanisms during emotion face perception: Evidence from functional neuroimaging', *Neuropsychologia* **45**(1), 174 - 194.
2. George et al. (2001), 'Seen Gaze-Direction Modulates Fusiform Activity and Its Coupling with Other Brain Areas during Face Processing', *NeuroImage* **13**(6), 1102 - 1112.
3. Kawashima al. (1999), 'The human amygdala plays an important role in gaze monitoring: A PET study', *Brain* **122**(4), 779-783.
4. Krolak-Salmon et al. (2004), 'Early Amygdala Reaction to Fear Spreading in Occipital, Temporal, and Frontal Cortex: A Depth Electrode ERP Study in Human', *Neuron* **42**(4), 665 - 676.
5. Moses et al. (2007), 'Dynamic neural activity recorded from human amygdala during fear conditioning using magnetoencephalography', *Brain Research Bulletin* **71**(5), 452 - 460.
6. Cornwell et al. (2008), 'Evoked amygdala responses to negative faces revealed by adaptive MEG beamformers.', *Brain Res* **1244**, 103--112.
7. Attal et al. (2009) 'Modeling and detecting deep brain activity with MEG and EEG IRBM, Biomed. Eng. & Res. 30, 133-3
8. Chupin et al. (2007), 'Anatomically constrained region deformation for the automated segmentation of the hippocampus and the amygdala: Method and validation on controls and patients with Alzheimer's disease', *NeuroImage* **34**(3), 996 - 1019.
9. Aggleton, ed. (2006), *The Amygdala*, Oxford University Press.
10. Washburn & Moises. (1992), 'Electrophysiological and morphological properties of rat basolateral amygdaloid neurons in vitro', *J. Neurosci.* **12**(10), 4066-4079.
11. Schumann & Amaral. (2005), 'Stereological estimation of the number of neurons in the human amygdaloid complex.', *J Comp Neurol* **491**(4), 320--329.
12. Pakkenberg, B. & Gundersen, H. J. (1997), 'Neocortical neuron number in humans: effect of sex and age.', *J Comp Neurol* **384**(2), 312--320.
13. Lundqvist, D., Flykt, A., & Öhman, A. (1998). *The Karolinska Directed Emotional Faces–KDEF.* CD-ROM from Department of Clinical Neuroscience, Psychology section, Karolinska Institutet (Stockholm, Sweden), ISBN 91-630-7164-9.
14. Okada and Papuashvili. (1996), 'Maximum current dipole moment density as an important physiological constraint in MEG inverse solution', *10th Int Conf on Biomag, Santa Fe NM*, 149.
15. Hillebrand & Barnes. (2002), 'A quantitative assessment of the sensitivity of whole-head MEG to activity in the adult human cortex.', *Neuroimage* **16**(3 Pt 1), 638--650.
16. Barbeau, et al. (2008), 'Spatio temporal dynamics of face recognition.', *Cereb Cortex* **18**(5), 997--1009.
17. Halgren et al. (1994), 'Spatio-temporal stages in face and word processing. I. Depth-recorded potentials in the human occipital, temporal and parietal lobes [corrected]', *J Physiol Paris* **88**(1), 1--50.

Author: Thibaud Dumas
Institute: CNRS, USR 3246, Centre Emotion
Street: 47, Bd de l'Hôpital
City: 75651 Paris Cedex 13
Country: France
Email: thibaud.dumas@gmail.com

Dynamic Brain Responses to Semantic Incongruencies in the Visually Presented Language Streams

Kouichi Sutani and Sunao Iwaki

Living Informatics Group, Institute for Human Science and Biomedical Engineering (HSBE),
National Institute of Advanced Industrial Science and Technology (AIST), Ikeda, Japan

Abstract— **Semantically inappropriate words embedded in a sentence elicits an event related potential (ERP) component called the N400 at latencies around 400 ms after the stimulus onset. In the present study, neuromagnetic measurements were used to investigate dynamic brain responses to semantic incongruencies in visually-presented language streams. Neural activities specific to the presentation of incongruent words in the language streams were observed in the left superior temporal, the parieto-temporal, and the inferior frontal/prefrontal regions, indicating the contribution of multiple information processing, including phonological and semantic language processing and working memory processes, in the generation of the N400 event-related component. The current results indicate that the working memory as well as the language-specific processes was involved in the generation of the N400 component.**

Keywords— **Word recognition, N400 response, LCMV spatial filter, semantic incongruency.**

I. Introduction

Semantically inappropriate words embedded in a natural language stream evoke an event related potential (ERP) component called N400 at the latencies around 400 ms after the stimulus onset [1]. Although the N400 component has been studied extensively, the profile of the N400 generator is still controversial, mainly due to its complex spatial distribution. The complexity of the scalp distribution of the N400 component indicates that there are multiple sources contributing to the generation of this component. The possible candidates for the generation of the N400 are neural activities in (a) the hippocampus [2,3], (b) the fusiform gyrus [4], (c) the left superior temporal gyrus [5,6,7], and (e) the dorsolateral prefrontal cortices [8].

In the present study, sentences consisting of six phrases, whose last phrase determines the semantic congruency of the sentence, were presented to normal adult subjects. Event-related brain activities evoked by the presentation of the last phrase were investigated using magnetoencephalography (MEG). The distribution of the neural activity corresponding to the N400 component was estimated using the linearly constrained minimum variance (LCMV) beamformer [9,10].

II. Metohds

Seventeen native Japanese speakers (2 females, 21 - 55 years of age, mean age 26.59±8.56, 2 left-handed male subjects) participated in the study as volunteers. None of the subjects had a history of neurological disorders and all gave informed consent to participate in the study, which was approved by the institutional review board of the National Institute of Advanced Industrial Science and Technology.

Japanese sentences each consisted of 6 phrases, each of which contained 2 - 6 characters, were displayed at 550 ms interval on a screen placed in front of the subject. The words were chosen from standard textbooks used in the Japanese public education system. The last (sixth) phrase was either a semantically congruent word (CONGRUENT condition as the control) or an incongruent word (INCONGRUENT condition as the target). The subjects were instructed to respond by lifting their index (CONGRUENT) or middle (INCONGRUENT) finger after the presentation of the reaction request cue. There were no significant differences in the average lengths of the last (sixth) phrase between the conditions (3.8 characters for the CONGRUENT and 3.6 characters for the INCONGRUENT).

Neuromagnetic responses were measured using a 122 channel whole-head planar-gradiometer system (Neuromag122^{TM}) [11] which was located in a magnetically shielded and air-conditioned room. The sensor array covered the whole scalp and contained 61 pairs of orthogonal planar gradiometers; each of these sensor pairs measured the two independent tangential derivatives of the magnetic field component normal to the helmet-shaped surface approximating the shape of the scalp. With this planar gradiometer configuration, the largest signal occurs just above a dipolar source.

Stimulus-related epochs of 1,200 ms including 200 ms pre-stimulus baseline were recorded with a pass-band of 0.03-100 Hz and sampled at 498 Hz. The event-related

MEG responses to correctly identified epochs were separately averaged over 50 times for each condition. The epochs with neuromagnetic signal change exceeding 3,000 fT/cm were rejected from signal averaging.

The time frequency representation (TFR) [12] of the recorded MEG raw data between 5 and 100 Hz was used to assess the event-related changes of the spontaneous brain activities. T-statistics with Bonferroni correction for multiple comparison was used to test if there was significant differences in the TFR power between the conditions.

Distribution of the N400 generator was evaluated using the LCMV beamformer [9,10]. The beamformers were implemented as a weighted sum of the data recorded at the MEG sensor array. These weights were determined so as to minimize the filter output power subject to a linear constraint, which forces the filter to pass a specific distribution of the measured signal generated from a brain electrical source at a specified location, while the minimization of power attenuates activity originating at other locations. In calculating the functional activation map using an LCMV spatial filter, the filter output power as a function of location is normalized by the estimated noise power (NAI: Neural Activity Index) [9] to compensate for the noise dominance at the locations where the sensitivity of the sensor array is low. The results were superimposed on magnetic resonance (MR) images for each subject.

III. Results

Figure 1 shows MEG responses to the presentation of the sixth phrases in INCONGRUENT and CONGRUENT conditions averaged over all subjects. Prominent differences between conditions in the averaged waveforms were observed in the left temporal and left frontal areas in 17 of 19 subjects around latencies of 400 ms.

Results of the TFR analysis were shown in Fig. 2, in which significant increase in the MEG signal power in the INCONGRUENT condition was tested. Increase in gamma-band (30 – 50 Hz) power was observed in the left temporal and the left frontal sensors.

Results of the beamformer analysis in the typical subject are shown in Fig. 3. The activities in the left middle temporal region and the left parieto-temporal regions including posterior superior-temporal area were observed between latencies of 380 and 420 ms (Fig. 3). Increased activities in the left inferior frontal, the left inferior prefrontal regions between latencies of 340 and 380 ms were observed as well.

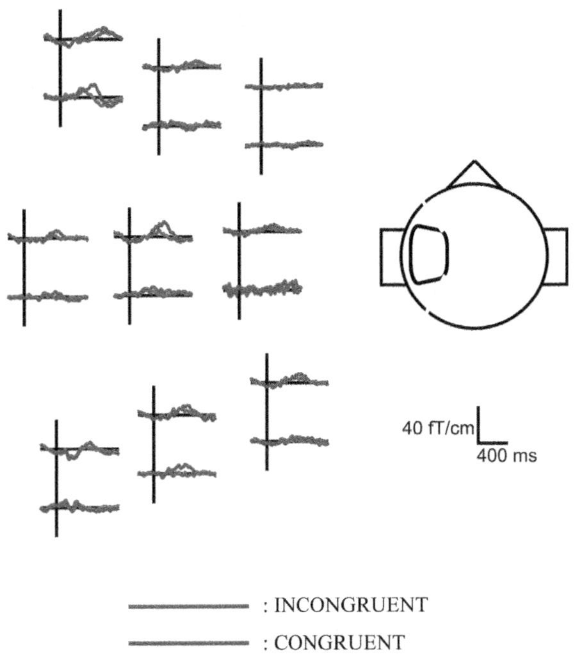

Fig. 1 MEG waveforms measured during the sentence recognition task under grand average of 17 subjects

IV. Discussion

The activity observed in the left parieto-temporal region is consistent with the results reported by Koyama [5]. These areas are known to be involved in human phonological and semantic language processing [13] including so-called Wernicke's area. Activities in the left frontal and prefrontal regions might be responsible for verifying semantic congruency of the presented phrase in the language stream [14]. Typically, the activities in these areas during word-recognition tasks are reported to be observed around latencies of 200 ms [15,16]. The longer activation latencies in these temporal language areas suggest that these areas might be activated in response to feedback input from other parts of the brain, where the semantic evaluation of the language streams takes place.

Previous studies suggest the contribution of the left frontal and inferior prefrontal areas to syntactic language processing or speech production [17,18] and the working memory [19,20], respectively.

The present results suggest that multiple brain areas including the left temporal, parieto-temporal, frontal and prefrontal regions are involved in the generation of the N400 component during the presentation of semantically incongruent words in the language stream.

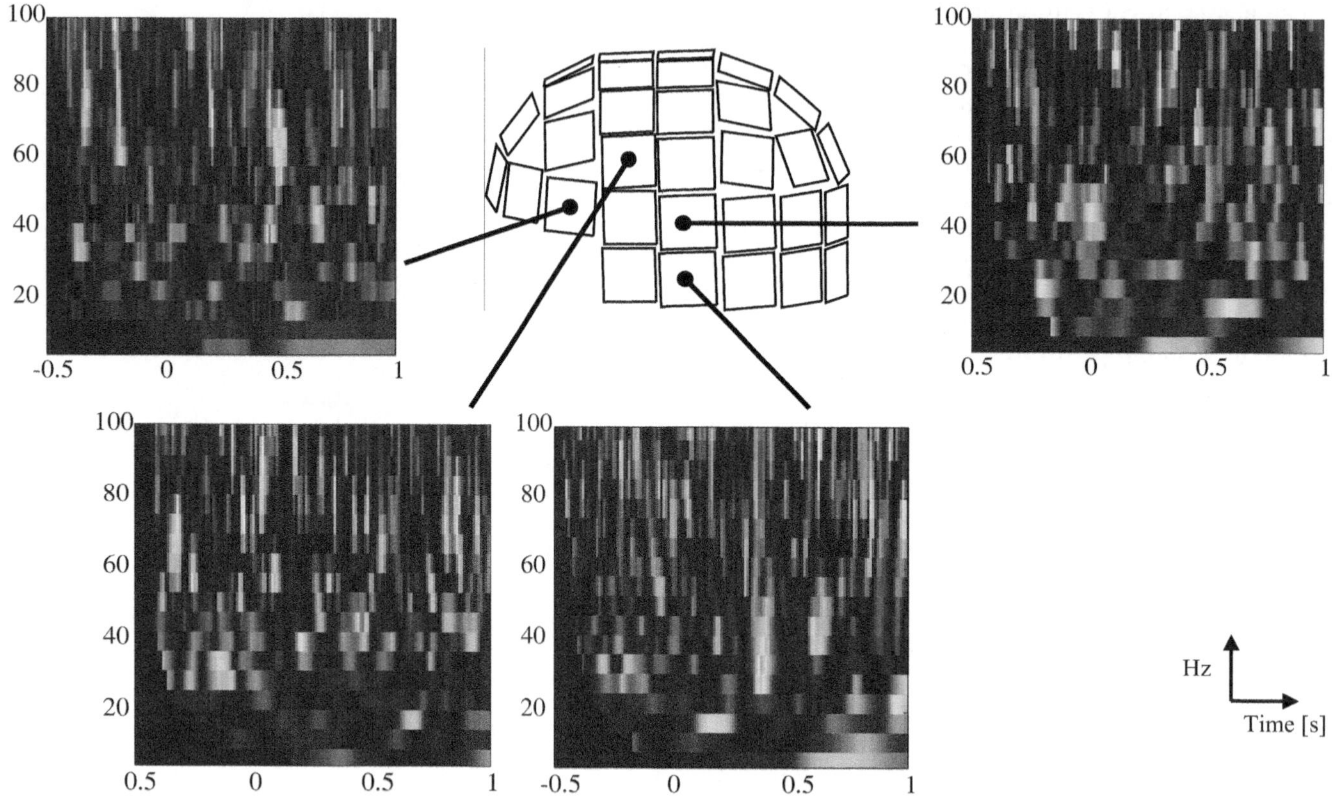

Fig. 2 Results of the TFR analysis of the MEG frequency power. Orange to red colors denote the sensor where statistically significant increase of signal power was observed in the INCONGRUENT condition

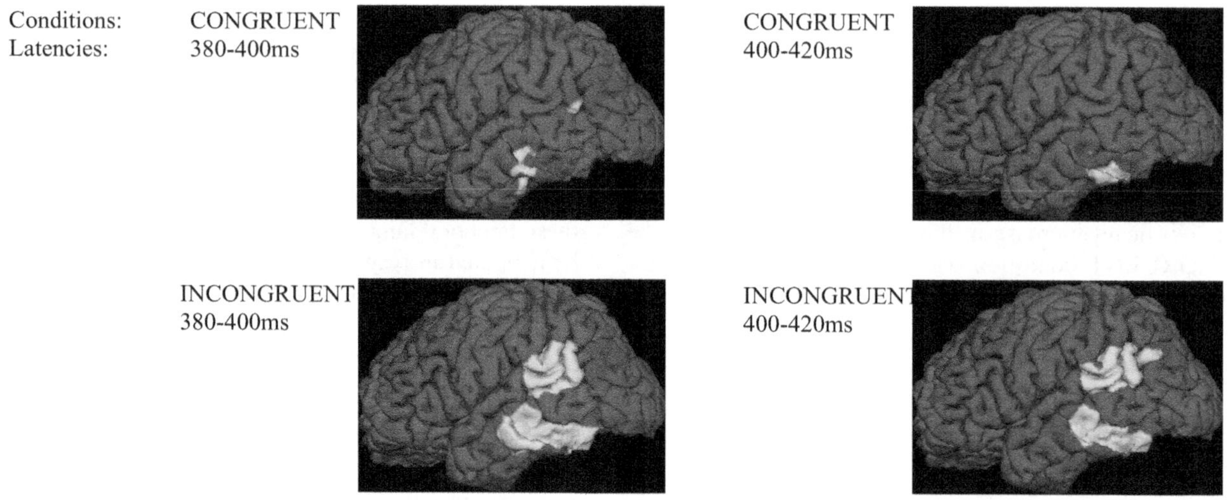

Fig. 3 Estimated neural activity distribution during sentence recognition task obtained by the beamformer analysis around the latencies of 400 ms

V. Conclusion

Brain responses during the recognition of semantic incongruencies in visually-presented language streams were investigated using neuromagnetic measurements. Neural activities specific to the presentation of incongruent words were observed in the left superior temporal, the parieto-temporal, and the inferior frontal/prefrontal regions, indicating the contribution of multiple information processing, including phonological and semantic language processing and working memory processes, in the generation of the N400 event-related component.

References

1. M. Kutas, S. A. Hillyard (1980) Reading senseless sentences: Brain potentials reflect semantic incongruity. Science 207:203-205
2. ME. Smith, JM. Stapleton, E. Halgren (1986) Human medial temporal lobe potentials evoked in memory and language tasks. Electroenceph clin Neurophysiol. 63:145-159
3. K. Hiramatsu, M. Fukuda, K. Itoh et al. (1995) MEG study of N400 component in semantic processing. Clinical Electroencephalography 37:293-297
4. AC. Nobre, T. Allison, G. McCarthy (1994) Word recognition in the human inferior temporal lobe. Nature 372:260-263
5. S. Koyama, R. Kakigi, Y. Nageishi (1994) N400 : Areview on measurement, clinical findings and its generator. Clinical Electroencephalography 36:30-34
6. S. Koyama, Y. Nageishi, M. Shimokochi. (1988) Contextual effect on word recognition and event-related potentials (N400). Clinical Electroencephalography 8:496-500
7. P.G. Simos, L.F.H. Basile, A.C. Papanicolaou (1998) Source localization of the N400 response in a sentence-reading paradigm using evoked magnetic fields and magnetic resonance imaging. Brain Res. 762:29-39
8. E. Halgren, R.P. Dhond, N. Christensen et al. (2002) N400-like Magnetoencephalography Responses Modulated by Semantic Context, Word Frequency, and Lexical Class in Sentences. Neuroimage 17:1101-1116
9. BD. Van Veen, W. Van Drongelen, M. Yuchtman et al. (1997) Localization of brain electrical activity via linearly constrained minimum variance spatial filtering. IEEE Trans. on Biomed. Eng. 44:867-880
10. S. Iwaki, S. Ueno, T. Imada et al. (1999) Dynamic cortical activation in mental image processing revealed by biomagnetic measurement. NeuroReport 10:1793-1797
11. M. Hamalainen, R. Hari, R. Ilmoniemi et al. (1993) Magnetoencephalography-theory, instrumentation, and applications to nonivasive studies of the working human brain. Rev. Modern Phys. 65:413-497
12. C. Tallon-Baudry, O. Bertrand, C. Delpuech et al. (1996) Stimulus specificity of phase-locked and non-phase-locked 40 Hz visual responses in human. J. Neurosci. 16:4240-4249
13. M. Kutas, S. A. Hillyard (1984) Brain potentials during reading reflect word expectancy and semantic association. Nature 307:161-163
14. M. Iwata (1996) Brain science series 21 : brain and language, Tokyo, Japan, 1996, pp 68-69
15. T. Sekiguchi, S. Koyama, R. Kakigi (2001) The effect of stimulus repetition on cortical magnetic responses evoked by words and nonwords. NeuroImage 14:118-128
16. L. Pylkkanen, A. Stringfellow, A. Marantz (2002) Neuromagnetic Evidence for the timing of lexical activation: An MEG component sensitive to phonotactic probability but not to neighborhood density. Brain and Language 81:666-678
17. R. Wise, F. Chollet, U. Hadar, K et al. (1991) Distribution of cortical neural networks involved in word comprehension and word retrieval. Brain 114:1803-1817
18. Y. Naito, H. Okazawa, I. Honjo et al. (1995) Cortical activation with sound stimulation in cochlear implant users demonstrated by positron emission tomography. Brain Res. Cog. Brain Res. 2:207-214
19. J. Jonides, E. E. Smith, R. A. Koeppe et al. (1993) Spatial working memory in humans as revealed by PET. Nature 17:623-625
20. A. Bechara, H. Damasio, D. Tranel et al. (1998) Dissociation of working memory from decision making within the human prefrontal cortex. J. Neurosci. 18:428-437

Author: K. Sutani and S. Iwaki
Institute: National Institute of Advanced Industrial Science and Technology (AIST)
Street: 1-8-31 Midorigaoka
City: Ikeda
Country: Japan
Email: k-sutani@aist.go.jp

Time Course and Neural Network for Comparing Written and Spoken Words: A MEG and DTI Study

Lu Meng[1,2,*], Jing Xiang[1], Douglas Rose[1], Rupesh Kotecha[1], Jennifer Vannest[3], Anna Byars[3], and Ton Degrauw[3]

[1] MEG Center, Cincinnati Children's Hospital Medical Center, 3333 Burnet Avenue, Cincinnati, OH, 45220, USA
[2] Key Laboratory of Medical Image Computing of Ministry of Education, Northeast University, Shenyang, 110004, China
[3] Department of Neurology, Cincinnati Children's Hospital Medical Center, 3333 Burnet Avenue, Cincinnati, OH, 45220, USA

Abstract— **The present study was designed to investigate the spatiotemporal signature as well as the neural network of brain activation associated with word recognition. Twenty participants were studied with a whole head magnetoencephalography (MEG) system and a 3T magnetic resonance imaging (MRI) scanner. The word stimuli consisted of 100 matched words (the visually and acoustically presented words were the same) and 20 mismatched words (the visually and acoustically presented words were different). The time course of brain activation was analyzed with a virtual sensor technique at source levels. The spatial location and connection were estimated with wavelet-based beamformer and diffusion tensor imaging (DTI) tractography. To our knowledge, this is the first study focusing on the time course and neural substrates involved in comparing spoken and written words with both MEG and DTI. The visually and acoustically presented words evoked three responses named as PM1, PM2 and PM3, respectively. The results of magnetic source imaging showed that the PM1 was localized to the primary auditory and visual cortices. The PM2 and PM3 were localized to Wernicke's area and Broca's area. Interestingly, virtual sensor waveforms showed a clear response between PM1 and PM2 from the left temporal-occipital junction. This component was consistently identified around 162 ± 8.9 ms (M160) in 20 participants (100%, 20/20). MEG source-guided DTI tractography revealed neural fibers in this junction area. The results suggested that the M160 was one part of the neural network of auditory and visual word recognition. The results of the present study yield highly convincing evidence that the left temporal-occipital junction plays an important role in comparing visually and acoustically presented words.**

Keywords— **Word recognition, Magnetoencephalography, Virtual sensor, Auditory/Visual evoked magnetic fields, Diffusion tensor imaging.**

I. Introduction

Magnetoencephalography (MEG) is an imaging technique used to measure the magnetic fields produced by electrical activity in the brain [1]. Previous reports have shown that MEG reliably records the morphology, latency and amplitude of neuronal activities in real time with excellent spatial resolution [2,3]. Recent advances in signal processing technology have also made it possible to localize specific volumes brain activation and compute virtual sensors in the brain. Wavelet based beamformer is one of the newly developed methods for analysis of MEG data for this purpose [4]. In comparison to the conventional dipole modeling, wavelet-based beamformer can scan the entire brain for possible sources without making any assumption of the number of sources.

Diffusion tensor imaging (DTI) can provide information about white matter fiber tracts in the brain. DTI tractography is a procedure to demonstrate neural tracts by acquiring DTI in at least six non-collinear directions. The result is a 3D network formed by long and short connections among different cortical and subcortical regions, and the network discloses the neuronal connections within biological tissues. The combination of MEG with DTI has the possibility to track neural activation and propagation [5].

To our knowledge, the time course and neural network involved in comparing spoken and written words with both MEG and DTI have not been reported. The objective of the present study was to investigate the neural network associated with word processing with a specific focus on the interaction of the visual, auditory and language systems. The spatial location of word processing in the brain was investigated using wavelet-based beamformer. The time course and connectivity of word processing in the brain was investigated with virtual sensor technique and DTI tractography.

II. Materials and methods

A. Subjects and Language Stimuli

Twenty healthy, native English-speaking adults were recruited for the present study. In the twenty volunteers, all participants (age: 19-49, mean age: 30 years; 10 female and 10 male) qualified for the present study. A written informed consent approved by the Institutional Review Board (IRB) at Cincinnati Children's Hospital Medical Center (CCHMC),

S. Supek and A. Sušac (Eds.): Advances in Biomagnetism – BIOMAG2010, IFMBE Proceedings 28, pp. 338–341, 2010.
www.springerlink.com

was obtained from each participant prior to testing. All participants were right-handed as measured by a questionnaire based on the Edinburgh Handedness Inventory.

The stimuli consisted of 120 words. The words were presented both visually on a backlit screen positioned at a comfortable position in front of the participants and auditorily through earphones. One hundred stimuli consisted of matched words, which meant that the seen and heard word were identical; twenty stimuli consisted of mismatched words, which meant that the seen word was different from the heard word. All the auditory and visual words were presented simultaneously. If the words did not match, the subjects were asked to press a button; no response was required when the visual and auditory stimuli matched. The onset of the visual and auditory presentation was aligned. Word presentation and response recording were accomplished using BrainX software [1].

B. Data Acquisition

A 275-channel whole cortex CTF OMEGA MEG system was used for recordings (VSM Ltd., Port Coquitlam, Canada). The MEG measurements were performed in a magnetic shielded room (MSR) with a system white noise level below 10 fT/√HZ. All subjects were supine during the whole data acquisition procedure, their left and right arms rested on either side. The location of the subject's head relative to the sensor array was measured using three small coils affixed to the nasion and the left and right preauricular points. The three coils were used to measure the position of the sensor array with respect to the nasion-ear coordinate system by means of continuous head localization, which tracks head motion during the entire procedure.

MRI scans were recorded by research technologists for all participants with a 3T MRI scanner (Siemens, Erlangen, Germany). T1-weighted MR images were recorded with the resolution of 256×256 and 128 slices for each serial. Aiming at the co-registration of MEG data and MRI data, we placed three fiducial points at the same locations as the positions of the three coils used in the MEG recordings. Consequently, we used them as landmarks to co-register these two data sets.

All of the DTI images were acquired on a 3T Siemens Trio MR imaging scanner (Siemens, Erlangen, Germany). A 46-section, diffusion-weighted, spin-echo echo-planar imaging scan was acquired in the axial plane with the following parameters: TR/TE=6000/87 ms; FOV=25.6×25.6 cm; matrix=128×128; section thickness=2mm; b-value=1000 s/mm2; and 4 repetitions. Reference T2-weighted images (b=0) were also acquired. The duration of the DTI sequence was 5 minutes 48 seconds.

C. Data Analysis

MEG data were averaged and then filtered with a band-pass filter of 1-200 Hz. Neuromagnetic responses (or deflections) were then identified with the averaged data using DataEditor (VSM MedTech Ltd., Port Coquitlam, BC, Canada). Subsequently, the latencies and the amplitudes of each recognizable response were measured using DataEditor.

Magnetic sources were localized using wavelet-based beamformer. Wavelet-based beamformer consisted of creating a three-dimensional lattice of spatial filters at 2.5 mm resolution throughout the brain, where the output of each spatial filter was an estimate of the source strength of a single source at that location. Thus, the result of the estimation of the sources was a volumetric image. The intensity of each voxel was the strength of the brain activation at the designated location. Beamformer images were fused into MRI using Magnetic Source Locator (MSL) software based on the three matched fiducial points.

Virtual sensors were also computed with wavelet-based beamformer. In this study, the virtual sensors were placed within the functional areas that were localized by wavelet-based beamformer at the voxel locations that showed peaks in functional activation. MEG virtual sensor waveforms at these locations were used to analyze the time course of neuromagnetic activity. The latency and amplitude of the response peak of the virtual sensor waveform were measured with MEG Processor.

DTI tractography was analyzed and computed using Riemannian framework. Pennec X and colleagues developed a Riemannian framework for the space of 2nd order tensors and recognized the whole brain white matter as a Riemannian manifold. MEG results were coregistered with DTI scans to provide Regions of Interest (ROI) for evaluating tractography between those regions. Connections were found between sites of high coherence in each subject via white matter fiber tracts. We used the software toolkit named DTI track to perform the whole procedure.

III. Results

The waveforms corresponding to the matched words showed at least three clear reproducible responses. The three responses were named as PM1, PM2, PM3, respectively (See Fig.1). These three responses' latencies of averaged MEG physical waveform components were 104.10±5.20 ms, 198.80±6.00 ms, and 302.10±10.01 ms, respectively. Differences were tested for significance using a paired t-test ($p<0.05$).

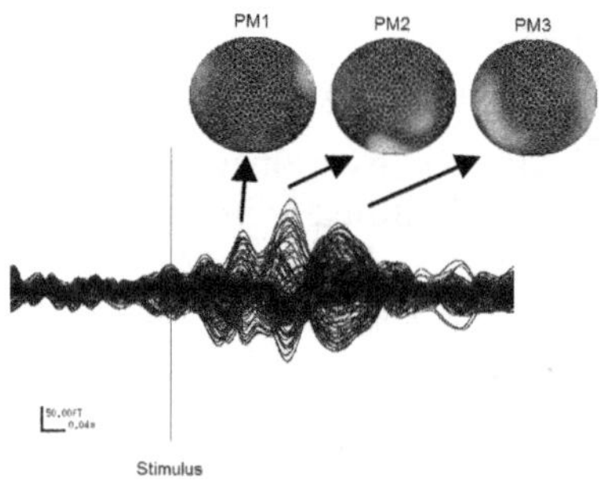

Fig. 1 MEG waveform and topographical map of neuromagnetic activation elicited by visual and auditory words. Three neuromagnetic responses are clearly identified

The neuromagnetic responses were localized using wavelet-based beamformer (See Fig.2). The PM1 was localized in the occipital and temporal cortices for all participants. The PM2 and PM3 were respectively localized in the left superior-posterior temporal cortex, namely Wernicke's area (18/20, 90%) and the inferior frontal cortex, namely Broca's area (18/20, 90%).

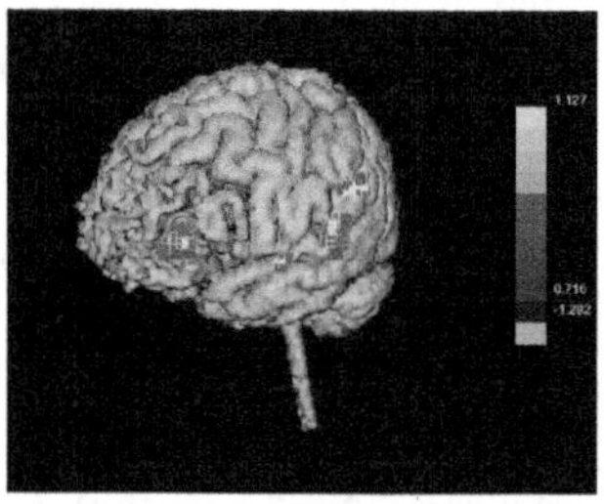

Fig. 2 Magnetic source imaging showing the brain activation elicited by simultaneously-presented written and spoken words. The Broca's area, Wernicke's area, and the junction of temporal and occipital area are clearly identifiable. The colour bar indicates the colour coding of the level of activation

To determine the time course of brain activation at source levels, virtual sensors were placed at the center of magnetic sources. The results showed that the peak latencies of the virtual sensor waveforms in the occipital, the left and right temporal cortices were 95 ± 3, 97 ± 2 and 98 ± 2 ms, respectively. The amplitude of the virtual sensor response in the occipital region was significantly stronger than the activation in temporal cortices ($p < 0.01$).

Virtual sensor waveforms also revealed activation in the left superior-posterior temporal cortex (Wernicke's area) and the inferior frontal cortex (Broca's area) at peak latencies of 320 ± 11 ms and 340 ± 9 ms, respectively. Although the amplitude of the virtual sensor response in Broca's area was larger than the response in Wernicke's area in 9 subjects, this difference was not statistically significant ($p > 0.05$).

The junction of the left temporal and occipital region was activated around 162 ± 8.9 ms in 20 participants. With spatial source localization and latency information, we considered that this component correlated with the responses in temporal and occipital cortices, and Wernicke's area.

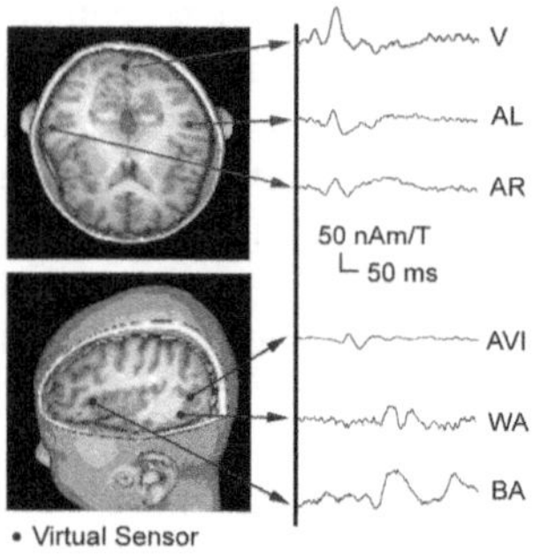

Fig. 3 Waveforms of virtual sensors showing the neuromagnetic activation at source levels. The neuromagnetic activation is clearly identifiable in the occipital and temporal cortices. In addition, there is clear activation in the posterior-superior temporal cortex (Wernicke's area, WA) and the inferior frontal cortex (Broca's area, BA). Waveforms of virtual sensors placed at the junction of the temporal and occipital regions show clear neuromagnetic activation, which is considered as a new finding

DTI tractography was computed based on the ROIs which were localized by magnetic source location and virtual sensor technology. In the left hemisphere, two DTI tractography pathways were found. One pathway connected Wernicke's area to Broca's area (20/20, 100%), which was named "pathway A"; the other pathway connected the junction of the left temporal and occipital region to Broca's area (18/20, 90%), which was named "pathway B" (See Fig.4). Noteworthy, there were also bunches of white fiber tract from the junction area to the temporal and occipital cortex. All these DTI tractography formed a white matter fiber network of language processing.

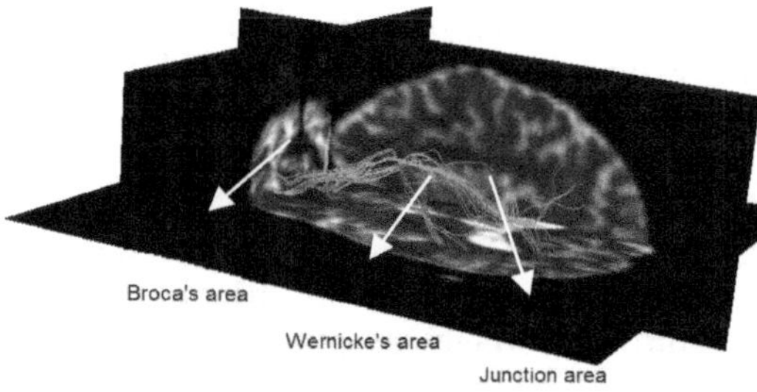

Fig. 4 DTI tractography showed the neural network of comparing visually and acoustically presented words. The color indicated the fibers' main directions as follows: red was left-right, green was anteroposterior, blue iwas superior-inferior. The ROI are the functional areas volumetrically estimated with MEG. There are clear connections (fibers) between Wernicke's area, Broca's area, and the junction of the left temporal and occipital area (we used arrows to depict the areas). There are also neural fibers connecting the junction area to the temporal and occipital cortices

IV. Discussion

The novel results of the present study are the time-course and neural network revealed by MEG and DTI. Studies of fMRI using audio-visual (AV) stimuli have suggested that there is a specific neural network for the integration of visual and auditory inputs [6]. However, the time course of the integration remains unclear. Recent EEG studies have reported the effects of crossmodal integration of different cortical levels [6-9]. However, EEG results fail to provide anatomical information showing the location of connections between visual, auditory and the language cortices in the human brain. To address this issue, we used both MEG and DTI to reveal the time course as well as the neural network. The results have demonstrated that there were clear connections between the visual cortex, Wernicke's area and Broca's area. Importantly, the M160 in the present study was just localized to the area which had connections to the visual and language cortices. Since M160 was identified in latency between visual/auditory activation and language activation, it is reasonable to consider that this component represents the integration of auditory and visual inputs for language processing. Thus, the results of the present study suggest that the bimodal temporal integration window occurs around 160 ms. This observation is consistent with the conclusion by van Wassenhove V and colleagues, who studied 43 normal participants and found that the bimodal temporal integration is around 200 ms.

The results of DTI tractography suggested that a neural network was involved in processing visually and acoustically presented words, and demonstrated that there was a brain area for integration of visual and auditory inputs in word processing. Our findings are in agreement with previous reports [10]. It has been suggested that the auditory input was conveyed via two possible pathways: 1) from the superior temporal gyrus to Broca's area if the information was to be repeated immediately or 2) relayed to the cortex below the superior sulcus for lexical semantic comprehension. The lexical-semantic area is located on the boundary between the auditory and visual association cortex, receiving inputs from both Wernicke's area and primary auditory cortex, as well as the visual cortex [10,11].The lexical-semantic area in their studies perfectly coincides with our virtual sensor results both in time course (162 ± 8.9 ms, after auditory-visual activation and prior to activation in Broca's area) and space localization (the junction of temporal and occipital cortices). The two pathways of conveying decoded information are vividly identifiable in the DTI tractography images.

V. Conclusions

The present study investigated the time course and neural network of visual-auditory word comparison using wavelet-based beamformer, virtual sensors and DTI tractography. The results suggested that the task of comparing written and spoken words activated the auditory, visual and language cortices (Broca's area and Wernicke's area) as well as the left junction of the temporal and occipital region. The activation in the left junction of the temporal and occipital region was detectable around 160 ms. The neural connection between the junction, Wernicke's area and Broca's area were identifiable in DTI tractography.

Acknowledgment

This study was partially supported by a Trustee Grant from CCHMC, OH, USA.

References

1. Xiang, J., Wilson, D., Otsubo, H., Ishii, R., & Chuang, S. (2001). Neuromagnetic spectral distribution of implicit processing of words. *Neuroreport, 12,* 3923-3927.
2. Robinson, S. E. (2004). Localization of event-related activity by SAM(erf). *Neurol.Clin.Neurophysiol., 2004,* 109.
3. Vrba, J. & Robinson, S. E. (2001). Signal processing in magnetoencephalography. *Methods, 25,* 249-271.
4. Kotecha, R., Xiang, J., Wang, Y., Huo, X., Hemasilpin, N., Fujiwara, H. et al. (2009). Time, frequency and volumetric differences of high-frequency neuromagnetic oscillation between left and right somatosensory cortices. *Int.J Psychophysiol., 72,* 102-110.
5. Mohamed, I. S., Otsubo, H., Shroff, M., Donner, E., Drake, J., & Snead, O. C., III (2007). Magnetoencephalography and diffusion tensor imaging in gelastic seizures secondary to a cingulate gyrus lesion. *Clin.Neurol.Neurosurg., 109,* 182-187.
6. Doehrmann, O. & Naumer, M. J. (2008). Semantics and the multisensory brain: how meaning modulates processes of audio-visual integration. *Brain Res., 1242,* 136-150.
7. Gallace, A. & Spence, C. (2006). Multisensory synesthetic interactions in the speeded classification of visual size. *Percept.Psychophys., 68,* 1191-1203.
8. Ghazanfar, A. A. & Schroeder, C. E. (2006). Is neocortex essentially multisensory? *Trends Cogn Sci., 10,* 278-285.
9. Lakatos, P., Chen, C. M., O'Connell, M. N., Mills, A., & Schroeder, C. E. (2007). Neuronal oscillations and multisensory interaction in primary auditory cortex. *Neuron, 53,* 279-292.
10. Price, C. J. (2000). The anatomy of language: contributions from functional neuroimaging. *J Anat., 197 Pt 3,* 335-359.
11. Hickok, G. & Poeppel, D. (2004). Dorsal and ventral streams: a framework for understanding aspects of the functional anatomy of language. *Cognition*, 92, 67-99

Neuronal Effects of the SpeechEasy Treatment for Stuttering

S.M. Bowyer[1,2,3], J. Peacock[1], N. Tepley[1], and J.E. Moran[1]

[1] Henry Ford Hospital, Detroit, USA
[2] Oakland University Rochester, USA
[3] Wayne State University, Detroit, USA

Abstract— **MEG imaging was used to localize brain regions activated during language processing in subjects who stutter with and without the use of the SpeechEasy device (compared to non-stuttering control subjects). This study was done to determine if the SpeechEasy device actually alters cortical processing when it helps relieve the symptoms of stuttering. Stuttering is a disruption in speech production, characterized by repetition, blocks or prolongation. The SpeechEasy is an in-the-ear auditory feedback device demonstrated to enhance fluency in people who stutter. It combines delayed auditory feedback with frequency altered feedback to create a choral effect, which occurs when people who stutter speak or sing in unison with others and the stutter is dramatically reduced or even eliminated. Eight stuttering subjects and five control subjects underwent MEG imaging to detect cortical activity during verb generation (VG) and single word speaking aloud (SA) tasks. MEG data analyses were performed utilizing a distributed source model (MR-FOCUSS) and displayed on a standard MRI brain image. These results indicated that VG task activates Wernicke's similarly regardless of the SpeechEasy device in the supramarginal gyrus (SMG). The SA tasks activated Broca's areas with a longer delay with the SpeechEasy device in place than without the use of the device (487 vs. 479ms). Greater activation in Broca's area was seen with the use of the SpeechEasy device. The main aim of this study is to establish the efficacy of specific MEG imaging techniques in determining the structure, activation sequence, and strength of neuronal interaction modulations during language processing with and without the SpeechEasy device. This study of MEG neuroimaging has increased our understanding of how choral effects, from the SpeechEasy device, impact the process of stuttering.**

Keywords— **Stuttering, MEG, MR-FOCUSS, SpeechEasy, Treatment.**

I. Introduction

Stuttering is a disruption in speech production that is primarily characterized by repetitions (sounds, syllables, words or phrases) or prolongations (audible or inaudible). There are many different theories about the anatomic basis for persistent stuttering, including those involving auditory perceptual deficits, hemispheral asymmetry, and deficits in motor planning or sequencing (Sommer, 2002; Foundas 2001 and 2004; Braun 1997; Salmelin 2000). To-date the underlying mechanism of stuttering has not been discovered and there is no cure for stuttering. One method for reducing the frequency of overt stuttering involves the presentation of auditory feedback. One type of treatment uses a device called the SpeechEasy. The SpeechEasy is an in-the-ear auditory feedback device that is reported to enhance fluency in people who stutter. It combines delayed auditory feedback (DAF) with frequency altered feedback (FAF) to create a choral effect. The choral effect occurs when people who stutter speak or sing in unison with others and the stutter is dramatically reduced or even eliminated.

Magnetoencephalography (MEG) imaging has been used to localize brain regions activated during language processing in normal subjects (Salmalin 2007, Bowyer 2004, Simos 1998). Since MEG can detect the activation of brain regions during langauge processing, it has been used in studies to explore cortical activation in people who stutter (PWS) and fluent speakers (Salmelin 2000, Salmelin 1998). These studies indicate differences in cortical organization between PWS and fluent speakers, but acknowledge that more detailed studies are required to determine the functional roles of the areas affected.

fMRI research has looked at the results of DAF on cortical functioning. Results suggest that there may be two biological subgroups of PWS, 1) those with anomalous anatomy of the auditory cortex, who improved more when using DAF than 2) those considered to have typical anatomy of the auditory cortex (Foundas 2004).

MEG was used in the present study to image the location of cortical processes of stuttering with and without the SpeechEasy device and to determine the latency and sequence of activation of the cognitive neural pathways involved in stuttering. These results will be of clinical use in determining if the SpeechEasy device actually alters cortical processing when it helps relieve stuttering.

II. Methods

A. Subjects

Nine patients who stutter and 5 non-stuttering control subjects were studied with 148 channel MEG (Magnes WH2500, 4D-Neuroimaging). Seven of the patients who

S. Supek and A. Sušac (Eds.): Advances in Biomagnetism – BIOMAG2010, IFMBE Proceedings 28, pp. 342–345, 2010.
www.springerlink.com

stutter had never used the SpeechEasy device. The other two has used the SpeechEasy for over 1 year.

The stuttering population was referred by the Henry Ford Hospital Division of Speech-Language Sciences and Disorders. The control subject population had no history of stuttering, drug use, psychiatric or neurological (including head injury leading to loss of consciousness) and was not on any active CNS medications. All subjects gave written informed consent prior to MEG study. The Institutional Review Board of Henry Ford Hospital approved the protocol.

B. MEG Data Collection Procedures

Neuromagnetic recordings were carried out to measure spontaneous and language evoked brain activity. Each subject was prepared for the MEG study in our standard way (Bowyer 2003). The subject then lay comfortably on the bed, inside the magnetically shielded room. The neuromagnetometer helmet containing the detector array was placed around the subject's head in close proximity to most of the cortical surface. The subject was asked to avoid excessive eye blinks and body movements during data collection. Each data collection run lasted 5-10 minutes. From start to finish, the MEG procedure lasted ~1.5 hours.

MEG data were sampled at a rate of 508.63 Hz with a low pass filter set to 0.1 Hz and the high pass filter set to 100 Hz. Changes in the subject position between the beginning and end of a study were detected by changes in magnetic fields from the coils on the forehead and ears. Runs during which the subject's head shifted position more than 0.5 cm were repeated. Epochs that contain large artifacts were eliminated. Each subject was monitored by video camera and two-way audio speaker system during the time he/she is in the shielded room.

Spontaneous (resting state) brain activity was recorded by MEG for 10 minutes, while the subject was lying quietly on his/her back, with eyes open. Keeping eyes open eliminated coherent activity in the occipital cortex corresponding to alpha activity (7-12 Hz range).

Language evoked brain activity was measured during two separate tasks:

1) A semantic language task involved Verb generation (VG) which depicted everyday objects as visually printed nouns. During each presentation the subject was asked to generate a verb that is linked to each noun (e.g. airplane-fly). A set of 60 nouns were randomly shown for 2 seconds each with a new word generated every 3 seconds.

2) The second task was a Speaking aloud (SA) task where the subjects read the word visually presented. Approximately 60 concrete nouns [every day objects, selected for concreteness, and high frequency] were randomly shown for 1-second. After a blank interval of 1 second, a question mark appeared for 2 seconds, prompting the subject to read the word aloud.

These language tasks were run first with out the SpeechEasy device (stuttering treatment device). Then they were repeated with the SpeechEasy device in place. The SpeechEasy device was taped the patient's shoulder and an 8 inch tube was used to connect it to a sponge ear insert in the right or left ear. The analyses was performed on the latency, location and amplitude of functional brain activity utilizing a standard MRI scan rescaled to their digitized head shape, collected during the MEG study.

C. MEG Data Analysis

All data processing was performed with MEG-Tools (Moran 2004) using MATLAB. Data were then forward and backward filtered 1-50Hz. Independent Component analysis (ICA) was used to remove heart artifact from the raw MEG data. Then singular valued decomposition (SVD) of MEG data was used if needed to eliminate and other noise components, such as dental artifact, not removed by the ICA and frequency filtering.

Analysis of language evoked data: Language tasks were recorded as continuous data. The timing of stimuli presentations were recorded as pulse codes on a trigger channel. These trigger events were used to segment the data in to epochs of 2 seconds. These activation epochs (N=60) were averaged with a 500 ms pre-stimulus baseline and 1500 ms post-stimulus time. Data analysis was performed utilizing MR-FOCUSS (Moran 2005). MR-FOCUSS is a whole brain current density imaging technique designed to image focal concentrations of cortical activity. The MR-FOCUSS technique employs a discrete model of approximately 4000 source locations matched to the distribution of cortical gray matter derived from the volumetric MRI. Wernick's area activations were studied using the verb generation task. Evoked responses in the time interval 200-270 ms after word onset were located and used as the time mark for determination of the amplitude and latency of this response. Broca's area was studied using the Speaking aloud task. Evoked responses in the time interval 390-480 ms after word onset were used to located the latency and amplitude of this response.

Analysis of the Spontaneous resting state MEG data: Coherence imaging analysis was performed on the 10 minutes of resting state MEG data to identify cortical sources that interacted strongly within each frequency. MEG data were analyzed using MR-FOCUSS-ICA. First the ICA signal separation was applied to the filtered data to obtain signals from distinct cortical sources and then Secondly MR-FOCUSS-ICA (Moran 2005) was used to image cortical activation corresponding to these ICA signals. Then for each frequency of the FFT spectra, the cross-spectral matrix of active brain sources was calculated then normalized and the absolute magnitude of each off-diagonal element was the coherence between 2 source locations. Finally, for each active cortical site, the average coherence with all other sources was calculated for each frequency. In addition, for each cortical site the connectivity was also calculated. The connectivity spectrum of an active site with other active sites was calculated by creating a histogram of the number of sources that are coherent with the target site for each of 20 levels of coherence between 0.05 and 1 for each FFT frequency component. Coherence images were generated for each consecutive 7.5 seconds of the 10 minutes of MEG data using MR-FOCUSS. These were averaged to obtain coherence magnitude and location of strongly interacting cortical sources. The variance across this set of images is a measure of the stability of the cortical network activity and allows changes in coherence across time to be assessed for statistical significance.

III. Results

All patients who stutter (PWS) completed the language tasks with and without the Speech Easy device and underwent a 10-minute continuous resting state MEG scan without the SpeechEasy device. One subject's data (#9) was contaminated with noise artifact. Control subjects completed the language tasks and had a resting state scan.

Visual activation was detected in all subjects at ~100ms after stimulus onset during the language task runs.

Results from the verb generation (VG) task indicated activation in Wernicke's area was similar in controls and patients who stutter regardless of the SpeechEasy device. Activation was located in the supramarginal gyrus (SMG) in the latency interval 230$\pm$ 20 ms across all subjects. These latencies are consistent with those found in our previous study using a Picture naming task in Epilepsy patients and control subjects (Bowyer 2003) where latency of activation in Wernicke's area was 239$\pm$31 ms for all subjects.

The speaking aloud (SA) task was divided into two epochs the reading of the word and the speaking out loud of the word. The reading of the word detected activation in Wernicke's area similar to the VG task. In the speaking out loud trials activation of Broca's areas (Brodmann's area 45) was detected with a longer delay in patients who stutter with the SpeechEasy device in place than without the use of the device (438ms vs. 403ms). Figure 1 displays the MEG traces from a subject who stutters. The top trace is the evoked response to reading the word out loud without the speech easy device in place. In the bottom trace the same

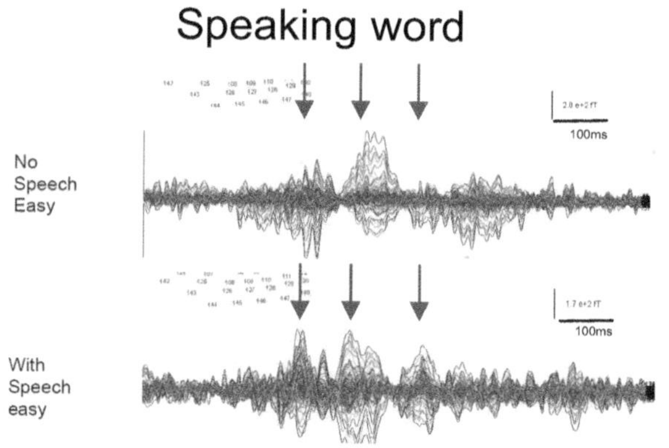

Fig. 1 148 MEG channel butterfly plots. MEG averaged evoked responses during speaking words aloud. Initial peak is visual processing, second peak is Wernicke's activation, and third peak is Broca's activation. Note in bottom trace Broca's activation is clearly seen with the use of the SpeechEasy device

subject is using the SpeechEasy device and a larger amplitude wave is seen under the third arrow.

Figures 2 displays brain activation during Broca's area activation without the use of the SpeechEasy device in one left handed subject who stutters. Figure 3 is the same subject with using the speech easy device during the same speaking aloud trial of the SA task. Note greater activation in the Right Inferior Frontal Gyrus (IFG) Brodmann's area (BA) 45 also known as Broca's area was detected with the SpeechEasy device in place (Fig 3 ~0.17 nAm) than without (Fig 2 ~ 0.08). The Talairach XYZ center for this subject was located at: 42.5 40.6 8.1 in mm and the MNI XYZ center was located at: 47.6 45.4 6.5 also in mm. Activation was detected in the Left IFG in right-handed subjects. Our previous study of picture naming to determine Broca's activation detected activation latencies of 436$\pm$40 ms across Epilepsy patients and control subjects. This study found similar latencies for Broca's activation.

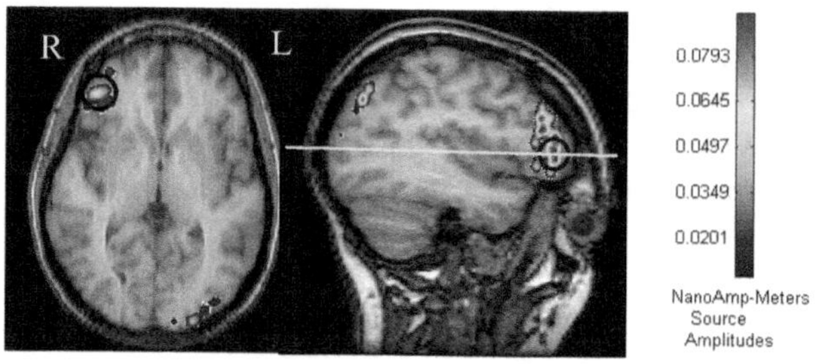

Fig. 2 MRI scan with cortical activation located in Broca's area (BA 45) seen at 403ms after onset of visual stimuli (question mark). This is a Left handed subject who stutters, but does not have the SpeechEasy device in his ear. Yellow line indicates axial slice

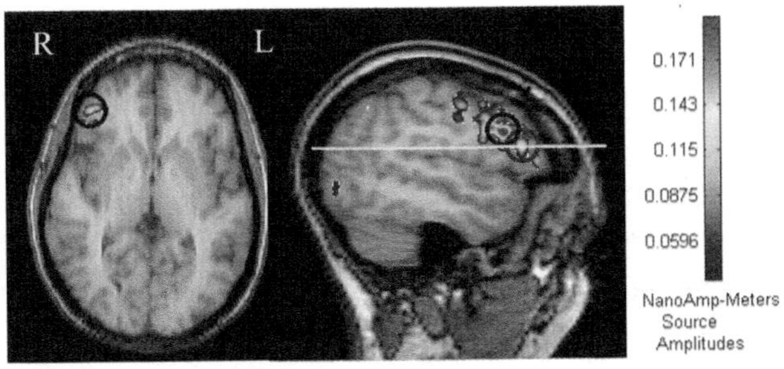

Fig. 3 MRI scan with cortical activation located in Broca's area (BA 45 and 46) seen at 438ms after onset of visual stimuli. This is the same subject in fig 2 while using the SpeechEasy device. Note higher source amplitudes and larger areas of activation in the IFG compared to figure 2. Yellow line indicates axial slice

Coherence imaging detected high activation in the inferior frontal cortex (Broca's area) compared to controls during the resting state (no speech). Figure 4 displays the IFG area of a patient who stutters while at rest. Activation in the right or left IFG and STG did not correlate with handedness. All PWS had high coherence in the IFG. Coherent levels were ~0.33 for the PWS group. The Control subjects had very low levels of coherent activity detected during the rest state MEG scan, all coherence levels were below 0.2.

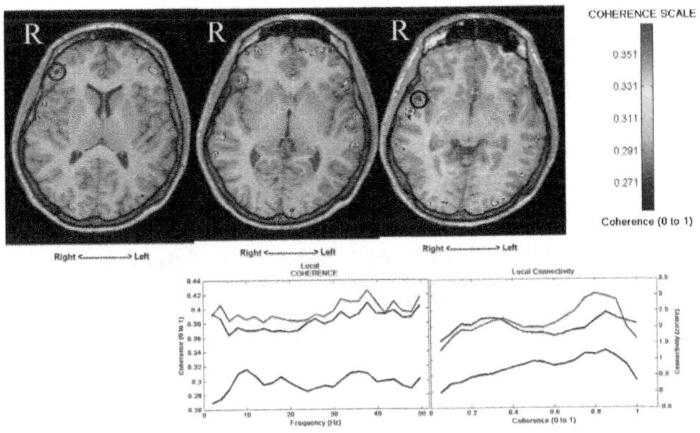

Fig. 4 Resting state data shows high levels of coherence in the inferior frontal gyrus (IFG) (Circles: blue BA 45, red BA 47) as well as auditory cortex (Circle: black BA 22), same subject as Fig 2 and 3

IV. Discussion

The main aim of this study was to determine the location, latency and and strength of neuronal interaction differences during language processing with and without the SpeechEasy device. We found increased cortical activity in each subject IFG (Broca's area) during the use of the Speech Easy Device compared to the MEG measurement without the SpeechEasy in place. The increased cortical activation with the SpeechEasy device in place (~0.200nAm) was still lower compared to control subjects (~0.380nAm).

Our findings are similar to other studies that noted activation differences in Broca's area between subjects who sutter and contol subjects. We also found high areas of coherent activity in the inferior frontal gyrus and auditory cortex during the resting state while the subject was not speaking. Further investigation is indicated as a result of the resting state finds.

V. Conclusions

This study of MEG neuroimaging has increased our understanding of how choral effects, from the SpeechEasy device, impact the process of stuttering.

Acknowledgment

Research supported by Henry Ford Research A30882 and NIH/NINDS Grant RO1-NS30914.

References

1. Sommer M, Koch MA, Paulus W, et al. (2002) Disconnection of speech-relevant brain areas in persistent developmental stuttering. Lancet 360(9330):380-3.
2. Foundas AL, Bollich AM, Corey DM, et al. (2001) Anomalous anatomy of speech-language areas in adults with persistent developmental stuttering. Neurology. 57(2):207-15.
3. Foundas AL, Corey DM, Hurley MM, et al. (2004) Verbal dichotic listening in developmental stuttering: subgroups with atypical auditory processing. Cogn Behav Neurol 4:224-32.
4. Braun AR, Varga M, Stager S, et al. (1997) Altered patterns of cerebral activity during speech and language production in developmental stuttering. An H2(15)O positron emission tomography study. Brain. 120:761-84.
5. Salmelin R, Schnitzler A, Schmitz F, et al. (2000) Single word reading in developmental stutterers and fluent speakers. Brain 123:1184-202.
6. Simos PG, Breier JI, Zouridakis G, et al. (1998) Identification of language-specific brain activity using magnetoencephalography. J Clin Exp Neuropsychol. 20:706-22.
7. Bowyer SM, Moran JE, Mason KM, et al (2004) MEG localization of language-specific cortex utilizing MR-FOCUSS. Neurology. 62:2247-55.
8. Salmelin R. (2007) Clinical neurophysiology of language: the MEG approach. Clin Neurophysiol. 118(2):237-54.
9. Bowyer SM, Mason KM, Tepley N, et al (2003) MEG Validation Parameters for Clinical Evaluation of Interictal Epileptic Activity. J of Clin Neurophys 20: 87-93.
10. Moran J. (2004). MEG Tools. www.megimaging.com
11. Moran JE, Bowyer SM, Tepley N (2005) Multi-Resolution FOCUSS: a source imaging technique applied to MEG data. Brain Topg18:1-17.

Author: Susan Bowyer
Institute: Henry Ford Health Systems
Street: 2799 West Grand Blvd
City: Detroit, Michigan
Country: USA
Email: sbowyer1@hfhs.org

MEG Detection of Attention and Memory Processes in Individuals with Dyslexia

S.M. Bowyer[1,2,3], L. Pawluk[4], A. Olszewski[4], M.L. Gallaway[4], A. Mansour[4], D. Jacobson[4], L. Erdodi[4], P. Lewandowski-Powley [4], J. Moran[1], N. Tepley[1], and R. Lajiness-O'Neill[1,4]

[1] Henry Ford Hospital, Detroit, MI, USA, [2] Wayne State University, Detroit, MI, USA, [3] Oakland University, Rochester, MI USA, [4] Eastern Michigan University, Ypsilanti, MI, USA

Abstract— **Atypical right and bilateral hemispheric activations in temporoparietal regions have been reported in individuals with dyslexia during nonword reading. We hypothesized similar aberrant activity would be evident during verbal and spatial working memory (WM) in individuals with dyslexia compared to controls as measured by MEG, suggesting inefficiencies in fundamental attentional processes necessary for reading. Nine controls and seven individuals with dyslexia underwent MEG to detect cortical activity during verbal and spatial working memory (VWM, SWM). Differential patterns of activation in acquired (AD) compared to developmental dyslexia (DD) were also examined in three 14-year-old males. Subject's MEG field responses were measured to visual presentations of a series of upper case letters (verbal) or squares presented in one of twelve locations around an imaginary circle (spatial). Brain activity was analyzed with MR-FOCUSS [1]. Results revealed that VWM and SWM activated earlier in dyslexia compared to controls, particularly in superior frontal (SF) regions during VWM and temporal regions during SWM. Individuals with dyslexia demonstrated higher amplitudes of activation in left parahippocampal regions during VWM. Additionally, lower mean amplitudes were observed in dyslexia in right superior temporal regions and the right precentral gyrus during VWM. During SWM, controls demonstrated later activation in precentral and middle and inferior temporal regions. During SWM, higher normalized mean amplitudes were noted in right superior temporal regions in controls relative to dyslexia. Results of WM performance in DD compared to AD suggest more effortful processing in left temporal regions relative to right in individuals with DD possibly suggestive of neural reorganization in AD. Overall findings suggest that children and young adults with dyslexia display aberrant working memory systems which may result in fewer resources available for the maintenance of information during reading and other verbally mediated tasks.**

Keywords— **MEG, Dyslexia, Verbal Working Memory, Spatial Working Memory.**

I. Introduction

Working memory (WM) has been shown to be closely associated with reading ability and literacy in typically developing children [2]. Atypical right and bilateral hemispheric brain activation in temporoparietal regions is reported during nonword reading in individuals with dyslexia during functional neuroimaging [3], although less is known about the relative contribution of WM for reading. While there have been important links found between WM, language, and reading in children and adults with dyslexia [1, 4] recent reports have not found WM to be predictive of reading ability in dyslexia [5]. This study sought to examine differences in WM during MEG in controls compared to those with dyslexia and examine differential effects of verbal compared to spatial working memory.

II. Methods

A. Participants

Nine controls (Males = 8; Mean Age = 26) (Mean (SD) Full Scale IQ (FSIQ) = 115 (14)) and seven individuals with dyslexia (Males = 5; Mean Age = 24) (Mean (SD) FSIQ = 112 (18)) underwent MEG to detect cortical activity during verbal and spatial working memory (VWM, SWM) [6]. Dyslexia was diagnosed based on a test of phonological decoding or reading if performance was below the 25th percentile.

Additionally, three right-handed, 14-year-old adolescent males were also compared on WM tasks to examine developmental trends. One boy was a neurotypical control (NC), one boy had developmental dyslexia (DD), and one boy had acquired dyslexia resulting from an early brain trauma (AD).

B. Measures

Subjects' MEG field responses were measured to visual presentations of a series of upper case letters (verbal) or squares presented in one of twelve locations around an imaginary circle (spatial). Each stimulus was presented for two seconds in three-second intervals. The paradigm was a 2 n-back task. 148 channel whole head MEG [4D

S. Supek and A. Sušac (Eds.): Advances in Biomagnetism – BIOMAG2010, IFMBE Proceedings 28, pp. 346–349, 2010.
www.springerlink.com

Neuroimaging, Magnes WH2500)] collected cortical activity. All data processing was performed with MEG-Tools (Moran 2005) using MATLAB. Data were forward and backward filtered 1-50Hz. Independent Component analysis (ICA) was used to remove heart artifact from the raw MEG data. Then singular valued decomposition (SVD) of MEG data was used if needed to eliminate and other noise components, such as dental artifact, not removed by the ICA and frequency filtering. Evoked magnetic field data from -0.2 - 0.65 seconds after the stimulus onset was analyzed with MR-FOCUSS [1], a current distribution technique.

An average normalized amplitude per active sources in 54 brain regions based on MNI coordinates was calculated. The left and right hemisphere regions of interest (ROI) were summed and averaged based on total active sources for that ROI to generate a normalized mean amplitude for the ROI over the 650 ms epoch to compare groups. The amplitude normalization procedure allows us to compare across brains by collapsing 4134 brain regions into 54 areas of interest defined by MNI/Talairach coordinates. Group differences were examined with a *t*-statistic. The *p*-values were not corrected for multiple comparisons given the small N and to decrease the probability of a Type II error.

III. Results

Verbal Working Memory. In individuals with dyslexia, significantly earlier activations were primarily noted in superior frontal (SF) regions. On average, individuals with dyslexia displayed a mean onset in SF regions at ≈210 ms, while controls demonstrate activation at ≈325 ms, $t(15) = 2.02$, $p = .05$. See Figure 1 (A and B). Subjects with dyslexia also demonstrated significantly higher amplitudes in the left parahippocampal regions, $t(15) = 2.18$, $p = .048$. See Figure 1 (C and D). Normalized mean amplitudes were significantly lower in the right precentral region, $t(15) = 2.45$, $p = .029$ and right superior temporal region, $t(15) = 3.341$, $p = .005$ in dyslexia. Figure 2 displays a comparison of control vs. subjects with dyslexia ROI amplitudes, note increased amplitude for control subjects in the superior temporal region.

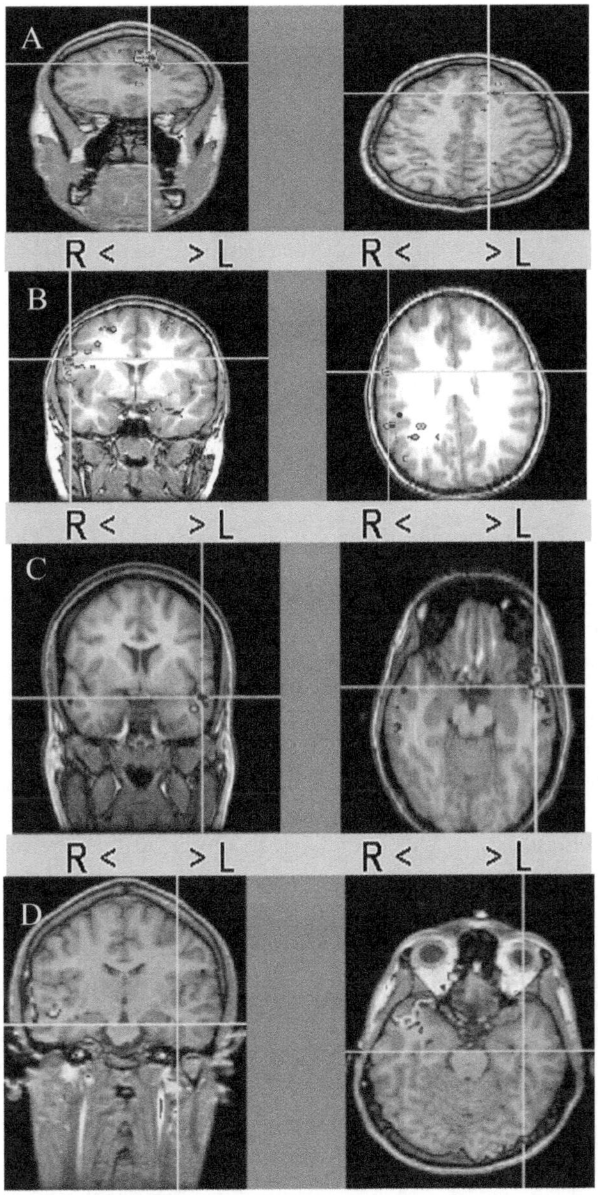

Fig. 1 Verbal Working Memory. Note pronounced left superior frontal activation in (A) control subject at about ≈300 ms compared to subject with (B) dyslexia who displays alternate, right inferior frontal activation at this time revealing possible reversed hemispheric processing during VWM. Note pronounced activation in left parahippocampal region in dyslexia (C) relative to control subject (D) at about ≈ 350ms possibly suggesting compensatory recruitment of inferior hippocampal memory regions rather than a frontoparietal working memory network when attempting working memory tasks

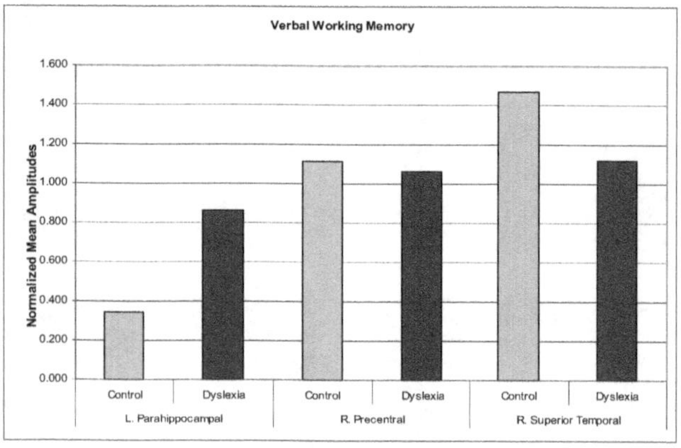

Fig. 2 Comparison of control subjects vs. subjects with dyslexia for 3 normalized ROI amplitudes during verbal working memory

Spatial Working Memory. During spatial working memory, significantly earlier activations occurred in bilateral precentral, $t(15) = 3.59$, $p = .008$, and middle and inferior temporal regions, $t(15) = 2.86$, $p = .02$, in dyslexia compared to controls. The onset of temporal activation was noted at ≈150 ms in subjects with dyslexia ≈150 ms while similar regions activated in controls at ≈350 ms. On average, over the 650 ms, more robust amplitudes were noted in right superior temporal regions in controls compared to those with dyslexia $t(15) = 2.579$, $p = .023$. See Figure 3. This increase was similar to that found in the VWM. See Figure 2

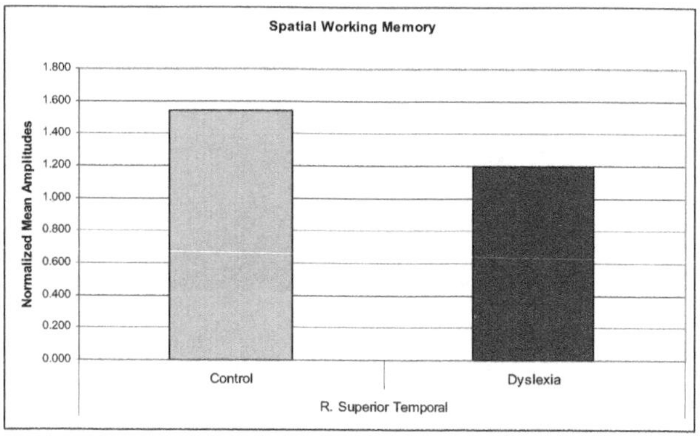

Fig. 3 Normalized amplitudes during spatial working memory reveal higher amplitudes in right superior temporal ROI during spatial working memory in controls compared to dyslexia

Developmental Trends. During onset of the VWM task, left parietal activation was observed in NC around ≈240 ms that was not observed in the DD or AD subjects. See Figure 4. The most robust activations in DD were bilateral temporal and frontal while predominately right temporoparietal activations were noted in AD. In general, during WM, the NC showed pronounced activation in the superior frontal parietal cortices. The clinical subjects displayed a higher number of significant amplitudes of activation and more variability in neural activity throughout the 650 ms epoch.

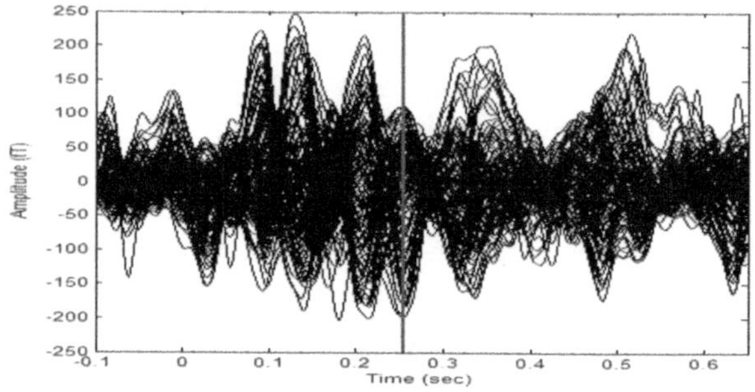

Fig. 4 Butterfly plot showing all 148 MEG channel waveforms during Verbal Working Memory task in Neurotypical Control (NC) at 240 ms arising from left parietal activation

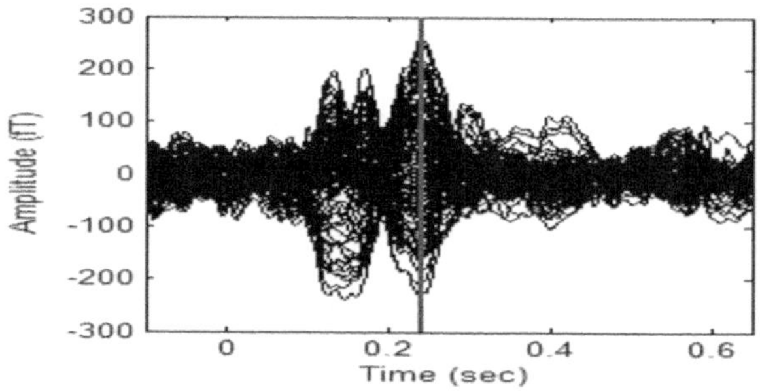

Fig. 5 Butterfly plot showing all 148 MEG channel waveforms during Verbal Working Memory task in Developmental Dyslexia (DD) at 244ms arising from left frontal activation

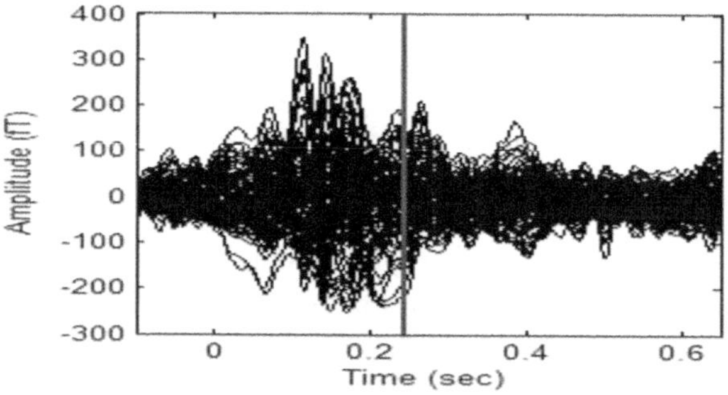

Fig. 6 Butterfly plot showing waveform from the 148 sensor array of MEG during Verbal Working Memory task in Acquired Dyslexia at 240 ms arising from right temporoparietal activation

During spatial WM, clear right hemisphere dominance was observed in NC. The opposite pattern characterized the DD. The AD produced symmetric bilateral activation throughout the task.

IV. Discussion

Individuals with dyslexia demonstrated a pattern of activation during both VWM and SWM suggestive of frontotemporal inefficiencies and or impairment, particularly in superior frontal and temporal regions. In general, individuals with dyslexia tended to demonstrate an earlier pattern of activation during WM tasks. Consistent with prior investigations of orthographic recall, structures in basal temporal regions were more heavily recruited during VWM tasks [7]. That is, individuals with dyslexia may be relying more heavily on ventral or basal temporal language areas and orthographic recall during performance of the VWM task rather than dorsal frontotemporal or frontoparietal regions and the phonological loop. A similar pattern of early bilateral frontal activation in dyslexia differentiated controls from those with dyslexia during SWM. In addition, middle and inferior temporal regions were found to be active significantly earlier in the dyslexic population.

Interestingly, individuals with dyslexia also displayed decreased activation in right superior temporal regions during SWM; a finding not previously reported. Increased amplitudes in right superior temporal regions during both the WM tasks in controls may be primarily due to the visual nature of the verbal working memory task.

In examining developmental differences in acquired compared to developmental dyslexia, we report a gradient of deviation from NC during VWM, with the AD being a closer approximation of the normative brain activity than the DD. This may suggest a time-related reorganization of neural activity post trauma versus a relatively persistent deviation associated with developmental dyslexia. In spatial WM, the aberrant lateralization observed in dyslexics also depicts the AD (bilateral) as an intermediate condition between NC (right dominance) and DD (left dominance). This pattern could be interpreted as evidence for plasticity in acquired neurological disorders.

V. Conclusion

While verbal memory inefficiencies have been a consistently reported finding in dyslexia, the literature is equivocal with respect to the relative importance of verbal working memory processes necessary for reading efficiency in dyslexia. In addition, limited studies have examined spatial working memory in dyslexia despite the visual nature of reading and common visual processing problems such as transpositions with orthographs. The current MEG study reveals differences in latency in frontal regions and amplitudes in bilateral superior temporal regions in controls compared to dyslexics. This suggests inefficiency in dorsal frontoparietal and superior temporal regions in dyslexia during working memory for which compensatory reversed hemispheric or bilateral ventral temporal activation may occur.

Acknowledgement

NIH/NINDS Grant R01-NS30914 to N. Tepley.

References

1. Moran J, Bowyer S, Tepley N. (2005) Multi-Resolution FOCUSS: source imaging technique applied to MEG data. Brain Topo 18:1-17.
2. Bayliss DM, Jarrold C, Gunn DM, Baddeley AD (2003). The complexities of complex span: explaining individual differences in working memory in children and adults. J Exp Psychol Gen. Mar;132(1):71-92.
3. Simos PG, Fletcher JM, Foorman BR, Francis DJ, Castillo EM, Davis RN, Fitzgerald M, Mathes PG, Denton C, Papanicolaou AC. (2002). Brain activation profiles during the early stages of reading acquisition. J Child Neurol. Mar;17(3):159-63.
4. McNamara JK, Wong B. (2003). Memory for everyday information in students with learning disabilities. J Learn Disabil. 2003 Sep-Oct;36(5):394-406.
5. Willcutt EG, Pennington BF, Olson RK, Chhabildas N, Hulslander (2005). Neuropsychological analyses of comorbidity between reading disability and attention deficit hyperactivity disorder: in search of the common deficit. J. Dev Neuropsychol. ;27(1):35-78.
6. Functional MRI studies of spatial and nonspatial working memory. (1998). D'Esposito M, Aguirre GK, Zarahn E, Ballard D, Shin RK, Lease J. Brain Res Cogn Brain Res. Jul;7(1):1-13.
7. Cohen L, Cohen S, Dehaene L, Naccache et al. (2000) The visual word form area: Spatial and temporal characterization of an initial stage of reading in normal subjects and posterior split brain patients. Brain 123:291-307.

Author: Susan Bowyer, Ph.D.
Institute: Henry Ford Hospital, Neuromagnetism Lab
Street: CFP 78/79, 2799 West Grand Blvd.
City: Detroit
Country: USA
Email: sbowyer1@hfhs.og or drsusan@umich.edu

Oscillatory Activity in Prefrontal Cortex during Implicit Letter-Location Binding

Claudia Poch[1], Pablo Campo[2], Fabrice B.R. Parmentier[3, 4], Jane V. Elsley[3], Francisco del Pozo[2], and Fernando Maestú[1]

[1] Center of Magnetoencephalography Dr. Pérez-Modrego, Complutense University of Madrid, Madrid, Spain
[2] Center of Biomedical Technology (CTB), Polytechnic University of Madrid, Madrid, Spain
[3] Clinical Neuropsychology Group, Department of Psychology and Health Sciences Institute, University of the Balearic Islands, Spain
[4] School of Psychology, University of Plymouth

Abstract— **In this study we demonstrated implicit verbal-spatial binding effects that were dependent on the task-relevant feature. We used MEG to measure brain activity underpinning the maintenance of verbal and spatial features in two recognition tasks, based on a letter-location paradigm previously used in binding studies. Both tasks were identical in terms of their perceptual characteristics and only differed with respect to the instructions given. Thus, in the verbal task participants attended consonants, while in the spatial task they attended locations. We observed that maintaining the identity of verbal information (consonants) arranged in a spatially distributed manner resulted in the concurrent processing of task-irrelevant location information. Critically, the reverse relationship does not hold true, supporting the notion of associative asymmetry. This implicit verbal-spatial binding was linked to a specific effect described as a greater oscillatory activity over prefrontal regions in "classical" frequency bands during the first half of the retention period and accompanied by greater activity in PPC and temporal regions.**

Keywords— **MEG, Time-Frequency, oscillations, cross-code binding.**

I. Introduction

Despite decades of research focusing on the functional distinction between verbal and visuo-spatial representations [1]surprisingly little attention has been devoted to the integration of verbal and spatial information. While evidence supports the notion of bound representations [2], the exact mechanisms underpinning binding in working memory (WM) remain relatively unspecified. Two issues have recently arisen that may constrain the development of a mechanistic model of binding: the role of attention and the asymmetrical nature of binding.

Whether binding occurs in an involuntary manner remains unclear. Several studies have shown that in the absence of attention visual features can be represented jointly [3]. Consequently, it has been proposed that attentional processes directed to task-relevant features of an object automatically spreads to other irrelevant features of the attended object leading to a bound representations of "spatiotemporally colocalized features" [4]. Thus, according to this position, attendance to a task-relevant object feature may entail the involuntary binding of all features of that object.

A second and related issue of importance is whether binding is best described by full integration of features or by asymmetric links between features [5]. There is indeed evidence suggesting that not all features are of equal importance in the formation of integrated representations [6]. These findings support asymmetry such that attending to an aspect of an objects' identity entails binding to its spatial location, while attending to spatial location does not necessarily entail its binding to the item's identity [7].

Results obtained in previous neuroimaging studies of visual-feature binding [8, 9] , highlight prefrontal and parietal regions as the neuroanatomical correlates of verbal-spatial binding [10, 11]. An alternative approach proposes that neural oscillations might be a potential mechanism by which separate features of an object are bound [12].

Accordingly, in the present study, we used magnetoencephalography (MEG) to measure brain activity underpinning the maintenance of verbal and spatial features in two recognition tasks, based on a letter-location paradigm previously used in binding studies [11]. Of critical importance to measure binding [13] positive-recognition probes could either be made up of a letter that was in location (intact probe) or a letter and a location that were both in the array, but were not seen together (re-combined probe). If verbal and spatial features are integrated in WM, accuracy and/or reaction time (RT) should show an advantage for intact probes as compared to re-combined probes [11, 13].

The aim of the current study was to elucidate whether neural activity specific to binding would be observed in conditions in which participants attend to one feature only (involuntary binding); and whether the brain activity underpinning binding processes varies as a function of the feature attended to by participants (binding asymmetry).

Thus, the predictions of the present study were as follows: From a behavioral perspective, if verbal and spatial binding is characterized by binding asymmetry, we should observe a binding effect (intact over re-combined probe advantage) when item (consonant) identity is task relevant, but not when spatial location is task-relevant. Considering previous neuroimaging results on feature binding [8-11], the

S. Supek and A. Sušac (Eds.): Advances in Biomagnetism – BIOMAG2010, IFMBE Proceedings 28, pp. 350–353, 2010.
www.springerlink.com

condition in which a binding effect occurred should be characterized by greater oscillatory activity within prefrontal and parietal regions.

II. Materials and Methods

A. Subjects

Eleven adult subjects (M=25.36 years, SD=3.13 years, six female), without any history of neurological or psychiatric illness, volunteered for the study.

B. Stimuli and Tasks

Experimental tasks were modelled on a letter-location paradigm developed by Prabhakaran et al [10] consisting of a visual memory array of four consonants displayed in four locations. The verbal stimuli comprised a set of eight consonants (Arial font; 48 pt), selected so as to differ in appearance between upper- and lower-case forms (D, F, H, J, N, Q, R, T). The spatial stimuli were presented within a set of eight spatial locations, marked by squares, placed equidistantly in a circular manner at an eccentricity of 3'63° from the center of the squares to fixation cross. Consonants, displayed in white (against a black background) and in uppercase, selected at random (without replacement) from the above set of eight. Each consonant appeared in a distinct location randomly selected (without replacement) from the possible set of eight. The letters were presented within a 1'87° x 1'87° white frame to reduce variations in spatial configuration caused by the consonants per se. Both tasks used identical stimuli, but differed with respect to the feature to be attended and memorized. In the verbal task, participants were asked to remember the identity of the consonants, irrespective of their location. In the spatial task, participants were instructed to remember the location of the consonants while ignoring their identity.

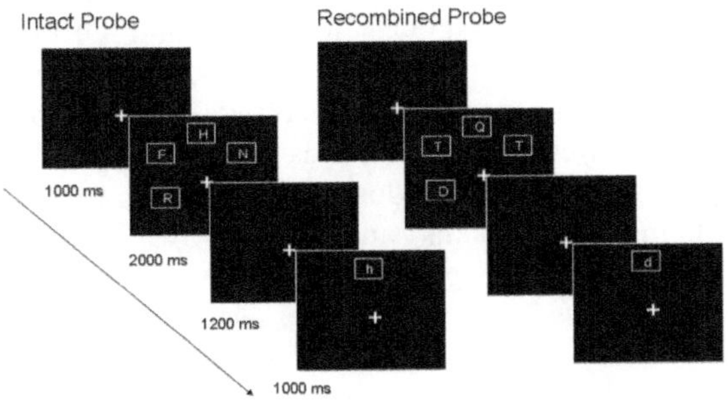

Fig. 1 Schematic illustration of both types of positive recognition probes

Two types of positive-recognition probes were presented, a consonant in its original location (intact probe), and a recombination of a consonant and a location presented in that trial but not together (recombined probe). The negative-recognition probes (requiring a "no" response) were as follows: a new consonant in a new location (in both the verbal and the spatial tasks); an old consonant in a new location (in the spatial task only); and an old location occupied by a new consonant (in the verbal task only). In both tasks, instructions emphasized the importance of both accuracy and speed. See Figure 1 for a scheme of the tasks.

C. Data Collection and Analysis

MEG scans were obtained during the verbal and the spatial tasks. A total of 480 trials were presented in each task. Tasks were counterbalancelly presented. Tasks were purpose-written using E-prime. The stimuli were projected through a LCD video-projector (SONY VPL-X600E), situated outside the shielded room on-to a series of in-room mirrors, the last of which was suspended approximately 60 centimeter above the subject's face (19'8° x 26'1°).

MEG recordings were carried out using a whole-head neuromagnetometer containing an array of 148 magnetometers (4-D WHS 2500®, San Diego). The data were collected using a sample rate of 254 Hz and band pass filtered between 0.1 and 50 Hz. MEG data were submitted to an interactive noise reduction procedure. Trials containing artifacts were removed using the automated artifact rejection algorithm implemented in the BESA 5.1 software suite. Only the retention period of trials with correct responses was included for subsequent analyses.

Digitized MEG data were imported into MATLAB Version 7.4 for analysis using the Fieldtrip software package. Time-frequency (TF) representation of MEG data was calculated on a single trial basis for a 1700 ms time window starting from 500 ms before the onset of the stimulus presentation, and 1200 ms starting from the beginning of maintenance period using a multi-taper Fourier transform approach applied to short sliding time windows (50 ms). We used adaptive Hanning tapers of four cycles, with frequencies ranging from 4 to 50 Hz, in 1-Hz steps.

In order to construct the source image of the TF data a minimum-norm estimation procedure (MNE) was applied to estimate the cortical origin of the retention period neuronal response. A Montreal Neurological Institute (MNI) dipole mesh (3004 nodes) was used to calculate the forward solution using a head model based on overlapping local spheres. The inverse solution was calculated by applying l2 MNE, with standard Tikhonov regularization. The underlying current source density of 'classical' frequency bands (θ, α, β

and γ) was estimated by calculating the MNE in the frequency domain [14]. Here, the real and imaginary parts of each Fourier component averaged within each of the four frequency bands was submitted to the MNE analysis, and then combined by using the root square of the sum of squares of the two Fourier parts as an estimate of absolute amplitude. The change in amplitude was calculated with respect to a baseline period before the beginning of each trial (between 500 and 0 ms before stimulus onset). The amplitude values at each dipole location of the brain surface mesh corresponding to the TF-values were statistically analyzed by using a Student's t-test to compare the verbal and spatial conditions. Associated p-values were thresholded at p < 0.001 (uncorrected). All reported p-values correspond to the maximum t-value observed within a significant source cluster.

III. RESULTS

A. *Behavioral Data*

Performance was assessed in terms of accuracy (% correct) and reaction time (RT) for both positive probe trials in each task.

In the verbal task, participants were more accurate for intact (M = 89.78 %; SD = 8.61) relative to recombined probe trials (M = 83.67 %; SD = 7.17) ($p < .01$). RTs in the verbal task were also significantly faster on intact probe trials (M = 908.36 ms; SD = 163.67) relative to recombined probe trials (M = 927.27 ms; SD = 158.21) ($p < .05$). In contrast, in the spatial task, accuracy measures for intact (M = 90.59 %; SD = 8.73) and recombined probe trials (M = 89.18 %; SD = 8.51) did not show significant differences (p >.20), and did not differ in terms of RT measures (intact: M = 859.36 ms; SD = 224.60; recombined: (M = 850.54 ms; SD = 224.85) (p >.30).

B. *Distributed Source Localization of Oscillatory Activity during Maintenance Period*

Theta frequency band (4 – 8 Hz): Higher T-F source amplitudes in the verbal task as compared to the spatial task were observed in left premotor area between 100 – 150 ms , left dorsolateral prefrontal cortex (DLPFC) between 350 – 400 ms, and posterior parietal regions (PPC) between 750-800 ms, 850-900 ms and 1150-1200 ms. Greater activity in the spatial task was also observed over PPC between 100-150 ms, 150-200 ms, 950-1000 ms and 1150-1200 ms (Figs 2 and 3 upper panels).

Alpha frequency band (8 – 12 Hz): Greater oscillatory activity in the verbal task was observed in left premotor area between 100-150 ms, medial PFC between 200-250 ms and DLPFC between 450-500 ms. Greater activation during the spatial task was observed in PPC, bilaterally, between 250 – 350 ms and between 500 – 650 ms, and in right premotor area between 650-700 ms (Figs. 2 and 3 mid-upper panel).

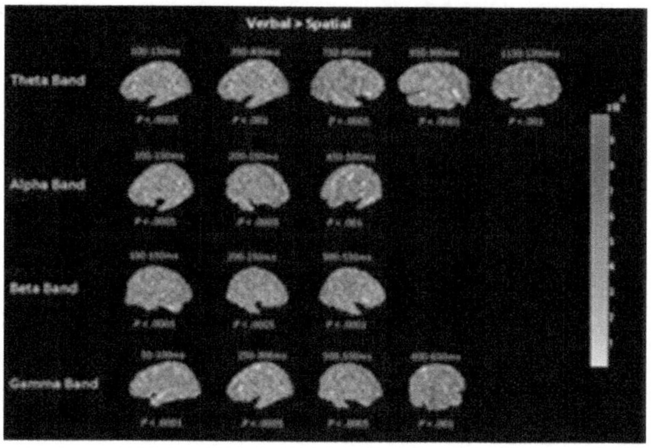

Fig. 2 Cortical source clusters indicating statistically significant increase in amplitude for the verbal task relative to the spatial task

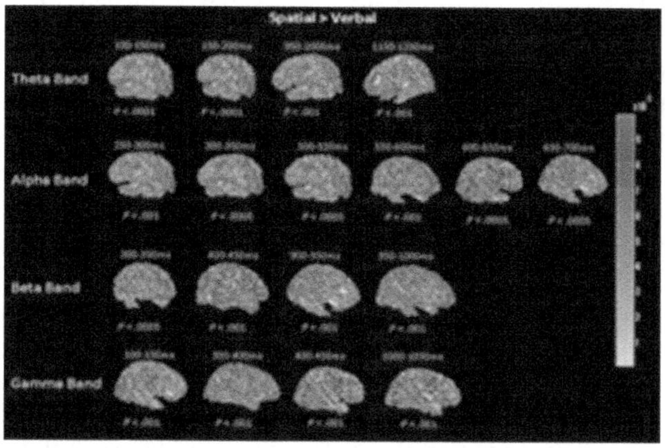

Fig. 3 Cortical source clusters indicating statistically significant increase in amplitude for the spatial task relative to the verbal task

Beta frequency band (13 – 30 Hz): Greater oscillatory activity in the verbal task was found in right DLPFC and VLPFC between 100-150 ms and 500-550 ms, and in parietal and temporal areas between 200-250 ms and between 500-550 ms. Greater activity in this frequency was observed mainly in premotor areas during the spatial task between 300-350 ms, 900-950 ms, and 950-100 ms. Additionally, the PPC also showed greater oscillatory activity between 400-450 ms (Figs 2 and 3 mid-lower panel).

Gamma frequency band (30 – 50 Hz): An enhancement over left VLPFC and anterior temporal regions between 50-100 ms, right posterior temporal region between 500-550 ms, and PPC between 600-650 ms was found during the verbal task. Parietal areas showed greater oscillatory activity during the spatial task between 100-150 ms, 400-450 ms and 1000-1050 ms, and also premotor areas between 350-400 ms (Figs 2 and 3 lower panels).

IV. Discussion

The results of the current study are twofold. From a behavioral perspective, our results indicated that verbal-spatial binding occurred when spatial information was irrelevant for task completion. Consequently, our findings are in agreement with the proposal of an asymmetrical contribution of the spatial dimension previously reported in visual WM studies [3]. These results indicate that in the absence of explicit processing, verbal and spatial features might be represented in an integrated format, although depending on the task-relevant feature [4].

Concerning T-F analyses, the most interesting finding is that although differences in activity in all analyzed frequency bands were observed in the verbal and spatial tasks in anterior regions, only activity in the prefrontal cortex, encompassing dorsolateral and ventrolateral regions, was greater during the verbal task. Prefrontal cortex has been proposed as a crucial area in integration processes since it is densely connected with sensory-specific cortices and heteromodal cortical regions [15]. Moreover, earlier functional neuroimaging [11] and neurophysiological studies [10] have revealed increased activation in prefrontal areas during maintenance of integrated verbal-spatial information. As verbal-spatial binding was observed only in the verbal task it can be inferred that the greater oscillatory activity found in PFC might be signaling neural processes specifically involved in cross-code binding. Thus, rather than functioning as a storage medium for bound representations, the PFC may function as a "pointer" indexing the features that form an object, while the features themselves are retained in more posterior regions.

We also observed activations common to both verbal and spatial tasks engaging distinct parts of the PPC, mainly in theta and gamma bands. The PPC activity during the retention period of the spatial task is compatible with the proposed role of this area in retaining spatial locations of stimuli [16]. Additionally, there is a large body of evidence showing that PPC is also engaged in different forms of visual feature integration and cross-modal binding [17].

V. Conclussions

Maintaining the identity of letters spatially distributed resulted in the concurrent processing of task-irrelevant location information. Critically, the reverse relationship does not hold true supporting the notion of associative asymmetry. This implicit verbal-spatial binding was linked to a specific effect in oscillatory activity over PFC in all analyzed frequency ranges during the first half of the retention period and accompanied by greater activity in PPC, matching previous neuroimaging feature binding results.

References

1. Smith, E.E., Jonides, J., (1998) Neuroimaging analyses of human working memory. Proc Natl Acad Sci U S A 95, 12061-12068.
2. Luck, S.J., Vogel, E.K., (1997) The capacity of visual working memory for features and conjunctions. Nature 390, 279-281
3. Meegan, D.V., Honsberger, M.J., (2005) Spatial information is processed even when it is task-irrelevant: implications for neuroimaging task design. Neuroimage 25, 1043-1055.
4. Melcher, D., Papathomas, T.V., Vidnyanszky, Z., (2005) Implicit attentional selection of bound visual features. Neuron 46, 723-729.
5. Maybery, M.T., Clissa, P.J., Parmentier, F.B., et al (in press) Binding of verbal and spatial features in auditory working memory. Journal of Memory and Language.
6. Hommel, B., Colzato, L., (2004) Visual attention and the temporal dynamics of feature integration. Visual Cognition 11, 483-521.
7. Jiang, Y., Olson, I.R., Chun, M.M., (2000) Organization of visual short-term memory. J Exp Psychol Learn Mem Cogn 26, 683-702.
8. Vogel, E.K., Machizawa, M.G., (2004) Neural activity predicts individual differences in visual working memory capacity. Nature 428, 748-751
9. Xu, Y., Chun, M.M., (2006) Dissociable neural mechanisms supporting visual short-term memory for objects. Nature 440, 91-95.
10. Campo, P., Maestu, F., Ortiz, T., et al (2005). Time modulated prefrontal and parietal activity during the maintenance of integrated information as revealed by magnetoencephalography. Cereb Cortex 15, 123-130.
11. Prabhakaran, V., Narayanan, K., Zhao, Z., et al (2000) Integration of diverse information in working memory within the frontal lobe. Nat Neurosci 3, 85-90.
12. Singer, W., Gray, C.M., (1995) Visual feature integration and the temporal correlation hypothesis. Annu Rev Neurosci 18, 555-586
13. Oberauer, K., Vockenberg, K., (2009) Updating of working memory: lingering bindings. Q J Exp Psychol (Colchester) 62, 967-987
14. Jensen, O., Vanni, S., (2002) A new method to identify multiple sources of oscillatory activity from magnetoencephalographic data. Neuroimage 15, 568-574.
15. Duncan, J., (2001) An adaptive coding model of neural function in prefrontal cortex. Nat Rev Neurosci 2, 820-829.
16. Corbetta, M., Shulman, G.L., (2002) Control of goal-directed and stimulus-driven attention in the brain. Nat Rev Neurosci 3, 201-215.
17. Macaluso, E., Driver, J., (2005) Multisensory spatial interactions: a window onto functional integration in the human brain. Trends Neurosci 28, 264-271.

Author: Fernando Maestú
Institute: Centro de Magnetoencefalografía, UCM
Street: Av/Comlutense s/n
City: Madrid
Country: Spain
Email: fmaestuu@psi.ucm.es

Spectral Power of Brain Activity Associated with Emotion — A Pilot MEG Study

George Zouridakis[1,2,3], Udit Patidar[2], Nikhil S. Padhye[4], Luca Pollonini[1], Antony Passaro[4], and Andrew C. Papanicolaou[4]

[1] Engineering Technology, [2] Computer Science, and [3] Electrical & Computer Engineering, University of Houston, Houston, TX, USA
[4] The University of Texas Health Science Center at Houston, Houston, TX, USA

Abstract— **In this study we analyzed broadband, whole-head MEG activity to investigate affective processing in the human brain and assess whether spectral power in nonoverlapping frequency bands correlates preferentially with specific anatomical locations and distinct emotions elicited by affective stimuli that differed in valence from unpleasant to pleasant and arousal from low to high. We analyzed recordings from 20 normal young adults using linear mixed effects models computed separately in each frequency band on a reduced set of recording sites represented by 8 principal components. Our results showed statistically significant differences between the pleasant and neutral conditions in all frequency bands and between the pleasant and unpleasant condition in the theta band.**

Keywords— **Affect; Emotion; MEG; Spectral power; Mixed effects models.**

I. Introduction

Several neuroimaging studies have focused on the neurophysiological substrates of affective processing by analyzing brain responses to affective stimuli that differ in valence from unpleasant to pleasant and arousal from low to high. The most recent evidence suggests that a highly interconnected network comprising both cortical and subcortical regions is involved in emotion. However, no specific brain area has been found to be activated consistently across individual emotions or activation tasks. Rather, separate brain regions would be involved with different aspects of emotion [1,2].

Most neurophysiological studies based on electroencephalography (EEG) rely on evoked potential analysis [3,4,5], computation of nonlinear measures, such as Kolmogorov entropy and Lyapunov exponents [6,7], functional connectivity [8], coherence analysis [9], and spectral content estimation [10,11], while those based on magnetoencephalography (MEG) have focused on dipole localizations to identify the various brain regions involved and the sequence of their activation [12,13].

In a recent EEG study [10] broad- and narrow-band spectral measurements were made from EEG obtained when subjects listened sentences with emotional underpinnings. The authors extracted power in predefined standard neurophysiological bands and found that higher delta power was elicited by the love emotion, while beta power showed changes when the subjects were exposed to frustration. In another study of narrow-band spectral measurements [11] higher power in the frequency range between 7.6 and 9.5 Hz was observed when stimuli with neutral emotional content were presented to subjects. In a study involving video clips containing emotionally valenced stimuli [6], lower alpha band activity was observed in both positive and negative clip categories compared to the neutral category at the C3, F3, and F7 electrode sites in the International 10-20 system. In a similar experimental setup [7], alpha and beta band activity was compared again across three distinct emotionally valenced film categories. The study found a higher power in the lower alpha band that was localized in left temporal areas for the neutral films when compared with emotional films. In the higher alpha band, negative films resulted in decreased power at the right temporal regions. More pronounced overall beta suppression was observed for the negative film category.

In the present study we analyze broadband, whole-head MEG recordings to investigate whether spectral power in the classical clinical frequency bands correlates preferentially with specific anatomical locations and the distinct emotions evoked by a series of affective stimuli that differ in valence from unpleasant to pleasant and arousal from low to high. This study is an extension of a previous report [13] that focused only on MEG dipole localizations and employs the same set of stimuli and experimental protocol but includes twice as many subjects. We deliberately disregard the time component of the MEG signals to reduce problem complexity.

II. Methods

A. Subjects and Recording Procedure

Data from 20 neurologically intact right-handed paid volunteers, 16 male and 4 female, were analyzed in this study. All subjects were medical students or fellows at the University of Texas Medical School at Houston, between the age of 25 and 55 years at the time of recording. Participants were screened, using self-report, for medication affecting the neurophysiological brain activity, as well as metallic implants, such as dental crowns, which adversely affect MEG recordings. The study protocol was approved by the appropriate committee for the protection of human subjects. All subjects gave written informed consent prior to the study.

Sixty picture stimuli were selected from the International Affective Picture System (IAPS) [14] based on normative affective ratings and organized into three affective categories, unpleasant (e.g., poisonous snakes, violent death), pleasant (e.g., happy babies, attractive nudes), and neutral (e.g., umbrellas, household items). The pictures were also classified according to their arousal with final classification as unpleasant (mean valence 2.75 and mean arousal 6.12), pleasant (mean valence was 7.35 and mean arousal 4.33), and neutral (mean valence was 5.18 and mean arousal 3.16).

MEG recordings were obtained in a magnetically shielded room using a 248-channel whole head MEG scanner (Magnes WH3600, 4D Neuroimaging Inc, San Diego, CA). A total of 160 single trial responses, all having the same duration, were obtained from each subject. Each single trial included 500 ms prestimulus (baseline) and 2 seconds of poststimulus activity. Data were digitized at a sampling rate of 290 Hz.

During preprocessing, any linear trends and the prestimulus mean values were removed from each single trial. In addition, channels showing absolute amplitude values higher than 6 times the standard deviation of that channel were replaced by the Gaussian average of the surrounding channels.

B. Data Processing

To compute signal power at each recording location, each single trial was first weighted using a unit-amplitude Tukey window with 5% tapering on each side to avoid the introduction of spurious frequencies due to discontinuities in the data. All single trials from a particular channel were then juxtaposed to form a single signal and then bandpass filtered between 0.1 and 40 Hz, using a third order Butterworth filter.

The power spectral density (PSD) of each channel was estimated using the periodogram method at discrete frequencies between 0.1 Hz and 145 Hz (half the sampling frequency f_s). The power P(f) at frequency f for a sequence of n discrete time points $x_1, \dots x_n$ was estimated using Welch' method [15] which is given by

$$P(f) = \frac{1}{f_s n}\left|\sum_{l=1}^{n} w_l x_l e^{-j2\pi f l}\right|^2 \Big/ \frac{1}{n}\sum_{l=1}^{n}|x_l|^2,$$

where $w_1 \dots w_n$ is the window used to weight the signal sequence.

Total power in the delta, theta, alpha, beta, and gamma bands, as well as total power of the full bandwidth signal was computed by summing the PSD values in the frequency range 0.1–4 Hz, 4–8 Hz, 8–13 Hz, 13–30 Hz, 30–40 Hz, and 0.1–40 Hz, respectively. This resulted in 248 power spectral values for each emotion category and each subject in a particular frequency band.

C. Statistical Analysis

The natural logarithm of the power spectral estimates was taken to improve departure from the Gaussian distribution [16]. Initially, analysis of variance (ANOVA) was performed using power as dependent variable and emotion and channel as independent categorical variables along with interaction between them. To test whether there were strong variations between subjects, Subjects was included as an independent variable in the ANOVA.

Furthermore, principal components were used to group the 248 channels into fewer ones, and the analysis was repeated. Finally, linear mixed models were used because of the advantage they offer over ANOVA in being able to handle correlated data that result from repeated-measures study designs. To account for correlations, a random intercept for channels nested within subject was added in the linear mixed model. A simultaneous advantage gained by this approach is that it can better accommodate inter-subject variability.

III. Results

The initial ANOVA showed that there were differences of power between emotions and between channels, but the interaction was not significant. Therefore channels do not behave differently in the emotions; they simply have slightly higher or lower power as a whole for each emotion.

Including Subjects as an independent variable in the ANOVA showed a significant interaction with emotion, indicating that response to emotion differs across subjects. Exploratory analysis using principal components separately in each frequency band was able to group the 248 channels into only eight, when the 90% variance method was used as component selection criterion. This channel reduction is credible, since it produces a spatial organization in the channels that map predominantly to one component, as Fig. 1 shows.

Linear mixed effects models computed on the reduced set with 8 principal components (down from 248 original channels) show that it is possible to distinguish between the pleasant emotion and the neutral emotion ($p < 0.05$). This is true for all frequency bands. The distinction is found in channel #1 of the reduced 8-channel system. Additionally, in the theta band, the power in the pleasant emotion was found to be greater than that in the unpleasant one ($p < 0.05$). However, the neutral and unpleasant emotions could not be distinguished from each other. Model coefficients and statistics for the theta band are shown in Table 1.

Furthermore, model predictions of greater power in the pleasant emotion over the neutral emotion resulted in accuracy of 65% (13 of 20 subjects) for the delta, beta, and gamma bands. The prediction accuracy for the theta and alpha bands was higher at 85% (17 of 20 subjects). The additional prediction in the theta band of greater power in the pleasant emotion over the unpleasant emotion had an accuracy of 60% (12 of 20 subjects).

Fig. 1 Channels that contribute predominantly to channels 1 (yellow), 2 (blue) and 3 (green) of the reduced 8-channel system. The black squares indicate channels that are not important

IV. Conclusions

The present study shows a strong relationship between the spectral power of the magnetic fields and the emotional content of affective picture stimuli. Analyzing the data separately in distinct well-established clinical frequency bands allowed the separation between the pleasant and neutral conditions, and between the pleasant and unpleasant one, which, otherwise, would not be possible.

Furthermore, a simple 8-channel reduced sensor space, identified through principal component analysis, delineated the brain regions involved in neural processing of emotional stimuli, and these regions were in agreement with a plethora of neuroimaging studies that have identified various cortical and subcortical structures, including the anterior cingulate and the medial prefrontal cortex.

Preliminary analysis of the same data after estimation of the intracranial neural generators using the MNE software package [17] to compute cortically-constrained L2 minimum-norm current estimates showed a similar pattern as the magnetic fields presented here (unpublished data). We are currently working on analyzing the data after preprocessing with an independent component analysis based method that minimizes the effects of extraneous signals on the MEG recordings [18].

Table 1 Linear mixed model coefficients for theta band. Emot(N) and Emot(U) are neutral and unpleasant emotions, respectively, while DF are degrees of freedom

	Value	Std. Error	DF	t-value	p-value
Intercept	17.37	0.86	304	20.1	<.001
Emot(N)	-3.51	0.97	304	-3.6	0.000
Emot(U)	-2.21	0.97	304	-2.3	0.023
Channel 2	-20.40	1.22	133	-16.7	<.001
Channel 3	-11.90	1.22	133	-9.8	<.001
Channel 4	-14.45	1.22	133	-11.8	<.001
Channel 5	-20.78	1.22	133	-17.0	<.001
Channel 6	-14.45	1.22	133	-11.8	<.001
Channel 7	-16.92	1.22	133	-13.9	<.001
Channel 8	-18.41	1.22	133	-15.1	<.001
Emot(N)*Chan 2	4.10	1.37	304	3.0	0.003
Emot(U)*Chan 2	1.12	1.37	304	0.8	0.414
Emot(N)*Chan 3	2.68	1.37	304	2.0	0.051
Emot(U)*Chan 3	0.25	1.37	304	0.2	0.857
Emot(N)*Chan 4	2.91	1.37	304	2.1	0.034
Emot(U)*Chan 4	0.75	1.37	304	0.5	0.583
Emot(N)*Chan 5	3.94	1.37	304	2.9	0.004
Emot(U)*Chan 5	1.59	1.37	304	1.2	0.245
Emot(N)*Chan 6	3.47	1.37	304	2.5	0.012
Emot(U)*Chan 6	2.44	1.37	304	1.8	0.075
Emot(N)*Chan 7	3.01	1.37	304	2.2	0.028
Emot(U)*Chan 7	1.85	1.37	304	1.4	0.176
Emot(N)*Chan 8	3.74	1.37	304	2.7	0.007
Emot(U)*Chan 8	2.20	1.37	304	1.6	0.108

Acknowledgment

This work has been supported in part by NSF grant 521527, the Institute for Space Systems Operations, and the Texas Learning and Computation Center at the University of Houston.

References

1. Berridge KC. Comparing the emotional brains of humans and other animals. In: Davidson RJ, Scherer KR, Goldsmith HH, editors. Handbook of affective sciences. New York: Oxford University Press; 2003. pp 25-51.
2. Rauch SL. Neuroimaging and the neurobiology of anxietj' disorders. In: Davidson RJ, Scherer KR, Goldsmith HH, editors. Handbook of affective sciences. New York: Oxford University Press; 2003. pp 963-975
3. Hajcak G, Dunning JP and Foti D (2009) Motivated and controlled attention to emotion: time-course of the late positive potential. *Clinical Neurohpysiology*, 120(3):505-10
4. Utama NP, Takemoto A, Koike Y, Nakamura K (2009) Phased processing of facial emotion: an ERP study. *Neuroscience Research*, 64(1):30-40
5. Vardi Y, Volos M, Sprecher E, Granovsky Y, Gruenwald I, Yarnitsky D (2006) A P300 event related potential technique for assessment of sexually oriented interest. Journal of Urology, 176(6 Pt 1):2736-40
6. Aftanas LI, Lotova NV, Koshkarov VI, Pokrovskaja VL, Popov SA, Makhnev VP (1997) Non-linear analysis of emotion EEG: calculation of Kolmogorov entropy and the principal Lyapunov exponent, *Neuroscience Letters*, 226(1):13-6
7. Aftanas LI, Lotova NV, Koshkarov VI, Makhnev VP, Mordvintsev YN, Popov SA (1998) Non-linear dynamic complexity of the human EEG during evoked emotions. *International Journal of Psychophysiology*, 28(1):63-76
8. Fingelkurts A, Fingelkurts A, Rytsälä H, Suominen K, Isometsä E and Kähkönen S (2007) Impaired Functional Connectivity at EEG Alpha and Theta Frequency Bands in Major Depression. *Human Brain Mapping*, 28(3):247-61
9. Hinrichs H and Machleidt W (1992) Basic emotions reflected in EEG coherence. *International Journal of Psychophysiolgy,* 13, 225-232
10. Marosi E, Rodríguez H, Yañez G, Bernal J, Rodríguez M, Fernández T, Silva J, Reyes A, Guerrero V (2001) Broad band spectral measurements of EEG during emotional tasks, *International Journal of Neuroscience,* 108(3-4):251-79
11. Marosi E, Bazán O, Yañez G, Bernal J, Fernández T, Rodríguez M, Silva J, Reyes A (2002) Narrow-band spectral measurements of EEG during emotional tasks, *International Journal of Neuroscience*, 112(7):871-91
12. Northoff, G., Richter, A., Gessner, M., Schlagenhauf, F., Fell, J., Baumgart, F., Kaulisch, T., Kötter, R., Stephan, K.E., Leschinger, A., Wagner, T., Barel, B., Witzel, T., Hinrichs, H., Bogerts, B., Scheich, H., Heinze, H.J., 2000. Functional dissociation between medial and lateral prefrontal cortical spatiotemporal activation in negative and positive emotions: a combined fMRI/MEG study. Cereb. Cortex 10 (1), 93–107.
13. Leon-Carrion J, McManis MH, Castillo EM, Papanicolaou AC (2006) Time-locked brain activity associated with emotion: a pilot MEG study, *Brain Injury*, 20(8):857-65
14. Lang PJ, Bradley MM, Cuthben BN. Intemational affective picture system (IAPS): Technical manual and affecrive ratings: Center for Research in Psychophysiology. Gainesville, FL: University of Florida; 1997.
15. Welch PD (1967) The Use of Fast Fourier Transform for the Estimation of Power Spectra: A Method Based on Time Averaging Over Short, Modified Periodograms, *IEEE Transactions on Audio Electroacoustics*, Vol. AU-15:70-73
16. Gasser T, Bächer P, Möcks J (1982) Transformations towards the normal distribution of broad band spectral parameters of the EEG, *Electroencephalography and Clinical Neurophysiology*, 53(1):119-24
17. Fischl B, Salat D, Busa E, Albert M, Dieterich M, Haselgrove C, van der Kouwe A, Killiany R, Kennedy D, Klaveness S, Montillo A, Makris N, Rosen B, Dale A.M.. Whole brain segmentation. Automated labeling of neuroanatomical structures in the human brain. Neuron. 2002; 33(3):341-355
18. Zouridakis G, Iyer D, Diaz J, Patidar U (2007) Estimation of individual evoked potential components using iterative independent component analysis, *Physics in Medicine and Biology,* 52(17):5353-68

Corresponding Author:

Author: George Zouridakis
Institute: University of Houston
Street: 218 Philip G. Hoffman Hall
City: Houston, TX 77204-3058
Country: USA
Email: zouridakis@uh.edu

Fair Play in the Brain – Cortical Activity in Response to Fair and Unfair Offers of a Fictitious Partner in a Gambling Design

Astrid C. Steffen[1], Gerald Schneider[2], and Brigitte Rockstroh[1]

[1] Department of Psychology, University of Konstanz, Konstanz, Germany
[2] Department Political Sciences, University of Konstanz, Konstanz, Germany

Abstract— **The present MEG study explored whether and how fair and unfair offers made by a fictitious teammate modulate cortical responses to decision-related cues.**

Electromagnetic activity analyzed in the source space (minimum norm estimate) disclosed larger (bilateral) fronto-temporal activity 170 - 250 ms after cue-onset for fair compared to unfair offers, irrespective of the subsequent decision to accept/reject. Bilateral posterior and frontotemporal activity 120-190 ms after feedback-cue onset was larger following 'reject'- compared to 'accept'-decisions. Frontotemporal activity 150 - 200 ms tended to be larger following satisfied rejection (1€ gain) compared to 'loss' feedback upon reject-decision. Brain activity patterns did not vary with the emotional expression of the fictitious teammate (first cue). Results suggest that evaluation of a fair or unfair offer in an experimental team-play situation modulates (fronto-)cortical activity related to decision-making, and that cortical responses to feedback reflect evaluation of risky (rejection may be rewarded by gain or punished by loss) decisions.

Keywords— **MEG, source analysis, decision-making, fairness.**

I. Introduction

EEG and fMRI have been used to study gambling within or without a social context, e.g. [1], [2], [3]. MEG allows monitoring of the involved processes with high temporal *and* spatial resolution. Neuroimaging studies substantiated the cortical correlates of involved processes like value and reward prediction and potential mediators like social context or fair/unfair bargaining [3].

The present study examined the impact of a fictitious teammate („neutral" or „angry" looking) and his fair or an unfair offer of how to share one Euro on participants' responses (accept / reject offer) and brain activation patterns.

Participants of this experiment were in a rewarding situation, and thus we expected reward-related temporo-parietal and frontal cortical areas to be involved in offer and feedback evaluation (compare [1], [4], [5]).

Based on the literature, we hypothesized that offer and feedback (success / failure) would have more influence than the teammate's facial expression on behavior and brain responses, and (e.g. [3]) that participants tend to be cooperative.

II. Methods

Participants: Ten right-handed [6] healthy student volunteers (5 female; mean age 28 ± 0.79 years, range 25-33 years). All subjects had normal or corrected to normal vision. Subjects received a financial bonus which varied between 40 and 51 € depending on their individual success in the experimental task (average reimbursement: 45,83 ± 1,49 € (mean ± standard error (SE)). Each participant was informed in detail about the MEG-measurement procedure and task and gave written consent.

Design and Materials: Each trial started with the fictitious teammate's face, followed by a cue indicating this teammate's offer of how to share one Euro (1 €), upon which the subjects had to decide by button press whether to accept or reject the offer. The face cues (neutral or angry) and the offer cues (fair (50 c : 50 c) or unfair (10 c : 90 c)) varied within a 2 x 2 repeated measures design (fig. 1). Participants were informed that accepting would result in a payoff of the exact amount of money that the teammate had offered; and that rejecting the offer, resulted in the possibility to win the whole Euro or nothing: The teammate could back down or take the whole Euro. The following feedback indicated how much the participant had gained, or whenever the subject did not press a button within 2 sec, that the trial was discarded (and no gain was made).

If participants rejected a fair offer, in half of the cases they gained nothing and they gained the whole Euro in the other half. The distribution of "reject"-feedback-probabilities was not disclosed to the participants: If participants rejected an unfair offer, they lost the Euro in 90% and gained it in only 10% of cases. Accordingly, presentation of feedback cues following "reject"-decisions was randomized.

Stimuli were presented on a monitor in a distance of 60 cm from the subject. The teammate was represented by a neutral or angry face (Ekman FACS database [7]). The offer was represented by a gray-shaded 1-€-coin-symbol, with the teammate's offer symbolized by blue colored parts on it. Subjects indicated their decision by pressing one of two adjacent buttons, defined as 'accept' and 'reject' button, with the index finger of their right hand. Feedback symbols indicating gain, no gain, or outrun of time were provided.

S. Supek and A. Sušac (Eds.): Advances in Biomagnetism – BIOMAG2010, IFMBE Proceedings 28, pp. 358–361, 2010.
www.springerlink.com

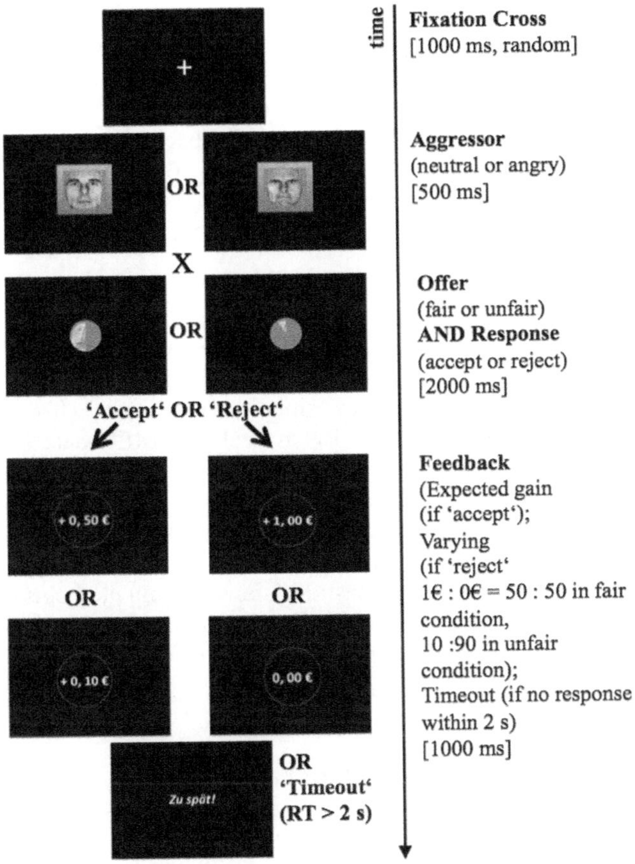

Fig. 1 Experimental Design including timing and variations

Each of the total 320 trials lasted on average 4580 ms including a 500-1500 ms fixation cross (baseline), a 500 ms face, a 2000 ms offer with the 2000 ms response interval (starting with the stimulus onset), and at last 1000 ms feedback stimulus. Trials were separated by 80 ms during which the screen remained black. 160 trials included fair and 160 trials unfair offers, within each category 80 trials were preceded by a neutral, 80 trials by an angry face. Presentation of face and offer stimuli was randomized.

Data Acquisition and Analysis: MEG was continuously recorded while subjects were in a prone position, using a 148-channel magnetometer (MAGNES™ 2500 WH, 4D Neuroimaging, San Diego, USA) with a sampling rate of 678.17 Hz and a bandpass filter of 0.1 to 200 Hz. The subject's nasion, left and right ear canal, and head shape were digitized with a Polhemus 3Space® Fasttrack prior to each session. Following noise reduction, MEG data were corrected for heartbeat-related artifacts. On average, 356 ± 2 (mean ± standard deviation) artifact-free trials per subject were available for the analyses of Face and Offer, 348 ± 57 trials for feedback-cue analysis ('win' vs. 'no-win', respectively feedback following 'accept' vs. 'reject' decision). Data epochs including 100 ms before (baseline) and 500 ms after onset of the respective stimulus were averaged and filtered with a 0.5 Hz (6dB/oct, forward, enabled for scan and averaging) high-pass and a 40 Hz (48dB/oct, zero phase, enabled for scan and averaging) low-pass filter using BESA software (Megis Software GmbH, Munich, Germany).

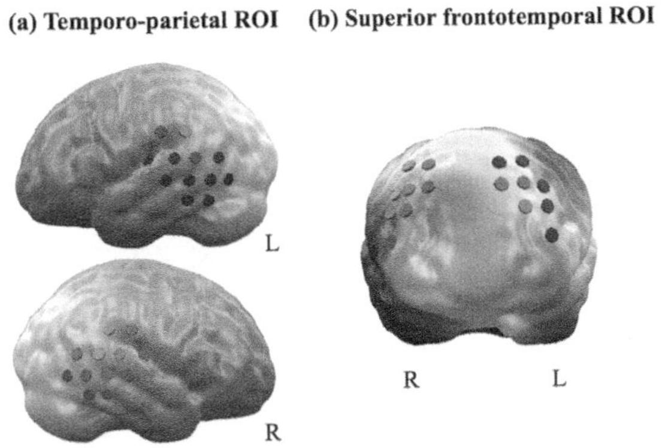

Fig. 2 Schematic positions of the dipole projections. Reflecting (a) temporoparietal and (b) superior frontotemporal ROIs

Source analyses involved minimum norm estimates (MNE; [8], [9], [10], [11]). Relying on EMEGS© custom software [12] written in Matlab® (MathWorks Inc., MA, USA), a spherical shell with 2 (azimuthal and polar direction) x 350 evenly distributed dipoles served as the source model. Point-wise repeated-measures analyses of variance were calculated separately for each dipole and identified spatio-temporal modulation as a function of stimulus type (face, offer, or feedback). Time windows and cortical regions of interest (ROI) were determined by plotting the statistical measures of activity differences (F-ratios) onto a spherical configuration of dipoles for time windows of significant stimulus effects. The two regions of differential brain activity were (see fig. 2): (a) A left- and right temporo-parietal ROI, scored for average activity among 12 dipoles in the left and 12 in the right hemisphere with distinguished dipole activity 120-190 ms after feedback stimulus onset, (b) a left- and right superior fronto-temporal ROI, scored for average activity among 8 dipoles per hemisphere with distinguished activity 170-250 ms after offer stimulus onset and 120-200 ms after feedback stimulus onset.

Separate repeated-measures ANOVAs each of which contained the factor Hemisphere (left ROI vs. right ROI), and moreover either the within-subject factors Face (neutral vs. angry) and Offer (fair vs. unfair), or Offer (fair vs. unfair) and Decision (accept vs. reject), or Gain-Feedback (win vs. no-win), or Decision-dependent Feedback (win vs. no-win) were calculated to derive statistical effects about dipole modulation in the ROIs for the respective time intervals. The alpha level was set at .05 for each comparison.

III. Results

Distribution of Decisions and Decision Time: Fair offers were accepted three times more often than rejected and unfair offers were accepted 2.3 times more often than rejected (difference between offers non-significant). Reaction times (RTs) were shorter for fair (688 ± 44 ms (mean ± SE)) than for unfair offers (638 ± 49 ms; F (1,7) = 8.2, p < .05), but neither Type of Decision (accept, reject) nor Face (neutral, angry) had significant effects.

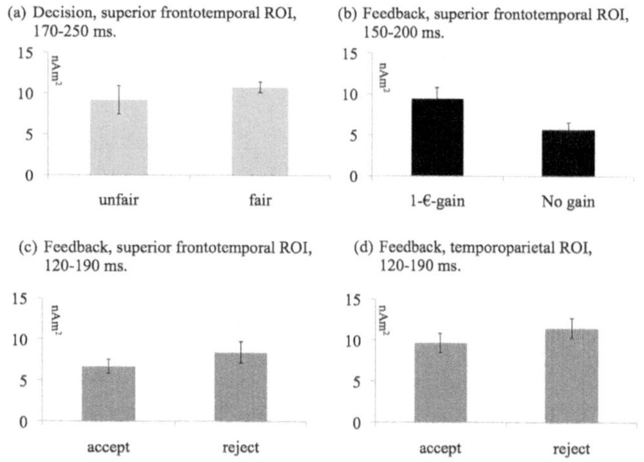

Fig. 3 Grand means of estimated source activity. Following the stimulus onset of (a) the offer stimulus in the superior fronto-temporal ROI 170-250 ms, and the feedback stimulus (b) in the superior frontotemporal ROI 150.200 ms, (c) in the superior frontotemporal ROI 120-190 ms and (d) in the temporoparietal ROI 120-190 ms

Dipole Activities in Response to Face: Dipole activity did not vary with the facial expression of the teammate (neutral, angry).

Dipole Activities in Response to Offer: Activity varied with the offer (fair, unfair) around 200 ms. 170-250 ms after cue-onset, fair offers (50 c: 50 c) evoked more pronounced superior fronto-temporal activity than unfair offers (10 c : 90 c; F (1,9) = 29.65, p < .005, fig. 2 (a)), independent of Decision (accept, reject).

Dipole Activities in Response to Feedback: Dipole activity to feedback cues varied with 1€-gain or no gain: Between 150-200 ms, 'gain'-feedback provoked more pronounced activity than that for 'no gain' in the superior frontotemporal ROI (F (1,9) = 5, p = .05; see fig. 2 (b)). Interestingly, reject-decisions evoked larger activity than accept-decisions 120-190 ms after feedback stimulus onset in (a) the temporoparietal ROI (F (1,9) = 6.76, p < .05) and (b) in the superior frontotemporal ROI (F (1,9) = 6.13, p < .05; see fig. 2(d) + (c)).

IV. Conclusions

Results suggest that evaluation of a fair or unfair offer in an experimental team-play situation modulates (fronto-) temporal cortical activity. Behavioral and MEG-data support the hypothesis that subjects focus on their own profit more than on their teammate's expressed emotion (maybe partly because they knew he was fictitious). Participants tended to cooperate without a rational need to do so: Future gains did not depend on past decisions. Augmented cortical responses to feedback may indicate evaluation of the risky reject-decisions reflected by augmented frontal reward-related activity.

Acknowledgments

This research was supported by the Deutsche Forschungsgemeinschaft and the Stiftung der Deutschen Wirtschaft. We thank Amra Covic and Sabine Scheermesser for helpful comments on the manuscript.

References

Knutson, B., Fong, G. W., Bennett, S. M., Adams, C. M., and Hommer, D. (2003). A region of mesial prefrontal cortex tracks monetarily rewarding outcomes: characterization with rapid event-related fMRI. NeuroImage, 18(2):263–272.

Daw, N. D., O'Doherty, J. P., Dayan, P., Seymour, B., and Dolan, R. J. (2006). Cortical substrates for exploratory decisions in humans. Nature, 441(7095):876–879.

Fehr, E. Fischbacher, U., Kosfeld, M. (2005). Neuroeconomic Foundations of Trust and Social Preferences: Initial Evidence. AmEconRev, 95(2): 346-351.

Smith, B.W., Mitchell, D.G., Hardin, M.G., Jazbec, S., Fridberg, D., Blair, R.J., and Ernst, M. (2009). Neural substrates of reward magnitude, probability, and risk during a wheel of fortune decision-making task. NeuroImage, 44(2): 600-609.

Bechara, A., Damasio, H., Tranel, D., and Damasio, A. (2005). The iowa gambling task and the somatic marker hypothesis: Some questions and answers. TrendsCognSci, 9(4):159–162.

Oldfield, R. C. (1971). The assessment and analysis of handedness: The Edinburgh Inventory. Neuropsychologia, 9:97–113.

Ekman P, Friesen WV (1976) Pictures of facial affect. Consulting Psychologists Press, Palo Alto

Baillet, S., Mosher, J.C., and Leahy, R.M. (2001). Electromagnetic brain mapping. IEEESignalProcMag, 18:14-30.

Haemaelaeinen, M. S. and Ilmoniemi, R. J. (1984). Interpreting measured magetic fields of the brain: Estimated of current distributions. Helsinki: Technical Report No. TKK-F-A559.

Hauk, O. (2004). Keep it simple: a case for using classical minimum norm estimation in the analysis of EEG and MEG data. NeuroImage, 21(4):1612–1621.

Hauk, O., Keil, A., Elbert, T., and Müller, M. M. (2002). Comparison of data transformation procedures to enhance topographical accuracy in time-series analysis of the human EEG. JNeurosciMeth, 113(2):111–122.

Junghoefer, M. and Peyk, P. (2004). Analysing electrical activity and magnetic fields in the brain. Matlab News & Notes.

Address of the corresponding author:

Author: Astrid Steffen
Institute: Department of Psychology
PO Box: D23
City: 78457 Konstanz
Country: Germany
Email: astrid.steffen@uni-konstanz.de

rtMEG: A Real-Time Software Toolbox for Brain-Machine Interfaces Using Magnetoencephelography

Gustavo Sudre[1], Wei Wang[1,2,4], Tao Song[3], Matti Kajola[3], Ramana Vinjamuri[2], Jennifer Collinger[2,7], Alan Degenhart[1,4], Anto Bagic[5,6], and Doug J. Weber[1,2,4]

[1] Center for the Neural Basis of Cognition, Carnegie Mellon University, Pittsburgh, US
[2] Department of Physical Medicine and Rehabilitation, University of Pittsburgh, Pittsburgh, US
[3] Elekta Oy, Helsinki, FI
[4] Department of Bioengineering, University of Pittsburgh, US
[5] Department of Neurology and Neurological Surgery, University of Pittsburgh, Pittsburgh, US
[6] Center for Advanced Brain Magnetic Source Imaging (CABMSI), University of Pittsburgh Medical Center, Pittsburgh, US
[7] Human Engineering Research Laboratories, Department of Veteran Affairs, Pittsburgh, PA, USA

Abstract— **Magnetoencephalography (MEG) is a non-invasive method to study brain functions with high temporal resolution. There is an emerging interest in studying the potential use of MEG for brain-machine interfaces (BMI) research. To date, the majority of studies have performed offline analysis to reveal detailed information about the spatial and temporal evolution of neural activity as it relates to a task, or to measure neuroplasticity resulting from an intervention. However, real-time MEG feedback could benefit many areas of research, including BMI. Currently there is no available method to capture the large amount of information from a 306-channel Elekta Neuromag® MEG system in order to provide real-time feedback. We have developed a toolbox that can stream in real-time MEG signals from this system to any computer. These signals can be processed with minimal delay (<30 ms) and used for various applications. Our MEG toolbox is integrated with BCI2000, a widely used open source software package for BMI research and development [1], and it can be easily configured to relay the real-time signal in binary format to any arbitrary host in the network. Preliminary results indicate that we can achieve an update-rate of approximately 35 Hz with 324 channels of data sampled at 1000 Hz, which is sufficient for many real-time BMI studies. This real-time software can be a valuable tool for real-time BMI research, including studies of neurofeedback training for stroke and spinal cord injury rehabilitation, and other general neuroscience research. The toolbox will be made available to the scientific research community as open source along with the BCI2000 software, and we hope that it can be used to support many new areas of real-time MEG research.**

Keywords— **real-time, Neuromag, brain-machine interfaces, magnetoencephalography.**

I. Introduction

Magnetoencephalography (MEG) has been used to study brain function in a wide variety of scenarios, from sensory mapping to motor functions [2, 3]. The ability to process and analyze MEG data in real-time creates new opportunities for neuroscience research and potentially useful clinical applications. For example, real-time neurofeedback can be used to train subjects to modulate their neural activity in order to perform a task via a brain-machine interface (BMI). This form of neurofeedback training may be useful for promoting neuroplasticity aimed at strengthening spared corticospinal pathways after stroke or spinal cord injury [4,5].

The area of adaptive paradigms (or optimal experiment designs [6,7]) can also benefit from the possibility of analyzing MEG signals in real-time. For example, in a cognitive state decoding task, one could figure out what types of stimuli to present next, or what classes to collect more repetitions for, in order to improve the model and therefore increase classification accuracies.

This work presents a software toolbox to acquire signals from an Elekta Neuromag® device in real-time (rtMEG). We first describe the system setup and the tests that were performed to assess the delay in the system. Then, we show results by measuring this delay and comparing the signal recorded by the toolbox with the one saved in the MEG FIF file.

II. Methods

A. System Description

The toolbox was developed to operate in parallel with the regular operations in a MEG center without affecting the standard MEG data acquisition workflow. Figure 1 shows how the computer running the toolbox interacts with the rest of the system. In a typical scenario, there is a computer responsible for running the Acquisition software and saving the data in the MEG system to FIF files. Although the communication between this acquisition computer and the digital signal processing (DSP) units in the system involves several steps, for the purpose of understanding how the toolbox works it is sufficient to imagine that the acquisition

S. Supek and A. Sušac (Eds.): Advances in Biomagnetism – BIOMAG2010, IFMBE Proceedings 28, pp. 362–365, 2010.
www.springerlink.com

computer connects to the machine using the network (as depicted by the network switch in Figure 1). In addition to the acquisition computer, there is usually a second computer providing stimuli to the subject. The stimuli can take different formats (auditory, visual, etc), and this computer also sends trigger pulses through the parallel port in order to mark the onset of stimuli in the recorded FIF files.

In this scenario, the toolbox can be installed on either the computer that presents the stimuli or a separate computer in the network. All that is required is to connect the computer running the toolbox to the external switch (usually located next to the acquisition computer), and set up the computer to be part of the network by configuring it with a static IP address.

BCI2000, a widely used open source software package for BMI research [1], has a modular implementation that has four components: an operator, source module, signal processing module, and application module. The rtMEG toolbox was written as a source module in the BCI2000 system. This way, any researchers already using BCI2000 in their experiments can simply re-use their module pipeline and exchange the source module with the one described here. The source code and the executables for the toolbox can be downloaded from the BCI2000 website, or by contacting the authors of this paper. There you can also find documentation and parameter files for different setup configurations (e.g. number of channels to stream).

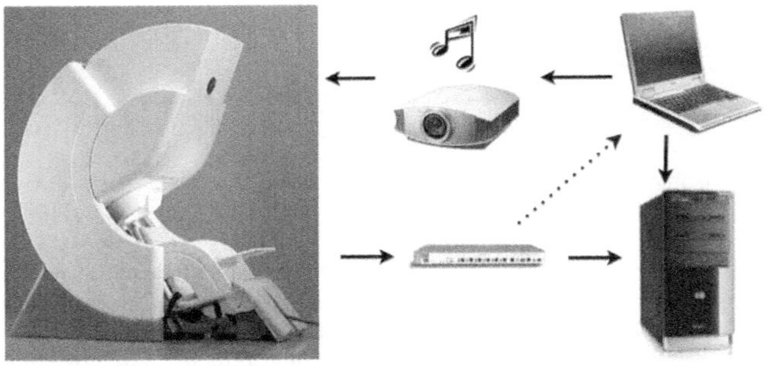

Fig. 1 Simplified illustration of the setup in a MEG center. A personal computer drives the experiment paradigm by providing stimuli to the subject, together with triggers to mark the MEG FIF file. The toolbox can be run in the stimuli computer or in any other computer connected to the external switch (dotted arrow)

Although the toolbox is well integrated with the BCI2000 system, the user is free to use his/her own applications to interface with the real-time MEG signals if so desired. That is accomplished by running an additional module also written by the authors that broadcasts the signals in binary format to any computer in the network the user requests.

The toolbox saves the data as 32-bit integers and interacts with the Elekta Neuromag® DSP units directly, by-passing the acquisition computer and any other links in between. The parameters set up in the acquisition machine still apply for the signals acquired by the module (e.g. filter values), and the stimulus codes can also be saved along with the data. Note that the toolbox was designed to work with systems that have DSPs employing Ethernet connection for data transfer, which may not apply to older systems that have a different type of data bus.

B. Testing Procedure

Minimizing the system delay is critical for real-time operations. Therefore, it is important to measure how much delay is added to the data when saved by the module. The setup used to perform this measurement is described in Figure 2.

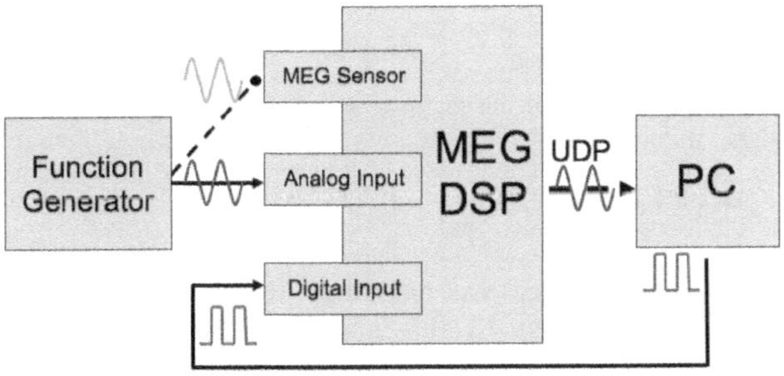

Fig. 2 Schematic illustration of the setup to measure the system delay. The toolbox is run on the PC, which marks the data recorded by the acquisition computer using parallel port pulses

A signal generator was used to produce a 5Hz sine wave, and the output of this generator was split to be recorded by the MEG system simultaneously in two different ways. One of the outputs was connected to the first auxiliary analog input channel in the MEG system. The other output was connected to a 30KΩ resistor positioned on the floor of the acquisition room, such that the channels in the MEG helmet were able to pick up the sinusoidal magnetic signal created by currents passing through the resistor. The toolbox received all the MEG and analog channels in real-time, and output to the parallel port a ‘0’ or ‘1’ depending on the polarity of the sine wave in the first analog channel, forming a square wave with value ‘1’ when the sine wave was positive and ‘0’ when the sine wave was negative. This binary output was sent to the MEG system's digital input/trigger channel so that the FIF file would save both the raw sine wave and the digital output from the rtMEG toolbox.

After the test, data recorded in the MEG FIF file was used to characterize the real-time delay between the square wave (when the PC "saw" the data) and the analog sine wave (e.g. the actual data). Figure 3 shows an example of the three signals in time (i.e. the analog input to the MEG machine and the signal captured by a magnetometer, both driven by the function generator, and the binary output of the PC running the rtMEG toolbox).

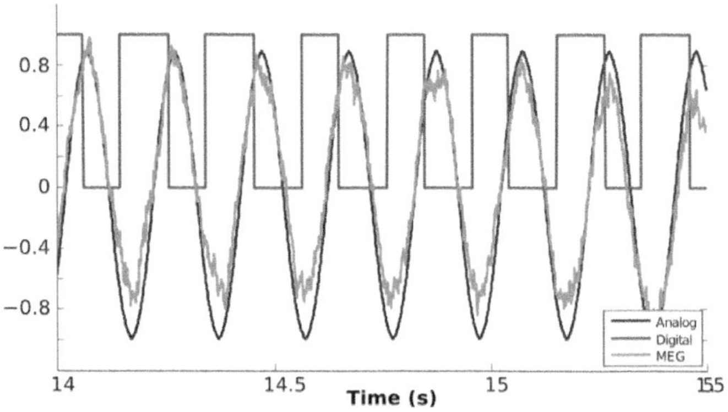

Fig. 3 Characterization of the real-time delay. Sine waves correspond to the signal from the function generator as seen by the MEG machine (connected to the analog input and sensed by a magnetometer), plotted with the pulse generated by the computer hosting the toolbox

The average delay was measured to be 30.5ms (±8ms), and was insensitive to the number of channels being transmitted simultaneously over the network. This results shows that the toolbox itself introduces very little delay to the measured signal, sufficient for most real-time MEG experiments.

The next step was to validate the data acquired by the rtMEG toolbox against the data recorded by the Elekta Neuromag® scanner. A task was developed such that a visual cue appeared on the computer screen for 10 seconds, and then disappeared for another 10 seconds. This process was repeated for ten trials while the experimenter manually modulated the frequency of the sine wave recorded by the MEG machine by changing the settings of the function generator in response to the visual cue. While the visual cue was present, an 18Hz sine wave was produced; a 2Hz sine wave was produced otherwise. The function generator was connected to the MEG machine using the two inputs (auxiliary analog input and resistor) as done before.

Figure 4 compares the signal as saved by the toolbox with the data recorded by the MEG machine. The top plot shows the signal of a given magnetometer over the whole length of the experiment. The middle plot zooms in a representative repetition of the task. The visual cue appears on the screen at zero, and after a reaction time of about 400ms the frequency of the sine wave changes simultaneously for the MEG machine and also for the toolbox. Finally, the bottom plot shows the power in the two bands used in the experiment (2 and 18Hz) for that same repetition. The conversion to frequency domain was obtained by running a Fast Fourier Transform in the signals from both sources, confirming the high degree of similarity in the results.

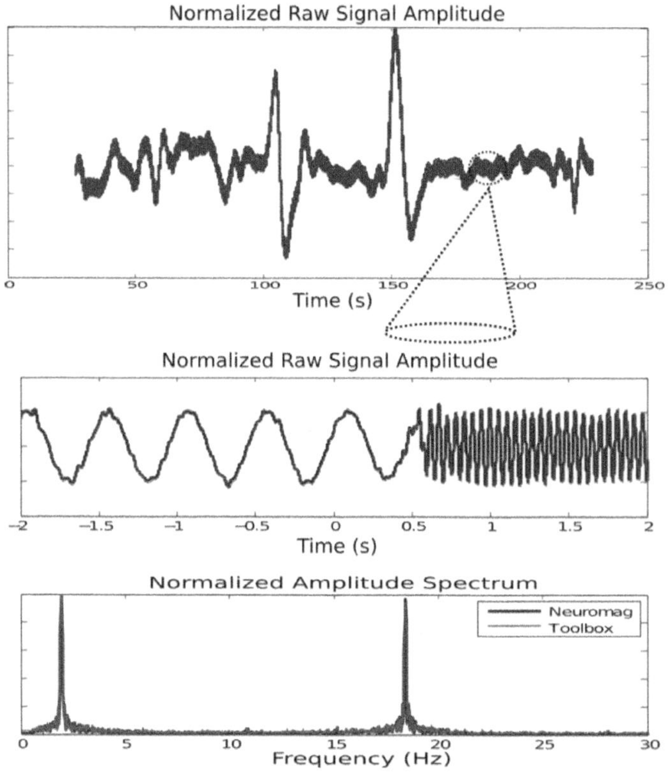

Fig. 4 Comparison between the data saved by the Elekta Neuromag® device (blue) and the toolbox (red) for a given magnetometer. Top plot shows the normalized amplitude of the raw signals for the whole period of the experiment. Middle plot shows the normalized amplitude of the data for a single representative repetition of the experiment, where zero correspond to the onset of the visual cue. Bottom plot shows the results of a FFT in the signals for the same repetition as the middle plot. Data in bottom two plots were band-pass filtered between 1-40Hz for better visualization

III. Conclusions

The ability to analyze MEG signals in real-time creates new possibilities for the study of brain function. These potential applications include:

- *Brain-machine interfaces:* the rtMEG toolbox could be used to support several areas of research in the BMI field. For example, researchers can test various neural processing, decoding, and user training paradigms "on the fly" within a single MEG session. Furthermore, one may be able to use

MEG as a pre-surgical tool to localize the optimal placement site for an ECoG grid for obtaining real-time BMI control.

- *Rehabilitation:* real-time neurofeedback training can be used to promote neuroplasticity [4,5]. Through the operation of a rtMEG-BMI system, users can learn to voluntarily modulate or change their brain activity, inducing neuroplasticity for recovery of motor function or to improve control of neuroprosthetics.
- *Research:* real-time MEG will open up the opportunity for dynamic or adaptive paradigms, which is a powerful alternative to the traditional way of doing neuroscience research. Dynamic paradigms will significantly speed up the common research cycle of having a hypothesis, designing a paradigm, running the study, going back to the lab to analyze the data, revising the paradigm based on results, and going back to the MEG center to collect more data. All those activities can potentially be set up through a dynamic paradigm and hence complete multiple research cycles in one MEG session. For example, in order to find signals tuned to particular features (e.g. human faces) which can only be described in a very high dimensional space, a dynamic paradigm may allow rapid identification of optimal feature(s) and collect data along that dimension instead of having to explore the whole feature space. Finally, this new functionality creates the possibility of producing new tools that further develop MEG research, such as real-time visualization of MEG in source space/brain surface.
- *Clinical:* clinicians may benefit from using MEG as a tool for real-time brain mapping. It enables a faster evaluation cycle, since the major elements of the offline analysis can be streamlined along with the toolbox, and this way provide advantages such as dipole localization in real-time. These features could potentially help clinicians more quickly determine treatment options. For example, the BCI2000 software package is currently being adapted for real-time brain function mapping [8], allowing clinicians to quickly locate functional brain areas without the use of cortical stimulation. Additionally, as the rtMEG toolbox presented here is a native BCI2000 module, clinicians will have access to the advanced signal processing and analysis features offered by BCI2000.

This paper describes a toolbox that relays the MEG signal directly from the MEG machine to any computer in the network with minimal delay. This toolbox is provided for free to the scientific community, with the hope to aid MEG and brain research in many different areas.

Acknowledgment

We thank the University of Pittsburgh Medical Center (UPMC) Center for Advanced Brain Magnetic Source Imaging (CABMSI) and Magnetic Resonance Research Center (MRRC) for providing the scanning time for MEG and MRI data collection. This work would not be possible without generous help from Elekta Neuromag®, Elekta Oy (Helsinki, Finland). We specifically thank Mrs. Anna Haridis at UPMC CABMSI for assistance in MEG set up and data collection. This work was partially supported by NSF Cooperative Agreement EEC-0540865, U.S. Army (TATRC) Agreement W81XWH-07-1-0716, and Grant Number 5 UL1 RR024153 from NCRR of NIH, and a special grant from the Office of the Senior Vice Chancellor for the Health Sciences at University of Pittsburgh, as well as a student travel fund from Center for Neural Basis of Cognition. Additional funding support was provided by NIH grants from the NIBIB (1R01EB007749) and NINDS (1R21NS056136) for Douglas Weber.

References

1. Schalk, G., McFarland, DJ, Hinterberger, T., Birbaumer, N., and Wolpaw, J. (2004) BCI2000: a general-purpose brain-computer interface (BCI) system. IEEE Transactions on Biomedical Engineering. 51 - 6.
2. Papanicolaou AC, Castillo EM, Billingsley-Marshall R, Pataraia E, Simos PG. (2005). A review of clinical applications of magnetoencephalography. Int Rev Neurobiol. 68:223-47.
3. Wheless JW, Castillo E, Maggio V, et al. (2004) Magnetoencephalography (MEG) and Magnetic Source Imaging (MSI). The Neurologist. 10(3):138-153.
4. Birbaumer N, Cohen LG. (2007) Brain-computer interfaces: communication and restoration of movement in paralysis. The Journal of physiology. 579:621-36
5. Wang, W., Collinger, J., Weber, D. et al. (2010) Neural Interface Technology for Rehabilitation: Exploiting and Promoting Neuroplasticity. Phys Med Rehabil. Clin. N Am, Feb; 21(1):157-78.
6. MacKay DJ. (1992) Information-Based Objective Functions for Active Data Selection. Neural Computation. 4(4):590-604.
7. Chaloner K, Verdinelli I. (2008) Bayesian Experimental Design: A Review. Statistical Science. 10(3):273-304.
8. Schalk G, Leuthardt EC, Brunner P, et al. (2008) Real-time detection of event-related brain activity. NeuroImage. 43(2):245-9.

Author: Doug Weber

Institute: Department of Physical Medicine and Rehabilitation
University of Pittsburgh
Street: 3471 Fifth Avenue, Suite 202
City: Pittsburgh, PA
Country: USA
Email: djw50@pitt.edu

Source Space Based Brain Computer Interface

Minkyu Ahn, Jun Hee Hong, and Sung Chan Jun

Department of Information & Communications, Gwangju Institute of Science and Technology, Gwangju, South Korea

Abstract— **Non-invasive Brain Computer Interface (BCI) using EEG or MEG is one of hot research areas nowadays. This technology gives a new way to communicate between human and computer or machine. BCI system enables persons to command a computer without spontaneous limb movement by measuring the brain signal and interpreting the intention of users in real-time basis. MEG/EEG signal at sensor should be combination of signals originated from infinitely many sources. So, signals at sensor are so noisy and are hard to discriminate different state information (feature), which contains implicitly user's intention. In general, it is too difficult to improve the performance of EEG based BCI without seeking better feature extraction. For seeking different features from the existing features, we try to do source localization in the belief that projection of EEG data into source space enables us to see unseen information in sensor space. Eigenbased minimum variance beamformer is introduced for this purpose in this work. It gives reconstructed source image on each voxel and is too robust to noise as well as fast (real-time reconstruction is tractable in some sense), which is well-suited to BCI system. Each trial for the given condition is projected into source space by beamformer localizer. Then in the source space feature extractions and classification are attempted.**

In this work, left-right hand motor imagery experimental paradigm was used, and common spatial pattern (CSP) and Fisher's linear discriminant analysis (FLDA) were applied for feature extraction and classification, respectively. In the specific regions of interest (for example, regions under the C3 and C4), voxel-based CSP is applied and compared with sensor-based CSP. Their BCI performance are investigated and compared for various datasets.

Keywords— **Brain Computer Interface (BCI), Localization, Beamformer.**

I. Introduction

Brain Computer Interface (BCI) is the method to communicate between human and computer. BCI does not need peripheral nervous system and it tries to understand the mental state or intention of human. Widely used experimental paradigm for BCI is left or right hand motor imagery. When human moves left and right hand some physiological changes occur and it can be measured by various ways using devices. For example, for real movement encephalography (EEG) power change can be detected on corresponding brain region since the physical movement leads increasing the number of interconnecting neurons so that the magnitude of brain activation becomes higher and a number of active neurons activate coherently [1]. The most BCI system uses such feature to discriminate mental states. Therefore BCI system requires to better figure out the intention of the user. Common Spatial Pattern (CSP) is the one of methods to successfully discriminate between two different conditions. Given two different state signals, CSP finds the vectors maximizing the variance for one signal while minimizing the variance for another signal. While CSP has been said to yield good performance there are some difficulties CSP is suffering. In many cases CSP might not work well because the data in feedback session can differ from the data in train session. Moreover the change of the mental state and condition of the subject contaminates the data, thereby resulting in bad performance. So many other methods to overcome this problem have been proposed such as invariant Common Spatial Pattern, Common Spectral Spatial Pattern (CSSP) etc. These methods were investigated and compared in [4]. To get higher performance BCI system better feature extraction methods are certainly to be developed. In this sense, efforts to seek other features in source space as well as in sensor space are made in this work. Beamformer is one of well known source imaging methods. It is a kind of spatial filter with good performance to localize sources and to reconstruct source magnitude in a voxelwise manner [2]. In this work, we apply beamformer technique to generate source imaging for EEG-based BCI system. We focus on finding the possibility to improve BCI performance from some features in source space rather than those in sensor space.

II. Methods

A. Common Spatial Pattern

CSP is the famous feature extraction technique used in the BCI field. It enables us to well discriminate multi-class components. CSP is a spatial filter maximizing the variance for one data set while minimizing the variance for another data set. The filters can be obtained by solving well known generalized eigenvalue problem below:

$$W'C_lW = \lambda_l\,,\; W'C_rW = \lambda_r = I - \lambda_l$$

W is a filter matrix of our interest and C_l and C_r are covariance matrices of left and right imagery datasets, respectively. Sometimes CSP method shows poor

S. Supek and A. Sušac (Eds.): Advances in Biomagnetism – BIOMAG2010, IFMBE Proceedings 28, pp. 366–369, 2010.
www.springerlink.com

performance for highly noisy data and its performance highly varies over subjects. Other variants of CSP method have been proposed for improving the classification of BCI system. Invariant CSP uses pre-knowledge information such as idle state signal for explaining invariant property. CSSP utilizes spectral information as well as spatial information. It was reported that these methods improved BCI system [4].

B. Adaptive Beamformer

Minimum-variance spatial filter is one of the representative adaptive spatial filters [2]. To estimates the source activity, the following spatial filter technique is applied.

$$\hat{Q}(\mathbf{r},t) = \mathbf{w}^T(\mathbf{r})\mathbf{m}(t), \quad (1)$$

where, $\hat{Q}(\mathbf{r},t)$ and $\mathbf{m}(t)$ are the estimated magnitude of the source activity at location $\mathbf{r}$ and spatiotemporal measurement with total number of S sensors and time T, respectively. Spatial filter $\mathbf{w}(\mathbf{r})$ is derived such that

$$\mathbf{w}(\mathbf{r}) = \arg\min \mathbf{w}^T(\mathbf{r})\mathbf{C}\mathbf{w}(\mathbf{r}), \text{ subject to } \mathbf{w}^T(\mathbf{r})\mathbf{l}(\mathbf{r}) = 1, \quad (2)$$

where, $\mathbf{C}$ is the covariance matrix of the measurement defined as $\mathbf{C} = \langle \mathbf{m}(t)\mathbf{m}^T(t) \rangle$ and $\langle \cdot \rangle$ indicates the time average over a certain time window. Lead-field vector is denoted $\mathbf{l}(\mathbf{r})$. This constraint minimization problem can be solved using the method of the Lagrange multiplier and explicit form of the weight vector can be represented by

$$\mathbf{w}(\mathbf{r}) = \frac{\mathbf{C}^{-1}\mathbf{l}(\mathbf{r})}{\mathbf{l}^T(\mathbf{r})\mathbf{C}^{-1}\mathbf{l}(\mathbf{r})}. \quad (3)$$

In general, the estimated source activity obtained from minimum-variance beamformer may include distortion or noise. This eigenspace-projection (ES) spatial filter enables to consider the reducing the noise effect by the projection onto the signal subspace. The projection matrix $\mathbf{E}_S$ can be obtained by singular value decomposition (SVD) of the covariance matrix $\mathbf{C}$ such that

$$\mathbf{C} = \mathbf{E}_S\mathbf{\Lambda}_S\mathbf{E}_S^T + \mathbf{E}_N\mathbf{\Lambda}_N\mathbf{E}_N^T \quad (4)$$

where, Λ indicates the diagonal matrix consisting of the eigenvalues which is composed of signal Λ_S and noise Λ_N part as decreasing order. Those corresponding eigenvectors are $\mathbf{E}_S$ and $\mathbf{E}_N$, respectively. When we consider about between lead-field vector $\mathbf{l}(\mathbf{r})$ and covariance $\mathbf{C}$, the term $\mathbf{C}^{-1}$ $\mathbf{l}(\mathbf{r})$ is orthogonal to each other in ideal conditions. In real case, the first term of Eq. (4) is used to calculate weight vector because the noise term enables to decrease the orthogonality. Accordingly, the following weight vector for ES is shown as

$$\mathbf{w}_{ES}(\mathbf{r}) = \mathbf{E}_S\mathbf{E}_S^T\mathbf{w}(\mathbf{r}). \quad (5)$$

III. Experiment and Performance Evaluation

A. Experimental Procedure

The experiment basically consists of measuring data when left-motor imagery and right-motor imagery which is well known paradigm for 2 class BCI. Simultaneous MEG/EEG datasets for four male subjects (age 23-30) from Eleckta Vectorview system (MEG: 306, EEG: 34 channels) at MEG center, Seoul National University Hospital were collected under left-right motor imagery movement BCI paradigm (Fig. 1) and all of data were digitized at a 1 KHz sampling rate; five experiment were carried out, 2 for 1st subject and 3 for other 3 subjects. Each trial data of our interest with 1 second long starting 0.5 seconds after appearance of arrow instruction (cue) was extracted from EEG. Finally 60 trials each condition (left or right) were obtained.

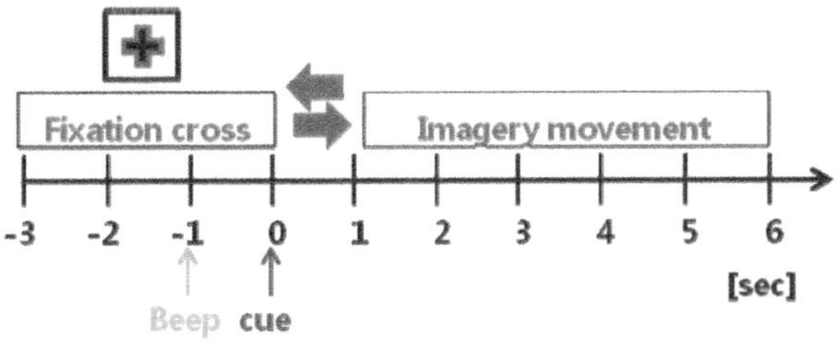

Fig. 1 Experimental procedure: resting before cue and main task after cue

B. Performance Evaluation

Every data set are band-pass (10–15 Hz) filtered just before evaluating the performance since this band is around SMR and is known to have significant discriminative information, being good for BCI. CSP is applied to the filtered data to extract features. This process makes a feature vector consisting of powers of the linearly combined channels by CSP spatial filters. In this work, first 5 and last 5 CSP spatial filters are used to make the power features. To construct classifier, 70% of data were categorized as a training set and this set were used to make FLDA classifier. Then remaining 30% of data were tested to compute this classifier performance. This procedure was several times repeated to cross validate success rate using uniform random number generator.

IV. Results

A. Spatial Filtering by Beamformer

Beamformer reconstructs the source imaging of the given spatiotemporal measurement data on the pre-defined voxel space. To make it, 3-shell 'ary' model and forward model in [5] was introduced. We focused on voxels near C3 and C4 area since these positions have been known as very discriminative region generating motor area dependent signal.

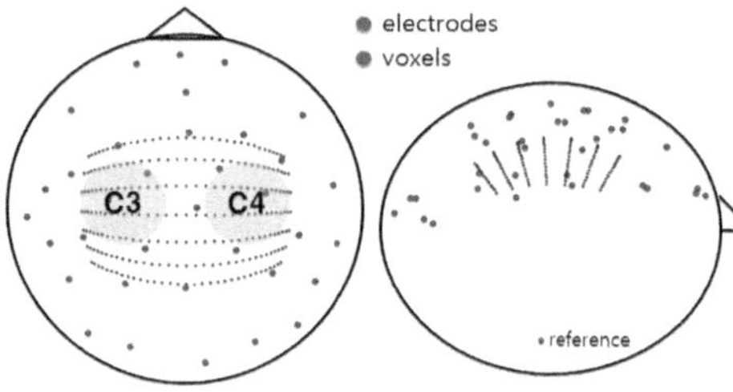

Fig. 2 Sensor location and scattered voxel space near C3 and C4

Figure 2 shows the description of sensor and source space. The data of each subject were projected by minimum variance beamformer from sensor space to source space. This source space signals have more (or different) information than sensor space since beamformer filters out some fraction of noise signal out of pre-defined source space. Therefore we expect to get more discriminative information in source space. To find this kind of feature in voxel signals, we investigated power changes over time in source space. But most of localization technique needs time-rocked data to average all of trials since it reduces the noise effect while increasing information in signals. However BCI system have trouble to use averaging over trials, each trial may differ in the meaning of phase and time so averaged data may lose the information by cancelling each other. Therefore we used single trial data to see power changes. The final image was constructed by accumulating voxel powers over trials after localization. It is illustrated in Figure 3.

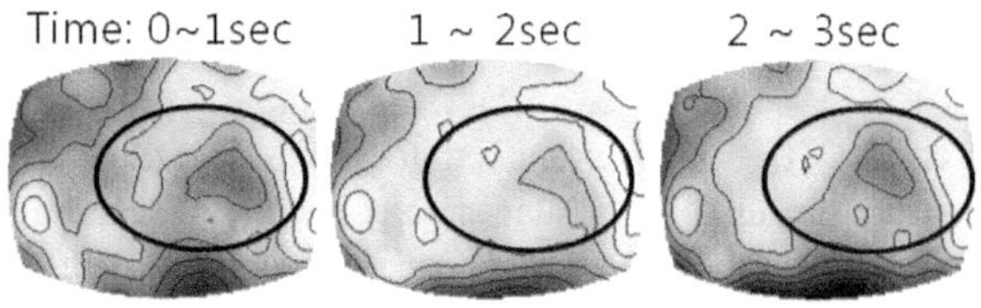

Fig. 3 Power changes over time in source space when right imagery; accumulated power image of every trial for each voxel

Figure 3 depicts the power distribution of voxels in Figure 2 after localization. As seeing, the power decrease is found near C4. This power decrease occurs from 1sec to 2sec and then it is recovered from 2sec. These decreasing and increasing of power are the good evidence that the subject brain generated some activity by event condition; left or right imagery.

B. Performance Test

As we found the power change features, it is expected that the performance would be better than sensor space. If the reconstructed voxel signals are used directly to BCI system, it may result in good success rate. Now we scatter different number of voxels (different spatial resolutions in source space) on same region that is near C3 and C4 as previous case. Four different numbers of voxels are chosen to evaluate its performance.

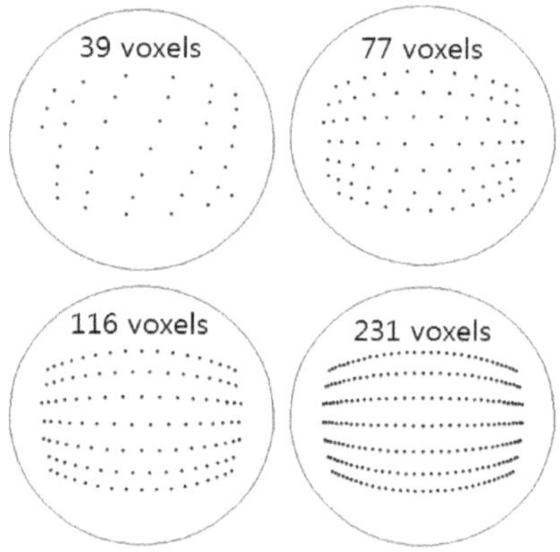

Fig. 4 Four source spaces: Different numbers of voxels, all of spaces are scattered near C3 and C4 that is different spatial resolution in source space

The data of each subject are projected onto 39, 77, 116, 231 voxels space. Considering that the number of EEG sensors is 34, the number 39 is relatively same but 231 of voxels are about seven times more than sensors. This may be advantageous in the sense of expanding the number of channels. But also it is worth to investigate that different session data from one subject may be used as expanded trials after projecting sensor space onto fixed voxels, although different sessions give un-consistent features varying over time and situation in sensor space. This kind of study to use previous data for same subject is under investigation.

In source space, source time series at each voxel was reconstructed. So, for each trial we got source time series which is represented as a matrix of the number of voxels times the number of time points. This means source reconstruction procedure transformed sensor matrix into source matrix. Then, source region of our interest was extracted with time window 0.5sec to 1.5sec after onset. We bandpass filtered them first and then tested it according to the way described in 'Performance evaluation' section. The result is shown in Figure 5. As we expected, beamformer filtered noisy sensor space signal to voxel and then it gives better success rate. In the case of 39 voxels which is almost comparable to the number of sensors, there is far more significant increase of the success rate than CSP of EEG sensor signals in sensor space. It tells that beamformer reconstructed original source signal well and filtered out brain noise as we expected. As the number of voxels increases (more refined in source space), it is found that up to 77 of voxels gives improvement of success rate while its improvement is not clear for other voxels. From these results we expect that there is a limitation and it may be related to the region of our interested source space. In this work we just used the broad area near C3 and C4. We believe that it is possible to get better performance if more valid region is chosen.

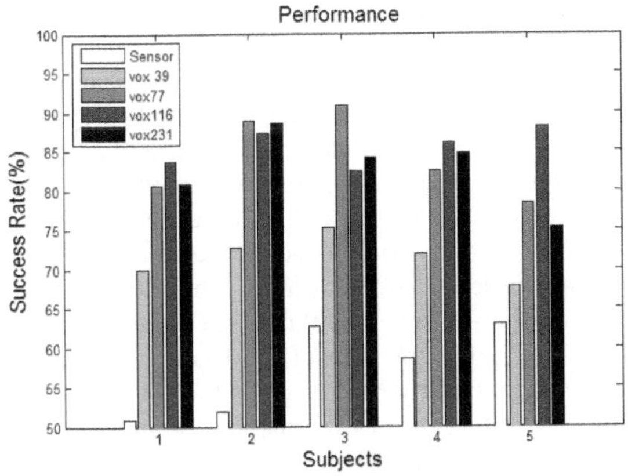

Fig. 5 Performance comparison over sensor and different source spaces

V. Discussion

The performance limitation may come from various issues; how to determine voxel space, how forward model is accurate and how this approach in source space is overfitted. This issue should be carefully investigated in the future study. Head modeling incorporating conductivity distribution in the head plays a crucial role in EEG source localization. More realistic modeling yields more accurate localization. MEG source localization is more robust to head modeling error. We expect MEG to yield the similar improvement in BCI system, which is under investigation.

VI. Conclustion

Most of BCI system has suffered from the low signal-to-noise ratio. Therefore many kinds of method to make BCI system robust to noise have been proposed. In this paper, BCI using minimum variance beamformer is used to reconstruct the clearer voxel signal in source space rather than sensor space. Voxels near C3 and C4 show the power change over time that is well known feature in BCI field. Different numbers of voxels (different spatial resolution) in source space were evaluated to see comparative performance of BCI system. In 37 voxels case, it results in significantly higher success rate than normal BCI using sensor signal. We expect that expanding source space could give enhanced accuracy. In addition, localization enables us to collect more trials or channels by simply merging different sessions without hassle, which is not likely in sensor space.

In conclusion, our proposed BCI using source imaging information by beamformer could improve its success rate significantly and it shows its applicability. This work is only for off-line BCI. BCI for on-line feedback is under investigation.

Acknowledgment

This work is supported by KRF-2008-531-C00012, the BioImaging Research Center at GIST, KOSEF/JSPS researcher exchange program, and the NIPA (National IT Industry Promotion Agency) (NIPA-2009-C1090-0902-0008).

References

1. G. Pfurtscheller, F.H. Lopes da Silva (1999) Event-related EEG/MEG synchronization and desynchronization: basic principles, Clinical Neurophysiology, 110, 1999 1842-1857.
2. K. Sekihara and S. S. Nagarajan (2008) Adaptive spatial filters for electromagnetic brain imaging. Berlin, Heidelber: Springer-Verlag
3. Benjamin Blankertz, Ryota Tomioka, Steven Lemm, Motoaki Kawanabe, Klaus-Robert Müller (2008) Optimizing Spatial Filters for Robust EEG Single-Trial Analysis, IEEE Signal Processing Magazine, Vol. XX, 2008.
4. S. Kang, M. Ahn, S. C. Jun (2009) Performances among various Common Spatial Pattern Methods for simultaneous MEG/EEG data, Proceeding of ICMIT2009.
5. Zhi Zhang (1995), a fast method to compute surface potentials generated by dipoles within multilayer anisotropic spheres, Phys. Med. Biol. 40 (1995) 335-349.

Author: S.C. Jun
Institute: Gwangju Institute of Science and Technology (GIST)
Street: 261 Cheomdan-gwagiro(Oryong-dong), Buk-gu
City: Gwangju
Country: South Korea
Email: scjun@gist.ac.kr

Exploiting Prior Neurophysiological Knowledge to Improve Brain Computer Interface Performance

Arne Ewald, Andreas Ziehe, Forooz Shahbazi, and Guido Nolte

Fraunhofer Institute FIRST, IDA Group, Kekuléstr. 7, 12489 Berlin, Germany

Abstract— **Most EEG/MEG based Brain Computer Interfaces (BCI) employ machine learning techniques to discriminate and classify the recorded data belonging to different classes. Usually, no neurophysiological knowledge is used within the classification algorithms. Here, a method is proposed that includes prior knowledge about the locations of sources of imagined movement of the left and the right hand by projecting EEG/MEG data onto a subspace defined by modeled sources at the corresponding locations in somatosensory areas.**

Three different source models are investigated. First, one radial dipole on each side is based on the assumption that both location and orientation are known. Hence, for two sides, a 2-dimensional subspace is selected. Second, three dipoles at each location span a 6-dimensional subspace assuming known locations but uncertain orientations. Third, we modeled the sources as multipoles up to quadrupolar order resulting in a 16-dimensional subspace. The multipole expansion systematically corrects for inaccuracies both in location and exact shape of the source.

After the projection onto respective topographies, feature extraction is performed on the reduced data by Common Spatial Filter (CSP) analysis. Finally, Linear Discriminant Analysis (LDA) is applied for classification. The projection of the data leads to a reduction of the dimensionality of the signal focusing on those parts of the signal which are generated or suppressed in the motor cortex during imagined hand movement. Since EEG/MEG data are strongly affected by various types of artifacts hampering the classification the proposed procedure leads to a removal of parts of the signal and therefore a reduction of artifacts. For EEG data it is shown that a projection with respect to source locations prior to CSP analysis leads to a gain of BCI performance when the sources are modeled as multipoles.

Keywords— **BCI, source modeling, multipoles, ERD.**

I. Introduction

Machine learning techniques have become the tool of choice for developing non-invasive Brain Computer Interfaces (BCI) as they take user specific variations of brain data into account [1]. Therefore the effort of learning is shifted from the user towards the machine. A well working technique for feature extraction within the Berlin Brain Computer Interface (BBCI) setup is Common Spatial Pattern (CSP) analysis [2]. It is used for spatial filtering as different regions in the brain are activated during real or imaginary movements of different extremities like the left or the right hand. Based on an optimized EEG analysis by advanced machine learning algorithms 12 out of 14 BCI-naive subjects could achieve a performance better than 70% [3], meaning that 70% of binary decisions, each made every few seconds, are correct. The difficulty of achieving higher performance rates is mainly due to physical limits of measuring brain currents with noninvasive techniques as the EEG signal is strongly affected by noise. Furthermore, many artifacts like eye blinking, breathing, or muscle activation disturb the classification.

Since no neurophysiological knowledge is applied to the BCI processing procedure and the classification only depends on the measured data, we propose a method of applying knowledge about the corresponding source locations to reduce artifacts. First, brain sources generated by movement of the left and the right hand are modeled at pre-defined locations (see III-A). Three different modeling techniques are examined. One single dipole at each source location, three dipoles at each source location and a multipole up to quadrupolar order at each location. Then, the induced electric field on the scalp is calculated via the electric lead field method [4]. Finally, the measured EEG data are projected in a subspace spanned by the result of the forward calculation to achieve a reduction of dimensionality and to reduce artifacts contained in the signal. The procedure is described in details in Section III-B. After the projection with respect to the modeled source locations is executed the machine learning techniques for discrimination and classification are applied on the reduced data. Section IV shows that the BCI performance can be improved by applying knowledge about source locations.

II. Background

The Berlin Brain Computer Interface employs event related desynchronisation (ERD) as the chosen neurophysiological paradigm. In a BCI online session the user performs imagined movement of the left and the right hand to steer any device, e.g. a spelling software or a pinball game. As the location of the ERD varies according to which hand was moved, the modulation of the sensory motor rhythm can be used to discriminate two classes. It is known that the

S. Supek and A. Sušac (Eds.): Advances in Biomagnetism – BIOMAG2010, IFMBE Proceedings 28, pp. 370–373, 2010.
www.springerlink.com

region around the electrode 'C3' which is positioned with respect to the standardized 10-20 system desynchronizes during phases of imagined movement of the right hand whereas the region of 'C4' desynchronizes during phases of movement of the left hand. Therefore, assumed brain sources beneath the electrodes 'C3' and 'C4' locations are used in this work as the prior neurophysiological knowledge (see Section III-A). To discriminate the spatial variation of the ERD according to the imaginary hand movement the spatial filtering technique Common Spatial Pattern (CSP) is used. It has been shown to be one of the best working algorithms for feature extraction in the BCI field [2, 3]. The aim is to decompose the recorded, sampled and preprocessed signal $\mathbf{X} \in \mathbb{R}^{C \times T}$ (C: Channels, T: sampled points in time) from the sensor space into a lower dimensional space of size J such that the data of the two given classes can be discriminated by a linear classifier. A projection matrix $\mathbf{W} \in \mathbb{R}^{C \times C}$ is determined by maximizing the variance of signals of one condition (e. g. left hand movement) and at the same time minimizing the variance of the signals of another condition (e.g. right hand). This leads to a high discriminability of the two conditions. The reduction of dimensionality is achieved by choosing J vectors from $\mathbf{W}$ resulting in $\mathbf{W}' \in \mathbb{R}^{J \times C}$.

After the data is projected into the CSP-space, Linear Discriminant Analysis is applied to separate the space into two different classes by a hyperplane. As the variance of the CSP-projected data is approximately chi-squared distributed and a linear classifier in general is optimal for Gaussian probability distributions, the logarithm is taken to achieve an approximately normal distribution. Hence, the linear discrimination is represented by

$$f(X:\{\mathbf{w}_j\}^J\{\beta_j\}^J) = \sum_{j=1}^{J} \beta_j \log(\mathbf{w}_j^T X X^T \mathbf{w}_j) + \beta_0$$

with $\mathbf{w}_j$, $j=1..J$ being the row vectors of $\mathbf{W}'$. The coefficients $\beta_0..\beta_j$ are calculated by a least squares regression. The separating hyperplane determined by the function f, is fitted into the projected data such that the sum of the squared distances between each data point and the hyperplane is maximized.

III. Methodology

Within this work an intermediate step is inserted before the discrimination procedure. To reduce the data containing information related to the sources of left and right hand movement, the discrete pre-processed data are projected onto a subspace defined by the known locations of the sources (see III-B). How the subspace is defined and how the sources are modeled is described in III-A.

A. *Source Modeling*

The concept of equivalent current dipoles is the basic approach for modeling brain sources as it is a convenient representation for coherent activation of a large number of pyramidal cells [5]. Three different cases are considered to model the sources of brain activity due to ERD. The results strongly depend on the type of the modeled source. The assumed location of the modeled sources is the same in all three cases. The pre-defined dipoles are located $\kappa = 2.5$ centimeters beneath the electrodes 'C3' and 'C4' in radial direction with respect to the curvature of the applied head model. Hence, the locations of the sources $\mathbf{d}$ can be calculated via $\mathbf{d} = \mathbf{e} - \kappa \mathbf{n}$, where $\mathbf{e}$ denotes the position of the electrode on the surface of the head and $\mathbf{n}$ the outward normal at 'C3' and 'C4'.

1) Radial Dipoles One single dipole is defined under each electrode 'C3' and 'C4'. The direction of each dipole is in radial direction according to the curvature of the head. To obtain the subspace in which the measured data are projected, the resulting scalp potential due to the modeled sources is calculated. This forward calculation is performed by the electric lead field method in a realistic volume conductor [4,6]. The results of the forward calculation $\mathbf{A}_N \in \mathbb{R}^{C \times N}$, with $N = 2$, describe a subspace with respect to the two modeled sources.

2) Rotating Dipole: The second case describes modeling a single source with three dipoles pointing along orthogonal axes. The technique differs from the previous by the direction and the number of dipoles. The locations of the sources due to left and right hand imagined movements are determined in the same way as before (see III-A). Instead of using a single dipole at each source location, three different dipoles are employed. They point along the three axes of an arbitrary Cartesian coordinate system. Therefore, they are orthogonal with respect to each other and serve as a basis for modeling a single brain source at the computed location. Applying forward calculation to each dipole results in a subspace $\mathbf{A}_N \in \mathbb{R}^{C \times N}$, with $N = 6$, as the two locations beneath 'C3' and 'C4' are considered.

3) Multipole Expansion: An extension to the model of equivalent current dipoles is modeling neuronal activity by multipoles. The approximation of brain activity by multipoles goes far beyond the dipole model as only the center and the magnitude is described. Multipoles are calculated by adding higher order coefficients which are obtained by an expansion (e.g. Taylor) of Green's function [7]. These higher order multipole coefficients can be related to other relevant properties like spatial extend or curvature. It is shown that as the sources increase in size, they become less accurately modeled by current dipoles [8].

Multipolar currents can be illustrated as two or more dipole currents which are infinitesimally shifted against each other at a distance δ [9]. When a current source is modeled by a dipole and a quadrupole at a given location, the direction in which the quadrupole components are pointing and the direction of the shift can be chosen arbitrarily. Therefore, in principal nine different cases have to be taken into consideration to model the quadrupole components of one source. By solving a singular value decomposition of the resulting scalp potentials of these nine different cases one can observe a linear dependency. Only five out of the nine combinations of direction and shift of the quadrupole components provide additional information. As a multipole is represented by a central dipole and the quadrupole components, each source is modeled by the five quadrupoles and three dipoles. For the two sources beneath the electrodes 'C3' and 'C4' the forward calculation leads to a subspace $\mathbf{A}_N \in \mathbb{R}^{C \times N}$, with N = 16.

B. Projection with Respect to the Source Locations

Considering the sampled and filtered EEG data as a matrix $\mathbf{X} \in \mathbb{R}^{C \times T}$, the projection is performed by a matrix $\mathbf{P} \in \mathbb{R}^{N \times C}$ via $\mathbf{Y} = \mathbf{PX} = ((\mathbf{A}_N^T \mathbf{A}_N)^{-1} \mathbf{A}_N^T)\mathbf{X}$. The resulting matrix $\mathbf{Y}$ is of size N×T, where N represents the dimension of the subspace in which the projection takes place. After the calculation $\mathbf{Y}$ contains the reduced data which is further processed by the common techniques CSP and LDA.

As a standard signal processing routine the Whitening Transformation is applied to decorrelate and further reduce the noise. The projection is extended to $\mathbf{Y}=\mathbf{P}\,\mathbf{\Sigma}^{-1}\mathbf{X}= ((\mathbf{A}_N^T \mathbf{A}_N)^{-1} \mathbf{A}_N^T)\,\mathbf{\Sigma}^{-1}\mathbf{X}$, with $\mathbf{\Sigma}$ being the estimated covariance matrix of the data.

IV. Results

To determine the BCI performance of a subject, data of advised imaginary movement of the left and the right hand are used. These EEG data sets are cut into two halves where the first half is used to train the classifier and the second half is used to evaluate the performance.

Considering the number of trials of the second half of the data N_t and the number of wrongly classified trials F, the performance in terms of an error ratio of one subject can be calculated as $e = F/N_t$. As data of 24 subjects are investigated, the error is averaged over all subjects and given in percentage by E.

A. Findings

Table I shows the results of the experiments considering the three source modeling techniques described in section III-A. The reference error that is obtained by not employing any projection with respect to source locations was E*=25.11%.

The results, given in Table I show that modeling the source locations by a single dipole for each condition leads to a fatal decrease of performance with $E^*- E_2$ =-13.36%-pts. The BBCI performance of only 4 of the 24 subjects improved slightly.

Table I Results

radial dip.(N = 2)	axial dip.(N = 6)	multipoles(N = 16)
E_2 =38.47 %	E_6 = 32.38 %	E_{16} = 23.03%

Modeling the sources of neuronal activity with three dipoles at each location leads to a better result in comparison to the modeling technique composed of only two dipoles. The error ratio decreases by E_2–E_6=6.09%-pts. Nonetheless, applying knowledge about the source locations by modeling them as dipoles impairs the performance compared to the reference value E*. This observation changes when the multipoles come into play. Table I illustrates that the BCI performance improves by modeling the sources with multipoles by $E^*- E_{16}$=2.08%-pts.

As improvement or decline of performance varies from subject to subject, Figure 1 shows the individual errors as a scatter plot. One can observe that the performance improves for 16 out of the 24 subjects, when the sources are modeled as multipoles as the error ratios lie above the diagonal, that is shown in red.

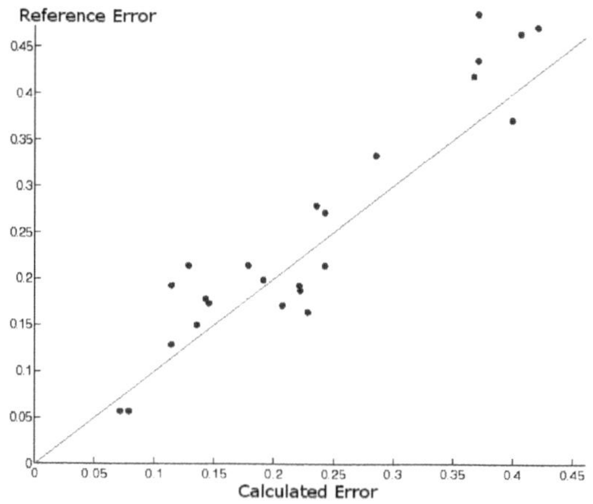

Fig. 1 The BCI performance relating the reference error E* and E_{16}.The performance of 16 out of 24 subjects improved by the applied method

B. *Interpretation*

Projecting the data into a subspace spanned by modeled source locations before the CSP method is applied leads to a loss of relevant information when the source locations are modeled as dipoles only. Since the exact locations of the sources vary a little from subject to subject, it is conceivable that the assumed locations do not exactly match with the real source locations. This effect can not be compensated by the reduction of artifacts. Discarding relevant information by the prior projection does not happen when the multipole technique is used to model the source locations, because the additive terms of the quadrupoles effectively introduce a spatial extent of the sources and diminish the effect of mislocation. This effect can be verified by a simulation that is shown in Figure 2. An activated patch of cortex is illustrated by the red region. This source of brain activity is modeled at the location indicated by the black dot which is one centimeter away from the original source. The misplacement and the fact that a distributed source is modeled by a point source introduce an error. This error e is given in the figure for all three modeling techniques and is defined as the relative error of the resulting scalp potential of the dipole/multipole model with given location and fitted moments. One can observe that the modeling error is nine times smaller when the source is modeled as a multipole instead of a single dipole.

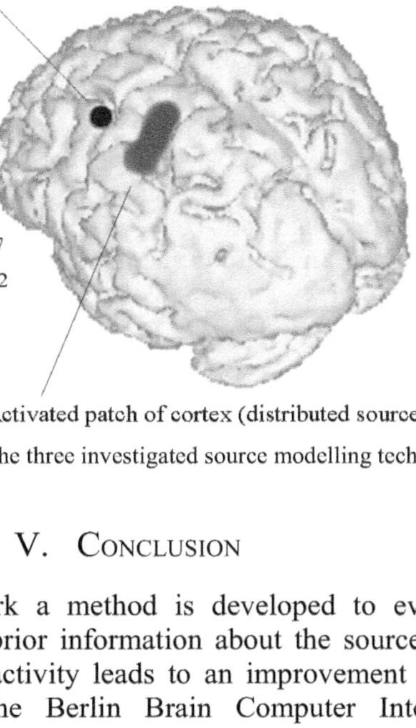

Fig. 2 Comparison of the three investigated source modelling techniques

V. Conclusion

Within this work a method is developed to evaluate whether applying prior information about the source locations of neuronal activity leads to an improvement of the performance of the Berlin Brain Computer Interface. The brain activity according to movement and imagined movement of the left and right hand can be distinguished by the locations of the activity. Therefore, the prior projection was designed to reduce the data to the part which is based on sources close to a known origin and hence to weaken artifacts and noise. Three different source modeling techniques were taken into consideration. In total, i.e. for both hemispheres, the sources were modeled by two dipoles in radial direction, six dipoles in axial directions and 16 multipoles consisting of dipoles and quadrupoles. It was observed that the performance of the BBCI could be improved by modeling the source locations by multipoles. Examining the recorded data of 24 subjects, the performance of 16 subjects led to better classification results. Applying dipoles, either two in radial direction or six in axial directions as the modeled sources, led to an extensive decrease of performance. The approach of applying prior knowledge to a BCI system where the classification is solely based on machine learning techniques still offers space for improvement and further research. As the sources vary from subject to subject, it would for example be useful to determine subject specific source locations in the area of the motor cortex that can be used for a prior projection.

Acknowledgment

This work was supported by the "Bundesministerium für Bildung und Forschung".

References

[1] K.-R. Müller, M. Kraudelat, G. Dornhege, G. Curio, and B. Blankertz, "Machine learning techniques for Brain Computer Interfaces," Biomed Tech, vol. 49, pp. 11–22, 2004.
[2] B. Blankertz, R. Tomioka, S. Lemm, M. Kawanabe, and K.-R. Müller, "Optimizing spatial filters for robust EEG single-trial analysis," IEEE Signal Processing Magazine, vol. 25, pp. 41–56, 2008.
[3] B. Blankertz, F. Losch, M. Krauledat, G. Dornhege, G. Curio, and K.-R. Müller, "The Berlin Brain Computer Interface: Accurate performance from first-session in BCI-naive subjects," IEEE Trans Biomed Eng, 2008.
[4] G. Nolte and G. Dassios, "Analytic expansion of the EEG lead field for realistic volume conductors," Phys Med Biol., vol. 50, pp. 3807 – 3823, 2005.
[5] S. Baillet, J. Mosher, and R. Leahy, "Electromagnetic Brain Mapping," IEEE Signal Processing Magazine, vol. 18, pp. 14–30, 2001. [Online]. Available: http://cogimage.dsi.cnrs.fr/publications/2001/BML01
[6] G. Nolte, "The magnetic lead field theorem in the quasi-static approximation and its use for MEG forward calculation in realistic volume conductors," Phys Med Biol., vol. 48, pp. 3637 – 3652, 2003.
[7] G. Nolte and G. Gurio, "Current multipole expansion to estimate lateral extent of neuronal activity: A theoretical analysis," IEEE. Trans. Biomed. Eng., pp. 1347–1355, 2000.
[8] K. Jerbi, S. Baillet, J. Mosher, G. Nolte, L. Garnero, and R. Leahy, "Localization of realistic cortical activity in MEG using current multipoles," Neuroimage, pp. 779–793, 2004.
[9] J. Wikswo and K. Swinney, "A comparison of multipole expansions," J. Appl. Phys., vol. 56, pp. 3039–3049, 1984.

Neural Dynamics of Alcohol Effects on Cognitive Control: Eriksen Flanker Task

Ksenija Marinkovic[1] and Sheeva Azma[2]

[1] University of California, Radiology, San Diego, USA
[2] Georgetown University, Neuroscience, Washington DC, USA

Abstract— **Alcohol intoxication interferes with the capacity to inhibit maladaptive reactions and execute correct responses. In order to investigate alcohol's effects on cognitive control, a group of healthy subjects (N = 20, 9 women) participated in both alcohol (0.6 g/kg ethanol) and placebo conditions in a color version of the Eriksen flanker task. MEG signal was recorded from the whole head and the noise-normalized minimum norm inverse estimates were anatomically constrained to each person's cortical surface reconstructed from MRI. Alcohol increased error rates in the response-conflict condition. Under placebo, the anterior cingulate (AC) was sensitive to the level of conflict, supporting an optimized response strategy. Alcohol blunted this conflict-induced differentiation in AC activity. This suggests that alcohol impairs cognitive control by affecting top-down regulation of response preparation and execution. Alcohol-induced impairment of executive functions may result in poor self-control, expressed as socially-inappropriate behavior and an inability to refrain from drinking.**

Keywords— **MEG, Eriksen flanker, cognitive control, alcohol, anterior cingulate gyrus.**

I. Introduction

Chronic alcohol abuse is associated with deficits in executive functions, with the frontal lobes being especially vulnerable to its adverse effects [1]. Similarly, acute alcohol intoxication disrupts executive abilities in situations of increased complexity as it interferes with cognitive assessment of the environment and the capacity to inhibit maladaptive responses and use an appropriate response strategy [2]. Indeed, the inability to maintain inhibitory control over harmful levels of drinking is considered to be fundamental to alcohol abuse [3]. Though subsuming diverse functions, "executive capacity" refers to the ability to sustain focus on a goal in the face of changing contextual demands [4].

Neuroimaging studies indicate that different executive tasks regularly engage highly overlapping regions centered primarily in the anterior cingulate (AC) and lateral prefrontal (LPF) cortices. Conflict-monitoring theory [5] describes how this strategy-driven regulatory function arises from their functional interplay. Alternatively, the "Reinforcement Learning" theory proposes that a dopamine-dependent signal acting through the basal ganglia permits the AC to engage in top-down regulation [6]. In both cases, the AC is essential to subserving goal-oriented actions especially those that are not well rehearsed or habitual, under contextually-challenging circumstances. Its anatomical connections include the spinal cord, motor and limbic cortices, and prefrontal association areas, supporting this multifaceted role [7]. The AC is sensitive to task difficulty, response conflict, and error processing [8] and its activity has been probed in a series of cognitive control tasks.

The Eriksen flanker task was devised to investigate the effects of compatibility between irrelevant information (flanker letters) and targets (central letter in a letter array) on decision making [9]. Event-related potential (ERP) studies suggest that response conflict evokes a N2 component at ~350ms after stimulus onset, and error-related negativity (ERN) at ~50ms after response, presumably generated in the AC [10]. The ERN is attenuated by alcohol intoxication [11], indicating impaired performance monitoring. In order to investigate spatio-temporal characteristics of alcohol's effects on cognitive control, we used anatomically-constrained MEG (aMEG) method. It combines whole-head high-density MEG and a distributed source modeling approach with high-resolution structural MRI to estimate the anatomical distribution of the involved neural networks with high temporal resolution [12, 13]. The task version employed in this study included comparison between the stimulus-level vs. response-level incongruity, permitting assessment of alcohol's effects on these two aspects of processing.

II. Methods

A. Subjects

Young, healthy volunteers (N=20, 9 women; age (mean ± st. dev) = 25 ± 3.4 years) served as their own controls as they participated in both alcohol and placebo conditions in a counterbalanced manner. They were all right-handed native English speakers and were prescreened for a negative history of alcoholism and drug use. They were non-smokers and reported moderately low alcohol use, imbibing 3 ± 1 drinks per occasion, 2 ± 1 times per week on average. They all came from non-alcoholic families. All participants gave written informed consent to participate in the study.

B. Task

A color version of Eriksen flanker task was employed in this study [14]. Two flanker squares were presented in green, red, blue, or yellow for 200ms, followed by a target square presented in central location (Fig. 1, left panel). Subjects were instructed to respond to the color of a target square by pressing left button to green or red and right button to blue or yellow. On congruent (CO, 50%) trials the target and the immediately preceding flankers were of the same color. The incongruent trials were of two types: on stimulus-level incompatible (SI, 25%) trials the flankers and targets were different but used the same response hand; on response-incompatible (RI, 25%) trials the flankers and targets differed in color as well as response mapping. A total of 512 trials were presented every 1600ms with Presentation® software (Neurobehavioral Systems).

C. Data Acquisition and Analysis

Continuous MEG signals were recorded from 204 channels (102 pairs of planar gradiometers) at 600 Hz with a whole-head Neuromag instrument (Vectorview, Elekta Neuromag), at the Martinos Center in Boston. Averages for each condition included only trials on which correct responses were given that were free of eyeblinks or artifacts. Digitizing positions of the main fiduciary points, magnetic coils, and a large array of random points spread across the scalp with 3Space Isotrak system (Polhemus Inc.) permitted co-registration with structural MRI images. For each participant a high-resolution structural 3D MPRAGE T1-weighted scan was obtained with a 3 T Trio body scanner (Siemens, Erlangen), with TR = 2.54 sec, TE = 3.25 msec, flip angle = 7^0, FOV = 256, 128 sagittal slices, 1.33 mm thickness, in-plane resolution 1 x 1 mm. Each person's cortical surface was reconstructed with an automatic segmentation, tessellation and inflation FreeSurfer tool (http://surfer.nmr.mgh.harvard.edu/), subsampled to ~2500 dipole locations per hemisphere and used to constrain inverse solution. The atomically-constrained MEG (aMEG) estimates were calculated with the MNE software (http://www.nmr.mgh.harvard.edu/martinos/userInfo/data/sof MNE.php). Averages contained only correct responses and were equated for the number of included trials across beverage conditions. Activity estimates were averaged to stimulus onset (Fig. 2) and to response execution (Fig. 3) in order to estimate alcohol effects on stimulus evaluation vs. response-preparation processing stages. The forward solution was based on a boundary element model and dipole strength power was estimated at each cortical location and each time point with a linear noise-normalized minimum norm estimation approach with no constraints on dipole orientation [15, 16]. The resulting "brain movies" consist of frames of statistical parametric maps displaying the statistical tests of the null hypothesis that, at each latency and location, there is no difference in the activity evoked by the condition and the baseline period. Group averages were obtained by averaging the inverse estimates using cortical surface alignment [17, 18]. Figures 2 and 3 show the group average estimates of the overall activity patterns evoked by each stimulus type at selected latencies for the stimulus-locked and response-locked averages respectively. Additionally, timecourses of estimated dipole-strength moments in the regions of interest in the cortical source space are presented in Fig 4.

D. Experimental Procedure and Design

All subjects first participated in an introductory session during which they were familiarized with the recording setup and provided detailed information about their medical status, history of alcohol and drug use, handedness, and dispositional profile. In each of the subsequent two sessions, scheduled 31 days apart on average, participants were given either alcohol or placebo in a counterbalanced order. Ten out of twenty subjects were given alcohol in the first session and placebo in the second. Prior to each session women were given a pregnancy test to ascertain that they were not pregnant. Blood alcohol concentration (BAC) was measured with a breathalyzer (Draeger, Inc.) on several occasions outside the magnetically shielded room. During the actual measurement, Q.E.D. Saliva Alcohol Test (OraSure Techn. Inc.) was used to estimate BAC. Alcohol was administered as a cocktail containing vodka (Grey Goose®, Bacardi), 0.6 g/kg of ethanol for men and 0.55 g/kg for women, mixed with orange juice in 20%v/v [19]. The average BAC just before the task reached 0.056 ± 0.01 and measured at 0.053 ± 0.01 upon the completion of the task. On a 1 to 5 Likert scale, the subjects reported being moderately intoxicated (2.7 ± 0.7) and rated the task as being moderately easy (2.7 ± 0.9). No gender effects were observed.

III. Results

A. Performance

Performance accuracy was affected by flanker-target compatibility, as indicated by the main effect of Condition, $F_{2,36} = 7.4$, $p < 0.005$ (Fig 1). Participants were least accurate on RI trials as compared to both CO, $F_{1,18} = 7.5$, $p < 0.05$, and SI conditions, $F_{1,18} = 7.7$, $p < 0.05$. The significant Beverage x Condition interaction, $F_{2,36} = 7.9$, $p < 0.005$, was due to more accurate responding under placebo than alcohol on RI trials, $F_{1,18} = 12.6$, $p < 0.005$. Reaction times (RTs) indicated that the condition difficulty increased from the CO to SI, $F_{1,18} = 114.6$, $p < 0.0001$, and from the SI to RI condition, $F_{1,18} = 64.0$, $p < 0.0001$, Fig 1. The Beverage x Condition interaction, $F_{2,36} = 4.2$, $p < 0.05$, reflected slower responses under alcohol than placebo on RI trials, $F_{1,18} = 3.3$, $p < 0.1$.

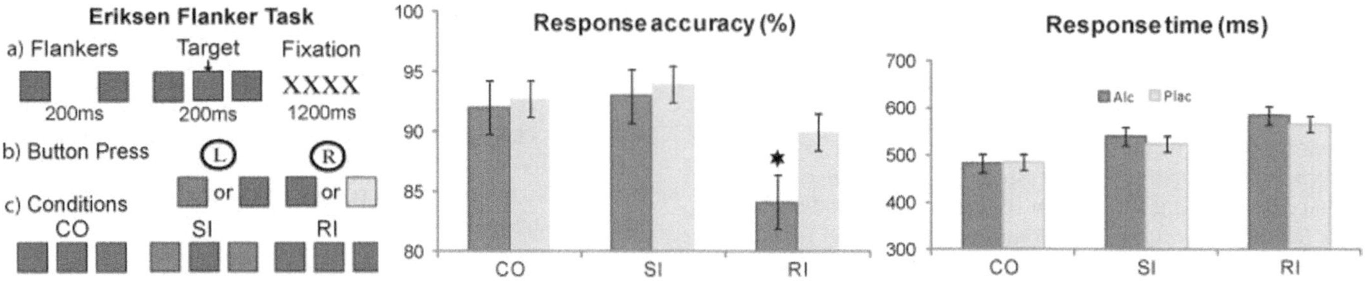

Fig. 1 left panel illustrates the task. a) Two flanker squares are presented for 200 ms. Target square is presented in the central location for 200 ms, followed by fixation for 1200 ms. b) Participants respond to the color of the target square by pressing left button to green *or* red and right button to blue *or* yellow. c) Examples of the Congruous (CO), Stimulus-level incongruous (SI), and Response-incongruous (RI) trials. Middle and right panels show response accuracy and RTs

Stimulus-level incompatibility increased task difficulty but that was offset by slower RTs. In contrast, response-mapping incompatibility was not mitigated by slower RTs as participants made the most errors on such trials. Furthermore, alcohol selectively affected performance in this condition, indicating deficits on neural systems supporting response preparation and execution. No gender effects were observed on any measure.

Even though all the participants reported drinking in moderately low doses, those with higher drinking levels had slower RTs overall, $r = 0.45$, $p < 0.05$. Self-reported impulsivity was related to response accuracy. Scores on Thrill and Adventure Seeking (TAS) sub-scale of the Sensation Seeking Scale [20] were negatively correlated with accuracy under both alcohol ($r = -0.62$, $p < 0.01$), and placebo conditions ($r = -0.52$, $p < 0.05$), suggesting a dispositional tendency to make errors.

B. aMEG Estimates

Stimulus-locked averages (Figs 2 and 4a) indicate that the AC is sensitive to conflict with strongest response to RI starting at 200ms after target onset under placebo, suggesting its importance in stimulus evaluation and top-down regulation. Even though the activity was stronger under alcohol at this stage overall, there was more conflict-related differentiation under placebo. Under placebo, RI selectively activated the AC after ~400ms, perhaps embodying AC contributions to executive regulation under the response-conflict condition. AC activation was also present under alcohol but did not differentiate strongly between conditions. Overall, the AC and motor/premotor cortex (MC) were the main foci of estimated activity.

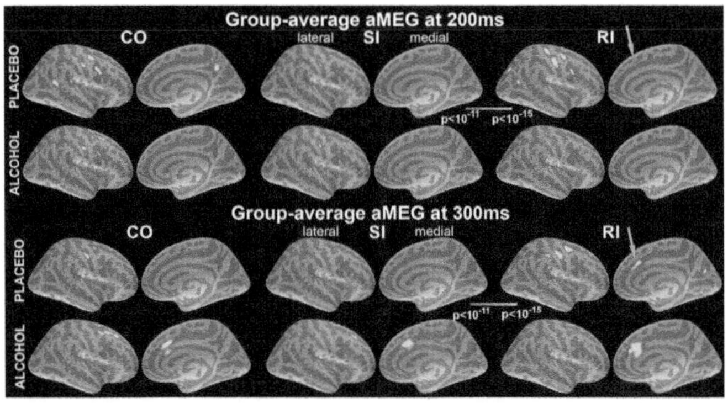

Fig. 2 Group average dynamic statistical parametric maps of estimated responses to all three trial types at 200 ms and 300 ms after stimulus onset under both placebo and alcohol conditions

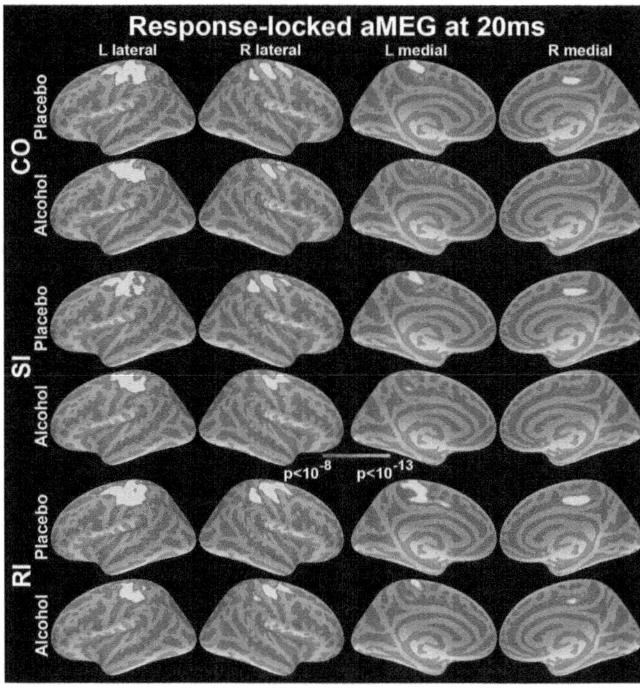

Fig. 3 Group average aMEG time-locked to button press responses

In order to explore response preparation and execution processing stages, response-locked aMEG estimates are presented in Fig 3. Together with the timecourses extracted from the MC and AC (Figs 4b,c), they indicate stronger activity under placebo overall. Conflict-related differentiation is expressed under placebo only, whereby the executive involvement of both the AC and MC are dependent on the level of incongruity between flankers and targets. Moreover, the preparatory increase in MC activity to SI and RI at ~80 ms before the response is visible only under placebo (Fig 4b), indicating optimized response strategy.

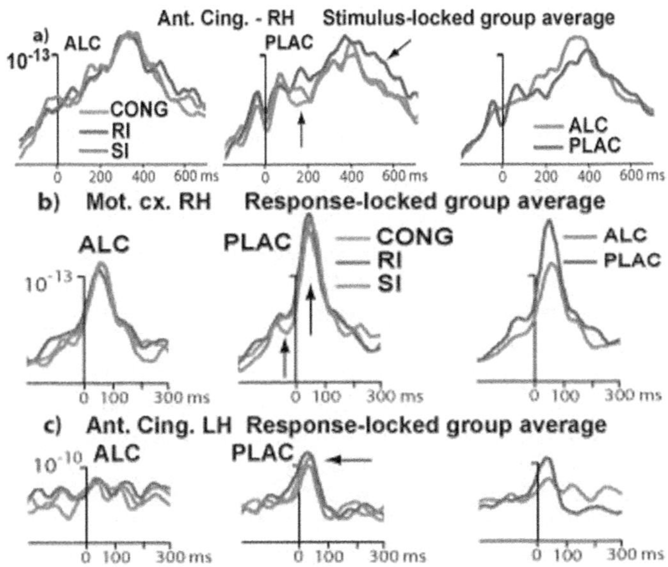

Fig. 4 Group average timecourses of the estimated dipole strengths in the AC and MC regions locked to stimulus onset (a) and button-press response (b,c)

IV. Conclusions

Moderately low intoxication affected executive capacity, particularly in the response-level incompatibility condition, wherein a primed response needed to be inhibited in favor of the opposite motor mapping. Whereas slower RTs successfully offset increased difficulty on SI trials, response-level incompatibility on RI trials resulted in significant increase in errors, particularly when intoxicated. Under placebo, the AC was sensitive to the level of conflict both during stimulus evaluation (~200ms) and response preparation (~400ms) processing stages, whereas alcohol blunted that conflict-induced differentiation. Moreover, increased activity during conflict under placebo during response preparation and execution in the AC and MC indicates a contribution to optimized response strategy. In contrast, alcohol intoxication impairs the regulative (AC) and motor executive circuitry, resulting in behavioral deficits under conditions of response conflict.

Acknowledgment

We are thankful to Elinor Artsy, Sarah Sheldon, and Sanja Kovacevic for their help. Supported by AA13402, AA016624 and Alcoholic Beverage Medical Research Foundation.

References

1. Oscar-Berman M, Marinkovic K (2007) Alcohol: effects on neurobehavioral functions and the brain. Neuropsychol Rev 17:239-57.
2. Finn P (2000) Acute effects of alcohol on cognition and impulsive-disinhibited behavior, in Rev of NIAAA's neurosci and beh res. A. Noronha et al. Ed., Bethesda, MD. 337-356.
3. Fillmore MT (2003) Drug abuse as a problem of impaired control: current approaches and findings. Behav Cogn Neurosci Rev 2:179-97.
4. Lezak MD (1995) Neuropsychological Assessment. New York: Oxford Univ. Press.
5. Botvinick MM et al. (2001) Conflict monitoring and cognitive control. Psychol Rev 108:624-52.
6. Holroyd CB, Coles MG (2002) The neural basis of human error processing: reinforcement learning, dopamine, and the error-related negativity. Psychol Rev 109:679-709.
7. Devinsky O, Morrell MJ, Vogt BA (1995) Contributions of anterior cingulate cortex to behaviour. Brain 118:279-306.
8. Wang et al. (2005) Responses of human anterior cingulate cortex microdomains. J Neurosci 25:604-613.
9. Eriksen BA, Eriksen CW (1974) Effects of noise letters upon the identification of target letters in a non-search task. Percept Psychophys 16:143-149.
10. Van Veen V, Carter CS (2002) The timing of action-monitoring processes in the anterior cingulate cortex. J Cogn Neurosc 14:593-602.
11. Ridderinkhof KR et al. (2002) Alcohol consumption impairs detection of performance errors in mediofrontal cortex. Science 298:2209-11.
12. Dale AM et al. (2000) Dynamic statistical parametric mapping: combining fMRI and MEG for high-resolution imaging of cortical activity. Neuron 26:55-67.
13. Marinkovic K et al. (2003) Spatiotemporal dynamics of modality-specific and supramodal word processing. Neuron 38:487-97.
14. Hazeltine E, Poldrack R, Gabrieli JD (2000) Neural activation during response competition. J Cogn Neurosci 12:118-129.
15. Dale AM, Sereno MI (1993) Improved localization of cortical activity by combining EEG and MEG with MRI cortical surface reconstruction: A linear approach. J Cogn Neurosci 5:162-176.
16. Hämäläinen MS, Ilmoniemi RJ (1994) Interpreting magnetic fields of the brain: minimum norm estimates. Med Biol Eng Comput 32:35-42.
17. Dale AM, Fischl B, Sereno MI (1999) Cortical surface-based analysis. I. Segmentation and surface reconstruction. Neuroimage 9:179-94.
18. Fischl B, et al. (1999) High-resolution intersubject averaging and a coordinate system for the cortical surface. Hum Brain Mapp 8:272-84.
19. Marinkovic K, Halgren E, Maltzman, I (2004) Effects of alcohol on verbal processing: An ERP study. Alcohol Clin Exp Res 28:415-423.
20. Zuckerman M (1971) Dimensions of sensation seeking. J Consult ClinPsychol 36:45-52.

Corresponding author:

Author: Ksenija Marinkovic, Ph.D.
Institute: Univ. of California, San Diego, Radiology
Street: 9500 Gilman Dr. 0841
City: La Jolla, CA 92093
Country: USA
Email: xenia@ucsd.edu

From Auditory Change Detection to Reading and Word Processing: Impairments in Children with Intractable Epilepsy

Milena Korostenskaja[1,2], Maria Pardos[1], Ki Heyeong Lee[1], Hisako Fujiwara[1], Teija Kujala[3], Jing Xiang[1], Jennifer Vannest[1,2], Yingying Wang[1], Nat Hemasilphin[1], Ton deGrauw[1], and Douglas Rose[1,*]

[1] Department of Neurology, Cincinnati Children's Hospital Medical Center, Cincinnati, Ohio, USA
[2] Pediatric Neuroimaging Research Consortium, Cincinnati Children's Hospital Medical Center, Cincinnati, Ohio, USA
[3] Cognitive Brain Research Unit, Department of Psychology, University of Helsinki, Finland

Abstract— **Objective evaluation of receptive language function is critical for children with intractable epilepsy, because deficient language function can be associated with poor surgery outcomes in these patients. The purpose of the present study was to evaluate the receptive language function in children with intractable epilepsy by means of magnetoencephalography (MEG). Six children with intractable epilepsy (M/F 3/3, MEAN±SD 14.5±2.07 yr.), undergoing evaluation for epilepsy surgery, and twelve healthy age- and gender-matched healthy controls participated in the study. Two paradigms were used to achieve our aim: (1) Multi-feature sound MMN paradigm (Naatanen et al. 2004), which makes it possible to register MMN for 5 different features (duration, frequency, intensity, location, and silent gap); and (2) word "match/mismatch" paradigm, which gives possibility to simultaneously register acoustically and visually presented words The words were presented simultaneously from visual (screen) and auditory (earphones) sensory inputs. The participants were asked to compare visually and acoustically presented words and to press the response button only if they did not match. There was a trend for longer latencies and smaller amplitudes in (1) all five MMN*m* responses and (2) in all M1-M5 components in patients as compared to healthy controls. For (1), the main effect of *GROUP* was significant for latencies and amplitudes of MMN*m* responses to all five deviant types in both left and right hemispheres with p = 0.004, and p = 0.035, respectively. For (2), after Bonferroni correction, the following effects remained significant: the main effect of *GROUP* was significant for M4 latency and amplitude with p = 0.005, and p = 0.01, respectively. These results show both early change detection mechanism and later stages of cognitive processing related to reading and word recognition are impaired in children with intractable epilepsy by reduced amplitudes and delayed latencies of event-related field components representing these processes.**

Keywords— **Auditory, Visual, Epilepsy, Language, Magnetoencephalography.**

I. INTRODUCTION

Adolescents with intractable epilepsy often possess global cognitive abnormality evident in memory and attention problems, decreased reading abilities, and language deficits. Objective identification of these abnormalities is highly important for making decisions regarding epilepsy surgery. This is especially important in children, who very often fail to respond well during neuropsychological examination due to their inability or unwillingness to communicate. Magnetoencephalography (MEG) is a non-invasive method, allowing objective detection of changes in neuromagnetic brain activity. Event-related fields, such as mismatch negativity response (MMN), recorded with MEG can provide invaluable information about information processing in intractable epilepsy patients. The MMN is a brain response to some repetitive change in the environmental stimulation, and represents ability of the individual to detect environmental changes (Naatanen et al. 2001). Therefore, MMN is often considered to reflect a "change-detection" mechanism in human brain. Importantly, some studies suggest that abnormal MMN generation represents a core cognitive dysfunction, leading to further cognitive changes, such as reading and language impairments (Kujala et al. 2007b). In this MEG study we aimed to register two types of event-related responses: (1) early MMN response to evaluate deficits in change detection mechanism, and (2) late "match/mismatch" response, to evaluate reading and word recognition. By registering both responses, we aimed to determine if complex cognitive abnormalities in reading and word processing in intractable epilepsy patients are preceded by early deficits in auditory change detection.

II. METHODS

We studied six children with intractable epilepsy (M/F 3/3, Mean/SD 14.5±2.07 yr., all right-handed), and twelve age- and gender- matched healthy controls. Patients had frontal and fronto-parietal lobe seizures (N = 3: N=2 with left hemisphere, N= 1 with right hemisphere) as well as left temporal lobe seizures (N=3). Two paradigms were used to achieve our aim: (1) Multi-feature sound MMN paradigm (Naatanen et al. 2004), which makes it possible to register MMN for 5 different features (duration, frequency, intensity, location, and silent gap); and (2) word "match/mismatch" paradigm, which gives possibility to simultaneously register acoustically and visually presented words (Wang et al. 2008).

S. Supek and A. Sušac (Eds.): Advances in Biomagnetism – BIOMAG2010, IFMBE Proceedings 28, pp. 378–380, 2010.
www.springerlink.com

During the first paradigm, we used a passive odd-ball approach for stimulus delivery with 5 deviant tones ("Optimum-1", also called multi-feature MMN paradigm, see Kujala et al. 2007b). All 5 deviants (P = 0.1, each) were presented in the same sequence so that tones alternated between standard (P = 0.5) and one of the 5 deviants. In each sequence, the first 15 tones were standards. The deviants were presented so that in an array of 5 deviants, each deviant category was presented once, and 2 deviants of the same category never followed each other. During this multi-feature MMN paradigm, the stimuli were presented at a stimulus-onset-asynchrony (SOA) of 500 ms in three 5 min sequences (1845 stimuli in total), with the total recording time for the 5 types of deviants thus being 15 min.

During the second paradigm, the words were presented simultaneously from visual (screen) and auditory (earphones) sensory inputs. Two types of stimuli were presented: (1) matched words (N = 100), for which the visually and acoustically presented words were identical; (2) mismatched words (N = 20), for which the visually and acoustically presented words were different. The participants were asked to compare visually and acoustically presented words and to press the response button only if they did not match. The latencies and amplitudes of neuromagnetic responses M1, M2, M3, M4 and M5 elicited during "match" condition were analyzed separately from the left and right hemispheres.

The amplitude and latency comparisons for MMN*m* components were carried out with a mixed ANOVA model with three fixed factors: *GROUP*: epilepsy patients *vs.* healthy controls, *HEMISPHERE*: left *vs.* right and *STIMULUS TYPE*: D1, D2, D3, D4, D5, and with one random factor: *SUBJECT*. The amplitude and latency comparisons for M1-M5 components were carried out with a mixed model analysis of variance (ANOVA) with two fixed factors: *GROUP*: epilepsy patients *vs.* healthy controls and *HEMISPHERE*: left *vs.* right and one random factor: *SUBJECT* nested within group. Statistically significant effects in this case were tested with Bonferroni correction for multiple ERF comparisons.

III. Results

There was a trend for longer latencies and smaller amplitudes in (1) all five MMNm responses and (2) in all M1-M5 components in patients as compared to healthy controls. In (1) case, the main effect of *GROUP* was significant for latencies and amplitudes of MMN*m* responses to all five deviant types in both left and right hemispheres with $p = 0.004$, and $p = 0.035$, respectively. In (2) case, after applying Bonferroni correction, the following effects remained significant: the main effect of *GROUP* was significant for M4 latency and amplitude with $p = 0.005$, and $p = 0.01$, respectively. There were no significant main effect of *HEMISPHERE* on either ERF latencies or amplitudes and no significant *GROUPxHEMISPHERE* interaction.

IV. Conclusions

These results show that both early change detection mechanism and later stages of cognitive processing related to reading and word recognition are impaired in children with intractable epilepsy, as indexed by reduced amplitudes and delayed latencies of event-related field components representing these processes. This is in line with previous reports, showing change detection (Boatman et al. 2008; Liasis et al. 2006) and reading deficits (Breier et al. 2005) in epilepsy patients. However, this is a first study to show that the abnormalities in change-detection mechanism are followed by later dysfunction in patients' ability to read and recognize words. Our results are supported by previous studies showing that the deficits in MMN generation, for example decrease in amplitude or delay in latency, represent a core neurophysiological dysfunction related to global impairments in everyday functioning (Light and Braff 2005); moreover, this global dysfunction can hinder language and other complex cognitive development (Kujala et al. 2007a). In the future, MMN test can be suggested as a useful assessment tool of core cognitive abilities for children undergoing evaluation for epilepsy surgery. However, before that, larger sample sizes of both patients and healthy subjects are needed in order to confirm and develop further the results of the current study.

References

1. Boatman DF, Trescher WH, Smith C, Ewen J, Los J, Wied HM, Gordon B, Kossoff EH, Gao Q, Vining EP (2008) Cortical auditory dysfunction in benign rolandic epilepsy. Epilepsia 49: 1018-26
2. Breier JI, Castillo EM, Simos PG, Billingsley-Marshall RL, Pataraia E, Sarkari S, Wheless JW, Papanicolaou AC (2005) Atypical language representation in patients with chronic seizure disorder and achievement deficits with magnetoencephalography. Epilepsia 46: 540-8
3. Kujala T, Aho E, Lepisto T, Jansson-Verkasalo E, Nieminen-von Wendt T, von Wendt L, Naatanen R (2007a) Atypical pattern of discriminating sound features in adults with Asperger syndrome as reflected by the mismatch negativity. Biol Psychol 75: 109-14
4. Kujala T, Tervaniemi M, Schroger E (2007b) The mismatch negativity in cognitive and clinical neuroscience: theoretical and methodological considerations. Biol Psychol 74: 1-19
5. Liasis A, Bamiou DE, Boyd S, Towell A (2006) Evidence for a neurophysiologic auditory deficit in children with benign epilepsy with centro-temporal spikes. J Neural Transm 113: 939-49
6. Light GA, Braff DL (2005) Mismatch negativity deficits are associated with poor functioning in schizophrenia patients. Arch Gen Psychiatry 62: 127-36

7. Naatanen R, Pakarinen S, Rinne T, Takegata R (2004) The mismatch negativity (MMN): towards the optimal paradigm Clin Neurophysiol, pp 140-4
8. Naatanen R, Tervaniemi M, Sussman E, Paavilainen P, Winkler I (2001) "Primitive intelligence" in the auditory cortex. Trends Neurosci 24: 283-8
9. Wang Y, Xiang J, Kotecha R, Vannest J, Liu Y, Rose D, Schapiro M, Degrauw T (2008) Spatial and frequency differences of neuromagnetic activities between the perception of open- and closed-class words. Brain Topogr 21: 75-85

* Corresponding author:
Author: Douglas Rose
Institute: Cincinnati Children's Hospital Medical Center
Street: 3333 Burnet Avenue
City: Cincinnati, OH, 45242
Country: USA
Email: douglas.rose@cchmc.org

Hemispheric Differences in Neural Activation during Gaze Cueing in Autism Spectrum Disorder (ASD) Measured by Magnetoencephalography (MEG)

R. Lajiness-O'Neill[1,2], A.M. Chase[1], A. Olszewski[1], M.A. Boyle[1], L. Pawluk[1], A. Mansour[1], D. Jacobson[1], M.L. Gallaway[1], P. Lewandowski-Powley[1], B. Gorka[1], J. Moran[2], and S.M. Bowyer[2,3,4]

[1] Eastern Michigan University, Ypsilanti, MI, USA, [2] Henry Ford Hospital, Detroit, MI, USA; [3] Wayne State University, Detroit, MI, USA, [4] Oakland University, Rochester, MI USA

Abstract— **Atypical hemispheric lateralization during language tasks has been reported in autism spectrum disorders (ASD). Six participants with ASD and eight neurotypical controls underwent MEG while performing a gaze cueing task. Participants indicated whether a character's gaze shifted congruently or incongruently to targets (asterisk), words, or faces. We hypothesized similar aberrant activity and contextual differences would be evident during joint attention, suggesting fundamental cognitive processing biases in regions necessary for either verbal or nonverbal communication. Data were analyzed with MR-FOCUSS [1]. Results revealed higher mean amplitudes in left occipital and parietal brain regions during gaze shifts to both targets and faces in ASD while higher mean amplitudes in right inferior temporal and medial orbitofrontal regions was evident in controls. Pronounced amplitudes were noted in left angular regions during gaze shifts to faces in ASD. Latency data further reveals contextual differences and a possible double dissociation in brain regions recruited for joint attention. While controls displayed an earlier onset in activation to words in frontal regions and later activation to faces in inferior temporal regions, ASD demonstrated the opposite pattern; an earlier activation in inferior temporal regions and later activation in frontal regions to words. Results suggest atypical hemispheric specialization for joint attention with ASD recruiting left posterior hemispheric regions during gaze shifts to social stimuli while controls recruit right inferior temporal and medial orbitofrontal regions. Consistent with aberrant connectivity and a "crowding" hypothesis, ASD subjects appear to be using regions that underlie reading such as posterior parietal and occipital regions for joint attention. This is the first study to reveal these deficits in ASD children and adolescents with neuroimaging.**

Keywords— **MEG, Autism, Gaze Cueing, Joint Attention, Social Cognition.**

I. INTRODUCTION

Socialization deficits are a major source of impairment in autism spectrum disorders (ASD), despite cognitive or language skills [2]. The research literature is replete in identifying the social impairments associated with ASD [3].

Early neurological models of the social deficits in ASD emphasized the presence of localized dysfunction within the social cognition network, with particular focus on the prefrontal cortex, temporal lobe, and the limbic lobe [4]. Functional neuroimaging studies of individuals with ASD have found altered brain activity among these regions in response to the perception of social stimuli [5]. To date, research has not identified a "smoking gun" in the identification of a potential social biomarker in ASD; however, the early behavioral deficit in joint attention appears a highly plausible candidate behavior for which an enduring neural marker is likely to be evident.

During social interaction, a person's eyes convey information about his or her direction of attention, emotion, and mental state [6]. A specific form of gaze following behavior is joint attention, a complex communicative mechanism that develops in early childhood. Joint attention is a method of nonverbal communication that occurs when one individual follows another's eye gaze or gesture to an object or third individual [7]. Joint attention is a fundamental part of social functioning as it has been linked to both social and language development [8]. However, this pivotal skill fails to develop appropriately in infants with autism, and almost nothing is known about the persistence of these gaze-following deficits into adolescence or adulthood and the impact on social functioning.

The goal of this study was to accelerate the discovery of the neural correlates underlying the profound social impairment in ASD using MEG. We attempted to demonstrate aberrant connectivity and differential activation in brain structures of individuals with ASD using a gaze cueing paradigm. We examined differential effects of stimuli attended to, such as faces and words, in children and adolescents with and without ASD. These stimuli were chosen because individuals with ASD do not appear to attend to faces any differently than other

S. Supek and A. Sušac (Eds.): Advances in Biomagnetism – BIOMAG2010, IFMBE Proceedings 28, pp. 381–384, 2010.
www.springerlink.com

"objects." Moreover, clinically, individuals with ASD tend to display an affinity for attending to orthographs (i.e., words) and often display accelerated word reading abilities in early toddlerhood, despite impaired language [9]. The mechanisms of these processes and implications for social functioning are poorly understood.

II. Methods

Participants: Six participants with ASD (Mean age (SD/range) = 16.6 (4.9/13); mean IQ = 120; Males = 4) and eight neurotypical controls (Mean age (SD/range) = 17.5 (2.9/8); mean IQ = 115; Males = 4) underwent MEG while performing a gaze cueing task. Results from the Autism Diagnostic Interview-Revised [10] confirmed diagnoses. All APA Ethical Guidelines were followed and Institutional Review Board approval was obtained.

Measures: Individuals viewed a digital photograph of a character whose gaze was shifted toward or away from (congruent vs. incongruent) targets (asterisk), printed words, or faces projected onto a screen (see Figure 1). There were a total of 60 targets, faces, or words (30 congruent and 30 incongruent in each condition); hence, over the three conditions there were a total of 180 stimuli. Each set (targets: congruent and incongruent; faces: congruent and incongruent; or words: congruent and incongruent) contained 30 responses to determine the averaged evoked response. The subject was asked to press the button when the subject was looking toward something. 148 channel whole head MEG (4D Neuroimaging, Magnes WH2500) collected cortical activity. Evoked magnetic field data was used from -0.2 seconds prior until 0.65 seconds after the stimulus onset. Brain activity was analyzed with MR-FOCUSS, a current density technique [1].

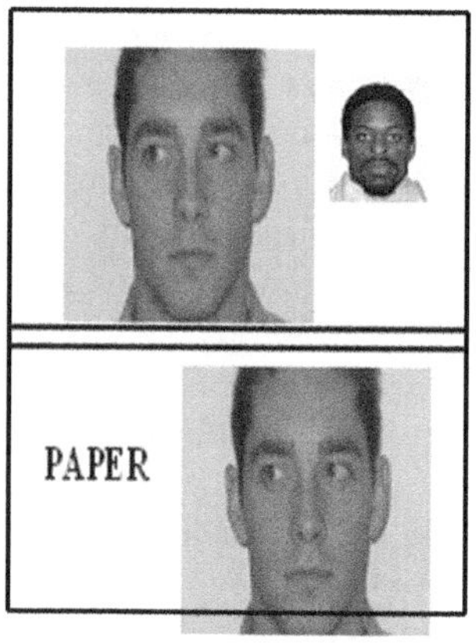

Fig. 1 Gaze cuing stimuli

An average normalized amplitude per active sources in 54 brain regions based on MNI coordinates was calculated for each subject. The left and right hemisphere ROIs were summed and averaged based on total active sources for that ROI to generate a normalized mean amplitude for the ROI over the 650 ms epoch to compare groups. Group differences were examined with a *t*-statistic. The *p*-values were not corrected for multiple comparisons given the small N and to decrease the probability of a Type II error.

III. Results

Joint Attention to Targets: There were no differences with respect to the onset of activation during gaze shifts to targets, although a significant number of differences were noted in the amplitudes of activation across different brain regions in ASD compared to controls, and a clear double dissociation was evident. Higher normalized mean amplitudes were noted in ASD, predominantly in the left posterior hemisphere including the left cuneus, $t(11) = 2.85$, $p = .02$, middle occipital, $t(11) = 2.56$, $p = .03$, poscentral gyrus, $t(11) = 2.66$, $p = .02$, superior parietal, supramarginal, $t(11) = 2.44$ $p = .02$, and right cuneus, $t(11) = 2.99$, $p = .01$. In contrast, higher amplitudes of activation were noted in the right inferior temporal, $t(11) = 2.59$, $p = .03$, and right middle orbitofrontal regions, $t(11) = 2.49$, $p = .04$, during gaze shift to targets in controls (see Figure 2).

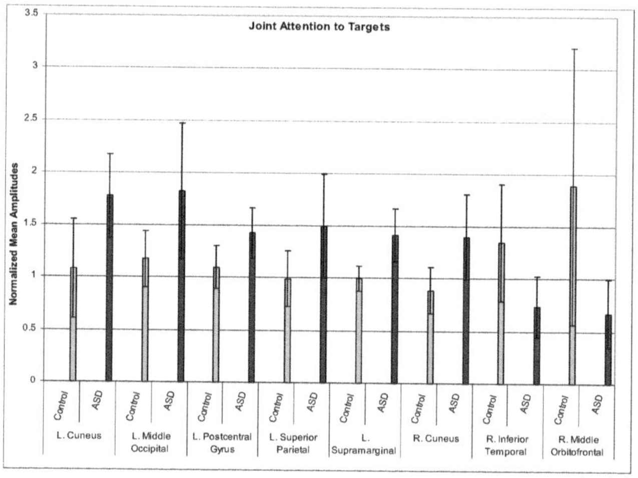

Fig. 2 Joint attention to targets. Note predominately left hemispheric activation in ASD and focal right frontotemporal activation in controls. All regions were at a $p < .05$

Joint Attention to Words: When ASD subjects engaged in gaze shifts to words, significantly higher normalized mean amplitudes were noted in the left superior occipital region in ASD, $t(11) = 3.53$, $p = .01$, while higher amplitudes were noted in the right putamen in controls, $t(11) = 3.12$, $p = .02$ (See Figure 3). Significant differences in the latency of activation were noted in middle and superior frontal regions, $t(11) = 2.15$, $p = .04$; $t(11) = 2.48$, $p = .03$, as well as in supramarginal regions, $t(11) = 3.24$, $p = .02$, with ASD subjects displaying a delay relative to controls with onset at ≈300-350 ms whereas controls displayed activation at about ≈200 ms. in frontal regions (see Figure 4).

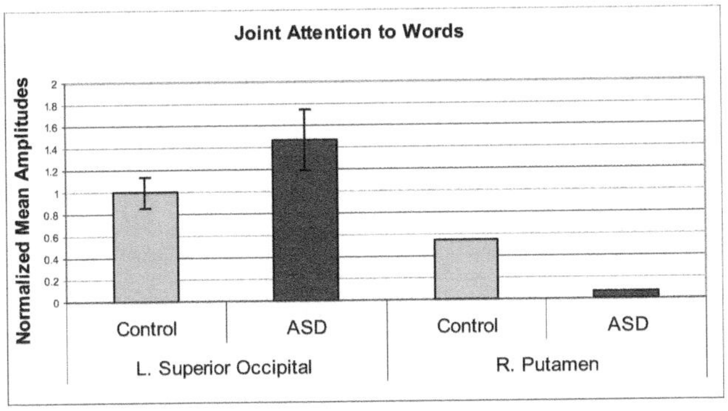

Fig. 3 Normalized amplitudes for joint attention to words with significantly higher left occipital activation noted in ASD. All regions were at a $p < .05$

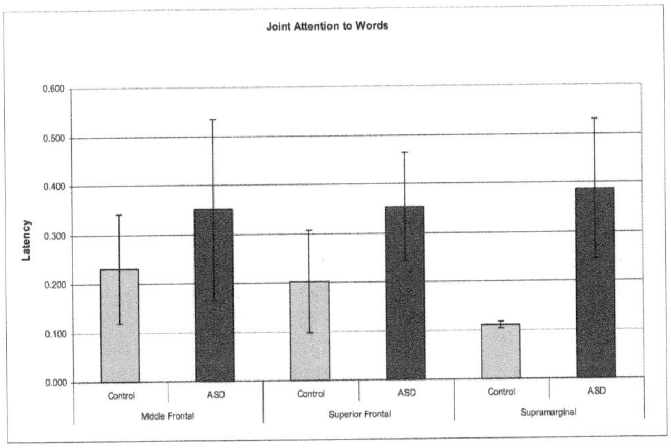

Fig. 4 Latency for joint attention to words. Note earlier latency in controls to words compared to later latency to faces in Fig. 6. The opposite pattern was noted in ASD. All regions were at a $p < .05$

Joint Attention to Faces: Significantly higher normalized mean amplitudes were noted in the left angular region in ASD compared to controls during gaze shifts to faces, $t(11) = 3.1$, $p = .01$ (see Figures 5 and 6). In contrast, a significantly earlier latency of activation was noted in inferior temporal regions in ASD compared to controls, with control subjects displaying activation in inferior temporal regions at ≈300 ms, while ASD subjects onset of activation was at ≈200 ms, $t(11) = 2.48$, $p = .03$ (see Figure 7).

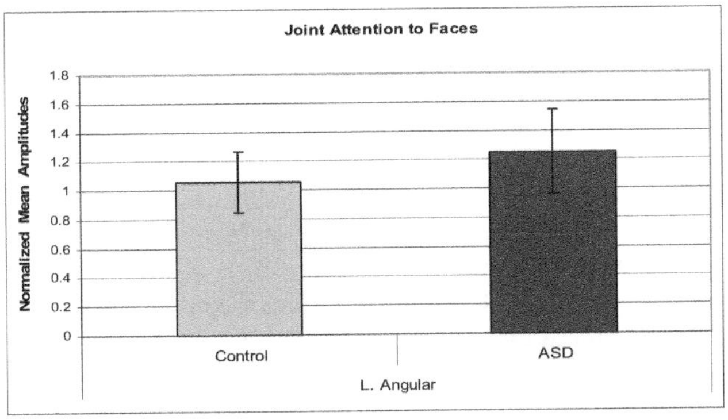

Fig. 5 Higher normalized mean amplitudes in left parietal regions during joint attention to faces in ASD relative to control. $p = .01$

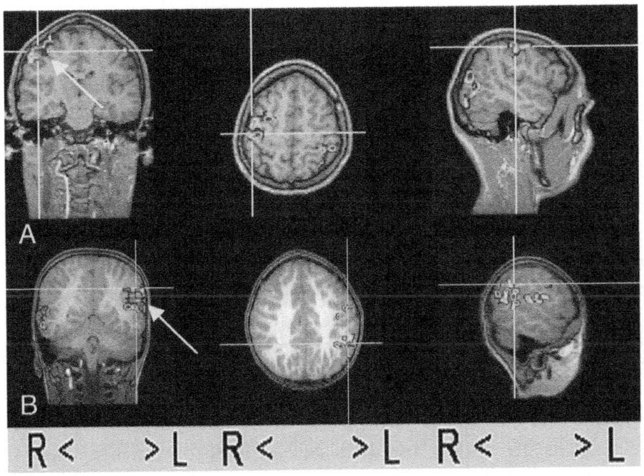

Fig. 6 Right frontoparietal activation in control subject (A) at 194 ms (above) while ASD subject (B) has significant activation in left parietal regions during face condition at 194 ms (below)

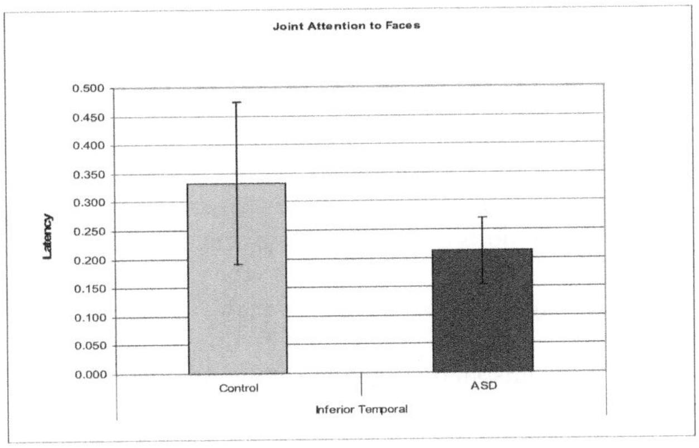

Fig. 7 Earlier latency of activation in ASD at ≈200 ms during joint attention to faces while controls activate later at ≈300 ms. $p = .03$

IV. Discussion

The neurobiological underpinnings of gaze cueing in children and adolescents with ASD are poorly understood and could have significant implications for understanding the etiology of social impairment. Results of this study revealed higher mean amplitudes in left occipital and parietal brain regions during gaze shifts to both targets and faces in ASD. In contrast, higher mean amplitudes in right inferior temporal and medial orbitofrontal regions was evident in controls, the latter of which has been implicated in the social cognitive network. Pronounced left angular amplitudes were also noted in left parietal, specifically angular, regions during gaze shifts to faces in ASD. Our preliminary findings suggest that there may be crowding in brain regions that underlie these processes which may be secondary to the recently reported atypical and rightward lateralization of language in ASD. Gaze cueing studies in ASD adults have revealed deficits in a number of brain regions believed to underlie social cognition, such as the superior temporal gyrus and sulcus (STG and STS), inferior parietal lobule (IPL), middle temporal gyrus (MTG), and inferior and middle frontal gyri [11]. This is the first study to reveal these deficits in ASD children and adolescents with neuroimaging.

Latency data further reveals contextual differences and a possible double dissociation in brain regions recruited for joint attention. While controls displayed an earlier onset in activation to words in frontal regions and later activation to faces in inferior temporal regions, ASD demonstrated the opposite pattern; an earlier activation in inferior temporal regions and later activation in frontal regions to words.

V. Conclusion

Results suggest atypical hemispheric specialization for joint attention with ASD recruiting left posterior hemispheric regions during gaze shifts to social stimuli while controls recruit right inferior temporal and medial orbitofrontal regions. ASD subjects appear to be using regions that underlie skills necessary for reading, including posterior parietal and occipital regions for joint attention to gaze shifts. This may be due to "crowding" or altered hemispheric specialization in regions that underlie both social and speech perception, such as the right inferior parietal and superior temporal lobe (e.g. IP and STS). Moreover, we believe the crowding occurs not only due to competition in regions that underlie social and "speech" perception, but both oral and written "language" perception in general.

Acknowledgement

The first author acknowledges Eastern Michigan University for a New Faculty Award and Faculty Research Fellowship which supported this project.

References

1. Moran JE, Bowyer SM, Tepley N. *Multi-resolution FOCUSS: A source imaging technique applied to MEG data. Brain Topography.* 2005; 18: 1-17.
2. Carter AS, Davis NO, Klin A, & Volkmar FR, editors. Social development in autism. Hoboken, NJ: Wiley; 2005.
3. Tse J, Strulovitch J, Tagalakis V, Meng L, Fombonne E. Social skills training for adolescents with asperger syndrome and high functioning autism. *Journal of Autism and Developmental Disorder.* 2007; 37(10): 1960-1968.
4. Adolphs R. Cognitive neuroscience of human social behavior. *Nature Reviews.* 2003; 4: 165-178.
5. Pelphrey KA, Morris JP, Michelich CR, Allison T, McCarthy G. Functional anatomy of biological motion perception in posterior temporal cortex: An fMRI study of eye, mouth and hand movements. *Cerebral Cortex.* 2005; 15(12): 1866-1876.
6. Frischen A, Bayliss AP, & Tipper SP. Gaze cueing of attention: Visual attention, social cognition, and individual differences. *Psychological* Functional anatomy of biological motion perception in posterior temporal cortex: An fMRI study of eye, mouth and hand movements. *Cerebral Cortex.* 2005; 15(12): 1866-1876. Visual attention, social cognition, and individual *Bulletin.* 2007; 133: 694-724.
7. Emery NJ. The eyes have it: The neuroethology, function and evolution of social gaze. *neuroscience and biobehavioral reviews. Neuroscience and Biobehavioral Reviews.* 2000; 25: 581-604.
8. Charman T. Why is joint attention a pivotal skill in autism? In: Frith U, Hill EL, editors. *Autism: Mind and brain* ; 2004. p. 67-87.
9. Newman TM, Macomber D, Naples AJ, Babitz T, Volkmar F, Grigorenko EL. Hyperlexia in children with autism spectrum disorders. *Journal of Autism and Developmental Disorders.* 2007; 37(4): 760-774.
10. Lord C, Rutter M, & Le Couteur A. Autism diagnostic interview-revised: A revised version of a diagnostic interview for caregivers of individuals with possible pervasive developmental disorders. *Journal of Autism and Developmental Disorders.* 1994; 24(5): 659-685.
11. Iacoboni M, Molnar-Szakacs I, Gallese V, Buccino G, Mazziotta JC, Rizzolaatti G. Grasping the intentions of others with one's own mirror neuron system. *PLoS Biology.* 2005; 3(3).

Author: Renee Lajiness-O'Neill, Ph.D.
Institute: Eastern Michigan University
Street: Dept. of Psychology, 537 Mark-Jefferson
City: Ypsilanti
Country: USA
Email: rlajines@gmail.com or rlajines@emich.edu

Cortico-Motor Coherence during Deep Brain Stimulation for Parkinson's Disease

Chun Kee Chung[1,2,3]

[1] MEG Center, Seoul National University Hospital
[2] Department of Neurosurgery, Seoul National University College of Medicine
[3] Interdisciplinary Program in Cognitive Science, Seoul National University, Seoul Korea

Abstract— **Parkinsonian motor symptoms can be relieved by deep brain stimulation on the subthalamic nucleus. However, the underlying neurophysiology has not been fully understood. In a magnetoencephalography study by Park et al., the authors measured the cortico-muscular coherence from three Parkinson's disease subjects with resting tremor. The coherence was measured while the deep brain stimulation was both on and off. Meanwhile, previous studies have shown that healthy humans performing isometric contractions show increased cortico-muscular coherence in the beta-band. Hypothesizing that this can be a measure of symptomatic improvement in motor performance after deep brain stimulation on the subthalamic nucleus, the authors compared the beta-band cortico-muscular coherence during the two states of stimulators. The results showed that the cortico-muscular coherence was elevated for the tremorous hand when the stimulator was on. The authors suggest that the result indicates that deep brain stimulation resulted in better cortico-muscular coordination.**

Keywords— **cortico-muscular coherence, magnetoencephalography, Parkinson's disease, subthalamic nucleus deep brain stimulation.**

I. Introduction

Parkinson's disease (PD) is a chronic degenerative disease of the central nervous system, and is one of the most well known of movement disorders. Major motor dysfunctions are bradykinesia, rigidity, and resting tremor. Today, deep brain stimulation (DBS) on the subthalamic nucleus (STN) or the globus pallidus interna has been an outstanding solution for these symptoms.

However, the neurophysiological mechanism underlying the remarkable alleviation of motor symptoms following DBS has not yet been fully understood. With the advantages of its precise temporal resolution up to millisecond scale and its non-invasiveness, magnetoencephalography (MEG) is a relatively new device in the field of neurophysiology which has growing potential in disclosing the dynamics of the human brain function. With all its virtue, the difficulties in utilizing this device is that since it has sensors which are extremely sensitive enough to pick up the magnetic field created by the electrical activities within the brain, it is also very sensitive to any kind of noise created by body movement or electrical devices nearby. Therefore, obviously it is a challenge to measure the magnetoencephalography of Parkinson's disease subjects who have had deep brain stimulation surgery, which leaves stimulators inside the brain generating electrical pulses and also a pacemaker, which is an electrical pulse generator inserted just beneath the chest skin. Also the involuntary movements caused by PD motor symptoms may cause serious contamination on the MEG data.

Park et al. [1] managed to measure the motor evoked field (MEF) of three PD subjects who underwent bilateral subthalamic nucleus deep brain stimulation surgery. The authors fully utilized the temporal signal-space separation method [3] provided as Maxfilter™ with the Neuromag software, which is expected to eliminate the artifact from the chest pacer and other body movements of the subject. Also by applying a novel averaging method previously introduced as being based on shifting the MEG epochs according to the simultaneously measured time lag between the spike times of the electromyogram signal [2], the authors have been able to localize the primary motor cortex of the thumb. Although the motor network is closely involved with the basal ganglia and therefore deep brain structures must be taken into account, it is expected that motor network research at the cortical level may be possible with magnetoencephalography data interconnected with the electromyogram using existing signal processing methods.

In the literature, many studies show that while performing moderate isometric muscle contraction, the electromyogram (EMG) signal obtained from the contracting muscle and MEG measured from the corresponding contralateral primary motor cortex has significant coherence in the beta-band [4-5]. This is thought to be a cortico-muscular oscillatory feature characteristic of a healthy person performing isometric contraction. With these backgrounds, Park et al. focused on the cortico-muscular coherence (CMC) within the beta-band (13–30 Hz) while the subject performed moderate strength isometric contraction. The authors compared the beta-band cortico-muscular coherence of stimulator on state and stimulator off state in order to reveal the effect on the cortical muscular coordination by deep brain stimulation.

S. Supek and A. Sušac (Eds.): Advances in Biomagnetism – BIOMAG2010, IFMBE Proceedings 28, pp. 385–388, 2010.
www.springerlink.com

II. Methods

A. Subject Information

All three subjects were diagnosed with idiopathic Parkinsons's disease. Subject 1 had right-hand dominant resting tremor, and both subject 2 and 3 were left-hand dominant. All three subjects had stereotactic bilateral subthalamic nucleus deep brain stimulation. DBS parameters for the subjects on the day of measurement are presented in Table 1.

Subject 1 and 3 were not on medication on the day of measurement. Subject 2 was on ropinirol 3mg/day, selegiline 5mg/day, Sinemet CR (Korea MSD, Seoul, Korea) (L-dopa/carbidopa 200/50mg) 300mg/day.

Table 1 Subject information and DBS parameters

	Age	DBS parameters (amplitude, pulse length, frequency)
Subject 1	63	Rt(1-), 2.0V; 60μs; 130Hz: Lt(1-), 2.5V; 60μs; 130Hz:
Subject 2	57	Rt(2-), 3.5V; 60μs; 130Hz: Lt(2-3-), 2.4V; 60μs; 130Hz:
Subject 3	69	Rt(2-3-), 3.3V; 60μs; 185Hz: Lt(2-3-), 2.9V; 60μs; 185Hz:

Numbers in the brackets in the DBS parameters indicate the number of electrode being stimulated. The minus sign means the electrodes were set as cathode.

B. Data Acquisition/Data Analysis

MEG was acquired with a 306 channel Elekta Neuromag Vectorview system (Elekta Neuromag Oy). All measurements were taken place n a magnetically shielded room. MEG data were analogue filtered between 0.1 and 200Hz and the sampling frequency was 3000Hz for subject 1, 1000Hz for subject 2 and 3. Subject 1's data was downsampled to 1000Hz before analysis. Head movements were tracked with four head position indicator (HPI) coils attached to the subjects' head.

Surface electrodes were placed on the abductor pollicis brevis of the hand performing the task to obtained electromyogram data. The EMG data were digitized at the same frequency as the MEG signals. Data were monitored for artifacts, tremor, and muscle contraction signals.

Raw MEG data were applied with the tSSS algorithm (Maxfilter^{TM} version 2.0). Then the MEG data were averaged with the phase-lock averaging method [2]. The primary motor cortex of each hand for each subject was localized with the averaged data. Dipoles were localized with the equivalent current dipole method with the Neuromag software [6]. The dipoles were overlayed on the structural MRI images of subject 2 and subject 3. Head movement was not tracked for subject 1 and thus MRI images were not used. Only those dipoles with goodness-of-fit above 80% were used.

Coherence, which is a measure of the phase consistency between two time signals, was measured between the MEG and EMG signals. The dipole coordinates was imported into BESA 5.1.6 (MEGIS Software, Gräfelfing, Germany) and the source activity was extracted. The cortico-muscular coherence was calculated with the MEG source activity signal and EMG signal.

In order to compare the coherence values when DBS was on and off, the authors created 10,000 surrogate data sets by randomly shuffling the order of a dataset created by connecting the MEG-EMG pair data for DBS OFF and DBS ON status. The half of the data was assigned DBS OFF state and the other DBS ON state for every 10,000 datasets. Coherence was calculated for each set and the coherence difference between DBS OFF and DBS ON were compared with the original coherence differences, frequency-wise. A $p<0.05$ criterion was used to threshold the significance of difference.

III. Results

A. MEG Measurement with DBS Subjects

Motor evoked field from three subjects were measured on different dates, between the period from February 2007 and January 2008. Subject 1 was post operation 7 months status when the MEG measurement took place, subject 2 was 3-half years, and subject 3 was 1 year and 9 months. Their symptoms varied, though all three had the diagnosis of resting tremor as the dominating motor dysfunction.

Figure 1 shows raw data and filtered data, averaged data before and after filtering, and the power spectrum comparing the raw and filtered data from the magnetoencephalographic signals of subject 1. (A) shows the raw signals before applying temporal signal-space separation algorithm and after, during DBS OFF and DBS ON. The gray signals are before tSSS, and black signals are after applying tSSS. Note the scale difference between DBS OFF (10^-10fT/cm) and DBS ON (10^-9fT/cm). The abnormal big fluctuations are evident for the raw data, especially when the DBS is ON, the signals are contaminated with fine regular electrical activities which may be due to the 130Hz deep brain electrical stimulation. The signals cleared with tSSS are much refined as shown superimposed with the raw signals. (B) shows a part of the left frontal and temporal signals of the full view of averaged data during DBS. The improvement of the signal-to-noise ratio is evident. The amplitude of the data for DBS ON was too big to be shown in the same scale

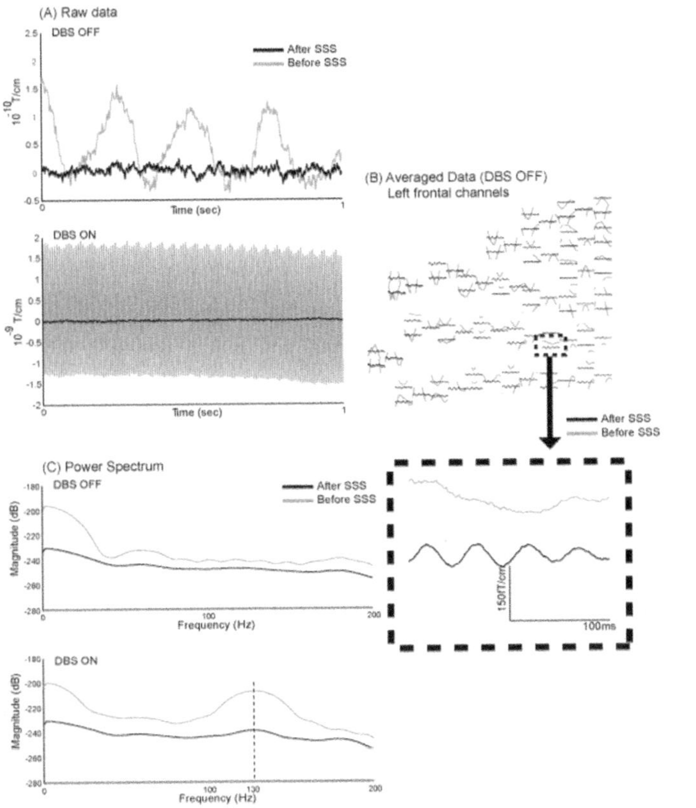

Fig. 1 (A) Raw data and data applied with SSS superimposed. Gray line is raw data and black line is filtered data. Data measured while DBS OFF (up) and DBS ON (down). (B) Averaged data of raw data and filtered data overlayed in full view. Only left frontal channels are shown. Gray line is raw data and black line is filtered data. (C) Power spectrum of raw data and filtered data for DBS OFF (up) and DBS ON (down) Gray line is raw data and black line is filtered data. Note the hump in the DBS ON raw data around 130Hz which is the stimulator frequency and the diminished power for the filtered data. All data in this figure were obtained from subject 1

of the filtered data for the full view plot, so is not displayed. (C) shows the power spectrum of filtered and non-filtered data. Overall power level is reduced when the tSSS algorithm is applied. For the DBS ON data, a considerable elimination of the power, presumably due to the activity coming from the chest's pacemaker at subject 1's stimulating frequency (130Hz) can be seen.

Figure 2 shows the dipole source modeling results of subject 2 and subject 3. All 8 dipoles are located on the hand knob of corresponding hemisphere, where the DBS ON state seems to show a slightly better result for all four hands.

B. Cortico-Muscular Coherence

Figure 3 shows the CMCs for all three subjects, for both left and right hands, while the stimulator was off and on.

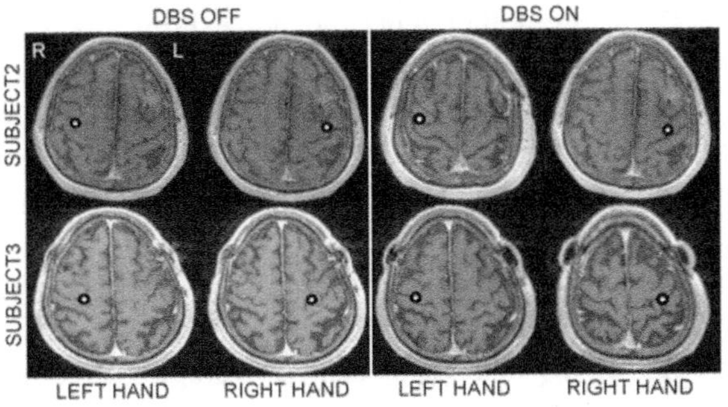

Fig. 2 Primary motor cortex localization for subject 1 and 2. The white circles indicate the localized dipoles. (Reproduced under permission. Park et al., 2009)

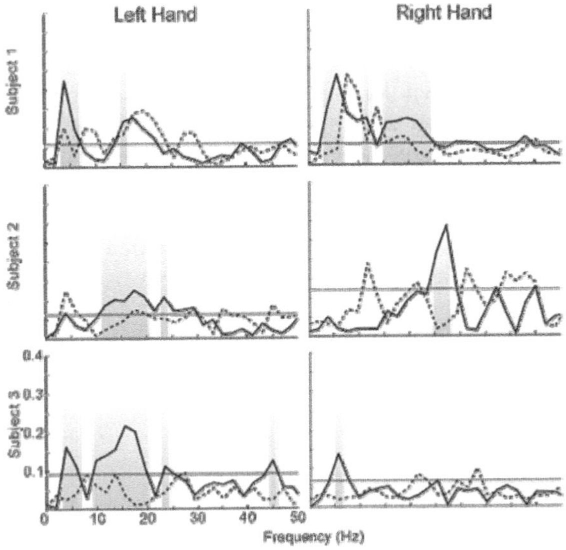

Fig. 3 Cortico-muscular coherences of all three subjects. According to the UPDRS tremor scores, subject 1 is right hand dominant, subject 2 is left hand dominant, and subject 3 is left hand dominant. DBS ON coherence values in the beta-band are significantly elevated for tremor dominant hands compared to that of the DBS OFF values for all three subjects. The solid graph indicates coherence values for DBS ON, dotted graph for DBS OFF, tightly dotted lines are 95% confidence level of the coherence values, shaded areas indicate frequencies where the DBS ON values were significantly larger than DBS OFF values. Range and scale of frequency and coherence are the same for all six figures. (Reproduced under permission. Park et al., 2009)

Data for DBS OFF and DBS ON are superimposed on each plot. The result shows that focusing on the beta- band frequency (13–30 Hz), the tremor dominant hand shows significantly elevated coherence; subject 1 right hand, subject 2 left hand, subject 2 right hand, and subject 3 left hand, during DBS ON. Other plots did not exhibit any kind of strong tendency of elevation or any differences.

IV. Discussion

Park et al. [1] have showed that magnetic fields from Parkinson's disease subjects can be measured with magnetoencephalography with the aid of noise and artifact removal algorithms and sophisticated averaging methods to increase the signal to noise ratio of the signal. This is the first study to localize the primary motor cortex with magnetoencephalography obtained from subjects with Parkinson's disease with deep brain stimulator implanted. Although there have been studies which involved measuring subjects with deep brain stimulation, studies which dealt with the motor evoked field obtained from DBS subjects itself was rare, if any.

It is well known that the motor network is highly associated with the deep brain structure such as the basal ganglia and also the cerebellum. Since the current study revealed the possibility to find the cortical level of the motor network, further investigation involving the other motor related cortices such as the supplementary motor area and the premotor cortex may proceed. This study provides the first guide for utilizing MEG in investigating the motor network of Parkinson's subjects with DBS.

References

1. Park H, Kim JS, Paek SH et al. (2009) Cortico-muscular coherence increases with tremor improvement after deep brain stimulation in Parkinson's disease. NeuroReport 20(16):1444-1449.
2. Kim JS, Chung CK (2007) Robust source analysis of oscillatory motor cortex activity with inherently variable phase delay. Neuroimage 37(2):518-29.
3. Taulu S, Hari R, (2009) Removal of magnetoencephalographic artifacts with temporal signal-space separation: Demonstration with single-trial auditory-evoked responses. Human Brain Mapping 30(5):1524-1534.
4. Andrykiewicz A, Patino L, Naranjo J et al. (2007) Corticomuscular synchronization with small and large dynamic force output. BMC Neurosci 8:101-112.
5. Salenius S, Portin K, Kajola M et al. (1997) Cortical control of human motoneuron firing during isometric contraction. J Neurophysiol 77:3401-3405.
6. Hämäläinen M, Hari R, Ilmoniemi RJ et al. (1993) Magnetoencephalography-theory, instrumentation, and applications to noninvasive studies of the working human brain. Rev Mod Phys 65:413-497.

Corresponding author:

Author:Chun Kee Chung
Institute:Seoul National University
Street:101 Daehang-ro, Jongro-ku, 110-744
City:Seoul
Country:Republic of Korea
Email:chungc@snu.ac.kr

MEG Slow Wave Dipole Density (SWDD) in Presurgical Evaluation of Epilepsy Patients: Preliminary Results of a Prospective Study

Martin Kaltenhäuser, Stefan Rampp, Tanja Ehrenfried, Marcel Heers, and Hermann Stefan

Epilepsy Center (ZEE), Department of Neurology, University Clinics, Erlangen, Germany

Abstract— **Main objective(s)**

The aim of this prospective study is to examine the potential contribution of SWDD localisation in clinical preoperative epilepsy focus localisation.

Methods

Preoperative MEG-SWDD localisations are calculated and overlaid on MR-images of pharmaco-resistant epilepsy patients undergoing presurgical focus localisation evaluation. For all 10 patients who have been operated within at least 6 months, postsurgical outcomes available up to now, are discussed. SWDD localisation results are compared to other preoperative findings and to MEG-spike results and are correlated to postoperative outcome (Engel) after 6 months or 12 months.

Results

7 of 10 patients showed significant local increase of SWDD. In these SWDD localisation results were always concordant with preoperative findings. All 7 patients were postoperatively seizure-free or had a worthwhile improvement of seizure situation. 3 of 10 patients showed no significant local increase of SWDD and, therefore, were without SWDD result.

Comparison to MEG spike results: Among these 7 patients, 6 showed MEG-spike concordant localisations, one had no spikes in the MEG.

Conclusions

SWDD is able to localise brain areas exhibiting pathological slow wave activity and might become a useful tool for preoperative epilepsy diagnosis.

Keywords— **Magnetoencephalography, slow wave, slow wave dipole density, epilepsy, postoperative outcome.**

I. Introduction

Magnetoencephalography (MEG) in epilepsy patients shows interictal epileptiform discharges (IED) in up to 70% of all patients [1]. It would be desirable, to get information about epileptic areas in the other 30% of patients, too.

One such alternative is the investigation of pathologic slow wave activity. In patients with focal epilepsy localisation of brain areas correlated to the epilepsy and exhibiting slow wave activity was shown in 2007, using slow wave dipole density analysis (SWDD) [2].

In the present prospective study, SWDD is compared to preoperative findings in epilepsy patients who underwent epilepsy surgery.

II. Methods

Between January 2007 and November 2007, 44 patients had magnetoencephalography (MEG) investigations in the course of preoperative epilepsy workup. A Magnes II system consisting of two dewars with 37 MEG channel each was used (4D-Neuroimaging, San Diego, CA). All MEG data were analysed by SWDD [2]. Eleven of 44 patients underwent epilepsy surgery. In 10 of these 11, a postoperative outcome (Engel, at least 6 months) was available and the 10 patients were included in this study.

SWDD analysis was done by custom Erlangen software, which includes some standard routines of the vendor's system software, other analysis steps were performed by custom programs.

SWDD data analysis consists of: filtering using a narrow band pass filter from 2 to 6 Hz, principle component analysis (PCA), dipole fit of time intervals selected automatically by PCA, dipole density calculation and visualization of SWDD localisation by projection on MR-images of the patient.

The following SWDD-parameter were defined:

"pswd_max" ("percentage of slow wave dipoles") describes the maximum value of SWDD within ccm-voxels in the patient.

"pswd_>1%" describes the sum of the SWDD contents of all ccm-voxels containing more than 1% of all slow wave dipoles.

Cut-off-values to determine, whether SWDD is increased in a patient are defined as follows in reference to healthy subjects data [2]:

If parameter "pswd_>1%" is more than 6% and parameter "pswd_max" is more than 1.5%, a significant increase in SWDD is defined.

S. Supek and A. Sušac (Eds.): Advances in Biomagnetism – BIOMAG2010, IFMBE Proceedings 28, pp. 389–392, 2010.
www.springerlink.com

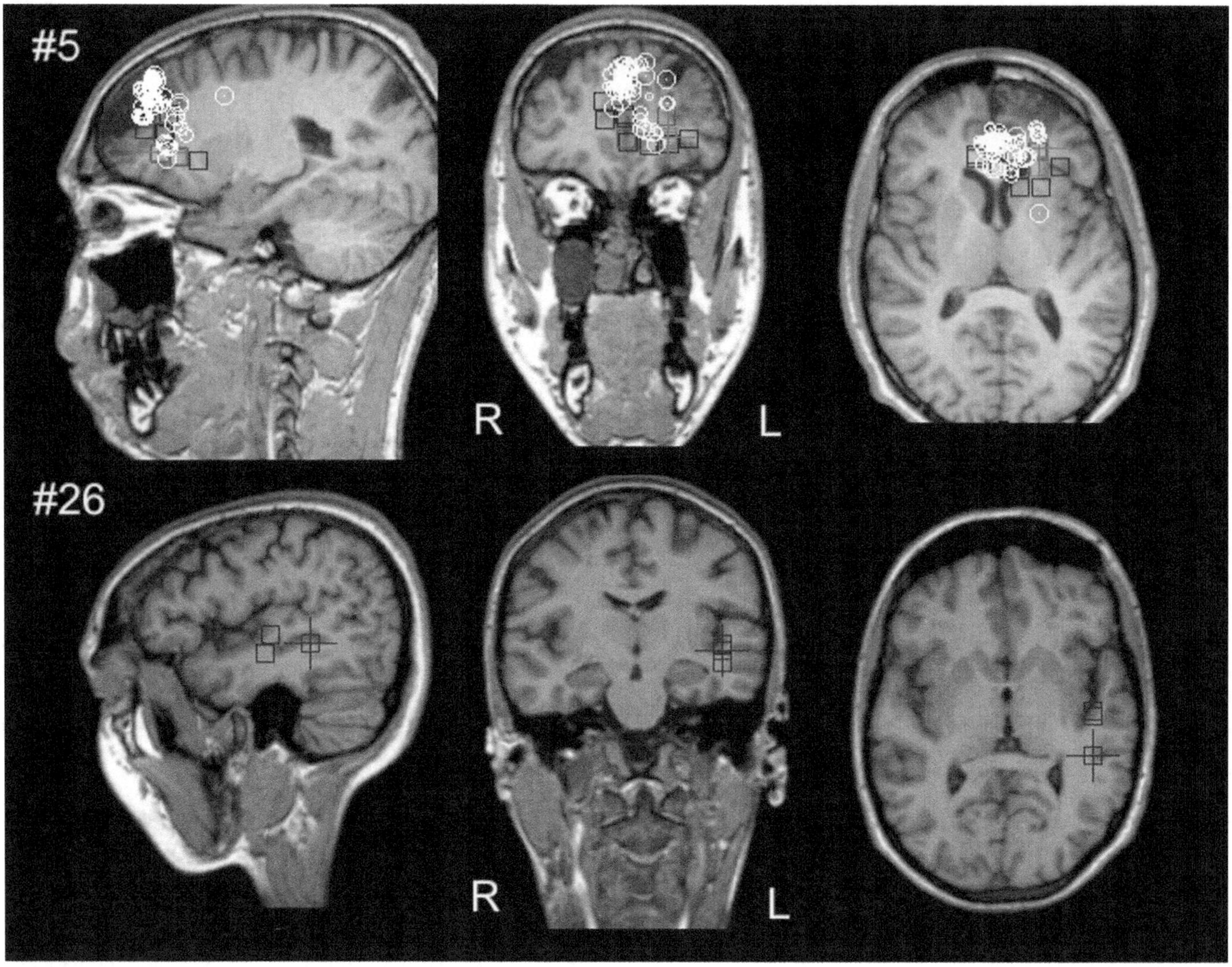

Fig. 1 MRI with SWDD-localisations; patient #5 with SWDD increase, patient #26 without significant SWDD increase; magenta square: pswd > 2.0%, red square: pswd > 1.5%, blue squares: 1.5%>pswd >1.0%; white circles: single dipole fit localisation results of interictal MEG spikes

SWDD-results were obtained without knowledge of other findings in the preoperative epilepsy workup. SWDD is an automated analysis routine that does not need input by the physicians or physicists doing the calculations. In this study, the results of SWDD analysis were not presented to the physicians working on other findings in order to not influence decision of target volume for epilepsy surgery.

III. Results

Seven out of 10 patients showed local SWDD increase, (Parameter "pswd_>1%": mean value 13,8%, range 6,5% - 22,3%) in contrast to 3 patients without SWDD increase (Parameter "pswd_>1%": mean value 1.4%, range 0% - 3%). In relation to other findings SWDD show:

1. Localisations of SWDD overlaid on MRI were concordant with resected brain areas on a sublobar level.
2. Six of the 7 patients with increased SWDD had MEG spike results. These were concordant, too.
3. Histological findings proofed pathological tissue damages in the resected areas.
4. Good outcomes (median 1A) show success of epilepsy surgeries.

Table 1 Patients' medical information and SWDD results; abbreviations: HS: hippocampal sclerosis; MST: multiple subpiale transsections; DNT: dysembryoblastical neuroepithelial tumor; FCD: focal cortical dysplasia; VH: vascular hamartoma; t.r.: tailored resection

patient	gender	outcome	pswd_max	pswd_>1%	SWDD increase	SWDD-localisation	interictal MEG spike	epilepsy classification	surgery	histology
#2	f	2B	2,5%	19,1%	Yes	left fronto-temporal	left fronto-temporal	symptomatic	left temp, t.r.	HS
#4	m	2B	1,5%	7,1%	Yes	right frontal	right frontal, pre-motor-cortex	cryptogenic	MST, right frontal	no histology (MST)
#5	m	1D	2,3%	17,9%	Yes	left frontal	left fronto-basal	symptomatic	left frontal	gliosis, after cranio-cerebral injury / trauma
#7	m	1A	2,0%	11,5%	Yes	right parieto-occipital	right parietal (only discrete finding)	symptomatic	right parietal, lesionectomy	VH
#23	f	1A	1,2%	1,2%	No	-----------	left frontal and left temporo-basal	symptomatic	left temporal, t.r.	HS
#24	f	3A	2,5%	22,4%	Yes	right parieto-occipital	right temporo-parietal	symptomatic	right parie-tal-occipital, lesionectomy	FCD type II b
#26	f	1A	1,0%	3,0%	No	-----------	-------------	symptomatic	left parietal-occipital	DNT degree I
#33	f	1A	0,9%	0,0%	No	-----------	-------------	symptomatic	left temp, t.r.	HS
#36	m	1A	1,6%	6,5%	Yes	left temporal	left temporo-basal (only discrete finding)	symptomatic	left temporal, lesionectomy	DNT degree I
#42	m	1A	1,9%	12,2%	Yes	right temporal	-------------	symptomatic	right temporal, t.r.	HS and FCD type I a

IV. Discussion

The relatively high rate of SWDD-results (7 out of 10 patients) and the concordance of SWDD localisation results with resected brain areas, verified by good outcomes, suggest that SWDD might be helpful in preoperative epilepsy diagnosis. Our prospective study started in 2007. The preliminary results presented here were restricted to address only operated patients. Most of them were MRI-positive. Concerning the possible value of SWDD analysis for preoperative epilepsy investigation, it might be objected that in MRI-positive cases no other additional measurements are necessary to decide and plan extent of epilepsy surgery, and thus, SWDD analysis might not be necessary. Nonetheless, the high sensitivity of SWDD of (7 out of 7 patients) shows the possible potential of this method. In future phases of the presented study, all 44 patients, especially those who won't undergo epilepsy surgery, will be discussed. Comparison of SWDD results and of all other findings will be done then.

SWDD results were not used for planning of epilepsy surgery. Additional, later post-operative outcomes of more patients might add information.

SWDD is able to localize functional areas of the brain exhibiting slow wave activity in theta and delta band. The pathophysiological mechanisms producing this activity in epilepsy patients is yet unclear. Using SWDD, Kamada et al. [3] showed areas of local slow wave activity in tumor patients, which were localized in the border zones of the tumors. He hypothesized that tumors or local edema lead to cell damages, which and might produce pathological slow wave activity there. It can not be decided yet whether in the 10 cases of epilepsy patients of our study, similar pathomechanisms produced the observed local slow wave activity. Another alternative hypothesis might be that local damages are induced due to epileptic processes in the epileptogenic network by numerous seizures over a long time.

Even if the involved pathomechanisms producing slow wave activity remain unclear, SWDD seems to have a potential to be helpful in determining the epileptic focus.

V. Conclusions

The preliminary results of this prospective study allow suggest that slow wave dipole density might be a useful tool in epilepsy surgery workup.

Acknowledgment

This work was supported by the ELAN (Erlanger leistungsbezogene Anschubfinanzierung und Nachwuchsförderung) under fund NL-06.11.20.1.

References

1. Stefan H, Hummel C, Scheler G, Genow A, Druschky K, Tilz C, Kaltenhäuser, M,Hopfengärtner R, Buchfelder M, Romstöck J (2003) Magnetic brain source imaging of focal epileptic activity: a synopsis of 455 cases. Brain 126:2396-2405
2. Kaltenhäuser M, Scheler G , Rampp S , Paulini A., Stefan H (2007) Spatial intralobar correlation of spike and slow wave activity localisations in focal epilepsies: A MEG analysis. NeuroImage 34:1466-1472
3. Kamada K, Saguer M, Möller M, Wicklow K, Kaltenhäuser M, Kober H, Vieth J (1997) Functional and metabolic analysis of cerebral ischemia using magnetoencephalography and proton magnetic resonance spectroscopy. Ann Neurol 42:539-41

Author: Martin Kaltenhäuser
Institute: Epilepsy Center Erlangen (ZEE)
Street: Schwabachanlage 6
City: Erlangen
Country: Germany
Email: martin.kaltenhaeuser@uk-erlangen.de

Brain Network Connectivity Dynamics during Voluntary Finger Movement in Right Handed Adults with Down Syndrome: Evidence for Contralateral and Ipsilateral Dominance

N. Moiseeva[1], A. Moiseev[1], I. Lott[2], R. Haier[2], K. Head[2], U. Ribary[1,3], and N. Virji-Babul[1,3]

[1] Down Syndrome Research Foundation, MEG Laboratory, Burnaby, B.C. Canada
[2] University of California, Irvine, USA
[3] Simon Fraser University, Burnaby, B.C. Canada

Abstract— **We previously reported that in a small group of right handed adults with Down syndrome (DS), voluntary finger movements were characterized by ipsilateral dominance. The aim of this study was to examine the dynamics of network connectivity during the same motion. Thirteen typical adults and twelve adults with DS participated in this study. Cortical responses were recorded with a 151 channel MEG system (VSM MedTech). Subjects pressed a button on the keypad using their right index finger. The SAM beamformer was used to obtain spatial distributions of the evoked activity in 0.2 – 50Hz, 0.2-12Hz and 12-50Hz frequency bands. Individual distributions were spatially normalized to a standard template brain using SPM and averaged. Locations with the highest evoked signal to noise ratio in the normalized averaged data were found, and transformed back to each subject's original brain coordinates.**

For the control group the most significant activation in the low frequency band was in the contralateral motor region. In the high frequency band, significant activations were found in bilateral motor regions. In the DS group, we observed two distinct patterns of activation: an ipsilateral dominance and a contralateral dominance in the low frequency band. DS individuals were subsequently divided into two subgroups based on laterality calculated in the broadband (0.2-50Hz). The ipsilateral group (n=6) showed largest activations in the right temporal region at low frequencies. At high frequencies the distribution was bilateral, involving both motor and temporal regions. The strongest peaks in the contralateral group (n=6) were found in the left motor area for both low and high frequency bands. PLV analysis showed significant differences in connected locations between the controls and the DS ipsilateral and contralateral groups. Further work is needed to better understand the emergence of these two distinct patterns of neural activity.

Keywords— **Down syndrome, motor control, phase locking, functional connectivity.**

I. Introduction

Down syndrome (DS) is a genetic condition resulting from the presence of an additional whole or part copy of chromosome 21 which disrupts many aspects of cognitive and sensory-motor development. Structural brain imaging studies have shown that there are significant differences in the structure of the brain in DS, however there are considerable gaps in our understanding of the functional significance of these structural differences and in particular how these structural differences relate to the dynamics of basic sensory-motor processes. Considerable behavioural research has shown a left hemisphere dominance in DS for the control of movements, similar to the typical population. However, we have recently shown that in a small group of right-handed adults with DS, voluntary finger movements were characterized by ipsilateral dominance [1]. Our working hypothesis is that some of the motor difficulties observed in DS may be due to altered sensory-motor processing. The objective of this work was to gain more insights into the underlying neural processing by studying both the spatial localization and the underlying functional connectivity during voluntary movements of the dominant right index finger in a larger group of adults with and without DS.

II. Methods

A. Study Participants

13 healthy right-handed adult subjects and 12 adults with Down syndrome (DS), aged 20-40 years participated in this study. All participants provided informed, written consent and had normal or corrected to normal vision. This study was approved both by the Simon Fraser University Research Ethics Board and the Down Syndrome Research Foundation Research Ethics Committee and conducted in accordance with the Declaration of Helsinki.

B. Procedure

Participants pressed a button on a response pad placed on the right armrest and depressed one button with their right index finger at their own pace, without counting. The nonmoving hand rested on the opposite armrest. An average of 80 movements were recorded.

S. Supek and A. Sušac (Eds.): Advances in Biomagnetism – BIOMAG2010, IFMBE Proceedings 28, pp. 393–396, 2010.
www.springerlink.com

C. MEG Data Acquisition

The MEG data was acquired using a 151-channel whole head MEG system (VSM Medtech, Coquitlam, Canada) installed within a magnetically shielded room. Approximately 300 seconds of continuous data were initially collected at a rate of 600 samples per second with a 150-Hz low pass filter, using third-order gradiometer noise cancellation. The data was manually inspected for artifacts and corresponding segments were marked as bad.

D. Data Processing

The event-related SAM beamformer spatial filtering algorithm [2, 3] was used to obtain spatial distributions of the evoked activity in 0.2 – 50 Hz frequency band. Data in [-.6, .5] s interval relative to the button press event were used for the analysis. Each subject's head shape was digitized at the end of the recording using a Polhemus digitizer. These data were used for head modeling. Individual distributions were spatially normalized to a standard template brain and averaged. Locations with the highest evoked signal to noise ratio in the averaged data were found, along with their prototypes in each subject's brain space.

We obtained the laterality index of the group-averaged images by using a 95% significance statistical threshold (obtained by bootstrap) to choose significant voxels in the image. The values of these pixels were integrated separately for left and right hemispheres, and the laterality index (LI) was estimated $LI = (left - right)/(left + right)$.

Synchronization between pairs of significant sources was computed by using phase locking value (PLV). PLV analysis was performed in multiple 4Hz-wide bands selected from 3 to 46 Hz frequency range. All PLVs were normalized on their standard deviation in the control interval [-.65, -.2] s relative to the event, then baselined using the mean value in that interval, and group-averaged. For a given frequency band and selected active time window (see fig. 2) two locations were considered connected if *both* of the following conditions where met: a) the group-averaged PLV time course showed high *sustained* event-related synchronization or desynchronization characterized by the ratio of rms PLVs in active and control windows exceeding specified threshold and b) the group-averaged (de)synchronization resulted from a statistically significant number of *individual* PLV time courses also showing rms PLV ratios above the threshold. The number of individuals was considered significant if it exceeded the 5% level based on the Bernoulli distribution.

III. RESULTS

Based on the laterality index calculated in the broadband (0.2 – 50 Hz), the control group showed highly significant contralateral dominance ($LI = 0.36$, p =0.005). In contrast, the DS group demonstrated almost ideal bilateral distribution ($LI = -0.01$). We then divided the DS group into two subgroups based on the sign of the laterality index: a contralateral group (6 participants) that had an average broad band LI of .44 and an ipsilateral group (6 participants) that had an average LI of -.46. Mean LI values for the DS subgroups in the low frequency band were .41 and -.23, respectively, with 2-sided *p*-values of .04 and 0.02. The fact that the LI values for the subgroups are statistically significant strongly suggests the presence of two distinct patterns of activation and not just a broad spectrum of possible LI readings for the DS individuals.

Figure 1 shows the location of evoked activity during voluntary self-paced right index finger movement averaged over the time interval starting 200 ms prior to the button press event and ending 200 ms following the event, in all three groups in the .2-12 Hz range. The top set of figures (A) shows the locations of peak activity in the control group. The most significant activations approximate the left primary motor and sensorimotor regions as has been reported in previous MEG studies [4; 5]. Smaller but significant activity was also observed in the ipsilateral motor region.

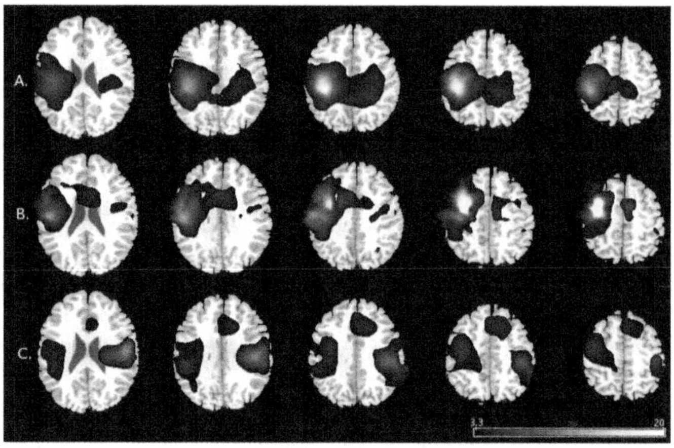

Fig. 1 Group averages of peak activity in the .2-12 Hz range in three groups: A) Control group, B) DS contralateral group and C) DS ipsilateral group

The middle row shows the activations for the DS contralateral group (B). The strongest peaks were observed in the left sensory-motor regions, similar to the control group. The locations of the activations however, were more anterior in comparison to the control group. The ipsilateral group (C) showed largest activations in the ipsilateral temporal region at low frequencies. At high frequencies the distribution was bilateral, involving both motor and temporal regions.

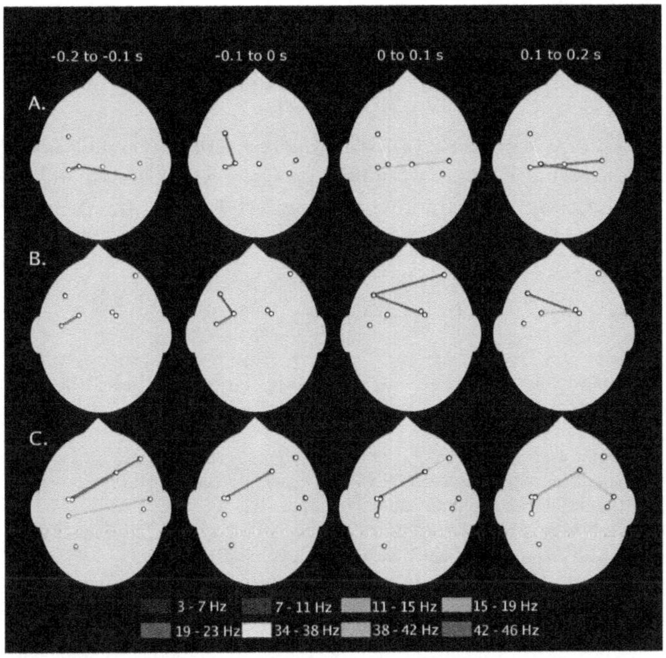

Fig. 2 Phase locking in the following 100ms windows (left to right): [-.2, -.1], [-.1, 0], [0, .1], [.1, .2] s in three groups: A) Control group, B) DS contralateral group and C) DS ipsilateral group. (Three-dimensional locations were projected on the plane of the figure)

PLV analysis showed significant differences in the spatial activation and time-frequency dynamics of the connectivity patterns between the controls and the DS ipsilateral and contralateral groups. Figure 2 shows the most significant phase locking between pairs of sources in four 100 ms time windows in the interval from -200 ms to 200 ms relative to the event. For the control group, significant phase locking was observed in the low frequency band (7-11Hz) between (L) Postcentral Gyrus/(L) Temporal Gyrus, and (L) Postcentral Gyrus/(R) Inferior Parietal Lobule at 200 ms prior to the onset of movement. At 100 ms prior to onset of movement, significant phase locking was observed in the high gamma range between 42-46 Hz in the left hemisphere between the (L) Postcentral Gyrus/(L) Temporal Gyrus and (L) Postcentral Gyrus/(L) Insula. 100 ms after movement onset, phase locking was observed in gamma in the 34-42 Hz frequency range between (L) Postcentral Gyrus/(L)Insula and (L) Temporal/(R) Postcentral Gyrus. Further phase locking was observed in the following frequencies: gamma, between (L) Temporal/(R) Postcentral Gyrus and (L) Postcentral Gyrus/(L) Insula and low alpha frequency range between (L) Postcentral Gyrus/(L) Temporal Gyrus, and (L) Postcentral Gyrus/(R) Inferior Parietal Lobule at 200 ms post movement. The pattern in the control group was characterized by phase locking in the low frequency band prior to the start of movement and moving to phase locking predominantly in the high frequency band following the movement.

In the DS contralateral group significant phase locking was observed in beta (19-23Hz) band between (L) Middle Frontal Gyrus/(L) Precentral Gyrus, followed by phase locking in the low alpha (7-11 Hz) band between (L) Mid Frontal Gyrus/(L) Middle Frontal Gyrus, (L) Mid Frontal Gyrus/(R) Cingulate Gyrus, and (L) Mid Frontal Gyrus/(R) Inferior Frontal Gyrus. At the termination of movement phase locking was observed in the gamma range between (L) Mid Frontal Gyrus/(R) Cingulate Gyrus. In this group the general pattern can be characterized as moving from high frequency phase locking to predominantly low frequency phase locking as the movement progressed.

In the DS ipsilateral group, the most significant phase locking was observed in the gamma band between (R) Precentral Gyrus/(L) Inferior Parietal Gyrus, (L) Insula/(R) Cingulate, and (L) Precentral Gyrus/(R) Mid Frontal Gyrus, followed by phase locking in the beta band between (L) Insula/(R) Cingulate. After the onset of movement, phase locking was observed primarily in the alpha/beta range between (R) Precentral Gyrus/(R) Postcentral Gyrus, (L) Insula/(R) Cingulate, (R) Cinglate/(R) Mid Frontal Gyrus, and (L) Inferior Parietal/(L) Precentral Gyrus. In this group the general pattern can be characterized as starting with high frequency phase locking to lower frequency phase locking at the end of movement.

Table 1 summarizes the PLV analysis for the movement preparation phase (-200 and -100 msec prior to movement onset) for all three groups.

Table 1 PLV analysis for the movement preparation phase (-200 and -100 msec prior to movement onset) for all three groups

Group	-200 ms prior to motion	-100 ms prior to motion
A. Control	7-11 Hz: (L) PCG/(L) TG (L) PCG/(R) IPL	42-46 Hz: (L) PCG/(L)TG (L) PCG/(L) Ins
B. DS Cont	19-23 Hz: (L)MFG/PreCG	7-11Hz: (L) MFG/(R)CG
C. DS Ipsilat	42-46Hz: (R) PreCG/(L) IPG 38-42Hz:(L) PreCG/(R)MFG	19-23Hz: (L) Ins/(R) CG

IV. Discussion

Our results show that in control subjects, predominantly contralateral as well as ipsilateral areas are selectively activated during the preparation and execution of simple unilateral finger movement, consistent with previous MEG results [5,6]. In addition, we show that phase locking occurs primarily in the low frequency bands during the movement preparation phase and predominantly in the high frequency band following the movement. Specifically, we observed significant synchronization between motor and temporal sources within the contralateral hemisphere, as well as cross hemisphere connectivity between the sensory-motor, temporal and parietal regions, during the movement preparation phase. During movement execution, long-range connectivity was observed in cross hemisphere sources between the sensory-motor area, temporal and parietal regions. These activated areas are consistent with previous observations on network activations during voluntary movement [7,8] and provide important insights of the underlying network dynamics of correlated local and distributed network activations across frequency and time.

Overall, individuals with DS showed altered network connectivity dynamics during right handed finger movement in relation to changes the activation of motor related brain regions or in relation to activation across low/high frequency scales. In the DS contralateral group, the strongest peaks were found in the (L) sensory-motor area for both the low and high frequency bands. PLV analysis showed significant synchronization primarily between sources in the (L) frontal and premotor areas with some cross-hemisphere connections observed following movement.

In contrast, the DS ipsilateral group showed the largest activation in the (R) temporal region in the low frequency band with ipsilateral and contralateral activation in the high frequency bands. PLV analysis showed significant cross-hemisphere synchronization between sources in the premotor and parietal regions as well as between premotor and frontal regions during the preparation phase. During movement execution significant synchronization was observed both within the ipsilateral and contralateral hemispheres between sources in the motor, frontal and parietal regions as well as between the insula and cingulate cortex.

Without any direct evidence of structural neuroanatomical differences or behavioral data, we suggest that the two distinct patterns of neural activation and the underlying network dynamics may reflect differences in sensorimotor organization. Further work is needed to chart the behavioral and neural trajectories in individuals with Down syndrome.

Acknowledgment

This work is supported by the Natural Sciences and Engineering Research Foundation (NSERC) Canada and the BC Leading Edge Endowment Fund (UR). We thank all the participants who took part in this study.

References

1. Virji-Babul, N., Moiseev, Cheung, Weeks, Cheyne, Ribary (In Press) Neural mechanisms underling action observation in adults with Down syndrome. AJIDD.
2. Robinson, S. & Vrba, J. (1999) Functional neuroimaging by synthetic aperture magnetometry (SAM). *Recent Advances in Biomagnetism,* Sendai, Japan, Tohoku Univ. Press, pp. 302–305
3. Robinson, S. (2004) Localization of event-related activity by SAM(erf). Neurol Clin Neurophysiol, vol. 2004, p.109.
4. Cheyne D, Weinberg H, Gaetz W, Jantzen KJ (1995) Motor cortex activity and predicting side of movement: neural network and dipole analysis of pre-movement magnetic fields. Neurosci Lett 188:81– 84.
5. Cheyne, D., Bakhtazad, L., & Gaetz, W. (2006) Spatiotemporal mapping of cortical activity accompanying voluntary movements using an event-related beamforming approach. *Human Brain Mapping, 27*(3), 213-229.
6. Huang M, Harrington D, Paulson K, Weisend M, & Lee R. (2004) Temporal dynamics of ipsilateral and contralateral motor activity during voluntary finger movement. *Human Brain Mapping, 23*(1), 26-39.
7. Volkman J, Joliot M, Mogilner A, Ioannides A, Lado F, Fazzini E, Ribary U and Llinas R (1996) Central motor loop oscillations in Parkinsonian resting tremor revealed by magnetoencephalography. Neurology, 46, 1359-1370
8. Schnitzler A and Gross J. (2005) Nature Rev Neuroscie 6, 285-296

Author: N. Virji-Babul
Institute: Down Syndrome Research Foundation
Street: 1409 Sperling Avenue
City: Burnaby, B.C.
Country: Canada
Email: naznin@dsrf.org

White Matter Abnormalities in Children with Temporal Lobe Epilepsy: A DTI and MEG Study

Lu Meng[1,2,*], Jing Xiang[1], Douglas Rose[1], Rupesh Kotecha[1], Jennifer Vannest[3], Anna Byars[3], and Ton Degrauw[3]

[1] MEG Center, Cincinnati Children's Hospital Medical Center, 3333 Burnet Avenue, Cincinnati, OH, 45220, USA
[2] Key Laboratory of Medical Image Computing of Ministry of Education, Northeast University, Shenyang, 110004, China
[3] Department of Neurology, Cincinnati Children's Hospital Medical Center, 3333 Burnet Avenue, Cincinnati, OH, 45220, USA

Abstract— **The widespread propagation of synchronized neuronal firing in seizure disorders may affect cortical and subcortical brain regions. The purpose of this study was to investigate the abnormal changes of white matter in children with focal temporal lobe epilepsy (TLE) using diffusion tensor imaging (DTI) and magnetoencephalography (MEG). Eight children with clinically diagnosed TLE and eight age- and gender-matched healthy controls were studied. DTI images were obtained with a 3T magnetic resonance imaging (MRI) scanner. The epileptic foci were localized with MEG. Fractional anisotropy (FA), trace, 1-volume ratio (VR), parallel ($\lambda_{\|}$) and perpendicular ($\lambda_{\perp}$) diffusivities in the genu and splenium of the corpus callosum, external capsule, anterior limbs of the internal capsule, and the posterior limbs of the internal capsule were calculated. The DTI parameters in the same brain regions between patients and controls were statistically compared. The DTI parameters of white matter between the hemisphere with seizure foci and the hemisphere without seizure foci were statistically compared. In comparison to controls, patients with epilepsy had significantly lower FA and higher trace, 1-VR, $\lambda_{\|}$ and $\lambda_{\perp}$ in white matter structures. There was no significant difference between the hemisphere with seizure foci and the hemisphere without seizure foci in TLE patients in terms of FA, trace, 1-VR, $\lambda_{\|}$ and $\lambda_{\perp}$. The results of the present study indicate that children with TLE had significant abnormalities in the white matter in the hemisphere with seizure foci. Furthermore, these abnormal changes could be identified in the hemisphere without seizure foci. Though further investigation and more patients' data are necessary, it seems that white matter abnormalities represent an underlying predisposing factor in the development of TLE.**

Keywords— **Diffusion tensor imaging, Temporal lobe epilepsy, Pediatric, Anisotropy.**

I. Introduction

Epilepsy is the fourth most common neurologic disorders in all ages [1]. Temporal lobe epilepsy (TLE) is the most common form and is the most frequent type of partial epilepsy refractory to medical therapy [2]. It has been found that TLE is associated with white matter abnormalities. The white matter abnormalities may represent sequelae of recurrent partial seizure activity, reflecting an abnormal network associated with epileptogenesis. The results of previous reports provide compelling evidence that a neural network of functional and structural reorganization may occur in patients with TLE both ipsilateral and contralateral to the epileptic brain tissue [3]. Diffusion tensor imaging (DTI) is a magnetic resonance imaging technique that can detect in vivo anisotropic diffusion properties in white matter by measuring water diffusion and its directionality in three directions [4]. DTI characterizes diffusive transport of water by an effective diffusion tensor D. This symmetric 3×3 tensor is very important, since it contains useful structural information about the tissue. The eigenvalues of D are the three principal diffusivities and the eigenvectors define the local fiber tract direction field. A previous study indicated the presence of bilateral limbic system DTI alterations in patients with unilateral TLE [5].

It has been suggested that the integration of multiple imaging modalities can elucidate functional brain dynamics and cortical connectivity [6]. The present study combined DTI and MEG to investigate the diffusion characteristics of white matter in children with TLE. MEG analysis was used to accurately localize the sources of interictal epileptiform discharges, and DTI analysis provided the unique in vivo information of the whiter matter architecture. To assess the integrity of white matter in the developing brain with epilepsy, the present study was designed to quantify the fractional anisotropy (FA), trace, 1-volume ratio (VR), parallel ($\lambda_{\|}$) and perpendicular ($\lambda_{\perp}$) diffusivity. The present study focused on the corpus callosum (GCC), splenium of the corpus callosum (SCC), external capsule (EC), anterior limbs of the internal capsule (AIC), and posterior limbs of the internal capsule (PIC).

II. Materials and methods

A. Participants

Eight children with TLE (3 girls and 5 boys) and eight age- and sex-matched healthy children (3 girls and 5 boys) were retrospectively studied. The mean age of pediatric TLE patients was 11.95 years (range: 5.5-21.0 years); and the mean age of normal subjects was also 11.87 years

S. Supek and A. Sušac (Eds.): Advances in Biomagnetism – BIOMAG2010, IFMBE Proceedings 28, pp. 397–400, 2010.
www.springerlink.com

(range: 5.5-21.0 years). The mean age of seizure onset was 6.7 years (range: 0.5-15 years) and the mean duration of seizures was 5.2 years (range: 1-12 years). This study was approved by the institutional Review Board (IRB) at the Cincinnati Children's Hospital Medical Centre (CCHMC).

B. *Image Acquisition*

The MR images were acquired on a 3.0T Siemens Trio MR imaging scanner (Siemens, Erlangen, Germany). A 46-section, diffusion-weighted, spin-echo echo-planar imaging scan was acquired in the axial plane with the following parameters: TR/TE= 6000/87 ms; FOV=25.6×25.6 cm; matrix= 128×128; section thickness=2 mm; b-value= 1000 s/mm2; and 4 repetitions. Diffusion-weighted scans were acquired in 12 optimized directions.

The MEG recordings were obtained in a magnetically shielded room (MSR) using a whole head CTF 275-Channel MEG system (VSM MedTech Systems Inc., Coquitlam, BC, Canada) in the MEG Center at CCHMC. Before data acquisition commenced, a small coil was attached to the nasion, left and right pre-auricular points of each subject. These three coils were subsequently activated at different frequencies for measuring subjects' head positions relative to the MEG sensors. Patients were deprived of sleep the night before MEG and MR imaging studies. All patients were drowsy and fell asleep during the recording sessions. Seizures were incidentally captured during MEG recordings.

C. *Image Processing*

All DTI images' reconstruction and post-processing were performed with DTIStudio 2.4 from John Hopkins University [7] in regions of interest. For DTI data, diagonalization of the diffusion tensor yielded three eigenvalues (λ_1, λ_2, and λ_3, and $\lambda_1>\lambda_2>\lambda_3$) and eigenvectors [4]. Eigenvalue λ_1 represents the diffusivity along the primary diffusion direction. In theory, the primary diffusion direction in a voxel coincides with the mean of the unit vectors that are parallel to the physical axes of the neurons included in that voxel. The second and third eigenvalues (λ_2 and λ_3) represent the diffusivities along directions that are perpendicular to the primary diffusion direction, and to each other[4,8]. Then five diffusion parameters were FA, $\lambda_1+\lambda_2+\lambda_3$, 1-VR, λ_1 and $(\lambda_2+\lambda_3)/2$.

D. *DTI Measurement*

In our ROI-based group comparison, a series of white matter regions were selected for each participant: genu of the corpus callosum (GCC), splenium of the corpus callosum (SCC), external capsule (EC), anterior limbs of the internal capsule (AIC), and the posterior limbs of the internal capsule (PIC). As shown in Figure 1, ROIs were outlined on FA maps following the approach described by Hermoye and colleagues [9]. We first used one control subject and manually drew all the ROIs as demonstrated in Figure 1(A). Because it is hard to distinguish the borders of AIC and PIC on the FA map where these two structures are continuous, we also produced color-coded maps for all subjects. The borders in color-coded map could be easily detected. Then, the identified ROIs were used as a guide to manually define ROIs for other subjects to maximize the reproducibility.

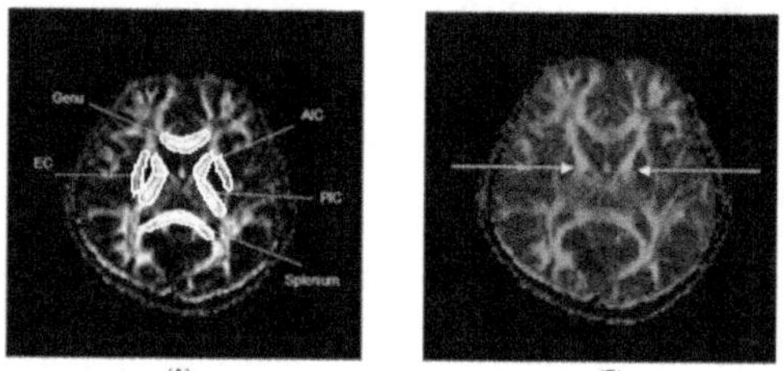

(A) (B)

Fig. 1 (A) FA map from a control. All selected structures (genu of the corpus callosum (GCC), splenium of the corpus callosum (SCC), external capsule (EC), anterior limbs of the internal capsule (AIC), and posterior limbs of the internal capsule (PIC)) are outlined on an axial slice. (B) Color-coded map of the same slice. Arrows illustrate the exact borders between AIC and PIC

E. *Localization of Epileptic Foci*

The epileptic foci were localized with wavelet-based beamformer. The beamformer algorithm implemented in the method was improved for performance and accuracy. The detailed mathematic algorithms have been described in previous reports [10]. One of the major features of the improved beamformer in this study was its ability to detect correlated sources. A 3D-head model was created with each subject's MR image. For the magnetic spatial filtering analysis, multiple local spheres were fitted to each subject's head model. A customer-designed program, MEG Processor, was used to estimate magnetic sources for signals in a frequency band of 20 - 70Hz for MEG data. The whole brain was scanned with a 2.5-millimeter resolution for sources.

F. *Statistical Analysis*

Statistical analyses were performed using SPSS version 15.0 for Windows Evaluation (SPSS Inc., Chicago, IL, USA). A paired t-test was used to compare data between the patient group and the control group. Subsequently, a paired t-test was used for hemispherical comparisons. The parameters used for the comparisons were FA, trace, 1-VR, $\lambda_{\parallel}$ and $\lambda_{\perp}$ of the GCC, SCC, EC, AIC and PIC. A p-value < 0.05 was considered as statistically significant.

III. Results

The dominant epileptic activities were localized to temporal cortex for all patients with MEG data. Two patients had epileptic foci in the right temporal cortex while six patients had epileptic foci in the left temporal cortex. The MEG results were in agreement with MRI and clinical semiology. However, wavelet-based beamformer revealed that there were clear rhythmic fast wave in the contralateral hemisphere for 4 patients. Based on the MEG source localization, we analyzed the differences between normal children and children with epilepsy in terms of FA and Trace. In comparison to the healthy control, the DTI data in both the ipsilateral and control lateral hemispheres were significantly altered. However, there was no statistically significant difference between seizure focus side and the contralateral side in terms of FA and Trace.

Figure 2 shows the results of the comparison of parameters between patient group and control group in the selected ROIs. In comparison to healthy children, children with TLE had a significant decrease of FA in the AIC, PIC and SCC ($p<0.05$), and a significant increase of trace, $\lambda_\perp$ in the EC, AIC and PIC ($p<0.002$), and a significant increase of 1-VR, $\lambda_\parallel$ in SCC ($p<0.05$).

For children with TLE, there was no statistically significant difference between seizure focus side and the contralateral side. For confirmation, we also made the comparison between the patients' contralateral side and the controls' ipsilateral. Moreover the results were very consistent with the comparison between patient group and control group.

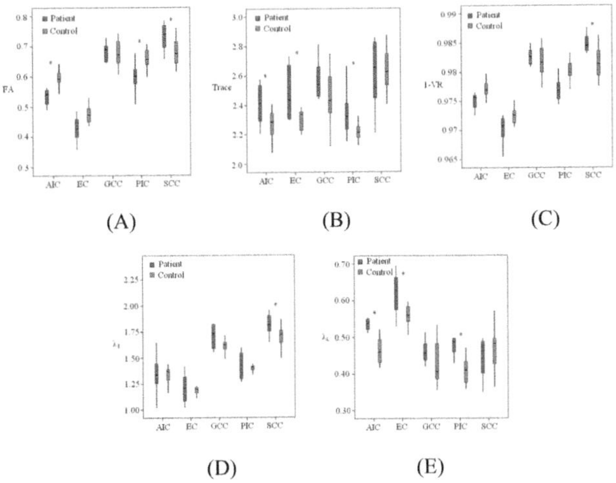

Fig. 2 Boxplots of the comparison of DTI parameters between patient group and control group in the selected ROIs. (A) FA value in AIC, PIC and SCC has significant difference between patient group and control group; (B) Trace value in AIC, PIC and EC has significant difference between patient group and control group; (C) 1-VR value in SCC has significant difference between patient group and control group; (D) $\lambda_\parallel$ value in SCC has significant difference between patient group and control group; (E) $\lambda_\perp$ value in AIC, PIC, and EC has significant difference between patient group and control group

IV. Discussion

For the first time, the present study combined MEG and DTI to investigate the white matter changes in the hemispheres with and without the seizure foci as well as the differences between children with TLE and healthy children. Our results have demonstrated that pediatric epilepsy is associated with abnormal white matter, and DTI can reveal abnormalities in pediatric patients with focal TLE. These findings are highly concordant with the results reported by Kimiwada and colleagues [11], which assessed the DTI parameters in children with TLE. They found decreased FA and increased ADC in the hippocampus ipsilateral to the seizure focus, and significantly decreased FA in the contralateral hippocampus.

In our study, the age range of the subjects was from 5.5 to 21.0 years. We decided to focus on children in this age range because the white matter in children younger than 5 years old has significant maturational changes. Hermoye and colleagues have reported that diffusion parameters comprised generally three phases: rapid changes during the first 12 months, slow modifications during the second year, and relative stability after 24 months [9]. According to their conclusion, we chose the youngest subject in our experiment to be older than five years to avoid the effect of age on the detection of the diffusion parameters in children with TLE.

The results of the present study indicated that pediatric epilepsy was associated with a decrease of FA in the AIC, PIC and SCC. In addition, pediatric epilepsy had a significant increase of diffusivity in the EC, AIC and PIC. All these white matter abnormalities reflected degeneration of fibers or edema, which may be one of the main reasons to develop epilepsy or cause seizure disorder. The white matter abnormality in epilepsy could be the sequel of repetitive seizures or a predisposing factor. According to the results of the present study, it seems that white matter abnormality is a predisposing factor. The reasons for considering that is our data also showed that the white matter in the contralateral hemisphere was significantly different from that of the normal controls. If the white matter was the sequel of seizures, the white matter abnormality should be in the ipsilateral hemisphere to the seizure foci but not in the hemisphere contralateral to the seizure foci. This observation was also supported by several previous studies[5,12].

Interestingly, there was no significant difference between the hemisphere with seizure foci and the hemisphere without seizure focus in any of the selected white matter structures. Since there were significant differences between patient and control group for DTI parameters in both hemispheres, it seems that the abnormal changes in the white matter is not just restricted to the seizure focus side of TLE, but might extend to the other brain hemisphere as well. Careful analyses of MEG data in a wide frequency range have revealed that almost half of the patients had strong rhythmic fast wave activities. Those rhythmic fast wave activities were probably the conventional "mirror epileptic activity". Therefore, according to the MEG and DTI data, it seems that white matter could be damaged by secondary epileptic activities.

Previously studies also have found that there are no significant differences in DTI parameters of the cingulum and of temporal lobe white matter on the seizure focus side versus the contralateral side in adults with TLE [5] or children with TLE [13]. In contrast, several studies in adult TLE patients showed significant diffusion abnormalities in the contralateral side white matter [14]. This discrepancy may be due to several reasons. First, the differences between adults and children could be caused by the differences in duration of TLE between adults and pediatric TLE patients. Second, people with different ages have different responses to seizures and different susceptibilities of white matter to injuries, which were induced by seizures. Third, adults and pediatric TLE are caused by different histopathological factors, generally, TLE in adults is caused predominantly by hippocampal sclerosis, but TLE in pediatric patients is caused by low-grad tumors, gliosis and cortical malformations [13]. Under the current circumstance, we consider that this discrepancy needs further investigation.

According to the results of the present study, we consider that the white matter abnormality is a predisposing factor for children with TLE. First, our data showed bilateral white matter abnormality in TLE. If the white matter abnormality was the sequel of seizures, the white matter in the hemisphere ipsilateral to the seizure foci should have prominence abnormality. Second, our data also showed that the white matter in the contralateral hemisphere was significantly different from that of the normal controls. The result implied that the white matter in the contralateral hemisphere was abnormal. Third, several previous reports also suggest that white matter abnormality is a predisposing factor for epilepsy.

V. Conclusions

The results of the present study indicate that children with TLE had significant abnormalities in the white matter in the hemisphere with and without seizure foci. The white matter abnormality in children with TLE might be a predisposing factor instead of sequel of seizures.

Acknowledgment

This study was partially supported by a Trustee Grant from CCHMC, OH, USA.

References

1. Hirtz, D., Thurman, D. J., Gwinn-Hardy, K., Mohamed, M., Chaudhuri, A. R., & Zalutsky, R. (2007). How common are the "common" neurologic disorders? *Neurology, 68,* 326-337.
2. Ahmadi, M. E., Hagler, D. J., Jr., McDonald, C. R., Tecoma, E. S., Iragui, V. J., Dale, A. M. et al. (2009). Side Matters: Diffusion Tensor Imaging Tractography in Left and Right Temporal Lobe Epilepsy. *AJNR Am.J Neuroradiol.*
3. Cascino, G. D. (2006). A cerebral network reflecting reorganization in medial temporal lobe epilepsy. *Epilepsy Curr., 6,* 46-48.
4. Basser, P. J., Mattiello, J., & LeBihan, D. (1994). Estimation of the effective self-diffusion tensor from the NMR spin echo. *J Magn Reson.B, 103,* 247-254.
5. Concha, L., Beaulieu, C., & Gross, D. W. (2005). Bilateral limbic diffusion abnormalities in unilateral temporal lobe epilepsy. *Ann.Neurol., 57,* 188-196.
6. Mohamed, I. S., Otsubo, H., Shroff, M., Donner, E., Drake, J., & Snead, O. C., III (2007). Magnetoencephalography and diffusion tensor imaging in gelastic seizures secondary to a cingulate gyrus lesion. *Clin.Neurol.Neurosurg., 109,* 182-187.
7. Jiang, H., van Zijl, P. C., Kim, J., Pearlson, G. D., & Mori, S. (2006). DtiStudio: resource program for diffusion tensor computation and fiber bundle tracking. *Comput.Methods Programs Biomed., 81,* 106-116.
8. Pierpaoli, C., Jezzard, P., Basser, P. J., Barnett, A., & Di, C. G. (1996). Diffusion tensor MR imaging of the human brain. *Radiology, 201,* 637-648.
9. Hermoye, L., Saint-Martin, C., Cosnard, G., Lee, S. K., Kim, J., Nassogne, M. C. et al. (2006). Pediatric diffusion tensor imaging: normal database and observation of the white matter maturation in early childhood. *Neuroimage., 29,* 493-504.
10. Xiang, J., Liu, Y., Wang, Y., Kirtman, E. G., Kotecha, R., Chen, Y. et al. (2009). Frequency and spatial characteristics of high-frequency neuromagnetic signals in childhood epilepsy. *Epileptic.Disord., 11,* 113-125.
11. Kimiwada, T., Juhasz, C., Makki, M., Muzik, O., Chugani, D. C., Asano, E. et al. (2006). Hippocampal and thalamic diffusion abnormalities in children with temporal lobe epilepsy. *Epilepsia, 47,* 167-175.
12. Gass, A., Niendorf, T., & Hirsch, J. G. (2001). Acute and chronic changes of the apparent diffusion coefficient in neurological disorders--biophysical mechanisms and possible underlying histopathology. *J Neurol.Sci., 186 Suppl 1,* S15-S23.
13. Nilsson, D., Go, C., Rutka, J. T., Rydenhag, B., Mabbott, D. J., Snead, O. C., III et al. (2008). Bilateral diffusion tensor abnormalities of temporal lobe and cingulate gyrus white matter in children with temporal lobe epilepsy. *Epilepsy Res., 81,* 128-135.
14. Lee, J. H., Chung, C. K., Song, I. C., Chang, K. H., & Kim, H. J. (2004). Limited utility of interictal apparent diffusion coefficient in the evaluation of hippocampal sclerosis. *Acta Neurol.Scand., 110,* 53-58.

Obtaining Auditory M100 Responses: Success Rates in Children with Autism Spectrum Disorders

Sarah Y. Khan[1], J. Christopher Edgar[1], Justin F. Monroe[1], Kate M. Cannon[1], Lisa Blaskey[1,2], Saba Qasmieh[2], Sue E. Levy[1,2], and Timothy P.L. Roberts[1]

[1] Lurie Family Foundations MEG Imaging Center, Department of Radiology, Children's Hospital of Philadelphia, Philadelphia, PA, United States

[2] Department of Pediatrics, Children's Hospital of Philadelphia, Philadelphia, PA, United States

Abstract— **Electrophysiological research demonstrates auditory abnormalities in children with autism spectrum disorders (ASD) and in children with language impairment. Although these auditory measures are promising as biomarkers [1], it is currently unclear how consistently these measures can be obtained in children. The present study examined the success in obtaining 50 ms (M50) and 100 ms (M100) auditory responses in 102 children: controls (N=35), ASD without language impairment (N=36), ASD with language impairment (N=22), and specific language impairment (N=9). Binaural 200, 300, 500 and 1000 Hz tones were presented, and left and right M50 and M100 superior temporal gyrus evoked activity modeled. In each hemisphere, a response was considered present when observed for at least 2 of the 4 tones, and with source strength significantly above background. Chi-squared analyses examined the presence or absence of M50 and M100 activity as a function of gender, hemisphere, age (young 6 to 10-years-old, older 11 to 15-years-old), and diagnosis.**

M50 responses were observed less often in the right (67.6%) than left hemisphere (85.3%, χ^2 = 8.83, p < 0.01). M100 responses were observed less often in the left (67.6%) than right hemisphere (87.3%, χ^2 = 11.2, p < 0.01), and less often in young (72.1%) than older participants (89.1%, χ^2 = 7.2, p < 0.01). M100 responses were also observed less often in patients than controls in the left (χ^2 = 8.39, p < 0.05) and right hemisphere (χ^2 = 8.82, p < 0.05). Simple-effect analyses indicated that patient and control differences were due to less frequent M100 responses in the young and older language impaired children (*p's* < 0.05).

In sum, M50 and M100 measures can be obtained in most subjects. In some cases, the lack of an auditory response appears to be of clinical significance, suggesting abnormal auditory cortex maturation. Present findings suggest that, within the examined populations, the most difficult responses to measure are the left-hemisphere M100 response in younger subjects, and left and right M100 responses in ASD subjects with language impairment.

Keywords— **Autism Spectrum Disorders, M50, M100, Magnetoencephalography, Language Impairment, Auditory evoked response.**

I. INTRODUCTION

Autism Spectrum Disorders (ASD) are a set of disabilities diagnosed in childhood that have significant impact throughout development and into adulthood. As a neurodevelopmental disorder, it is likely that a better understanding of the neurobiological abnormalities associated with ASD is needed to allow advances in the diagnosis and treatment of individuals included under this umbrella diagnosis. Specifically, whereas clinical diagnosis and ongoing assessment is presently made on the basis of observed behavioral characteristics, underlying structural and functional brain abnormalities may better characterize this heterogeneous disorder, allowing for more effective treatment and therapy monitoring [2].

In a recent MEG study, presenting 200, 300, 500, and 1000 Hz tones, Roberts et al. [1] observed a delayed right-hemisphere M100 response in children with ASD compared to typically developing (TD) children. This finding indicates that early auditory cortex responses could serve as a biomarker for ASD. For a measure to serve as a clinical biomarker, however, it must be possible to easily obtain the measure in most subjects [3]. An interesting aspect of Roberts et al. [1] was that responses were observed much less often in the left than right hemisphere. The present study examined this hemisphere difference in a larger sample in order to determine the success rate of obtaining a measurable M100 as a function of gender, age, and diagnosis.

Aside from serving as an autism biomarker, there are other reasons why early auditory measures are of clinical interest. Several studies have observed a relationship between age and M50 and M100 latencies in TD children. For example, Roberts et al. [4] and Kotecha et al. [5] showed significant associations between M100 latencies and age, with earlier latencies in older children due, in part, to increased acoustic radiation myelination in older children [4]. Such findings indicate that the M50 and M100 auditory cortex measures can be used to assess normal as well as pathological auditory cortex development. The present study examined the success of obtaining these neuromagnetic measures in a relatively large population of TD and ASD children.

S. Supek and A. Sušac (Eds.): Advances in Biomagnetism – BIOMAG2010, IFMBE Proceedings 28, pp. 401–404, 2010.
www.springerlink.com

II. Methods

A. Participants

Subjects with ASD were recruited from the Regional Autism Center of The Children's Hospital of Philadelphia (CHOP), the Neuropsychiatry program of the Department of Psychiatry of the University of Pennsylvania School of Medicine, and from local and regional parent support groups such as ASCEND (Asperger Syndrome Information Alliance for Southeastern Pennsylvania), Autism Society of America – Greater Philadelphia Chapter, and local chapters of Autism Speaks. Children screened for inclusion in the ASD sample had a prior ASD diagnosis made by an expert clinician, typically a developmental pediatrician in the Regional Autism Center at the Children's Hospital of Philadelphia. The original diagnosis was made after an extensive clinical interview, documentation of DSM-IV criteria for ASD, and use of various ASD diagnostic tools, such as the Childhood Autism Rating Scale and, in many cases, the Autism Diagnostic Observation Schedule. Subjects with typical development (TD) were recruited through local newspaper advertisements and from pediatric practices of the CHOP primary care network. Research participants made two visits to CHOP. Clinical and diagnostic testing was performed to confirm a prior ASD diagnosis, to administer psychological tests, and to ensure that TD children met study inclusion criteria. ASD diagnosis was confirmed with standard diagnostic tools, including direct observation with the *Autism Diagnostic Observation Schedule (ADOS*; [6]) and parent report on the *Social Communication Questionnaire (SCQ*; [7]). To confirm presence or absence of language impairment (LI), all subjects were evaluated with the *Clinical Evaluation of Language Fundamentals-4th edition* (*CELF-4*; [8]). The ASD group with language impairment (ASD+LI) was comprised of subjects with a *CELF-4* Core Language score below the 16th percentile. The TD and the ASD group without LI (ASD–LI) performed at or above the 16th percentile on the CELF-4.

B. Auditory Stimuli

Auditory stimuli were presented using Eprime v1.1 (Psychology Software Tools Inc., Pittsburgh, PA). Auditory stimuli were delivered via a sound pressure transducer and sound conduction tubing to the subject's peripheral auditory canal via eartip inserts (ER3A, Etymotic Research, Illinois, USA). Prior to the MEG exam, each participant's hearing threshold was determined, and the auditory stimuli were presented 45 dB SPL above threshold. During the MEG exam, 200, 300, 500 and 1000 Hz sinusoidal tones of 300 ms duration were binaurally presented (digitized at 1041.7 Hz with a 10 ms rise time). Tones were randomly presented, with a 1s interstimulus interval (jittered +/- 100 ms). Over approximately 10 min of recording time, 105 tones at each of the 4 frequencies were presented.

C. MEG Data Analysis

M50 and M100 source localization was done blind to participant group. Epochs 500 ms pre-stimulus to 500 ms post-stimulus were obtained from the continuous recording. Data was corrected for eye blink and heartbeat activity. Scanning the eye blink and heartbeat corrected raw data, epochs with artifacts other than blinks were rejected by amplitude and gradient criteria (amplitude > 1200 fT/cm, gradients > 800 fT/cm/sample). Non-contaminated epochs were averaged according to stimulus type and a 1 Hz (6 dB/octave, forward) to 40 Hz (48 dB/octave, zero-phase) bandpass filter was applied. To determine the source strength and latency of M50 and M100, a standard source model [9-11] was applied to each subject's raw MEG surface activity to transform the sensor data into brain space. The standard source model included left and right superior temporal gyrus (STG) dipole sources (placed at Heschl's gyrus), nine fixed regional sources that modeled background brain activity and served as probe sources, and eye blink and heartbeat source vectors to remove eye blink and heartbeat activity. Orientation of the STG sources was optimized by orienting dipoles to the maximum of the M50 and M100 in each subject and in each hemisphere. To measure M50 (40-90 ms) and M100 (90–180 ms) STG amplitude and latency, prestimulus baseline activity (-400 to -100 ms) was subtracted, and left and right M50 and M100 STG peak latency and peak source strength (measured in nano-Ampere-meters, nAm) were calculated from the largest point in the M50 and M100 scoring windows using in-house MatLab software. Slightly extended M50 and M100 latency ranges allow for the M50 and M100 latencies observed in younger children and for low frequency stimuli. In each hemisphere, a M50 and M100 response was considered present when observed for at least 2 of the 4 tones, with source strength significantly above background, and with a latency between 40-90 ms (M50) and 90-180 ms (M100). Chi-squared analyses examined the presence or absence of M50 and M100 activity as a function of gender, hemisphere, age, and diagnosis.

III. Results

Data from 102 subjects were analyzed: 36 children with ASD-LI, 22 children with ASD+LI, and 35 TD children.

M50. No gender differences were seen. As shown in Figure 1, an overall effect of hemisphere was observed, such that M50 was less frequently observed in the right (67.6%) than left hemisphere (85.3%, $\chi^2 = 8.83$, $p < 0.01$). Marginally significant diagnostic group differences were observed for the younger children in the left hemisphere ($\chi^2 = 5.03$, $p = 0.08$), and the older children in the right hemisphere ($\chi^2 = 5.23$, $p = 0.07$). Specifically, in the younger subjects, left-hemisphere group differences were due to more frequent M50 responses in the ASD-LI group (96%) than the ASD+LI (73.7%) and TD groups (75%). In the older subjects, right-hemisphere group differences were due to more frequent M50 responses for the ASD-LI (81.8%) and ASD+LI (100%) groups than the TD group (46.7%).

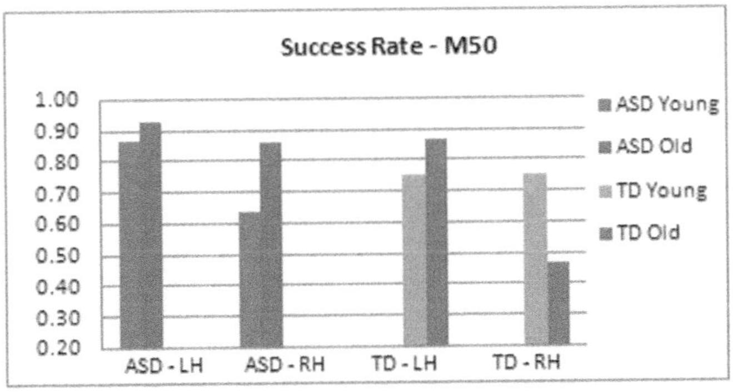

Fig. 1 M50 obtained less often in the right hemisphere, especially in older typically developing (11-15 yrs)

M100. No gender differences were seen. As shown in Figure 2, M100 responses were observed less frequently in the left (67.6%) than right hemisphere (87.3%; $\chi^2 = 11.2$, $p < 0.01$). Although the presence or absence of M100 did not differ as a function of age in the right hemisphere ($\chi^2 = 1.77$, *ns*), left-hemisphere M100 responses were observed less frequently in younger (60%) than older participants (84.4%; $\chi^2 = 5.96$, $p < 0.05$). Finally, diagnostic group differences were observed in the left hemisphere ($\chi^2 = 6.82$, $p < 0.05$), with M100 responses appearing less frequently for younger ASD+LI children (36.8%) than the younger ASD-LI (68%) and TD groups (75%). Diagnostic group differences were also observed in the right hemisphere ($\chi^2 = 8.98$, $p < 0.05$), with M100 responses appearing less frequently in the older children with ASD+LI (66.7%) than the older children with ASD-LI (100%) and the older TD group (100%).

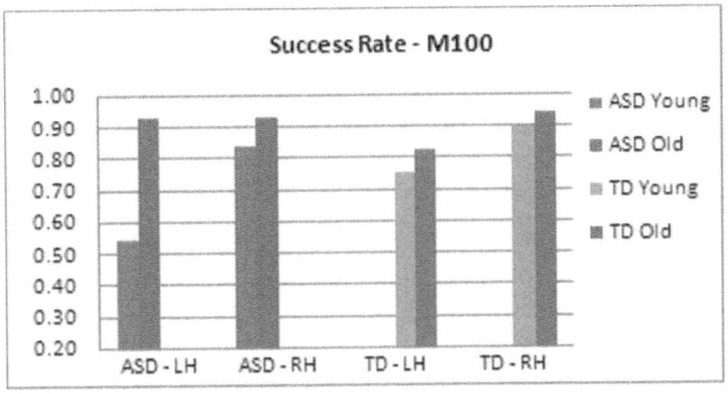

Fig. 2 M100 obtained less often in the left hemisphere, especially in younger children with ASD (6-10 yrs)

IV. Discussion

Present findings indicate that M50 and M100 auditory cortex measures can be successfully obtained in most children. For example, in the older TD sample, M50 measures were obtained in 87% of the subjects in the left hemisphere and 47% of the subjects in the right hemisphere. In the older TD sample, M100 measures were obtained in 87% of subjects in the left hemisphere and 100% of the subjects in the right hemisphere.

The relatively low right-hemisphere M50 success rate (47%) in the older TD sample was somewhat unexpected. It is likely, however, that difficultly recording this auditory component reflects a normal asymmetry in left- and right-hemisphere auditory cortex development, with the right-hemisphere auditory cortex developing at a faster rate than the left-hemisphere auditory cortex. For example, previous studies indicate more prominent M50 responses in younger than older children, and older children display a less prominent M50 and a more prominent M100 [5, 12]. As such, in the present study, the more mercurial right-hemisphere M50 as well as the very distinct right M100 in the older TD group may simply reflect a more mature auditory cortex in the older TD controls. The fact that M100 responses were observed more often in the right than left hemisphere also indicates a more slowly developing left hemisphere.

Deviations from this pattern of normal M50 and M100 development were observed in the patient group. For example, in the right hemisphere, M50 responses were observed more frequently in the children with ASD than the TD subjects, suggesting slowed or incomplete maturation of the right auditory cortex in ASD. In addition, the ASD+LI subjects had significantly fewer M100 responses bilaterally, again suggesting slowed or incomplete maturation of the auditory cortex.

In sum, M50 and M100 measures can be obtained in most subjects. In some cases, the lack of an auditory response appears to be of clinical significance, suggesting abnormal auditory cortex maturation. The most difficult responses to measure are the left-hemisphere M100 response in younger subjects, and left and right M100 responses in ASD subjects with language impairment.

REFERENCES

1. Roberts, T.P., et al., *MEG detection of delayed auditory evoked responses in autism spectrum disorders: towards an imaging biomarker for autism* Autism Research, in press.
2. Edgar, J.C., et al., *Psychophysiology in research on Psycopathology*, in *Handbook of Psychophysiology*, L.G. Tassinary, Cacioppo, J.T., Bernston, G.G., Editor. 2007, Cambridge University Press: New York.
3. Turetsky, B.I., et al., *Neurophysiological endophenotypes of schizophrenia: the viability of selected candidate measures.* Schizophr Bull, 2007. **33**(1): p. 69-94.
4. Roberts, T.P., et al., *Developmental correlation of diffusion anisotropy with auditory-evoked response.* Neuroreport, 2009. **20**(18): p. 1586-91.
5. Kotecha, R., et al., *Modeling the developmental patterns of auditory evoked magnetic fields in children.* PLoS One, 2009. **4**(3): p. e4811.
6. Lord, C., et al., *The autism diagnostic observation schedule-generic: a standard measure of social and communication deficits associated with the spectrum of autism.* J Autism Dev Disord, 2000. **30**(3): p. 205-23.
7. Rutter, M., et al., *SCQ: Social Communication Questionnaire.* 2003, Los Angeles, CA: Western Psychological Services.
8. Semel, E.M., et al., *Clinical Evaluation of Language Fundamentals (CELF-4).* 2003, San Antonio, TX: The Psychological Corporation.
9. Scherg, M., *Fundamentals of dipole source potential analysis*, in *Auditory evoked magnetic fields and electric potentials. Advances in audiology*, M.H.G.L.R. Gandori, Editor. 1990, Karger: Basel, Switzerland. p. 40-69.
10. Scherg, M. and P. Berg, *New concepts of brain source imaging and localization.* Electroencephalogr Clin Neurophysiol Suppl, 1996. **46**: p. 127-37.
11. Scherg, M. and D. von Cramon, *A new interpretation of the generators of BAEP waves I-V: results of a spatio-temporal dipole model.* Electroencephalogr Clin Neurophysiol, 1985. **62**(4): p. 290-9.
12. Oram Cardy, J.E., et al., *Prominence of M50 auditory evoked response over M100 in childhood and autism.* Neuroreport, 2004. **15**(12): p. 1867-70.

Early ERF Responses to Final Words in a Sentence Context during Reading in Individuals with Autism Spectrum Disorders: An MEG Study

Banu Ahtam, Sven Braeutigam, and Anthony J. Bailey

Department of Psychiatry, University of Oxford, Oxford OX3 7JX, United Kingdom

Abstract— **This study used magnetoencephalography (MEG) to study the neural basis of abnormalities in sentence context effects in individuals with an autism spectrum disorder (ASD) and typically developing (TD) individuals. Early ERF responses (~150ms) to three different types of final words [dominant homonyms (DH), subordinate homonyms (SH), unambiguous words (UW)] were analyzed. In both participant groups, strong responses were observed at word response latency over left-occipito and mid-parietal regions. There were no significant differences between the participant groups in any of the final words at 150ms. The neural responses, however, were significantly modulated by the final word category within groups. In TD, the response to UW was significantly different than both DH and SH predominantly over the left hemisphere. There were no differences between DH and SH conditions in TD individuals. On the other hand, in ASD, there was a significant difference between each paired final word condition, where significant responses were observed mainly over the right hemisphere. These results are consistent with the argument that early semantic access takes place as early as 150ms after the onset of a word. Moreover, these findings suggest atypical semantic representations for more and less frequent meanings of the homonyms in individuals with ASD.**

Keywords— **autism, language, context, homonym, MEG.**

I. Introduction

Autism is a neurodevelopmental disorder that is associated with impairments in reciprocal social interaction and communication with stereotyped, restricted, and repetitive patterns of interests and behaviors [1, 2]. It has been observed that individuals with autism show anomalous semantic processing during reading. Studies of homograph (i.e. words with several different meanings whose pronunciations may differ) reading demonstrate that individuals with autism are less likely to use sentence context spontaneously for the correct pronunciation of homographs. Instead, affected individuals show a tendency to pronounce the dominant meaning of the homograph regardless of sentence context [3, 4, 5].

Ambiguous words such as homonyms (i.e. words with several different meanings whose pronunciations and spellings are the same) are very useful for examining cognitive processes of lexical ambiguity in context because they have more than one semantic representation matching with a single physical representation [6]. To successfully understand a sentence, the reader must activate the correct meaning of each word and also integrate the meaning of individual words. For an ambiguous word the reader must use sentence context to activate the appropriate meaning and to suppress the contextually inappropriate meaning [7].

Recent EEG and MEG study results provide evidence that access to semantic information (i.e. mental lexicon) can take place as early as 100-200ms after visual word onset [8]. Differences in the early ERF responses were found between words and pseudowords [9], between content and function words [10], between nouns and verbs [11], between congruous and incongruous words [8, 12] and among words from different semantic categories [13, 14]. Moreover, semantic expectedness of a word in a sentence context is also found to affect ERP responses as early as 100-200ms [8, 15]. Semantic ambiguity effects were also shown in ERPs around 150-200ms [16].

This is the first study to look at early ERP/ERF responses to ambiguous words in ASD and TD groups to examine the semantic access process for dominant and subordinate homonyms as well as unambiguous words.

II. Methods

A. Participants

Fifty individuals participated in this experiment. Because of technical reasons the data from three individuals in each participant group had to be excluded. As a result, a total of forty-four data sets were analyzed. In the ASD group (n=22) there were two females and twenty males; in the TD group (n=22) there were three females and nineteen males. Participants were aged between 14 and 43 years (the second oldest participant was 33 years old). In each participant group there were 6 left-handed individuals (1 female). TD and ASD groups were matched on age, gender, handedness, as well as verbal and performance IQ. For participants with ASD, the clinical diagnosis was confirmed using the ADI-R

S. Supek and A. Sušac (Eds.): Advances in Biomagnetism – BIOMAG2010, IFMBE Proceedings 28, pp. 405–408, 2010.
www.springerlink.com

and the ADOS. Participants had normal or corrected-to-normal vision. This study has been approved by the local NHS (UK) Ethics Committee and all participants gave written informed consent.

B. Task

Participants read single English words back-projected sequentially (300ms word display, 400ms separation) onto a translucent screen to form 300 sentences. The length of the sentences was 7.5 ± 1.7 words. Each sentence ended in either a homonym (H) or an unambiguous word (UW). In the homonym condition, there were two types of sentences, one in which the homonym was used with its dominant meaning (DH), and the other in which the homonym was used with its subordinate meaning (SH). Each sentence was followed by a probe word. Participants were instructed to decide whether or not the probe word was semantically related to the meaning of the sentence, and to indicate their choice using one of two response pads after a visual cue. ERF responses to probe words are not reported in this paper.

Example sentences without the probe words:

The thief ran out of the bank. (dominant homonym)
The fisherman sat by the bank. (subordinate homonym)
The child ate a pizza. (unambiguous word)

C. Measurement

All measurements were taken at the Oxford Neurodevelopmental Magnetoencephalography Centre using a Neuromag-306 VectorViewTM system, providing a helmet-shaped array of 102 pairs of gradiometers. HPI measurement was taken at the start of each testing session, in order to locate the position of the participants' head under the helmet. It was ensured that all participants' heads were approximately in the same position with respect to the helmet and any head movements were monitored (<5mm). Electro-oculogram (EOG) and electro-cardiogram (ECG) responses were recorded during the experiment to control for artefacts. The data were sampled at 1000 Hz (0.03 to 330 Hz anti-alias filter).

D. Analysis

Prior to data analysis, each data set was filtered using Neuromag® MaxFilter 2.0 software. MaxFilter's *MaxST* function was used to reduce interferences coming from sources close to the brain such as eyes or dental work. MaxFilter was also used to transfer MEG signals of each participant to a reference set of head coordinates chosen separately for TD and ASD groups to ensure standardization within each participant group, thereby increasing the robustness of grand-mean estimates, and decreasing variation caused by subjects' different head positions.

For each participant, evoked responses were averaged separately for each of the three final word types at the long SOA: dominant homonym (DH), subordinate homonym (SH), unambiguous word (UW). Average signals were filtered using a 30 Hz low-pass filter (roll-off 0.5Hz).

The output of each pair of sensors is most sensitive to the tangential current flow in the region below the detectors. The local root-mean-square (RMS) signal summed over each detector site is a measure of the strength of the current. Differences in the evoked responses to different final word categories were sought using a time-dependent, chi-square-based measure $P(t)$ [17], which takes into account the data from all sensors and does not require *a priori* identification of peaks in the evoked responses.

$$P(t) = probability(\chi^2) \quad (\chi^2) = -2\sum_{i=1}^{N} \ln[f_i(t)]$$

N denotes the number of channels. For each channel, $f_i(t)$ denotes a Friedman (non-parametric) analysis of variance used to compare the responses to DH, SH, and UW final word conditions in each sentence. Then, $P(t)$ describes significance across the whole head of the channel differences (as a function of time). For each such interval with $P(t) < 0.01$, the set of all values f_i provided the spatial distribution of the significance of the differences between evoked response: these distributions were used to identify significant regions at the group level. For each identified significant latency, spatial maps of local RMS-signals (amplitude maps) were calculated.

III. Results

All participants completed the task with few errors and were comfortable using the keypads.

A. Typically Developing Participants

According to the global rms-curves, all final word conditions peaked at 150ms in the TD group. The strongest amplitude was for the DH condition followed by UW and SH (Table 1). Local rms-maps showed activity over posterior parietal and left occipito-temporal regions (Fig 1).

Table 1 Values taken from the global rms-curves. Peak latency (in milliseconds) and peak amplitude strengths [fT/cm] for all three final word conditions within the word interval in the TD group

	TD (n=22)		
	DH	SH	UW
Latency	147.87	147.71	151.01
Amplitude	12.19	11.13	11.67

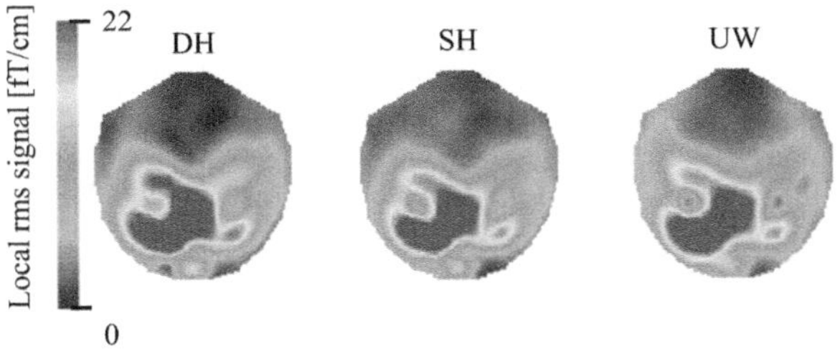

Fig. 1 Local maps of activity (rms-maps) at 150ms for all three final word conditions in the TD group

The Friedman analysis revealed significant differences among the final words between 150-160ms. *Post hoc* pairwise Wilcoxon analyses were conducted to follow up this result (Fig 2). Significant differences were observed over the left occipito-temporal regions between DH and UW and also between SH and UW conditions. Between SH and UW, significant differences were also observed over left anterior temporal cortex. There were no significant differences between DH and SH conditions.

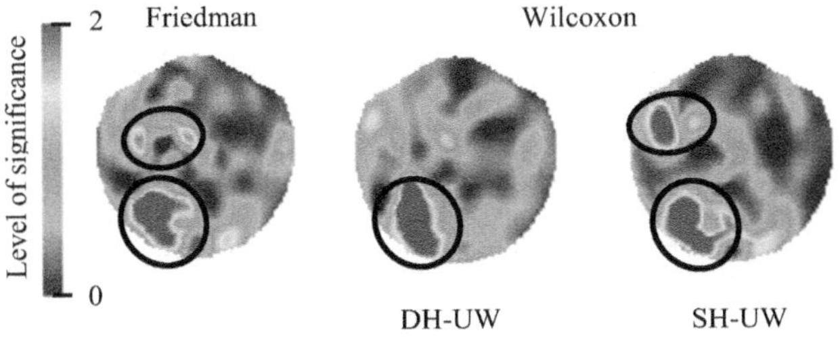

Fig. 2 Probability maps of Friedman and Wilcoxon analyses at 150ms in the TD group. Wilcoxon probability maps corresponding to significant differences between DH and UW and between SH and UW final word conditions. Probability map for Wilcoxon analysis for the difference between DH and SH final word conditions is not presented due to lack of significance

B. Participants with ASD

The word response peaked at 160ms in the ASD group for all final words. The strongest response was for the DH condition, followed by UW and SH (Table 2). The local rms-maps showed the main activity over left occipito-parietal and reduced activity over left posterior temporal regions for all three conditions (Fig 3).

Table 2 Values taken from the global rms-curves. Peak latency (in milliseconds) and peak amplitude strengths [fT/cm] for all three final word conditions within the word interval in the ASD group

	ASD (n=22)		
	DH	SH	UW
Latency	157.7	158.44	158.44
Amplitude	9.34	8.45	8.76

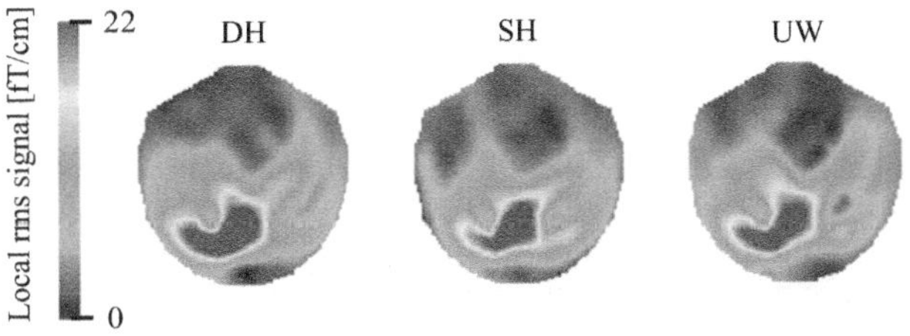

Fig. 3 Local maps of activity (rms-maps) at 150ms for all three final word conditions in the ASD group

The Friedman probability spectrum indicated significant differences among the three final words between 150-170ms. According to *post hoc* Wilcoxon pairwise analyses there were significant differences between each paired final word conditions (Fig 4). The distribution of significance between DH and SH and between SH and UW were mainly over right occipito-temporal regions. The significant difference between DH and UW was over the right superior temporal-posterior frontal cortices.

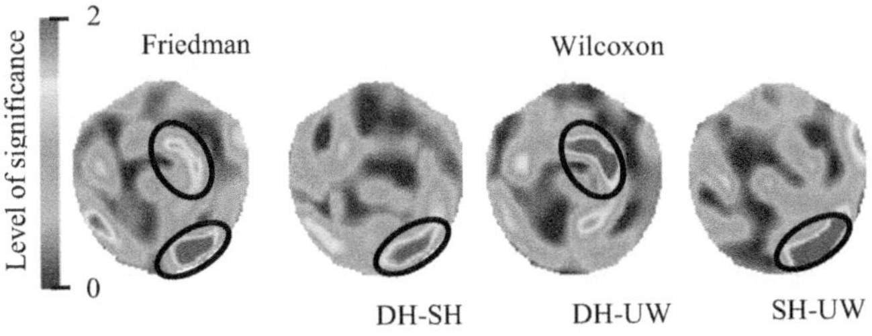

Fig. 4 Probability maps of Friedman and Wilcoxon analyses at 150ms in the ASD group. Wilcoxon probability maps corresponding to significant differences between DH and SH, between DH and UW, and between SH and UW final word conditions

IV. Discussion

The data for early ERF responses to final words are in keeping with suggestions that semantic information is accessed as early as 150-180ms after visual presentation of a word [8, 12]. Moreover, the difference between the three final word conditions provides evidence that sentence context has already influenced the meaning activation of final words by 150ms [8]. The fact that the ordering of the amplitude strengths of the final words was the same in both TD and ASD suggests that both participant groups showed a similar degree of sensitivity to context. Nevertheless, individuals with ASD may activate atypical neural networks during early stages of resolution of semantic ambiguity.

It is currently unresolved whether such anomalous processes imply a general delay and/or under-activation when accessing the less frequent meaning of a word within a given context.

V. Conclusions

This study provides further evidence for distinguishable stages associated with semantic processing well before the well-documented N400 responses. These early latency processes may be altered in certain disorders linked anomalies in language semantics and pragmatics.

Acknowledgments

This work was supported by the Clarendon Bursaries and the University of Oxford.

References

1. American Psychiatric Association (APA) (1994). Diagnostic and Statistical Manual of Mental Disorders (DSM-IV) (4th ed.). Washington, DC: American Psychiatric Association.
2. World Health Organization (WHO) (1992). The ICD-10 classification of mental and behavioral disorders: Clinical descriptions and diagnostic guidelines. Geneva.
3. Frith, U. & Snowling, M. (1983). Reading for meaning and reading for sound in autistic and dyslexic children. British Journal of Developmental Psychology, 1, 329-342.
4. Happe, F. (1997). Central coherence and theory of mind in autism: reading homographs in context. British Journal of Developmental Psychology, 15, 1-12.
5. Jolliffe, T. & Baron-Cohen, S. (1999). A test of central coherence theory: linguistic processing in high-functioning adults with autism or Asperger's syndrome: is local coherence impaired? Cognition, 71(2), 149-185.
6. Van Petten, C., & Kutas, M. (1987). Ambiguous words in context: An event-related potential analysis of the time course of meaning activation. Journal of Memory & Language, 26, 188-208.
7. Gernsbacher, M. A., & Faust, M. E. (1991). The mechanism of suppression: a component of general comprehension skill. J Exp Psychol Learn Mem Cogn, 17(2), 245-262.
8. Penolazzi, B., Hauk, O., & Pulvermuller, F. (2007). Early semantic context integration and lexical access as revealed by event-related brain potentials. Biol Psychol, 74(3), 374-388.
9. Hauk, O., Davis, M.H., Ford, M., Pulvermuller, F. & Marslen-Wilson, W.D. (2006) The Time Course Of Visual Word-Recognition As Revealed By Linear Regression Analysis of ERP Data, Neuroimage, 30(4), 1383-1400.
10. Pulvermuller, F., Lutzenberger, W., & Birbaumer, N. (1995). Electrocortical distinction of vocabulary types. Electroencephalogr Clin Neurophysiol, 94(5), 357-370.
11. Pulvermuller, F., Lutzenberger, W., & Preissl, H. (1999). Nouns and verbs in the intact brain: evidence from event-related potentials and high-frequency cortical responses. Cereb Cortex, 9(5), 497-506.
12. Shtyrov, Y., & Pulvermuller, F. (2007). Early MEG activation dynamics in the left temporal and inferior frontal cortex reflect semantic context integration. J Cogn Neurosci, 19(10), 1633-1642.
13. Hauk, O., & Pulvermuller, F. (2004). Effects of word length and frequency on the human event-related potential. Clin Neurophysiol, 115(5), 1090-1103.
14. Pulvermuller, F., Assadollahi, R., & Elbert, T. (2001). Neuromagnetic evidence for early semantic access in word recognition. Eur J Neurosci, 13(1), 201-205.
15. Sereno, S. C., Brewer, C. C., & O'Donnell, P. J. (2003). Context effects in word recognition: evidence for early interactive processing. Psychol Sci, 14(4), 328-333.
16. Federmeier, K. D., Segal, J. B., Lombrozo, T., & Kutas, M. (2000). Brain responses to nouns, verbs and class-ambiguous words in context. Brain, 123 Pt 12, 2552-2566.
17. Braeutigam, Rose, P., Swithenby, S., and Ambler, T. (2004). The distributed neuronal systems supporting choice-making in real-life situations: differences between men and women when choosing groceries detected using magnetoencephalography. European Journal of Neuroscience, 20, 1, 293-302.

The address of the corresponding author:

Author: Banu Ahtam
Institute: University of Oxford
Street: Warneford Hospital, Headington
City: Oxford OX3 7JX
Country: United Kingdom
Email: banu.ahtam@psych.ox.ac.uk

MEG Abnormalities in Creutzfeldt-Jakob Disease, a Case Report

Eero Pekkonen[1], Juha Wilenius[2], Jukka Lyytinen[1], Anders Paetau[3], Maarit Palomäki[4], and Jyrki P. Mäkelä[2]

[1] Department of Neurology, University of Helsinki, Helsinki, Finland
[2] BioMag laboratory/HUSLAB, Helsinki, Finland
[3] Department of Pathology, University of Helsinki, Helsinki, Finland
[4] Department of Radiology, University of Helsinki, Helsinki, Finland

Abstract— **Creutzfeldt-Jakob Disease (CJD) is a rare neurodegenerative disease, which belongs to a group of spongiform encephalopathies termed prion diseases.**

Main symptoms of CJD are progressive dementia, pyramidal and extrapyramidal symptoms, and myoclonus. Currently there is no effective therapy for CJD. Definite diagnosis is based on autopsy findings.

Here we report a male patient with a rapidly progressive memory disorder, ataxia and myoclonus. EEG showed slow wave activity, and head MRI revealed increased basal ganglia and left-dominant increase in cortical signal intensity. In cerebrospinal fluid (CSF) 14-3-3 test was positive. The autopsy confirmed the sporadic CJD.

A 306-channel MEG measurement was performed to the patient. Spontaneous brain activity and auditory evoked fields (AEF) were recorded.

Both hemispheres had slow wave activity and periodic sharp waves, activity being more pronounced in the left, more damaged, hemisphere. The source of auditory N100m were more lateral in the left than in the right hemisphere. P200m was larger and more posterior in the left hemisphere compared with the right hemisphere.

This is the first MEG-recording of a patient with sporadic CJD. The more damaged left hemisphere in MRI had more pronounced slow and sharp waves in MEG. The sources of auditory responses and amplitude of later P200m deflection were different between the more and less damaged hemispheres. MEG appears to be suitable to investigate and locate brain activity in prion diseases.

Keywords— **Creutzfeldt-Jakob Disease, MEG, spontaneous activity, auditory N100m and P200m, MRI.**

I. INTRODUCTION

Creutzfeldt-Jakob Disease (CJD) is a rare neurodegenerative disease with incidence of approximately one patient/million people/year. CJD belongs to a group of spongiform encephalopathies termed prion diseases, which appear to be caused by abnormal metabolism of cellular prion protein. The exact function of prion protein, which is abundant in neurons, is not known [1].

Although early symptoms of CJD may appear insidiously, the disease finally progresses rapidly. Main symptoms are gradual dementia, pyramidal and extrapyramidal symptoms, ataxia, and myoclonus. Few patients have peripheral neuropathy, aphasia or seizures [1]. Death occurs approximately within one year after the onset of symptoms. Currently there is no effective therapy available for CJD.

Clinical diagnosis of CJD is based on typical symptoms, generalized periodic triphasic sharp wave complexes in EEG, cortical and basal ganglia abnormalities detected in MRI, and appearance of 14-3-3 protein in CSF. 14-3-3- is a neuronal death signaling protein in CSF, and it is usually elevated in CJD. 14-3-3 is not, however, specific to CJD [1]. Definite diagnosis is based on neuropathological examination [1].

No MEG study on CJD has been reported. Here we describe MEG-findings of one male patient, who had sporadic CJD.

II. MATERIAL AND METHODS

A. Patient

Young male patient (33 years old) was admitted to our hospital because of progressive forgetfulness and motor clumsiness that had appeared within one year. Regular alcohol usage was reported.

On clinical examination patient had clear amnesia and ataxia. Head MRI examination was normal. Electromyography (EMG) revealed peripheral polyneuropathy. Ethanol related amnesia and polyneuropathy were diagnosed, and intravenous thiamine substitution was introduced. Patient was moved to the district hospital for rehabilitation.

Despite treatment, patient's clinical symptoms progressed rapidly, and he was readmitted to the university hospital. Clinical examination revealed severe amnesia, desorientation, and occasional limb myoclonus. New head MRI displayed abnormal intensity in the basal ganglia, thalami and mainly in the left frontotemporal cortex (Fig 1).

S. Supek and A. Sušac (Eds.): Advances in Biomagnetism – BIOMAG2010, IFMBE Proceedings 28, pp. 409–412, 2010.
www.springerlink.com

Increased level of Tau-protein, and positive 14-3-3 was detected in the CSF supporting the diagnosis of CJD. In EEG, general slow wave activity was prominent in the left hemisphere in line with the MRI. Periodic triphasic sharp wave complexes, which are typical for CJD, were not detected on serial EEGs.

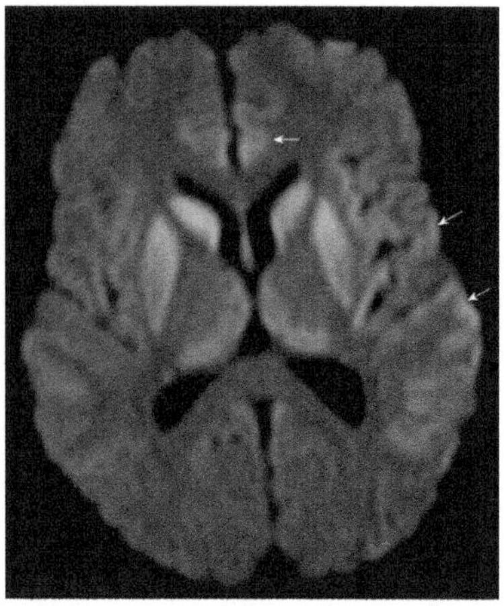

Fig. 1 Head MRI

Diffusion weighted imaging revealed abnormal hyperintensity in the thalami and striatum bilaterally. Cortical involvement was mainly in the left frontotemporal cortex (arrows).

Clinical diagnosis of CJD was based on patient's rapidly progressed symptoms and combined findings on MRI, EEG and CSF. Only one MEG recording was performed before the patient suddenly died from probable pulmonary embolism. Neuropathological examination confirmed sporadic CJD (Figs 2a and 2b).

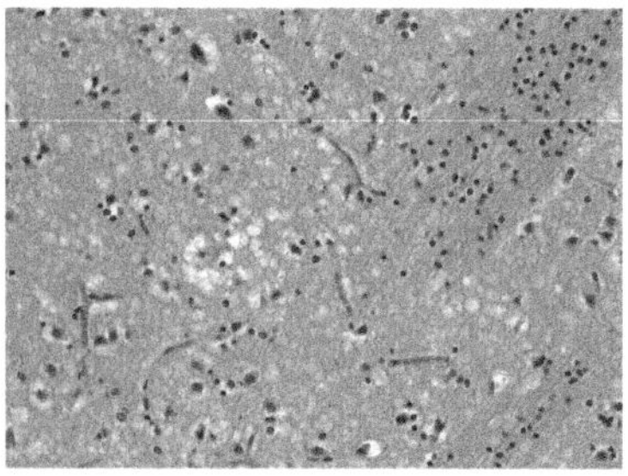

Fig. 2a Putamen

The texture of the neuropil in putamen is filled with small round vacuoles, seen also clustering focally. At the right upper corner a typical myelinated putaminopallidal stria is seen. Paraffin section, He, original magnification x200.

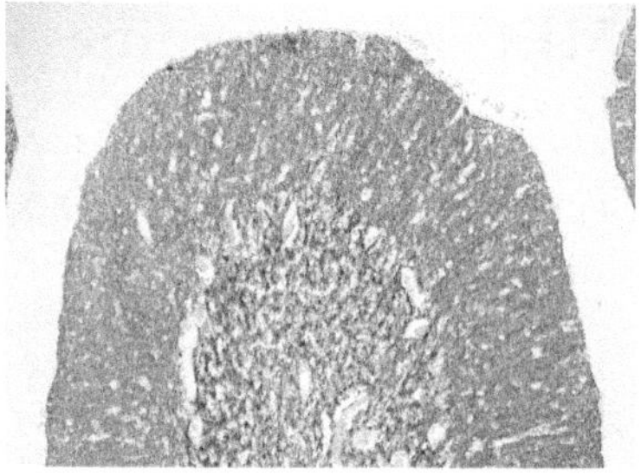

Fig. 2b Cerebellum

Cerebellar cortical folium shows dense granular mainly synaptic brown PrP-immunoreactivity both in the molecular and granular layers. Purkinje cells are not highlighted. Paraffin section, PrP immunohistochemistry (monoclonal 12F10 antibodies), original magnification x100.

B. MEG

The MEG was recorded with a 306-channel Elekta Vectorview^{TM}. The patient was in supine position and vertical and horizontal EOG were recorded. Head position information was obtained by using standard HPI-coils, and head position was registered continuously during the whole measurement. Spontaneous brain activity was registered for 16 minutes. Binaural auditory stimuli (1 kHz, 50 ms duration and 1.5-2.5 s random ISI) were presented during the first seven minutes of the recording (n=207). Median nerve stimulation was not performed because of muscle jerks, which were partly provoked by external stimuli.

Before the analysis, the data were filtered by tSSS with movement compensation (Elekta Maxfilter). The spontaneous brain activity abnormalities were located by visual inspection. AEF sources were found by equivalent current dipole (ECD) analysis using standard Elekta analysis software.

III. Results

MEG revealed general slow wave activity over both hemispheres that was more prominent in the left, more affected hemisphere (Fig 3).

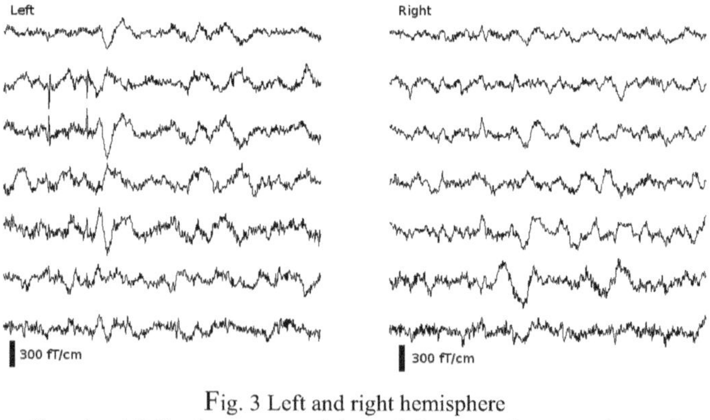

Fig. 3 Left and right hemisphere

Prominent 3 Hz slow wave activity and sharp transients are observed on seven selected MEG channels over the left and right temporal region at the same time period. Length of the data segments are 3.3 s.

Clear auditory P50m, N100m and P200m responses over both hemispheres were detected, the P50m and P200m being larger in amplitude in the more damaged left hemisphere than in the right hemisphere (Figs. 4a and 4b).

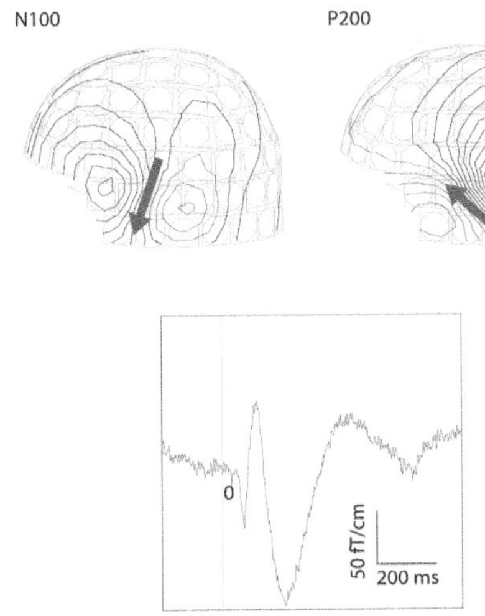

Fig. 4a Left hemisphere

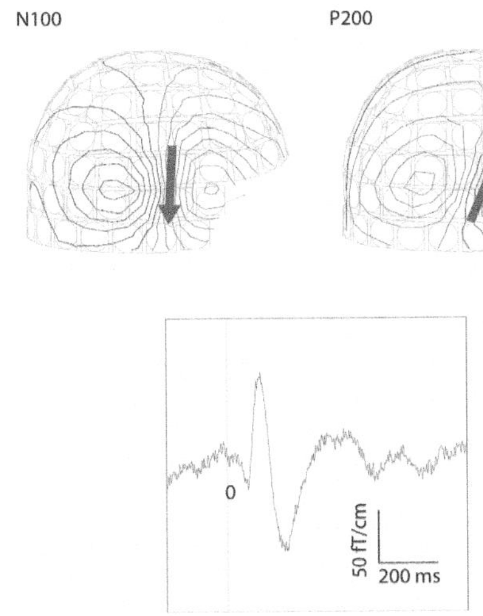

Fig. 4b Right hemisphere
Field maps of AEFs over the left (3a) and right (3b) hemispheres. Estimated dipoles are shown for N100m and P200m (top left and right). Below and AEF from sensor displaying maximum activity in each hemisphere.

In ECD analysis, the location of the N100m source was more lateral in the more damaged left than right hemisphere (Fig. 5).

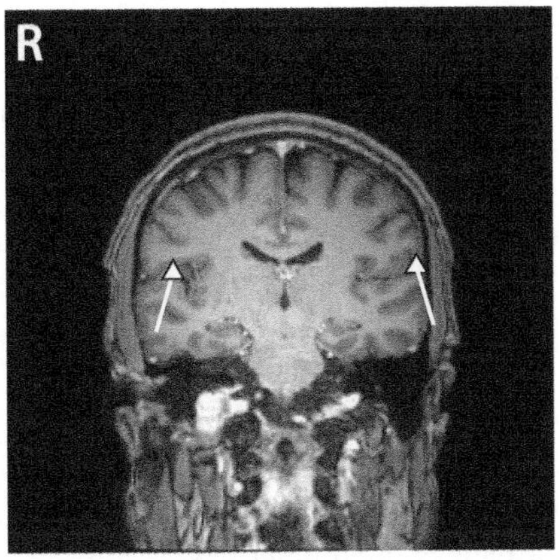

Fig. 5 N100m source modeling
Source location of N100m overlaid on the patient's coronal MR-image. R indicates right hemisphere.

The location of the P200m source was more posterior in the left than in the right hemisphere (Fig 6).

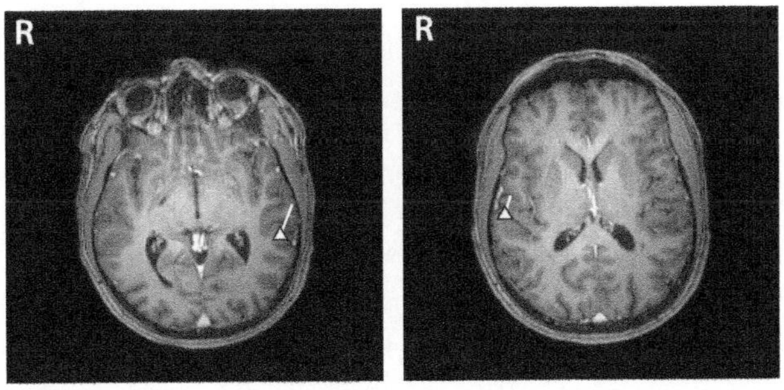

Fig. 6 P200m source modeling
Source location of P200m on axial MRI slices.

IV. Discussion

This is the first MEG-recording on a patient with neuropathologically verified sporadic CJD. MEG revealed abnormal slow wave activity that was more prominent in the more affected left hemisphere. This finding correlated with the cortical abnormality detected by MRI.

Previous EEG studies have shown either normal [2] or deteriorated [3] brainstem auditory evoked potentials in CJD. Present results suggest that cortical auditory processing is not diminished in CJD. In contrast, there appears to be excitation or disinhibition of the P50m and P200m generators.

Spontaneous MEG of this patient did not have typical 0.5-2 Hz slow wave activity observed in CJD [1]. According to the literature, typical spontaneous activity is not detected from all patients and it is not present constantly [1]. Serial MEG-recording could have revealed additional information about effects of disease progression on brain function.

Our patient was quite young and had thalami hyperintensity in MRI raising suspicion of variant CJD. Neuropathological examination verified, however, sporadic CJD.

V. Conclusion

MEG is suitable to measure cortical activity in CJD. Follow-up studies of CJD are needed to evaluate how disease progression affects cortical functions.

References

1. Mastrianni JA (2006) Prion Disease: Transmissible Spongiform Encephalopathies. In: Neurological Therapeutics Principles and Practice. The future of modern genomics. Informa Healthcare, Oxon.
2. Aguglia U, Farnarier G, Regis H et al. (1990) Sensory evoked potentials in Creutzfeldt-Jakob disease Eur Neurol 30:157-61
3. Pollak L, Klein C, Giladi R et al (1996) Progressive deterioration of brainstem auditory evoked potentials in Creutzfeldt-Jakob disease: clinical and electroencephalographic correlation Clin Electroencephal 27:95-99

The address of the corresponding author:

Author: Eero Pekkonen
Institute: Department of Neurology, University of Helsinki
Street: Haartmaninkatu 4
City: Helsinki
Country: Finland
Email: eero.pekkonen@hus.fi

Nondipolar Content of Cardiac Magnetic Field Maps in Patients after ST Elevation Myocardial Infarction

Peter Van Leeuwen[1], Birgit Hailer[2], Gregor Eiling[2], and Dietrich H.W. Grönemeyer[1]

[1] Dept. Radiology and Microtherapy, University Witten/Herdecke, Bochum, Germany
[2] Dept. of Internal Medicine, Philippusstift, Essen, Germany

Abstract— **Electrophysiological changes occurring in the context of coronary artery disease (CAD) have been investigated using magnetocardiography. Various approaches have been developed and studied which quantify different aspects of the magnetic signals and reconstructed magnetic field maps (MFM). In this work, we compare the coherence of MFM by determining their relative nondipolar content on the basis of the Karhunen-Loeve transformation in healthy subjects and patients with acute ST elevation myocardial infarction (STEMI). Magnetocardiograms were obtained for 3 minutes in 31 healthy subjects and 75 STEMI patients using a 61 channel biomagnetometer. For each subject, the signals of ~50 beats of similar duration were averaged. Integral maps were calculated for 6 different segments of the QRST interval. For each segment 12 eigenmaps were constructed using Karhunen-Loeve transform. Nondipolar content was determined as percentage of the contribution of all eigenvectors beyond the third. The results showed that the nondipolar content of the MFM was generally increased in the patient group. Discriminatory power of each segment was assessed by the area under the curve (AUC) of the receiver operating characteristic. Results showed that the AUC was highest during repolarization: ST segment 74%, T wave 78% , terminal T wave 77% (p<0.001). Depolarization segments discriminated poorly (QRS, R peak: ~50%). Furthermore, maps with dipolar content (maps 1-3) permitted excellent discrimination with AUCs > 90% for the T wave (p<0.001). The results indicate that the nondipolar content of MFM in patients with acute symptoms of CAD increases particularly during repolarization. However T wave coherence in the dipolar maps differed even more so between healthy subjects and patients with myocardial infarction. Principal component analysis of cardiac MFM using the Karhunen-Loeve transform may thus be helpful in the surveillance of patients with CAD.**

Keywords— **Magnetic field maps, coronary artery disease, myocardial infarction, nondipolar content, Karhunen-Loeve transform.**

I. Introduction

Coronary artery disease (CAD) induces changes in the electrophysiological processes in the heart. These changes are routinely clinically documented using electrocardiography (ECG) but they may also assessed on the basis of the magnetic component of the electrical activity [1]. Magnetocardiography (MCG) permits the acquisition of cardiac magnetic signals and various approaches have been developed which quantify different aspects of these signals. Apart from the examination of signal amplitude and duration during ventricular de- and repolarization [2] [3], magnetic field maps (MFM) can be reconstructed and analyzed. One of the most widely examined MFM characteristics is its orientation [4] which has been shown to be altered under ischemic conditions. The overall spatial structure of cardiac MFM has also been examined in terms of nondipolar content in patients with relevant stenosis of the coronary arteries [5] and in post myocardial infraction (MI) patients with and without ventricular tachycardia (VT) [6].

The principal spatial patterns present in MFM obtained during depolarization and repolarization can be identified on the basis of the first 12 eigenvectors of these maps obtained by singular value decomposition [7]. Nondipolar content is estimated by examining the relative contribution of the 4th - 12th maps to the total information content of the first 12 maps. This approach has been successfully applied to ECG and MCG QRST maps particularly in the context of risk analysis [6] [7]. In this work, we compare the coherence of MFM in selected QRST subintervals by determining both the relative dipolar and nondipolar content of these subintervals on the basis of the Karhunen-Loeve transformation in healthy subjects and patients with acute ST elevation myocardial infarction (STEMI).

II. Methods

A. Subjects

Seventy-five patients (aged 59.5±11.9 years, 60 male) who had suffered acute STEMI and received successful interventional therapy were included. Anterior infarction was present in 37, posterior in 38 patients; 31 had single vessel, 23 2-vessel and 21 3-vessel disease. The right coronary artery was involved in 53, left anterior descending in 56 and circumflex in 31 patients. Ejection fraction (EF) was 49.2%±10.0%. Wall motion disturbances were present in 59. The normal group consisted of 31 healthy subjects (aged 55.3±9.4 years, 22 male) with no history of cardiovascular disease and normal resting ECG and echocardiograms.

S. Supek and A. Sušac (Eds.): Advances in Biomagnetism – BIOMAG2010, IFMBE Proceedings 28, pp. 413–416, 2010.
www.springerlink.com

B. Data Acquisition

MCG were obtained in all subjects using a 61 channel biomagnetometer (Magnes 1300C, 4D Neuroimaging, San Diego), in the patients 5.7±2.4 days after infarction and immediate successful intervention. The sensor with a coverage of ca. 800 cm^2 was placed over the frontal thorax. Data were recorded for 3 minutes at a sampling rate of 1 kHz with a 200 Hz low pass filter. To reduce the effects of external noise, data were registered inside a standard 3-layered magnetically shielded room (Vakuumschmelze, Hanau).

C. Data Analysis

For each subject, all cardiac beats were identified on the basis of correlation to a representative QRS template from the subject's data. Beats within ±10 ms from the modal RR interval were selected for averaging (~50 beats/subject). In the signal averaged data, Q onset, R peak, QRS end, T onset, T apex and T end were determined. The following intervals were extracted for analysis: QRS, ST-segment, T wave, terminal T wave (Tmax-Tend) as well as 20 ms around R peak and 40 ms around Tmax. Singular value decomposition was performed for each of these intervals on the basis of the Karhunen-Loeve transform (KLT). Nondipolar content was determined as percentage of the contribution of the eigenvectors 4 to 12. Beside the nondipolar content, we also examined the relative proportion of dipolarity as found in maps 1-3.

We compared the %-content between the normal and STEMI groups on the basis of descriptive statistics and the Mann-Whitney U-test (2-sided). Further, the capacity of the maps to distinguish between the two groups was examined on the basis of the receiver operating characteristic (ROC) where the area under the curve (AUC) was used as a measure for sensitivity/specificity.

III. Results

For all subjects and intervals examined, most of the information content was found in eigenmaps 1, 2 and 3. This is demonstrated in Fig. 1, which shows the first 12 maps of the T wave for a healthy subject and a STEMI patient with 3-vessel disease. In both subjects, maps 1-3 display a dipolar structure and contain >99% of the information. The remaining maps are increasingly nondipolar and unstructured but, in the patient, the content is higher than in the healthy subject (0.48% vs. 0.03%).

Overall, the nondipolar content (eigenvectors 4-12) was similar for QRS, R peak and Tmax but significantly higher in the STEMI group for ST, T wave and terminal T wave (Table 1). Based on the AUC (see Fig. 2), the discriminatory power was best for the T wave (79%) and terminal T wave (80%) and somewhat less for the ST segment (73%).

Furthermore, examination and comparison of the eigenvectors of the dipolar maps 1, 2 and 3 showed differences between the subject groups. In particular there was a shift of content from map 1 to map 2 in the STEMI group. Again, this was clearest in the intervals ST, T wave and terminal T wave (Fig. 3): in the healthy subjects the relative information content was high in the first map and dropped

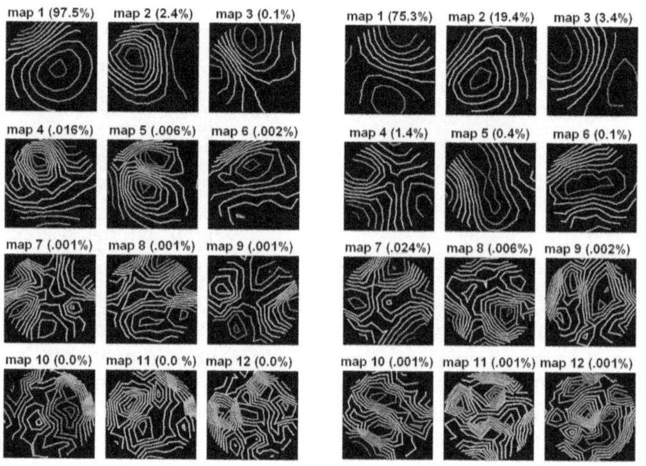

Fig. 1 KLT eigenmaps of the T wave
The first 12 maps of a healthy subject (left) and a STEMI patient (right). The % information content is given for each map. Maps 1-3 show dipolar spatial features whereas the maps 4-12 are increasingly nondipolar and unstructured. In maps 4-12, the STEMI patient reveals more content.

Table 1 Nondipolar content of the MFM for the different intervals. N = healthy group; STEMI = patient group, Values given in quantiles, p-value based on a Mann-Whitney U-test (p<.05 in bold format)

Interval	group	min	1.Q	median	3.Q	max	p
QRS	N	0.19%	0.68%	0.98%	1.38%	2.23%	0.709
	STEMI	0.14%	0.52%	0.97%	1.41%	4.25%	
Rpeak	N	0.06%	0.12%	0.18%	0.34%	1.34%	0.810
	STEMI	0.02%	0.12%	0.20%	0.31%	1.97%	
ST	N	0.02%	0.04%	0.10%	0.17%	0.53%	**0.000**
	STEMI	0.00%	0.13%	0.21%	0.47%	2.05%	
T	N	0.01%	0.01%	0.03%	0.04%	0.07%	**0.000**
	STEMI	0.00%	0.03%	0.07%	0.12%	1.92%	
Tmax	N	0.02%	0.05%	0.10%	0.15%	0.70%	0.680
	STEMI	0.01%	0.05%	0.11%	0.25%	2.36%	
terminal	N	0.00%	0.00%	0.01%	0.01%	0.02%	**0.000**
T	STEMI	0.00%	0.01%	0.02%	0.04%	0.36%	

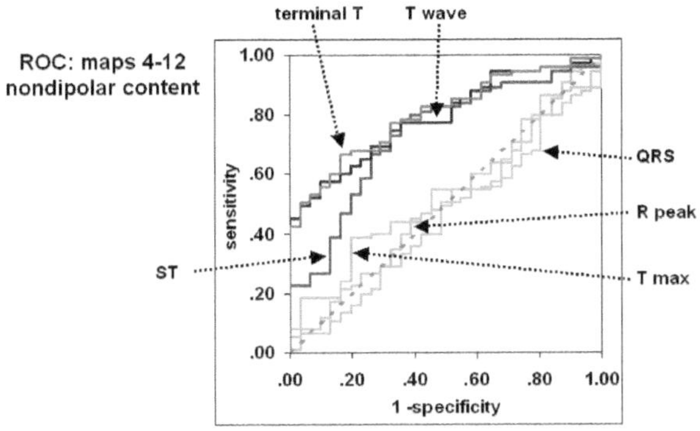

Fig. 2 ROC curves for nondipolar content for the different intervals. (gray curves: no statistical significance)

dramatically in the second (and third) map. In the STEMI patients the content was not as high in map 1 and not as low in map 2 ($p<0.001$ for ST segment, T wave and terminal T wave).

The ROC analysis showed that the discrimination between group was better using the dipolar maps than the nondipolar: the AUC for map 1 were: T wave 90%, terminal T wave 78%, ST segment 83%; for map 2: T wave 90%, terminal T wave 77%, ST segment 81% (all $p<0.001$). The values for QRS, R peak and Tmax were <60% and not significant (see also Fig. 4).

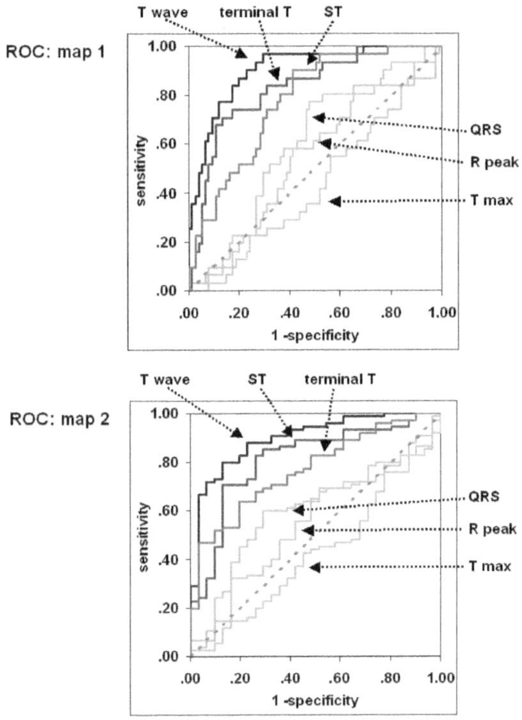

Fig. 4 ROC curves for maps 1 & 2 for the different intervals. (gray curves: no statistical significance)

IV. Discussion

The results indicate that the information content of MFM in patients who have recently suffered myocardial infarction is changed compared to that of healthy subjects. Overall there seems to be a shift away from coherent spatial structures to more disorganized patterns. The comparison between groups is appropriate as the subjects in the healthy and patient groups were relatively well-matched with respect to age and gender.

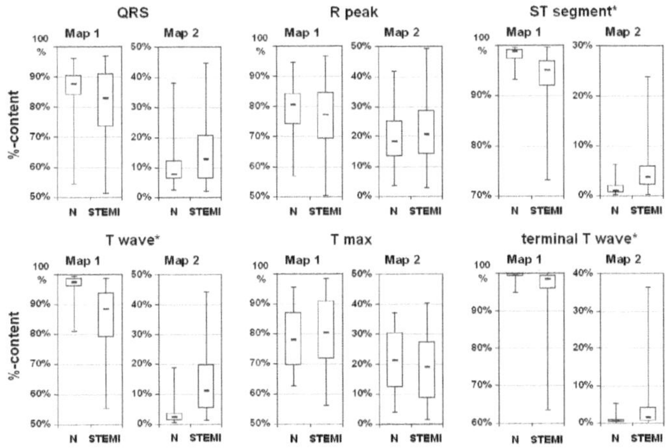

Fig. 3 Boxplots of %-content of map 1 and map 2 for the different intervals. * = $p<0.001$ for comparisons between N and STEMI, abbreviations as in Table 1

Nondipolarity has been presumed to be captured well by the KLT eigenvectors beyond the third [7]. This was confirmed in our examination of MFM in specific intervals between Q onset and the end of the T wave, with maps 1-3 showing dipolar patterns and maps 4-12 generally being multipolar. In the STEMI patients the presence of increasingly complex patterns was documented by more nondipolarity in particular in the MFM during repolarization. This will most likely have been due to myocardial damage associated with repolarization disturbances in the context of acute myocardial infarction. The fact that the depolarization intervals did not show such changes to the same degree was surprising. However, STEMI subjects with wall motion disturbances had significantly more QRS nondipolar content (maps 4-12) than patients without wall motion disturbances (1.28±0.96% vs. 0.65±38%, p=0.002; cf. Table 1). This indicates that subjects with preserved contraction of the ventricles as a result of a successful and early interventional therapy displayed relatively normal information content of the QRS map whereas less structured MFM will be associated with hypo- or akinesia of portions of the ventricular walls as a result of myocardial scar.

We note with interest that examination of the maps singly – in particular maps 1, 2 and 3 – indicated that there were substantial changes even in the more structured, dipolar maps. Whereas the MFM of the healthy subjects had a very high proportion of information in the first map, this was reduced in the maps of the patients, again primarily during repolarization. The loss of information in the patients' first map was compensated by an increase of content in their second map. This complementarity can be seen for single intervals in the inverse relationship of sensitivity and specificity in the ROC curves for map 1 and map 2 (Fig. 4). These results indicate that, in the patients, not just the nondipolar but also the dipolar content of the MFM was affected by myocardial infarction. In fact, the high AUC values, in particular for the T wave, suggest that the shift of information within the dipolar maps may be more relevant to the pathological state than that from dipolar to nondipolar. E.g. at a threshold of 95% information content for map 1 of the T wave, we obtained a sensitivity of 87% with a specificity of 81%.

The loss of spatial structures during repolarization due to myocardial damage induced by infarction can thus be adequately quantified on the basis the loss of information content of the first eigenvector map of T wave activity. This measure may be useful in patients who have suffered myocardial infarction in assessing success of treatment in the acute phase immediately following the event as well as the myocardial recovery in the follow-up period. Whether it has value as a predictor of restenosis or further possible events remains to be seen. Furthermore it might be of interest to examine dipolar content of repolarization in the broader context of CAD, i.e. in patients who have not had myocardial infarction but in whom ischemia and stenosis of the coronary arteries are present.

V. Conclusions

The results indicate that the nondipolar content of MFM in patients with acute symptoms of CAD increases particularly during repolarization. However T wave coherence in the dipolar maps differed even more so between healthy subjects and patients with myocardial infarction. Principal component analysis of cardiac MFM using the Karhunen-Loeve transform may thus be helpful in the surveillance of patients with CAD.

Acknowledgment

The authors would like to thank Silke Lange for helpful discussions of the methods and results.

References

1. Stroink G, Hailer B, Van Leeuwen P (2006) Cardiomagnetism. In: Andrä W, Nowak H, eds., Magnetism in Medicine. Wiley-VCH, Weinheim, pp 164–209
2. Hänninen H, Takala P, Mäkijärvi M et al. (2001) Recording locations in multichannel magnetocardiography and body surface potential mapping sensitive for regional exercise-induced myocardial ischemia. Basic Res Cardiol 96: 405–414 DOI 10.1007/s003950170049
3. Hailer B, Van Leeuwen P, Lange S et al. (1998) Spatial dispersion of the magnetocardiographically determined QT intervals and its components in the identification of patients at risk for arrhythmia after myocardial infarction. Ann Noninvasive Electrocardiol 3: 311-318 DOI 10.1111/j.1542-474X.1998.tb00039.x
4. Van Leeuwen P, Hailer B, Lange S et al. (2008) Quantification of cardiac magnetic field orientation during ventricular de- and repolarization, Phys Med Biol 53:2291-2301 DOI 10.1088/0031-9155/53/9/ 006
5. Van Leeuwen P, Stroink G, Hailer B et al. (1999) Rest and stress magnetocardiography in coronary artery disease, Med Biol Eng Comput 37:1482-1483.
6. Stroink G, Meeder RJ, Elliott P et al. (1999) Arrhythmia vulnerability assessment using magnetic field maps and body surface potential maps, Pacing Clin Electrophysiol 22:1718-1728 DOI 10.1111/j.1540-8159.1999.tb00403.x
7. Hubley-Kozey CL, Mitchell LB, Gardner MJ et al (1995) Spatial features in body-surface potential maps can identify patients with a history of sustained ventricular tachycardia, Circulation 92:1825-1838

Use macro [author address] to enter the address of the corresponding author:

Author: Peter Van Leeuwen
Institute: Dept. Biomagnetism, Chair of Radiology and Microtherapy, University of Witten/Herdecke
Street: Universitätsstr. 142
City: 44799 Bochum
Country: Germany
Email: petervl@microtherapy.de

Magnetic Reconstruction of Circulating Excitation

Taysia Sosnytska[1], Pavlo Sutkovyy[2], Volodymyr Sosnytskyy[2], and Ekaterina Koronovska[2]

[1] P.L. Shupik Medical Academy for Postgraduate Education of UH Ministry, Kyiv Ukraine
[2] MCG Laboratory, Glushkov Institute of Cybernetics, Kyiv, Ukraine

Abstract— **Cardiac arrhythmia re-entry is most important mechanism of fatal tachy arrhythmias such as ventricular fibrillation. This arrhythmia re-entry may be provoked by such anisotropic factor as circulating fiber orientation. Our hypothesis is that the circulating excitation can be modeled by a set of electrical sources with a different orientation. The purpose of this study was to investigate how well two dimensional current density vectors (CDV) map reflects real distribution of the circulating excitation of the set of electrochemical "cells". A simple realistic torso model and source model (SM) suitable for magnetic measurements was constructed by layers of discrete electrochemical "cells" mounted inside a torso phantom with saline solution. The continuous and discontinuous nature of active circulating excitation were simulated by a set of electrical sources arranged in a circle. Cells were switched on stepwise one after the other. Magnetic signals were recorded by means of 4x3–channel MCG system in an unshielded setting. After averaging the beat at rectangle grid 6x6, magnetic field distribution maps were plotted in the course of an SM activity. CDV maps were plotted by Magnetic Moments Method of the " inverse problem solution " Results show a good correspondence between current density vector maps determined from the actual measured magnetic maps and active circulating excitation. Pattern of activation and regions of earliest and latest activation were observed with acceptable accuracy. Our study confirmed the good correlation between CDV and circulating distribution of current flow sources.**

Keywords— **MCG, electrical sources, current density vector.**

I. Introduction

Cardiac arrhythmia re-entry is most important mechanism of fatal tachy arrhythmias such as ventricular fibrillation. This arrhythmia re-entry may be provoked by such anisotropy factors as circulating fiber orientation and dispersion of cardiac repolarization.

During cardiac excitation its magnetic field as well as the surface electrode derived ECG are based on events taking place at cellular level at multiple sites in the myocardium. Yet differences in the mode of charge transport and in the anatomical structure relationships may affect the magnetic signal mapping in a different way than is the ECG, derived from surface electrodes. In particular, discontinuities in the cardiac muscle may affect the magnetic field such that they may be detected by the MCG whereas the voltage measurements from the standard ECG may not do so. Our focus of this work was on the ionic mechanisms that contribute to magnetic signals mapping. The "inverse solution" with spatial-temporary distributions of sources in the heart was carried out with derivation of some parameters available for diagnostic conclusions. Even so, it remains a moot point, as stated by "how well such a current density distributions map reflects reality" [1]. To evaluate these potentials uncertainties, a number of experiments were carried out. The aim of this study was to visualize directly discontinuous circulating circuits as prototype of the re-entrant mechanism of activation using standard MCG mapping.

II. Materials and Methods

We hypothesized that the electrical heterogeneity can be modeled by a set of discrete electrical sources with a activated current flow of a characteristic shape and duration. An extended physical source model (SM) suitable for magnetic measurements was constructed by discrete electrical "cells" (Fig.1) with dimension of 10 mm x 10 mm x 30 mm.

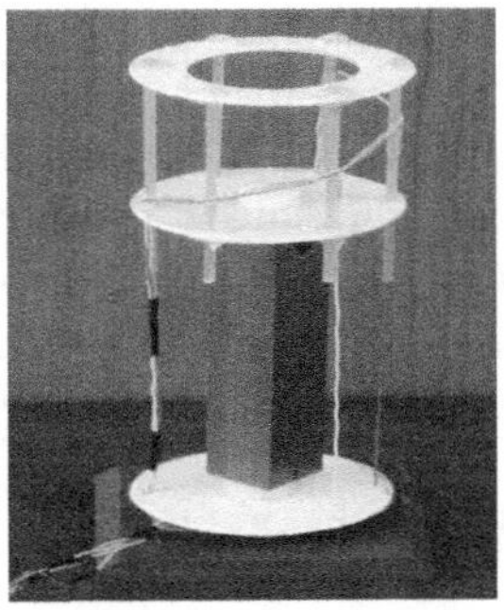

a

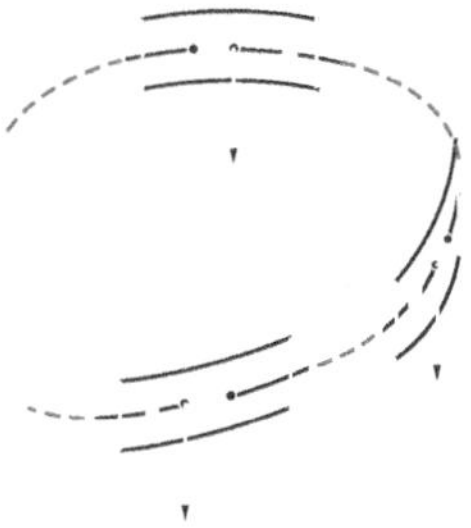

b

Fig. 1 Physical source model suitable for magnetic measurements (a) and schematic representation of the poison of the electrochemical "cells"

Each "cell" consists of a small electrochemical battery, with the negative electrode directly connected to an exterior copper electrode and the positive one connected through an electronic "key" to the other exterior copper electrode. The continuous and discontinuous nature of active circulating excitation were simulated by a set of electrical sources

S. Supek and A. Sušac (Eds.): Advances in Biomagnetism – BIOMAG2010, IFMBE Proceedings 28, pp. 417–419, 2010.
www.springerlink.com

arranged in a circle. Cells were switched on stepwise one after the other. Also, the direction of the "cells" was not perfectly parallel to the measurement plane (ie, they form an angle of attack at the torso model surface). The "cells" were finally covered and carefully isolated by epoxy resin, so that no fluid could leak. The source model was assumed to be located in a homogeneous volume conductor in an anatomically realistic torso model of adult male subject. Our procedures for building the heart-torso model and the magnetic field profiles for computing the current distribution are described elsewhere [2].

The torso model was filled with a 0,9% saline solution and both copper electrodes were immediate connected with a fluid. In order to simulate a shift in the direction of ion currents, each of the 3 "cells" was turned by 90° in a plane, parallel to the measurement plane. Each discrete "cell" had a characteristic current flow duration and electrode direction assigned to it. A key feature of our model was that the electrical sources act as independent variables. In order to simulate a shift in the position of ion currents, the potential impulses with different durations and rectangular shape were supplied to the electronic "key" electrodes with a frequency of 1.0 Hz. The SM behaves as a set of sources that generate a potential in the surrounding volume conductor. The SM voltage gradient that develops as a result of imposition of a current sources layer in the "heart" region of the torso model plays a role in the inscription of the electro potential signal. The current inside the SM can be interpreted as an impressed (i.e., applied) current. Magnetic signals were obtained with a 4x3-channels MCG system (Kiev Medical Group Ltd, Ukraine) in an unshielded setting. The magnetic field component normal to a plane that is tangential to the torso model has been recorded sequentially at 36 points forming rectangular measuring grid in a planar surface located 1.0 cm in front (anterior) of the torso and 4.0 cm in front of SM. The 36 sites with grid space of 4 cm in both directions resulted in a total grid dimension of 24 x 24 cm. Magnetic data were recorded while the "cells" were switched on stepwise one after the other.

After averaging the beats at each of the 36 measurement points, equiinductional maps of magnetic field distribution were plotted in the course activation interval with 10 ms steps. Current density vector (CDV) maps were plotted by Magnetic Moments Method of the "inverse problem solution" [3]. The "inverse problem" can be formulated as follows: the field source is a 2D system of currents situated in the plane parallel to measurement plane from which the spatial configuration of the currents producing the magnetic field is derived.

III. Results

Figure 2 (a,b,c) displays the changes of distribution and direction of a current density vectors in a frontal view for current sources (cells) after activation. A color–coded CDV maps reflect the magnitude of the current density at three different time instant of the cells activity. Pattern of activation (propagation of the wave front) and regions of earliest and latest activation can be observed from CDV maps.

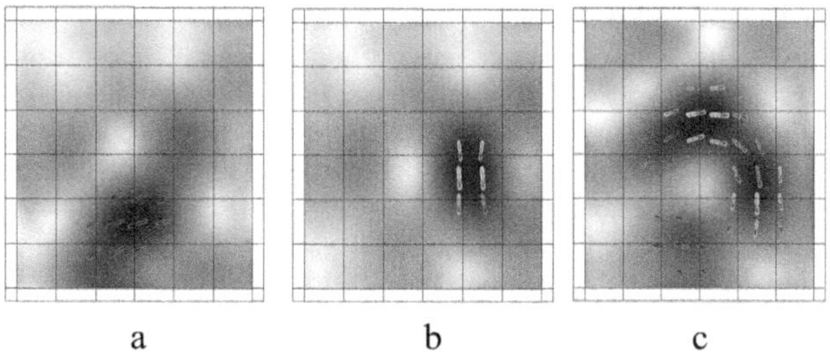

Fig. 2 CDV maps reflect patterns of the "cells" electrical activity and position at the actual time instant (a,b,c)

Current density distribution exhibits a clockwise rotation for activation originating from SM and reflects the underlying current dipole direction.

IV. Discussion

Sachse et al.[4] demonstrated in a realistic impedance model of the human torso the electric potentials to be sensitive to heterogeneous and isotropic conductivity but not to anisotropy. In such important aspects as tissue anisotropy and discontinuities in tissue structure the myocardium is different from the generic excitable media typically considered in the theory of vortex currents. To simplify the situation we can use the fact, that the activating cardiac cells act just as elementary current sources, which produce two fields, E and H. The CDV map as constructed from magnetic field depends on a number of factors: the quality of the signal, the reduction in the signal to noise ratio (of the residual ambient magnetic field), the algorithm in the preliminary signal processing and, most important of all, the variability in the biochemical origin of the ionic driving force. Ramon et al. [5] and Plonsey [6] have furthermore argued that the helical geometry and arrangement of cardiac fibres, provide a structural reason for specific current flow patterns. Taccardi et al. [7] identified the elongated shape of the myocardial fibres, the epi-endocardial rotation in the fibre direction as contributing factors to electrical heterogeneity. Kleber [8] has specifically pointed out that adjacent muscle layers are electrically insulated and that activation as well repolarization use direct muscle bridges. Kazataxa and Xoral [9] have reported on a model study to elucidate these influences. Based on this latest knowledge we can assume that the heterogeneity of an action potential is caused by an intramural heterogeneity of a current flow and for each time instant a result the "inverse problem solution" can display a resulting current density vectors, whose quantity and direction correspond a quite particular allocation of action current densities at a certain level (depth) and reflect the architecture of the fibre through which the current propagates. In our model study only a two-dimensional reconstruction was carried out, whereas in reality current flow is three-dimensional at various depths. The quantity of parameters in a magnetic field of the heart and the character of its allocation depend basically on the intensity of these sources, their spatial arrangement and the geometrical shape. One of the key differences of the phantom experiment from a conducting myocardium is that the phantom ions moved only as a reflection of the electromotive force. In the human myocardium the driving force for ion current flow consists of 2 components: electromotive and thermodynamic driving forces. The real influence of the geometrical factors and the exact nature of a ionic driving force should be demonstrated by measuring data from living intact organ.

V. Conclusions

With the help of physical phantom experiments we quantified the influence of active circulating circuits on the strength of magnetic signals and we tested the ability of standard source localization algorithms to reconstruct circulating currents. Our results should be considered as preliminary ones. This idealized representation of the circulating activation is just used for demonstration the ability of the MCG-technology to visualize directly discontinuous circulating circuits as prototype of the re-entrant mechanism of excitation.

Acknowledgment

This work was supported by a grant N4719 from STCU.

References

1. Koch H (2001) Magnetocardiography: Status and Perspectives,IEEE Trans. on applied superconductivity 11:49-50
2. Sosnytskyy V, Hugenholtz P, Sutkovyy P, Awolin B (2005) The Magnetocardiogram for the Assessment of Current Density Heterogeneity. A Torso Model Study USiM.N3:25-28
3. Romanovych S, Steinberg F, Sosnytsky V, Ivanova S (1999) Imaging of heart biomagnetic sources by current lines in a plane using the magnetic moments method. Proceedings of the EMBEC99.Part 1: 410-411
4. Sachse F, Werner C, Meyer-Waarden K, and Dossel O (1997) Comparison of solution to the forward problem in electrophysiology with homogeneous, heterogeneous and anisotropic impedance models. Biomedizinische Technik Band 42:277-280
5. Ramon C, Wang Y, Haueisen J et al.(2000) Effect of Myocardial Anisotropy on the Torso Current Flow Patterns, Potentials and Magnetic Fields, Phys.Med.Biol. 45:1141-1150
6. Plonsey R (1974) An Evaluation of Several Cardiac Activation Models, J. Electrocardiol. 7:234-244
7. Taccardi B, Lux RL, Ershier PR et al. (1997) Anatomical Architecture and Electrical Activity of the Heart.Acta Cardiol. 52(2):91-105
8. Kleber AG, Janse MJ, Fast VG (2001) Normal and Abnormal Conduction in the Heart, Handbook of Physiology, vol.1:The Heart, Oxford University Press
9. Kazataxa G, Xoral R (2002) Ionic Current Basis of Electrocardiographic Waveforms. A Model Study. Circulation Research 3:889-896

Author: V.Sosnytskyy
Institute: Glushkov Institute of Cybernetics
Street: 40,prosp.Glushkova
City: Kiev
Country: Ukraine
Email: sosna1@voliacable.com

Comparison of Cardiac Activation Models Using 3D MCG Data

Massimo De Melis and Yoshinori Uchikawa

Graduate School of Advanced Science and Technology, Tokyo Denki University, Tokyo, Japan

Abstract— **In the study of the inverse problem in biomagnetism, the measured signals are used to localize the source that generates a given magnetic field distribution. The problem has no unique solution, since the same magnetic field map (MFM) can be generated by an infinite number of possible source configurations. To restrict the set of possible solutions, physiological information can be used. In magnetocardiography (MCG), the knowledge of the heart functionality may help in making a suitable choice to model the current source in the human heart. A possible choice for the current model is a single dipole, with varying position, orientation and amplitude inside the heart volume, the so-called source space. An alternative solution can be a multiple dipole model, characterized by a limited number of dipoles, with varying positions and orientations, or a limited range of variability for the position inside the source space. A more general description of the wavefront propagation can be provided by a distributed sources model, which considers an elevated number of dipolar sources, located in fixed positions inside the heart volume region, having varying amplitudes and directions, to model the current flow during the heart cycle. A single dipole model is a suitable one only for short segment of the MCG signals, whereas a distributed model allows estimating the current density in the source space. We performed the source localizations of MCG signals using different source models on the same data sets, to evaluate whether the differences in the source models descriptions can be minimized in the MCG data sets, or in some of them. The analyses are repeated on normal subjects and WPW patients, using normal components data and vector component data of the measured MFMs.**

Keywords— **MCG, inverse problem, source model, volume conductor.**

I. Introduction

In MCG analyses, the signals recorded by multi-channel systems can be displayed in the form of a magnetic field map (MFM) that changes periodically during the cardiac cycle. The time evolution of the MFMs follows a well-known pattern in healthy subjects, and deviations from the normal sequence of the MFMs are an indicator of a possible cardiac pathology.

A typical application of MCG signal analysis is to use the recorded fields to reconstruct the electrical source that generates the MFMs [1]. Because of the non-uniqueness of the biomagnetic inverse problem [2], it is common to makes assumptions on the source configuration and on the volume conductor model to reduce the complexity of the problem. In case that the source is mainly localized in a limited region of the source space, a single dipole can be used as source model. When the heart activation is spread over a large portion of the myocardium, a distributed source model provides a more generalized approach to describe the heart activity.

In this study MCG source localizations were performed using single and distributed source models on normal subjects and WPW patients to evaluate if the differences between the source models can be reduced, to obtain a more general model that describes the heart excitation. All analyses were repeated using only the normal components and 3D components data, to study the effect of the information provided by the additional tangential components. The analyses were performed using the experimental software MFIView of the BMDSys Gmbh (Jena, Germany).

II. Methods

A. Source Models

A simple solution for the source model, suggested by the dipolar pattern of the MFM, is to model the electrical activity by a single current dipole, having varying position, orientation and amplitude in the source space region. Non-linear minimization algorithms, as the downhill-simplex method [3], are used to localize the solution.

An alternative and more general description of the heart activation can be provided by a distributed source model, characterized by a great number of current dipoles, each one having a fixed position in the heart volume, and varying orientations and amplitudes. The equation to be solved is in the form of:

$$[L][J]=[B] \tag{1}$$

If M is the number of sensors and N the number of dipoles, J is a 3xN parameters vector of unknown positions, B is a M values vector of measured field points, and L is a Mx(3xN) transfer matrix of the measurement system which expresses the sensors sensitivities.

Since the number of unknown is generally much higher than the number of known values, minimum norm

S. Supek and A. Sušac (Eds.): Advances in Biomagnetism – BIOMAG2010, IFMBE Proceedings 28, pp. 420–423, 2010.
www.springerlink.com

techniques are used to find a plausible solution [4, 5], which is expressed as:

$$J = L^T \left(LL^T\right)^{-1} B \qquad (2)$$

The solution is stabilized using Tikhonov regularization [6], which is finding a modified solution of eq. (2) in the form of:

$$\left(L^T L + \alpha I\right) J = L^T B \qquad (3)$$

where I is the identity matrix and α the regularization parameter used to stabilize the solution.

To compare the results of single dipole and distributed source models, an equivalent current dipole can be calculated for the distributed sources model [7], having as amplitude and position the values defined by eq. (4) and (5):

$$\vec{J}(t) = \frac{1}{N_D} \sum_{i=0}^{N_D} \vec{J}_i(t) \qquad (4)$$

$$\vec{P}(t) = \frac{1}{\sum_{i=0}^{N_D} \left|\vec{J}_i(t)\right|} \sum_{i=0}^{N_D} \left(\left|\vec{J}_i(t)\right| \vec{P}_i\right) \qquad (5)$$

where J(t) is the average of all amplitudes, P(t) is the center of mass, i is the dipole index, N_D is the total number of dipoles, $|J_i(t)|$ is the amplitude of the i-th dipole at the time t, and P_i its position.

B. *Conductor Model*

The volume conductor model used is a Boundary Element Method (BEM) model, which is a common choice in MCG localization studies [8].

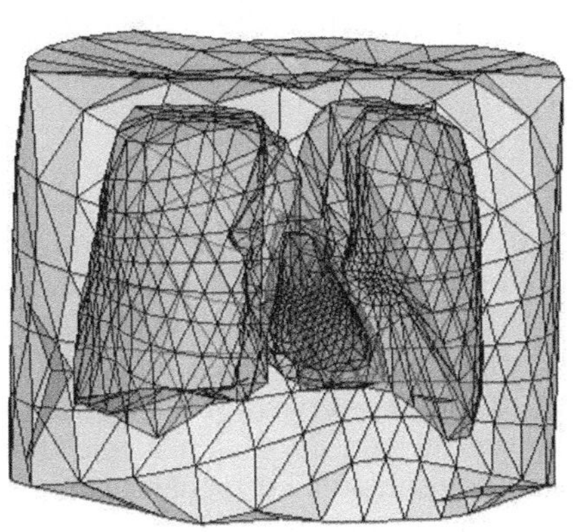

Fig. 1 BEM model obtained from CT data

The conductor is approximated by a series of triangulated surfaces, and the potential is supposed to vary linearly over each triangle. The conductivity is supposed homogeneous and isotropic inside each surface.

C. *Sensors System*

The MCG data were recorded using a vector second order gradiometer, orthogonal wound with Nb-Ti-Cu wire on a rectangular solid 3x3x6cm, which is able to record simultaneously three-dimensional magnetic fields [9]. The frequency for data sampling is 1 kHz, and a band pass filter was used in the range 0.1-80 Hz. The system consists of the 16 vector sensors shown in fig. 2a. Four scans are performed to cover the whole chest area, so that at the end 48 measurement points having 4 cm interval are available, and averaged data over 4 min of measurement are used for the analysis. These 48 points are the locations of the sensors in fig. 2b.

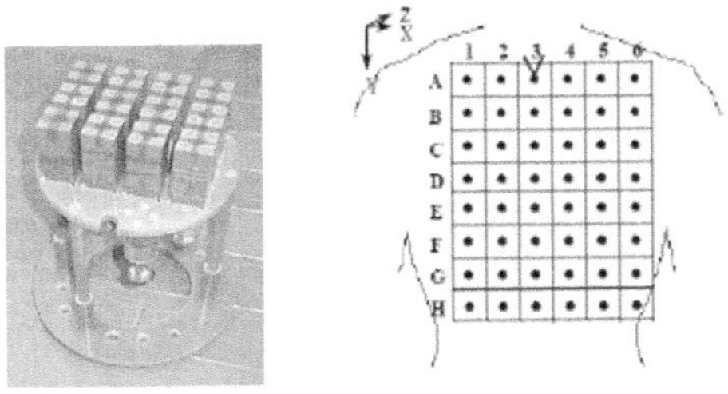

(a) (b)

Fig. 2 A 3D second order gradiometer (a) and coordinate system for MCG (b)

III. Results

The current source reconstructions were performed on characteristic time segments of the MCG signals, namely the P, Q, R, S and T waves.

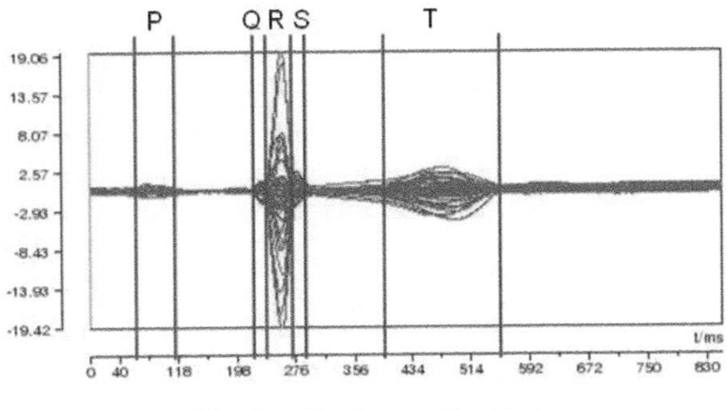

Fig. 3 MCG of a normal subject

After calculating the solution with single dipole and distributed source models, the equivalent dipole was calculated for the distributed model, as defined by eq. (4) and (5), and the differences in the dipole positions and orientations (fig. 4-7) with the single dipole were evaluated for all intervals of interest. Five normal subjects and one WPW patient data were examined.

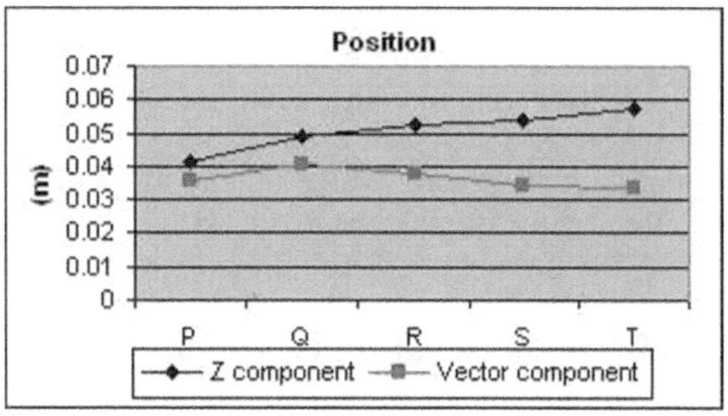

Fig. 4 Positions differences for normal subjects on planar and 3D data

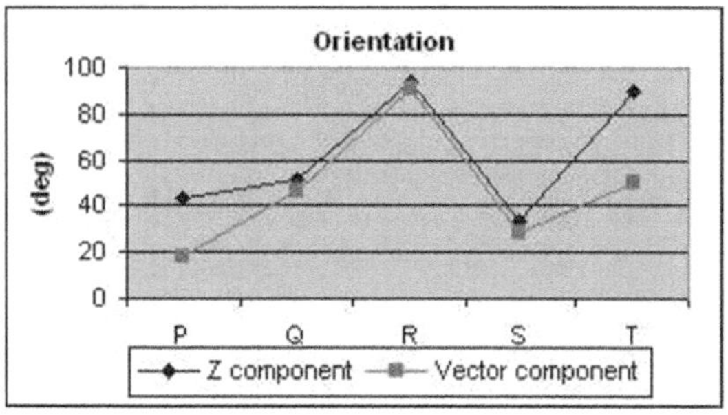

Fig. 5 Orientations differences for normal subjects on planar and 3D data

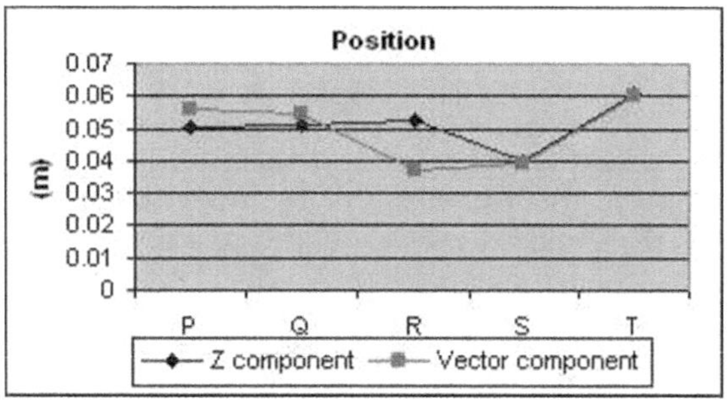

Fig. 6 Positions differences for WPW patient on planar and 3D data

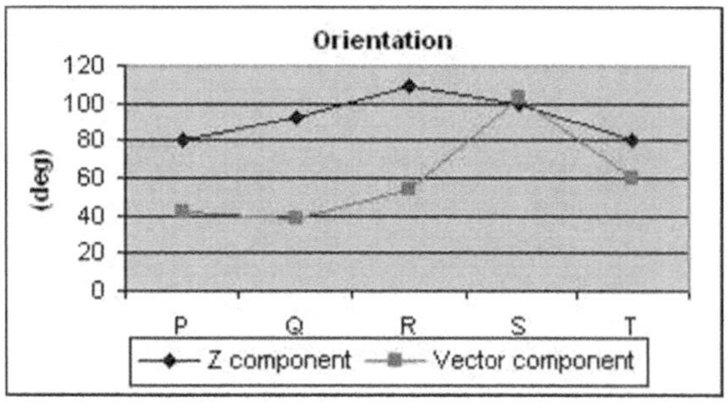

Fig. 7 Orientations differences for WPW patient on planar and 3D data

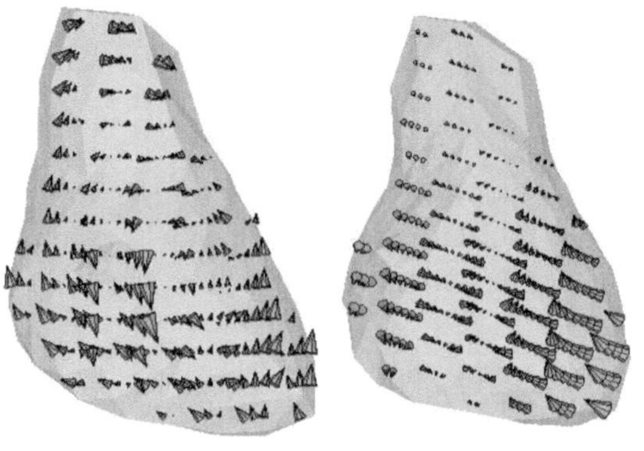

(a) (b)

Fig. 8 Distributed sources of a normal subject using normal (a) and vector (b) component data

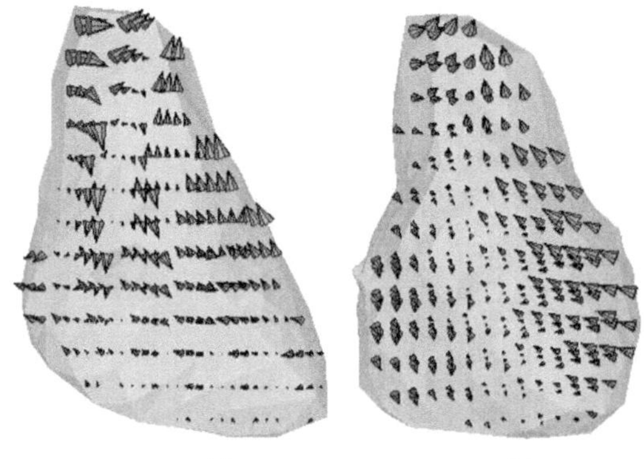

(a) (b)

Fig. 9 Distributed sources on WPW patient data using normal (a) and vector (b) component data

The fig. 4 and 5 show the comparison between Z component and 3D data for the positions and the orientations respectively, in case of normal subject's data. The fig. 6 and 7 illustrate the difference in positions and orientations for a WPW patient data. The results of the source reconstruction using a distributed model for normal subjects and WPW data are shown in fig. 8 and 9 respectively at the beginning of the QRS complex, where the effect of the additional pathway results in a deformed signal pattern.

IV. Discussion

If the source reconstruction is performed on normal subject's data, the differences in the localizations between single dipole and equivalent dipole models, obtained using vector data, are smaller than when using only the normal components. The center of mass of the activation represents the same concept behind the single dipole model: a main activation source in a confined region inside the heart volume; therefore the smaller difference given by 3D data is not a surprising result, since three component data provide more complete information about the heart stimulation process.

The results change slightly if WPW data are used. In this case the existence of an extra conduction pathway between the atria and the ventricles is responsible of a pre-excitation of the ventricles, which may lead to tachycardia or chest palpitation. The effect of this activation disturbance can be recognized in ECG or MCG time signals, specifically in the first part of the QRS complex. The pre-excitation is the result of a congenital heart malformation, and is usually localized in a limited region of the source space. The differences in the localization may depend on the time segment where the source is reconstructed, but the differences in the dipoles orientations remain smaller when using 3D data.

The CSR with a distributed source model allows visualizing how the excitation wavefront spreads over the myocardium. If the CSR is performed using vector data, the higher number of sensors used allows a better S/N ratio, but simulation studies performed using a limited number of dipoles [10] or a distributed current model [11] showed that, on equal S/N values, vector data give results that are numerically better as well as less dependent on the specific value of the regularization parameter. A visualization of the distributed sources solutions obtained considering the first part of the QRS complex, when the activation begins propagating to the ventricles, reveals clear differences between normal subjects and WPW patient data. In case of a normal subject, the excitation wavefront is mainly localized in the lower part of the heart, that is, the ventricular region. In case of WPW data, and specifically when using vector data, the activation looks prominent in the left part of the cardiac volume. The considered sensor type, used to acquire 3D WPW data, allowed localizing the Kent bundle also when using a two-source model [12].

V. Conclusions

The choice of the current source model to solve the inverse problem in biomagnetism is usually done on the basis of physiological considerations. In this study we presented a comparison of the results of current source reconstruction using single dipole models and distributed sources models. The direct comparison is performed calculating an equivalent average dipole for the distributed model and evaluating the positions and orientations errors. If these differences become reasonably small, a general current source model can be used to describe the heart electrical activity. For normal subjects, three component data give smaller differences than normal component data. For WPW data, the error positions have a larger variability, since the WPW syndrome is characterized by a localized accessory conduction pathway, but it is interesting to note that the orientations errors are still smaller using 3D data.

If the source is reconstructed by means of a distributed model, localized sources are harder to detect, but a visual inspection of the reconstruction results obtained using vector data allows identifying the area where the most significant activity is concentrated, thanks to the sensitivity to the additional tangential components.

References

1. Hämäläinen M.S., Haarii R., Ilmoniemi R.J., Knuutila J. and Lounasmaa O. (1993) "Magnetoencephalography – Theory, instrumentation and applications to noninvasive studies of the working human brain" Rev. Mod. Phys., 65, num. 2: 413-497.
2. Helmutz H. (1953) "Über einige Gesetze der Vertheilung elektrischer Ströme in körperlichen Leitern, mit Anwendungen auf thierisch elektrischen Versuche" Ann. Phys. Chem., 89: 211.
3. Nelder J.A. and Mead R. (1965) "A simplex method for function minimization" Comp. J., 7: 308-313.
4. Hämäläinen M.S. and Ilmoniemi R.J. (1994) "Interpreting magnetic fields of the brain: minimum norm estimates" Med. Biol. Eng. Computing, 32: 35-42.
5. Nenonen J.T., Hämäläinen M.S. And Ilmoniemi R.J. (1994) "Minimum norm estimation in a boundary element torso model" Med. Biol. Eng. Computing, 32: 43-48.
6. Tikhonov A.N. and Arsenin V.Y. (1977) Solution of ill-posed problems, W. H. Winston&Sons, Wiley, New York.
7. De Melis M., Kim B.S., Uchikawa Y., Müller H.P. and Erné S.N. (2008) "Discussion of current distribution of exercise 3D MCG using an inhomogeneous human volume conductor" J. Magn. Soc. Jap., 32: 195-200.
8. Ferguson A.S., Zhang X. and Stroink G. (1994) "A complete linear discretization for calculating the magnetic field using the boundary element method" IEEE Trans. Biomed. Eng., 41, num. 5: 455-460.
9. Kobayashi K. and Uchikawa Y. (1999) "Development of a three dimensional biomagnetic measurement system with 39-channel SQUIDs" Recent advances in biomagnetism, ed. by Yoshimoto T., Kotani M., Kuriki S., Karibe H. and Nakasato N., Tokyo: 35-38.
10. Arturi C. M. Di Rienzo L. and Haueisen J. (2004) "Information content in single-component versus three component cardiomagnetic fields" IEEE Trans. Magn., 40, num. 2: 631-634.
11. De Melis M. and Uchikawa Y. (2009) "Analysis of standard and irregular patterns in a simulated human magnetocardiogram" J. Magn. Soc. Jap., 33: 179-184.
12. Kobayashi K., Uchikawa Y., Nakai K. and Yoshizawa M. (2002) "Analysis and estimation of excitation conduction with Wolff-Parkinson-White syndrome patients based on a 3D magnetocardiogram" IEEE Trans. Magn., 38, num. 5: 3338-3340.

Corresponding author:

Author: De Melis Massimo
Institute: Graduate School of Advanced Science and Technology, Tokyo Denki University.
Street: 350-0394 Ishizaka, Hiki-gun.
City: Hatoyama, Saitama.
Country: Japan
Email: demelis@ff.f.dendai.ac.jp

Development of Pattern Recognition Method for Diagnosis of Myocardial Ischemia and Noncoronarogenic Myocardial Diseases Based on Current Density Distribution Maps

I.A. Chaykovskyy[1,2], M.M. Budnyk[1], M.A. Najafian[2], S.S. Martynenko[3], A.S. Dovbysh[3], and O.S. Kovalenko[2]

[1] Glushkov Institute for Cybernetics/Department of Sensors and Technologies for Non-contact Diagnostics, Kyiv, Ukraine
[2] International Research and Training Center "IT and Systems"/Department of Medical Informatic Systems, Kyiv, Ukraine
[3] Sumy State University/Department of Informatics, Faculty Electronics and IT, Sumy, Ukraine

Abstract— **Aim of study is to develop advanced pattern recognition method based on current density distribution (CDD) maps. Sets of typical maps within ST-T interval of healthy volunteers, patients with CAD, myocarditis and maps with high level of noise were selected. These images were classified with help of information-extreme intellectual technology (IEIT) based on Kullback measure in parametric Hemming space. Faultless error-free decision rules were obtained in learning population. Decision rule was examined by test Healthy–CAD. Examination group consisted of 203 healthy volunteers and 256 CAD pts. Sensitivity and specificity equal to 93% and 87 %, respectively, have been achieved.**

Keywords— **pattern recognition, information-extreme intellectual technology (IEIT), current density distribution.**

I. BACKGROUND

Magnetocardiography (MCG) is the modern method for registration and analysis of the heart magnetic field. It is a noninvasive and safe method which allow receives principle new information about the heart electrophysiology in norm and at pathological changes. MCG examination is fully risk-free for a patient.

Current density distribution maps (CDD) are the most valuable form of MCG data representation. However, visual estimation of above maps is too complex and it is required considerable experience and high qualification from a doctor-cardiologist for recognition of MCG data. Such estimation is especially difficult at the analysis of data registered in unshielded environment due to noisy effects. Significant efforts are recently undertaken on development of methods of automatic classification of MCG data [1-9]. However, all these methods allow maps division only into 2 classes – normal and pathological (ischemic), i.e. they do not possess sufficient specificity.

Aim of study is to develop advanced pattern recognition method for classification of CDD maps into 4 classes, which are typical for daily clinical practice and very frequently occur at unshielded environment. These classes are the following: normal, ischemic, non-coronarogenic heart disease (myocarditis), maps with high level of noise.

II. NON-FORMAL DESCRIPTION OF TYPES OF CDD MAPS

Researchers paid special attention to ventricles repolarization, i.e. ST-T interval. The most informative MCG images are vector maps of CDD. Examples of maps of 4 classes at T peak apex are shown at Fig.1-2 based on visual analysis of large MCG database.

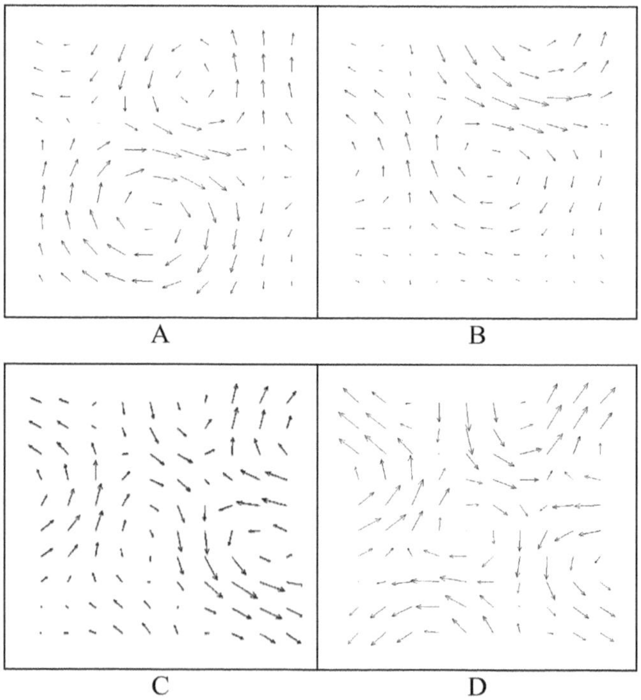

Fig. 1 Typical maps of healthy volunteers (a, b) and patients with CAD (c, d)

S. Supek and A. Sušac (Eds.): Advances in Biomagnetism – BIOMAG2010, IFMBE Proceedings 28, pp. 424–427, 2010.
www.springerlink.com

Normal maps demonstrate a dipole structure having direction to the left-bottom, and two nearly equal current vortexes are usually take place. There is only single area of vectors with maximum length, located usually in the map center (Fig. 1a), and sometimes it can be displaced up- or downward. On Fig. 1 c-d is represented two different variants of maps of CAD pts. These maps are characterized by non-dipole structure, presence of few pathological areas of vectors of large length forming a few variants of special structures.

Maps of patients with the heart pathology of non-coronarogenic origin are represented on the Fig. 2 a,b. They differ from the maps of CAD pts because pathological areas of vectors are more extended and do not form structures, typical for CAD pts. Maps of 4th group (Fig. 2 c,d) are the maps with low signal-to-noise ratio (SNR). They also have non-dipole structure, but areas of maximal vectors are located at the map edge out of the heart area.

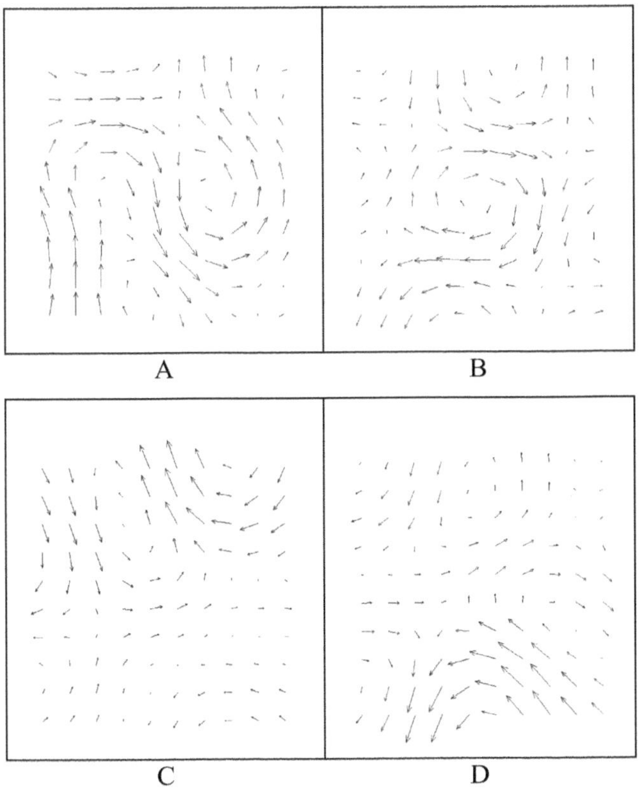

Fig. 2 Typical maps of patients with myocarditis (a, b) and noisy maps (c, d)

III. METHOD OF PATTERN RECOGNITION

4 samplings of 40 color CDD maps of above classes were selected by an experienced cardiologist with help of the originally developed software package MAGWIN [7]. Direction, length and brightness of vectors are considered as recognition signs.

Single typical CDD map was selected from set of maps within ST-T interval of one person, i.e. 40 persons were used to form the learning dataset in each group. Aim of such preliminary selection is decreasing of amount of learning data. In result, procedure of synthesis of decision rules was essentially simplified and accelerated.

After map processing in polar co-ordinates it was formed multi-dimension learning matrix in which every line involves 400 signs describing the brightness of red and blue color, and 100 signs describing the map topology.

Here it is presented the first results based on information-extreme intellectual technology (IE IT) based on maximization of ability of the pattern recognition during learning process by introducing of additional restrictions [10,11].

Learning was implemented on the basis of IE IT with optimization of control tolerances for recognition signs. The basic idea of learning consists in transformation from a priori fuzzy fragmentation of sign space to non-fuzzy fragmentation of classes of equivalence by possible optimal transformations of input mathematical description in sub-paraceptual parametric Hemming space. As a optimization criterion the modified Kullback measure was used [10,11].

$$E_m^{(k)} = \log_2 \frac{P_t}{P_f} * (P_t - P_f) = \begin{bmatrix} P_t^{(k)} = 0{,}5D_1^{(k)} + 0{,}5D_2^{(k)}, \\ P_f^{(k)} = 0{,}5\alpha^{(k)} + 0{,}5\beta^{(k)} \end{bmatrix} =$$

$$= 0{,}5\log_2\left(\frac{D_1^{(k)} + D_2^{(k)}}{\alpha^{(k)} + \beta^{(k)}}\right)\left[(D_1^{(k)} + D_2^{(k)}) - (\alpha^{(k)} + \beta^{(k)})\right] =$$

$$= \log_2\left(\frac{2 - (\alpha^{(k)} + \beta^{(k)})}{\alpha^{(k)} + \beta^{(k)}}\right)\left[1 - (\alpha^{(k)} + \beta^{(k)})\right] \quad (1)$$

where P_t, P_f – full probabilities of true and false decision-making, respectively; $D_1^{(k)}, D_2^{(k)}, \alpha^{(k)}, \beta^{(k)}$ – 1st and 2nd reliability, errors of the 1st and 2nd kinds, respectively; obtained at the k-th step of learning.

On Fig. 3 it is shown the dependence of criterion of functional effectiveness (CFE) (1) from a parameter δ, characterizing the width of the symmetric field of tolerances, which is obtained in the process of parallel optimization of control tolerances for the recognition signs. The work (acceptable) zones of CFE (1) are marked by the dark areas. Analysis of Fig. 3 shows that maximal value of CFE is reached at optimum value of parameter $\delta^* = \pm 51$.

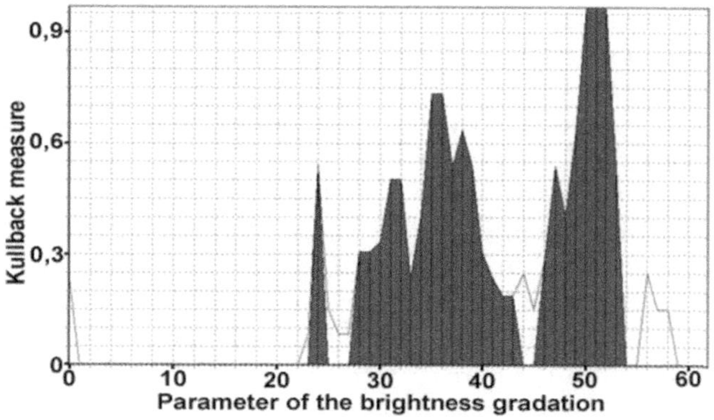

Fig. 3 Dependence (1) of Kullback criterion from the parameter of the brightness gradation δ

Let us consider setting of task of synthesis of MCG recognition system. Let us have an alphabet of classes of recognition $X_m^o, m = \overline{1,M}$ ($M \geq 2$ – for a general case), efficiency of study is characterized by CFE E_m. Learning matrix $\|y_{m,i}^{(j)}\|, i = \overline{1,N}, j = \overline{1,n}$ is known, where N, n – amount of recognition signs and tests, respectively. It is necessary to find optimum coordinates of vector of learning parameters $g_m =< g_{m,1},...,g_{m,q},...,g_{m,Q} >$ with the known limitations $R_q(g_1,...,g_Q) \leq 0$, by organization a lot of iteration procedures in order to obtain the optimum values of vector coordinates g_m, which will provide maximum CFE for learning of the recognition system in the tolerance area of learning parameters G [11]

$$E_{\max}^{*} = \max_{G} E_m \tag{2}$$

In the examination mode the recognition system must make (with reliability close to asymptotical) decision about belonging the image to the proper class from the alphabet, formed at the learning stage.

Let us consider the recognition on the example of images of healthy volunteers (Fig. 1a) and CAD patients (Fig. 1c). Recognition signs for RGB color components formed according to (3)

$$\Theta_j = \frac{\sum_{i=1}^{N_j} \theta_i}{N_j} \tag{3}$$

where Θ_j is spectral value at j-th radius; $j = \overline{0,R}$; Θ_i is a value of color component in i-th pixel; N_j is a total amount of pixels within j-th circle; R is a radius of container, i.e. circle, in which pixels are sampled.

Images processing within polar co-ordinates system does them invariant to the shift, rotation and scaling. The results of image processing are RGB-spectrums appropriated to every realization of recognition class (see Fig. 4).

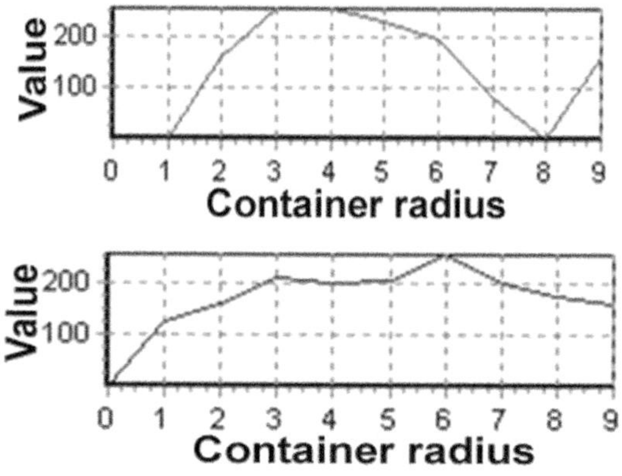

Fig. 4 Dependence (3) of spectral value from the container radius: red component for healthy (a); blue component for the CAD patient (b)

Additionally the dictionary of recognition signs includes geometrical characteristics of CDD vectors. Forming of rectangles, area of which depends on length and direction of CDD vectors it is shown on Fig. 5. Thus, 100 numbers was additionally added to learning matrix for each class describing the rectangle areas.

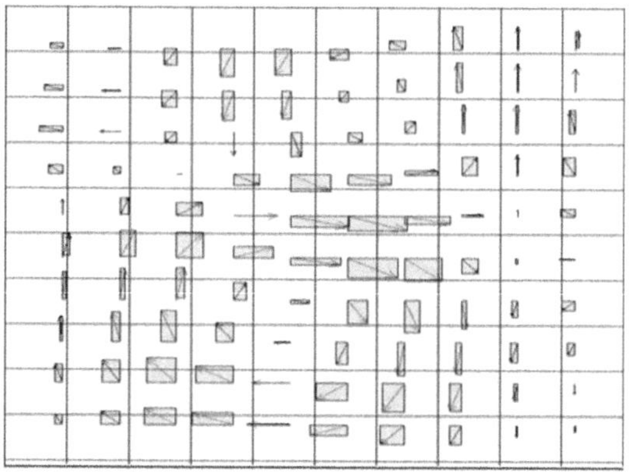

Fig. 5 Additional signs of recognition

Each class of recognition has 40 realizations enabled to build a representative learning sampling. After forming of learning matrix the algorithms of learning and parallel optimization of the control tolerances system were realized, that enabled to build faultless error-free decision rules. The resulted values of CFE are shown on Fig. 6.

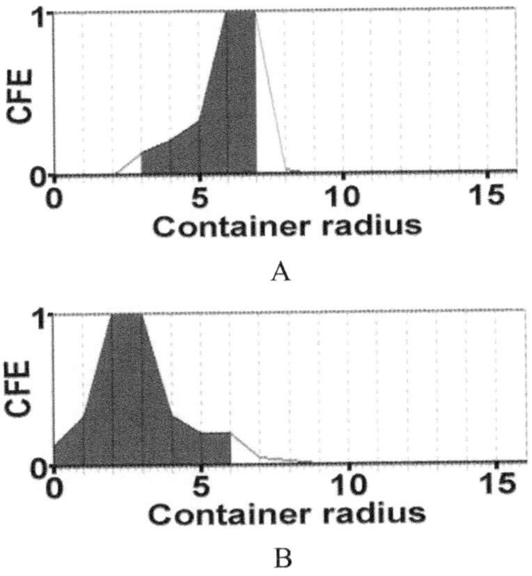

Fig. 6 Dependence of CFE from the container radius for norm (a) and CAD (b)

To examine the obtained decision rule Healthy vs. CAD, 203 healthy volunteers and 256 CAD patients were studied. For each person, set of CDD maps within ST-T interval were analyzed. Sensitivity and specificity equal to 93% and 87 %, respectively, have been achieved. Further examination of above approach in other patients group would be done. Besides, increase of number of map classes to be planned.

IV. Conclusions

- Software, developed based on IE IT, allows recognize CDD maps with high accuracy.
- Processing of magnetocardiograms in polar co-ordi-nates provides the recognition algorithm which is invariant to the shift, rotation and scaling of CDD maps.
- Error-free decision rules were obtained by optimization of parameters of learning matrixes.
- CFE approach demonstrated high reliability of decisi-on-making under uncertainty of the procedure of MCG signal registration, i.e. absence of information about precise link between placement and depth of heart and measurement grid.
- IEI technology allows use as learning dataset a relatively small amount of data. High accuracy of recognition was achieved in a larger examined group based on this approach.

Acknowledgment

This work was performed with financial support of the Science and Technology Center in Ukraine (STCU) under project No 4719.

References

Chaikovsky I, Sosnytsky V, Budnyk M et al. (2001) Detection of CAD in patients with normal or unspecifically changed ECG on the basis of magnetocardiography, Proc, 12th Int. Conf. on Biomagnetism, Espoo, Finland, 2000, pp 565–568

Hailer B, Chaikovsky I, van Leeuwen P et al. (2005) The Value of Magnetocardiography in Patients with and without Relevant Stenoses of the Coronary Arteries using an Unshielded System. PACE 28:8-16.

Chen J, Thomson P, Nolan V, Clarke J (2004) Age and sex dependent variations in the normal magnetocardiogram compared with changes associated with ischemia. Ann Biomed Eng 32:1088-1099

Park J, Hill P, Hugenholtz P et al. (2005) Magnetocardiography predicts coronary artery disease in patients with acute chest pain. Ann Noninvasive Electrocardiol 10:312-323

Kim K, Kwon H, Lee Y et al. (2005) Clinical Parameter Assessment in Magnetocardiography by Using the Support Vector Machine. Int J Bioelectromagnetism 7(1):224-227

Chaikovsky I (2005) Magnetocardiography in unshielded location in coronary artery disease detection using computerized classification of current density vectors maps, Dissertation of Doctor of medicine, Univ Duisburg-Essen

Fainzilberg L, Chaikovsky I, Hailer B et al. (2007) Sensitivity and specificity of magnetocardiography, using computerized classification of current density vectors maps, in ischemic patients with normal ECG and echocardiogram, Int Congress Series 1300, New Frontiers in Biomagnetism, Elsevier, Rotterdam, pp 468-471

Van Leeuwen P, Hailer B, Lange S et al. (2008) Quantification of cardiac magnetic field orientation during ventricular de- and repolarization. Phys Med Biol 53:2291-2301

Tantimongcolwat T, Naenna T, Isarankura-Na-Ayudhyu Ch et al. (2008) Identification of ischemic heart disease via machine learning analysis on magnetocardiograms. Computers in Biol and Med 38(7):817-825

Krasnopoyasovskyi A (2002) Classification data analysis, Sumy State Univ, Sumy, Ukraine

Krasnopoyasovskyi A (2004) Informatic synthesis of intellectual control systems: approach based on method of functional-statistical tests, Sumy State Univ, Sumy, Ukraine

Address of the corresponding author:

Author: I.A. Chaykovskyy
Institute: International Research and Training Center "IT and Systems"/Department of Medical Informatic Systems
Street: 40 Glushkov Avenue
City: Kyiv
Country: Ukraine
Email: illya_6@hotmail.com

Evaluation of Electrophysiological Alteration Created by Local Autonomic Dysfunction Using Magnetocardiography

Taysia Sosnytska[1], Leonyd Stadnyuk[1], and Volodymyr Sosnytskyy[2]

[1] P.L. Shupik Medical Academy for Postgraduate Education of UH Ministry, Kyiv Ukraine
[2] MCG Laboratory, Glushkov Institute of Cybernetics, Kyiv, Ukraine

Abstract— **The pattern of local neurocircuitry is likely to play a critical role in influencing electric heterogeneity of repolarization (EHR), a fundamental factor in arrhythmogenesis. Magnetocardiography (MCG) has revolutionized the investigation of cardiac cellular physiology and advanced our understanding of basic mechanisms of electrical activity. In our study we investigated whether MCG mapping could expose the electrophysiological alteration created by local autonomic dysfunction and which are sensitive to myocardial cooling. 24-h Holter monitoring, ECG, Tissue Doppler echocardiograms (TDE) and MCG records were performed in 4 male patients affected by ventricular extrasystolia Patients were without cardiac structural abnormalities, history of myocardial infarction and sign of ischemia on resting ECG. Current density vector (CDV) maps were plotted in the course of ventricle repolarization interval by the magnetic moments method of the "inverse problem solution". The measurements were made before (baseline) and shortly (15 minutes) after 4 patients ingested 150 ml of cold (6 degree C) water. Standard ECG patterns and TDE were unaffected when the patients drank cold water. CDV directions differed significantly between 2 measurements. MCG examinations before ingestion showed additional local area with high level of regional EHR. The changes in the CDV maps indicated decreasing of the heterogeneity of the repolarization by cooling the surface of the apex of the ventricle which was located closer to the stomach. Our findings can be understood in terms of the known temperature sensitivities of the kinetics of the membrane ion channels responsible for recovery and suggest a potential role of MCG in the evaluation of electrophysiological alteration created by local autonomic dysfunction.**

Keywords— **MCG, heterogeneity, repolarization, current density.**

I. Introduction

The pattern of local neurocircuitry is likely to play a critical role in influencing electric heterogeneity of repolarization (EHR), a fundamental factor in arrhythmogenesis. Magnetocardiography (MCG) has revolutionized the investigation of cardiac cellular physiology and advanced our understanding of basic mechanisms of electrical activity. The electrocardiographic effects of ingesting cold liquids were fist recorded by Wilson and Finch in 1923 [1], and repeated by Dowling and Hellerstein in 1951 [2]. Maximal the effect of drinking ice water on T waves in the electrocardiogram occurred within five minutes following ingestion of the water and subsided completely within 25 minutes. In view of the electrocardiographic changes demonstrated in these studies concerning the effects of drinking iced water, a procedure such as MCG mapping would be expected to demonstrate more marked alterations. The electrophysiological mechanism for the development of significant electric heterogeneity of repolarization in heart disease is not well understood. The heterogeneity of the repolarization should be measured in diseased myocardial segments under physiological conditions by means of non-invasive imaging of cardiac electro genesis. In our study we investigated whether MCG mapping could expose the electrophysiological alteration created by local autonomic dysfunction and which are sensitive to myocardial cooling.

II. Materials and methods

Standard clinical examination (e.g. Imaging and Doppler echocardiograms (ECHO, Kontron medical, Imagic 5500), 12-channel ECG, 24-h ECG monitoring (Philips, Zymed) and resting MCG were performed in 4 male patients (age 27±1.9) affected by ventricular extrasystolia with EHR an identical location to an area of the heart near major sympathetic trunk. Patients were without cardiac structural abnormalities, history of myocardial infarction and sign of ischemia on resting ECG. Echocardiography images were obtained in the apical 4- chamber views with using tissue Doppler imaging (TDI) in pulsed regime on the lateral part of the mitral and tricuspid valve annulus: early diastolic mitral (Em) and tricuspid (Et) velocities, end diastolic mitral (Am) and tricuspid velocities, systolic mitral (Sm) and tricuspid (St) velocities were measured. MCG was recorded by means of 4x3–channel MCG system (Kiev Medical Group Ltd, Ukraine) within a 20×20 rectangular grid with a 4 cm pitch over the precordial area in an unshielded setting. Recording time was 60 s for each position with an acquisition rate of 1000 Hz. Recording bandwidth was 0.01–80 Hz. Lead II of the surface ECG served as reference channel. After averaging the beat at each of the 36 measurement points, equiinductional maps of magnetic field distribution were plotted in the course of an ventricular

S. Supek and A. Sušac (Eds.): Advances in Biomagnetism – BIOMAG2010, IFMBE Proceedings 28, pp. 428–430, 2010.
www.springerlink.com

repolarization interval with 10 ms steps. Current density vector (CDV) maps were plotted by Magnetic Moments Method of the "inverse problem solution" [3]. The "inverse problem" can be formulated as follows: the field source is a 2D system of currents situated in the plane parallel to measurement plane from which the spatial configuration of the currents producing the magnetic field is derived.

MCG based indices, namely direction of the largest CDV, its deviation and shift at several time instants of ST-segment and T-wave were automatically computed [4]. The presence of additional regions with pathologic direction of CDV served for further analysis. Measurements were made before and at frequent intervals after 4 patients ingested about 150 ml of cold (6 degree C) water.

III. Results

Standard ECG patterns and TDI were unaffected when the subjects drank cold water. CDV orientation differed significantly between the 2 measurements. In our patients the first MCG examinations (baseline) showed additional regions with pathologic direction of CDV (red marked vectors) located in the right area of the apex of the heart (Fig.1). The main maximum and second (red marked vectors) maximum peak values are: 0.64 and 0.48 μA cm^{-2} respectively.

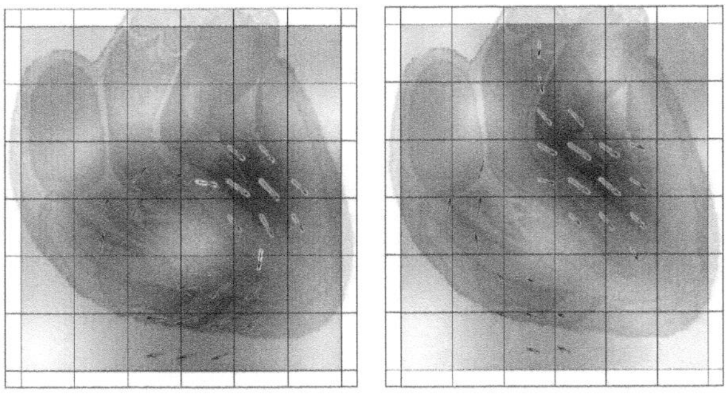

a b

Fig. 1 The changes in the CDV maps occurred within 15 minutes following ingestion of the cold water

Figure 1 displays the typical changes of the magnitude, distribution and direction of a current density vectors in a frontal view for patient before (a) and after (b) ingestion of cold water. A color–coded CDV map (a) reveals the occurrence of abnormal repolarization process (70 ms after T-apex) current distribution due to local autonomic dysfunction. The changes in the CDV maps occurring within 15 minutes following ingestion of the cold water have been associated with a decrease in the complexity of the CDV maps. They indicated lessening of the heterogeneity of the repolarization by cooling the surface of the apex of the ventricle which had been in contact with the stomach and was cooled more than other parts of the ventricles.

IV. Discussion

Hypothermia increases the duration of the ventricular action potentials and prolongs the ventricular effective refractory period [2,5,6]. This effect has been related to slowing of conduction, a heterogeneous increase in ventricular repolarization, and the dispersion of refractoriness [7]. On the other hand, myocardial cooling has also been used to modify different cardiac rhythm disturbances. Such interventions have been made for the protective effect of mild hypothermia against ventricular arrhythmias in experimental models [8], and myocardial cooling has been used to terminate reentrant arrhythmias induced in isolated atrial preparations [9]. The aim of the present study was to analyze the modification of CDV patterns induced by the reduction of myocardial temperature in hearts with newest non-invasive imaging tool. The presence of regions with pathologic direction of CDV associated with zones where the time of repolarization and refractoriness were shorter. We used local interventions (epicardial cooling) to remove regions with increased dispersion of myocardial repolarization. They were removed when the time of repolarization has been prolonged by low temperature. These findings can be understood in terms of the known temperature sensitivities of the kinetics of the membrane ion channels responsible for recovery. The mechanism for the decreased complexity CDV maps at low temperature is probably related primarily to temperature-induced changes in membrane ion channels. The action potential duration is determined by a critical balance between the inward and outward ionic currents and all these currents might be sensitive to temperature changes. The prolongation of repolarization is due primarily to a slowing in the activation of the delayed rectifier channel, which is relatively temperature sensitive. In addition, the recovery kinetics of the delayed rectifier channel must also be slowed. It is well known that lowering myocardial temperature slows the kinetics of the membrane ion channels responsible for activation and recovery, and their relative temperature sensitivities vary [7,10]. Because of these differences in the temperature sensitivity of various channels and the fact that slowing their kinetics may have either opposing or synergistic effects dependent on how they contribute to a given electrophysiological phenomenon.

V. Conclusions

Our findings can be understood in terms of the known temperature sensitivities of the kinetics of the membrane ion channels responsible for recovery. This study was limited in several ways. First, the results must be viewed with caution as the group of the patients is rather small. Second, although the MCG is a powerful tool for understanding spatial time–variant activation our reconstructed current density distributions may be used to identify changes in local cardiac electric activity although the non-uniqueness of the inverse problem requires that results must be viewed with caution. An experimental model for documenting the actual temperature of the heart muscle would be required for a future study. While our results are preliminary because of the above limitations, they are important in terms of understanding the potential role of MCG in the evaluation of electrophysiological alteration created by local autonomic dysfunction. MCG can be a useful tool in elucidating mechanisms of local myocardial disturbances.

Acknowledgment

This work was supported by a grant N4719 from STCU.

References

1. Wilson FN, Finch R (1923) The effect of drinking iced-water upon the form of the T deflection of the electrocardiogram. *Heart.* 10:275–278.
2. Dowling CV Hellerstein HK.(1951) Factors Influencing the T Wave of the Electrocardiogram.II.Effect of Drinking Iced Water, *Am Heart J.* 41-50
3. Romanovych S, Steinberg F, Sosnytsky V, Ivanova S (1999) Imaging of heart biomagnetic sources by current lines in a plane using the magnetic moments method. Proceedings of the EMBEC99. Part 1:410-411
4. Sosnytkaya T, Boroday A, Sosnytskyy V (2008) Detection of patients with ventricular extrasystolia using cardiac magnetic field mapping at rest, *European Heart Journal,(Abstract Supplement)* 29:169
5. Herve JC, Yamaoka K, Twist VW, Powell T, Ellroy JC, and Wang LC.(1992) Temperature dependence of electrophysiological properties of guinea pig and ground squirrel myocytes. *Am J Physiol Regul Integr Comp Physiol* . 263:R177-R184
6. Hoffman BF and Cranefield PF (1976) Electrophysiology of the Heart Mount Kisco. NY: Futura, 95-98
7. Kiyosue T, Arita M (1989) Late sodium current and its contribution to action potential configuration of guinea-pig ventricular myocytes. *Circ. Res.* 64:389-397
8. Gambassi G, Cerbai E, Pahor M, Capogrossi MC, Carbonin P, and Mugelli A.(1994) Temperature modulates calcium homeostasis and ventricular arrhythmias in myocardial preparations.*Cardiovasc Res.* 28: 391-399
9. Kiyosue, T, Amrita M, Muramatsu H, Spindler A J, Noble D (1993) Ionic mechanisms of action potential prolongation at low temperature in guinea-pig ventricular myocytes. *J Physiol.(Lond.)*, 468:85-106
10. Kirsch GE, Sykes J S (1987) Temperature dependence of Na currents in rabbit and frog muscle membrane.. *Gen. Physiol.* 89:239-251

Author: V. Sosnytskyy
Institute: Glushkov Institute of Cybernetics
Street: 40,prosp.Glushkova
City: Kiev
Country: Ukraine
Email: sosna1@voliacable.com

Magnetocardiography Study of Cardiac Anomalies

K. Gireesan[1], S. Sengottuvel[1], C. Parasakthi[1], Rajesh Patel[1], M.P. Janawadkar[1], T.S. Radhakrishnan[1], K.S.R. Murthy[2], and A. Vijaya[2]

[1] Condensed Matter Physics division, Indira Gandhi Centre for Atomic Research, Kalpakkam – 603102, India
[2] Department of Atomic Energy Hospital, Kalpakkam – 603102, India

Abstract— **Magnetocardiography (MCG) recordings were carried out on 40 subjects comprising of both normal and those with cardiac anomalies, referred by the collaborating hospital, with the view to compare with the routine 12 lead ECG. An indigenously built four channel Low Tc - SQUID system inside a magnetically shielded room was used for the MCG. In this paper, results on a few specific subjects are presented. In the case of a 68 years old male with ECG indicating Left Bundle Branch Block characterized by wide QRS and a barely split R peak, the MCG showed two well separated RR' waves making a M pattern over several chest positions. A 51 years old male with normal sinus rhythm in the ECG, exhibited fragmented QRS (<120 ms) in MCG in 21 of the 36 chest positions recorded. Such very well resolved features were also obtained in MCG for subjects diagnosed with myocardial infarction on the basis of ECG. As a control study, some subjects with normal ECG exhibited regular and symmetric MCG patterns across the chest. In the MCG, several cardiac cycles were signal averaged to one cycle by appropriate time epoching. The magnetic field map at the instant of the R wave peak indicated the position of the equivalent current dipole to be in the region of the left ventricle. While the data could be fitted to one dipole in the case of the normal subject, those with fragmented QRS and other anomalies needed more than one dipole to fit the data. Limiting the bandwidth of the MCG data in representative cases by appropriate filters to the lower bandwidth employed in routine ECG did not suppress the major features, indicating that the resolved features seen in MCG are intrinsic to the technique. We conclude that MCG is inherently more sensitive and gives highly resolved information as compared to the ECG.**

Keywords— **Magnetocardiography, SQUID, Cardiac anomalies, Fragmented QRS, Magnetic Field Mapping**

I. INTRODUCTION

Magnetocardiography (MCG) is the measurement of magnetic field due to the flow of elementary currents inside the myocardial muscle of the heart and is a non-invasive and non-contact method. Although MCG has not yet been established as a routine clinical tool, diagnosis on MCG is reported to be successful and advantageous [1, 2] over ECG in several clinically important problems, including the risk assessment of life-threatening arrhythmias, detection and characterization of coronary heart diseases, noninvasive localization of cardiac activation such as arrhythmia causing regions or ischemic currents, and studies of fetal heart function. This paper presents the MCG studies on a few subjects to illustrate the effectiveness of MCG in probing cardiac abnormalities as a part of a study carried out on 40 subjects (both normal subjects and those with cardiac dysfunctions) in collaboration with cardiac specialists.

II. THE FOUR CHANNEL MCG SYSTEM

The MCG system developed indigenously consisted of four Nb based DC SQUID channels incorporated inside a FRP liquid helium dewar and housed inside a two layer magnetically shielded room (MSR). The MSR itself was erected inside a special building constructed using grade 316 stainless steel rods as reinforcement in concrete (SS-RCC) to minimize the influence of transient magnetic noise. The immunity to transient disturbances was expressly checked by moving a tractor outside this construction; the noise recorded inside the SS-RCC building was about 10 times lower than that in a conventional steel reinforced building. Further, the MSR was erected on a 50Tonne SS-RCC foundation block to reduce vibrations, measured with an accelerometer to be below 10μg. Each SQUID was coupled to a superconducting first order axial gradiometer with a 15mm loop diameter and 50mm base-line. The gradiometers were seated at 56mm PCD in the recesses provided in the bottom plate of the dewar reducing the stand off distance to about 15mm. The outputs of the SQUIDs were linearised using a SQUID electronics module located close to the dewar. The cable carrying this linearised output was routed through a waveguide to a radio frequency shielded room (RFSR) where the output of each channel was simultaneously digitized with a resolution of 24bits at a sampling rate of 1 kHz. The digitized data was transmitted for data storage outside the RFSR using optical fibre. This elaborate noise suppression arrangement has been provided essentially as a part of a Magnetoencephalography system, where the signals are typically about two orders of magnitude smaller than the MCG signals. The measured system noise floor was 12 femto Tesla/√Hz in each channel.

S. Supek and A. Sušac (Eds.): Advances in Biomagnetism – BIOMAG2010, IFMBE Proceedings 28, pp. 431–435, 2010.
www.springerlink.com

The output from the four channels can be simultaneously recorded.

III. Experimental

The subjects for the MCG study were referred by the Department of Atomic Energy Hospital associated with our Research Centre. About 40 subjects have so far been studied, comprising of both normal (without any known cardiac dysfunction) and with cardiac anomalies inferred based on their ECGs recorded at the hospital. In all cases, the MCG recordings were carried out within two hours of the ECG recordings. The MCG recordings were made over the chest of the subject on a grid at 36 equispaced positions, the grid spacing being 42mm which is the distance between two adjacent gradiometers in the four channel MEG system. The recordings were carried out by placing a paper grid on the subject's chest and moving the subject to cover a total of nine configurations, each configuration simultaneously measuring from four spatial positions.

The as recorded time traces (magnetic field in pico Tesla on the Y-axis and time in seconds on the X-axis) are prone to noise and artifacts, the major ones being the interference of the line frequency and its harmonics and the breathing of the subject. During the measurement, the time traces were recorded with the hardware filter set at 300Hz and subsequently band pass filtered to limit the signals to the frequency band of 0.01 to 120 Hz (ECGs are normally recorded in the frequency band 0.05 - 35 Hz). The artifacts were removed by the application of the wavelet transform. The eliminated component distinctly showed the breathing rhythm of the subjects at the breathing rate of about 17 to 20 cycles per minute. The cleaned time traces (about 20 cardiac cycles in each position) were then averaged to further suppress the effects of random noise. For this, the traces in each configuration were time referenced to the R wave peak and then epoched; thus, the averaging of a number of cardiac cycles at a given location was performed. The pre and post R-peak durations for epoching and averaging purposes were suitably selected for each subject so that all the features of the cardiac cycle were covered. The resulting time traces, thus averaged to one cardiac cycle are designated as SAMCG (signal averaged MCG).

IV. Results and discussions

We present here in some detail the results of a few specific cases from among the 40 subjects studied, in order to bring home the efficacy of the MCG technique over the routine ECG. Fig 1 shows the signal averaged MCG of a 68 years old male, diagnosed with left bundle branch block (LBBB) on the basis of ECG. The ECG shows a widened QRS as well as a barely seen notch at the top of the R wave (highlighted with a circle in the figure). In contrast, the MCG shows well split R waves in several positions, in the region between the 4th and 5th intercostal spaces on both the left and right side of the heart. Two cardiac cycles from the unaveraged MCG time trace of a position in this region are also shown in the inset, where the enhanced splitting in the R wave is highlighted with a circle. The iso-magnetic field contour plot (magnetic field mapping or MFM for short) for the instant of the R wave peak, has contributions from at least two equivalent current dipoles. The result of a least square fit shows one dipole to be located in the 4th intercostal space near the V3 lead position and above V4 of the ECG and the other dipole to the left of the sternum in the third intercostal space.

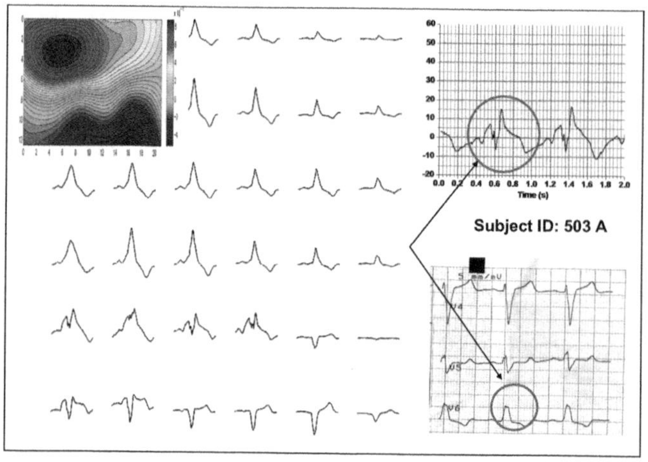

Fig. 1 Signal averaged MCG at the spatial positions recorded over the chest. The top left panel is the corresponding magnetic field map. The ECG at lead V6 showing features of LBBB, highlighted with a circle is at bottom right. The inset at the top right shows the time trace of two cardiac cycles in MCG at a position above V4 of ECG. The split in the R wave is highlighted with a circle. The maximum y-magnitude is 80 pT

It is illustrative now to discuss the case of a normal subject. Fig. 2 shows the signal averaged MCG of a 48 years old male having a normal ECG. The MCG is regular in all the spatial positions and symmetric over the chest. The magnetic field map at the instant of the R wave peak shows one dominant dipole in the region of the left ventricle. The field data could be least square fitted to a single dipole located at the intersection of the mid clavicular line, with the 4th intercostal space.

MCG has shown very distinguishing features in the case of a subject, a 51 years old male having normal ECG. The MCG traces have shown intra QRS fragmentation (designated as fQRS) in at least 21 of the 36 positions recorded. The QRS itself is not large, being less than 120msec duration. Fig. 3 shows the signal averaged MCG where this feature can be seen in an inset. The magnetic field map of the as recorded data clearly shows two dipoles, one deeper as compared to the other, judging from the distance between the minima and maxima of the corresponding dipolar field pattern.

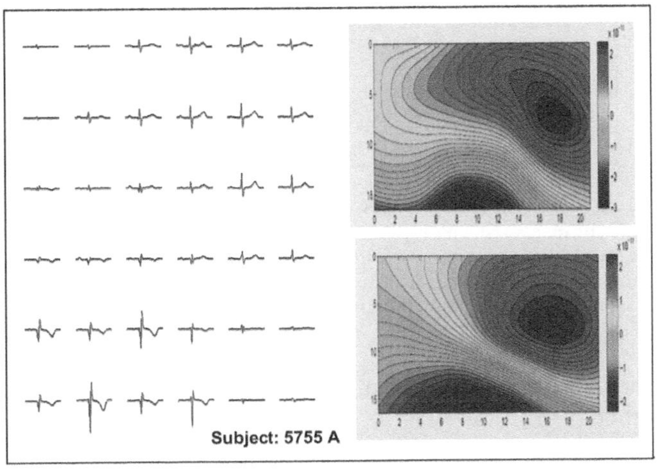

Fig. 2 Signal averaged MCG of a normal subject, a 48 years old male. The magnetic field map of the data is on the top right and the least square fit to the data is at the bottom right. The maximum y-magnitude is about 50pT

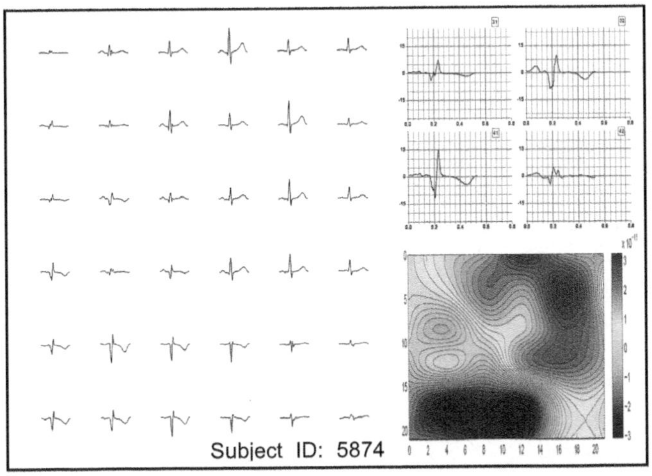

Fig. 3 Signal averaged MCG of a 51 years old male exhibiting fragmented QRS in several spatial positions. The inset at top right shows the enlarged traces at four of the positions covering the third intercostal space on the right side of the heart. The magnetic field map of the as recorded data is shown at the bottom right. The maximum y-magnitude is about 35pT

Indeed, the fragmentation in QRS is observed in MCG in a few subjects with normal ECG. It appears that it is often seen, but not always, in the MCG at "zero-line" (upper right to lower left across the thorax); the signals measured at these sites will tend to be ambiguous, i.e. they often do not show a clear R peak but have several inflections during the QRS complex, which may not be ascribed to cardiac anomalies. However this aspect has not been discussed extensively in the literature. There are albeit other cases, where the fragmentation could be observed outside of the zero line. Fig. 4 shows the MCG trace recorded at a location in the 3rd intercostal space at the right side of the heart of a 52 years old male exhibiting fQRS in a few spatial positions. Fig 4 also shows the MCG trace in the same position of another subject, a 69 years old female also with normal ECG. The fragmentation is more marked in the latter. The magnetic field maps at the R wave peaks of the as recorded data in both cases show clearly the presence of two dipoles. In both the cases, the least square fits performed have better figures of merit for a fit with two dipoles compared to that of a single dipole (not shown).

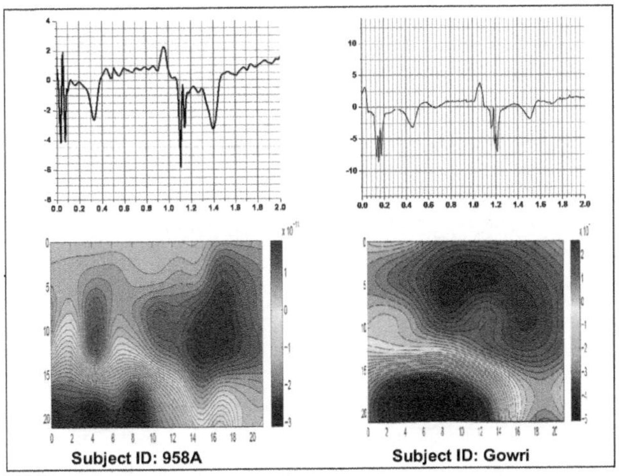

Fig. 4Top left: MCG trace at the third intercostal space at the right side of the heart of a 52 years old male with a normal ECG (not shown). Bottom left: The MFM of the as recorded data at the R peak, over the 36 spatial positions. Top right and bottom right: MCG at the same position and the MFM at the R peak for a 69 years old female, also with a normal ECG. The fragmentation is more marked in the latter

The occurrence of fQRS where the QRS duration is itself less than 120 msec (different from typical bundle branch blocks) is the focus of a number of recent investigations [3,4,5]. A study [4] considered the fQRS as markers of altered ventricular depolarization owing to a prior myocardial scar and postulated that the presence of fQRS might improve the ability to detect a prior myocardial infarction compared with detection of Q waves by ECG alone. Another study [5] associated a computed fragmentation score to MCG and using case histories over a prolonged duration correlated this MCG fragmentation score to the risk of death or arrhythmic events. The fQRS may be thought to arise as a result of transmural voltage gradients in action potentials similar to the occurrence [6] of Brugada syndrome from several causes including those ascribed to molecular genetics.

Intrafragmentation in QRS has also been seen in ECG in a few investigations, albeit in signal averaged ECG (designated as SAECG). Such a study [7] saw a reduced manifestation of fragmentation when the bandwidth of measurement was limited to 0-25 Hz from 0-150 Hz where a somewhat larger effect was seen. This study also hypothesized that fQRS could indicate a substrate for ventricular fibrillation.

In contrast to the SAECG studies, it is significant that fQRS is readily seen even in unaveraged MCG traces. The fragmentation components are often prominent and well resolved, akin to the manifestation of LBBB in MCG (compared to the ECG) contrasted in fig. 1. To probe the issue whether the higher resolution seen for LBBB as well as the prominent observation of fQRS in MCG as compared to ECG, are related to the higher bandwidth employed in MCG, the representative MCG data were bandwidth limited.

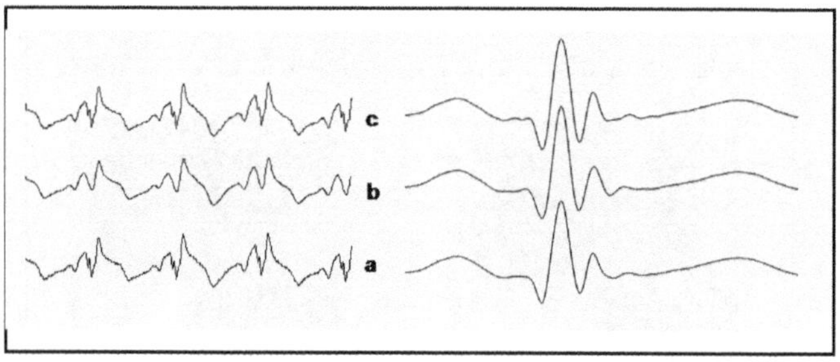

Fig 5 MCG time traces of two subjects, one with LBBB (left traces) and the other exhibiting fQRS (right traces). (a): as recorded data; (b) data subjected to a low pass filter with cut off at 35 Hz; (c): cut off at 45 Hz

Fig. 5 shows the as recorded time traces of the cardiac cycles for two subjects, one showing LBBB (left traces in the figure) and the other showing fQRS (right traces). The figure also shows the traces in each case, after subjecting the data to low pass filters with cut off frequencies at 35 and 45 Hz. In both the cases (LBBB and fQRS), the marked resolution of the MCG is seen to survive and the effects are not suppressed by limiting the bandwidth.

As another example, Fig.6 shows the MCG of a 59 years old male, diagnosed with myocardial infarction with characteristic tombstone like ST elevation in the ECG, shows these features more prominent and enhanced. The Magnetic field map at the instant of the R peak clearly

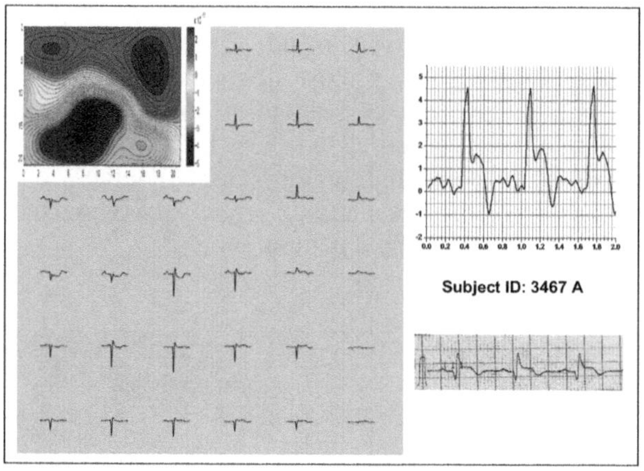

Fig. 6 The signal averaged MCG at several spatial positions of a 59 years old male diagnosed with myocardial infarction. The ECG at lead III and an enlarged MCG trace recorded just above the lead V5 of the ECG at the fourth intercostal space, are also shown. The magnetic field map at the R peak shows clearly the presence of two dipoles. The maximum y-magnitude is about 50 pT

shows two dipoles, one inferred to be deeper lying than the other.

V. Conclusions

In the several cases discussed, the MCG has revealed more resolved features as compared to the routine ECG and even compared to the signal averaged ECG. Particularly significant is the ability to reveal finer features like the intrafragmentation in QRS, which is the topic of many recent investigations. The Fragmented QRS is not readily seen in routine ECG and is only barely seen in the SAECG. The inherent higher resolution in the MCG should be related to the reduced sensitivity of MCG to the conductivity distribution of intervening tissues and skin surfaces as well as to their inhomogeneous and anisotropic electrical properties.

Acknowledgement

The work has been carried out as part of a project sponsored by the Department of Science & Technology, Government of India, with the expertise on SQUIDs [8] and skills developed over the years in the Department of Atomic Energy. We thank Dr. Baldev Raj and Dr. C.S.Sundar for encouragement and N.Mariappa for support in computation. We thank the referee for critical comments that has led to the revision of the paper in some aspects, particularly, the discussion pertaining to the nature of the QRS complex across the zero line.

References

1. Koch H (2001) IEEE Trans. Appl. Supercon. 11:49-59.
2. Ikefuji H et al. (2007) J. Med. Investigation, 54:116.
3. Jonathan S et al. (2006) J. Electrophysiol., 17:1-3.
4. Das M.K et al. (2006) Circulation 113:2495-2501.
5. Karhonen P et al. (2006) J. Cardiovas Electrophysiol. 10:1540-8167.
6. Scheiman M et al.(2007) J. Am. Coll. Card., 49:2061-2069.
7. Morita H et al. (2008) Circulation 118: 1697-1704.
8. Janawadkar M P et al. (1999) Current Science 77:759-769.

Magnetic Field Map Orientation in Patients after ST Elevation Myocardial Infarction

Peter Van Leeuwen[1], Birgit Hailer[2], A. Beck[1], Waleid Sherifa[2], Gregor Eiling[2], and Dietrich Grönemeyer[1]

[1] Dept. Radiology and Microtherapy, University Witten/Herdecke, Bochum, Germany
[2] Dept. of Internal Medicine, Philippusstift, Essen, Germany

Abstract— **Magnetic field map (MFM) orientation has been successfully applied in a number of studies investigating patients with coronary artery disease. In these patients, the degree of deviation of MFM orientation from normal values is dependent of the severity of the disease. In this work we examine MFM orientation in a large sample of patients who had just suffered acute ST Elevation Myocardial Infarction (STEMI) with successful interventional therapy. Ejection fraction (EF) and the presence of microvascular obstruction was assessed with cardiac MRI. Magnetocardiograms were obtained for 3 minutes in 83 STEMI patients and 31 age- and gender-matched healthy subjects using a 61 channel biomagnetometer. For each subject, the signals of ~50 beats of similar duration were averaged. MFM orientation was calculated for 420 points in the time-normalized QT interval and the percent deviations from the normal course, defined as the 95% confidence interval of the orientation values of the healthy subjects, was determined. The relative number deviations from MFM orientation was substantially higher for the STEMI subjects (35±18% vs. 4±7%, $p<0.001$). Sensitivity / specificity values were high, with an area under the receiver operating characteristic curve of 95% ($p<0.001$). Within the patient group, the % deviation was significantly higher in subjects with low EF ($p<0.05$), with three affected vessels ($p<0.005$) and with microvascular obstruction ($p<0.05$). Also, comparing patients with anterior and posterior infarction, orientation deviated less in the positive and more in the negative direction for the former group ($p<0.001$, $p<0.05$). The results demonstrate that MFM orientation is significantly changed after acute STEMI and that these changes can be associated with severity of the infarction. Furthermore infarct localization affected the direction of the deviation from normal orientation. MFM orientation may thus be useful in monitoring patients after acute myocardial infarction.**

Keywords— **Magnetic field map orientation, coronary artery disease, myocardial infarction.**

I. Introduction

In patients with coronary artery disease (CAD), several methods are used to investigate cardiac electrophysiological processes due to ischemia. Apart from routine electrocardiography (ECG), a further approach in this context is magnetocardiography (MCG). MCG measures the magnetic component of the cardiac electrical activity. Different aspects of the MCG signal have been examined and the effects of CAD have been described in numerous studies [1]. Among others, ischemia-induced changes in magnetic field map (MFM) orientation have been found. The majority of these studies were performed on CAD patients without myocardial infarction (MI). Some studies also included patients with chronic MI and found that deviations from normal orientation increased with increasing overall severity of CAD [2].

In order to better understand the effect of ischemia on structural myocardial changes, we examined patients who had just suffered an acute ST-elevation myocardial infarction (STEMI) and had undergone successful interventional therapy. Aim of this work was to investigate the discriminatory power of MFM orientation in STEMI patients. We also associated MFM orientation to a number of patient characteristics including infarct localization, ejection fraction, affected vessels and microvascular obstruction as assessed by cardiac magnetic resonance imaging (MRI). The overall goal is the assessment of MCG as a marker for the long-term benefit of revascularization therapy after successful reperfusion therapy.

II. Methods

A. Subjects

Eighty-three patients (aged 59.9±11.7 years, 66 male) who had suffered acute STEMI and received successful interventional therapy were included. Anterior infarction was present in 41, posterior in 42 patients; 35 had single vessel, 27 had two-vessel and 21 had three-vessel disease. The right coronary artery (RCA) was involved in 56, left anterior descending (LAD) in 62 and circumflex (RCX) in 34 patients. Ejection fraction (EF) was 49.7%±9.8%. Wall motion disturbances were present in 67. Microvascular obstruction was found in 36 patients. The normal group consisted of 31 healthy subjects (aged 55.3±9.4 years, 22 male) with no history of cardiovascular disease and normal resting ECG and echocardiograms.

B. Data Acquisition

MCG were obtained in all subjects using a 61 channel biomagnetometer (Magnes 1300C, 4D Neuroimaging, San

S. Supek and A. Sušac (Eds.): Advances in Biomagnetism – BIOMAG2010, IFMBE Proceedings 28, pp. 436–439, 2010.
www.springerlink.com

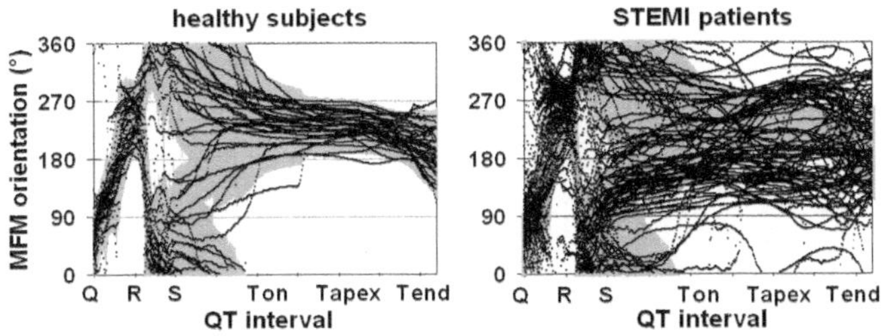

Fig. 1 MFM orientation during the QT interval for the healthy subjects and STEMI patients. The gray shading shows the 95% confidence interval of the courses of the healthy subjects

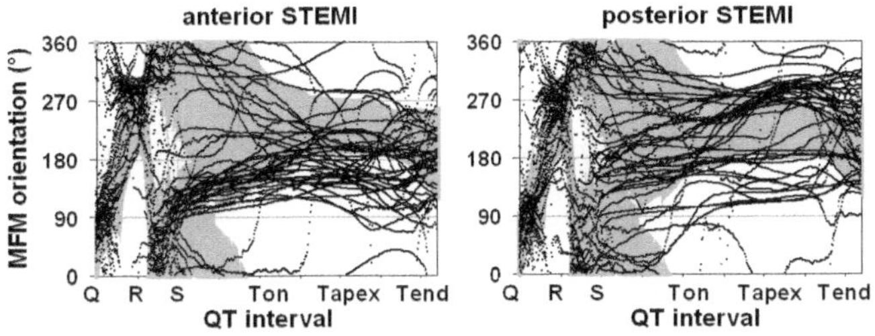

Fig. 2 MFM orientation during the QT interval for patients with anterior and posterior infarction. The gray shading shows the 95% confidence interval of the courses of the healthy subjects

Diego), in the patients 5.7±2.3 days after infarction. The sensor with a coverage of ca. 800 cm^2 was placed over the frontal thorax. Data were recorded for 3 minutes at a sampling rate of 1 kHz with a 200 Hz low pass filter. To reduce the effects of external noise, data were registered inside a standard 3-layered magnetically shielded room (Vakuumschmelze, Hanau). EF, wall motion disturbances and microvascular obstruction were assessed using 1.5 Tesla Twinspeed Excite MRI system (GE Deutschland, Munich).

C. Data Analysis

For each subject, all cardiac beats in the MCG data were identified on the basis of correlation to a representative QRS template from the subject's data. Beats within ±10 ms from the modal RR interval were selected for averaging (~50 beats/subject). In the averaged signal of each subject, Q onset, R peak, QRS end, T apex and T end were semi-automatically determined [3]. To enable rate independent, inter-individual comparison of MFM orientation for the complete QT interval, the sub-intervals separated by Q onset, R peak, QRS end, T onset, T apex and T end were time-normalized to give 420 time steps. For each time step, MFM orientation was quantified as the angle between 1) a left-right line across the thorax and 2) a line connecting positive and negative centers of gravity of the magnetic field [2]. This was done for a channel configuration of 46 channels, as the discriminatory power of this configuration has been shown to be higher than that of a 61 channel configuration [4].

On the basis of the MRI examination, EF (in %) was calculated on the basis of ventricle volumes. The presence of wall motion disturbances was determined visually. The existence of microvascular obstruction was identified in contrast-enhanced images as early hypo-enhancement.

D. Statistics

In all subjects, the number of deviations from the 95% confidence interval of the normal MFM orientation of the healthy subjects over the QT interval were determined. We compared the deviations from normal MFM orientation between the normal and STEMI groups on the basis of descriptive statistics and the Mann-Whitney U-test (2-sided). Further, the capacity of the deviations to distinguish between the healthy subjects and patients was examined on the basis of the receiver operating characteristic (ROC) where the area under the curve (AUC) was used as a measure for sensitivity/specificity.

The number of deviations from normal orientation in sub-groups of patients based on infarct localization, wall motion disturbances, ischemia, and microvascular obstruction were tested with the Mann-Whitney U-test, the number of affected vessels with the chi-square test (both 1-sided). EF was examined on the basis of regression analysis. A p-value <0.05 was considered statistically significant.

III. Results

In the healthy subjects, MFM orientation displayed a similar course over the QT interval with the values lying within a narrow range of about 40°-70° except for the ST interval (Fig. 1, left panel). Deviations from the normal course were obvious in the STEMI group, particularly during the T wave (Fig. 1, right panel). The healthy subjects had, on average, 4±7% deviations from the normal course whereas the patients had 35±18% ($p<0.001$). The ROC analysis showed a 94.9% area-under-the-curve ($p<0.001$), indicating that MFM orientation was able to discriminate extremely well between the two subject groups. At a cut-off of 10% deviations from the normal course, sensitivity was 88% and specificity was 90%.

Examination of patients sub-groups showed that although the number of deviations from the normal course was not influenced by infarct localization (anterior 35±19%, posterior 35±17%, $p=0.85$), the direction of the deviation from normal orientation was strongly affected. Particularly during repolarization in the subjects with anterior infarction, there was a strong trend for orientation to shift clockwise

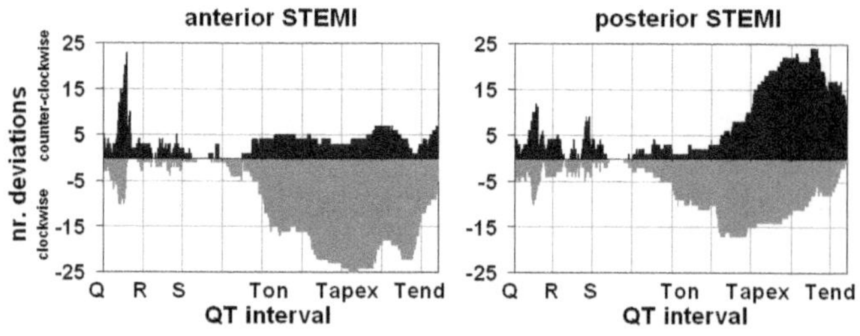

Fig. 3 Number of positive and negative deviations from the normal course of MFM orientation during the QT interval for patients with anterior and posterior infarction (cf. Fig. 2)

away from the normal values (Fig. 2, left panel). For most patients with posterior infarction, such a trend was not as clear, although the majority shifted counter-clockwise during the T wave while the others shifted clockwise (Fig. 2, right panel). For almost all patients, however, the tendency was to rotate away from the normal orientation values of ~ 230° around T apex.

The number of negative (clockwise) deviations from the normal course differed between these two patient subgroups: anterior –27±20%, posterior –17±20% ($p=0.024$). This was also true for the positive (counter-clockwise) numbers: 9±10% vs. 18±14% ($p<0.001$). Fig. 3 shows the distribution of these differences during the QT interval, which are obvious during repolarization but which also indicate some smaller differences during depolarization, particularly during QR upstroke.

In the patient group, regression analysis showed a negative relationship between the number of deviations from normal orientation and overall left ventricular function: patients with low EF had a more abnormal MFM orientation values (slope –0.44, $p=0.031$). Fig. 4 shows that of the 34 patients with an EF < 50% all but three patients had more than 25% deviations from normal orientation.

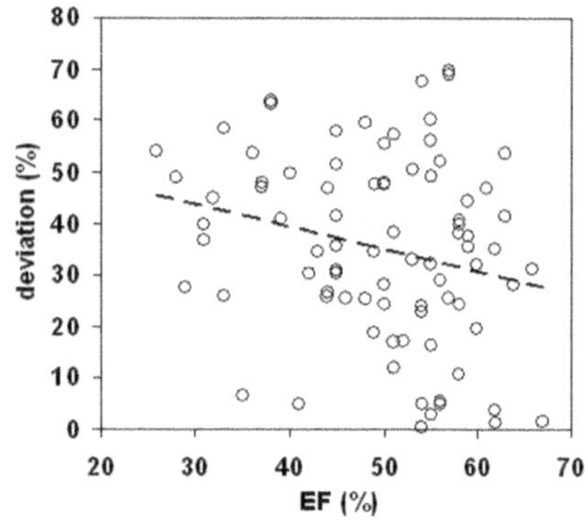

Fig. 4 Relationship between ejection fraction (EF) and the percent deviations from the normal course of MFM orientation in the STEMI patients. The dashed line shows the slope of the regression

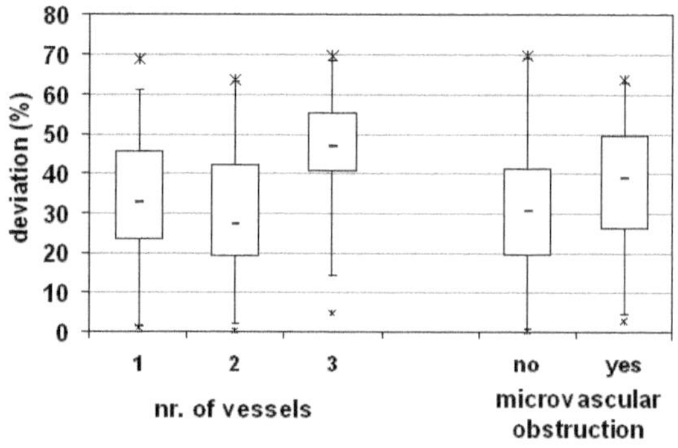

Fig. 5 Boxplots of the percent deviations from the normal course of MFM orientation in the STEMI patients with respect to the number of affected vessels (left) and the presence of microvascular obstruction (right)

Abnormal orientation was also associated with the number of affected vessels: patient with 3-vessel disease had significantly more deviation form normal orientation than those with 1- or 2-vessel disease ($p=0.003$; Fig. 5, left). Also, in patients in whom microvascular obstruction was detected, there was a tendency to more deviations than in those patients without such obstruction ($p<0.05$; Fig 5, right). The degree of deviations from normal orientation did not differ between patients with or without wall motion disturbances or ischemia.

IV. Discussion

The results demonstrate that MFM orientation is significantly changed after acute STEMI. The myocardial damage inflicted by infarction leads to changes in electrophysiological properties of the heart and these changes are visible not only in the ECG but also the MCG. Multichannel MCG offers the possibility of mapping the distribution of the cardiac magnetic field and following the changes in these fields over ventricular de- and repolarization with high temporal and spatial resolution. Of the various approaches that have been developed to efficiently condense the large amount of information captured in the MFM, the orientation of the field has been frequently used [5] [6]. Only few reports exist of the examination of MFM orientation in patients with previous [2] [7] or acute [8] myocardial infarction.

Here we compared the orientation of patients who had suffered acute STEMI with an age- and gender-matched group of healthy subjects with no history of cardiovascular disease. We could show that the discriminatory power of MFM orientation was high. On the one hand, this is not surprising, as the patients had suffered significant damage to their cardiac tissue. On the other hand, the MCG measurement was performed after successful percutaneous coronary intervention (PCI). Thus the degree of deviation from normal orientation may be an indicator of the extent of the damage and may help in the prognosis for these patients. Indeed, the results indicate that the severity of the infarction as reflected in ventricular function (EF), the number of affected vessels and the presence of microvascular obstruction can also be associated with the extent of abnormal orientation.

We also note that, not only does the deviation, as such, indicate functional electrophysiological disturbances, but also that the direction of the deviation reflects morphological structural changes. The localization of the infarcted tissue led to shifts in the distribution of magnetic field strength and these were dependent on the localization of the infarction. This was most clearly seen in patients with anterior infarction, almost all of whom showed a clockwise rotation away from normal orientation during the ST segment and in particular during the T wave. To a lesser degree and in the opposite direction, this was also true for patients with posterior infarction. Tracking the changes in MFM orientation after PCI may thus aid in the assessment of recovery and normalization of microvascular perfusion and ventricular function. Furthermore, the method could serve as a prognostic tool for identification of those patients who might not profit from early reperfusion and who consequently develop large myocardial scars.

V. Conclusions

The results demonstrate that MFM orientation is significantly changed after acute STEMI and that these changes can be associated with severity of the infarction. Furthermore infarct localization affected the direction of the deviation from normal orientation. MFM orientation may thus be useful in monitoring patients after acute myocardial infarction.

References

1. Stroink G, Hailer B, Van Leeuwen P (2006) Cardiomagnetism. In: Andrä W, Nowak H, eds., Magnetism in Medicine. Wiley-VCH, Weinheim, pp 164–209
2. Van Leeuwen P, Hailer B, Lange S, Grönemeyer D (2006) Identification of patients with coronary artery disease using magnetocardiographic signal analysis. Biomed Tech 51: 83-88 DOI 10.1515/BMT.2006.015
3. Van Leeuwen P, Geue D, Poplutz C, Klein A, Lange S, Grönemeyer D (2003) Reliability of automated determination of QRST times in ECG and MCG. Biomed Tech 48(suppl 1): 370-371
4. Van Leeuwen P, Klein A, Matil K, Geue D, Poplutz C, Lange S, Hailer B, Grönemeyer D (2003) Influence of area of coverage on cardiac magnetic field map orientation. Biomed Tech 48 (suppl 1): 238-239
5. Gapelyuk A, Wessel N, Fischer R, Zacharzowsky U, Koch L, Selbig D, Schutt H, Sawitzki B, Luft FC, Dietz R, Schirdewan A (2007) Detection of patients with coronary artery disease using cardiac magnetic field mapping at rest. J Electrocardiol 40: 401-407 DOI 10.1016/j.jelectrocard.2007.03.013
6. Van Leeuwen P, Hailer B, Lange S et al. (2008) Quantification of cardiac magnetic field orientation during ventricular de- and repolarization, Phys Med Biol 53:2291-2301 DOI 10.1088/0031-9155/53/9/ 006
7. Hänninen H, Takala P, Korhonen P, Oikarinen L, Makijarvi M, Nenonen J, Katila T, Toivonen L (2002) Features of ST segment and T-wave in exercise-induced myocardial ischemia evaluated with multichannel magnetocardiography. Ann Med 34: 120-129. DOI: 10.1080/07853890252953518
8. Lim HK, Kim K, Lee YH, Chung N (2009) Detection of non-ST-elevation myocardial infarction using magnetocardiogram: New information from spatiotemporal electrical activation map. Ann Med 41: 533-546. DOI: 10.1080/07853890903107883

Author: Peter Van Leeuwen
Institute: Dept. Biomagnetism, Chair of Radiology and Microtherapy, University of Witten/Herdecke
Street: Universitätsstr. 142
City: 44799 Bochum
Country: Germany
Email: petervl@microtherapy.de

AC Biosusceptometry as a Tool for Monitoring the Gastrointestinal Transit of Multiparticulate Drug Delivery System

G.F. Oliveira[1], P.C. Ferrari[1], L.A. Corá[2], U. Andreis[2], J.R.A. Miranda[2], and R.C. Evangelista[1]

[1] UNESP – School of Pharmaceutical Sciences, Araraquara, Brazil
[2] UNESP – Biosciences Institut, Botucatu, Brazil

Abstract— **Drug delivery systems based on natural polysaccharides, such as chitosan (CS) and pectin (PC), rather than on synthetic polymers, have been widely studied. Some reasons for that are low toxicity and costs and high biodegradability of the formers. A multiparticulate system based on CS and PC was developed in our laboratories, including the addition of an enteric polymer, cellulose acetate phtalate (CAP). Such improvement promoted stronger gastric and enteric resistances, as assessed *in vitro*, making the systems more selective to enzymatic degradation in the colon. Although *in vitro* dissolution tests can simulate some properties concerning the gastrointestinal transit (GT), collaborating to characterize the systems behavior in the biological fluids, frequently they do not result in satisfactory *in vitro/in vivo* correlations. The objective of this work was to follow *in vivo* the GT of the particles developed by means of AC biosusceptometry (ACB), a non-invasive and of low cost methodology. The particles containing ferrite in powder form were prepared by complex coacervation using an ideal 3:1:1 mass ratio for PC:CS:CAP. The magnetic particles were administered to healthy volunteers by oral route. The GT was monitored by using multi-sensor ACB system and the signal acquisition was performed every 15 min until the colonic region was reached. By means of ACB technique, it was possible to acquiring images generated by the magnetic particles within the whole gastrointestinal tract including the colonic region. Variable particles transit times were observed among the volunteers, but without interference on the mapping of the particles until the colonic region. The particles were able to produce magnetic field strong enough to generate signals adequate for mapping the particles. The results suggest that integral particles reached the colon, after they resisted against gastric and enteric media. Studies associating transit time and *in vivo* drug release are in development in order to confirm the efficiency of the systems.**

Keywords— **multiparticulated systems, chitosan, pectin, AC biosusceptometry, colonic delivery systems.**

I. INTRODUCTION

Colon-selective drug delivery systems have been the focus of increasing interest for the last decades. This is mainly due to the recognized importance of this region of the gastrointestinal tract (GIT), not only for local but also for systemic therapy [1]. CS is a cationic polysaccharide made from alkaline N-deacetylation of chitin, the second most abundant natural polymer on earth [2]. CS is non-toxic, biocompatible, and biodegradable. PC is an anionic polysaccharide that provides colon-specific delivery for several drugs [3]. Amidated PC is low-methoxyl PC in which some of the carboxylic acid groups are amidated. They are more tolerant against pH variations and calcium level than conventional PC, which could make them useful in colonic delivery systems. The properties of such amidated PC particles may be altered by the formation of polyelectrolyte complex membrane around the particle using cationic polymers such as CS [4]. Various systems have been developed for colon specific drug delivery. These include systems developed using pH-sensitive polymers (enteric coating polymers), based on transit time or increasing of luminal pressure within the GIT, and enzymatically controlled delivery systems. The method by which the drug release will be triggered by the colonic flora appears to be more interesting with regard to the selectivity (Rubinstein, 1990). Multiparticulated systems have several advantages in comparison to the conventional single unit types for controlled-release technology, such as more predictable gastric emptying and fewer localized adverse effects (Mehta et al., 2001). The use of naturally occurring polysaccharides is attracting a lot of attention for drug targeting to the colon since these polymers are found in abundance, are inexpensive and available in a variety of structures with varied properties (Hovgaard and Brondsted, 1996).

A multiparticulate system based on CS and PC was developed in our laboratories, including the addition of an enteric polymer, cellulose acetate phtalate (CAP). Such improvement promoted stronger gastric and enteric resistances, as assessed *in vitro*, making the systems more selective to enzymatic degradation in the colon. Although *in vitro* dissolution tests can simulate some properties concerning the gastrointestinal transit (GT), helping to characterize the systems behavior in the biological fluids, frequently they do not result in satisfactory *in vitro/in vivo* correlations. For this reason the aim of this work was to follow *in vivo* the GT of the particles developed by means of AC biosusceptometry (ACB).

II. EXPERIMENTAL METHODS

A. Particles Preparation

The particles were prepared by a complex coacervation method. The ideal ratio between the polyelectrolytes was

S. Supek and A. Sušac (Eds.): Advances in Biomagnetism – BIOMAG2010, IFMBE Proceedings 28, pp. 440–442, 2010.
www.springerlink.com

applied, i.e., PC:CS=3:1. The CAP solution and Ferrite were incorporated into the pectin (PC) solution. Then, CS links to free carboxylic groups, hardening the system. After 3 h under stirring, the particles were filtered and freeze-dried. Four different formulations used in this study are shown in the Table 1.

Table 1 Particles prepared

Samples	Ca^{+2} %	PC 1,5%	QS 0,5%	CAP 0,5%	CAP 1%	TC mg	F g
(A)	2	15	40	5		200	1
(B)	2	15	40		5	100	1
(C)	2	15	40	5		200	1
(D)	2	15	40		5	200	1

B. Morphological and Granulometric Studies

Morphological and granulometric analyses were performed by optical microscopy using Leica Qwin software.

C. In vivo Studies

The *in vivo* investigation was approved by the Ethics Committee in Research of the Medical School, Universidade Estadual Paulista (UNESP). The *in vivo* study was carried out in four healthy volunteers. All volunteers fasted at least 12 h prior to the administration of magnetic particles. Each volunteer, in an orthostatic position in the measurement system, swallowed magnetic particles with 200 ml of water. The single sensor ACB system (Fig.1) was positioned on the abdominal surface (Fig. 2), and the magnetic signals were acquired each 15 min until they reached the colon.

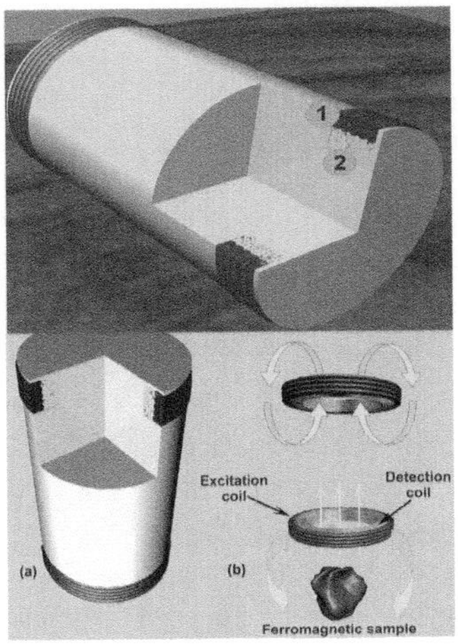

Fig. 1 Single-sensor AC Biosusceptometry system with the pair of excitation (1) and detection coils (2) coaxially arranged in a first-order gradiometric configuration

The single sensor ACB has two pairs of coils (φ = 3.0cm) separated by a fixed distance (baseline), where each pair of coils are composed of an excitation coil (external) and a detection coil (internal), in the first-order gradiometric configuration (Fig. 1). This system working as a double magnetic flux transformer with an air nucleus, in which the pair (excitation/detection), located more distant from magnetic material (ferrite), acts as reference [5]. Due to this configuration, when no magnetic material is near to the measurement system, the output signal is minimized. When a magnetic mass is approaching, an unbalancing in the magnetic flux of the gradiometric system occurs, and the magnetic material is detected [6].

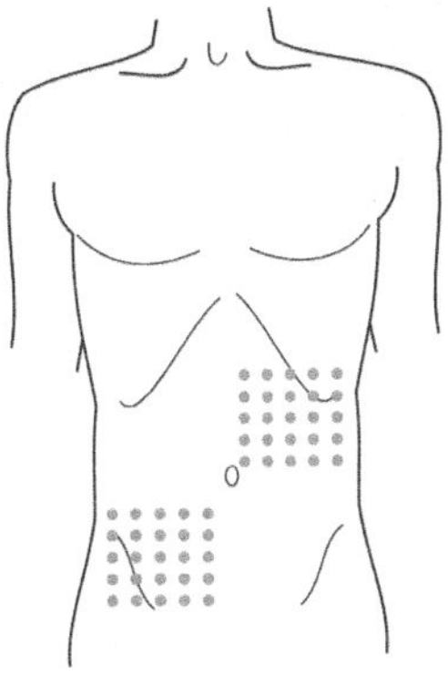

Fig. 2 Schematic representation of square point matrix (gray circles) drawn around gastric and colonic regions

III. Results and Discussion

A. Morphological and Granulometric Analyses

The incorporation of ferrite lead to significant size variation among the particles types, the mean diameter being 1.5-2.0 mm (n = 220) for particles without ferrite while the mean diameter for the particles with ferrite was 3.93 ± 0.66 (Fig. 3).

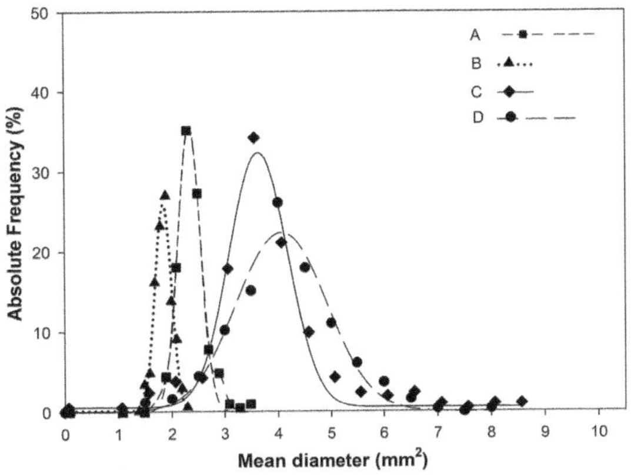

Fig. 3 Particles size distribution

B. *In vivo Studies*

Figure 4 show the sequence of magnetic images of the multiparticulated system in the colonic region. White outlines are the ribs on the top and the iliac crests on the botton.

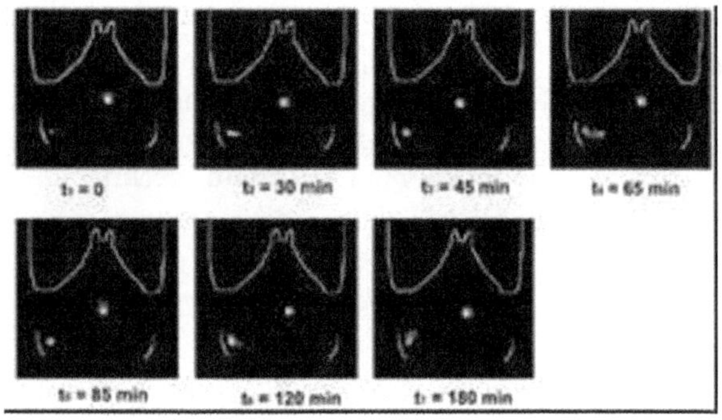

Fig. 4 Sequence of magnetic images of magnetic particles on the colonic region

Figure 5 shows the area of magnetic field detected *versus* time and exemplifies the gastric emptying (GE) and colon arrival (CA) profiles for a subject after ingestion of particles of samples (a) FCAP1TC (D) and (b) FCAP05TC200 (C) in fasted conditions.

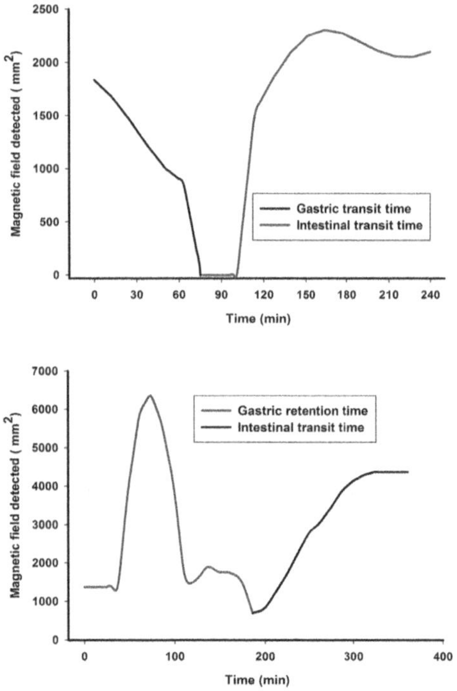

Fig. 5 Gastric emptying (line red) and colon arrival (line blue) in fasted conditions of voluntary after ingestion of the particles (a): with 1% of CAP end (b) with 0,5% of CAP

IV. Conclusion

The particles prepared were able to produce enough magnetic field to generate consistent magnetic signal, thereby allowing their mapping. The generation of magnetic images from the capsules containing the system under study allowed multiple-view the process of arrival at the colon, which suggests that the particles resisted against disintegration in the stomach and small intestine and achieved the colonic region.

Acknowledgments

The autors would like to thanks the Brazilian agencies FAPESP and CAPES for financial support.

References

1. G.S. Macleod, G.S. Fell, et al. (1999) An in vivo investigation into the potencial for bimodal drug release from pectin/chitosan/HPMC-coated tablets. Int. J. Pharm., v.188, p.11-18,
2. V.K. Gupta, T.E. Beckert, et al. (2001) A novel pH- and time-based multi-unit potential colonic drug delivery system. I. Development. International Journal of Pharmaceutics, v, 213, p. 83-91.
3. L. Yang, J.S. Chu, et al. (2002) Colon-specific drug delivery: new approaches and in vitro/in vivo evaluation. Int. J. Pharm., v.235, p.1-15.
4. N. Shimono, T. Takatori, et al. (2002) Chitosan dispersed system for colon-specific drug delivery. Int. J. Pharm., v.245, p. 45-54
5. L. A. Corá, F. G. Romeiro, M. Stelzer, et al. (2005) AC biosusceptometryin the study of drug delivery. Adv. Drug Deliv.**57:**1223-1241
6. L.A. Corá, U. Andreis, F.G. Romeiro et al. (2005). Magnetic images of the disintegration processo f tablets in the human stomach by ac biosusceptometry. Phys. Med. Biol. 50: 5523-5534

Author: Giselle Faria Oliveira
Institute: School of Pharmaceutical Science
Street: Rod. Araraquara-Jaú, Km1
City: Araraquara - SP
Country: Brazil
Email: gi.arena@hotmail.com

A New Technique for Magnetic Nanoparticle Imaging Using Magnetoencephalography Frequency Data

Teresa Cheung[1,3], Karen L. Kavanagh[1], and Urs Ribary[2,3]

[1] Simon Fraser University/Physics, Burnaby, Canada
[2] Simon Fraser University/Psychology, Burnaby, Canada
[3] Down Syndrome Research Foundation, Burnaby, Canada

Abstract— **Nanoparticles are objects of nanometer dimensions. Magnetic nanoparticles consist of an inorganic core of iron oxide. An outer coating of bioactive molecules such as peptides or antibodies surrounds this core, which can be chemically functionalized to conjugate with specific molecules in cells. There is potential to utilize the magnetic properties of these particles for internal imaging of living organisms using magnetoencephalography (MEG). While it is clear that bulk movement of a collection of particles should be detectable by MEG either as a flow or via rigid mechanical motion, it would be far more useful to detect the particles in situ. We have taken the first step towards this goal by demonstrating that MEG can detect these particles at fixed locations by focusing on the magnetic noise generated from random rotations of the particles' magnetic moments. Using a 151-channel MEG device, a glass vial of Fe3O4 in a colloidal suspension was measured at stationary locations within the MEG helmet region. The data was collected at 300, 1200 and 12000 Hz with up to 4000 Hz bandwidth. Between 10 and 50 ten-second trials were collected. Fast Fourier transforms for each MEG channel was generated. The net effect of these fluctuations was not detectable in the time domain. However, these fluctuations appeared as an increase in noise in the frequency domain. The character of the noise had a 1/f^n slope where 0<n<1 with n approaching 0 (= white noise) as the concentration of the sample was reduced. Spatial contour maps of the frequency data showed a distinct peak near the location of the sample. These results suggest that MEG is sensitive enough to detect and potentially localize a stationary vial of magnetic nanoparticles in a colloidal suspension.**

Keywords— **nanoparticle, imaging, MEG, ferrofluid.**

I. Introduction

Nanoparticles are objects of nanometer dimensions. They consist of an inorganic core, such as gold, silicon, cadmium selenium or iron oxide. This core provides the functionality of the particle such as being fluorescent or magnetic. An outer shell with a coating of bioactive molecules such as peptides or antibodies surrounds this core. These surface molecules can be chemically functionalized to conjugate with specific molecules in cells such as proteins. For example, semiconductor fluorescent nanoparticles have been synthesized to conjugate to neurons and glia [4]. Using fluorescent microscopy, the movement of neuroreceptors can be visualized and tracked [1]. This process, however, has limited application in vivo where the presence of opaque tissue may prevent the ability to perform fluorescent microscopy. Magnetic nanoparticles can be detected in opaque tissue. Coupled with the highly sensitive magnetic field detectors called SQUIDs (superconducting quantum interference device), there exists exciting research and clinical possibilities to create a new technique for internal medical imaging.

MEG is a device consisting of multiple SQUID channels for the purpose of non-invasively detecting biological magnetic signals. MEG has been used mainly to detect natural biological signals such as the magnetic fields produced by intracellular currents of depolarizing neurons in the brain [9]. There is no restriction technologically to expand its usage to other sources of magnetic field generators such as the movement of magnetic nanoparticles. Used in this way, the technique would be akin to PET for magnetic rather than radioactive particles.

There have been a small number of studies involving nanoparticles and MEG. Romanus [5] and Jia [2,3] have demonstrated that SQUID sensors can detect magnetic nanoparticles subjected to pre-magnetization or in motion. Saligram [6] showed that these particles could be detected in the MCG (magnetic cardiogram) of rats when the particles were introduced into the bloodstream or heart. These studies showed changes in magnetic field signals over time. There have been no studies examining the frequency and thereby noise characteristics of magnetic nanoparticles.

Ferrofluids are colloidal suspensions of magnetic nanoparticles in a liquid carrier. The particles are coated with a shell that keeps the ferrofluid grains from agglomerating. Magnetic nanoparticles with core sizes between 10 nm and 100 nm are predominantly single domain in their magnetic structure [7]. A single particle can be described as a magnetic dipole and the magnetic field equations are derived using Maxwell's equations. These particles can be made to be either superparamagnetic or ferromagnetic. Particles in a colloidal liquid have moments that change randomly according to two processes. Brownian rotation refers to the random rotation of the whole particle itself,

S. Supek and A. Sušac (Eds.): Advances in Biomagnetism – BIOMAG2010, IFMBE Proceedings 28, pp. 443–446, 2010.
www.springerlink.com

which causes a change in moment. Néel rotation refers to the random rotation of the moment within the particle [7]. There is also Brownian translational movement of the particle. The two rotations contribute to the net moment change for each particle and together with translational movement give the measured magnetic field at any point in space and time. When bounded to the target molecule, the magnetic nanoparticle movement and whole particle rotation will be dictated by the properties of the target molecule. However, Néel rotation is hypothesized to remain unchanged.

By looking in the frequency domain, it may be possible to observe these random rotations if the moments are large enough to be detected. Our study examines the potential for utilizing frequency data to detect and localize a colloidal sample of magnetic nanoparticles.

II. Methods

A. *Sample Preparation*

A sample of ferrofluid manufactured by Ferrofluidics Corporation (Burlington, Massachusetts) was used for our data collections. The sample was a water based colloidal suspension of Fe_3O_4. A 0.5 ml volume of the ferrofluid was placed in a non-magnetic cylindrical glass vial. The empty container was checked for magnetic contamination before usage. A second diluted sample was prepared using 5 parts de-ionized water to 1 part ferrofluid.

B. *TEM Measurements*

To check the particle size and confirm the particle's chemical composition, the diluted sample was scanned using transmission electron microscopy (TEM) using a Tecnai G2 STEM (200 keV, 10 log column vacuum). A carbon grid was used to hold a dried sample of the diluted ferrofluid. Bright field images (with objective aperture), high magnification images (no objective aperture) and selected area diffraction patterns were collected for the sample (Figure 1).

C. *MEG Measurements*

The sample was placed on a non-magnetic phantom stand. The stand has calibrated locations for placement of fiducial markers to track the sample's location during MEG measurement relative to the sensor coordinates. Polhemus was used to digitize the sample's location relative to these calibrated fiducials. We estimate the sample's measured location error was between 1 to 3 mm.

A 151-channel CTF MEG device within a 4-layer Vacuumschmelze magnetically shielded room (MSR) was used for the measurements.

Each colloidal sample mounted on the phantom stand was measured at stationary locations within the MEG helmet region. To minimize magnetization effects, head localizations, which involve the generation of magnetic fields within the helmet, were done only when necessary and only at the end of a data collection. The data was collected at 300, 1200 and 12000 Hz. There is a permanent anti-aliasing filter set at ¼ sample rate during all collections. The data was collected with synthetic third gradient cancellation enabled. Between 10 and 50 ten-second trials were collected. To check stability of the sample, data was collected over a period of hours without disturbing the sample.

D. *Data Processing*

The particle sizes were manually measured from the bright field images. Spatial Fourier transforms were performed on the fringe patterns observed in the high magnification images. The diffraction rings were measured by generating intensities along a line that intersected the ring centre as a function of distance. Peak pairs were marked to identify ring boundaries. The diffraction ring diameters were defined to be the distance between these peak pairs. The ratios between radii pairs were computed. These were matched to known patterns to identify the chemical composition.

There were no offline filters applied to the MEG data. The bandwidth was therefore 0 to ¼ sample rate. Fast Fourier transforms for each MEG channel was generated and averaged (stacked) across the trials. Topological maps were also generated in the frequency domain. Selected frequencies were examined in the alpha, beta, and gamma bands. An amplitude was generated for each sensor by integrating across approximately 1 Hz frequency range at the selected frequency.

III. Results

A. *TEM Results*

TEM revealed the particle core sizes were between 10 – 50 nm. Diffraction pattern analysis confirmed the core was made of iron oxide, FE_3O_4 and probably amorphous. Some particles seemed to have aligned to form lines. There was no evidence of the materials that surrounded the ferrofluid droplets.

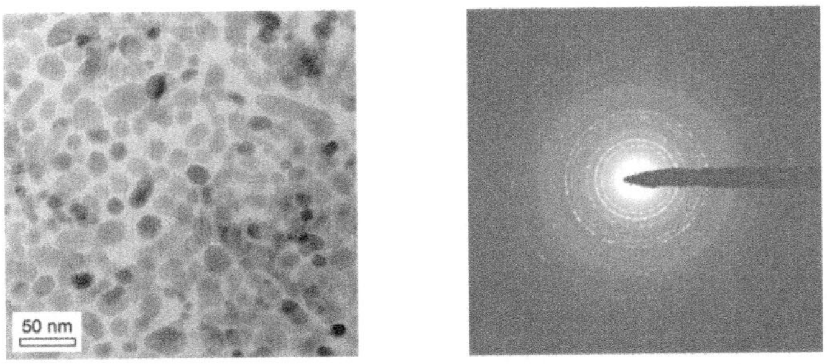

Fig. 1 Left – Bright field TEM image. Right – Diffraction pattern

B. MEG Results

The sensor noise floor is unique to each channel and does not generally change over time while the MEG system is cold and operational. Data from noise taken at the beginning and end of collections confirmed this stability. In the time domain, differences between the data with and without the sample were difficult to detect. Although some channels were noisier, the information was difficult to quantify in the time domain. However, in the frequency domain, these fluctuations appeared as an increase in the noise floor. Figure 2 (left) shows a log-log plot of one of the sensor channels closest to the ferrofluid sample. The slope of the noise has a $1/f^n$ characteristic where n is between 0 and 1. Figure 2 (right) shows a diluted sample placed at the same location for the same sensor channel. The slope of the noise is nearly white (n=0) but the data with the sample still shows an increase in the noise floor.

Figure 3 shows spatial contour maps of the frequency data in the bandwidth 1.98 to 2.94 Hz. In the left map, the sample was placed at the back of the helmet (occipital). In the right map, no sample was present. Comparison between the two maps shows a distinct increase in frequency power at the location of the sample. The same results were observed for different bandwidths until the noise floors of the two datasets merge at higher frequencies. It should be noted that noise data was also collected for the empty container, a sample of liquid water and the empty phantom stand. These datasets showed no significant difference from the right plot in Figure 3.

IV. Discussion

The MEG frequency data analysis suggests that a liquid colloidal suspension of magnetic particles with sufficiently large moments can be detected. The diluted sample showed a nearly white noise profile with an increase in noise floor when the sample was present. It is not clear if the undiluted sample included interactions beyond random rotations although the noise profile would suggest this is probable. Further investigation is underway to examine the effects of concentration.

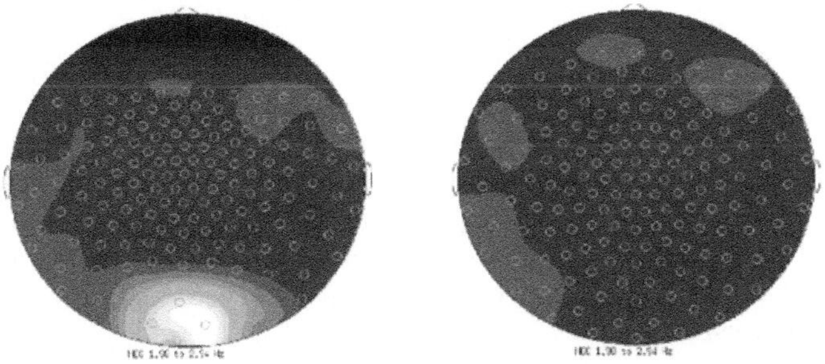

Fig. 3 Fourier transform spatial contour maps with (Left) and without (right) sample. Bandwidth = 1.98 – 2.94 Hz

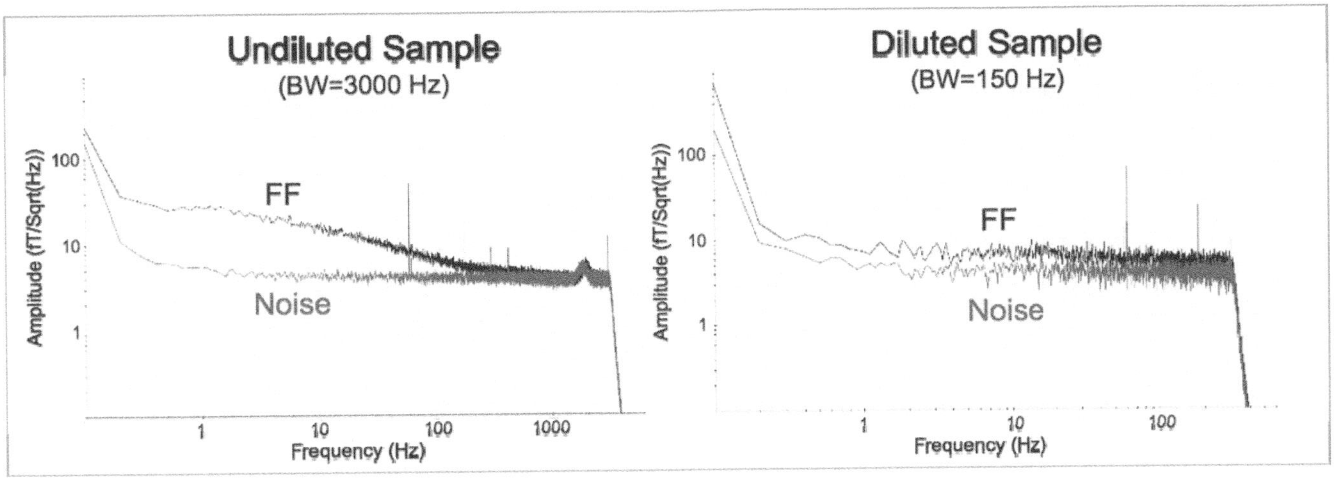

Fig. 2 Sensor noise compared to ferrofluid noise for a single channel. Left – undiluted sample. Right – diluted sample

There was a spatial correlation between location of the sample and increase in sensor power. These results support the possibility the MEG can be used to localize the sample spatially. A technique such as synthetic aperture magnetometry beamformers may be adaptable for such a purpose.

We have used a specific ferrofluid for these collections. Further tests to examine other magnetic nanoparticles are needed to determine which are suitable for this technique.

Whether the particles are superparamagnetic or ferromagnetic also plays a role in detectability. The non-zero magnetic field within the MSR allows superparamagnetic particles to be magnetized. The susceptibility of the particle will dictate the size of moments detected. In addition it is possible that introducing an external magnetic field may enhance the delectability of these particles. For example, the magnetic fields generated by a magnetic dipole such as those used for head localization could be used to introduce a field small enough not to unlock the sensors but perhaps large enough to affect the particles. These studies are also underway.

V. Conclusions

MEG can detect a colloidal suspension of magnetic nanoparticles particularly in the frequency spectrum. The spectrum approaches white when the concentration is lowered. The MEG topological maps clearly indicate a spatial correlation between sample location and sensor response. These results support the possibility of utilizing MEG to develop a tool to image magnetic nanoparticle within living organisms.

Acknowledgment

The authors would like to thank the support of Michael Smith Foundation for Health Research (Cheung); the BC LEEF Chair related Rix Student Award (Cheung) and the BC Leading Edge Endowment Fund (Ribary).

References

1. Dahan, M., Levi, S., Luccardini, C., et al. (2003) Diffusion Dynamics of Glycine Receptors Revealed by Single-Quantum Dot Tracking. Science, 302:Oct, 442-444.
2. Jia W, Xu G, et al. (2008) Nanoparticle-Enhanced MEG Detection for in vivo Imaging. Biomagnetism: Interdisciplinary Research and Exploration, 89-91.
3. Jia W, Xu G, et al. (2008) Detection of magnetic nanoparticles with magnetoencephalography. Journal of Magnetism and Magnetic Materials, 320:1472-1478.
4. Pathak, S., Cao, E., Davidson, M. C., Jin, S., Silva, G. A. (2006) Quantum Dot Applications to Neuroscience: New Tools Probing Neurons and Gia. Journal of Neuroscience, February 15, 26(7):1893-1895.
5. Romanus, E., Huckel, M., et al. (2002) Magnetic nanoparticle relaxation measurement as a novel tool for in vivo diagnostics. Journal of Magnetism and Magnetic Materials, 252:387-389.
6. Saligram, U., Moran, J.E., Tepley, N. (2000) MCG Studies in Rats Using Ferrofluids. In: Aine, C., Okada, Y., Stroink, G., Swithenby, S., Wood, C.C., editors. Biomag 96: Advances in Biomagnetism Research. New York: Springer-Verlag, pp 675-678.
7. Scherer, C., Figueiredo, N. (2005) Ferrofluids: Properties and Applications. Brazilian Journal of Physics, 35;3A:718-727.
8. Stok, C.J., et al, (1987) Inverse solutions based on MEG and EEG applied to volume conductor analysis. Phys. Med. Biol. 32:99-104.
9. Vrba, J., Anderson, G., Betts, K., Burbank, M.B., Cheung, T., Cheyne, D., et al. (1999) 151-Channel whole-cortex MEG system for seated or supine positions. Recent Advances in Biomagnetism, eds: Yoshimoto, T. et al, 93-96.

Author: Teresa Cheung
Institute: Simon Fraser University
Street: 8888 University Drive
City: Burnaby
Country: Canada
Email: tpc5@sfu.ca

Magnetic Detection of the Sentinel Lymph Node in Ex Vivo Tissue with Colorectal Cancer

B. ten Haken[1], M. Visscher[1], J.J. Pouw[2], J.M. Klaase[2], Q.A. Pankhurst[3], J. Galindo-Millan[4], A.H. Velders[4], and H. Rogalla[1]

[1] University of Twente, MIRA Institute for biomedical technology and technical medicine, Department of Low Temperatures - Biomagnetism, Enschede, The Netherlands
[2] Medisch Spectrum Twente, Department of Surgery, Enschede, The Netherlands
[3] The Royal Institution of Great Britain, Davy-Faraday Research Laboratory, London, UK
[4] University of Twente, MESA+ Institute for nanotechnology, Department of Supramolecular Chemistry and Technology, Enschede, The Netherlands

Abstract— **A new method for sentinel lymph node detection was investigated by ex-vivo experiments in colorectal cancer. A dispersion of superparamagnetic iron oxide (SPIO) nanoparticles was injected in the resected tissue around the tumor. A magnetic probe with a 2nd order gradiometer configuration was used to localize the sentinel lymph node with accumulated SPIO nanoparticles. Additional analysis of SPIO content was performed on the resected lymph nodes. High-field MRI and conventional microscopy revealed the presence and distribution of SPIO inside the lymph nodes. Magnetometry confirmed SPIO content in lymph nodes and could give an estimate of the amount of SPIO uptake.**

Keywords— **sentinel lymph node, superparamagnetic iron oxides, magnetometry, MRI.**

I. INTRODUCTION

Since detection of magnetic particles in clinical practice is presently mainly reserved for magnetic resonance imaging (MRI), it can be valuable to explore the possibilities for detection outside MRI. For example, there are serious problems to combine surgery and MRI at the same time. Therefore new technology to detect magnetic particles also used as contrast agent in MRI can be an important addition to the clinical tools for surgeons in the operating room. In this research a new approach for the detection of superparamagnetic nanoparticles in sentinel lymph nodes is investigated.

Tumor metastases occur primarily in the lymph nodes that drain the tumor area [1]. These so called sentinel lymph nodes (SLNs) are therefore important to investigate at tumor resection. In mamma carcinoma SLN detection is currently performed using a radioactive dye which is injected in or around the tumor [2]. The radioactive particles will accumulate in the lymph nodes that drain the area of injection. Using a gamma camera and a hand-held gamma probe the sentinel node can be localized for resection [3]. In addition to radioactive dye a blue dye is often used for localization of the sentinel node by visual inspection. Hence sentinel nodes are frequently defined as lymph nodes that have a level of radioactivity relative to background and/or lymph nodes that have a blue afferent lymphatic vessel directly coming from the tumor [4-6].

Since penetration depth of optical detection of the SLN with blue dye is limited, another non-radioactive additional sentinel node approach with larger penetration depth is welcome. The logistical complex and harmful radio-isotope based technique [7] might be replaced by a much safer magnetic approach using a magnetic dye and magnetic probe. A small clinical study on magnetic detection of the SLN in breast cancer presented promising results [8]. Also for other carcinoma SLN detection can improve diagnosis and treatment strategy. Recent ex vivo studies using blue dye, propose to introduce SLN identification for colorectal cancer in order to identify high-risk patients who may benefit from adjuvant therapy [9-10]. A combined, complementary approach of SLN identification with magnetic nanoparticles and blue dye might be promising.

The currently investigated magnetic approach for SLN detection depends on the application of superparamagnetic nanoparticles, which are already in use as contrast agent for MRI [11]. Resected tissue with colorectal cancer was injected with these magnetic particles and blue dye. SLN

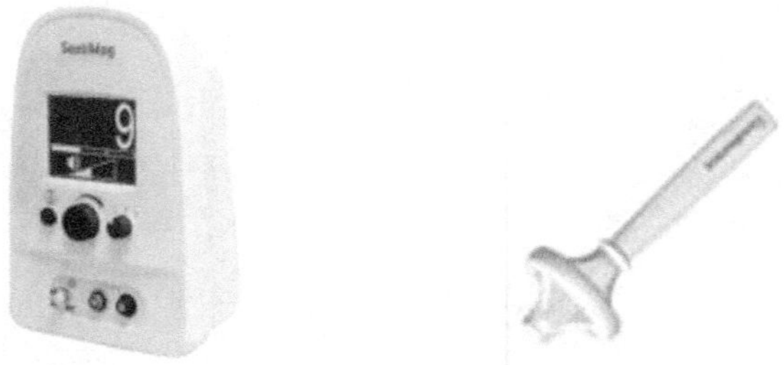

Fig. 1 The SentiMag® system

S. Supek and A. Sušac (Eds.): Advances in Biomagnetism – BIOMAG2010, IFMBE Proceedings 28, pp. 447–449, 2010.
www.springerlink.com

identification is investigated using the SentiMag®, a medical hand held device that detects magnetic particles non-invasively. Ex vivo injection of superparamagnetic iron oxide (SPIO) particles in cancerous colon resections is followed by analysis of the distribution of these particles through the tissue and especially the lymphatic system.

II. Methods

Patients with colorectal cancer were included after informed consent and subjected to surgical resection and lymphadenectomy. Colorectal resection of the tumor was performed following the normal procedure.

Immediately after resection of the colonic segment, the tissue sample was brought to a separate field and 2 mL of a SPIO nanoparticle dispersion (Endorem®, Guerbet Nederland BV, Gorinchem, The Netherlands; diameter: 80-150 nm, 11.2 mg Fe mL^{-1}) was injected subserosally around the tumor at multiple injection sites. At the same sites 2 mL of a blue dye (Patent Blue V, Guerbet Nederland BV, Gorinchem, The Netherlands) was injected, followed by a gentle massage for 5 minutes to enhance lymphatic flow mechanically. The SLN was identified by classical lymph node mapping, which includes tracing afferent blue lymph vessels draining Patent Blue V from the tumor area. This blue SLN was marked with a suture. Cutting the mesenteric tissue superficially was in some cases necessary to uncover the lymphatic pattern.

After fixation in formalin, the sample was fixed on a plate. Using both a T1- and T2-weighted sequence, the resected tissue was imaged in a 1.0 T MRI system (Philips Intera) to image the distribution of SPIO in the tissue sample.

For a precise localization of the lymph nodes containing SPIO the SentiMag® (Endomagnetics Ltd., London, UK) was used [12]. This device (figure 1) comprises a probe with an excitation coil that applies an alternating magnetic field (~ 2 kHz) to the sample. Sensing coils are arranged in a 2nd order gradiometer configuration enabling localization of magnetized lymph nodes by moving the probe over the area of interest. Using a magnetic field with certain frequency, the signal coming from the induced magnetization of the sample has the same frequency and noise sources can be easily suppressed.

The resected tissue was systematically scanned for SLNs by moving the the SentiMag® probe over the plate where the sample was placed on. The signal intensity of the SentiMag® was recorded according to the probe position on the glass plate. The localized SLNs were resected from the specimen by a pathologist. During resection the SentiMag® was also used with free access to the sample for a precise localization of the lymph nodes and to check if lymph nodes were missed in the systematic scan over the plate. After resection the signal intensity of the single SLN was measured, in order to prevent background contribution of SPIO. Vibrating sample magnetometry (VSM) was applied to

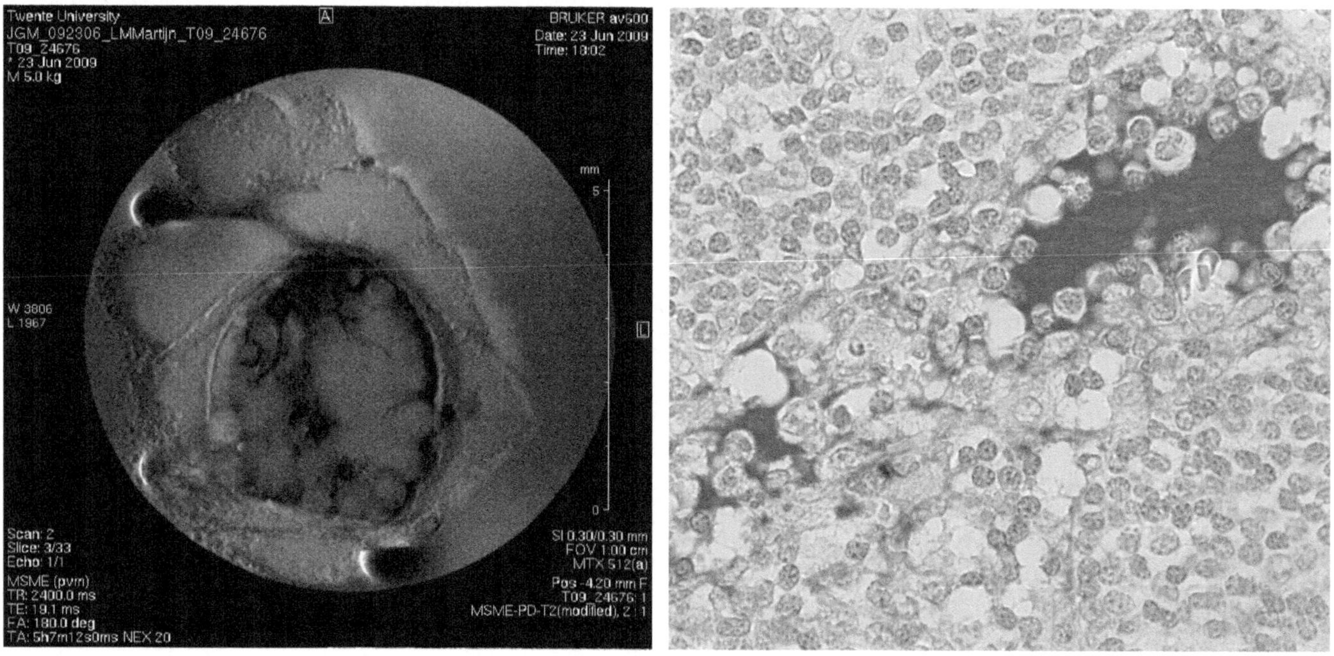

Fig. 2 *Left*: High field MRI of a single lymph node surrounded by fat tissue. The diameter of the circular image is ± 8 mm. Regions with low intensity in the lymph node indicate the presence of SPIO nanoparticles. - *Right:* Microscopic analysis with Pearls Prussian Blue confirmed the presence of iron in the lymph node

separate SLNs to determine the amount of SPIO content quantitatively.

Individual lymph nodes were imaged in a 14T MRI system (Bruker AvanceII 600 MHz system) in a special NMR-glass tube (Ø 8.2 mm). Both T1- and T2-weighted images were made to reveal the distribution of SPIO inside SLNs.

Finally the SLNs were subjected to normal microscopic analysis by a pathologist. Pearls Prussian Blue was added to SLN slices to prove the presence of iron in the lymph nodes. Furthermore, metastases in lymph nodes were investigated by H&E and immunohistochemical staining.

III. Results

In a subset of the patients with colorectal cancer the SLN could be found in the resected tissue sample after local administration of 2 mL of SPIO dispersion. Also lymph node mapping identified the SLNs filled with Patent Blue V. MRI of the specimens showed the distribution of Endorem® through the tissue from the injection site to the lymph nodes. Subsequently, the SentiMag accurately revealed the location of the SLNs filled with SPIO nanoparticles. The signal strength depends on the distance between the probe and the lymph nodes containing SPIO. Combining the 1.0 T MRI results with the localizations of the SentiMag a correlation study could be done, providing an estimate of the distance between the SentiMag probe and the SLN.

T2-weighted high field MRIs of some of the individual SLNs showed decreased intensity of the inner lymph node caused by the T2-shortening effect of SPIO (figure 2, left). This indicates distribution of Endorem® through lymph nodes. Using magnetometry the amount of iron present in the SLNs could be determined in the order of 10-100 µg.

Microscopic analysis with Pearls Prussian Blue confirmed the presence of SPIO by iron staining in the detected lymph nodes (figure 2, right). Endorem® was present in metastatic as well as non-metastatic lymph nodes.

IV. Discussion and conclusion

This ex vivo experiment showed the possibility of magnetic detection of lymph nodes in colorectal cancer. This case shows that clinical use of magnetic particle detection beside MRI can be regarded as promising. Further clinical studies to the applicability of magnetic SLN identification have to be done, including its impact on treatment and survival rates.

Although the distribution mechanisms of living tissue were not taken into account, this experiment indicated the usability of the SentiMag® to localize the SLN intra-operatively. Since only commercially available Endorem® is used as magnetic dye in this experiment, particle size is probably not optimized for SLN localization [13]. For that reason it is recommended to study the optimal particle size and size distribution for magnetic SLN identification.

References

1. M. A. Swartz, "The physiology of the lymphatic system," *Advanced Drug Delivery Reviews,* vol. 50, pp. 3-20, 2001.
2. H. Schraffordt Koops, *et al.*, "Sentinel node biopsy as a surgical staging method for solid cancers," *Radiother. Oncol.*, vol. 51, pp. 1-7, 1999.
3. N. P. Alazraki, *et al.*, "Lymphoscintigraphy, the sentinel node concept, and the intraoperative gamma probe in melanoma, breast cancer, and other potential cancers," *Seminars in Nuclear Medicine,* vol. 27, pp. 55-67, 1997.
4. M. H. E. Doting, *et al.*, "Lymphatic mapping with intralesional tracer administration in breast carcinoma patients," *Cancer,* vol. 88, pp. 2546-2552, 2000.
5. I. Bedrosian, *et al.*, "Accuracy of sentinel lymph node biopsy in patients with large primary breast tumors," *Cancer,* vol. 88, pp. 2540-2545, 2000.
6. S. H. Estourgie, *et al.*, "Lymphatic Drainage Patterns from the Breast," *Ann. Surg.*, vol. 239, pp. 232-237, 2004.
7. A. H. Strickland, *et al.*, "Sentinel node biopsy: an in depth appraisal," *Crit. Rev. Oncol. Hematol.*, vol. 44, pp. 45-70, 2002.
8. T. Joshi, et al., "Magnetic nanoparticles for detecting cancer spread," 30th Annual San Antonio Breast Cancer Symposium – December 13–16, 2007 - Breast Cancer Research and Treatment, vol. 106, p. S129, 2007.
9. E. S. Van der Zaag, *et al.*, "Improving staging accuracy in colon and rectal cancer by sentinel lymph node mapping: A comparative study," *Eur. J. Surg. Oncol.*, vol. 35, pp. 1065-1070, 2009.
10. P. M. Van Schaik, *et al.*, "Ex vivo sentinel lymph node "mapping" in colorectal cancer," *Eur. J. Surg. Oncol.*, vol. 33, pp. 1177-1182, 2007.
11. C. Corot, *et al.*, "Recent advances in iron oxide nanocrystal technology for medical imaging," *Advanced Drug Delivery Reviews,* vol. 58, pp. 1471-1504, 2006.
12. S. R. B. K. Hattersley, GB), Pankhurst, Quentin Andrew (Hertfordshire, GB), Brazdeikis, Audrius (Missouri City, TX, US), "Apparatus and method for determining magnetic properties of materials," United States Patent, 2009.
13. I. R. Jimenez, *et al.*, "Particle sizes of colloids to be used in sentinel lymph node radiolocalization," *Nucl. Med. Commun.*, vol. 29, pp. 166-172, 2008.

Author: B. ten Haken

Institute: University of Twente, MIRA Institute for biomedical technology and technical medicine, Department of Low Temperatures - Biomagnetism
Street: Drienerlolaan 5
City: Enschede
Country: The Netherlands
Email: B.tenHaken@utwente.nl

Effects of Magnetic Fields on Nervous Cells

J.C. Hernandez-Pavon[1,2,3,4,5], T. Cordova[1], S. Solorio[3], G. Barbosa-Sabanero[4], M. Sabanero-Lopez[5], and M. Sosa[1]

[1] University of Guanajuato/ Department of Engineering, Leon Guanajuato, Mexico
[2] Aalto University/ Department of Biomedical Engineering and Computational Science, Espoo, Finland
[3] Hospital of High Specialities/ Clinic T-1 IMSS, Leon Guanajuato, Mexico
[4] University of Guanajuato/Department of Medical Sciences, Leon Guanajuato, Mexico
[5] University of Guanajuato/Department of Biology, Guanajuato Guanajuato, Mexico

Abstract— **The effects of extremely low frequency magnetic fields (ELF-MF's) on neuroblastoma cell cultures (SK-NSH) were studied. The cells were exposed for 1 hour to magnetic fields of 19 mT at the rate of 8 Hz. Three groups of cells were studied: control, sham, and stimulated. Studies of protein expression and immunofluorescence were carried out to assess whether the ELF-MF's induce alterations in cytoskeletal structure. The results suggest that the groups exposed to MF's present higher protein expression than the control group. Furthermore, the samples exposed to MF's show a cytoskeletal morphology differing from that of the control groups. These findings suggest that a magnetic dose (MD) with these characteristics produces alterations in nervous cells. This study provides important information that describes some bioeffects underwent by neuronal systems under magnetic stimulation.**

Keywords— **Nervous cells, magnetic stimulation, protein expression, morphology.**

I. Introduction

In the past few decades, there has been an increasing effort by the scientific community to understand the possible effects of extremely low frequency electromagnetic fields (ELF-EMF's) on biological systems. In particular, the exposure of human beings to different sources of ELF-EMF's, that is, EMF's with frequencies less than 100 Hz, has motivated the increasing interest in understanding the effects of ELF-EMF's on biological systems and promoted the development of several kinds of investigations. Neuroblastoma is an extra cranial solid cancer in childhood. However, an undifferentiated neuroblastoma cell line is an excellent model for studying nervous cells. There are a large number of papers that have described the effect of ELF-EMF's at 50 – 60 Hz on neuroblastomas since that range coincides with household line frequencies. Some of these findings show that the exposure to magnetic fields (MF's) of 1 – 2 mT does not produce apoptosis (programmed cell death). In contrast, the MF's induce an increase in proliferation and changes in neuronal morphology [1, 2].

Other studies have indicated that the exposure to static and variable MF's do not produce a change in resting potential and action potential in human neuroblastoma cells, which suggests that the cellular mechanisms responsible for generating such potentials is not altered [3, 4]. Also, there is evidence that the MF's stimulate the mobilization of calcium ions, which can be seen as a mechanism to avoid a possible differentiation produced by ELF-EMF's [5].

In this study, the magnetic stimulation with ELF-MF's at a rate of 8 Hz on the neuroblastoma cell line SK-NSH is presented, which is a frequency range that is not covered in most publications, but is part of the alpha waves generated by the brain [6]. Experiments were carried out to evaluate whether this kind of stimulation produces alterations in the protein expression and morphology of the nervous cells.

II. Materials and Methods

A. Cell Cultures

The human neuroblastoma cell line SK-NSH was used. The cells were grown (1 x 10^5 cells/mL) at 37 °C in an atmosphere of 5% CO_2 in DMEM "Dulbecco's Modified Eagle Media" supplemented with 10 % of Fetal Calf Serum.

B. Exposure Conditions

The exposure system used to stimulate the neuroblastoma cells was composed of two Helmholtz configurations with a diameter of 20 cm. One pair of coils produced a magnetic field, whereas in the other pair, the magnetic field was cancelled according to the direction of the electrical current. The purpose of using two Helmholtz configurations was to have similar conditions in both the samples placed between the coils. One important physical issue was the control of thermal effects due to the presence of the electromagnetic field. It is well known that this same exposure system may produce some heating as a result of the joule effect, due to the electrical currents flowing in the copper coils. The coils were connected in parallel and each coil was powered by a chopped DC current of 10 V that produces a variable magnetic field. In each coil, the resistance was found to be R = 600 ± 50 mΩ and the inductance was L = 3.3 ± 0.2 mH. The system was fed with four (15 V, 30 A) DC power supplies from Steren Electronics (San Diego, USA). A separate power supply was used for each coil. The resulting current waveform was a typically square pulse that lasts 15 ms, giving dB/dt ≈ 1.2 T/s. The cells were exposed for one hour to a magnetic field of 19 mT and were stimulated by square waves produced by the DC current with a duty cycle of 12 % to rate 8 Hz. During operation, variables such as the temperature and the magnetic flux were monitored inside the coils. The magnetic field was measured using a Gaussmeter 2100 from Magnetic Instrumentation Inc.

Three groups of cells were studied: control cells (C), cells that were not magnetically stimulated; magnetically stimulated cells (MS), cells stimulated by the Helmholtz coil that produces magnetic field; and sham cells (S), sample placed in a Helmholtz coil where the magnetic field was cancelled. All the samples were maintained at room temperature outside the incubator. The stimulated and sham samples were placed in the homogenous region of the Helmholtz coil where the magnetic field was perpendicular to the samples.

C. Electrophoresis Studies

After stimulation time, the samples were prepared for carrying out gel electrophoresis studies. Proteins pellets of each cell group were fractionated in 10% SDS-PAGE gel (sodium dodecyl sulfate polyacrylamide gel electrophoresis) [7]. The gels were charged with the extract of the proteins (20 – 40μL) to be analyzed. The samples were run in a Mini-Gel Systems vertical electrophoresis chamber (C.B.S Scientific Company, Inc MGU-402, USA) fed with a power supply (BIO- RAD) at 80 Volts during 2 hours. In order to evaluate whether the proteins were of high, medium, or low molecular weight, molecular markers were used in one lane for reference and the samples were placed in the other lanes. The gels were stained with a 0.25 % solution of Coomassie blue to observe the protein profiles of the samples.

D. Immunofluorescence

To carry out the Immunofluorescence staining, the cells were grown in slides approximately at a concentration of 1 x 10^4 cells / slide. The cells were fixed with 4% p- formaldehyde and 0.05% glutaraldehyde for 30 minutes. The cells were then washed with PBS three times and permeabilized with 0.1% Triton X-100 for 3 minutes. Later, cells were incubated with a primary monoclonal antibody (anti-β-tubulin, Zymed, dilution 1:40) to study the microtubules and a polyclonal antibody (anti-goat-mouse, IgG coupled to FITC (Zymed), dilution 1:30). The time of incubation was one hour for each antibody. Finally, the cells were mounted in cover glasses by using DAPI (Vector) to observe the nuclei. The samples were analyzed in an epifluorescence microscope (Exc 400-420 nm).

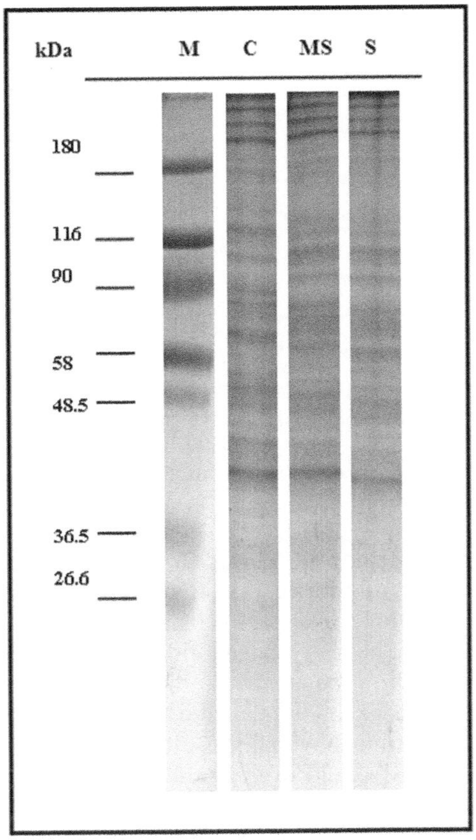

Fig. 1 Electrophoresis gel, (C), (MS) and (S) represent the control, stimulated and sham groups respectively, whereas (M) represents the molecular weight markers in Kilodaltons (kDa)

III. Results

Fig. 1 shows the electrophoresis gel with the profiles of proteins of the human neuroblastoma cell line SK-NSH. The peptides of high, medium, and low molecular weight are presented to compare with the molecular markers. The band definitions were similar in the C, MS, and S groups. No qualitative differences were evident between the different groups, so it was not possible to identify alterations in the synthesis of proteins using only the electrophoresis gel. In order to further investigate the effects of magnetic fields on the protein expression, a quantitative

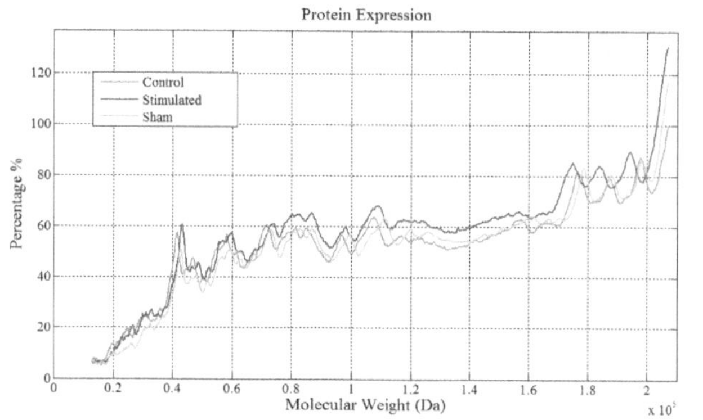

Fig. 2 The molecular weight versus the percentage of the protein concentration is shown for the control, stimulated, and sham groups. The stimulated group showed higher concentrations of proteins in comparison with the control group

analysis was performed. In Fig. 2, a comparison between the percentage of proteins and molecular weight is presented for the three groups of cells studied. The stimulated group shows a higher concentration of proteins than the control and sham groups. Due to the fact that the area and intensity of the bands is related to the protein expression, Fig. 2 was determined with the results of Fig. 1.

Fig. 3 shows a field of view of the morphology of the groups studied. In the control group, there is presence of well-defined neurons, as shown in Fig. 3A. The nuclei also show uniform structures, as shown in Fig. 3A'. For the stimulated cells, the morphology undergoes some changes. Magnetic stimulation seems to induce spherical morphologies, as shown in Fig. 3B. However, the nuclei do not show variations respect to the other groups Fig. 3B'. The sham cell group shows morphology more similar to the control group, shown in Fig. 3C. In this group, there were no alterations in the nuclei and likewise in the other groups, as shown in Fig. 3C'.

No evidence of alterations in the microtubules was observed in the cytoplasm of the studied cells. Nevertheless, the alterations were evident in the morphology of the cells, mainly in the cells exposed to magnetic stimulation. This suggests that the magnetic stimulation produces physiological alterations (electrical and chemical) in the neuronal processes.

IV. Conclusions

In this study, magnetic stimulation was applied to neuroblastoma cell cultures to detect the possible effects produced by MF's. The results demonstrate that the exposure to magnetic fields of 19 mT with square pulses at a rate of 8 Hz during one hour can induce alterations in the cells.

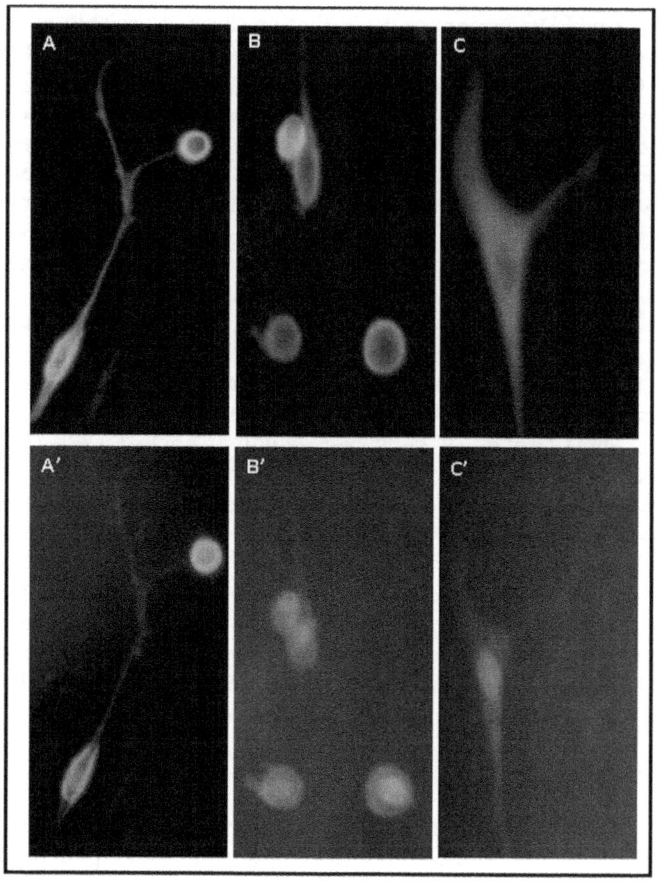

Fig. 3 The results of the immunofluorescence studies are shown; control cells (C), magnetic stimulated cells (MS), and sham cells (S). A, B, and C correspond to the microtubules of the C, MS, and S groups, respectively, whereas A', B', and C' correspond to the nuclei of the CC, MSC, and SC groups, respectively. The MS showed alterations in cytoskeletal morphology. It seems that MF's induce spherical morphology. However, the nuclei do not show alterations

The cells exposed to magnetic stimulation exhibit a higher percentage of protein expression with respect to the control and sham groups, which coincides with studies reported in the literature [8]. That is, EMF's activate the cellular stress response to increase the levels of protein expression. However, in our study it was not possible to determine which proteins are increased. Nevertheless, the most remarkable differences are observed for molecular weights above 60kDa.

The cytoskeleton showed morphological alterations in the stimulated cells mainly in the axons. The stimulated cells show reduced length of the axons or processes in comparison to the other groups. These alterations could be related with physiological changes in the cells, such as polarization and depolarization of the membrane and transport phenomenon. Since, the biochemical compounds in living cells are composed of charges and dipoles that can interact with electrical and magnetic fields by various mechanisms, more studies should be carry out in order to determine whether the morphological alterations are related to magnetoporation or electrorotation.

Finally, the results presented in this paper shows that a magnetic field of 19 mT at a rate of 8 Hz can produce alterations in neuroblastoma in various processes such as protein expression and morphological changes.

Acknowledgment

We would like to thank Christine Wallace for the review of this paper. Studies presented in this paper were supported by CONACYT J50182, and CONCYTEG 09-16-K662-08, as well as, IMSS with its research scholarship. We also want to thank Professor Risto Ilmoniemi, Aalto University and the Department of Biomedical Engineering and Computational Science, Finland.

References

1. Pirozzoli MC, Marino C, Lovisolo GA et al. (2003) Effects of 50 Hz Electromagnetic Field Exposure on Apoptosis and Differentiation in a Neuroblastoma Cell Line. Biolectromagnetics 24:510–516.
2. Pozzi D, Grimaldi S, Ledda M et al. (2007). Effect of 50 Hz Magnetic Field exposure on Neuroblastoma Morphology. IJIB 1:12–16.
3. Sonnier H, Kolomytkin OV, Marino AA. (2000). Resting potential of excitable neuroblastoma cells in weak magnetic fields. Cellular and Molecular Life Sciences 57:514–520.
4. Sonnier H, Kolomytkin O, Marino A. (2003). Action potentials from human neuroblastoma cells in magnetic fields. Neuroscience Letters 337:163–166.
5. Tonini R, Baroni MD, Masala E et al. (2001). Calcium Protects Differentiation Neuroblastoma Cells During 50Hz Electromagnetic Radiation. Biophysical Journal 81:2580–2589.
6. Cohen D. 1968. Magnetoencephalography: Evidence of Magnetic Fields Produced by Alpha-Rhythm Currents. Science 161:784-786.
7. Laemmli UK. (1970). Cleavage of Structural Proteins During the Assembly of the Head of the Bacteriophage T4. Nature 227:680–685.
8. Blank M, Goodman R. (2009). Electromagnetic fields stress living cells. Pathophysiology 16:71-78.

Corresponding author:

Author: Julio Cesar Hernandez-Pavon
Institute: Department of Engineering, Guanajuato University
Street: Loma del Bosque 103, Loma del Campestre, 37150
City: Leon, Guanajuato.
Country: Mexico.
Email: julius.fis@gmail.com

Author Index

GPSR Compliance

The European Union's (EU) General Product Safety Regulation (GPSR) is a set of rules that requires consumer products to be safe and our obligations to ensure this.

If you have any concerns about our products, you can contact us on ProductSafety@springernature.com

In case Publisher is established outside the EU, the EU authorized representative is:

Springer Nature Customer Service Center GmbH
Europaplatz 3
69115 Heidelberg, Germany

Batch number: 10755813

Printed by Printforce, the Netherlands